საქართველოს სსრ მეცნიერებათა აკადემია * აბასთუმნის ასტროფიზიკური ობსერვატორია
АКАДЕМИЯ НАУК ГРУЗИНСКОЙ ССР * АБАСТУМАНСКАЯ АСТРОФИЗИЧЕСКАЯ ОБСЕРВАТОРИЯ
ACADEMY OF SCIENCES OF THE GEORGIAN SSR * ABASTUMANI ASTROPHYSICAL OBSERVATORY

PROCEEDINGS OF THE THIRD EUROPEAN ASTRONOMICAL MEETING

TBILISI, 1—5 JULY, 1975

ТРУДЫ ТРЕТЬЕЙ ЕВРОПЕЙСКОЙ АСТРОНОМИЧЕСКОЙ КОНФЕРЕНЦИИ

ТБИЛИСИ, 1—5 ИЮЛЯ, 1975

მესამე ევროპული ასტრონომიული კონფერენციის შრომები

თბილისი, 1—5 ივლისი, 1975

STARS AND GALAXIES FROM OBSERVATIONAL POINTS OF VIEW

ЗВЕЗДЫ И ГАЛАКТИКИ В НАБЛЮДАТЕЛЬНОМ АСПЕКТЕ

ვარსკვლავები და გალაქტიკები დამზერით ასპექტში

Springer-Science+Business Media, B.V.

УДК 52

Сборник состоит из докладов и кратких сообщений, заслушанных Третьей Европейской Астрономической Конференцией,проведенной в Тбилиси 1-5 июля 1975года под эгидой Международного Астрономического Союза. Доклады и сообшения, также как и Конференция в целом, посвящены проблемам строения и эволюции звезд и галактик в наблюдательном аспекте.

ISBN 978-94-017-5308-1 ISBN 978-94-017-5306-7 (eBook)
DOI 10.1007/978-94-017-5306-7

Softcover reprint of the hardcover 1st edition 1975

Редактор Е.К.Харадзе

Т $\frac{20605}{М\ 607(03)-76}$

Ursprünglich erschienen bei АБАСТУМАНСКАЯ АСТРОФИЗИЧЕСКАЯ ОБСЕРВАТОРИЯ 1975

EDITOR'S PREFACE

The volume contains the papers presented at the Third European Astronomical Meeting devoted to the problem : ''Stars and galaxies from the observational points of view'' and sponsored by the International Astronomical Union and the Astronomical Council of the Academy of Sciences of the U.S.S.R., organized by Abastumani Astrophysical Observatory of the Georgian SSR Academy of Sciences, held in Tbilisi (Georgia, U.S.S.R.) on July 1-5, 1975.

The assistance of the Scientific Organizing Committee of the Conference and of the Astronomical Council of the Academy of Sciences of the U.S.S.R. is gratefully acknowledged.

The 3rd EAM had 11 sessions with 31 leading papers , 77 short communications and three lectures.

The number of the registered participants reached 361– representatives of 19 European countries and two participants from the U.S.A. The mid-age of the 3rd EAM participants was a little less than 40. This fact is emphasized with a certain satisfaction, since one of the aims of this new institution – European Astronomical Meetings – is to encourage the establishment of good scientific contacts between young astronomers and to involve them into discussions.

The papers and communications appear in the order they were delivered de facto.

No attempt was made to change the style of the English language of the authors.

The urgency to publish the Proceedings without sensible delay led the publishers to issue the book in the mimeographic way, thus, without sending the proofs to the authors.

Misunderstandings have been possibly avoided thanks to the tape recordings which were at the editor's disposal.

The authors have been timely informed on the subject.

Many people have helped in making this volume ready for publication. Of these a few stand out as having made a major contribution and the editor would like to thank especially Mrs.A.R.Azo, Dr.R.M.West, the Director of the Printing House of the Georgian Academy of Sciences R.M.Grigolia and Dr. T.M.Borchkhadze.

Abastumani Astrophysical Observatory

E.K.Kharadze, editor

LIST OF PARTICIPANTS

AUSTRIA

SCHNELL A.	Univ. – Sternwarte, Wien
WEISS WERNER W.A.	Univ. Obs., Figl–Obs. für Astrophys., Wien

BULGARIA

DOBRICHEV V.	Department of Astron. and National Obs., Ac. of Sc., Sofia
IVANOV G.	Department of Astron., Univ. of Sofia
KALINKOV M.	Department of Astron. and National Obs., Ac. of Sc., Sofia
KANEVA I.	Department of Astron., Ac. of Sc., Sofia
KOLEV D.	Department of Astron., Ac. of Sc., Sofia
NIKOLOV N.	Univ. of Sofia, Faculty of Physics
NIKOLOVA M.	Univ. of Sofia, Faculty of Physics
POPOVA M.	Department of Astron. and National Obs., Ac. of Sc., Sofia
RADOSLAVOVA TS.	Department of Astron. and National Obs., Ac. of Sc., Sofia
STAVREV K.	Department of Astron., Ac. of Sc., Sofia

CZECHOSLOVAKIA

ANDRLE P.	Astron. Inst., Prague
BAHYL V.	Astron. Inst., Slovak Ac. of Sc., Skalnaté Pleso
BOUSKA J.	Department of Astron., Charles Univ. and Journal "Říše hvězd", Prague
CHOCHOL D.	Astron. Inst., Slovak Ac. of Sc., Skalnaté Pleso
HLAD O.	Obs. of Prague-Petřín
KOUBSKY P.	Astron. Inst. ČSAU, Ondřejov
KŘÍŽ S.	Astron. Inst. ČSAU, Ondřejov
ONDERLIČKA B.	Department of Astron. Purkyně Univ., Brno

DENMARK

GYLDENKERNE K.	Copenhagen Univ. Obs.
NISSEN P.E.	Inst. of Astron., Univ. of Aarhus

FINLAND

JAAKKOLA T.	Helsinki Univ. Obs. and Astrophys. Lab.
MARKKANEN T.	Helsinki Univ. Obs. and Astrophys. Lab.
MATTILA K.	Helsinki Univ. Obs. and Astrophys. Lab.
PIIROLA V.	Helsinki Univ. Obs. and Astrophys. Lab.
TUOMINEN I.	Helsinki Univ. Obs.
VILHU O.	Helsinki Univ. Obs.

FRANCE

BALKOWSKI CH.	Obs. de Meudon
BOTTINELLI L.	Obs. de Meudon
BOULESTEIX J.	Obs. de Marseille
BRIOT D.	Obs. de Paris
COURTÉS G.	Obs. de Marseille

DAVOUST E.	Obs. de Besançon
DUVAL M.F.	Obs. de Marseille, Provence Univ.
GOUGUENHEIM L.	Obs. de Meudon
GUIBERT J.	Obs. de Meudon, Department de Radioastron.
HEIDMANN J.	Obs. de Meudon
HEIDMANN N.	LAT Inst. d'Astrophys., Paris
KUNTH D.	Obs. de Paris - Meudon
LACENTEL J.M.	Obs. de Paris
LE DENMAT	Obs. de Paris
OCHSENBEIN F.	Centre de Données Stellaires, Strasbourg
PELAT D.	Obs. de Paris
QUERCI F.	Obs. de Meudon
QUERCI M.	Obs. de Meudon
TULLY R.B.	Obs. de Marseille
VALTIER J.C.	Obs. de Nice

F. R.G.

BERKHUIJSEN E.M.	Max-Plank-Inst. für Radioastron., Bonn
BREDOW K.	Hamburger Sternwarte
BROSCHE P.	Sternwarte der Univ., Bonn
GARDNER F.	Max-Plank-Inst. für Radioastron., Bonn
GLIESE W.	Astron. Rechen - Inst., Heidelberg
MATERNE J.	Hamburger Sternwarte
MEZGER P.G.	Max-Plank-Inst. für Radioastron., Bonn
PREUSS E.	Max-Plank-Inst. für Radioastron., Bonn
RUF K.	Astron. Inst., Frankfurt
SCHMIDT-KALER TH.	Astron. Inst., Bochum Univ.
SMITH L.	Max-Plank-Inst. für Radioastron., Bonn
WIELEN R.	Astron. Rechen - Inst., Heidelberg

G.D.R.

BECK H.G.	Carl Zeiss Obs.
GÜRTLER J.	Univ. - Sternwarte, Jena
GUSSMANN E.A.	Zentralinstitut für Astrophysik, Potsdam
KÜHN L.	Friedrich-Schiller-Universitat, Jena
MARX S.	Zentralinstitut für Astrophysik, Potsdam
NOTNI P.	Zentralinstitut für Astrophysik, Potsdam
OETKEN L.	Zentralinstitut für Astrophysik, Potsdam
OLEAK H.	Zentralinstitut für Astrophysik, Potsdam
PAUL H.	Zentralinstitut für Astrophysik, Potsdam
RICHTER G.	Zentralinstitut für Astrophysik, Porsdam
RICHTER GOT.	Zentralinstitut für Astrophysik, Potsdam
SCHMIDT K.H.	Zentralinstitut für Astrophysik, Potsdam
SCHOLZ G.	Zentralinstitut für Astrophysik, Potsdam
WENZEL W.	Zentralinstitut für Astrophysik, Potsdam
ŽELWANOWA E.	Zentralinstitut für Astrophysik, Potsdam
ZIENER R.	Zentralinstitut für Astrophysik, Potsdam

G R E E C E

CONTOPOULOS G.	Department of Astron., Univ. of Thessaloniki
MERTZANIDES C.	Department of Astron., Univ. of Thessaloniki

HUNGARY

BALÁZS B.	Department of Astron., R, Eötvös Univ., Budapest
BALÁZS L.	Konkoly Obs., Budapest
JANKOVICH I.	Konkoly Obs., Budapest
KANYO S.	Konkoly Obs., Budapest
PAÁL G.	Konkoly Obs., Budapest
SZABADOS L.	Konkoly Obs., Budapest

ITALY

BENVENUTI P.	Asiago Astrophys. Obs.
BERTOLA F.	Astron. Obs., Padova
CRIVELLARI L.	Astron. Obs., Trieste
D'ODORICO S.	Astrophys. Obs. of the Univ. of Padova
SELVELLI P.	Astron. Obs., Padova

THE NETHERLANDS

BLAAUW A.	Sterrewacht - Huygens Lab., Leiden
STALLER R.P.A.	Astron. Inst., Univ. of Amsterdam
STROM R.G.	Leiden Obs.
TAKENS R.	Astron. Inst., Amsterdam
VAN DER HUCHT K.	Space Research Lab. of the Astron. Inst., Utrecht
WESSELIUS P.R.	Kapteyn Inst., Department of Space Research, Groningen

POLAND

BASINSKA-GRZESIK E.	Inst. of Astron., Ac. of Sc., Toruń
BRANCEWICZ H.	Astron. Obs., Jagellonian Univ., Kraków
BURNICKI A.	Astron. Inst., N.Copernicus Univ., Toruń
CUGIER H.	Astron. Obs. of the Wrocław Univ
FLOWER PH.	Inst. of Astron., Ac. of Sc., Wrocław
IWANISZEWSKA C.	Astron. Inst., N.Copernicus Univ., Toruń
IWANOWSKA W.	Astron. Inst., N.Copernicus Univ., Toruń
JUCHNIEWICZ J.	Warsaw Univ. Obs.
KLIMEK Z.	Astron. Obs., Jagellonian Univ., Kraków
KOZAR	Astron. Obs. of the Wrocław Univ
KREINER J.M.	Astron. Obs., Kraków
KRELOWSKI J.	Astron. Inst., N.Copernicus Univ., Toruń
KUBIAK M.	Astron. Obs. Warsaw
KUROCZKIN D.	Astron. Inst., N.Copernicus Univ., Toruń
MICHALEC A.	Astron. Obs., Jagellonian Univ., Kraków
RUDNICKI K.	Astron. Obs., Jagellonian Univ., Kraków
STEPIEN K.	Astron. Obs., of Warsaw Univ.
SZUMIEJKO E.	Astron. Obs. of the Wroclaw Univ.
URBANIK M.	Astron. Obs., Jagellonian Univ., Kraków

RUMANIA

TANASESCU E.	Politechnical Inst., Buchurest

SWEDEN

ARDEBERG A.	Astron. Inst., Lund
ELVIUS A.	Stockholm Obs.
ELVIUS T.	Lund Obs.

ERIKSSON K.	Astron. Obs., Uppsala
KINNANDER A.	Astron. Obs., Uppsala
LINDBLAD P.O.	Stockholm Obs.
LINDGREN H.	Lund Obs.
LUNDSTRÖM I.	Lund Obs.
NORDLUND A.	Inst. of Theoret. Physics, Stockholm Univ
SANDQVIST A.	Stocknolm Obs.
SHARMER G.	Stockholm Obs.
STENHOLM B.	Lund Obs.
WELIN G.	Astron. Obs., Uppsala

SWITZERLAND

TAMMANN G.	Astron. Inst., Binningen
WEST R.M.	ESO, Geneva

U. K.

BROWNRIGG D.R.K.	Cumputer Science Department, Reading Univ.
DAVIES R.D.	Univ. of Manchester, Nuffield Radioastron. Lab.
DODD R.J.	Royal Obs., Edinburgh
HARGRAVE P.J.	Mullard Radioastron. Obs., Cavendish Lab.,Cambridge
LONGAIR M.S.	Mullard Radioastron. Obs., Cavendish Lab.,Cambridge
MORGAN D.H.	Royal Obs., Edinburgh
NANDY K.	Royal Obs., Edinburgh

U.S.A.

HOUK N.	Department of Astron., Univ. of Michigan
SNOW TH.P.	Princeton Univ. Obs.

U.S.S,R.

ABBASOV G.I.	Shemakha Astrophys. Obs.
ABRAMIAN M.G.	Byurakan Astrophys. Obs.
AKIMOV L.A.	Astron. Obs., Khar'kov State Univ.
ALANIA I.F.	Abastumani Astrophys. Obs.
ALKSNIS A.K.	Baldone Radioastrophys. Obs., Riga
AMBARTSUMIAN V.A.	Byurakan Astrophys. Obs.
AMNUEL P.R.	Shemakha Astrophys. Obs.
ANOSOVA ZH. P.	Astron. Obs., Leningrad State Univ.
ARAKELYAN M.A.	Byurakan Astrophys. Obs.
ARKHIPOVA V.P.	Sternberg Astron. Inst., Moscow
ARTAMONOV B.P.	Sternberg Astron. Inst., Moscow
ASLANOV I.A.	Shemakha Astrophys. Obs.
AZO A.R.	Abastumani Astrophys. Obs.
BARANOV V.B.	Inst. for Space Research, Moscow
BARTAYA R.A.	Abastumani Astrophys. Obs.
BASKO M.M.	Inst. for Space Research, Moscow
BATCHIKOVA M.A.	Astron. Council, Acad. of Scien. of the USSR,Moscow
BELOKON E.T.	Astron. Obs., Leningrad State Univ.
BERULIS I.I.	Lebedev Phys. Inst., Moscow
BESHENOVA L.D.	Alma-Ata Astrophys. Inst.
BIBILEISHVILI TS.V.	Abastumani Astrophys. Obs.
BYSTROVA N.V.	Special Astrophys. Obs., Zelenchuk
BORCHKHADZE T.M.	Abastumani Astrophys. Obs.

BOTCHKAREV N.G.	Sternberg Astron. Inst., Moscow
BOYARCHUK A.A.	Crimean Astrophys. Obs.
BOYARCHUK M.E.	Crimean Astrophys. Obs.
BRONSHTEN V.A.	USSR Astron. and Geodes. Society, Moscow
BURDYUZHA V.V.	Inst. for Space Research, Moscow
CHAVUSHIAN O.S.	Byurakan Astrophys. Obs.
CHENTSOV E.L.	Special Astrophys. Obs., Zelenchuk
CHERNIN A.D.	Phys. Techn. Inst., Moscow
CHERNOMORDIK V.V.	Phys. Techn. Inst., Moscow
CHKHIKVADZE Ya.N.	Tbilisi State Univ.
CHUADZE A.D.	Telavi Pedagogical Inst.
CHUGAY N.N.	Astron. Council, Acad. of Scien. of the USSR, Moscow
CHUMAK Z.N.	Alma-Ata Astrophys. Inst.
CHUNAKOVA N.M.	Special Astrophys. Obs., Zelenchuk
CHUVAEV K.K.	Crimean Astrophys. Obs.
DAGKESAMANSKIJ R.D.	Lebedev Phys. Inst., Moscow
DANILOV V.M.	Ural State Univ., Sverdlovsk
DENISYUK E.K.	Alma-Ata Astrophys. Inst.
DEPADEL' D.M.	Abastumani Astrophys. Obs
DEUTSCH A.N.	Pulkovo Obs., Leningrad
DIBAY E.A.	Sternberg Astron. Inst., Southern Station, Crimea
DLUZHNEVSKAYA O.B.	Astron. Council, Acad. of Scien. of the USSR, Moscow
DOLIDZE M.V.	Abastumani Astrophys. Obs.
DOROSHENKO V.T.	Sternberg Astron Inst., Southern Station, Crimea
DOROSHKEVICH A.G.	Inst. of Applied Mathematics, Moscow
DUBYAGO I.A.	Kazan' Astron. Obs.
DZIGVASHVILI R.M.	Abastumani Astrophys. Obs.
DZHAPIASHVILI V.P.	Abastumani Astrophys. Obs.
DZHIMSHELEISHVILI G.N.	Abastumani Astrophys. Obs.
EFIMOV YU.S.	Crimean Astrophys. Obs.
EFREMOV YU.N.	Sternberg Astron. Inst., Moscow
EINASTO J.E.	Tartu Inst. of Astrophys. and Atm. Phys.
EMINZADE T.A.	Shemakha Astrophys. Obs.
ERGMA E.V.	Astron. Council, Acad. of Scien. of the USSR, Moscow
ESIPOV N.I.	Sternberg Astron. Inst., Moscow
FESENKO B.I.	Pskov Pedagogical Inst.
FINKEL'SHTEIN A.M.	Special Astrophys. Obs., Zelenchuk
FOMENKO A.E.	Special Astrophys. Obs., Zelenchuk
FROLOV M.S.	Astron. Council, Acad. of Scien. of the USSR, Moscow
GARIBJANYAN A.T.	Byurakan Astrophys. Obs.
GASANALIZADE A.G.	Shemakha Astrophys. Obs.
GASPARIAN K.G.	Garny Space Astron. Lab., Erevan
GEDEVANISHVILI L.D.	Tbilisi State Univ.
GERSHBERG R.E.	Crimean Astrophys. Obs.,
GLAGOLEVSKIJ JU.V.	Special Astrophys. Obs., Zelenchuk
GLUSHKOV YU.I.	Alma-Ata Astrophys. Inst.
GLUSHNEVA I.N.	Sternberg Astron. Inst., Moscow
GOSACHINSKIJ I.V.	Special Astrophys. Obs., Zelenchuk
GRIGOREVSKIJ V.M.	Odessa State Univ.
GUBANOV A.G.	Astron. Obs., Leningrad State Univ.
GYUL'BUDAGIAN A.L.	Byurakan Astrophys. Obs.
GURZADIAN G.A.	Garny Space Astron. Lab., Erevan
ISKANDARIAN S.G.	Byurakan Astrophys. Obs.
KAAZIK A.A.	Tartu Inst. of Astrophys. and Atm. Phys
KADLA Z.I.	Pulkovo Obs., Leningrad
KALANDADZE N.B.	Abastumani Astrophys. Obs.
KALLOGLIAN A.T.	Byurakan Astrophys. Obs.

KANAEV I.I.	Pulkovo Obs., Leningrad
KAPLAN S.A.	Gorkij Inst. of Radiophys.
KARACHENTSEV I.D.	Special Astron. Obs., Zelenchuk
KARACHENTSEVA V.E.	Special Astrom. Obs., Zelenchuk
KAKHNIASHVILI N.V.	Acad. of Scien. of the Georgian SSR,Tbilisi
KARIMOVA D.K.	Sternberg Astron. Inst. Moscow
KARITSKAYA E.A.	Astron. Council, Acad. of Scien. of the USSR, Moscow
KASUMOV F.K.	Shemakha Astrophys. Obs.
KAZANASMAS M.S.	Odessa Astron. Obs.
KAZIMIRCHAK-POLONSKAYA H.I.	Leningrad Inst. of Theor. Astron.
KEVANISHVILI G.F.	Tbilisi State Univ.
KHACHIKIAN E.E.	Byurakan Astrophys. Obs.
KHARADZE E.K.	Abastumani Astrophys. Obs.
KHETSURIANI TS.S.	Abastumani Astrophys. Obs.
KHODYACHIKH M.F.	Astron. Obs., Khar'kov State Univ.
KHOLOPOV P.N.	Sternberg Astron. Inst., Moscow
KIPPER M.A.	Tartu Inst. of Astrophys. and Atm. Phys.
KOGOSHVILI N.G.	Abastumani Astrophys. Obs.
KOLESNIK I.G.	Goloseevo Astron. Obs., Kiev
KOLESNIK L.N.	Goloseevo Astron. Obs., Kiev
KOLKHIDASHVILI M.G.	Tbilisi State Univ.
KOLOTILOV E.A.	Sternberg Astron. Inst., Moscow
KOMAROV N.S.	Odessa Astron. Obs.
KOMBERG B.V.	Inst. for Space Research, Moscow
KOPYLOV I.M.	Special Astrophys. Obs., Zelenchuk
KOROTKEVICH G.V.	"Priroda", Editorial board, Moscow
KOZLOVA K.I.	Special Astrophys. Obs., Zelenchuk
KRASNOBABTSEV V.I.	Crimean Astrophys. Obs.
KUMAJGORODSKAYA R.N.	Special Astrophys. Obs., Zelenchuk
KUMSIASHVILI M.I.	Abastumani Astrophys. Obs.
KUMSISHVILI YA.I.	Abastumani Astrophys. Obs.
KURCHAKOV A.V.	Alma-Ata Astrophys. Inst.
KUVSHINOV V.M.	Crimean Astrophys. Obs.
KUUZIK I.KH.	Tartu Inst. of Astrophys. and Atm. Phys.
KUZNETSOV V.I.	Goloseevo Astron. Obs., Kiev
KVIRKVELIYA G.D.	Tbilisi State Univ.
LEBEDEV V.S.	Special Astrophys. Obs., Zelenchuk
LEVIN B.YU.	Astron. Council, Acad. of Scien. of the USSR, Moscow
LIPOVKA N.M.	Special Astrophys. Obs., Zelenchuk
LOMINADZE J.G.	Acad. of Scien. of the Georgian SSR, Tbilisi
LOZINSKAYA T.A.	Sternberg Astron. Inst., Moscow
LYUTYJ V.M.	Sternberg Astron. Inst., Southern Station, Crimea
LUUD L.S.	Tartu Inst. of Astrophys. and Atm. Phys.
LYUBIMKOV L.S.	Crimean Astrophys. Obs.
MAGALASHVILI N.L.	Abastumani Astrophys. Obs.
MAGNARADZE N.G.	Tbilisi State Univ.
MALUMIAN V.G.	Byurakan Astrophys. Obs.
MALYUTO V.D.	Tartu Inst. of Astrophys. and Atm. Phys.
MAROCHNIK L.S.	Rostov State Univ.
MARTYNOV D.YA.	Sternberg Astron. Inst., Moscow
MASSEVICH A.G.	Astron. Council, Acad. of Scien of the USSR,Moscow
MATVEYENKO L.I.	Inst. for Space Research, Moscow
MEDVEDEVA G.I.	Astron. Council, Acad. of Scien. of the USSR, Moscow
MEGRELISHVILI T.G.	Abastumani Astrophys. Obs.
MESTIASHVILI Z.D.	Abastumani Astrophys. Obs.
METREVELI M.D.	Abastumani Astrophys. Obs.

MIKHAILOV A.A.	Pulkovo Obs., Leningrad
MIKIRTUMOVA G.G.	Abastumani Astrophys. Obs.
MIRZOYAN L.V.	Byurakan Astrophys. Obs.
MOROZ V.I.	Inst. for Space Research, Moscow
MUSTEL E.R.	Astron. Council, Acad. of Scien. of the USSR, Moscow
NIKONOV V.B.	Crimean Astrophys. Obs.
NICOGHOSSIAN A.G.	Byurakan Astrophys. Obs.
NOVIKOV I.D.	Inst. for Space Research, Moscow
NOVRUZOVA KH.I.	Shemakha Astrophys. Obs.
OHANESYAN O.V.	Garny Space Astron. Lab., Erevan
ORLOV M.YA.	Goloseevo Astron. Obs., Kiev
OZERNOY L.M.	Lebedev Phys. Inst., Moscow
PAVLOVSKAYA E.D.	Sternberg Astron. Inst., Moscow
PETROSIAN M.B.	Byurakan Astrophys. Obs.
PETROV P.P.	Crimean Astrophys. Obs.
PETROVSKAYA I.V.	Astron. Obs. of Leningrad State Univ.
POLYAKOVA G.D.	Pulkovo Obs., Leningrad
POPOV V.S.	Pulkovo Obs., Leningrad
PRONIK I.I.	Crimean Astrophys. Obs.
PRONIK V.I.	Crimean Astrophys. Obs.
PUGACH A.F.	Goloseevo Astron. Obs., Kiev
PURTSKHVANIDZE A.	Tbilisi Pedagogical Inst.
PUSTYLNIK I.B.	Tartu Inst. of Astrophys. and Atm. Phys.
PUSTYLNIK A.S.	Special Astrophys. Obs., Zelenchuk
PYATUNINA T.B.	Special Astrophys. Obs., Zelenchuk
RACHKOVSKAYA T.M.	Crimean Astrophys. Obs.
RADLOVA L.N.	Research Inst. of Scien.-Techn. Information,Moscow
RAKHAMIMOV SH.J.	Shemakha Astrophys. Obs.
RAZIN V.A.	Gorkij Inst. of Radiophys.
ROMANOV YU.S.	Astron. Obs. of Odessa Univ.
RUSTAMOV YU.S.	Shemakha Astrophys. Obs.
RYABCHIKOVA T.A.	Astron. Council, Acad. of Scien. of the USSR, Moscow
SABASHVILI SH.A.	Tbilisi State Univ.
SAKHIBULIN N.A.	Kazan' Astron. Obs., Kazan' Univ.
SALUKVADZE G.N.	Abastumani Astrophys. Obs.
SANAMIAN V.A.	Byurakan Astrophys. Obs.
SAPAR A.A.	Tartu Inst. of Astrophys. and Atm. Phys.
SEIDOV Z.PH.	Shemakha Astrophys. Obs.
SHAKIR-ZADE A.A.	Shemakha Astrophys. Obs.
SHANDARIN S.F.	Inst. of Applied Mathematics, Moscow
SHANIN G.I.	Tashkent Astron. Inst.
SHATSOVA R.B.	Rostov Pedagogical Inst.
SHAVRINA A.V.	Goloseevo Astron. Obs., Kiev
SHCHERBANOVSKI A.L.	Special Astrophys. Obs., Zelenchuk
SHCHERBINA-SAMOJLOVA I.S.	Research Inst. of Scien. Techn. Inform., Moscow
SHITOV YU.P.	Lebedev Phys. Inst., Moscow
SHIUKASHVILI M.A.	Abastumani Astrophys. Obs.
SHKLOVSKY I.S.	Inst. for Space Research, Moscow
SHOLOMITSKIJ G.B.	Inst. for Space Research, Moscow
SHUL'MAN L.M.	Goloseevo Astron. Obs., Kiev
SITSKA A.E.	Tartu Inst. of Astrophys. and Atm. Phys.
SLYSH V.I.	Inst. for Space Research, Moscow
SMIRNOV A.S.	Inst. for Space Research, Moscow
SOLOV'OVA L.B.	Kazan' Astron. Obs.
SOMOV B.V.	Lebedev Phys. Inst., Moscow
STARIKOVA G.A.	Sternberg Astron. Inst., Moscow

STRAUME YA.K.	Baldone Radioastrophys. Obs., Riga
SUCHKOV A.A.	Rostov State Univ.
SULTANOV G.F.	Shemakha Astrophys. Obs.
SYROVOJ V.V.	Ural State Univ., Sverdlovsk
SYUNYAEV R.A.	Inst. for Space Research, Moscow
TARASHASHVILI I.	Tbilisi Polytechn. Inst.
TEREBIZH V.Yu.	Sternberg Astron. Inst., Southern Station, Crimea
TOKHTAS'EV S.S.	Dept. of Astron., Kazan' Univ.
TOROSHELIDZE T.I.	Abastumani Astrophys. Obs.
TOTOCHAVA A.G.	Abastumani Astrophys. Obs.
TOVMASSIAN H.M.	Byurakan Astrophys. Obs.
TROITSKY V.S.	Gorkij Radiophys. Inst.
TUTUKOV A.V.	Astron. Council, Acad. of Scien. of the USSR, Moscow
UNT V.A.	Tartu Inst. of Astrophys. and Atm. Phys.
VILKOVISKIJ E.YA.	Alma-Ata Astrophys. Inst.
VOROSHILOV V.I.	Goloseevo Astron. Obs., Kiev
VSEKHSVYATSKIJ S.K.	Kiev State Univ.
YAANISTE I.A.	Tartu Inst. of Astrophys. and Atm. Phys.
YAKOVLEVA V.A.	Leningrad State Univ.
YATSKIV YA.S.	Goloseevo Astron. Obs., Kiev
YUNGEL'SON L.R.	Astron. Council, Acad. of Scien. of the USSR, Moscow
YUSIFOV I.M.	Shemakha Astrophys. Obs.
ZAITSEVA G.V.	Sternberg Astron. Inst., Southern Station, Crimea
ZAKHAROVA P.E.	Ural State Univ., Sverdlovsk
ZASOV A.V.	Sternberg Astron. Inst., Moscow
ZEL'DOVICH YA. B.	Inst. of Applied Mathematics, Moscow
ZHILYAEV B.E.	Goloseevo Astron. Obs., Kiev
ZHUKOV G.V.	Kazan' Astron. Obs.
ZVEREVA A.M.	Crimean Astrophys. Obs.

VATICAN

COYNE G.V., S.J.	Vatican Obs.
TREANOR S.J.P.J.	Vatican Obs.

The Program of the Third European Astronomical Meeting
STARS AND GALAXIES FROM OBSERVATIONAL POINTS OF VIEW
(Tbilisi, 1-5 July 1975)

Tuesday, July, 1

Opening of the Meeting

Prof. E.K.Kharadze acting as a Chairman opens the Meeting
Address to the Third European Astronomical Meeting on behalf of the Government of the Georgian SSR by the Deputy Chairman of the Council of the Ministers O.E.Tcherkezia
Address by the General Secretary of the International Astronomical Union Prof. G.Contopoulos
Address by the Chairman of the Astronomical Council of the USSR Academy of Sciences, Vice-President of the IAU Prof. E.R.Mustel
Address by the Rector of Tbilisi State University Prof. D.I.Chkhikvishvili
Speech of the Chairman of the Scientific Organizing Committee and Local Organizing Committee, Vice-President of the Academy of Sciences of the Georgian SSR, Prof. E.K.Kharadze
Performance of Chamber ensemble

First Scientific Session. "STARS"

Chairman: Prof. A.Blaauw

E.K.Kharadze and R.A.Bartaya(U.S.S.R.)Some results of two-dimensional MK classification of stars

R.M.West (European Southern Observatory). Multidimensional automatic classification of stellar spectra; present state and future programs

G.A.Gurzadian (U.S.S.R.) Ultraviolet spectra of stars and nebulae ("ORION-2")

P.R.Wesselius, J.W.G.Aalders, T.S.van Albada, C.D.Andriesse, K.S.De Boer, J.Borgman, R.J. V.Duinen, J.Koornneef, S.R.Pottasch, J.P.Vader and C.C.Wu (The Netherlands). First results obtained with the Dutch UV instrument on-board ANS

Short communications by:

L.D.Beshenova, A.V.Kharitonov; N.Houk, N.I.Irvine I. and D.Rosenbush; K.A. van der Hucht ; K.Nandy, E.Kontizas; O.V.Ohanesyan

Wednesday. July, 2 — Second Scientific Session. "GALAXIES". (Lecture)

Chairman: Prof. E.K.Kharadze

V.A.Ambartsumian(U.S.S.R.)The role of the nuclear activity in the overall evolutionary processes in galaxies.

Third Scientific Session. "STARS"

Chairman: Prof. A.G.Massevich

K.Rudnicki (Poland). Certain statistical problems of the supernovae
E.R. Mustel(U.S.S.R.)Some problems of physics of supernovae
L.V.Mirzoyan(U.S.S.R.)Observational aspects in studying the early stages of the evolution of stars
A.A.Boyarchuk (U.S.S.R.) Chemical composition of stars

Short communications by:

O.Kh.Gusseinov, F.K.Kasumov, I.A.Yusifov; M.A.Lozinskaya; N.N.Chugaj; E.V.Ergma, A.V.Tutukov; Th.P.Snow; L.Crivellari, P.Selvelli; P.Petrov, A.Scherbakov; D.H.Morgan, K.Nandy, G.I.Thompson; E.Zelvanova; I.A.Aslanov, Yu.S.Rustamov, V.M.Khalilov, A.A.Shakir-zade

Fourth Scientific Session. "GALAXIES"

Chairman: Prof. G.Contopoulos

B.A.Vorontsov-Vel'yaminov(U.S.S.R.)Fragmentation of galaxies and the birth of dwarf satellites

Short communications by:

D.Alloin, P.Benvenuti, S.D'Odorico; A.Ardeberg; C.Balkowski, P.Chamaraux; E.K.Denisyuk; V.Ye.Karachentseva; V.I.Korchagin, L.S.Marochnik; K.Mattila; A.A.Suchkov, Yu.A.Shekinov; Th. Schmidt-Kaler; T.Jaakkola; L.Bottinelli, R.Duflot, L.Gouguenheim, J.Heidmann; C.Casini, J.Heidmann; D.Alloin, D.Pelat, A.Bijaoui

Fifth Scientific Session. "STARS"

Chairman: Prof. R.E.Gershberg

R.E.Gershberg and L.S.Luud(U.S.S.R.)Emission lines in stellar spectra:observations and their interpretation

A. Sapar(U.S.S.R.)Models of stellar atmospheres in comparison with observations

Short communications by:

D.Briot; Ju.V.Glagolevskij, K.I.Kozlova, V.S.Lebedev, N.S.Polosukhina; N.Heidmann, S.Dumont, R.N.Thomas; G.V. Coyne; I.M.Kopylov, R.N.Kumajgorodskaya, L.I.Sneshko, V.V.Sokolov, N.M. Chunakova, D.Kuroczkin; T.Markkanen; P.E.Nissen; J. I. K.Straume.

Thursday, July, 3

Sixth Scientific Session. "GALAXIES"

Chairman: Prof. P.G.Mezger

V.A.Ambartsumian, L.V.Mirzoyan, M.B.Petrosian, R.K.Shahbazian (U.S.S.R.) Compact groups of compact galaxies

N.G.Kogoshvili (U.S.S.R.) Some results of computer investigations of the Catalogue of bright galaxies compiled on magnetic tape. 1. On Probable Bright Compact galaxies

T.M.Borchkhadze (U.S.S.R.) Some results of computer investigations of the Catalogue of bright galaxies compiled on magnetic tape. 2. Studies of apparent directions of Galaxy spiral arms' coiling

J.Heidmann(France).Morphology in tight extragalactic systems

M.S.Longair (U.K.) Radio and X-ray observations of clusters of galaxies

P.J.Hargrave and M.S.Longair (U.K.). Observations of extragalactic radio sources with the Cambridge 5-km telescope

Short communications by:

M.A.Arakelyan; R.G.Strom, J.P.Hamaker, A.G.Willis; B.P.Artamonov; F.Börngen, A.T.Kalloghlian

Short communications by:

Chairman (after the break): Prof. G.Courtes

M.Kalinkov, K.Stavrev, II.Kaneva, V.Dermenjiev; M.Kalinkov, K.Stavrev, II.Kaneva, V.Dermenjiev and K.Rudnicki; V.A.Sanamian, Gopal-Krishna; H.M.Tovmassian, R.Sramek; I.I. Pronik, L.P.Metik; M.F.Duval-Cheriguene; J.Boulesteix; I.Pauliny-Toth, E.Preuss, A.Witzel;L.Bottinelli, R.Duflot, L.Gouguenheim

Seventh Scientific Session. "GALAXIES".(Lecture)

Chairman: Prof. G.Courtés

I.S.Shklovsky (USSR). Galactic X-ray sources

Friday, July, 4

Eighth Scientific Session. "GALAXIES"

Chairman: Prof. H.Oleak

A.B.Blaauw, C.D.Garmany (The Netherlands). Space distribution and kinematics of F-stars

W.Iwanowska (Poland). Kinematics and structure of subsystems of different ages and contents

D.K.Karimova and E.D.Pavlovskaya(U.S.S.R.)The kinematical properties of some star subsystems of Galaxy

P.G.Mezger and L.F.Smith (F.R.G.). Formation of O-Stars and the rate of star formation in the Galaxy

P.O.Lindblad (Sweden). Birth-places of stars and galactic shocks

G.Contopoulos and P.Grosbøl (Greece). The past positions of the spiral arms of our Galaxy

Short communications by:

G.Courtes,P.Cruvellier,J.M.Deharveng,J.Maucherat,G.Monnet,M.Pellet,M.Simien;P.Benvenuti,S.D Odorico, P.Vettolani; Yu.I.Glushkov, E.K.Denisyuk, Z.V. Koryagina; J.Einasto, M.Jôeveer, A.Kaasik; R.Wielen; S.Collin-Souffrin, M.Joly; M.D.Metreveli; G.R.Ivanov, N.S.Nikolov; Yu.N.Efremov; Th.Schmidt-Kaler; N.B.Kalandadze, L.N.Kolesnik, V.I.Voroshilov

Ninth Scientific Session. "GALAXIES"

Chairman: Prof. P.O.Lindblad

F.Bertola and G.Di Tullio (Italy). Missing mass in galaxies

J.Einasto, M.Jôeveer, A.Kaasik, J.Vennik(U.S.S.R.)The missing mass around galaxies

I.D.Karachentsev(U.S.S.R.)Hidden mass-problem for double galaxies

H.Oleak (G.D.R.). Is there any missing mass?

Short communications by:

J.Materne, G.A.Tammann; W. Gliese; L.M.Ozernoy; G.Paál; R.B.Tully, J.R.Fisher; B.I.Fesenko; T.Jaakkola

Concluding remarks on the missing mass

R.D.Davies; L.M.Ozernoy; J.Einasto; B.I.Fesenko

Tenth Scientific Session. "GALAXIES"

Chairman: Prof. W.Iwanowska

R.D.Davies (U.K.). Gas and stars in the outer regions of galaxies

F.Ochsenbein (France). The present status of data available at the Stellar Data Center

D.R.K.Brownrigg (U.K.). Observation of an Sc type spiral in a computer model of a Galaxy

E.A. Dibay (USSR). The nuclei of Seyfert galaxies

Short communications by :

J.Guibert, F.Viallefond; R.M.West; O.B.Dluzhnevskaya, A.E.Piskunov; M.V.Dolidze; I.V.Petrovskaya; W.Weiss

Saturday, July, 5

Eleventh Scientific Session. "GALAXIES"

Chairman: Prof. A.A.Mikhailov

Short communication by:

Aa.Sandqvist, P.O.Lindblad, K.P.Lindroos

Ya.B.Zel'dovich(U.S.S.R.)Formation of galaxies in Friedmanian Universe (Lecture)

Closing Ceremony

Prof. W.Iwanowska (Poland). (Speech on behalf of the IAU)

Prof. A.Blaauw (The Netherlands). (Speech on behalf of foreign participants)

Two facultative communications:

Chairman: Prof. Ya. B.Zel'dovich

P.N.Kholopov(U.S.S.R.).On the nature of stellar associations

S.K.Vsekhsvyatskij(U.S.S.R.).Cometary matter in interstellar medium and probably in other galaxies

TUESDAY, JULY, 1

FIRST SCIENTIFIC SESSION. "STARS"
CHAIRMAN: PROF. A.BLAAUW

SOME RESULTS OF TWO-DIMENSIONAL MK CLASSIFICATION OF STARS

E.K.KHARADZE AND R.A.BARTAYA
ABASTUMANI ASTROPHYSICAL OBSERVATORY, U.S.S.R.
PRESENTED BY E.K.KHARADZE

НЕКОТОРЫЕ РЕЗУЛЬТАТЫ ДВУМЕРНОЙ МК-КЛАССИФИКАЦИИ ЗВЕЗД

РЕЗЮМЕ

На основе материала Абастуманского каталога, содержащего данные двумерной МК-классификации для звезд в площадках Каптейна №№ 2-43, построены и исследованы функции плотности и светимости звезд, а также рассмотрены некоторые вопросы пространственного распределения выявленных в ходе классификации Ap, Am звезд.
Отмечается рост процентного содержания гигантов на высоких галактических широтах. Звезды высокой светимости типов F, G, K и почти типа M, не обнаруживают тенденцию столь сильной концентрации к галактической плоскости, как это считалось до сих пор. Вид функции светимости для средних и высоких галактических широт явно отличается от того, что мы имеем для окрестности Солнца. Звездам Ap и Am не свойственна столь резкая концентрация к галактической плоскости, как это характерно для нормальных звезд соответствующих спектральных интервалов..

ABSTRACT

Abastumani Catalogue containing the data on two-dimensional MK- classification for stars in KA Nos 2-43 provided material to plot density and luminosity functions and to discuss some questions of space distribution of Ap, Am stars.
High galactic latitudes are marked with the increase of giant percentage. High luminosity stars like F, G, K and partly M do not show tendency of such a strong concentration towards the galactic plane as it has been recognized hitherto. Luminosity functions for the intermediate and high galactic latitudes are obviously different from what we have for the solar vicinity. Ap, Am stars are not distinguished with that high concentration towards the galactic plane as normal stars of the corresponding intervals.

The current development of researches of astrophysical and stellar astronomy interests based on spectral classification of stars is closely related to the use of moderately dispersed objective-prism spectra classified after the widely known Morgan-Keenan (MK) two-dimensional system.

Due to a continuous perfecting of that system and general rise of its accuracy, one has started recently to complement MK classification by discriminating those of the classified spectra, which are characterized by certain peculiarities.

That, indeed, enriches significantly the spectral data.

The well-known Michigan work on the reclassification of Henry Draper Southern Stars Catalogue (Houk, Cowley 1973) is an illustration to the above-said.

In Abastumani, where our experience has demonstrated the effectiveness of the meniscus telescope in low and moderate dispersion stellar spectroscopy, we have realized an extensive classification of about one hundred thousand stars.

Leaving aside the details, I allow myself to state that the accuracy of our data is close to the Michigan level and, what is very essential for what follows, these data are quite homogeneous, uniform in accuracy and in the penetration rate into the space, the limit being higher than that of Henry Draper Catalogue, namely, close to the 12th magnitude.

That must be considered as very important for the statistics if we expect reliable results from the latter; only that can lead us to authentic conclusions of general character in stellar astronomy.

I would like to submit for consideration two results of general importance based on the discussion of our classification data.

One of them concerns rather an old problem of luminosity function, and the other - a relatively new problem of peculiar star frequencies and their space distribution.

In spite of the great importance of luminosity function, it cannot be stated that its shape and character are firmly established.

One of the reasons of such a failure lies in the fact that scarcely homogeneous data have ever been used for plotting the function; the second not less important reason is that the luminosity classes have been determined directly, with application of rough methods of dividing stars merely into giant and dwarf groups; without the large scale using the second parameter of classification.

A comment must be made for better understanding of this statement.

The luminosity effect in spectra, i.e. the classification criteria very weak by themselves gradually disappear while advancing from giants to dwarfs (the late type stars are being considered).

It follows then that if the quality of spectra even slightly drops or the intensities in them deviate from the optimal one, the luminosity effect might completely disappear earlier than it should by its nature and the classificator attaches the star, say, of the luminosity class III to the class V, and the subgiant of class IV, for which the effect is much fainter by itself, to the same class V.

Consequently, the percentage of the fifth class is presented as being higher than it is in reality.

Something more could be said: when studying statistical problems, frequently, one considers rather arbitrarily the stars being at the limit of brightness and not possessing the determined luminosity classes as belonging to the fifth class (Upgren, 1962; 1963a).

Undoubtedly, that introduces considerable errors into the plotting of luminosity functions.

May I now draw your attention to stars of F-G5 types. As is known, in this spectral interval the luminosity effect on the spectra is very faint. Therefore, usually, one does not group these stars even merely into giant and dwarf groups and accounting for the presence of the so-called Hertzschprung gap, i.e. lack of giants in that spectral interval, one admits that we are dealing with dwarfs only (Upgren, 1963b).

Let us consider now what we do have in reality, when we try to make a conclusion on a much firmer basis of thorough two-dimensional classification. In this particular case, our classification comprises about eleven thousand stars of the Kapteyn Selected Areas Nos. from 2 to 43 (Kharadze, Bartaya, 1973) situated at different galactic latitudes and grouped into 5 groups according to the latitudes from that equal to zero to one reaching 70 degrees.

I invite you to look at Fig.1 plotted for F stars. Here we have the luminosity classes as abscissae and the visible numbers of stars (percentage) as ordinates for five groups differing in the latitudes.

At the low latitudes the apparent percentage of the III class stars is appreciably lower than that of the V class stars. The Hertzschprung effect is displayed here. But, it is important that at higher latitudes that percentage noticeably approaches that of the V class. That is important to notice.

Worth of attention is the next Fig.2 plotted for K stars. Here mainly the III class stars are present. That happens because the difference in absolute magnitudes for the III class and V class K stars is particularly great and correspondingly the same is for the penetration into the space; for the V class stars it is significantly limited. Their percentage is equal to the percentage of the I and II class stars.

At the same time a slight tendency to an increase of the III class stars with the increasing latitude is obvious. That also deserves our attention.

So we become convinced that in stellar statistics it is not competent to consider F and G type stars as the main sequence stars so freely, at random, without having determined their luminosity classes.

This statement becomes more valid when the polar regions are considered, where our data have demonstrated that the III and V class F and G type stars practically occur in equal proportions. The observed number of dwarfs among faint K and M stars could be increased indeed, but it is impossible to say to what extent they prevail, if the luminosity class determinations are not at our disposal.

Now a question arises: what may be a reason of a change of the percentage of the III and V class stars when we advance to the high latitude stars?

Perhaps, no appreciable decrease of the apparent number of giants occurs at the high latitudes, because of absence there of interstellar absorption. But at the same time neither could we expect a decrease of dwarf amount, since their distribution is considered as a uniform one at least in the solar vicinity.

Fig. 3 shows the apparent distribution. The number of the III class stars is nearly equal at all the latitudes. Meanwhile, the number of the V class stars sharply drops at the high latitudes.

The same may be seen for other type stars. That means the following: the visible number of the V class stars diminishes with the increasing latitude at a higher rate than it is in the case with the III class stars. That is the reason of the change of percentage of giants and dwarfs depending on the latitude.

Anyhow, the picture shows that a certain galactic concentration is characteristic for dwarf stars too.

May I remind that Weistrop has found in his recent work, (Weistrop, 1972) based on the U, B, V photometry that red dwarfs are closely concentrated at the galactic plane.

As far as we know, this problem is being studied for the Southern Sky in European Southern Observatory.

The observations which are in progress in Abastumani might be applied to the study, in the same line, of the Northern Sky.

The next diagram in Fig.4 shows that high luminosity stars of types F, G, K and almost M are not so strongly concentrated to the galactic plane as it has been thought hitherto.

The diagram in Fig.5 shows that the space distribution of K supergiants is characterized by groupings discretely at different distances from the galactic plane.

Here we could emphasize once more the importance, say - the urgent necessity - to base the conclusions on the knowledge of the luminosity classes determined as the second parameter of the MK classification.

Now I invite you to look at some diagrams, the first of which in Fig.6 represents general luminosity function, plotted for the stars situated quite close to the galactic equator.

The open circles correspond to Van Rhijn standard luminosity function calculated for the solar vicinity (Van Rhijn,1965).

As for certain deviations for the stars with M = +3 and fainter, the recent determinations by Luyten (1968) and Wanner (1972) are in good agreement with ours.

The next diagram in Fig.7 shows the luminosity functions separately for stars at different galactic latitudes: 1 - in the galactic equator, 5 - the most distant from it.

The curves at the low latitudes differ insignificantly from each other, if we neglect the regular decrease of highly luminous stars here. On the contrary, there is an evident difference at the high latitudes especially for the stars in the interval of absolute magnitudes between minus 1 and plus 2.

It may be assumed that this effect is the consequence of the mixture of the II population stars. We could remember that Upgren (1963b) has recently pointed out a similar effect.

No doubt, the structural complexity of our Galaxy implies the existence of a lot of versions of luminosity function curves, different in different directions or depths of the Galaxy. To study this problem more completely, one should go on with the mass determinations of the accurate luminosity classes of a majority of stars advancing to fainter stars as much as possible.

Now, a few words on the second problem involved in our paper.

The thorough MK classification of stellar spectra obtained at good seeing and good guiding conditions gives the classificator a possibility of revealing peculiar Ap and Am stars.

Nearly 200 peculiar stars have been revealed in Abastumani only in the Kapteyn Areas under consideration (Bartaya, Kharadze, 1970 ; Bartaya, 1974). The homogeneity of the data permits us to make some statistics. So much that there exist very poor data of such kind.

We have come to the following conclusions:

1) Although the distribution of these stars in the sky is not uniform, they do not show any tendency to gather up into stellar clusters and associations;

2) Having considered the mean percentage of Ap and Am stars in the whole quantities of stars in the intervals of B8-AO and A2-A5, we have found that the Am stars significantly prevail over the Ap stars;

3) Ap stars occur mainly at the galactic latitudes less than 20 degrees; but Am stars - up to 60 degrees.

Let us demonstrate now the diagram in Fig.8 where the percentage of Ap, Am stars increases with the latitude.

The galactic concentration is not so sharply characteristic for Ap, Am stars as it is for the normal B8-A5 stars (Fig.9).

Fig. 10 where the distribution of stars is considered along the Z-axis confirms this last statement. We meet Ap stars at the distances not farther than 200 parsecs from the galactic plane; Am stars are met at 400 ps.

These facts may have some far going consequences regarding the age and evolutionary development of Ap, Am stars.

In the association Orion I c, which is 10^6 years old, Smith (1972) has recently revealed several Am stars among A type stars on the evolutionary track before the main sequence. If so, remembering the presence of Am stars in older stellar clusters, it turns out to be that among Am stars we have representatives of different subsystems. That gives some source for speculations for cosmogonists or for people occupied with evolutionary research.

We are going on with our work on MK classifications of a bulk of stars and the introduction of the third parameter (that is the metal abundance criterion). Dr West (1970, 1972) has shown that it is quite accessible to determine it quantitatively using Abastumani spectra but that is the topic of his own talk.

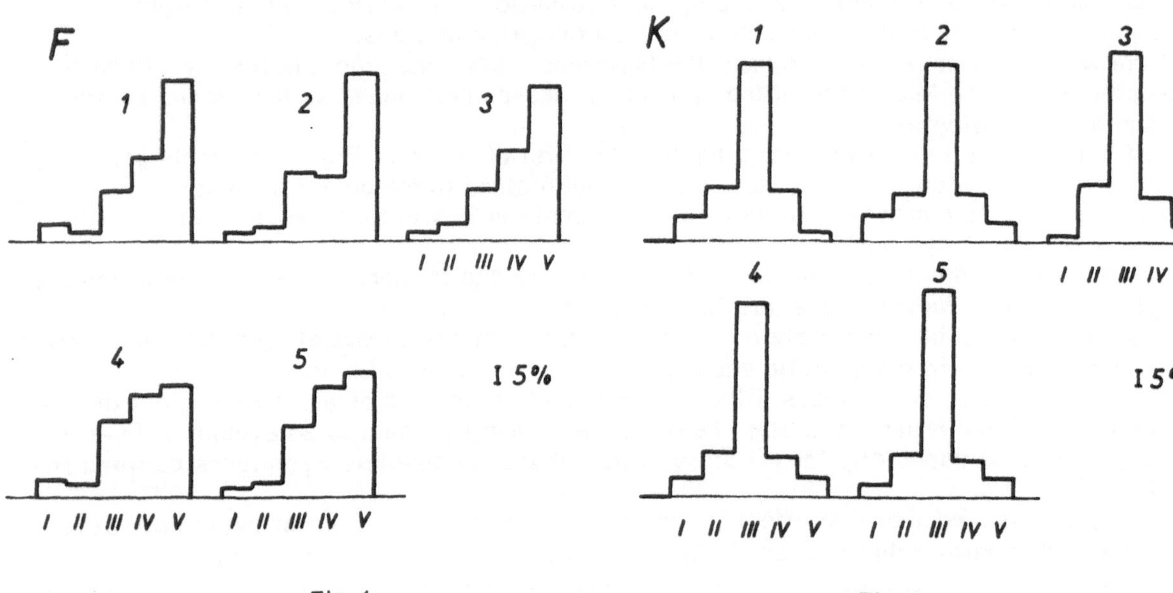

Fig. 1.

Fig. 2.

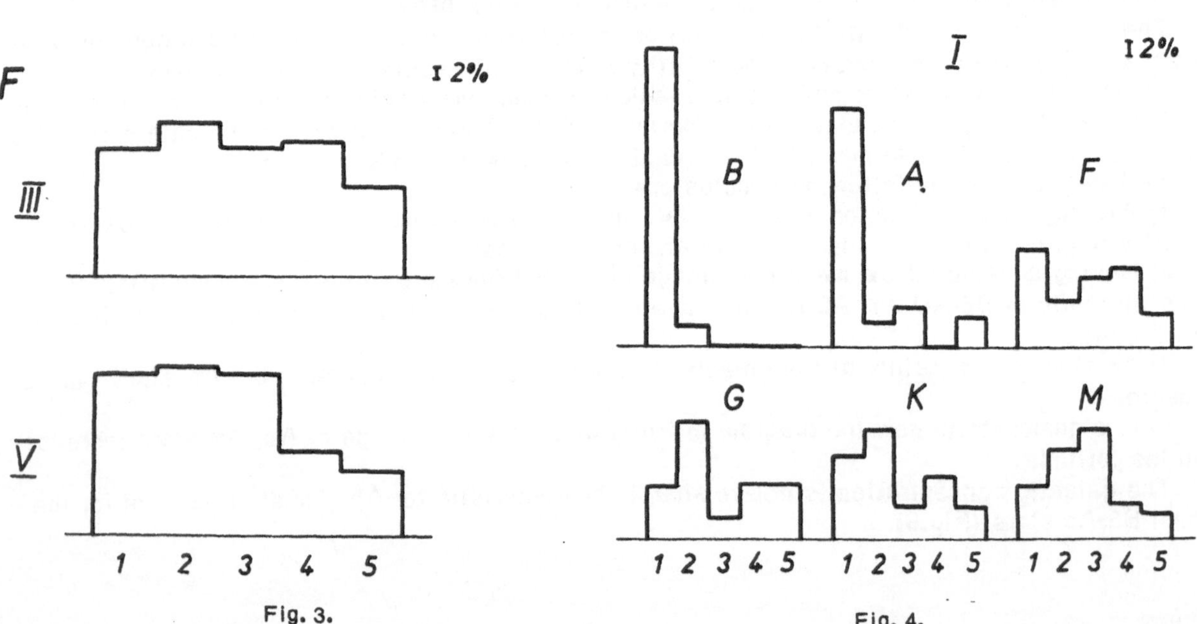

Fig. 3.

Fig. 4.

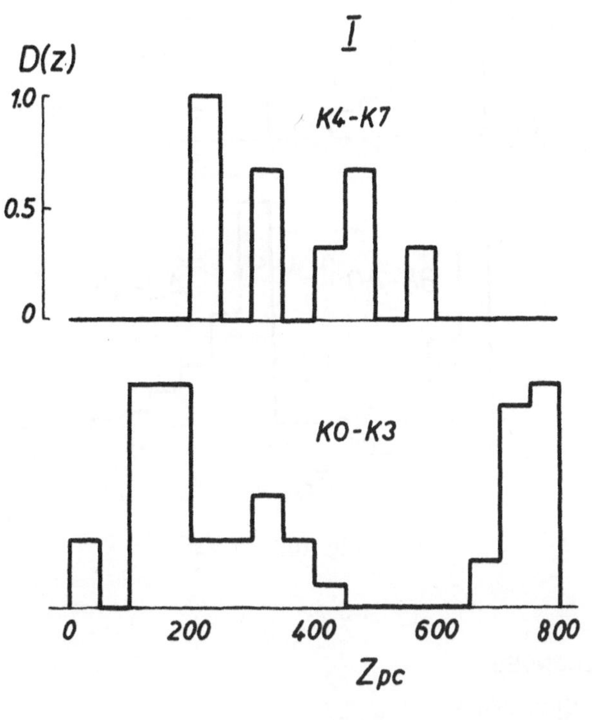

Fig.5.

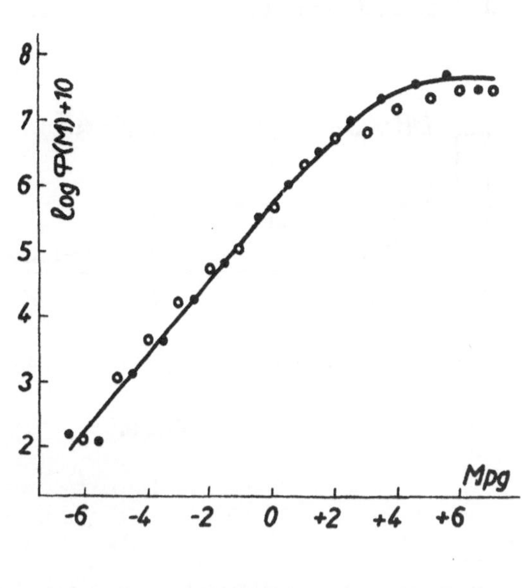

Fig.6.

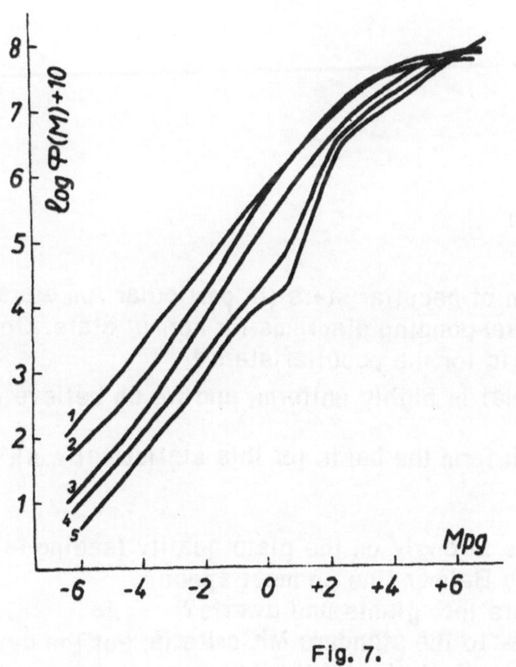

Fig. 7.

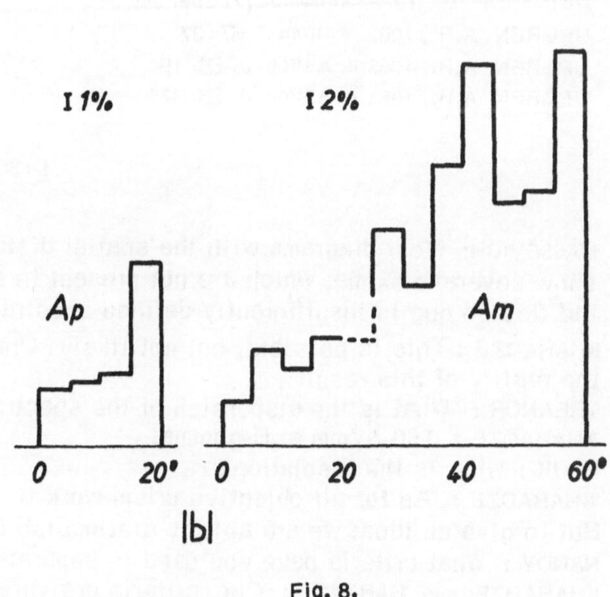

Fig. 8.

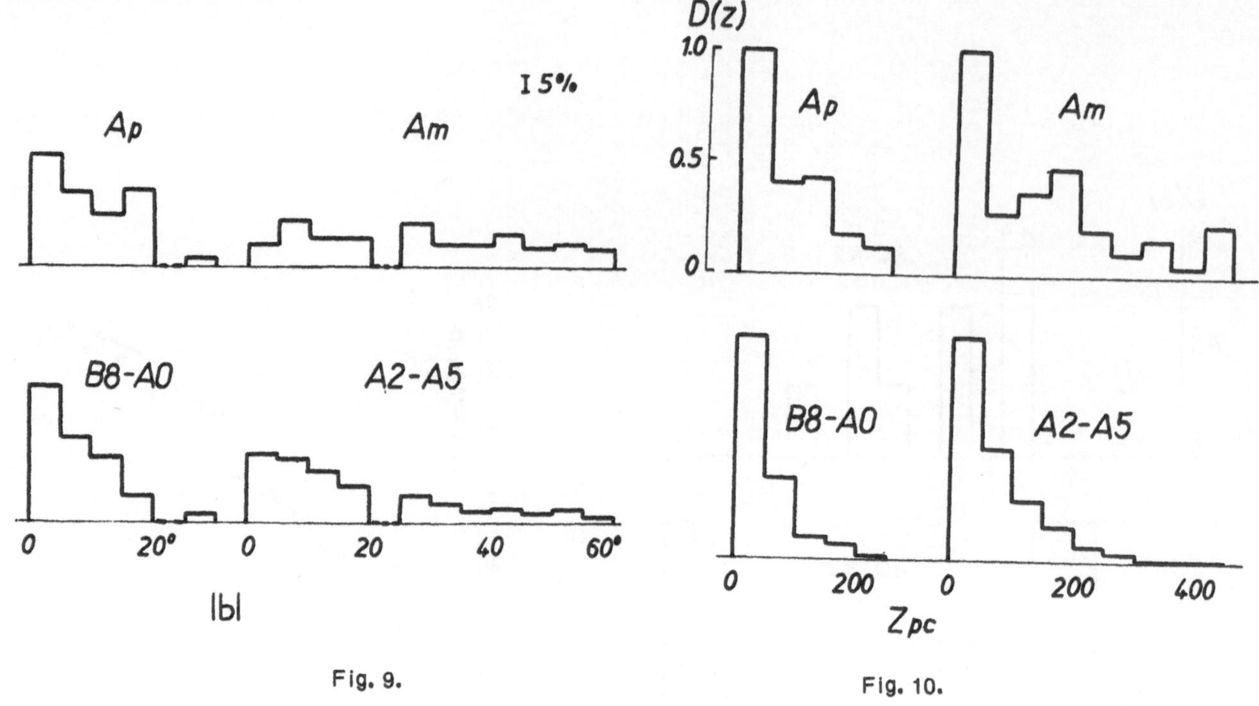

Fig. 9.

Fig. 10.

REFERENCES

BARTAYA R.A., KHARADZE E.K., 1970, Soviet Astron. Circ. 583, 5

BARTAYA R.A., 1974, Soviet Astron. Circ. N 845, 3

HOUK N., COWLEY A., 1973, Spectral Classification and Multicolour Photometry. IAU Symp. N 50, 70

KHARADZE E.K., BARTAYA R.A., 1973, Spectral Classification and Multicolour Photometry. IAU Symp.N 50,91

LUYTEN W., 1968, Mon. Not. R. Astr. Soc., 139, 221

SMITH M., 1972, Astrophys. J. 175,765

VAN RHIJN P.J., 1965, Galactic Structure, ed.Blaauw and M.Schmidt, 27

WANNER J., 1972, Mon. Not, Astr. Soc., 155, 463

WEST R.M., 1970, Bull. Abast. Astrophys. Obs. 39, 29

WEST R.M., 1972, Bull. Abast. Astrophys. Obs. 43, 109

WEISTROP D., 1972, Astron.J. 77, 366; 849

UPGREN A.R., 1962, Astron. J. 67, 37

UPGREN A.R., 1963 a, Astron. J. 68, 194

UPGREN A.R., 1963 b, Astron. J. 68, 475

DISCUSSION

MASSEVICH: Your diagrams with the spatial distribution of peculiar stars (in particular Am stars) show several maxima, which are not present in the corresponding diagrams for normal stars. Could this be due to insufficiently defined spectral criteria for the peculiar stars?

KHARADZE : This is possible, but not likely. Our material is highly uniform, and we do believe in the reality of this result.

TREANOR : What is the dispersion of the spectra which form the basis for this statistical work?

KHARADZE : 166 Å/mm at H-gamma.

HOUK : What is the resolution?

KHARADZE : As for all objective prism work it depends strongly on the plate quality (seeing etc.) But to give an idea: we are able to distinguish the 13th Balmer line on most spectra.

NANDY : What criteria have you used to separate M stars into giants and dwarfs?

KHARADZE and BARTAYA : Our criteria are very similar to the standard MK criteria, but the details may be found in our previous papers, published in the Bulletins of the Abastumani Astrophysical Observatory.

ROMANOV : Is your dispersion sufficiently high to well distinguish between Ap and Am stars?

KHARADZE : Yes, we think it is.

MULTIDIMENSIONAL AUTOMATIC CLASSIFICATION OF STELLAR SPECTRA – PRESENT STATE AND FUTURE PROGRAMS

R.M.WEST
ESO SKY ATLAS LABORATORY

МНОГОМЕРНАЯ АВТОМАТИЧЕСКАЯ КЛАССИФИКАЦИЯ ЗВЕЗДНЫХ СПЕКТРОВ – СОВРЕМЕННОЕ СОСТОЯНИЕ И ПРОГРАММА БУДУЩЕГО

РЕЗЮМЕ

Рассмотрено состояние автоматической классификации звездных спектров, в частности,-спектров, получаемых с предобъективными призмами. Недавний опыт применения двухкоординатного растрового сканирования означает шаг вперед в реализации автоматических спектральных обозрений. Опытная программа инициирована Европейской Южной Обсерваторией.

ABSTRACT

The present state of automatic classification of stellar spectra – in particular objective prism spectra – is reviewed. Recent experience with two-coordinate raster scanning implies an important step forward for the realization of automatic spectral surveys. A pilot project has been initiated at the European Southern Observatory.

1. Introduction. The concept of automatic classification of stellar spectra was first discussed in some detail at the IAU Symposium No. 50 in 1971. Progress in microphotometer-computer classification was reported by various authors and a summary of the general problems connected with the realization of automatic classification of objective prism spectra was given by the present author (West, 1973) Attention to the possible introduction of automatic means in the field of spectral classification was given by Commission 45 at the 1973 IAU General Assembly in Sydney, cf. the President's report (Westerlund,1973) and the resolution about "Priorities and Expected Developments" adopted by the Commission.

During the past four years, most of the effort in this field has been directed towards efficient software for the reduction of digitized spectra, which have been registered by means of conventional one-dimensional slit microphotometers. Many of the computer programs were discussed during a meeting on "Image Processing Techniques in Astronomy" held in Utrecht, the Netherlands in March 1975. Details may be found in the Proceedings of that meeting (de Jager and Nieuwenhuijzen, 1975) and in the summary given by Nieuwenhuijzen(1975).

New software has been developed for the reduction of high-dispersion (slit) spectra as well as objective prism spectra. It comprises noise-smoothing by truncation in the power (Fourier-transformed) spectrum, calibration to intensity (e.g. Ardeberg and Virdefors, 1972), classification of objective prism spectra (Diaz-Santanilla, 1972) and partially automated curve-of-growth analysis (Packett et al., 1973). Progress may also be noted in the computation of synthetic spectra at various levels of resolution, even for late-type stars (cf. Bell (1973) and Bell and Gustafsson (1975)), and good accuracy has been achieved in automated radial velocity determinations (Martini and Perinotto (1973), Weiss (1972)).

The results obtained by the above cited authors and others working in this field have greatly facilitated routine reductions of stellar spectra. This is particularly true for high-dispersion slit spectra, whereas the automated reduction of objective prism spectra still has to gain momentum in order to prove convincingly its usefulness in astronomical research.

2. Automatic classification of objective prism spectra. It is not the intention of this paper to discuss the reduction of high-dispersion spectra from digital registrations involving one spectrum at a time for which (more or less coarse) spectral parameters are already known. The first part of the analysis of a high-dispersion spectrum mainly consists in a straightforward calculation of equivalent widths which then form the basis for the interpretation of the spectrum in terms of stellar parameters, including abundances. Contrarily, a completely automatic classification of low-dispersion objective prism spectra presupposes no knowledge about the spectra to be classified and therefore

presents a number of basic problems which are not encountered with slit spectra.

Although there have been important advances in the reduction of digital spectral registrations, as described above, only a limited effort at a few observatories has centered on automatic classification during the recent years. This is probably not so much due to lack of interest as because of the complexity of the problem. However, a program, the aim of which is to achieve a completely automatic classification of objective prism spectra, has now been started at the European Southern Observatory (ESO) and will be discussed in some detail below. But let us first consider, in general terms, the feasibility of automatic classification, lest it is believed that it is not worth the effort.

The value of an automatic classification scheme in astronomical research must be judged in terms of the *quality* (i.e. the accuracy) of the classification and the *quantity* of spectra that can be classified within a reasonable time, as compared with conventional visual classification.

It has repeatedly been demonstrated that the application of quantitative criteria to spectral classification at intermediate to low dispersions permits an increase in classification accuracy above that of visual classification relying on subjective criteria ("this line is barely visible", "that line is somewhat stronger than this line", etc.). This is most obvious for the second classification parameter, the luminosity. The visual estimation of a third parameter, the "metallicity", is normally only possible in extreme cases. In particular, criteria that depend on more than one parameter (e.g. the CN-bands in late-type stars) can only be fully exploited when the effects of the parameters can be well separated, i.e. quantitatively.

The density resolution of the human eye varies with density, and is around 0.02 – 0.03 D (ASA diffuse) at the normal density levels for optimally exposed objective prism spectra. The practical density resolution for most microphotometers is 0.01 - 0.02 D. Due to the capability of the eye to disregard disturbing factors like plate faults and graininess, it has sometimes been stated that the eye is superior to the machine in recognizing very faint spectral features, and that accordingly the determination of the absence/presence of a faint line is more consistent in visual classification than in automated work. This is true when a wrong choice of slit width and step length for the microphotometer is made. Moreover, plate faults in a spectrum may create false lines if the spectrum is scanned with a one-dimensional slit, covering most of the width. However, the problem is overcome by two-dimensional scanning with a square aperture of optimal size (taking the grain noise and the resolution of the emulsion into account). -This method permits the elimination of those areas of the spectrum that are influenced by plate faults and to establish the reality of even very faint spectral features, provided they extend across most of the width of the spectrum. Moreover, the differential accuracy of a microphotometer is retained at high densities, where the human eye fails to notice weak features.

The speed in visual classification is roughly 1 spectrum/min for a trained classifier, strongly depending on the classification system and the number of peculiarities it includes. The ultimate speed of a completely automated system was estimated at 1 spectrum/sec. by West (1973). Recent experience shows that whereas this speed may still be achieved with future machines (possibly within the next decade), the presently available machines in connection with minicomputers will probably not be much faster than 1 spectrum/20 sec. Nevertheless, the endurance of the machine, when compared to the human classifier, still ensures that a vastly larger number of spectra may be classified automatically per unit time over a reasonably long period than visually, already with present-day hardware.

On the basis of past experience, and with a strong belief in the astronomical value of automated spectral surveys, a program has been started at the European Southern Observatory. The methods that are employed do not pretend to be radically different from those utilized at other observatories, but this project is, to the best of our knowledge, the first time the whole complex of fully automated classification of objective prism spectra is being attacked on a broad front.

3. The ESO project. In order to perform automatic classification of objective prism spectra, three main components are necessary. These are, on the hardware side, a *telescope* equipped with an *objective prism* and a *measuring machine* for scanning the spectra, and thirdly the *software* for the registration and classification of the spectra.

The hardware is available within the European Southern Observatory in the form of a 1 m Schmidt telescope on La Silla, Chile, and an Optronics S-3000 14" x 14" two-coordinate computer controlled measuring machine at the ESO Sky Atlas Laboratory in Geneva. Much of the classification software has now been written for the ESO-HP 2100 computer system at the ESO-TP Division Geneva. In what follows, details are given about a project that aims at an automated spectral survey of selected areas of the Southern Sky. In view of the complexity of the involved problems it is expected that the automatic classification system will be fully operational about a year from now.

3.1 The telescope. The ESO 1 m Schmidt telescope has been described by West (1974a). It is presently being used for direct surveys of the Southern Sky on 11a-0 and 098-04 plates (the ESO (B) and the ESO (R) Surveys of the Southern Sky). The telescope is equipped with a 4° UBK 7 objective prism which gives a dispersion of 450 Å/mm at H_{γ}. Test plates with this prism are very satisfactory (cf. Fig. 1; 30 min, widening 0.3mm) and the limiting magnitude of 11a-0 plates for 1 hour exposure, 0.2 mm widening is not far from 15^{m}. It is expected that the telescope will be used for limited spectral surveys of certain areas of the southern sky from 1976, and possibly, if the preliminary results are successful enough to justify the more extensive use of the telescope for this purpose, a survey of a larger part of the sky visible from La Silla (cf. West, 1974 b).

3.2 The measuring machine. The Optronics S-3000 two-dimensional measuring machine was installed at the ESO Sky Atlas Laboratory in Geneva late 1972. The machine and its performance in measurements of direct Schmidt plates has been discussed by West (1975). It now regularly measures ESO Schmidt plates and produces accurate positions and magnitudes for stellar images ($\pm$ 0.8 μm in X and Y, $\pm 0^{m}.08$ from the radius and the profile) at a rate of about 1 star/4 sec. The S-3000 is controlled by an 8K ALPHA-16 minicomputer and density data (0-4 D) are stored on magtape at 2 kHz rate. The magtape is then taken to the ESO HP-2100 computer system where the data are reduced.

In order to increase the efficiency, it has been decided to establish a direct 0.5 Mbit/sec data link between the S-3000 and the HP-2100 system (distance 50 m) and to install a 32 K HP 2100 instead of the ALPHA-16 computer. At the same time the data acquisition rate will be increased to at least 5 kHz. The new hardware configuration is shown in Fig. 2. The present program for the reduction of scans of direct images has been conceived in such a way that when this hardware change takes place (in October 1975), the software for the initial data reduction (detecting the stellar images) will be moved to the HP 2100 control computer. At the same time, a new and more efficient control program is being doveloped that will allow "active" scanning, i.e. guiding the scan aperture in a strategic way, in response to what is detected on the plate. This feature will be of particular importance for the scanning of objective prism spectra, since it allows an efficient economizing of the scan time by reducing the time spent on idling over the clear areas on the plate.

The question of whether the classification program will reside in the HP-2100 that controls the S-3000, or in the ESO HP-2100 system has yet to be investigated. However, with the very high capacity of the data link, it may be more economical to perform a preliminary reduction of the spectral registration in the control HP-2100 computer and free it for the subsequent extensive classification calculations by transferring the entire density arrays for the spectra to the large ESO HP-2100 system, e.g. equivalent to the method in direct scanning.

3.3 The software. A scheme for automatic classification of objective prism spectra was proposed by West (1973). It consists of five separate operations:

1. Registration of (a part of) the photographic plate
2. Recognition of the spectra
3. Transformation (e.g. to intensities)
4. Criteria evaluation
5. Classification

It was decided to adopt the general structure of this scheme for the ESO project.

At the time of writing (June 1975) functions 1), 3) and 4) are available in the ESO software, function 2) partially available and 5) has not yet been specified in detail. The system of interrelated programs (the software) is known under the name CLASS.

Since spectra of standard stars are not yet available from the ESO Schmidt telescope, a series of spectral plates (unbacked IIa-0), taken with the Abastumani 70 cm Menisc telescope has so far served as the basis for the software development. Spectra of late-type standard stars were available from an earlier investigation (West, 1972) and new plates were taken in October 1974 of a field at (2h, +13°), in which about 60 stars (B3-GO, 5^{m}- 11^{m}) were observed photoelectrically in the uvby-β system with the 50 cm ESO telescope on La Silla.

The dispersion of the Abastumani spectra, 166 Å/mm at Hγ, is of course much higher than that of the ESO spectra, 450 Å/mm at Hγ, for which the system will later be used but the software has been written in such a way that the preliminary reduction - until the point where the specific classification criteria are employed - may be easily applied to any dispersion between about 100-600 Å/mm.

The length of an Abastumani spectrum (5200 - 3600 Å/) is 10.5 mm; in most cases the width is 0.2 mm. A registration consists of four scans of the adjacent background (two on each side) and five scans along the spectrum itself. As optimal parameters we have found a step length of 2μm (i.e. 5250 points along one scan line) with a square aperture of 15 x 15μm². At the present 2 kHz data rate, this scan procedure takes about 25 sec per spectrum including carriage return. The density data are registered on magtape in blocks of 500points, in all about 46.000 points/spectrum. During the process of setting up the system, we have so far refrained from a "random" scanning and each spectrum is manually centered to within about 0.2 mm on the center of the (H + Hε) line. An "active" scanning mode, in which the approximate position is automatically sensed, will be introduced after the installation of the HP-2100 control computer.

Following the registration by the S-3000, a "recognition" of the spectrum takes place. The four background scans are compared and zones of possible overlapping with other spectra are identified. The five scans along the spectrum are compared point by point and discrepancies, which may be due to plate faults, are filtered out. The scan is then compressed to a one-dimensional array and the background is subtracted. The (H + Hε) line, which normally is very well visible for all spectral types later than about BO,is identified, and a quick search is made for "strong" lines in the spectrum. This is actually a coarse, preliminary classification. With the positions of the line centers - if the first guess is unsuccessful, another line is tried as possible candidate for the (H + Hε) line - a Hartman-fit ($\lambda = \lambda_0 - C/(d_0 - d)$) is made and improved by the method of least squares. The fit is best for late-type spectra with many lines, but even for AO stars, in which practically only the H-lines are visible,r.m.s. values are normally better than ± 1.0 Å. The exact value of the constant C depends on the position of the spectrum on the plate, due to the geometrical properties of the Menisc telescope.

Each Abastumani spectral plate is accompanied by a calibration plate on which 18 calibration spectra have been exposed. The present software includes a determination of the D (λ) p› log I (λ) transformation by the fitting of a Baker-curve ($\log I = A \ln (T/(1-T)) + B$)at 50 Å intervals (4900 - 3800 Å). The A, B (λ) values are stored together with the values of $D_{max}(\lambda)$ and $D_{min}(\lambda)$, i.e. the density interval in which the calibration is valid,and are used for the transformation of the entire spectrum array (5000 - 3700 Å) into intensities. This takes several seconds and is probably not necessary, once the specific spectral criteria for the classification are known. The log I (λ) arrays of the standard spectra are the basis for the next step, the identification of the classification criteria, and they are stored in the computer disc memory for easy access.

For the establishment of an optimal classification, various types of spectral criteria may be considered. It should of course be stressed that due to the different dispersions, the best criteria for the ESO spectra may be different from the best Abastumani-criteria, and that the final decision on the ESO criteria will have to await the observations of standard spectra with the ESO Schmidt, which are expected to take place later this year.

Experience with the CLASS program has shown that a B-magnitude is readily obtainable by the application of a simple gaussian-shaped mathematical filter that simulates the standard B-filter; the obtained accuracy is about ± 0.^{m}06. G colour index may be found in a similar way, or by a calculation of the strength of those spectral lines that are sensitive to the temperature. Both the "equivalent width" (defined as the line area below a reasonably traced continuum) or an Hβ - type filter as used by Furenlid (1971) and Clausen (1973) are useful color indicators, cf. Fig. 3. In the Abastumani spectra the K-line, the G-band and the 4427 Ca I line may serve as color indicators over a wide spectral range, if the luminosity effects are removed.

The luminosity is more difficult to evaluate, especially for F-type stars. Previous experiences with the Abastumani spectra (Shiukashvili (1969), West (1972)) has shown that an accuracy of about ± 1^m can be arrived at (FO-MO) when using a combination of line-depth ratios. It will undoubtedly be very difficult, if not impossible, to keep a similar accuracy for the 450 Å/mm ESO spectra, but a separation of late type stars in two, hopefully three luminosity groups should be possible.

What concerns the third parameter, investigations by Vasilevsky (1972), West (1972) and Malyuto (1974, 1975) have demonstrated that reasonably accurate determination may be made from the Abastumani spectra for F-G dwarfs and K dwarfs and giants. The work by Samson (1969) strongly supports the hope that a metallicity parameter may also be determined for the ESO spectra.

Most of the rare stellar types (Wolf-Rayet, P Cyg, carbon stars, etc.) are easily recognized at 450 Å /mm. Peculiar types (Ap, Am, etc.) may be more difficult to distinguish as such, except in extreme cases.

One of the most important advantages of quantitative, automatic classification is that the basic stellar parameters (T_e, g, abundances) may be found from the spectral criteria through a direct calibration by means of standard stars. This step will be taken as soon as an automatic classification in terms of standard photometrical indices and spectral classes has been arrived at.

3.4 The automated pilot spectral survey. As a pilot programme for classification of faint stars in the southern hemisphere, we plan to investigate four to six 5°5 x 5°5 fields near the South Galactic Pole with the above described automatic method. This exercise will first of all serve to test the system in practical large-scale work, but it should also give valuable astronomical information about the distribution of stellar types in these fields, down to about 15^m. It must of course be underlined that the classification accuracy will depend on the magnitude, and, since many of the criteria can not be applied for the faintest stars seen on the plates (the intensity is zero in the centers of the strongest lines and in the long and short wavelength areas), only a rough temperature class can be given for these stars. Some preliminary experiments indicate, however,that automatic assignment of the first parameter is possible for even very under-exposed stellar spectra.

4. Conclusion. The present state of automatic spectral classification, in particular of objective prism spectra, has been discussed. An ESO project has recently been started, which aims at the automation of classification of 450 Å/mm spectra taken with the ESO 1 m Schmidt telescope. The basic software has been written and the initial results are promising. There is every indication that the system will be sufficiently accurate and reasonably fast. Modifications of the S-3000 measuring machine (cf. 3.2) will facilitate the efficient registration of the spectra, and the recognition of the spectra on the plate poses no special problems. Plate faults in the spectra are efficiently eliminated through two-dimensional scanning and calibration to intensities is accurately performed. After a rigid wavelength calibration, all features in the spectrum may be assessed. The decision on exactly which criteria to be used in the calibration still awaits the observations of a sufficient number of standard stars. It is the intention to carry out automated surveys in a limited number of fields from 1976.

5. Acknowledgements. I am much indebted to Dr. R. Bartaya, Abastumani, for helping to take plates with the Menisc telescope and to Dr. Ch.Tolbert, Charlottesville, for photometry with the ESO 50 cm telescope. Mr. F. Middelburg of the ESO-TP Division Geneva, has been a most valuable collaborator in the development of the classification software (CLASS).

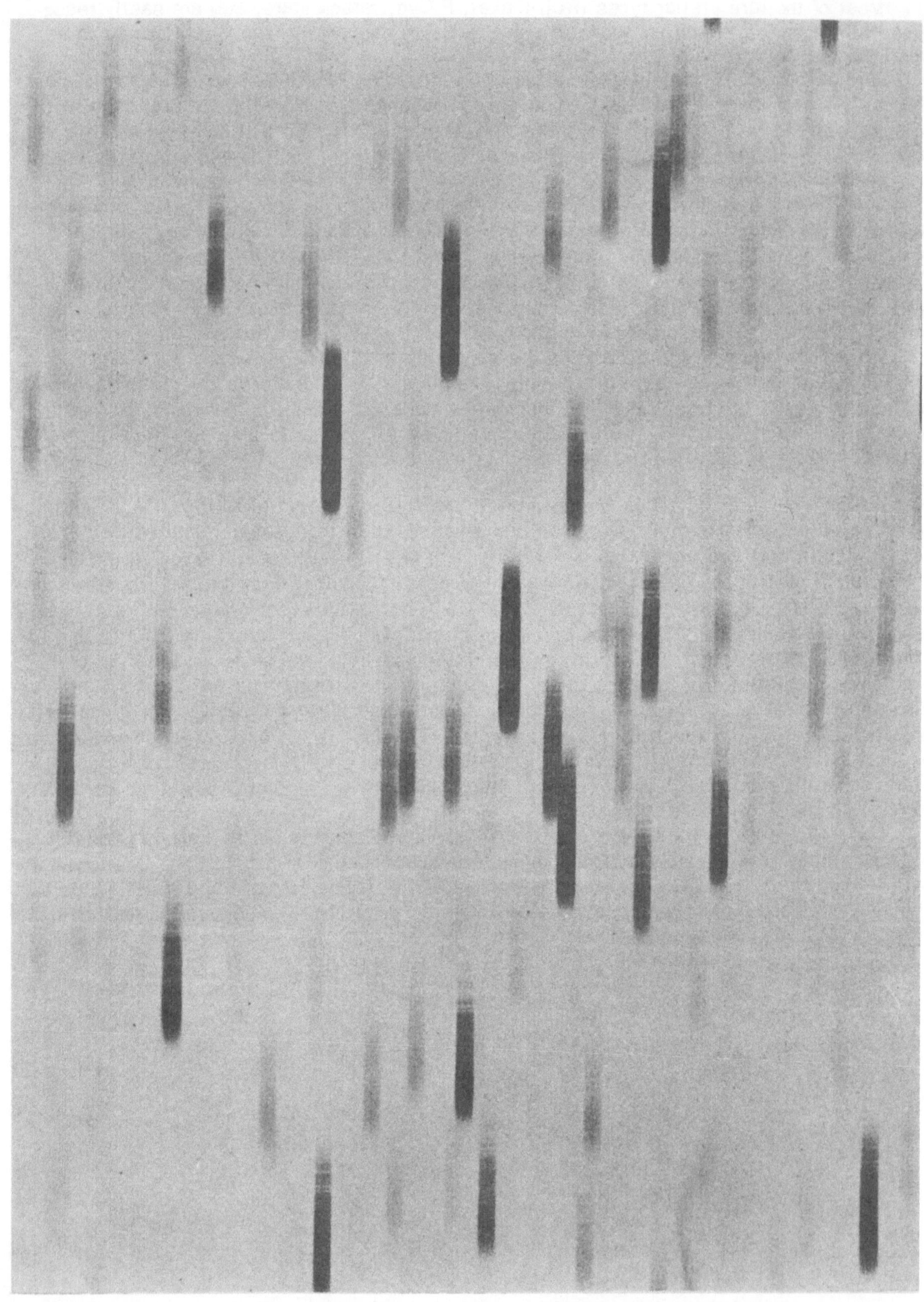

Fig. 1 Spectra taken with the ESO 1 m Schmidt telescope; IIa-O, 30 min, widening 0.3 mm

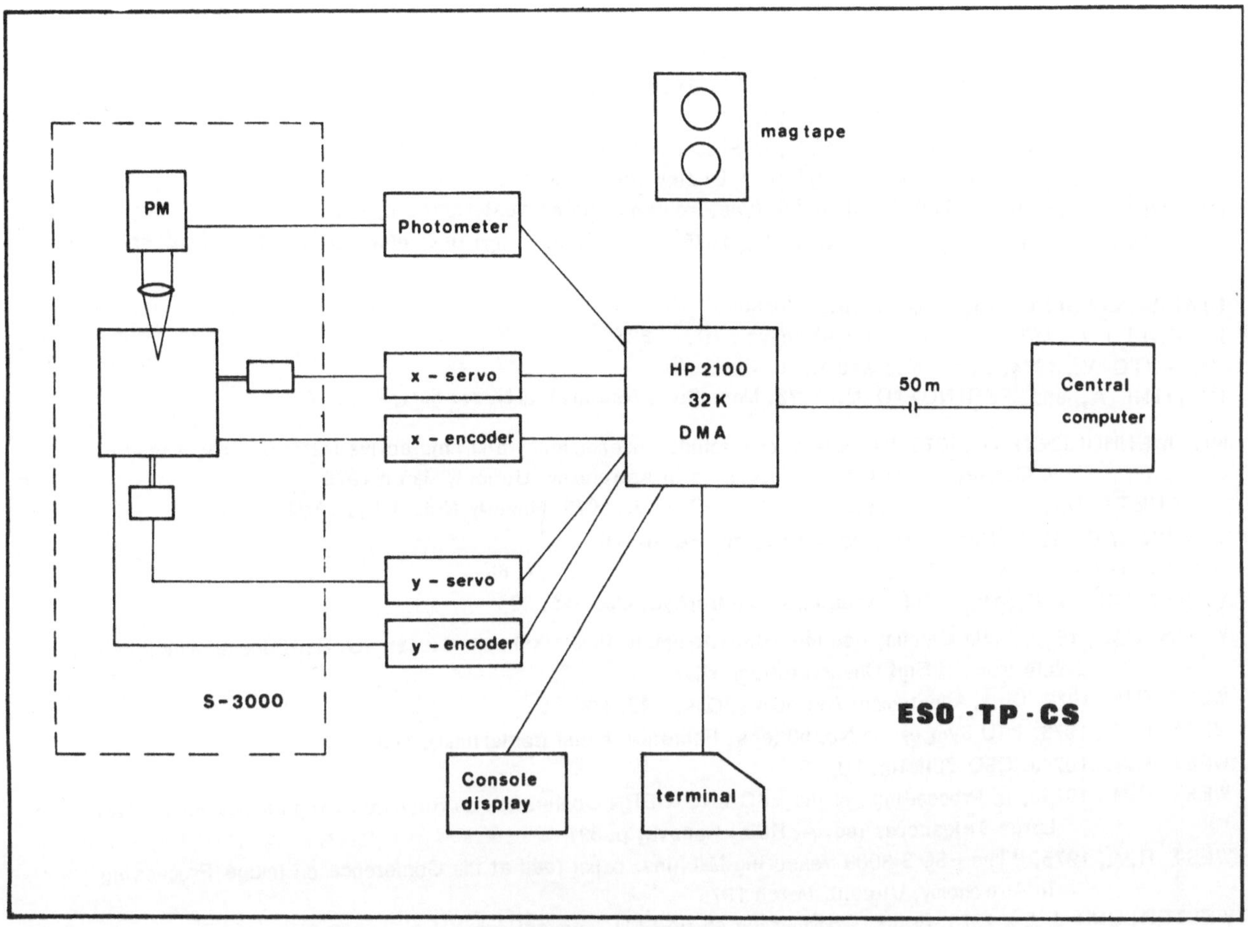

mag tape
PM
Photometer
x - servo
x - encoder
y - servo
y - encoder
S-3000
HP 2100
32 K
DMA
50 m
Central
computer
ESO · TP · CS
Console
display
terminal

Fig. 2 The new configuration for the S-3000 measuring machine

2.9
ß
2.8
2.7
2.6
3a
.1
.2
.3
.4
Iγ(30,150)
.4
(b-y)
.3
.2
.1
3b
.1
.2
.3
.4
.5
IK(3,12)

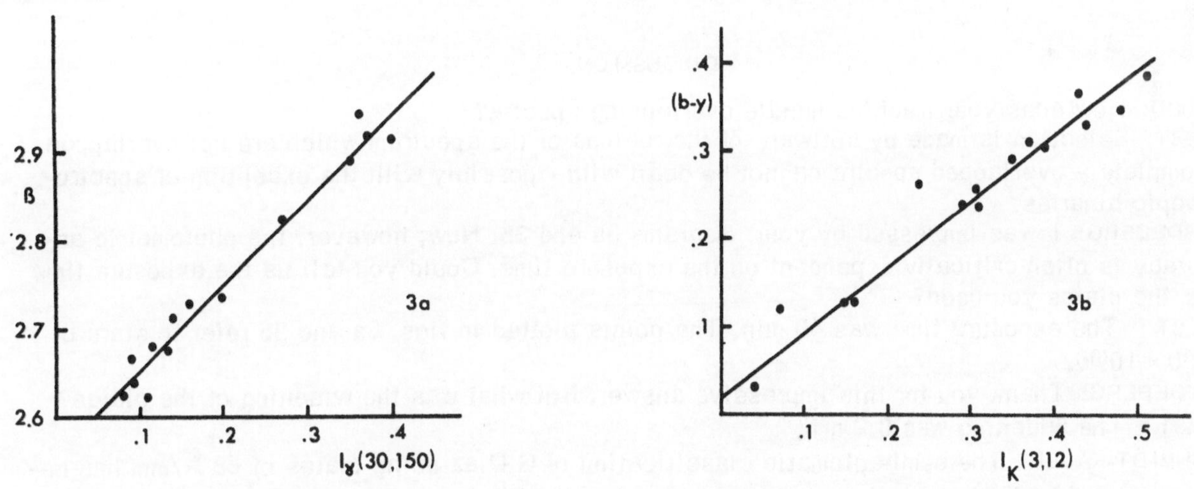

Fig. 3 Examples of the use of spectral criteria, comp uted by CLASS, for the determination of photometric indices a) the β - index vrs. I_{γ} (30, 150) as defined by Furenlid (1971). b) the (b-y) colour vrs. I_K (3, 12) for A-F stars.

REFERENCES

ARDEBERG A. and VIRDEFORS B., 1972, Astron.and Astrophys. **20**, 177

BELL R.A., 1973, Monthly Not. R.A.S., **164**, 197

BELL R.A., and GUSTAFSSON B., 1975, in Proceedings of "Multicolour Photometry and the Theoretical HR-diagram" (eds. Philip and Hayes), Dudley Obs. Report, **9**, p. 319

CLAUSEN J.V., 1973, IAU Symposium **50** (eds. Fehrenbach and Westerlund), p. 134.

DE JAGER C., and NIEUWENHUIJZEN H., 1975, Proceedings of Conference on Image Processing in Astronomy, Reidel, Dordrecht

DIAZ-SANTANILLA G., 1972, Thesis, Bochum

FURENLID I., 1971, Astron. and Astrophys., **10**, 321

MALYUTO V., 1974, Pub. Tartu Astrophys. Obs., **43**

MARTINI A., and PERINOTTO M., 1973, Mem. Soc., Astron. It., Nuova Ser., **44**, 247

NIEUWENHUIJZEN H., 1975, "Problems of Machine Independency and Interactive Matters", paper read at the Conference on Image Processing in Astronomy, Utrecht, March 1975

PATCHETT B.E., McCALL A., and STICKLAND D.J., 1973, Monthly Not. R.A.S., **164**, 329

SAMSON W.B., 1969, Publ. Roy. Obs. Edinburgh, **6**, No. 10

SHIUKASHVILI M.A., 1969, Bull. Abastumani Astrophys. Obs. **37**, 68

VASILEVSKY A.E., 1972, Bull. Abastumani Astrophys. Obs. **43**, 29

WEISS W.W., 1972, "Die Eignung des Mikrodensitometers PDS 1000 zum Messen von Radialgeschwindigkeiten", Note from L. Figl Observatorium, Wien

WEST R.M., 1972, Bull. Abastumani Astrophys. Obs. **43**, 109

WEST R.M., 1973, IAU Symposium No. 50 (eds. Fehrenbach and Westerlund), 109

WEST R.M., 1974a, ESO Bulletin, **10**, 25

WEST R.M., 1974b, in Proceedings of the ESO/SRC/CERN Conference on Research Programmes for the New Large Telescopes (ed. A. Reiz) Geneva, p. 321

WEST R.M., 1975, "The ESO S-3000 Measuring Machine, paper read at the Conference on Image Processing in Astronomy, Utrecht, March 1975

WESTERLUND B.E., 1973, Report Commission 45, in IAU Transactions XV A, p. 697

DISCUSSION

DODD: How does your machine handle overlapping spectra?

WEST: Selection is made by software of the regions of the spectrum which are not overlapped. Completely overlapped spectra cannot be dealt with - possibly with the exception of spectroscopic binaries.

ARDEBERG: I was impressed by your diagrams 3a and 3b. Now, however, the photometric accuracy is often critically dependent on the exposure time. Could you tell us the exposure time for the plates you used?

WEST : The exposure time was 40 min. The points plotted in figs. 3a and 3b refer to stars of $9^m_.0$ - $10^m_.5$.

ARDEBERG: Thank you for this impressive answer. But what was the widening of the plates?

WEST : The widening was 0.2 mm.

SCHMIDT-KALER: The semi-automatic classification of G.Diaz using plates of 88 Å/mm has been extended by H.Unger to lower dispersions (about 600 Å/mm) and towards complete automation. The correlations of $W(H_\gamma + H\delta)$ to H_β are of about the same quality as yours. However, we feel that the faintest spectra and the lowest dispersions still leave some ground for visual classifiers like Nancy Houk!

WEST: I shall be interested in learning the details of the new Bochum work.

HOUK: I think that for fainter stars - say fainter than 10^m - automatic classification is absolutely essential. I am sure that automatic classification is the way of the future and I wish you the best of luck.

WEST : Thank you, we need it! And let me please express my deep appreciation of those people like you, who still do the hard work in this fundamental field in astronomy.
COYNE : How accurately can you determine the metal abundance?
WEST : The [Fe/H].- ratio may be determined from late-type Abastumani spectra with a formal accuracy of about ± 0.2.
IWANOWSKA: Dr. A.Strobol at the Torun Observatory is attempting to establish photometric classification with objective prism spectra (250 Å/mm). The work is being automated up to the transformation of the density scale into magnitudes or intensity scale. His experience leads to two remarks: first, we are badly lacking reliable standards for the chemical composition. This severely limits his [Fe/H].- accuracy. Second, since almost all lines depend on all three parameters, would it not be reasonable to determine in the quantitative multiparametric classification the physical parameters, T_e, g and [Fe/H] rather than the traditional Sp, Mv and metallicity?
WEST: Yes, the possibility of determining directly the basic physical parameters is indeed one of the greatest advantages of automatic classification.
ARDEBERG: Another piece of information available from objective prism spectra is the (absolute) spectral energy distribution. Why do you only use step-wedge calibrations of the plates? Don't you intend to extend the calibration method?
WEST: As soon as we have the manpower and time available, we shall investigate the "Ardeberg" calibration method (Ardeberg and Virdefors, 1972).

ULTRAVIOLET SPECTRA OF STARS AND NEBULAE ("ORION-2")

G.A.GURZADIAN

GARNY SPACE ASTRONOMY LABORATORY, EREVAN, U.S.S.R.

УЛЬТРАФИОЛЕТОВЫЕ СПЕКТРЫ ЗВЕЗД И ТУМАННОСТЕЙ ("ОРИОН-2")

РЕЗЮМЕ

В статье приведено краткое изложение некоторых научных результатов, полученных астрофизиками Гарнийской лаборатории космической астрономии на основе обработки и измерений ультрафиолетовых спектральных снимков звезд в диапазоне длин волн 2000–3800 Å. Спектральные снимки были получены с помощью широкоугольного менискового телескопа с объективной призмой; сам телескоп был установлен на космической обсерватории "Орион-2".
Круг задач, доступных решению по материалам "Ориона-2", оказался довольно широкий ; они относятся как к непрерывным спектрам звезд в ультрафиолете, так и к коротковолновым линиям поглощения. Конкретно в настоящей статье рассматриваются следующие вопросы: характер распределения энергии в ультрафиолете непрерывных спектров горячих звезд и сравнение с теорией; эффект блокировки непрерывного излучения звезд ультрафиолетовыми линиями поглощения; депрессия непрерывного спектра на 2800 Å; структура непрерывного спектра γ Cas в ультрафиолете; эффект блендирования ультрафиолетовых эмиссионных линий ; закон межзвездного поглощения в ультрафиолете; распределение энергии в ультрафиолете горячей звезды $\varkappa$ Ori ; необычная группировка высокотемпературных звезд низкой светимости в Возничем; поведение дублета ионизованного магния 2800 Mg II в звездных спектрах; аномальные звезды A2; обнаружение хромосфер у холодных звезд и энергетическая мощность этих хромосфер; природа одной странной звезды; характер ультрафиолетовых спектров звезд промежуточных классов (F, G, K); спектральная классификация звезд по их ультрафиолетовым спектрограммам; аномально богатая кремнием газовая оболочка у одной высокотемпературной звезды; получение и расшифровка первой ультрафиолетовой спектрограммы планетарной туманности; характер коротковолновых спектров звезд класса O в ультрафиолете и т.д.

1. Introduction. One and a half year have passed since the time when an astrophysical experiment in space conditions was carried out by means of "Orion-2". The space observatory "Orion-2" set up in the outer part of the spaceship "Soyuz-13", inside a special astronomical cupola and manned by cosmonauts on board the ship, has made it possible to obtain, in five days, in the shady part of the orbit, ultraviolet spectrograms of more than 3000 stars up to the 13-th magnitude in the wavelength range 2000-5000 Å. The total number of spectrograms fit for measurements has exceeded 7000. This result is due, in particular, to the use of a wide-angle telescope of the meniscus system with an objective prism, as well as to the application of a highly precise tri-axial system of astrostabilization using two reference guiding stars.

The spectral pictures of "Orion-2" were destined from the very start for a study of the continuous spectra of stars in the above wavelength range. However, the quality of the pictures taken was so satisfactory that in some cases detection and even the measurements of particular spectral lines has been effected (Fig.1). This fact has widened the scope of scientific tasks awaiting solution, based on "Orion-2" data.

In the period under review J.B.Ohanesyan, R.Kh.Ohanesyan, O.V.Ohanesyan, R.S.Assatryan, S.S. Rustambekova, R.A. Ephremyan, A.A.Akopyan and other astrophysicists of the Garny Laboratory of Space Astronomy have elaborated and measured the spectrograms of about 600 stars, for various tasks, and over 2000 microphotometric recordings have been made for the spectral classification of 2000 faint stars alone (fainter than 10^m) by their ultraviolet spectrograms. The results have been generalized and formulated as separate scientific papers,part of which are already out of print, others - still in print in various journals. Some of those results have been set forth in review articles (Gurzadian, 1975, 1975a, 1974). Particular mention has also been made of the observatory "Orion-2" itself, its optical, kinematic, electronic and other systems, design, the basic points in operating and guiding it,matters of ballistic supply, ground and space tests and other issues (Gurzadian et.al, 1975).

The present paper, another general review, dwells basically on the latest results obtained as the material of "Orion-2" was further processed.

2. Ordinary hot stars. Let us start considering the peculiarities of the ultraviolet spectra in ordinary hot stars, of the B-A class and, in particular, the energy distribution in their continuous spectrum. Here and further the radiation flux F_λ will be presented in the wavelength scale and in stellar magnitudes, Δm_λ, taking the intensity of 3200 Å as the unit i.e.

$$\Delta m_\lambda = -2,5 \, Lg \, (E_\lambda / F_{3200}) \tag{1}$$

If the star under study is located far enough from us or its colour excess E(B-V) differs appreciably from zero, the magnitude Δm_λ resulting from direct measurements should be corrected to the impact of interstellar selective absorption. As a matter of fact this has been done in each particular case by one of the following means - depending on the circumstance – which aspect of the star under study we are familiar with: colour excess E(B-V) or its distance.

In the first case the corrected magnitude Δm°_λ is reached by the following expression:

$$\Delta m^\circ_\lambda = \Delta m_\lambda - (X_\lambda - X_{3200}) \, E(B-V), \tag{2}$$

where $X_\lambda = E(\lambda - V) / E(B-V)$. The average magnitudes of X_λ for the ultraviolet have been taken from Bless and Savage (1972).

In the second case,when E(B-V) is unkown but we are familiar with the distance of the star r , the magnitude Δm°_λ is reached by the following expression:

$$\Delta m^\circ_\lambda = \Delta m_\lambda - (\alpha_\lambda - \alpha_{3200}) \, r, \tag{3}$$

where α_λ is the coefficient of interstellar selective absorption per kps; its numerical values for a number of magnitudes of wavelengths is tabulated in (Gurzadian and R.Kh.Ohanesyan,1975).

The methods described above have been used in obtaining energy distribution in the ultraviolet of continuous spectra of more than one hundred hot stars, mostly of the B-A class. We should like to give a number of illustrations.

Fig. 2 illustrates energy distributions, according to the results of "Orion-2" spectrograms, in the ultraviolet of continuous spectra, corrected to interstellar absorption, in the spectra of two stars, one of which, HD 79186 of the B31 class, is in Vela constellation, in the southern hemisphere of the sky (Gurzadian, R.Kh.Ohanesyan,1975), while the other HD 33853 of the B9 class is in Auriga(J.B.Ohanesyan,1975).Theoretical distributions, according to Mihalas' model (Mihalas, 1965) at various effective temperatures are also given there as unbroken lines. In the given case, as we can see, the spectra observed have been in good agreement with the theory.

A similar result is likewise noticeable in the case of two stars of the same class B3V-121 Tau and 114 Tau, in Taurus (Rustambekova, 1975)(Fig. 3; here and further the circles indicate the observed fluxes, Δm_λ, while the dots show the fluxes corrected to interstellar absorption Δm°_λ).

Fig. 4 gives some idea of the internal agreement or disagreement in the nature of the energy distribution in the spectrum of stars of the same class, where the measurement results for stars, of the B8-B9 (J.B.Ohanesyan,1975) type, are compared. Here the observed magnitudes Δm_λ for all the cases proved to be much lower- sometimes as much as a whole stellar magnitude - from the theoretical curves. However, the corrected magnitudes Δm°_λ , turned out to be, nearly in all four cases, in good agreement with the theory.

Another instance: HD 32446, of the B3 class, is produced in Fig. 5. According to measuremen (J.B.Ohanesyan,1975) the observed points proved to be heavily deflected- towards the smaller fluxes - from the theoretical curve corresponding to the effective temperature of the star of this class 20000°K. However, after the correction to the effect of interstellar absorption, corresponding to the distance of this star r = 1000 ps (M=-1.6), the agreement with the theory has been complete.

Fig. 5 produces also data obtained by other observers for a number of stars of the same class B5V, especially, the stars HD 129116, 125823, (Bottemiller, 1972), HD 6588 (Sudbury, 1971), also η U Ma (Stecher, 1969); all of those stars are quite bright (3 - 5^m) and unreddened therefore the radiation fluxes obtained for them need no corrections. The good agreement of the obtained energy distribution in the spectrum of the star HD 32446 with the data for other stars of the B5V class indicates its belonging to the main sequence - the luminosity class of V.

Fig. 6 shows the results of the "Orion-2" measurements of two stars - HD 39319 and HD 66624, of the B9 class, one of which displays a peculiar feature (a strong line of silicon) (Ohanesyan, and Akopyan, 1975). Though both stars are of the same class, the agreement with the theory is obtained, in the first case, at an effective temperature of 11200°K, and 10000°K - in the second.

Other examples can also be given. Yet, the above suffices for arriving at the basic conclusion: in ordinary hot stars the macrostructure of their continuous spectra in the ultraviolet (up to 2000° Å) is on the whole in good agreement with the theoretically pre-calculated models at a given effective temperature of the star (in Mihalas' model only the continuous absorption due to hydrogen and helium is taken into account, while the absorption in spectral lines is disregarded). On the other hand, the comparison of ultraviolet observations with the theory has allowed in a number of cases to specify or determine some parameters of stars under investigation - distance, luminosity class, effective temperatures and so on; those cases are marked in (Curzadian, Ohanesyan R.Kh,1975; Ohanesyan J.B.,1975; Rustambekova, 1975; Ohanesyan and Akopyan, 1975).

3. Blocking effect by absorption lines. The case is different when we come to considering the microstructure of the continuous spectra of hot stars in the ultraviolet. Fig.7 illustrates the observed curve, corrected to interstellar absorption, of energy distribution in the continuous spectrum of the star HD 36589 of B6V class in the wavelength range 2000 - 3700 Å, along with the theoretical curve corresponding to the model when T = 15700°K (Rustambekova, 1975). Here the general structure of the observed spectrum is in a good agreement with the theory. On the other hand, distinctly pronounced depressions in the continuous spectrum at 2350 Å, 2250 Å, 2050 Å and 2800 Å are marked out.

We have reason to believe that the microstructure of the continuous spectra of hot stars in the ultraviolet is inconstant and can vary in the range of the given spectral class as we take up one star or another. As an example Fig. 8 pictures a group of five stars of the same class (A2) and nearly of the same brightness ($8^m.2$), selected by J.B. Ohanesyan (Ohanesyan, 1975). That group proved to be exclusively uniform in all respects (including the conditions for obtaining spectral pictures and their microphotomeasurements), save for the short-wave measurement boundaries that were dissimilar for the spectrograms of those five stars. Now comparing the spectrograms of those stars with one another we can see that the microstructure of their continuous spectrum in the range of wavelengths, shorter than 3000 Å, is inconstant and changes from one star to another (within the same class).

Fig. 9 presents another group of four stars of the same class AO, in which the above-noted spread in the microstructure of the continuous spectrum has been still more pronounced (Ohanesyan, J.B.,1975). Thus, for instance, if in the case of the star HD 34380 the energy distribution in the continuous spectrum follows with almost absolute accuracy the Fowler's (1974) theoretical curve for Sirius, especially, with two powerful depressions occurring at 2800 Å and 2400 Å, in the case of the succeeding stars a constant decrease of deviation from that "ideal" model is observed, as much as the complete disappearance of both depressions in the star HD 35848. The deep yet rather narrow depression at 2400 Å in the spectrum of HD 34399 deserves attention.

It should be noted that the observed spread in the microstructure of the continuous spectra of the classes of stars under consideration cannot result from interstellar absorption because, firstly, all of those stars are at a distance of 300 ps from us and, secondly, for those stars, the colour indices obtained as a result of our ground-based observations approached zero (Ohanesyan O.V.,1975). In view of the great degree of homogeneity of the selected groups of stars, it is equally hard to trace the cause of the above spread in experimental nuances or in elaborated method of the spectrograms. Despite all this we need some further evidence before arriving at a final conclusion on a genuine difference in the microstructure of the ultraviolet spectra of those classes.

At present it is beyond doubt that a consideration of the blocking effect by spectral lines in the theory of radiation transfer in the photospheres of stars leads to the results that differ essentially both in quantity and quality from those available in the case of ordinary theories which disregard that effect. The blocking of continuous emission by a very large number of absorption lines, pertaining mainly to neutral and ionized metals (titanium, chrome, iron, cobalt, nickel, etc.) distorts beyond recognition the general view of the continuous spectrum in the wavelength range shorter than 3000 Å. This effect, predicted by Elst nearly ten years ago (Elst, 1966), has been first confirmed experimentally by the observations of OAO-2 (Underhill, 1972) and OAO-3 (Underhill, 1974), during which the dissimilarity in the action of this effect in stars of different classes has been established: the depression in the range of 2000 - 3000 Å is quite strong in star A5-A6 with maximum depression at 2400 Å, amounting to 1^m.

Thus, it can be asserted with fair assurance that the above noted depressions are due to blocking by absorption lines. The point is that according to our data the blocking effect itself acts, due to some reason, to a different degree in stars of the same spectral class.

At present several attempts are being made aimed at developing a theory of stellar photospheres that takes into account the blocking effect. Most significant among them is the attempt of Kurucz et al,(Kurucz and Peytermann, 1972), considering in their calculations the effect of 1,760,000 spectral lines.

It follows from what has been stated above that the existing classification of stars (at least of the A class) is inadequate for describing their continuous spectra in the ultraviolet. If subsequently a confirmation of the reality of variety in the spectral microstructure in the ultraviolet comes, their spectral classification will apparently need a new parameter (for the usual designation of the sub-class) that will take into account one peculiarity or another in the spectral microstructure in the ultraviolet (2000 - 3000 Å).

4. Depression of the continuous spectrum at 2800 Å. Since depression in the continuous spectra results from the blocking effect of numerous absorption lines, the density and power of which depends in turn on the effective temperature of the star, it would be natural to expect some dependence between the energy power of the depression and the spectral class of the star. An attempt has been made to plot such a dependence empirically for the depression at 2800 Å.

Depression at 2800 Å is interesting for a number of reasons. Firstly, it is observed almost in all spectral classes - from the early B to the early K. Fig. 10 quoted from J.B.Ohanesyan (1975), indicates the change sequence of the depression power as we move from B3 class stars to FO. Secondly, the depression width almost does not depend on the spectral class of the star and is of the order of 250-300 Å, and the depression depth alone seems to change during transition from one class to another. Finally the spectrum area around 2800 Å is readily determined even in faint stars during moderate exposures, which enables us to use more observational material for the required dependence.

Let us call the ratio equal to

$$q = \frac{E_{depr}\ (2650 - 2900)}{E_{cont}\ (2650 - 2900)} \qquad (1)$$

the "power" depression, where E_{cont} (2650-2900) is the total continuous energy in the wavelength interval 2650-2900 Å when depression lacks at 2800 Å, while the E_{depr} (2650-2900) is the summary depression power, i.e. the missing energy in the continuous spectrum in the same wavelength interval.

Relying on the method described by R.A.Ephremyan (1975) the numerical values of q for 32 stars of the spectral class B9-KO, scattered in the Auriga and Cassiopeia constellations, have been obtained. Fig.11 shows the graphic dependence of q on the spectral class. As we see, the existence of a similar dependence is real and in future, plotting it in a more refined manner, making use of more extensive observational material, it can probably be used for the spectral classification of stars.

5. Short-wave spectrum of γ Cas. One of the prominent representatives of the B class stars with a gaseous envelope, γ Cas, is as before in the focus of attention of astrophysicists. Since 1970 a number of successful attempts have been made to obtain the ultraviolet spectrograms of this star (BO.5 IV re, V=2.2) in space conditions (Bohlin, 1970; Code and Bless, 1970; Lillie et.al., 1972; Boksenberg et.al. 1972).

A new attempt has been made with "Orion-2", giving us several spectrograms of γ Cas in the range 2000-3700 Å; Fig. 12 illustrates the microphotometric recordings of one of them. The energy distribution in the continuous spectrum of γ Cas, revealed as a result of processing the spectrograms, is given in Fig.14 (for details see Gurzadian, 1975 e). The solid line shows the theoretical spectrum of the star with the effective temperature 28000°K. A reservation should be made in advance to the effect that every comparison of that curve with observations will be of a formal character for the simple reason that the theoretical curve takes no account of the emission of the gaseous envelope. Meanwhile, the effect of the latter on the general emission of the star seems to be fairly appreciable, which follows from the Balmer discontinuity; it proved negative and equal to $D = Lg(F_{3646}^{+}/F_{3646}^{-}) \approx -0.08$, nearly twice as much as for normal BO.5 stars (+0.09). That is why, though the distribution found from observations agrees well with the values T =28000°K, this temperature will reflect neither the effective temperature of the star nor, all the more, of its envelope.

A peculiar feature of Fig.14 is the occurrence of two protrusions in the observed energy distribution in the spectrum of γCas,near 2200 Å and 2600 Å. Apparently, those protrusions are real and have emerged following the blending of a large number of emission lines(see section 6).

We are inclined to believe that the most interesting feature of the ultraviolet spectrum of γCas, is the occurrence of a large and fairly deep depression in its continuous spectrum at 2800 Å; it is seen well in Fig. 12 and is neglected in Fig.14. Reservation should be made, however, that the emulsion of the Kodak 103-OU-V film, on which all of the spectrograms of "Orion-2" were taken, is of low sensitivity especially at ~ 2800 Å. But the depression depth, due to the low sensitivity of the film, is much less (as it has been established by means of special laboratory measurement), than we have in the case of γ Cas. One can even notice a rather strong line 2800 Mg II on fragments of densitometric recordings from four spectrograms of γ Cas (Fig.13.). It should be noted that this line has nothing in common with the interstellar line, very narrow and very deep, discovered by Boksenberg et al. (1972), in the spectrum of that star. On the other hand,the above depression width has not been revealed in the case of Boksenberg et.al. (1972). for the reason that the working range of their spectrograph (2740 - 2870 Å, Fig. 13) turned out to be twice smaller than the depression width (~ 300 Å).

By most careful estimations the equivalent width of the line 2800 MgII on the "Orion-2" spectrogram of γ Cas is more than 10 Å. This value exceeds by more than one order the expected value W(2800) for the BO.5 class stars (Mihalas, 1972). It follows that the above depression can neither be of stellar origin.

It is very likely that the source of the origin of the depression at 2800 Å in the case of γ Cas is the circumstellar cloud surrounding the star. However, this assumption requires a special treatment.

6. Short-wave spectra of emission stars. Blocking effect of emission lines. The data of "Orion-2" have been treated and the ultraviolet spectrograms of the following B-type emission stars have been derived:

Star	Class	V	Reference
SAO 040183	B2e III-V	$6^{m}.9$	(Gurzadian, 1975c)
γ cas	BO,5 IV pe	2,2	(Gurzadian, 1975d)
HD 37967	B3 ne III-V	6,1	(Rustambekova, 1975)
ξ Tau	B2 n e	2,97	"
SAO 077308	BI V nn e	8,87	"
HD 38191	BI n e	8,73	"

The results of processing the spectrograms of the star SAO 040183 are presented in (Gurzadian,, 1975c; Gurzadian, 1975a). Energy distribution in the continuous spectrum of that star has been revealed; again, an attempt has been made to isolate about two dozen emission lines and an equal number of absorption lines. A point worthy of notice is the lack of depression at 2800 Å in the star (Fig.14) that was so typical of γCas (see the previous section).

Figs. 14 and 15 present the results of treating the shortwave spectrograms of the remaining four emission stars. The plotting of the theoretical curves, in those figures, corresponding to one effective temperature or another, is, as noted above, somewhat of a formal nature,since those curves had been plotted for pure photospheres, with no gaseous envelope. Nevertheless, the peculiar feature in those spectra is self-implied, viz. the presence of protrusions or local elevations in certain parts of the spectrum, more frequently at 2200 Å and 2600 Å.

If we leave alone the assumption that those local elevations can ensue systematic or casual errors in measuring the spectrograms themselves, then the question arises: how are we to account for the phenomena referred to above? As a matter of fact we observe additional emission in certain narrow parts of the stellar spectrum. This does not mean, however, that it is due in fact to an additional radiation of the star: the occurrence of some unknown sources is not ruled out either. Rather, the matter is simpler; namely, we observe *a blending of emission lines* - the reverse of blocking effect in the case of the absorption line (section 3). A large number of emission lines located quite close to one another and pertaining mainly to neutral and ionized metals, can be given off (omitted) in the interval of wavelengths (2000 - 3000 Å) of the gaseous envelope of the star. With inadequately high spectral exposure, as is our case (about 10 Å and 15 Å at 2000 Å and 2500 Å respectively), those emission lines will not be resolved; as a result, we mark out the total effect in the form of a "protrusion" of the continuous spectrum.

It could not have been otherwise, i.e. the true physical blending of the emission lines with each other, due to the dispersion of the thermal velocities of atoms and ions exciting those lines. As to the blending effect of the emission lines we have in mind precisely this phenomenon. However, we cannot as yet answer the question to what extent this effect is not "false" in the above case, i.e. is not exactly due to the physical blending of the emission lines. On the other hand , we have no guarantee that the solution of the problem will be eased in the case of spectrograms with very high resolution. For the isolation of "protrusions" against the general background of the continuous spectrum will become very complicated in view of the fact that the working region of the spectral apparatus will be sharply reduced,depriving us of the chance of simultaneous determining the location of the "protrusion" and its "flanks" i.e. the continuous areas of the normal level of the continuous spectrum.

In the light of the above statement the need arises for the elaboration of the theory of spectra of stars with a gaseous envelope, taking into account the radiation of emission lines, excited in the envelope, as well as with absorption of the photosphere radiation during its transfer through the gaseous envelope of a given thickness.

7. Interstellar absorption according to the "Orion-2" data. When confronted with the task of precise measurements, the "Orion" spectrograms attained the limit of 2000 Å not too frequently. Nevertheless the data at hand permit to make some conclusions concerning the nature of interstellar selective absorption in the ultraviolet ranging from 3000 Å to 2000 Å.

Circumstantial results on the study of interstellar absorption in the ultraviolet are formulated by Bless and Savage (Bless and Savage, 1972; Savage, 1975). The gist of their method of determining the magnitude of interstellar absorption on the given wavelength consists in comparing the spectra of two stars of the same class with each other, one of which has a colour index considerably differing from zero, while the other is nearly zero, which means that practically it has undergone stellar absorption.

In our case we acted somewhat differently, namely, the comparison was made between the observed spectrum of a given star and the theoretical spectrum corresponding to its effective temperature.

Fig. 16 gives the dependence curves of the absorption values $X_\lambda = E(\lambda - V)/E(B-V)$ from $1/\lambda$ found out by S.S.Rustambekova (1975), rated on photographic rays (4400 Å), and determined by the measurement results of O-B stars, two of which are emission stars. In the light of the reasons set forth above, O and B type stars do not suit that purpose and therefore the results quoted should be approached with reservation.

An additional confirmation of that comes from the results obtained following the processing of the spectrogram of another emission star - SAO 040183, of the B2e class; they are given in Fig.17(dots), along with the dependence curve of total absorption a_λ from $1/\lambda$ discovered by Bless and Savage, (1972),relying on the data of star ξ Per, of the class O7.5. With the results obtained in both cases fully coinciding, inadequacy of absorption at ~ 2200 Å is still noticeable in the case of SAO 040183 the fact due, in our opinion, to the blending effect of the emission lines.

Despite the shortcomings referred to above, the results obtained allow us anyhow to draw a number of conclusions concerning the nature of interstellar absorption, viz:

a). Interstellar absorption in the range of wavelengths 2000 - 3000 Å is caused, in line with the current concept, mainly by the graphical particles, almost spherical in form;

b). The law of interstellar absorption can be presented in the form of λ^{-1}, at least up to 2700 Å (from the side of long waves);

c). The application of stars with gaseous envelope in determining the law of interstellar absorption renders the results uncertain. especially in the range of~2200 Å, because of the blending effect of the emission lines.

8. Energy distribution in the ultraviolet of star $\varkappa$ ORI. It is a bright star (V=2.06) of the early spectral class BO.51a and as such it happened to be one of the first stars the shortwave spectral measurements of which have been made available, when outer space astronomy was still in its incipient stage.

The first short-wave spectrophotometric measurements of this star were made by Bogges and Kondo in 1968 in the interval of wavelengths 3000 - 2300 Å. Next, papers dealing with the observation of the short-wave spectrum of that star appeared - Ste cher (1969), Navach et.al.(1973), Evans (1972), Morgan et.al.(1975); the latter review compares all the previously published data on the short-wave spectrum of $\varkappa$ Ori in the range of wavelengths 2000 - 3700 Å.

Two dozen spectrograms have been obtained for $\varkappa$ Ori by means of "Orion-2"; seven of them have been worked out. As a result, the energy distribution in the continuous spectrum of that star in the range 2000 - 3700 Å has been found and corrected to the effect of interstellar absorption, corresponding to the colour excess of $\varkappa$ Ori, it is given in Fig. 18 (black dots). Tabulated in the same figure are also measurement results of the above observations and the curve corresponding to Mihalas' theoretical model when T_{eff} = 28000°K and Lgg=4. As we can see the above observations are in a good agreement with the theory at least up to 2400Å, upon which depression begins - energy deficit in the continuous spectrum as compared to the theoretically expected level, This depression increases as we move closer to the short wavelength side as far as the limit of our observations - 2000 Å.

It is interesting that our measurements completely coincide with those of Stetcher and Morgan et al (in the latter case the short-wave limit of observations is at 2650 Å). On the other hand, there is a considerable discrepancy with Evans' observations - toward newer emission flows in the ultraviolet, those of Navach et al - toward the greater flows; in the latter case we are kept on the alert by almost complete coincidence of the results of observations with the theoretical curve also in the region of wavelengths 2400 - 2000 Å, i.e. without signs of depression.

If we assume that all the six observations quoted in Fig. 18 hold, we shall be led to the conclusion concerning the reality of the scattering of the energy distribution in the ultraviolet of the star in time. Is that really so?It is difficult to say. At any rate the assumption that $\varkappa$ Ori is variable in the ultraviolet, but a relatively stationary star in the optical range,needs some additional basis. In principle such variability in the ultraviolet can, of course, be accounted for in terms of a gaseous envelope surrounding the star, variable in the power. In this case the emission lines pertaining to the neutral and ionized metals - they are in large numbers in the ultraviolet - can fill in the depressions of photospheric origin - in some cases or, blending with one another, form a "protrusion" - in others.

Apparently the assumption of the existence of a gaseous envelope or extented atmosphere, or,finally of a circumstellar ionized cloud around $\varkappa$ Ori is not very unlikely. This is indicated,in particular, by the discovery of the line H_α in the emission in the spectrum of that star (Rosendhal, 1973). This assumption becomes more likely in view of the fact that the disappearance of gaseous matter at a great velocity is a phenomenon characteristic of supergiants of the O-B class, judging by the results of space spectral observations (Morton, 1967).

9. "Ultraviolet" stars in the Auriga. Stellar association of a new type? The films of "Orion-2" revealed a group of more than 20 "ultraviolet" stars weaker than $9^m.5$ and up to $11^m.3$ in the constellation Auriga, in the vicinity of Capella. They were scattered in the region of the sky with an area less than one square degree. By the name "ultraviolet" we imply stars, the brightness of which lie in the abovementioned range and the continuous spectra of which are rather strong, at least up to 2500 Å. None of those stars figures in the existing catalogues. Their numbers in our observations, the photovisual magnitudes and the colour indexes B-V and V-B, obtained by O.V.Ohanesyan(1975) from data derived from ground-based observations, are quoted in Table 1.

The above group of stars was first described by Gurzadian,(1975b), who also provided an identification map for most of them. The average location of the group corresponds to the coordinates $\alpha = 0.5^h\ 10^m$, $\delta = +44°30'$.

The reproductions of "Orion-2" spectrograms of some stars from the above group were also listed in the paper of Gurzadian,(1975b).For the purpose of comparison the "Orion-2" spectrograms of two stars of the classes - AO and B9 - were also given there; they were in proximity to that group and selected in such a way that the photographic identity of the image in the longwave part of the spectrum was comparable to the blanketing density of the spectra of stars under consideration.A comparison of the spectrograms with those of our ultraviolet stars shows that the latter cannot be the objects of the class later than AO and B9; rather, they must pertain to the class of early B.

A more precise evaluation of the spectral class of those stars can, of course be, given either by the spectral lines - to this end we must have their slit spectrograms at our disposal - or by the measurement results of their continuous spectra.

Table 1

Photovisual magnitudes and colour indices of a group of "ultraviolet" stars in Auriga

No of star	V	B-V	U-B	No of star	V	B-V	U-B
50	$11^m_.16$	$+0,^m10$	$-0^m_.43$	61	$10^m_.75$	$+0.^m31$	$+0,^m18$
51	10,62	+0.01	-0.11	62	9,86	+0,30	+0,19
52	10,96	+0,26	+0,01	63	9,59	+0,01	-0,29
53	9,09	+0,06	-0,13	64	9,35	+0,36	+0,07
54	9,95	-0,04	+0,22	65	10,36	-0,01	-0,05
55	9,70	+0,03	+0,09	1511	10,50	+0,04	-0,40
56	9,93	+0,01	+0,23	1518	9,51	+0,04	-0,40
57	10,47	+0.02	+0,24	1533*)	12,84	+0,12	-0,04
58	11,01	+0,20	+0,11	1665	10,72	+0,03	-0,14
59	10,47	+0,06	-0,24	1670	11,29	+0,09	-0,09
60	10,96	+0,19	+0,22	1700	9,05	-0,01	-0,36

*) The star pertaining to the group is uncertain

The latter method was applied, as an illustration, in respect of one star from this group - No 50 with the magnitude $11^m_.2$ in the photographic rays. The measurement results for the continuous spectrum of this star are presented in Fig.19 (circles).

Plotted in the figure are also the theoretical curves corresponding to the various effective temperatures, from 10000°K to infinity. As we can see, the energy distribution already observed in the spectrum corresponds to the effective temperature much higher than 10000°K and comes close to 20000°K. This is obtained without taking into account the effect of interstellar selective absorption.

The appearance of the "Orion-2"spectrograms of star No 50 does not differ from those of the remaining groups. It can therefore be expected that we shall also have effective temperatures of the same order of magnitude for the remaining stars of the group as well.

We have no idea of the distance or the absolute luminosity of those stars. Assuming that they, including star No 50, are at a distance of 1000 pc from us, and making a corresponding correction to interstellar absorption, we come to the conclusion that its effective temperature must be of the order of 50000° K (Fig. 19). But this is the temperature of the O type star or, at any rate, of the early subclasses of B. It should be pointed out that according to the existing catalogues in the region of the sky around Capella, no star of a class earlier than B2 and brighter than 10^m occurs in the region 5° in diameter.

Thus although we cannot determine the exact effective temperatures in the group of stars under consideration by the energy distribution in the continuous spectrum,we can assert, however,that these temperatures are higher than 10000°K, i.e. they all form the essence of stars of earlier classes. This is sufficient for those stars to attract our attention; first of all, because the number of such stars in the area under review is great (considerably greater than illustrated in Table 1) and they form an entire group. Secondly, those stars are presumably not typical hot stars with an absolute luminosity of -3^m or -4^m, since in that case their distance will attain 5 - 10 kps, which is unlikely if we realize that they are in the direction of the anticentre of the Galaxy($l \approx 173°$, $b \approx +7°$). The impression is that M>O for those stars.

It is not a simple thing to combine such low absolute luminosities with high temperatures among the well-known categories of stars. The Humason - Zwicky objects or the former nuclei of planetary nebulae can be exception. But, in the first place, we are well aware of the groupings of Humason - Zwicky stars and, in the second, if we rely on the existing statistics of the Humason-Zwicky stars, their number in the region of the sky under study will be smaller by two stars and in any case not more than twenty.

Thus, the nature of the stellar group under review and that of the grouping itself is not clear to us. It is early as yet for us to make reference to the angular and linear dimensions of the grouping, all the more so - to the total number of stars that form part of it. However, even without such reference it is clear that the dimensions of the grouping are large enough to form an open cluster and inadequately large to form a stellar association. Nevertheless, we are inclined to believe that in the given case we have to deal with a *stellar association possibly of a new type, viz. a stellar association consisting of the stars of low luminosity.* However, much effort is required before we arrive at a final conclusion on the matter. We believe that the most important problem should be handled first: if the existence of the above grouping is real or it involves some misunderstanding.

10. Behaviour of the doublet 2800 MgII in the stellar spectra. As early as in 1956, long before the first ultraviolet spectrograms of stars in space conditions were obtained, a prominent expert of stellar spectra Merrill(1956) noted the great significance of the doublet of ionized magnesium 2800 MgII (2803 Mg II and 2795 MgII)for astrophysics in general. This doublet, which is an analog of the well-known lines H and K of ionized calcium, displays an obvious advantage as an indicator, rather sensitive, in deciphering the physical conditions in the atmospheres of stars as well as the interstellar medium.

To begin with cosmic abundance of magnesium is at least 20times higher than that of calcium. Owing to this, the range of the spectral classes of stars, where the doublet 2800 MgII can still be detected, will be much broader than in the case of H and K lines of CaII. Further, in difference from the lines of H and K of CaII,which are in an exclusively disadvantageous spot-among the hydrogen lines of higher numbers, the doublet 2800 MgII is located in an exclusively "calm" spot; at any rate free from a large accumulation of strong lines, the fact considerably contributing to its isolation and measurements.

At the beginning of 1974 our knowledge of the ultraviolet doublet of the ionized magnesium 2800 MgII in the stellar spectra was restricted to about twenty stars, bright ones brighter than 4^m (Doherty, 1971; Kondo et.al., 1972; Kondo et.al., 1970; Gurzadian and Ohanesyan, 1972; Lamers et.al. 1973; Aller et. al. 1955). The picture changed sharply after the experiment with "Orion-2" when ultraviolet spectrograms of an immensely large number of faint stars - up to 10^m and weaker were available . Even with the spectral resolution 20-25 Å at 2800Å the "Orion-2" material, due to its high degree of homogeneity and mass character, proved to be quite a valuebale source for studying the doublet 2800 MgII in the stellar spectra of different classes.

The first results derived from processing the "Orion-2" spectrograms for 51 stars (Gurzadian, 1975d) have enabled us to establish a number of regularities relatedto the behaviour of the doublet 2800 MgII in the stellar spectra.

At present the ultraviolet doublet of ionized magnesium 2800 MgII is distinguished and measured in the "Orion-2" material in the spectra of more than 300 stars (Gurzadian, 1975d; Assatryan 1975,), encompassing the spectral range O - M and up to 10^m. The results of analogous investigations are also expected, again from the "Orion-2" data, of another two hundred fainter stars - up to $12 - 13^m$. In this way, a possibility is not ruled out that in the near future our notion of the doublet 2800 MgII will involve over 600 stars.

The equivalent widths of the doublet 2800 MgII were obtained first of all for more than 300 stars of those; those data are tabulated in the papers of(Gurzadian, 1975d ;Assatryan, 1975).As the analysis of the obtained data shows the structure of the continuous spectra of stars near 2800 Å is essentially different in the stars of different spectral types and is determined generally by the strength of the doublet 2800 MgII. However, the character and structure of the continuous spectra near 2800Å as well as the doublet itself are more or less stable for a given spectral type. Sometimes derivation from the "mean" norm,as a rule of anomalous character, is also possible.

At present the available observational data about the doublet 2800 MgII in the stellar spectra enable us to make the following conclusions:

1. In the bright stars of BO-A9 types the observed magnitudes of W(2800 MgII) lie between the limits 1-5 Å in accordance, in general with the theoretically predicted values (Mihalas,1972).

2. In the faint stars of B-type scattered far away for 1000 pc and more the observed equivalent widths of 2800 MgII are determined completely by the interstellar absorption in the clouds of interstellar ionized magnesium. The mean eqiuvalent width for the interstellar line 2800 MgII is equal, according to our determinations, to 3.8Å per 1 kpc (for the three regions of the sky - α Aur , β Aur and γ Cas).

3. The equivalent width of 2800 MgII is increased from the stars of early types to later, ones and after reaching the greatest strength in stars of spectral classes F5-GO, is deccreased again. For F5-GO stars W(2800 MgII) is of the order of 50Å and more. In these stars the doublet of ionized magnesium is so powerful and extensive that actually a strong and wide depression of the continuous spectrum around 2800Å is formed. In Fig.20, as an illustration, the fragments from the "Orion-2" spectrograms for five stars of FO-GO type, are given. near 2800 Å, In these stars this depression can sometimes cause an abrupt declining or a precipice of the continuous spectrum at 2800 Å (Fig. 21). Incidentally, this circumstance (i.e. the breaking of the spectrum at 2800 Å and practically its disappearance on the shortwave side) may serve as a sufficient cri-

terion for identification of any unknown star as belonging to class F.

4. In stars of the type K2 and later the doublet 2800 MgII may, in some cases, be present in the form of a weak emission line (see section 12).

5. There exists a definite correlation between the behaviour of the doublet 2800 MgII and the multiplet, No 5, of single ionized titanium with a mean wavelength 3080 TiII in stellar spectra: they are both present simultaneously - either in emission or in absorption.

Table 2

Mean equivalent width of the ultraviolet doublet of ionized magnesium W(2800 MgII) in the stars of different spectral types

Spectra	W(2800 MgII) Å	Spectra	W(2800 MgII) Å
B2	0,8	F2	29
B5	1,8	F5	35
B8	3,5	F8	38
AO	5,5	GO	40
A2	8,3	G2	38
A5	14	G5	30
A8	21	G8	20
FO	26	KO	13

6. The "Orion-2" data confirm the existence of the empirical regularity, discovered earlier (Gurzadian, 1972; Gurzadian, J.B.Ohanesyan, 1975 ; Gurzadian, 1975d), between the equivalent width of 2800 MgII and the spectral class of the star. This relationship - it is shown in Fig.22 and Table 2, based on the wavelengths (R.S. Assatryan, 1975) - is very stable in form, it may be even used for spectral classification, i.e. having the observed magnitude of W (2800 MgII) to find the class of the star. The characteristic peculiarity of this relationship is the presence of the maximum of W(2800 MgII) approximately at the classes F5-GO. Towards later types from this, the equivalent width declines because of the decrease of the amount of magnesium ions, Mg^+, at low temperatures, while towards earlier classes it again declines but in this case because of the transition of Mg^+ into Mg^{++}.

11. Anomalous stars of A2-type. By the structure of continuous spectra the star HD 32296 of A2-type does not differ from the usual stars of the same type. But from one point of view it differs strongly; the absorption line 2800 MgII is extremely strong in its spectrum (Ohanesyan J.B., 1975). Fig. 23 , with two microphotometric recordings from two spectrograms of this star, confirms this fact. The equivalent width of 2800 MgII, according to these spectrograms, equals 30 Å(!) - more than three times as large as we have in usual A2-type stars (see Table 2); this magnitude is characteristic only of the stars FO and later. We have no reason to put under doubt the correctness of the spectral classification of this star; in all cases its belonging to the A2-type star is also confirmed by the results of energy distribution measurements, carried out by J.B. Ohanesyan (1975).

In the feature noted - usually powerful line 2800 MgII - we see the anomaly of the star HD 32296. Such stars apparently do not compose a great rarity. In all cases we are able to find also such a star among "Orion-2" spectrograms - HD 33332, of A2-type, for which W(2800MgII)= 20 Å.

More probable explanation of this fact may be the overabundance of magnesium in the photospheres of those stars. As the quantitative analysis shows the total amount of magnesium ions in those stars must be at least an order as large as in the usual A2-type stars. At the same time another possible explanation is not excluded either. We have in view the specific conditions in structure and physical processes, the conditions of the energy transfer in the photosphere of those anomalous stars and so on. At last the influence of the circumstellar cloud which may surround the star is not clear.

12. S t e l l a r   c h r o m o s p h e r e s   a n d   t h e i r   r a d i a t i v e   p o w e r. By means of "Orion-2" the ultraviolet spectrograms of late-type stars were also obtained. Some of these spectrograms proved to be of certain interest for the possible detection of stellar choromospheres. In Figures 24, 25 and 26 the microphotometric recordings for three of such stars of spectral types G5, K5 and M3 are given. The doublet of ionized magnesium, 2800 MgII, is clearly seen in emission (Gurzadian, 1975f).

Previously, the presence of 2800 MgII in emission was observed in the spectra of cool stars by Doherty, (1971), Kondo(1972). But, in the present case we encounter new facts according to which the presence of the doublet 2800 MgII in emission covers a wide spectral range - from G5 up to M3.

The star SAO 040296, of the K5 type, is the faintest (V=8.3) compared with the other two stars and perhaps this is the reason why the emission line 2800 MgII in its spectrum is not strongly pronounced.

We have an impression that the line 3080TiII is also present in emission in the spectra of these stars as a constant satellite of the line 2800 MgII. This line (multiplet No 5 of ionized titanium) is clearly visible in spectra of K5 and M3 stars and more vaguely in the spectrum of G5 star.

The emission lines 2800 MgII and 3080 TiII, however, are excited in the *chromosphere* of the star. Hence, their presence in emission provides a method of detecting the chromospheres of cool stars. The strength of these lines can be expressed as a fraction of the total radiative output of the stellar chromosphere. We can even derive some estimation of the lower limit of relative power for, the so-called, "magnesium" chromosphere for a given star with the known effective temperature, T_{eff}, it is the magnitude of the equivalent width of the line itself, W(2800). Under the "relative" power" we imply the ratio of the energy, E_{ch}, emitted by the chromosphere in the line 2800 MgII, to the total photospheric radiation, E_*. This ratio can be calculated from the observed equivalent width W(2800) in the following way :

$$\frac{E_{ch}}{E_*} = \frac{W(2800)\, B_{2800}(T_{eff})}{\sigma T_{eff}^4}$$

where $B_{2800}(T_{eff})$ is the Planck function, and σ - is the Stefan's constant.

In the subsequent quantitative applications we shall also include the stars α Boo (K2), α Tau(K5), and α Ori (M2), the OAO-2 ultraviolet spectrometric recordings of which are given by Doherty (1971), from which we have obtained very approximate values of W(2800) for those stars.

The final results for the measurements of W(2800) and calculations of the E_{ch}/E_* for the stars under examination are given in Table 3. It follows that the relative power of "magnesium" chromosphere in the late type stars is too large - of the order of 10^{-5} in the M2-M3 stars. It must be noticed that in the case of the Sun, the relative power of "Layman - alpha" chromosphere is of the order of 10^{-6} - considerably less than in the "magnesium" chromospheres of cool stars.

T a b l e 3

Equivalent widths, W(2800), of the emission line 2800 MgII, and its relative radiation power, Ech/E_*, in the "magnesium" photosphere of some late-type stars, from G5 to M3

Star	Spectra	T_{eff}	W(2800) Å	E_{ch}/E_*
SAO 040769	G5	5700°	14	70×10^{-5}
α Boo	K2	4850	20	40×10^{-5}
α Tau	K5	4400	15	13×10^{-5}
α Ori	M2	3500	18	2×10^{-5}
GC 7554	M3	3300	22	1×10^{-5}

It is clear that the total chromospheric emission power in these stars must be considerably greater, probably for one or two orders of magnitude, than in the case of solar chromosphere.

Particular attention should be paid to the star SAO 040769; here we encounter an abnormally powerfull chromosphere.

The resonance line 2852MgI, induced by *single* ionized magnesium, seems to be completely lacking in the spectrum of stars under examination. Thus we can conclude that magnesium in the chromospheres of these stars is practically found in a twice ionized state. Then we obtain for the minimum chromospheric temperature the value of the order of 10000°K.

The results obtained, despite their preliminary character, permit us to make at least one very important conclusion: the physical conditions in the chromosphere, its relative emissivity, the very fact of the existence of the chromosphere itself are independent of the spectral type or effective temperature of the star. In other words, the excitation and heating of the chromosphere in stars is not conditioned by thermal processes.

13. S t r a n g e s t a r No 1. We have found this faint star not far from Capella during the examination of the "Orion-2" material. Later, obtaining a ground-based picture, its brightness and colour characteristics were obtained: they turned out to be equal to : V = 12.45, B-V = -0.17 and U-B = -0.84(Ohanesyan O.V.,1975).The guiding map for this star is given in the paper by Gurzadian,(1975g.).Despite the faintness the shortwave boundary of the spectrum of this star was traced up to 2500Å which shows its high temperature. Actually, the measurement of its continuous spectrum revealed that the effective temperature was not lower than 20000°K (Fig. 27), even without the correction for the interstellar absorption. Considering this effect and assuming that star No 1 is at a distance of 1000 pc, we come to the conclusion that its effective temperature should be of the order of 50000°K, even higher. At the same time, this star cannot be simply an ordinary hot giant ($M \approx -3^m$ or -4^m), because of the same reason as considered above (section 9),in the case of the group of hot stars in Auriga. The requirement to be in Galaxy (from the side of the anticenter) results in $M > 0$ for these stars.In this we see the first strangeness of the star No 1.

The second and more essential strangeness of the star No1 is connected with its spectrum. As follows from the microphotometric recording (Fig.28), many emission like formations are present in its spectrum. Some of those formations, very probably, are not caused by functions of photographic density and are real. We have even tried to identify these lines (Gurzadian, 1975 g) which,certainly,will be reconsidered later. But even this endeavour has given some interesting results.

First of all, the forbidden lines are completely absent.This fact shows that in the case of the star No1 we have too dense stellar envelope.

Further, in the spectrum of the star No1 the emission lines even very strong ones of ionized magnesium, 2800 MgII, and ionized titanium, 3080 TiII, are present. By the way, this is the first case when the indicated lines are present in emission in the spectrum of a *hot* star.

At last, as we think, in the spectrum of the star No1 the emission lines of ionized and neutral helium are also present: 3205 HeII, 2733 HeII, 2945 HeI, 2830 HeI and so on. This fact also may be taken as an argument speaking in favour of the high temperature of this star.

The emission spectrum of the star No1 does not resemble the spectra of known to us objects-symbiotic stars, the Wolf-Rayet stars, Be or Of, the former nuclei of planetary or other peculiar objects. We do not know a single case when the lines of multiplet 3080 TiII, found at the limit of transparency of the Earth's atmosphere, were revealed in emission, although cases are known when one of the components of this multiplet, 3073 TiII, was discovered in the star spectrum as an absorption line of interstellar origin, (Dunham and Adams, 1939).

We can take another extreme case assuming that the star No 1 is a usual white dwarf. In this case it must be not far from us, not more than at 10-20 pc. Then how must we explain its emission spectrum? Perhaps as a white dwarf with a gaseous envelope? A subdwarf with the powerful chromosphere? Or no less powerful gaseous envelope around the star No 1?

The nature of the star No 1 is not known to us. The ground-based possibilities of astronomy, certainly, may be useful for elucidation of this problem earlier than a repeated attempt can be made to obtain new shortwave spectrograms of this star in space conditions.

14. Ultraviolet spectra of stars (F, G, K) of the intermediate type. From the abundant material of "Orion-2" we have also got a lot of information about the intermediate type stars - F, G, K, many aspects of the nature and structure of continuous spectra in ultraviolet of which are not clear for us.

The ultraviolet spectra of Capella, a GOIII type star, were processed first of all, (Gurzadian, 1974). The continuous spectrum of Capella extends to about 2100 Å. In the interval 2850-2100 Å nearly 10 absorption lines (bands) can be distinguished with more or less certainty. An "elevation" in the continuous spectrum at 2350-2430 is discovered also in the spectrum of this star. The impression is that there is a large-zone of depression on the both parts of the elevation.

In Fig. 29 the results of the processing of the continuous spectra for a group of F-G-K type stars are given, according to R.A.Ephremyan (1975) in the form of a graphical relationship of the fluxes, F_{λ}, from λ, taking $F_{3200}= 1$. All these stars are not far from us (100-130 ps) and, hence, the observed fluxes sometimes represent the true distributions of the energy in the spectrum, not affected by the interstellar absorption. The mean error of those measurements equals : 20% at 2400 Å and 12% at 3600 Å.

A comparison of the observed energy distributions in the spectrum of these stars with the theoretical distributions for model atmospheres, according to Parsons' calculations (Parsons, 1965), has been carried out. In these model-atmospheres the continuous absorption by hydrogen and helium was taken into account only without the influence of absorption in spectral lines. Despite this fact, the observed distributions have been in good agreement with the theoretical ones (Ephremyan, 1975). At the same time the considerable discrepancies have been found between observations and the theory in *certain parts* of the spectrum, which are a result, as we think, of blocking-effect of absorption lines, which are so numerous just in the F-G stars.

Apparently the divergences in the fine structure of the continuous spectra of different stars of one and the same spectral type are real. As an illustration, in Fig. 30 the observed energy distributions for three FO-type stars are shown. The most pronounced feature in these spectra is the presence of two wide and powerful depressions at 2800 Å and 2550 Å; the widths of each of these depressions is of the order of 200Å. In the first two stars - SAO 040256 - and 040226 - these depressions exist separately, in the case of the star SAO 011551 both these depressions merge and, as a result, a giant drop with the enormous extension from 2950Å down to 2450Å - is found in the continuous spectrum of this star.

The same as the depressions in the continuous spectra appear as a result of physical blending of single absorption lines, the observed scatter of the power and extensions of depressions must be interpreted as scatter of the density and power of the spectral lines themselves.

Such a variety of powers and extensions of the depression is also observed in the case of other subclasses- F2, F5, F8 and so on.

According to our measurements the best agreement between observations and the theory is found at the following values of T_{eff} and Lgg:

Class	T_{eff}	Lgg
FO	6900 °K	2.0
F2	6750	1.8 - 2.0
F5	6600	1.8
F8	6000	1.8

The results of "Orion-2" observations for the G-type stars are of certain interest. For such a star - SAO 021693- the results obtained are in good agreement with the OAO-2 observations for a G3V-type star, HD 53705 (Savage and Caldwell, 1974), and even for the Sun, a G2V-type star (Fig. 31). Nevertheless, the star SAO 021693 is classified in SAO catalogue as a G5 type. The colorimetric data also show an earlier type than G5 for this star (B-V =+1.08 , U-B =+1.8). The absence of the theoretical models for G-type and later stars makes impossible to make any conclusions concerning this question.

For another G5-type star, SAO 011491, the "Orion-2" results are in good agreement with the OAO-2 data for a G5V type star, HD 188376 (Fig. 31).

The observed energy distribution in the spectrum of the star SAO 021855 is shown in Fig.32; this star is classified in SAO catalogue as Ko type. But most probably it must be a G-subclass star; nearly complete coincidence of the continuous spectrum of this star - from "Orion-2" data, with the HD 180711, a G9III type star , - from OAO-2 data (circles in Fig.32), as well as the relatively strong line 2800 MgII in the spectrum of this star may be used as a confirmation of this point of view.

15. Spectral classification of stars by their ultraviolet spectrograms. The ultraviolet spectra give us a possibility to develop in principle new methods for the spectral classification of stars. At present, at least four such methods may be suggested in which one or another feature of ultraviolet spectra is used. These methods are the following:

1. *Classification by the continuous spectra in the ultraviolet.* This method is described in the paper of Gurzadian (1974a). Its first practical application for 2000 faint stars - from 10^m to 13^m - has been carried out by O.V.Ohanesyan (1975).

II. *Classification by the equivalent width of 2800 MgII.* This method was considered above, in section 10. Its practical application may be realized having the empirical relationship between W(2800 MgII) and the spectral type (Fig. 22).

III. *Classification by magnitude of depression, q, at 2800 Å.* This method was also considered above, in section 4. Its practical application is realized by means of the empirical relationship between q and the spectral class (Fig. 11).

IV. *Classification by the ultraviolet iron lines.* This method, suggested by R.A. Ephremyan (1975), is based on the fact that the ratio of the equivalent widths of the lines 2755 FeII and 2967 FeI, that is, the magnitude

$$Q = \frac{W(2755\ \mathrm{FeII})}{W(2967\ \mathrm{FeI})}$$

depends on the spectral type of the star. The empirical relationship between Q and spectra, constructed according to the "Orion-2" data for 64 stars, of FO-K2 type, is shown in Fig.33. Despite not very high accuracy of the measurements this relationship may be used for the spectral classification of the stars of intermediate types.

It does not exclude a possibility also to use the relationship "Q-Spectra" for the solution of different problems related to the physics of stellar atmospheres, for example at the determination of the effective temperature of the medium beyond the frequency of the ionization of iron.

16. Silicon rich stellar envelope. Among the data of "Orion-2" we also encounter a hot star with a gaseous envelope - SAO 077308, of the B1e type, and nearly 9-th magnitude, interesting due to a very intense emission line ~2520Å in its spectrum. After a series of attempts we have identified this line with the resonance sixtet of neutral silicon with the adopted wavelength 2520 SiI (Gurzadian and Rustambekova, 1975). The components of this sixtet are: 2507, 2514, 2516, 2519, 2524 and 2528 Å, moreover, the second of these lines is a resonance, the others are quasiresonances ($E_{exc} \approx 0.01 - 0.03$ eV). This identification, however, is not assumed to be final and is open to examination. But independently of possible changes in this identification the fact of the presence of a powerful emission line in the spectrum of this star remains incontrovertible and it is just the thing that we consider to be important. In other words, the question of abnormal abundance of the unknown element exciting line 2520 Å in the envelope of this star cannot be overlooked.

The emission line 2520 Å is the strongest line in the spectrum of the star under examination at least in the interval of wavelengths from 5000 Å to 2300 Å (Fig. 34).

In the " Orion-2 " material we were also able to find the spectrograms for the three hot stars with gaseous envelope and nearly of the same class. One of these stars is ξ Tau - a well-known emission star, of B2e-type, the others - HD 37967, of B3ne type, and HD 38191, of B1(V) ne type. The comparison of the microphotometric recordings of all these stars shows that the line 2520 SiI peaks out strongly in comparison to the neighbouring emission lines in the case of SAO 077308 and sometimes, in the case of the other three stars this line has the same intensity as the neighbouring lines, (2473 FeI, 2573 AlI and so on).

An approximate estimation shows that in the gaseous envelope of SAO 077308 silicon is nearly an order of the magnitude as abundant as in usual Be-type stars. As to the question in what degree this is also characteristic of the photosphere of SAO 077308,it can be the subject of another investigation, to which end it is necessary to search for and discover first of all the silicon *absorption* lines in the spectrum of the star under examination. Also the question is not clear what place this star takes among the so called silicon stars in general (peculiar B-stars), some characteristics of which have been studied, in particular, by Eggen (1974).

17. Ultraviolet spectra of a planetary nebula. With the help of "Orion-2" the ultraviolet spectra of the planetary nebula, IC 2149, and of its nucleus, an 07 type star have been obtained for the first time in astrophysical practice. The integral brightness of this nebula is $9^m.9$ in photographic rays and the brightness of the nucleus is $10^m.2$-in photovisual.

Only one successful spectrum was obtained for this nebula and its nucleus. The densitometric recording of this spectra is given in Fig. 35 (the central part of this spectrum, from 3300 Å to nearly 2800 Å, has a superimposed long wavelength spectrum of a nearby star of $11^m.5$ magnitude). In spite of extremely small dispersion at 5000 Å (≈ 3000 Å/mm) the lines N_1+N_2 [OIII], H_β and some of the higher lines of the Balmer series of hydrogen are visible here .
The 3727 [OII] line, the shortest line registered in the spectrum of this nebula by ground-based observations (Aller, 1951), is distinctly pronounced. Towards shorter wavelengths down to 2400 Å, many emission lines are present, some of which may be caused by fluctuations of photographic density, but others are real emission lines of the nebula. The suggested identification of some of these lines is given in Table 4. The details of the identification analysis of these lines are given in the papers of Gurzadian (1975a, 1975h). Here we confine ourselves to the general review of the results obtained. At the same time we shall give some applications connected with the physics of planetary nebulae.

Table 4

Ultraviolet emission lines in the spectrum of the planetary nebula IC 2149

Wavelength Å	Relative intensity *)	Identification
2440	20:	2440 + 2442 [NeIV]
2500 ?	-	-
2530?	-	-
2550	-	2545 HeI
2670	20	2669 AlI
2730?	-	-
2750 (?)	-	2750 FeI
2800	10	2795+2802 MgII
2830?	-	2829 HeI
2850	10	2852 MgI
2950?	15	2945 HeI
3080	13	3080 TiII
3135	10	3135 OII
3200	-	3200 FeI, 3200TiI
3280	-	-

* the Intensity of 3727 [OII] is taken as 10.

First of all, the discovered ultraviolet lines appear to belong to the well-known groups of emission lines, usually observed in planetary nebulae, that is, the resonance (2800 MgII, 2852MgI), the forbidden (2440 [NeIV]), and recombination lines (2445 HeI, 3135 OII). Further, nearly all the discovered lines have been predicted earlier - the forbidden lines by the author (Gurzadian,. 1969; Gurzadian, 1965). the fluorescete lines - by Aller (1961). These lines are, for example: the forbidden lines 2440+2442 NeIV, the resonance lines 2669 AlII and 2800 MgII, recombination line 3135 OII and so on.

A confirmation is available for the first time of the presence of two chemical elements-aluminium and titanium in the planetary nebula assuming of course, that our identifications of the resonance line 2669 Al II and multiplet No5 of ionized titanium at ~3080Å are correct). We also got a confirmation of the presence of magnesium, through the line 2800 MgII, in the planetary nebula (earlier, a very faint forbidden line 4571 [MgI]of neutral magnesium has been found only in one case - in NGC 7027 (Aller, 1955). It is interesting to note that only 16 chemical elements have been found in planetary nebulae at least during 50/60 year period of their studies.

The presence of the forbidden doublet of three times ionized neon, 2440 + 2442 [Ne IV], an analog of the doublet of single ionized oxygen, 3726+3729 [OII], in the spectrum of such a low excited nebula as IC 2149, is a curious thing. As the quantitative analysis shows (Gurzadian,, 1975h), in the high excited nebulae (for example, in NGC 7027) the doublet 2440+ +2442 [NeIV] must be almost the third emission line by its strength after the lines $N_1 + N_2$[OIII] and H_α.

The doublet 2440+2442 [NeIV] may be of a particular interest at deciphering physical conditions in planetary nebulae. The thing is that this doublet can be excited only in the central parts of planetary nebulae,where,according to the theory (Gurzadian, 1969), the electron temperature must be high in its "envelope"(twice - higher than the electron temperature in the main body or nebula). Hence, from the observed strength of this doublet we can get some information about the magnitude of the electron temperature, T_{el}, in the central parts of nebula. It is possible to solve this problem in practice using the following relationship (see Gurzadian, 1975 h).

$$\frac{E_{2440} + E_{2442}}{E_{4720}} \approx 8.4 \exp \frac{30500}{T_{el}} \qquad (1)$$

As to the electron temperature, T_e, in the main "envelope" of the nebula, it is obtained by the usual method - with the help of the lines $N_1 + N_2$ [OIII] and 4363 [OIII] .

The ratio E(2440 + 2442)/ E(4720) is very sensitive to the electron temperature T_{el} (this ratio equals, for example, 170 and 38 at T_{el}=10000° and T_{el} = 20000° respectively. Hence, it may be even enough to know very approximate magnitude of this ratio from observations to determine T_{el}.

The presence of strong emission lines of neutral (2852 MgI) and ionized (2800 MgII) magnesium opens interesting possibilities for deriving a relation between their intensities and some parameters of nebulae, including the effective temperature of the nucleus, T_*, and the factor of dilution W.

We can write for the emission of the nebula in the resonance lines 2800 MgII and 2852 MgI:

$$E(2800) = n^{++} n_e \alpha_t (Mg^{++}) p(Mg^+) A^+_{21} h\nu^+ \; ; \qquad (1)$$

$$E(2852) = n^+ n_e \alpha_t (Mg^+) p(Mg) A_{21} h\nu \; , \qquad (2)$$

where n^{++} and n^+ are the concetrations of twice and single ionized magnesium; $\alpha_t(Mg^+)$ - complete recombination rate for Mg^{++}; $p(Mg^+)$ - the relative part of transitions corresponding to the resonance level of neutral magnesium; A^+_{21} - the coefficient of spontaneous transition from the resonance to the normal level for Mg; all these notations with an additional cross are related to the ionized magnesium (or to the line 2800 Å),

The condition of ionization equilibrium between the single and twice ionized magnesium atoms will be written in the form:

$$Wn^+ \int_{\nu_0}^{\infty} \alpha\nu \frac{B_\nu (T_*)}{h\nu} d\nu = n^{++} n_e \alpha_t (Mg^{++}) \; , \qquad (3)$$

where $B_\nu (T_*)$ is the Planck's function at the effective temperature of the nuclei, T_*; ν_0 - the frequency of double ionization of magnesium; $\alpha\nu$ - the coefficient of the continuous absorption for single ionized magnesium.

From (1), (2), (3), we find, putting $\alpha_\nu = \alpha_o (\nu_o / \nu)^2$:

$$\frac{E(2800)}{E(2852)} = \frac{2\alpha_o \nu_o^2 \kappa}{hc} \frac{A^+_{21}}{A_{21}} \frac{W}{n_e} \frac{T_*}{\alpha_t (Mg^+)} \exp(-h\nu_o / \kappa T_*) =$$

$$9{,}3 \cdot 10^{14} \frac{W}{n_e} T_* \exp(-176000/T_*) \qquad (4)$$

where : $\alpha_o = 0{,}24 \cdot 10^{-18}\,cm^2$; $A^+_{21} = 52.\,10^7\,sec^{-1}$; $A_{21} = 46.10^7\,sec^{-1}$ (Allen, 1974) $\nu_o = 3.66 \cdot 10^{15}\,sec^{-1}$; $\alpha_t(Mg^+) = 1.77\,10^{-13} sec^{-1}$ at $T_e = 10000$°K (Tarter, 1971).

For the planetary nebula IC 2149 we have: $T_* = 50000$°K (the mean from the estimations by Gurzadian, (1969) and O'Dell (1962) $n_e \approx 2.5\,10^3 cm^{-3}$ (Gurzadian, 1969) and $E(2800)/E(2852) \approx 1$ (from Table 4). With these data we obtain from (4): $W = 1.2 \cdot 10^{-15}$. At the distance of this nebula ~1000 pc (Gurzadian, 1975 h) and the angular diameter~10'' (Perek and Kohoutek, 1967) we shall have for its linear diameter $D=3 \cdot 10^{17}$cm. Hence, we have for the radius of the nucleus(the central star) $R = DW^{1/2} \approx 10^{10}$cm $\approx 0.14\,R_\odot$.

The possibilities of the use of emission lines 2800 MgII and 2852 MgI apparently are not exhausted by a just considered example.

The role of the group of the emission lines of ionized titanium with the mean wavelength ~3080Å for the physics of planetary nebulae is not completely clear. This group consists of nine lines but only four of them - 3073, 3075, 3078 and 3088 Å - are the strongest, and two of them - 3066 TiII and 3073 TiII - are even resonances. However, before making any definite conclusions on this question, it is neccessary to have the atomic parameters for the corresponding transitions for the single ionized titanium.

The last remark also concerns the resonance line of ionized aluminium, 2669 Al II.

The microphotometric recording also shows the presence of other emission lines (Fig. 35), but the reality of these lines and especially their identification need more careful examination.

18. Ultraviolet spectrum of the nuclei of planetary nebula. The opportunity of making measurements of the spectra of both the nebula and of its central star on one and the same image is very rare; since then it is necessary to have almost the same photographic densities of each component. This circumstance may be used, in particular, for the presentation of the intensities of nebulae emission lines in the absolute energy units with the far-reaching aims. However, leaving the practical application of this possibility for the future, we shall present here the results of the measurements of the continuous spectra of nuclei in the region of the wavelengths 2400-3800 Å.

The level of the continuous spectrum of the nucleus has been traced in the microphotometric recordings with the allowance made for the presence of the emission component of the nebula itself and also of the near star (a broken line in Fig.35); it is considered that the level traced represents a lower limit. The results of the measurements are presented in Fig. 36 where Δm is plotted against λ (open circles), putting $\Delta m_\lambda = 0$ at $\lambda = 3500$ Å. The theoretical spectrum, according to Mihalas(1965) model for the effective temperature of the nucleus of IC 2149 (Gurzadian, 1969),(O'Dell, 1962)- is also given in Fig. 36 by a solid line.

The difference between the theoretical and observed data is great (see Gurzadian, 1975 h) and arises, of course, from the interstellar absorption. We introduced a correction for this effect under the following assumptions: (a). The low interstellar absorption at least up to 2400 Å has the form λ^{-1} (see section 6); (b). The value of the interstellar absorption is 2^m in photographic rays (4400Å). The final result - the corrected distribution of energy in the continuous spectrum - is presented in the same Fig. 36 by dots. The agreement with the theory, as we see, is quite satisfactory at least from 3800Å to 2700Å. At shorter wavelengths there seems to be an *excess* of the radiation from the nucleus in comparison to the theoretical model distribution. At the same time, as mentioned above, this estimate is considered as a lower limit.

19. U l t r a v i o l e t   s p e c t r a   o f   s t a r s   o f   O-t y p e. Two stars of O-type the nuclei of the planetary nebula IC 2149, of O7 type star, and HD 36879, of O6-type (Rustambekova, 1975), have been included by observation program of "Orion-2". The energy distribution in their continuous spectra is presented in Fig.36. In both cases the energy distribution is in a good agreement with the theoretical model at least to 2600-2700Å. Further, towards the shortwave region, an excess of the observed radiation appears.We think that this excess is real and cannot be explained by errors of measurements.

Of course, it is impossible to make any conclusions only by the data of these two stars. Nevertheless we connect the origin of this excess with the same effect of blending of emission lines, which we have examined in the case of Be stars (see section 6). If such an explanation may be true then we come to the conclusion about the existence of a faint gaseous envelope, around these typical O-type stars with the absorption spectrum. These envelopes cannot be discovered - due to some unknown causes - in the optical region of observations. In the case this assumption is confirmed,we shall come to the conclusion about the absence of O-type stars with absorption characteristics in general, the stars of OA-type do not exist.

The question of reality of the blending effect in O-type stars of course needs more careful examination. However, an important conclusion may be drawn even now; the stars of O-type cannot be used,without having in view their emission features,as indicators for finding the magnitude and character of the interstellar absorption in ultraviolet (2000 - 3000 Å) especially about 2200 Å where the excitation curve reaches the sharp maximum. This conclusion is in harmony with that of Haramundanis(1973)who came to the similar conclusion on the basis of OAO-2 observation data. Perhaps in this one should search for an explanation of the scattered nature of the results by Bless and Savage (1972) in all the cases when the general characteristics of the interstellar absorption were derived from the comparison of two O-type stars, one of which has large absorption, the other – small.

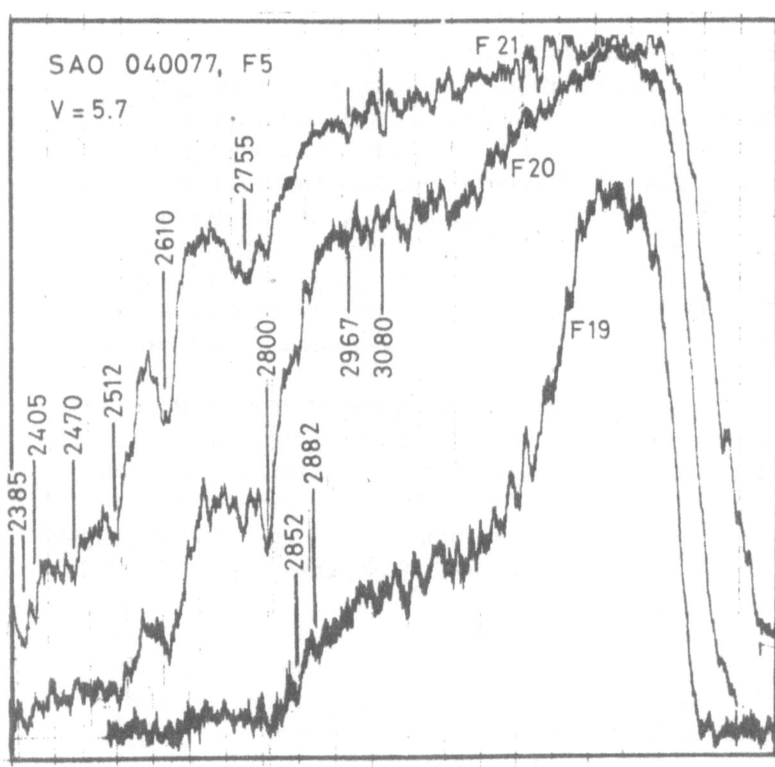

Fig. 1. Microphotometric recordings from three shortwave spectrograms of the star SAO 040077, type F5, obtained by "Orion-2" at the exposure times 18 min (F21), 1.5 min (F20) and 15 sec (F19). Some absorption lines are shown.

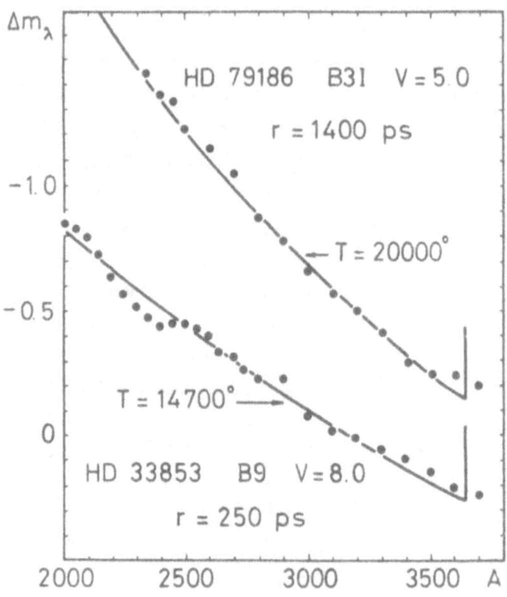

Fig.2. Distribution of the energy in the ultraviolet of the continuous spectrum of HD79186, type B3, and HD33853, type B9 corrected to the interstellar absorption (dots). Full lines are the theoretical distributions at different effective temperatures.

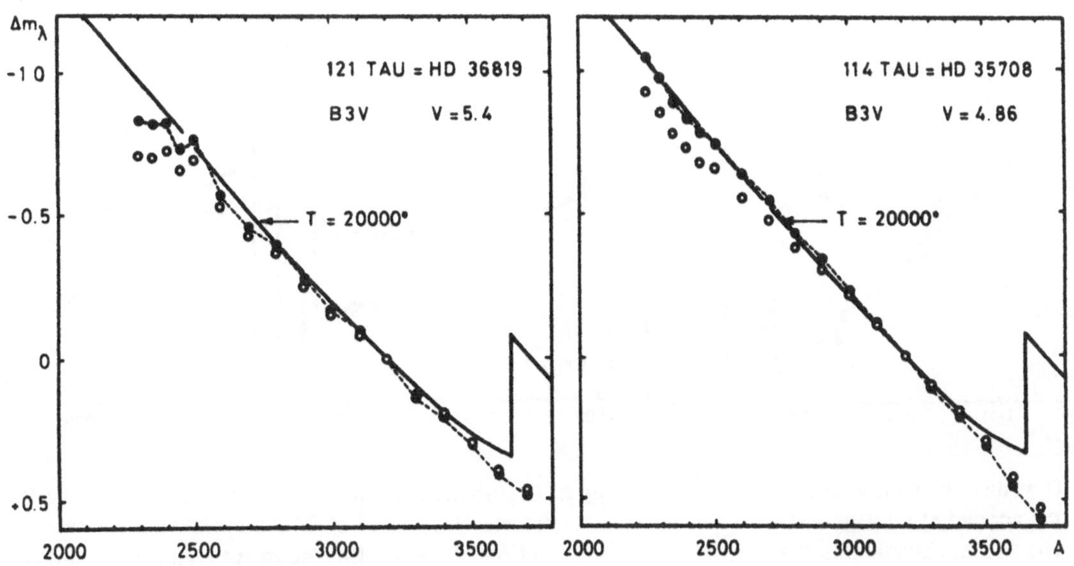

Fig. 3. Distribution of the energy in ultraviolet of two B3 type stars, 121 Tau and 114 Tau; observed (circles) and corrected for interstellar absorption (dots). Full lines - theoretical distributions at T_{eff} = 20000°K.

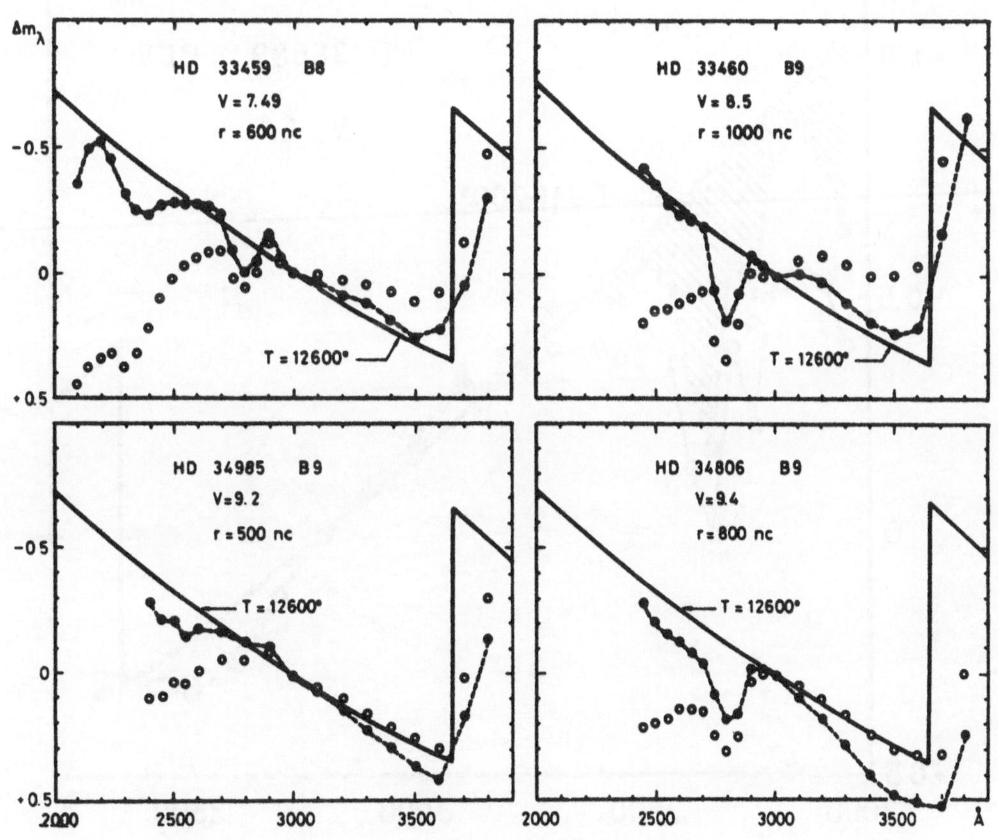

Fig. 4. Graphical dependence between the observed (circles) and corrected for the interstellar absorption fluxes(dots), Δm_λ on λ , for four B-type stars according to "Orion-2" data. Full lines-theory.

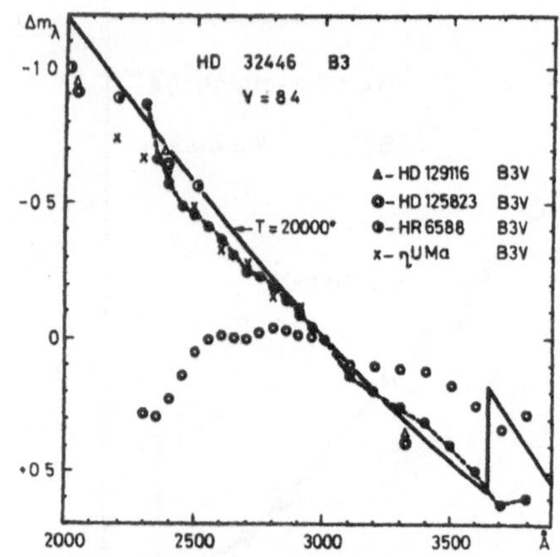

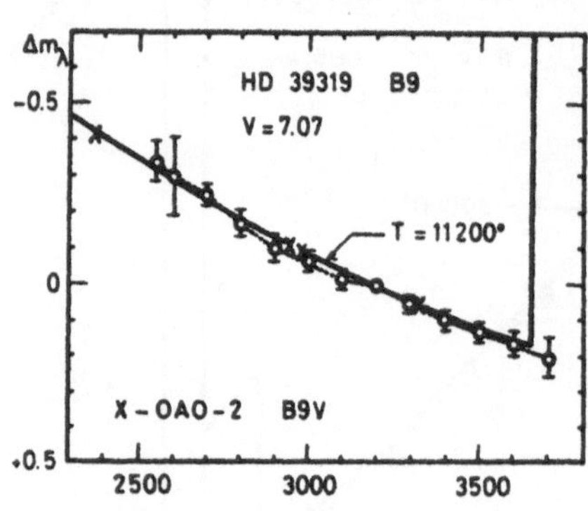

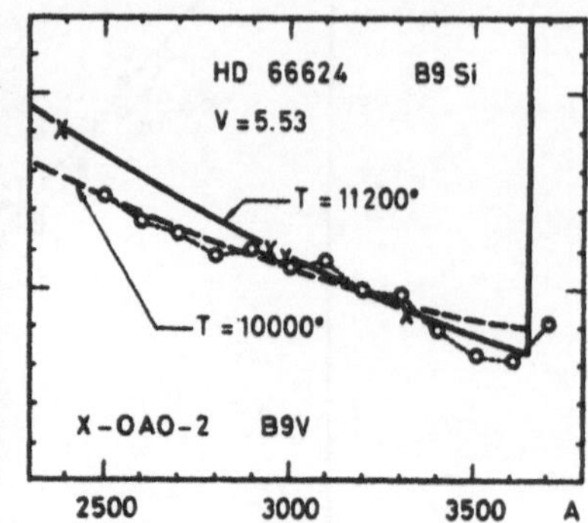

Fig.5. Distribution of the energy in the spectra of a B3-type star, HD 32446, according to "Orion-2" (circles and dots). The results of another observations for four B3V type stars are also shown(see [7]).

Fig. 6. Distribution of the energy in the spectrum of the B9 type stars. The results of OAO-2 observations (the mean values of Δm_λ from two B9V stars, 14 CVn and HD 4622) also are shown. Full lines - theoretical models.

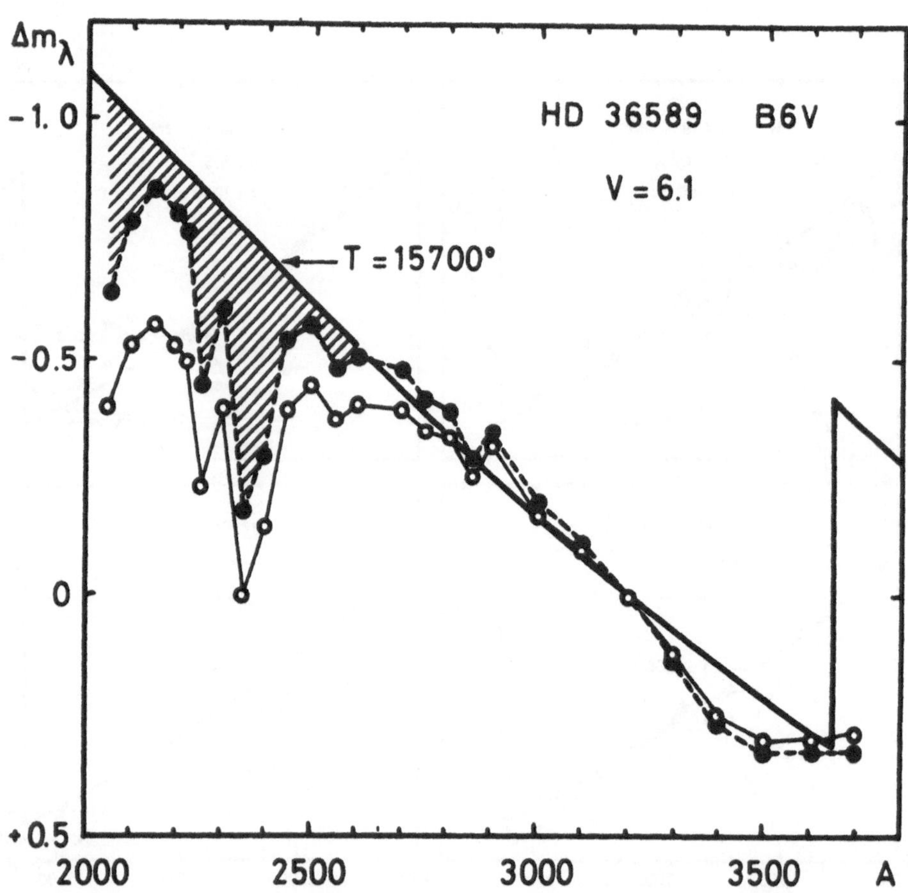

Fig. 7. Blocking effect by absorption lines in the spectrum of HD 36859, a B6V type star. The depressions in continuous spectra on 2350Å, 2250Å and 2050Å are visible.

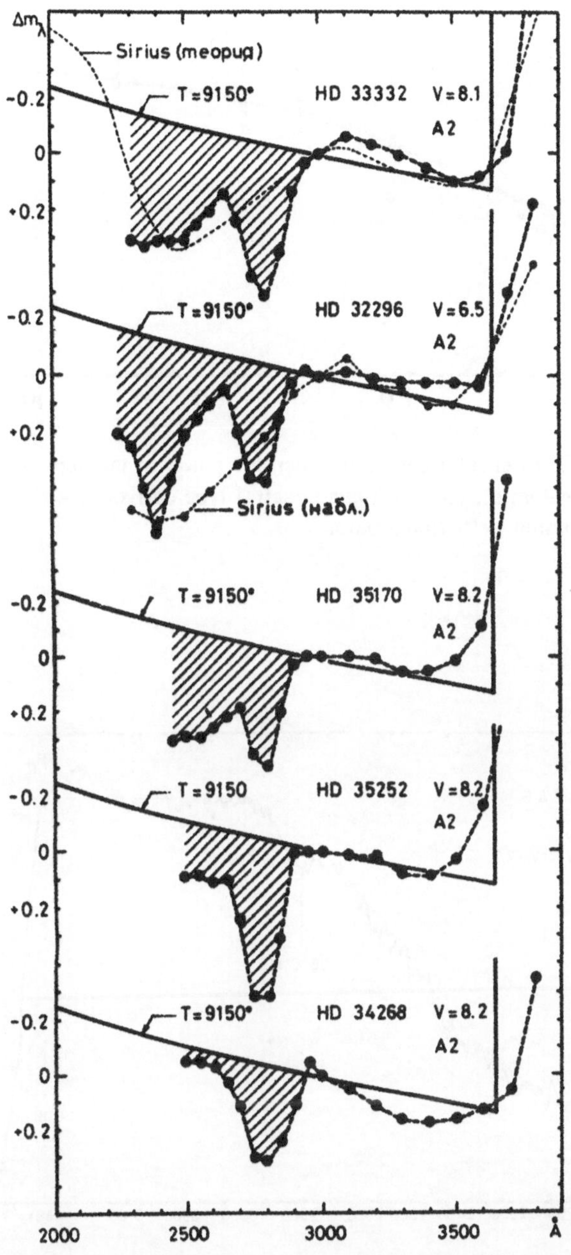

Fig. 8. Sequence of ultraviolet spectra of few A2 type stars with varying microstructure - the different powers of the depressions on 2400Å and 2800Å according to "Orion-2" data (dots).

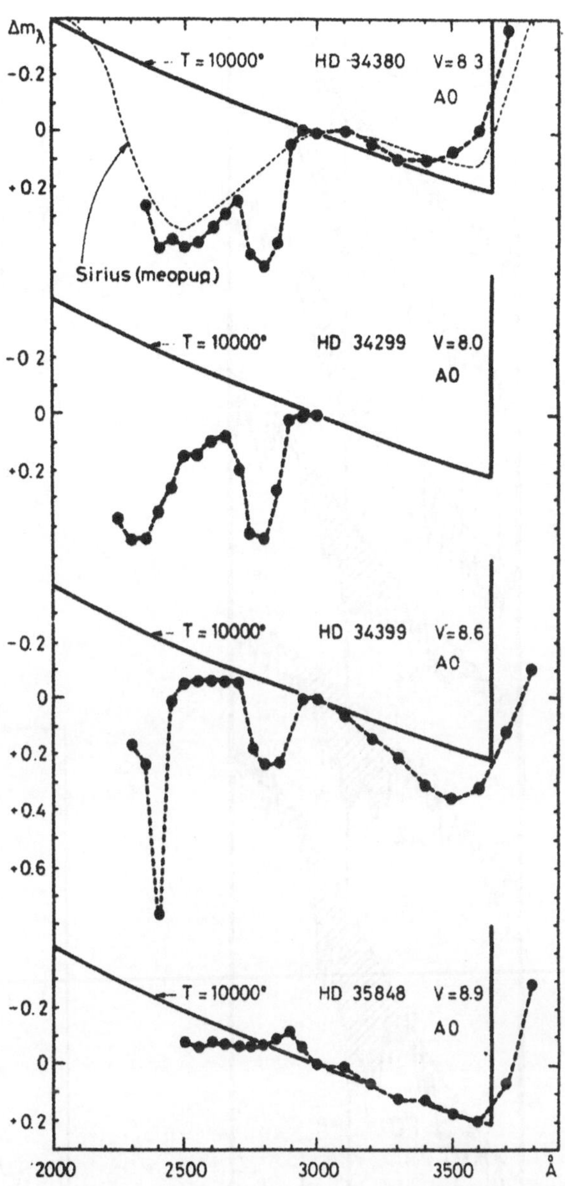

Fig. 9. Microstructure of ultraviolet spectra of a group from four AO type stars according to "Orion-2" data (dots). The stars with the extremely powerful depressions on 2800 Å and 2400Å(HD3480) as well as with nearly unnoticably depressions (HD 35848) are presented.

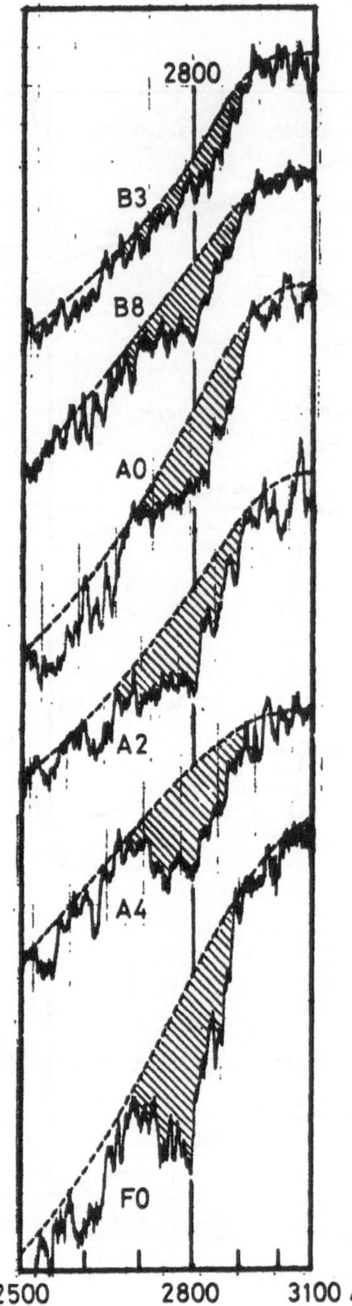

Fig. 10. Sequence of the fragments of microphotometric recordings from the spectrum of B3 - FO type stars, which illustrates the dependence of the depression on 2800Å from the spectral type. All spectrograms are from "Orion-2" and relate to the stars: B3 - HD 32446; B8 - HD 33542; AO - HD 34380; A2 - HD 33332 ; A4-HD32619; FO - HD 34331.

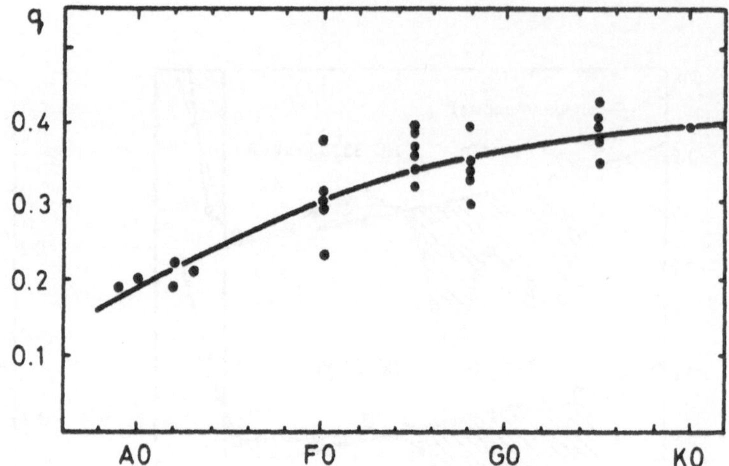

Fig. 11. Empirical relationship between the power of the depression, q , on 2800Å and spectral type derived according to "Orion-2"data.

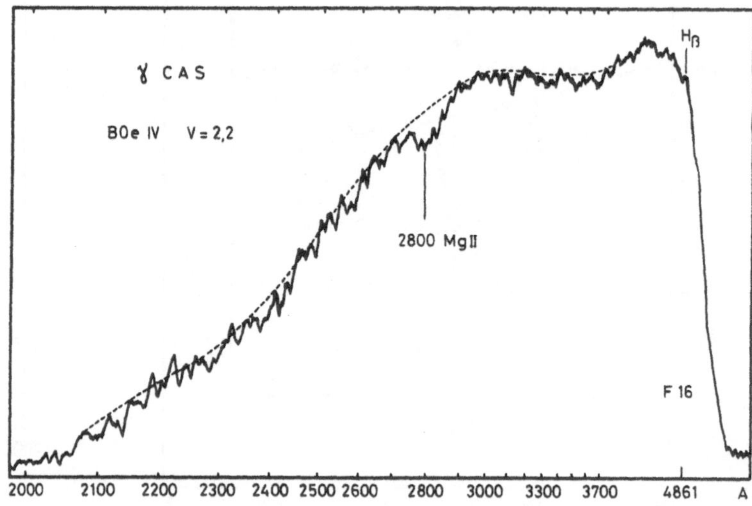

Fig. 12. Densitometric recording from the ultraviolet spectrogram (frame F16) of γ Cas obtained by "Orion-2". The dotted line is adopted continuous level. The strong depression on 2800 Å is visible.

Fig. 13. Fragments from the densitometric recordings of four spectrograms of γ Cas obtained by "Orion-2". The depression on the continuous spectrum, from ~ 2670 Å to 2970 Å (dotted vertical lines) and with the maximum depth near 2800 Å, is visible. The interval 2740 - 2870Å is the working area in the experiment of Boksenberg et al. [43]

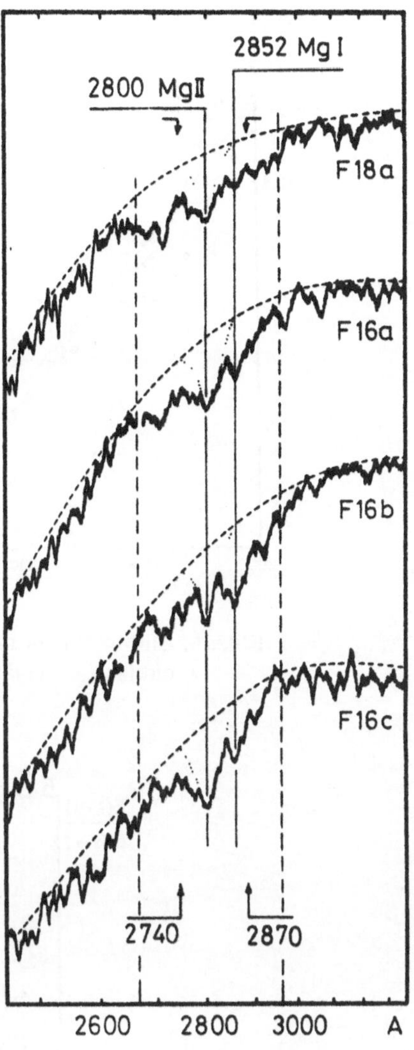

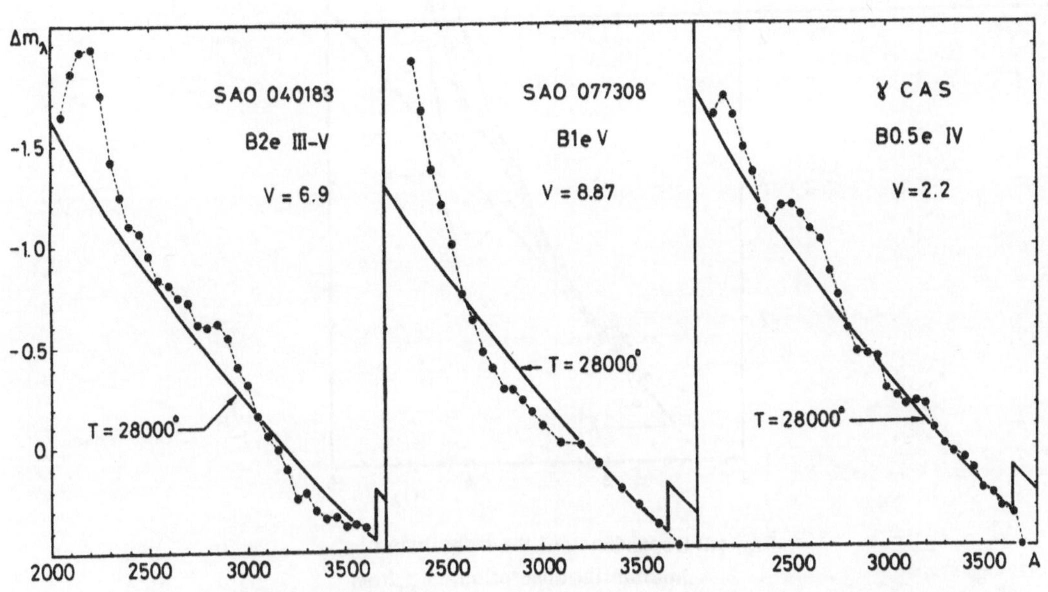

Fig. 14. Distribution of the energy in ultraviolet of continuous spectra for three emission stars (dots, corrected for interstellar absorption). Two protrusions on ~2200 Å and 2600 Å are visible.

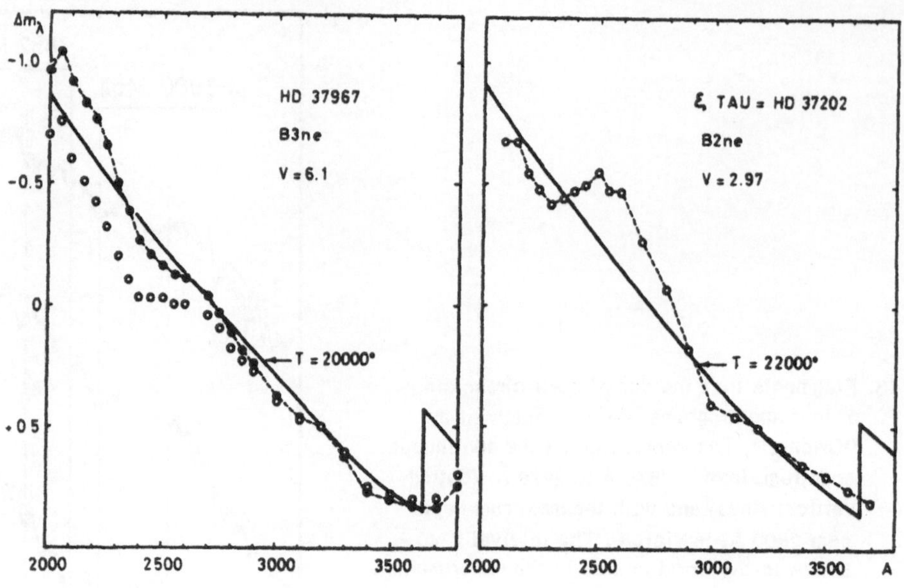

Fig. 15. Distribution of the energy in ultraviolet spectrum of two emission stars: circles - observations; dots - corrected for interstellar absorption.

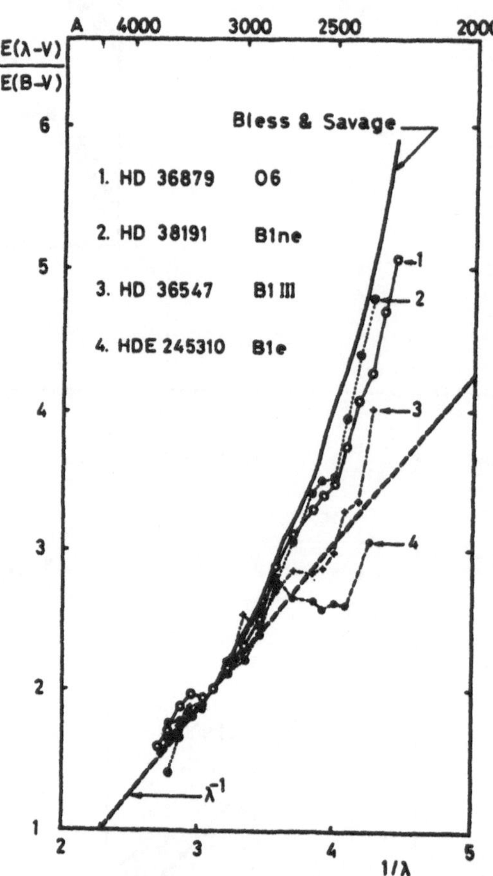

Fig. 16. Dependence of the magnitude of interstellar absorption, X_λ, from $1/\lambda$ according to "Orion-2" data, for four stars: HD 36547, 36879, 38191 and E 245310. Four comparisons, the mean OAO-2 data from [5] as well as the last λ^{-1} also are shown.

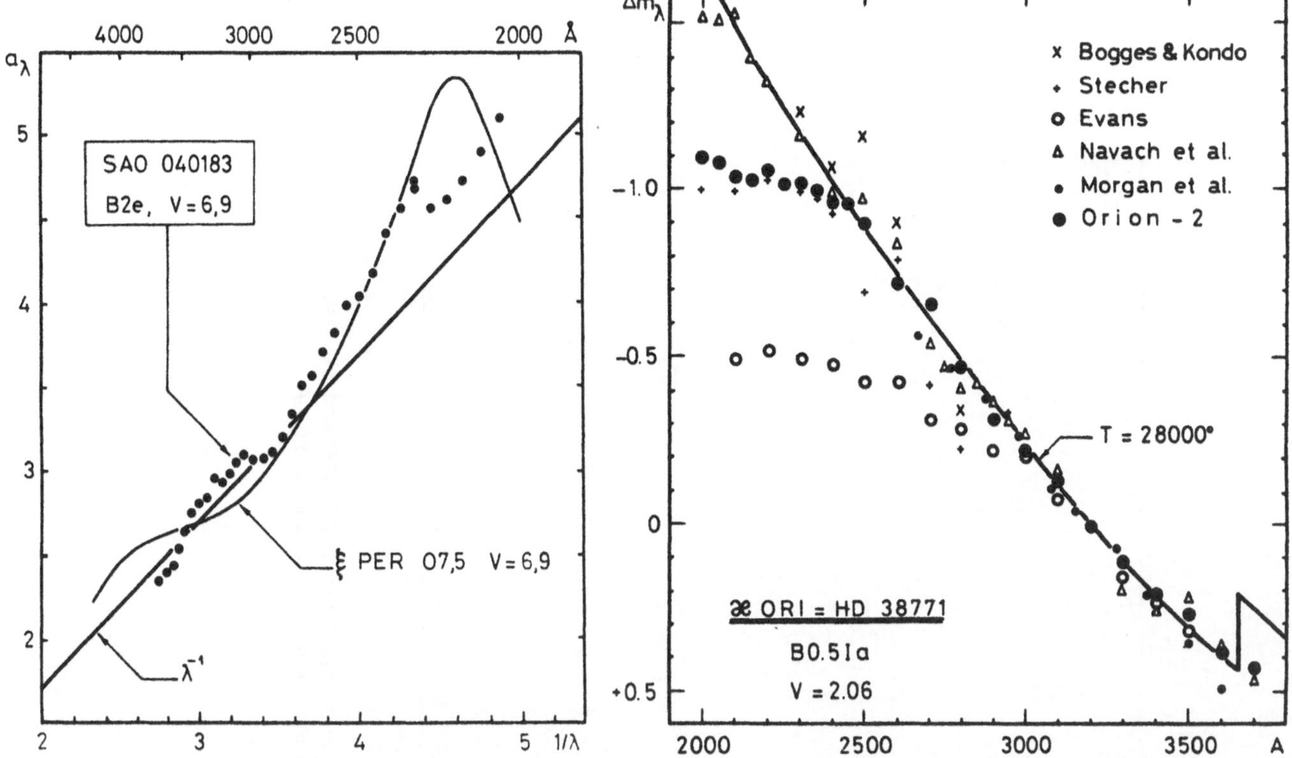

Fig. 17. Dependence of interstellar (and stellar) absorption, a_λ, from wavelength, λ, for the emission star SAO 040183 (points), according to "Orion-2" data. The interstellar absorption curve for ξ Per, according to OAO-2 data [5] also is shown. a_λ is in stellar magnitudes.

Fig. 18. Distribution of the energy in the ultraviolet of the star ϰ Ori, type BO, 5 1a, according to "Orion-2" (dots) and other observations. Full lines - theoretical model at $T_{eff} = 28000°K$ and lgg =4.0.

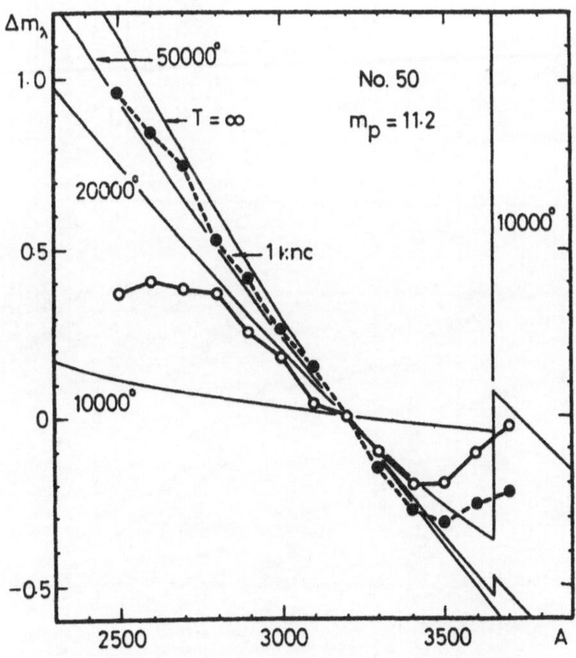

Fig. 19. Distribution of the energy in the spectra of an "ultraviolet" star No 50 in the wavelength region 2500 - 3700 Å. Circles - direct observations, dots - corrected for interstellar absorption, curves - theoretical distributions at various temperatures.

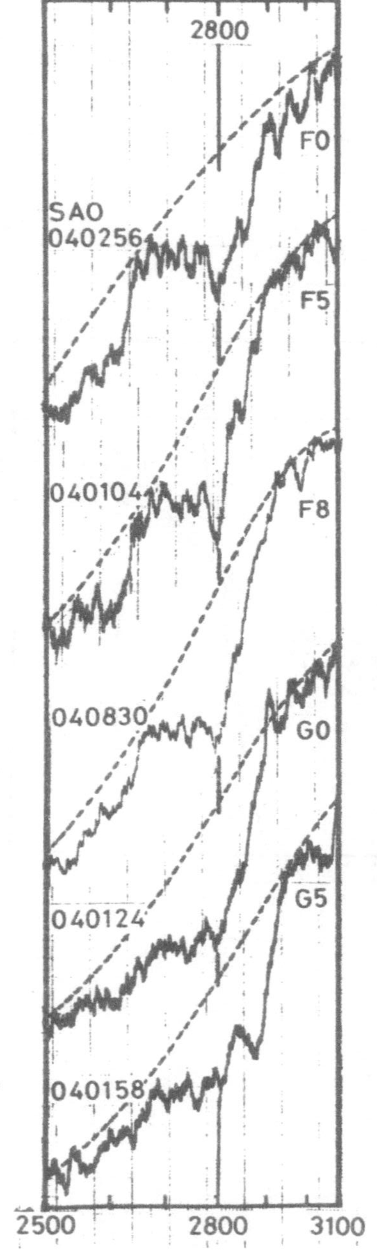

Fig. 20. Fragments from the microphotometric recordings of the spectra of few FO-G5 type stars near 2800Å.

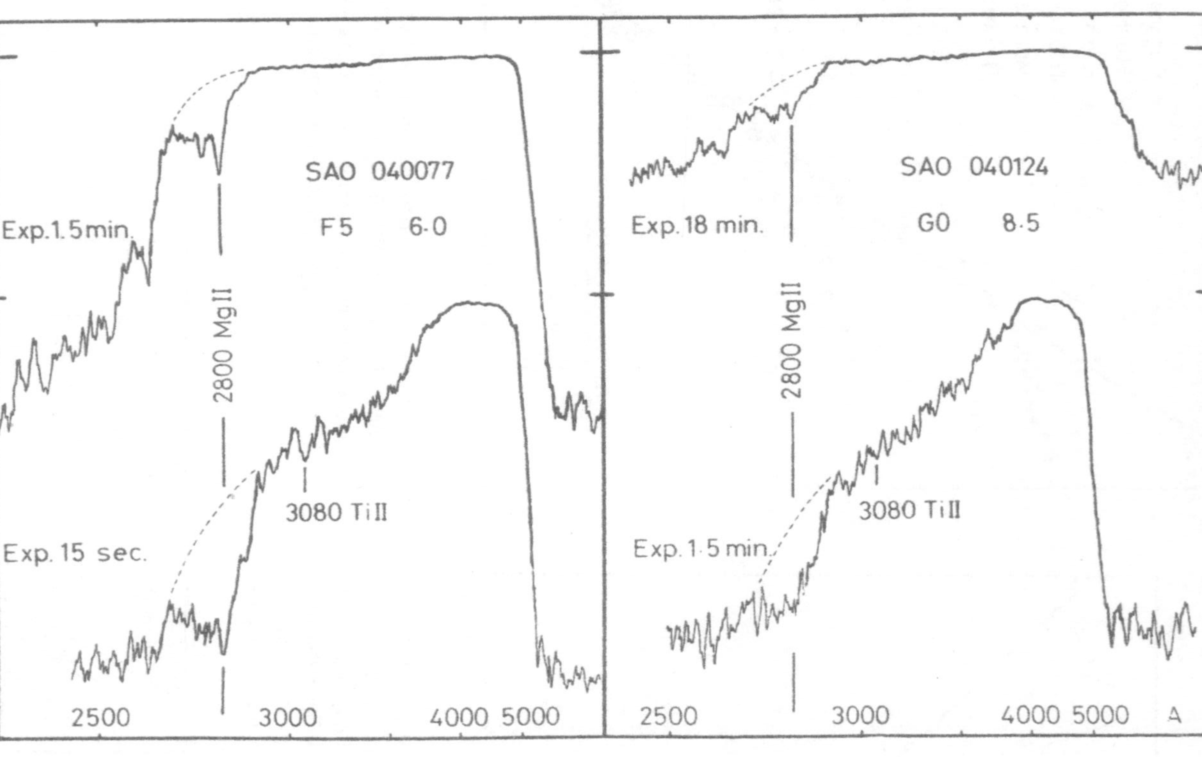

Fig. 21. Microphotometric recordings of the spectra of SAO 040077 and SAO 040124, F5 and Go type stars. A broad depression in the continuous spectrum at 2800Å is visible, as well as its break off at short exposures.

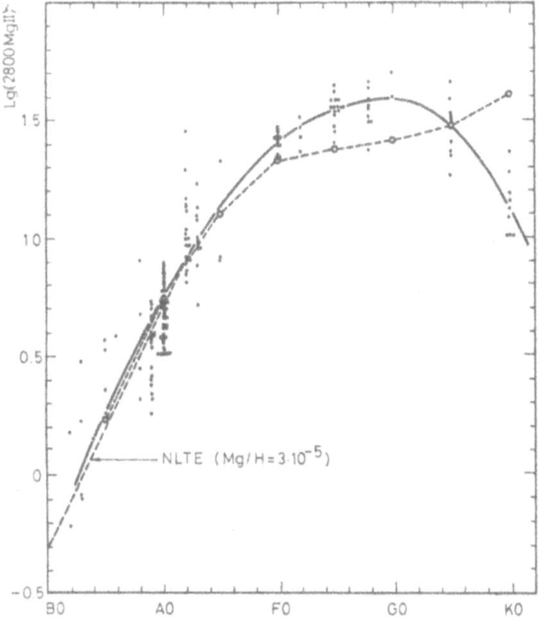

Fig. 22. Empirical relationship (continuous line) between the equivalent width of the resonance doublet of ionized magnesium 2800 MgII and the spectral class of the star according to the data of "Orion-2" data (dots). The results of Non LTE (dotted line) and approximate calculations (circles with dotted line) are also plotted.

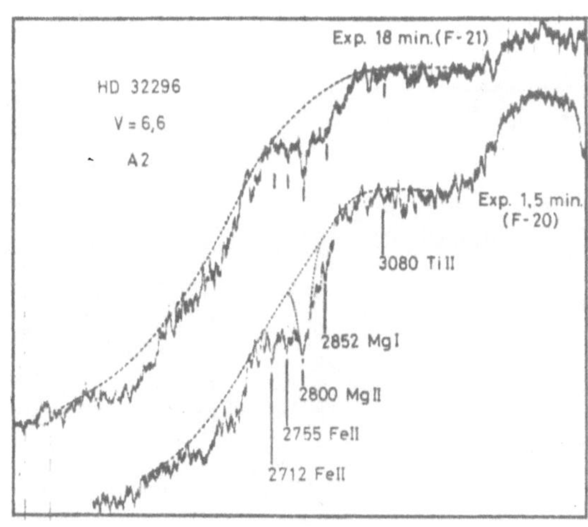

Fig.23. Microphotometric recordings of the spectrograms of an anomal star HD 32296, type A2, obtained with the exposure times 1,5 min (F20) and 18 min (F21). The anomaly strong line ionized magnesium on 2800 Å and line of neutral magnesium on 2852 Å are visible.

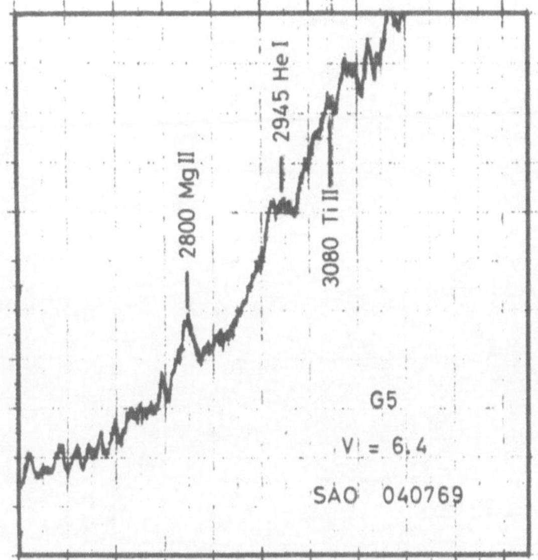

Fig. 24. A microphotometric recording of the ultraviolet spectrum of a G-type star, SAO 040769. The chromospheric emission line 2800 MgII is visible clearly, the line 3080 TiII - less confidently. The line 2945 HeI is of doubtful reality.

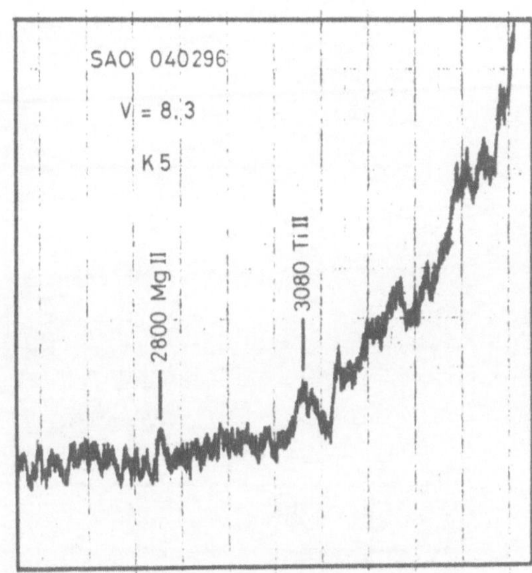

Fig. 25. Chromospheric emission lines 2800 MgI and 3080 TiII are visible clearly in the spectrum of a KS type, relatively faint, star, SAO 040296.

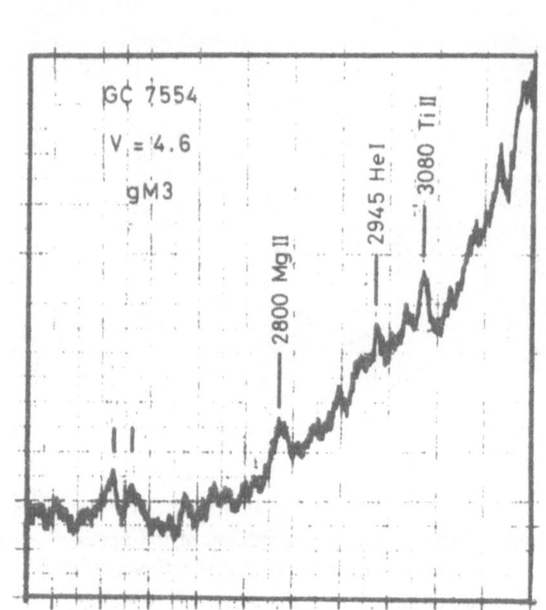

Fig. 26. A microphotometric recording of the ultraviolet spectrum of a late-type M3 star, GC 7554. The chromospheric emission lines 2800 MgII and 3080 TiII are visible, the line 2945 HeI is doubtful.

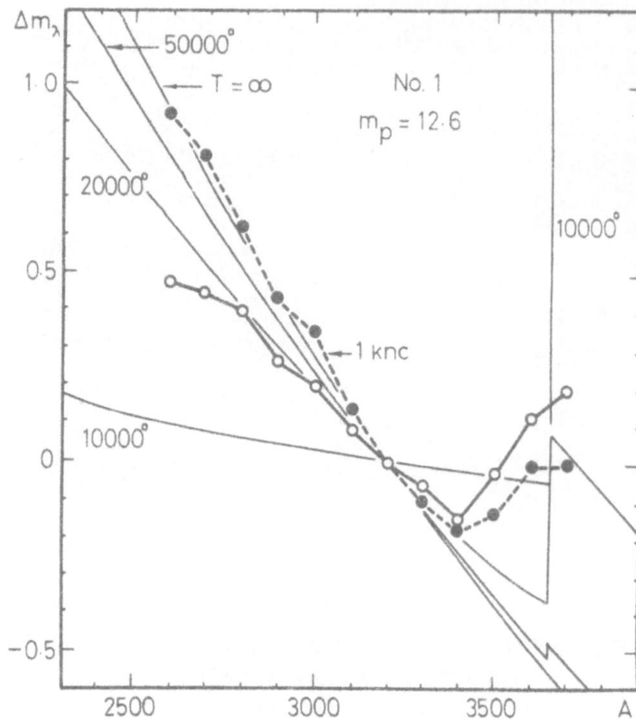

Fig. 27. Distribution of the energy in continuous spectrum of the remarkable star No 1, observed (circles) and corrected for interstellar absorption (dots). Full lines are the theoretical models for differer temperatures.

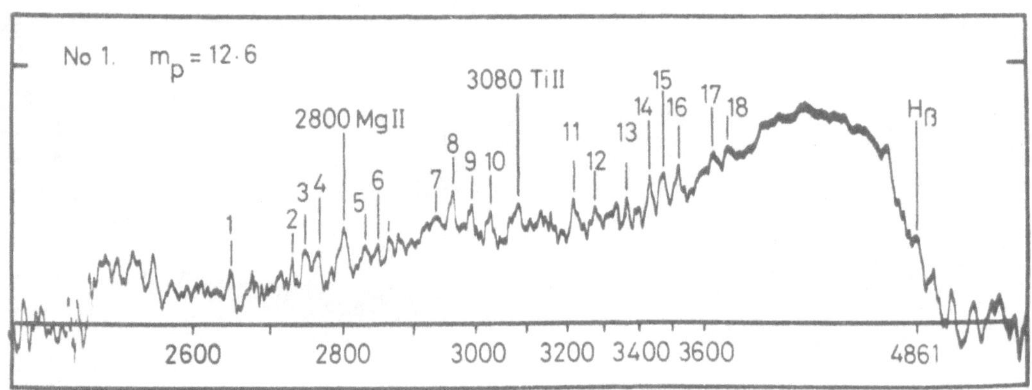

Fig. 28. Microphotometric recording from the ultraviolet spectra of the star No 1. The ultraviolet emission lines are visible.

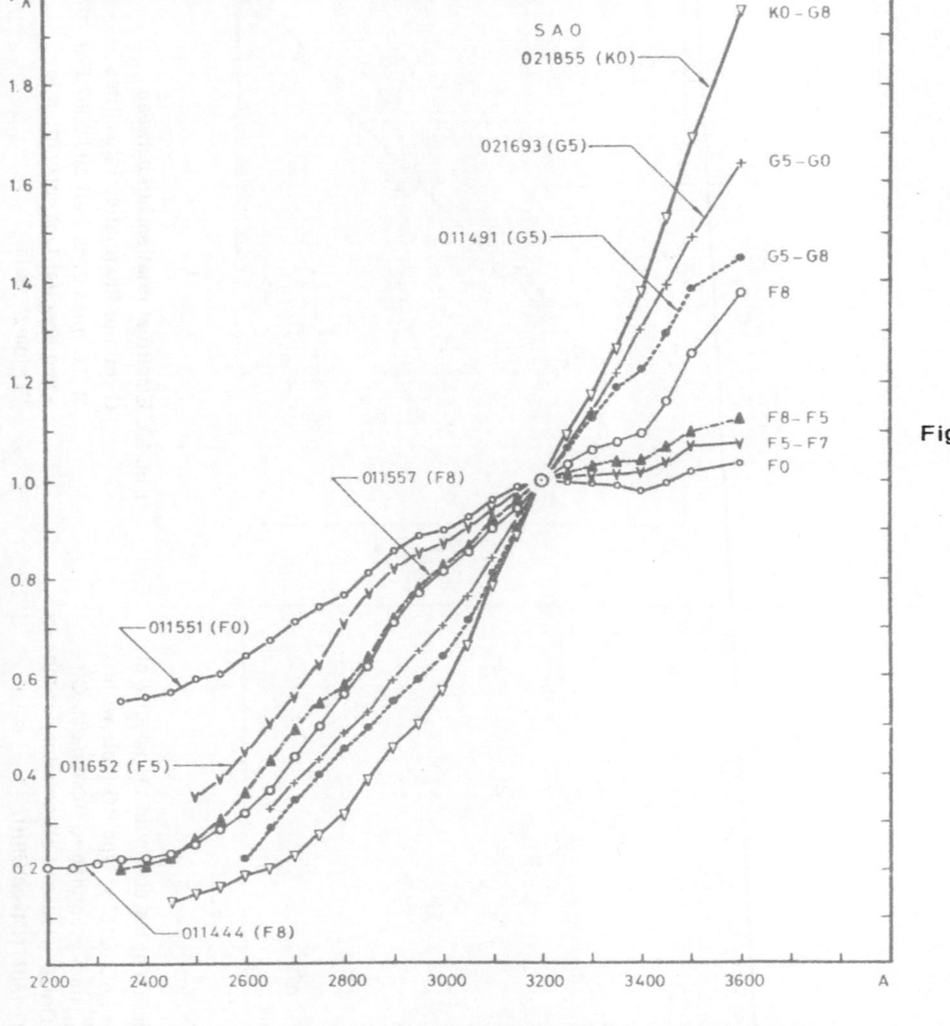

Fig. 29. Curves of the relative distribution of the energy in the ultraviolet for a group of FO-KO type stars scattered around γ Cas. It is taken $F_{3200} = 1$.

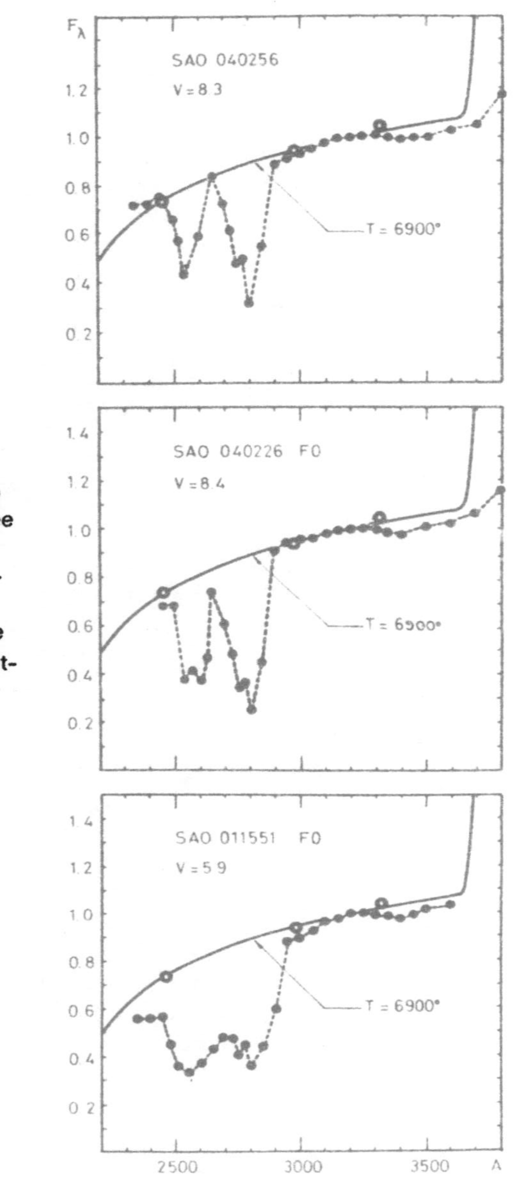

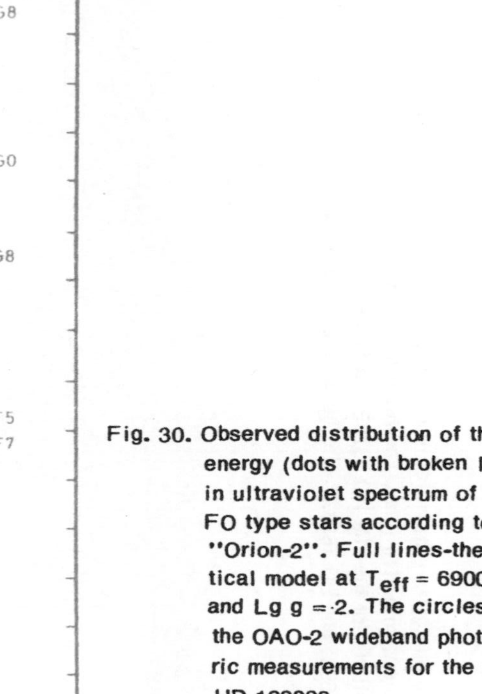

Fig. 30. Observed distribution of the energy (dots with broken line) in ultraviolet spectrum of three FO type stars according to "Orion-2". Full lines-theorêtical model at $T_{eff} = 6900°K$ and Lg g = 2. The circles are the OAO-2 wideband photometric measurements for the star HD 128898.

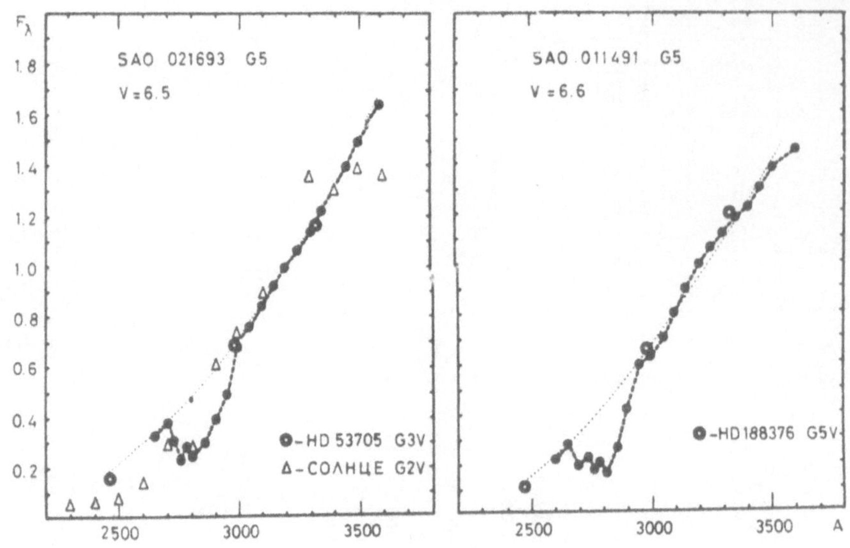

Fig. 31. Distribution of the energy in ultraviolet in the spectra of two G5 type stars (dots with dotted line). The results of short wavelength observations for the star HD 53705 and Sun - in the case of SAO 021693, and HD 188376 - in the case of SAO 011491, are shown also.

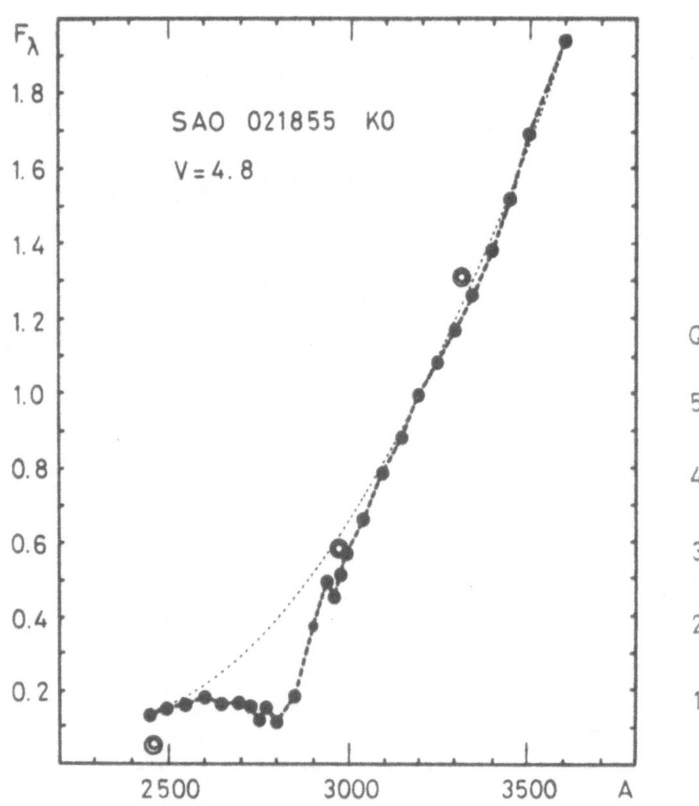

Fig. 32. Distribution of the energy in the spectra of the star SAO 021855, type KO (dots with broken line) according to "Orion-2". Circles - OAO-2 wideband observation for the star HD 180711, type G9III.

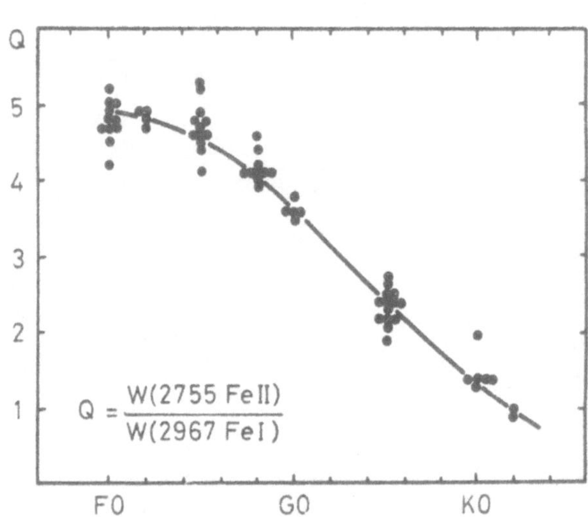

Fig. 33. Empirical relationship between Q-ratio of the relative intensities of the lines 2755 FeII and 2967 FeI - and spectral type, according to "Orion-2" data.

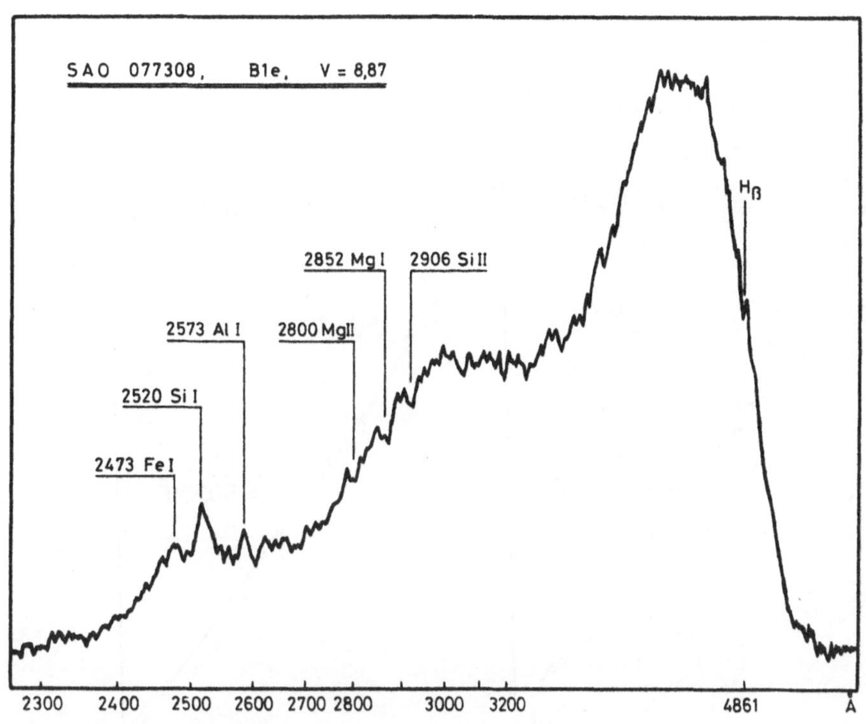

Fig. 34. The strongest emission line 2520 SiI in the spectrum of B1e type star SAO 077308 in the wavelength region from 4800 Å to 2400 Å.

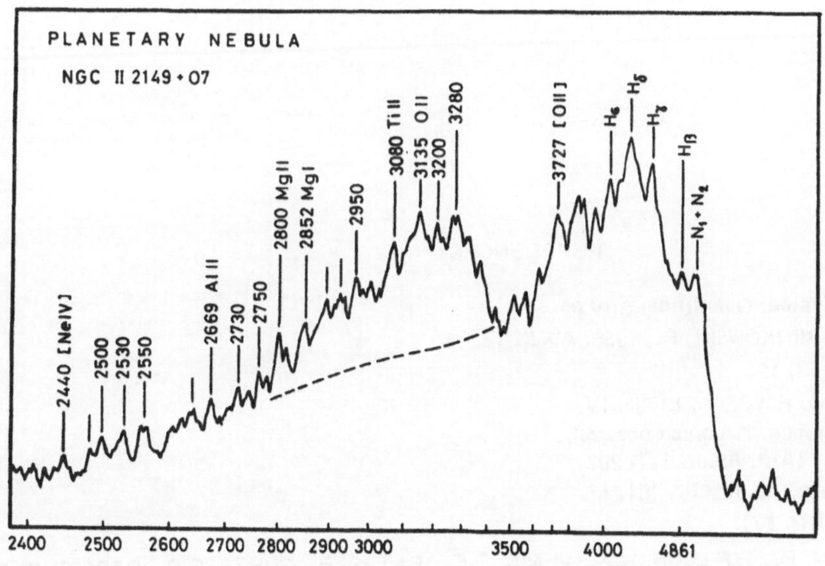

Fig. 35. The microphotometric recording obtained from the spectrogram of the planetary nebula Ic 2149 and its nuclei. The emission lines of the nebula are shown. The broken line shows the boundaries of the part of the spectrum affected by the nearby faint star.

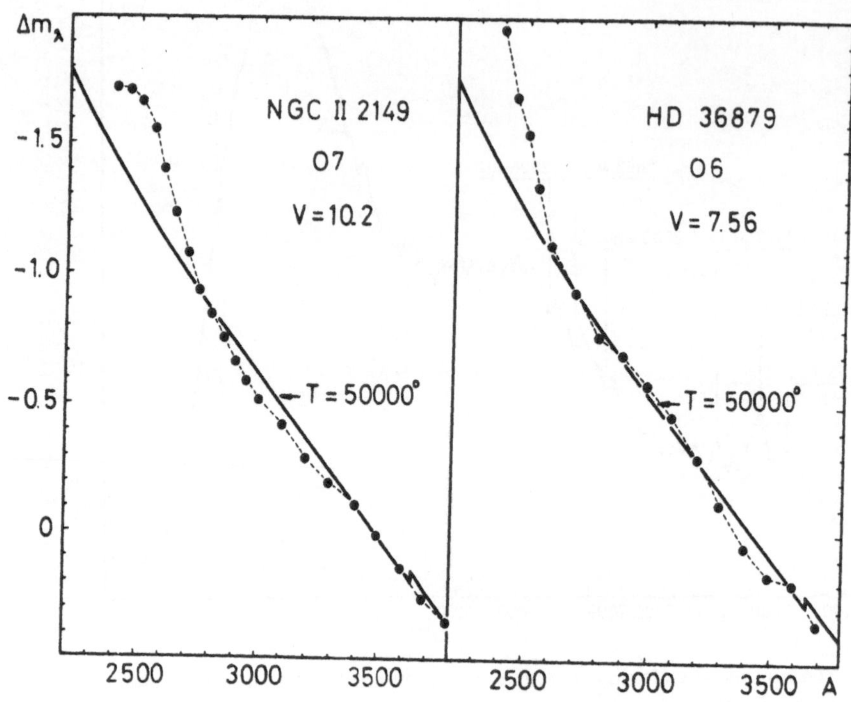

Fig. 36. Distribution of the energy in ultraviolet of the spectra of two O type stars - nuclei of planetary nebula IC 2149 and HD 36879. The excess of the energy on ∽ 2200 Å is visible in both cases.

REFERENCES

ALLEN C.W., 1974, Astrophysical Quantities, 3-rd ed.

ALLER L.H., BOWEN J.S., MINKOWSKI R., 1955, Ap.J., 122, 62

ALLER L.H., 1951, Ap.J., 113, 125

ALLER L.H., 1961, Mem. Soc. Roy. Sci., Liege, IV

ASSATRYAN R.S., 1975, Commun. Byurakan obs., 48,

BLESS R.C., SAVAGE B.D., 1972, Ap.J., 171, 293

BOGGES A., KONDO Y., 1968, Ap.J.Lett., 151, L5

BOHLIN R.C., 1970, Ap.J., 162, 571

BOKSENBERG A., KIRKHAM B., TOWLSON W.A., VENIS T.E., BATES B., CURTS G.R., CARSON P.P.D., 1972, Nature Phys. Sci., 240, 127

BOTTEMILLER R.L., 1972, The Sci. Results from OAO-2. Ed.A.Code.NASA Sp.-310, p.505

CODE A.D., BLESS R.C., 1970, Proc. IAU Sympos., No36, p.173

DOHERTY L.R., 1971, Phil.Trans.Roy.Soc.London, A270, 189

DUNHAM T., ADAMS W.S. 1939, PASP, 9, 5

EGGEN O.J., 1974, PASP, 86, 241
ELST E.W., 1966, Bull. Astron. Inst. Netherlands, 19, 90
EPHREMYAN R.A., 1975, Commun, Byurakan Observ. 48
EVANS D.C., 1972, The Sci. Results from OAO-2. Ed.A.Code. NASA SP-310, p.347
FOWLER J.B., 1974, Ap.J., 188, 295
GURZADIAN G.A., 1975, Vestnik Akad. Nauk SSSR, January
GURZADIAN G.A., 1975,(a) Space Sci. Rev., 18
GURZADIAN G.A., 1975(b) Obs., 94, 293
GURZADIAN G.A., 1975 (c) Astr. and Astrophys., 39, 213
GURZADIAN G.A., 1975(d), PASP, 87, 289
GURZADIAN G.A., 1975(e), Astr. and Astrophys., 40, 447
GURZADIAN G.A., 1975(f), Mon.Not. Roy. Astr. Soc., 172
GURZADIAN G.A., 1975(g)Astrofizika, 10, 379
GURZADIAN G.A., 1975(h) Mon.Not.Roy.Astr.Soc., 172, 249,
GURZADIAN G.A., KASHIN A.L., KRMOYAN M.N., OHANESYAN J.B., 1974, Astrofizika, 10, 177
GURZADIAN G.A., KASHIN A.L., KRMOYAN M.N., JARAKYAN A.L., LORETZYAN G.M., OHANESYAN J.B, 1975, Astrophys. and Space Sci., (in press)
GURZADIAN G.A., OHANESYAN R.Kh., 1975, Astrofizika, 11, No.3,
GURZADIAN G.A., OHANESYAN J.B., 1972, Astron. and Astrophys., 20, 321
GURZADIAN G.A., 1972, Sky and Telescope, 43, 350
GURZADIAN G.A., OHANESYAN J.B., 1975 Astrofizika, 11, No.4
GURZADIAN G.A., 1974, Nature, 250, 204
GURZADIAN G.A., 1974(a) Astron, and Astrophys., 35, 493
GURZADIAN G.A., RUSTAMBEKOVA S.S., 1975, Nature, 254, 322
GURZADIAN G.A., 1969, Planetary Nebulae. Gordon and Breach, New York
GURZADIAN G.A., 1965, Astrofizika, 1, 91
HARAMUNDANIS K.L., 1973, Ap.J.Lett., 185, L87
HONEYCUTT R.K., 1972, A.J., 77, 24
KONDO Y., GIULI T., MODISETTE J.L., RIDGREN A.E., 1972, Ap.J., 176, 153
KONDO Y., HENIZE K.G., KOTILA C.L., 1970, Ap.J., 159, 929
KONDO Y., 1972, Ap.J., 171, 605
KURUCZ R.L., PEYTERMANN E., AVERTT E.N., 1972, SAO Reprint, 209-201
LAMERS H.J., van der HUCHT K.A., SHIJDERS M.A., SAKHIBULLIN N., 1973, Astron. and Astrophys.,25,105
LILLIE C.F., BOHLIN R.C., MOLNAR C.A., LANE A.L., 1972, Science , 175, 321
MERRILL P.W., 1956, Lines of Chemical Elements in Astrophysical Spectra. Washington
MIHALAS D., 1965, Ap.J.Suppl., IX, No.92, 321
MIHALAS D., 1972, Ap.J., 177, 115
MORGAN T.H., SPEAR G.C., KONDO Y., HENIZE K.G., 1975, Ap.J., 197, 371
MORTON D.C., 1967, Ap.J., 144, 1, 1966; 147, 1017
NAVACH C., LEHRMAN M., HUGUENIN P., 1973, Astr. and Astrophys., 22, 371
O'DELL C.R., 1962, Ap.J., 135, 371
OHANESYAN J.B., 1975, Commun. Byurakan Obs., 48
OHANESYAN R.Kh., AKOPYAN A.A., 1975, Commun. Byurakan Observ., 48
OHANESYAN O.V., 1975 Commun, Byurakan Observ., 48
PARSONS B., 1965, Ap.J. Suppl., 18, 159
PEREK L., KOHOUTEK L., 1967, Catalogue of Planetary Nebulae Acad.Publ. House, Prague
POSENDHAL I.P., 1973, Ap.J., 186, 909
RUSTAMBEKOVA S.S., 1975, Commun. Byurakan Obs., 48
SAVAGE B.D., 1975, Ap.J., 199, 92
SAVAGE B.D., CALDWELL I.I., 1974, Ap.J., 187, 197
STECHER T.P., 1969, A.J., 74, 96
STECHER T.P., 1969, A.J., 74, 98
STRIGANOV A.R., SVENTISKIJ N.S., 1968, Table of Spectral Lines of Neutral and Ionized Atoms (New York: IFI/ Plenum.)
SUDBURY G.C., 1971, Mon. Not.Roy. Astr.Soc., 153, 241
TARTER C.B., 1971, Ap.J., 168, 313
UNDERHILL A.B., 1972, The Scien. Results from OAO-2. Ed.A.Code.NASA SP-310, p.367
UNDERHILL A.B., 1974, Ap.J., Suppl., 27, 359

DISCUSSION

NANDY: How do you calibrate the spectra to obtain the flux distributions and what is the accuracy?

GURZADIAN: Our spectra cover the ultraviolet region up to about 3800 Å. It is therefore possible to connect our measurements directly to optically determined absolute fluxes. For this purpose we use stars around spectral class AO. The accuracy is about 10 - 15%, but we hope to reach 10% or better in the future.

BLAAUW: Could you please say a little more about how the observations were obtained?

GURZADIAN : To do this in detail would take too much time so I limit myself to some words about the control of the "Orion-2" telescope. First, the cosmonaut-engineer fixes the first photoguider on the primary guide star and the observatory is stabilized in two axes. However, the field of "Orion-2" is large, about 5°. A second system of photo-guiders therefore fixes on a secondary guide star. When the observatory has been stabilized in three axes with the foreseen accuracy (5-7 seconds of arc), the cosmonaut pushes a button and the on-board control system takes over for the rest of the observation, without any interaction by the cosmonaut.

PRELIMINARY RESULTS OBTAINED WITH THE UV SPECTROPHOTOMETER, ONBOARD ANS

P.R.WESSELIUS, J.W.G.AALDERS, T.S. VAN ALBADA, C.D.ANDRIESSE, K.S. DE BOER, J.BORGMAN, R.J. VAN DUINEN, J.KOORNNEEF, S.R.POTTASCH, J.P.VADER AND C.C.WU

KAPTEYN ASTRONOMICAL INSTITUTE, DEPARTMENT OF SPACE RESEARCH, THE NETHERLANDS
PRESENTED BY P.R.WESSELIUS

ПРЕДВАРИТЕЛЬНЫЕ РЕЗУЛЬТАТЫ, ПОЛУЧЕННЫЕ С УЛЬТРАФИОЛЕТОВЫМ СПЕКТРОФОТОМЕТРОМ, РАБОТАВШИМ НА БОРТУ СПУТНИКА ANS

РЕЗЮМЕ

Изложены некоторые из предварительных результатов, полученных пятиканальным ультрафиолетовым спектрофотометром, установленным на борту Голландского Астрономического Спутника (ANS), запущенного 30 августа 1974 г. Описание прибора см. (Van Duinen et al., 1975) ; характеристики прибора в Табл.1.

ABSTRACT

In this paper we report some of the preliminary results obtained by the five channel ultraviolet spectrophotometer, onboard the Netherlands Astronomical Satellite (ANS), launched on August 30th, 1974.

1. I n t r o d u c t i o n. In this paper we report some of the first results obtained by the five channel ultraviolet spectrophotometer, onboard the Netherlands Astronomical Satellite (ANS), launched on August 30th, 1974. A description of the instrument may be found elsewhere (Van Duinen et al, 1975). Relevant instrument characteristics are summarized in Table 1. All data reduction was based on quick-look data. Preliminary results of six observational programs are discussed: hot subluminous stars, planetary nebulae, globular clusters, the extinction laws in the Large Magellanic Cloud and in Carina, and the reflection nebula Merope.

2. H o t s u b l u m i n o u s s t a r s. Data on twelve subdwarfs and white dwarfs are presented. They have been selected on the basis of groundbased colorimetric and spectroscopic results.

One of our selection criteria was, that these subluminous stars occurred at high galactic latitudes, where almost no reddening seems to be present (Philip and Tifft, 1971). In fact, none of the twelve objects shows a dip at 2200 Å in its energy distribution. Therefore, no reddening correction has been applied.

As a first approximation we have fitted the results assuming the atmospheres of such hot stars radiate as blackbodies (BB). Models of Hummer and Mihalas (1970) seem to indicate that the effective temperatures of subluminous stars are slightly overestimated using the BB approximation. Because of the large uncertainties in these models (Hummer, private communication) we have sticked to BB models for simplicity reasons, varying just one parameter - the effective temperature of the star - to fit the flux ratio, F_λ (1550 Å)/F (3300 Å). This ratio was chosen because it is most sensitive for the determination of high values of the temperature.

For all objects the measured flux ratios at 1800, 2200 and 2500 Å were lying satisfactorily close to the corresponding blackbody curve. For three objects the measured flux ratios are shown in figure 1. Some relevant information on the twelve objects is presented in Table 2.

The internal accuracy of the data is very good; e.g. for AGK2 + 81°266 eight sets of six consecutive samples of 32 seconds were available. The 8 averages gave for F(1550/F3300) a value of 15.23 ± 0.05 (one sigma). These one sigma-values are given in the second line of column 6 of table 2.

Unfortunately a few systematic errors are much larger than the statistical error. The most important one is the uncertainty with respect to the absolute calibration. There is a systematic uncertainty in F(1550)/F(3300) of about 10 percent. A detailed comparison of the energy distribution of a few early B stars - as measured by ANS-with models can decrease this uncertainty to a few percent.

In Table 2, column 8 values of the temperature are presented determined from ratios of F(1550) and F(3300) increased and decreased by 10 percent of the measured values.

These results are important for two reasons.

The highest temperatures for sdO result in a bolometric correction of 5 to 8 magnitudes versus the value of 3.5 mag- proposed by Sargent and Searle (l.c.). I.e. some of the sdO stars occupy another region in the HR diagram than previously thought. Also a few old novae and some planetaries were measured in the UV by ANS and it is our aim to study the possible relation between these categories of objects by determining their place in the HR diagram more accurately than possible from the ground.

The influence of these very hot subdwarfs and dwarfs on the interstellar medium may be considerable. According to Hills (1973) the cloud - intercloud structure of the interstellar medium could be due to overlapping Strömgren spheres of numerous hot pre-white dwarf-stars. Ten UV stars at 100 000 K are needed within a sphere having a radius of 100 pc. At a distance of 100 pc. such stars would have a visual magnitude of $11^m.1$. In our sample of 12 stars there are three stars having a temperature of about 10^5 K and visual magnitudes of 10.5, 12.1 and 12.3. There probably are more hot subluminous stars close to the Sun because our sample was rather incomplete. Thus it may well be that such hot pre-white dwarfs ionize the hydrogen in a large fraction of interstellar space.

3. Planetary nebulae. At present, positive measurements have been made of 17 nebulae, while upper limits are available for several others. The results are more difficult to interpret than most other ultraviolet measurements, since the 2½ arcminute diaphragm measures not only the central star, but the nebular continuum as well.

Furthermore, many of these nebulae are reddened, some considerably, and a correction must be made for this reddening. The reddening correction was made by assuming that the measured dip in the 2200 Å band is entirely due to interstellar extinction. The observations were then corrected, using the ANS reddening curve so that the corrected measurements formed a smooth curve. This method turns out to be quite a sensitive one, and the reddening obtained in this way is well correlated with the reddening obtained by measuring the ratio of the $H\beta$ flux to the radio emission of the nebula. In a more careful analysis, based on more data, we will be able to determine the value of the absolute extinction coefficient R with reasonable accuracy.

The spectrum of the nebula is under study at present by measuring those nebulae which are large so that the central star can be excluded, or have such a weak central star that only the nebula is measured. Preliminary results indicate that the two-quantum continuum is more important than is generally accepted (i.e. X = 1, thus all recombinations go through the 2S level). The spectrum of the nucleus can then be determined. The emission can be fitted either to a black-body curve or to a model atmosphere. The resultant temperatures vary considerably, from about T = 40,000°K or slightly lower for IC418 and NGC7662 to over 200,000°K for IC 1360. Finally, comparison of observations with the narrow band filter and broad band filter at 1550 Å indicates in several cases that the CIV line, presumably originating from the nebula, is observed in emission. A more detailed paper will be published by Pottasch, Wesselius and Wu.

4. Globular clusters. Earlier OAO-2 observations of 6 Globular Clusters show interesting results for these objects in the UV. The most surprising is the "upturn" of the spectrum in the 1500 Å range.

So far, ANS has observed 10 clusters; 3 in common with OAO. The spectrum of these 10, based on the ANS laboratory preflight calibration, is shown in fig. 2.

There is no clear indication for the blue upturn. The 1500 - 3300 colours correlate well with the relative population of the blue side of the horizontal branch (see e.g. van Albada and Baker, 1973).

For a more detailed analysis a comparison of the photometric systems of the OAO and the ANS is required.

A more detailed paper will be published by van Albada and de Boer.

5. The extinction law in the large Magellanic cloud. An article on this subject has been published in Astronomy and Astrophysics (Borgman et al, 1975). The reader is referred to this original paper.

6. The ultraviolet extinction law in Carina. The Carina region is very well suited for studies of large scale structure of the galaxy, mainly because the extinction is generally low and seems to increase only slowly with distance.

The space distribution of OB stars in Carina has been investigated by Graham (1970). Distances to 438 stars were obtained by using observations in Walraven's five color photometric system for spectral classification augmented by $H\beta$ photometry for luminosity determination.

To provide additional data to the original groundbased photometry we have attempted to obtain ultraviolet data on a sample of 130 stars in the region. In principle the ultraviolet data could be used to confirm and possibly improve the groundbased classification and therefore to improve the distance estimated, provided the corrections due to extinction can be made reliably. Here we present the result to an investigation of the behaviour of the ultraviolet extinction law in this region.

As is customary, the results of Graham's investigation are obtained by the application of a single reddening line and a fixed ratio for total to selective extinction. This may be a valid procedure in the visual spectral range, in the ultraviolet however it is questionable whether such a scheme is permissible. Especially the occurrence of the pronounced extinction feature at 4.6 μ^{-1} may or may not be universal, while in the far ultraviolet, there is a large variation of extinction for different stars (Bless and Savage, 1972).

It should be emphasized that the existence of the 4.6 μ^{-1} feature is only confirmed for relatively local stars. In the Carina region the behaviour of the extinction law can be tested out to much larger distances. The results described above on associations in LMC have demonstrated that the "hump" at 4.6μ^{-1} is virtually absent in the LMC local extinction law.

130 objects in the spectral range 09 to B3 were selected from Graham's list with distances up to 6 kpc, equally distributed in different distance bins (distances as obtained by Graham). We obtained data on 86 stars. Results which showed variation in the background countrate resulting in standard deviations exceeding twice the expected standard deviation were rejected. To investigate the extinction behaviour we computed the quantity defined as

$$\rho = -2.5 \log_{10} \frac{C_{2200}}{C_{1800} \times C_{2500}}$$

where C_λ represents the countrate in channel λ.

This parameter represents the depth of the extinction feature. The standard error of the mean was computed from the standard deviation of a series of observations on each star. In case only one or two observations were available we computed the error from the statistics of an individual observation. From a comparison of the statistical error with the error distribution for multiple observations we found a factor of 2 to be a conservative multiplier for the internal statistical error, which we attribute to the residual background countrate fluctuations.

The results are presented in figure 4. The parameter ρ is plotted as a function of E_{B-V} as obtained from Graham photometry. The crosses are obtained from nearby stars not in the Carina region. An unweighted least squares parabolic fit to the data is presented by the broken line. The curvature of the distribution is caused by saturation in the parameter ρ which results from the gradually increasing residual influence of the bump in the 1800 and 2500 Å channels. The deviations from the parabola are rather large; they could not be correlated with any quantity we tried: countrates, (U-B)' index (Graham, 1968), luminosity class and distance.

From synthetic colors in the ANS system as obtained by Peytremann we have studied the effect on the ρ parameter for stars of different temperature and gravity. ρ varies 0.12 mag in the spectral range 09 - B3.

Clearly only gross classification errors in the photometry could cause scatter effects as observed in the diagram. We do not believe such large errors to be present in Graham's classification.

We are led to conclude that generally the existence of the 4.6μ^{-1} feature is confirmed even if the observations are extended out to larger distances from the sun. The magnitude of the feature, however, seems to depend only loosely on the reddening in the visual. At least part of the observed scatter is caused by variations in the extinction law in the Carina region. A more detailed analysis based on a very much larger sample of reddened objects in the solar vicinity as observed by ANS is presently under way.

At the same time we attempt to investigate the consistency of the ANS intermediate band photometry with ground-based spectral type assignments for a number of unreddened and slightly reddened stars.

The results of these investigations together with the present results of the Carina investigation will learn whether the observed variation is a local effect in the Carina region or is more universal in character.

An article on this subject will be written by van Duinen and Vader.

7. The reflection nebula of Merope. 5 positions in the nebula and the star itself have been observed, with the intention to derive the ultraviolet albedo A and the scattering phase factor g of dust particles in the Nebula. The flux ratios to the star show a minimum that progresses from 4 to 5.5μm^{-1} for increasing distances. Studying these ratios one finds more subtle but also systematic effects of the distance. A preliminary analysis learns that the albedo is relatively low around 4.6μm^{-1}. The full analysis will include a computer study of

a model with 5 independent variables, viz, A, g, particle density, depth of nebula and line-of-sight distance of star to nebula.

A more extensive paper on this subject is being prepared by Andriesse and Witt.

8. Concluding remarks. A discussion of preliminary reductions on only a few hundred objects observed with ANS has been presented above. None of the results given is definitive. On all data a more refined reduction scheme will be applied. At the moment ANS has already observed 3.000 individual objects, and a few thousand will probably have been added when these proceedings are published.

Table 1

ANS-UV instrument "characteristics"

Telescope : 22 cm Ritchy Chretien

Field of view : 2.5 x 2.5 arcminute

Bandpasses :	central wavelength (nm)	bandwidth (nm)
	154.9	14.9
	154.5	5.0
	179.9	14.9
	220.0	20.0
	249.3	15.0
	329.4	10.1

Table 2

ANS UVX DATA						SUBLUMINOUS STARS		
NAME	V (mag)	SP.TYPE	B-V	U-B	$f = \frac{F_\lambda(1550)}{F_\lambda(3300)}$	T (°K)	T±	REMARKS
HZ 29	14.2	DBp	-0.23	-1.01	4.72 ±0.34	23.800	25.100 22.500	other name AM CVn
F 46	13.2	sd O	-0.25	-1.16	6.58 ±0.17	30,000	31,600 27.300	may be DO_s
F 36	12.5	sd B	-0.25	-1.02	7.52 ±0.09	32.600	35.400 30,000	
HZ 44	11.7	sd 08p	-0.29	-1.19	8.45 ±0.08	36.100	39.700 32.900	
HD 127493	10.1	sd 08p	-0.28		8.67 ±0.05	37.000	40.800 33.600	
HD 149382	9.1	sd B			9.30 ±0.05	39.700	44.300 35.800	
BD +25°2534	10.5	sd Op	-0.26	-1.19	9.43 ±0.02	40.300	45.000 36.200	other name: F 66
BD +25°4655	9.8	sd 06			9.53 ±0.02	40.800	45.600 36.600	hydrogen-poor
HZ 43	12.9	DA wk	-0.10	-1.14	13.10 ±0.16	64.800	80.100 53.800	
AGK2 +81°266	12.1	sd O			15.23 ±0.05	93.500	133.800 71.200	
F 24	12.3	DA wke	-0.23	-1.25	15.37 ±0.17	96.300	139.800 72,600	close binary: white dwarf + cool red dwarf
BD +28°4211	10.5	sd O7 p			16.1 ±0.7	113.200	181.700 81.300	

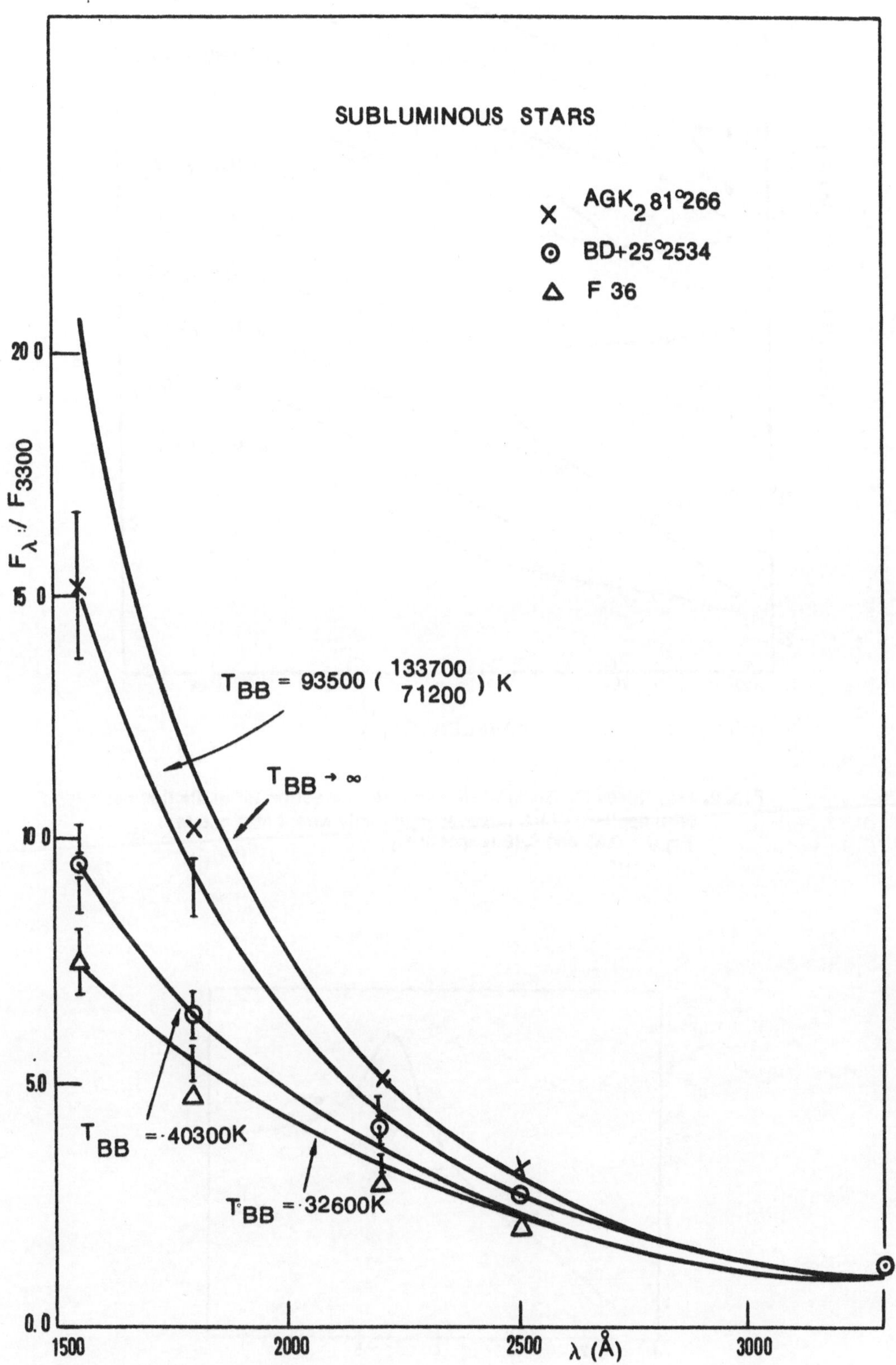

Fig. 1. The energy distributions in the UV of three subluminous stars.

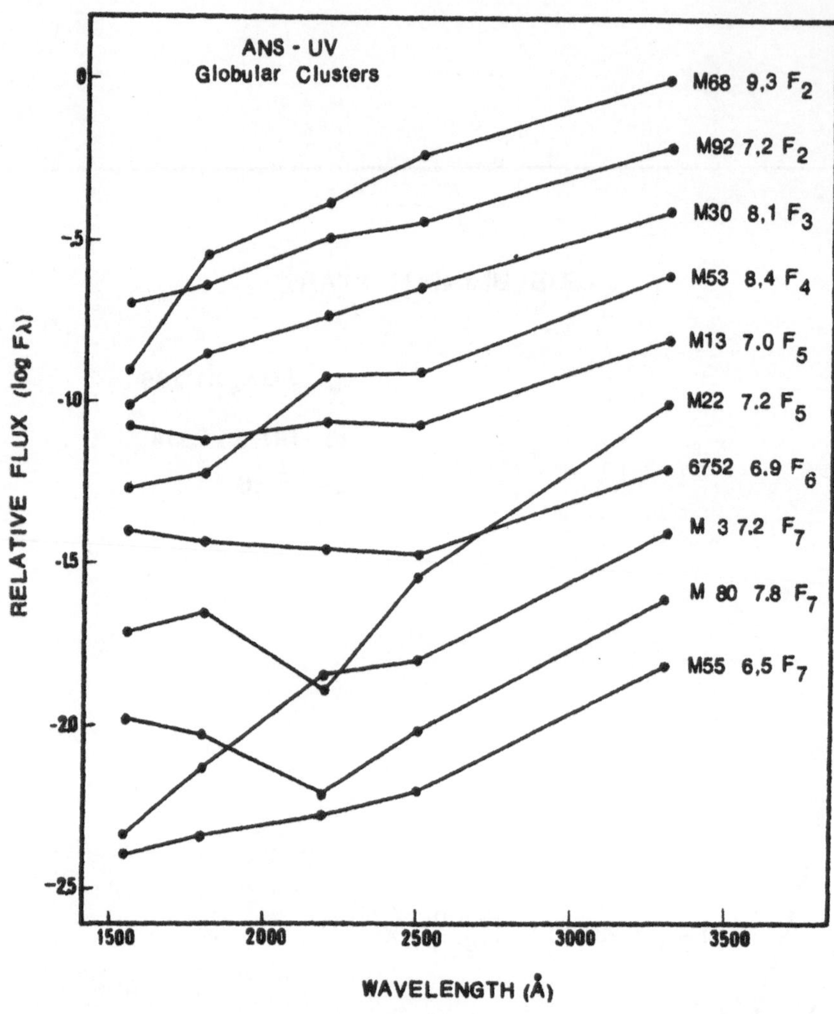

Fig. 2. ANS fluxes for Globular Clusters. No correction for extinction has been applied, which however would only affect M22 and M80 $E_{B-V} = 0.35$ and 0.18 respectively.

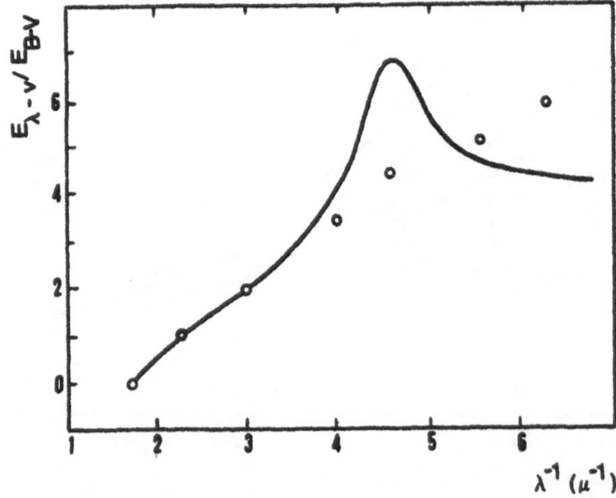

Fig. 3. The average extinction law of Bless and Savage (1972) compared with the proposed extinction law in the LMC associations (open circles).

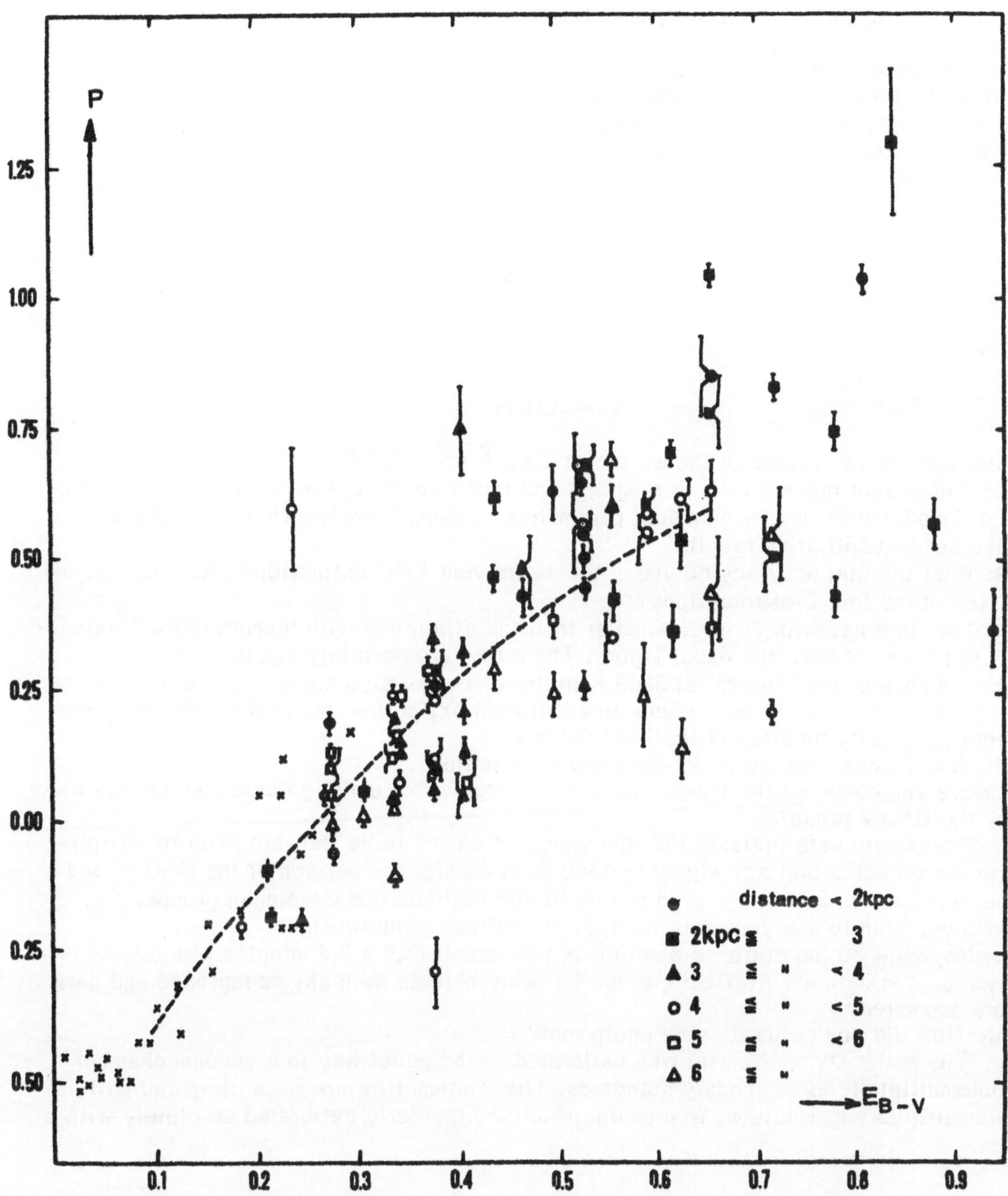

Fig. 4. The extinction parameter ρ versus the color-excess E_{B-V}. The crosses represent O9-B3 stars not in Carina, which have also been observed by ANS; they have been added to fill the lower left part of the diagram. The dashed curve is a parabolic fit (crosses not included).

REFERENCES

ALBADA T.S. VAN, BAKER N. 1973, Astrophys. J. 185, 477.

BLESS R.C., SAVAGE B.D. 1972, Astrophys. J. 171, 293.

BORGMAN J., DUINEN R.J. VAN, KOORNNEEF J. 1975, Astron. and Astrophys. 40, 461.

DUINEN R.J. VAN, AALDERS J.W.G., WESSELIUS P.R., WILDEMAN K.J., WU C.C., LUINGE W., SNEL D. 1975, Astron. and Astrophys. 39, 159

GRAHAM J.A. 1968, Bull. astr. Inst. Netherl. Suppl. 2, 397.

GRAHAM J.A. 1970, Astron. J. 75, 703.

HILLS G.G. 1973, Astron. and Astrophys. 26, 197.

HUMMER D.G., MIHALAS D. 1970, JILA Report No. 101.

PHILIP A.G.D., TIFFT L.E. 1971, Astron. J. 76, 567.

SARGENT W.L.W., SEARLE L. 1968, Astrophys. J. 152, 443.

DISCUSSION

NANDY: How low are the values of the albedo at 2200 Å and 1500 Å?

WESSELIUS: The reductions are still in progress and I can not give any figures yet. A computer study of a model with five independent parameters - among them the albedo - will hopefully soon give some quantitative results.

ARDEBERG: What internal accuracy do you estimate for your LMC extinction curve from the intercomparison of the four O-associations?

WESSELIUS: The internal error is much smaller than the difference with respect to the "galactic bump", but I do not have the exact figure. The errors are certainly small.

NANDY: The absence of the "bump" at 2200 Å in the LMC was also noted from the observations with the TD-1 satellite. It may be due to a different extinction law in the LMC, but also to a different luminosity function of the LMC O-stars.

WESSELIUS: Yes, I know that we prefer different explanations.

MATTILA: Have you observed the UV-upturn in the energy spectrum of galaxies which was announced by the OAO-2 group?

WESSELIUS: Preliminary data indicate that the spectrum seems to be flat, but even for elliptical galaxies we do not detect any signal at 1550 Å. A careful comparison of the OAO-2 and the ANS observations is now undertaken jointly by the Madison and Groningen groups.

MATTILA: Do you plan to use your data to study the diffuse galactic light?

WESSELIUS: No, because the entrance aperture is very small (2,5 x 2,5 minutes of arc.). In principle, however, the data are available, since for many objects both sky background and dark current were measured.

GURZADYAN: How did you calibrate your equipment?

WESSELIUS: The entire UV instrument was calibrated in the usual way in a vacuum chamber against photomultipliers as secondary standards. The photomultipliers were compared with a set of photomultipliers at Madison, Wisconsin, which are regularly calibrated absolutely with a synchrotron.

SPECTROPHOTOMETRY AND QUANTITATIVE SPECTRAL CLASSIFICATION OF M-GIANTS

L.D.BESHENOVA, A.V.KHARITONOV
ASTROPHYSICAL INSTITUTE OF KAZAKH ACADEMY OF SCIENCES, U.S.S.R.
PRESENTED BY L.D.BESHENOVA

СПЕКТРОФОТОМЕТРИЯ И КОЛИЧЕСТВЕННАЯ СПЕКТРАЛЬНАЯ КЛАССИФИКАЦИЯ М-ГИГАНТОВ

M-stars are studied worse than the representatives of other spectral classes. Apart from having but scarce data on the energy distribution in their spectra and the existence of certain complications in the process of model computing due to extreme insufficiency of the latter, we confront the lack of homogeneous data on star magnitude and colour indices in UBV system even for relatively bright stars (up to 6^m) and what is more important, on spectral classification in the MK system. Thus, for example, according to the scientific literature data, only seventy three stars from one hundred twenty six stars selected for our research work have magnitude V and colour indices U-B and B-V. But even these data are not homogeneous in photometric respect: they are scattered over eighteen original sources.

What concerns spectral classification, the state of things here is even worse. Only 44 stars from our list are classified in the MK system. That is why the main task of our present research is to work out the procedure of quantitative spectral classification of M-giants and to apply it to the stars of our scheme.

In the period of years from 1972 to 1974 the Astrophysical Institute of the Kaz SSR Academy of Sciences performed on its 50 cm reflector photoelectric spectra registration of 126 giants of M class ranging from 3200 to 8000 Å and obtained energy distribution E(λ) in the spectra of these stars expressed in absolute units ($erg/cm^2 \cdot sec \cdot cm$). Values of E(λ) are related to the integral spectrum, that is to the continuum, all lines and bands including. Averaging of intensities was made every 50 Å.

For qualitative classifications were used bands TiO. The minimum for each depression falls on the following wave lengths: 4950, 5450, 5900, 6200, 6700, 7100 Å. For each star were taken eleven criteria which makes it possible to increase the accuracy of the classification and to estimate its errors.

In the first group of criteria (defined by D) we ascertained the spectrum intensity bound close to the head of absorption band (fig.1a). In the other group of criteria (group T) we ascertained the amount of radiation flux coming in the intervals of wave length close to the bandhead to the radiation in the two fields of comparison situated on either side of the bandhead (fig. 1 b).

Calibration of functions D(Sp) and T(Sp) was made on 44 stars of M class from our list and on the group of stars of KI-K5 class for which we can find MK system classification in the scientific literature. According to these functions we found spectral subclasses Sp_{A-A} for all stars of our program.

The spectral classes obtained in this work we compared with spectral classes in the MK system and spectral classes obtained by Jamashita and Pedoussaut. The slope differences of the spectral curve between our classification and the MK system are practically none. As to the methodical difference with Jamashita and Pedoussaut, it is insignificant.

The method of definition of spectral subclasses cumulating the values D and T is highly sensitive and precise, mean square error makes from 0.1 to 0.2 of spectral subclasses. Owing to the fact, that M dwarfs are extremely weak and inaccessible for our telescope, and supergiants are rare stars, we did not manage to investigate the division criteria of the stars of different luminosity classes.

For 67 stars of our list which have the parallaxes in BS catalogue(III edition) the spectra-absolute magnitude diagram was plotted. The star magnitude values were considered by the energy distribution in the spectra and Sp_{A-A} was taken according to our classification. Fig. 2 shows this diagram, different stars were represented by different tokens depending on the parallax limits.

We paid attention to star isolated group lying in the lower part of R-H diagram with the mean absolute magnitude about +2.8. The name of each star from this group is pointed out in diagram (fig.2). Here prevailed the stars with relatively large and therefore sufficiently reliable parallaxes, so it seems their absolute magnitudes are true.

Probably this group is the continuation of the sequence of subgiants (IV class of the luminosity) in the field of M class which is limited near the class at KO in the summary diagram H-R. This limitation seems to be caused by insufficient accuracy of the classification of M-stars and very diffident separation them by classes of the luminosity.

Besides spectrophotometric gradients were drawn on the points of truncated continuum. Under the truncated continuum we understand a broken line, connecting the points of maximum intensity between the bands. Spectrophotometric gradients are found for three fields for red with λ_{ef} =6670 Å, for yellow with λ_{ef}= 5000 Å and for blue with λ_{ef} = 4350 Å. According to these gradients we ascertained spectrophotometric temperatures for the unreddened stars of our program. It would be interesting to know that for the red field the temperatures fall with increasing spectral subclass, for the yellow field - remain practically unchanged and for the blue field - rise. The illustration of these variation is in figs.3 and 4.

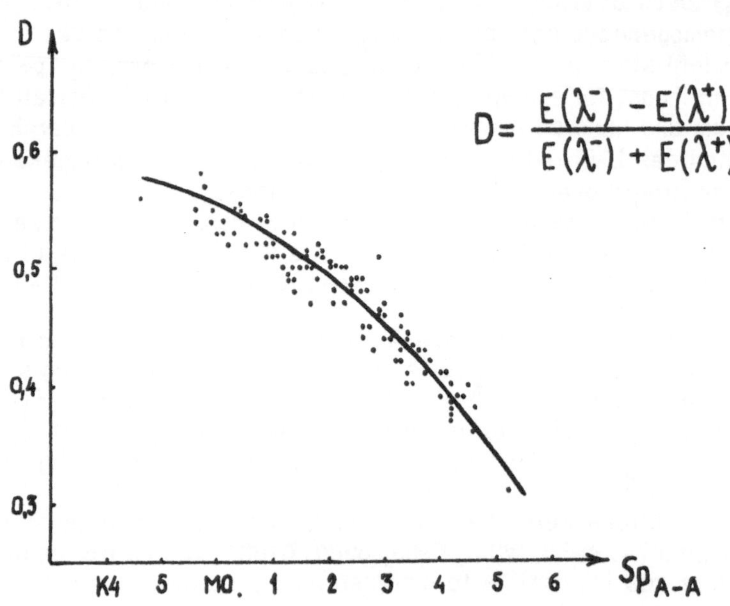

Fig. 1a. The relationship between the intensity jump D and the spectral subclass for depression with λ = 6700Å.

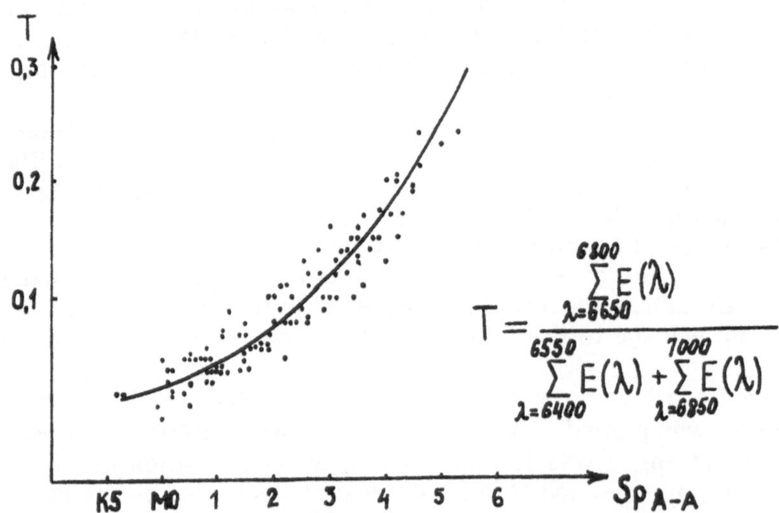

Fig. 1b. The relationship between the magnitude T and the spectral subclass for depression with λ = 6700Å.

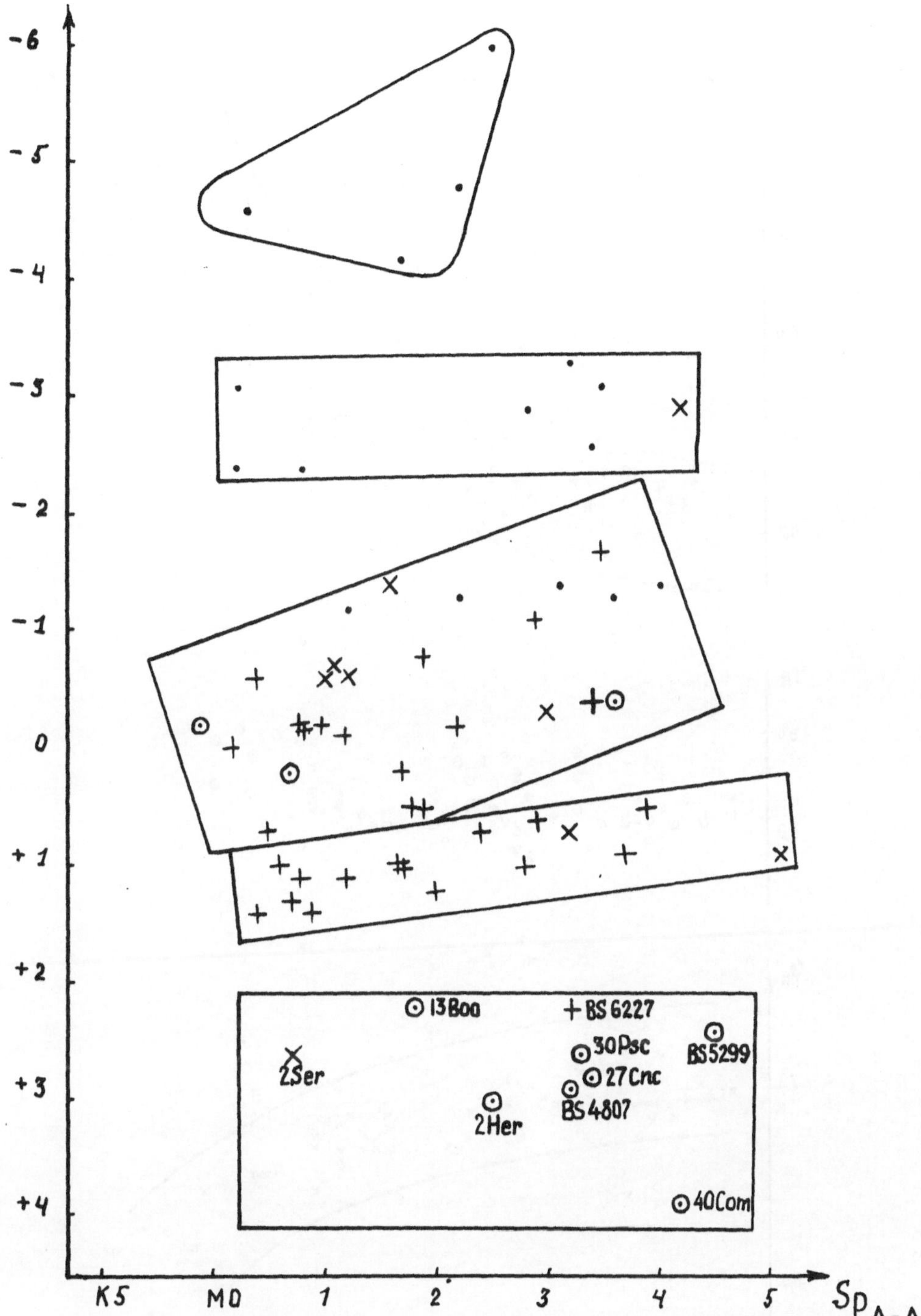

Fig. 2. The spectrum-absolute magnitude diagram for giants (by MK).
Here we marked: $\pi \leqslant 0.005$; × $0.005 < \pi \leqslant 0.010$; + $0.010 < \pi \leqslant 0.020$; ⊙ $0.020 < \pi$.

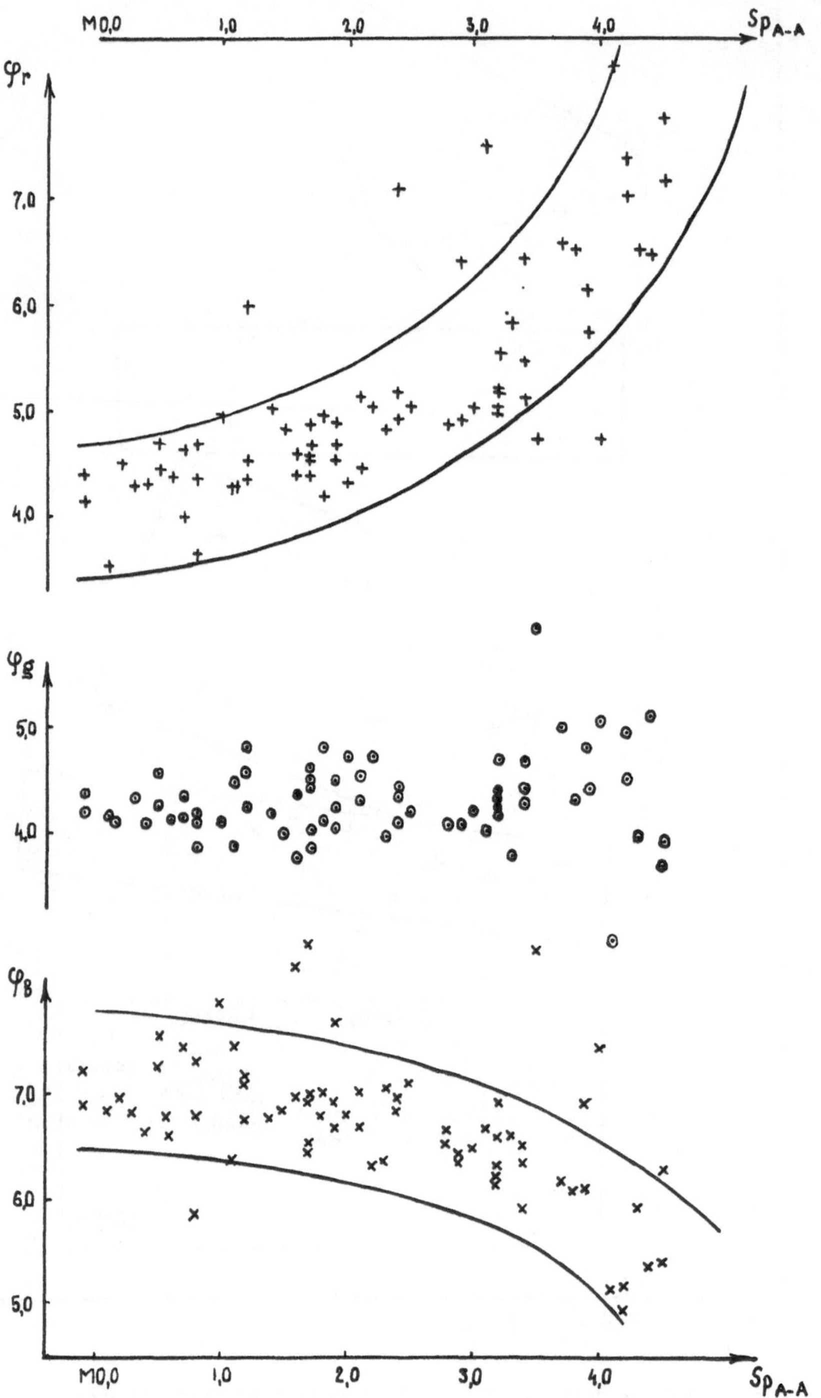

Fig. 3. Spectrophotometric gradients for unreddening M-giants:
φe - for red range, φg - for green range, φb - for blue.

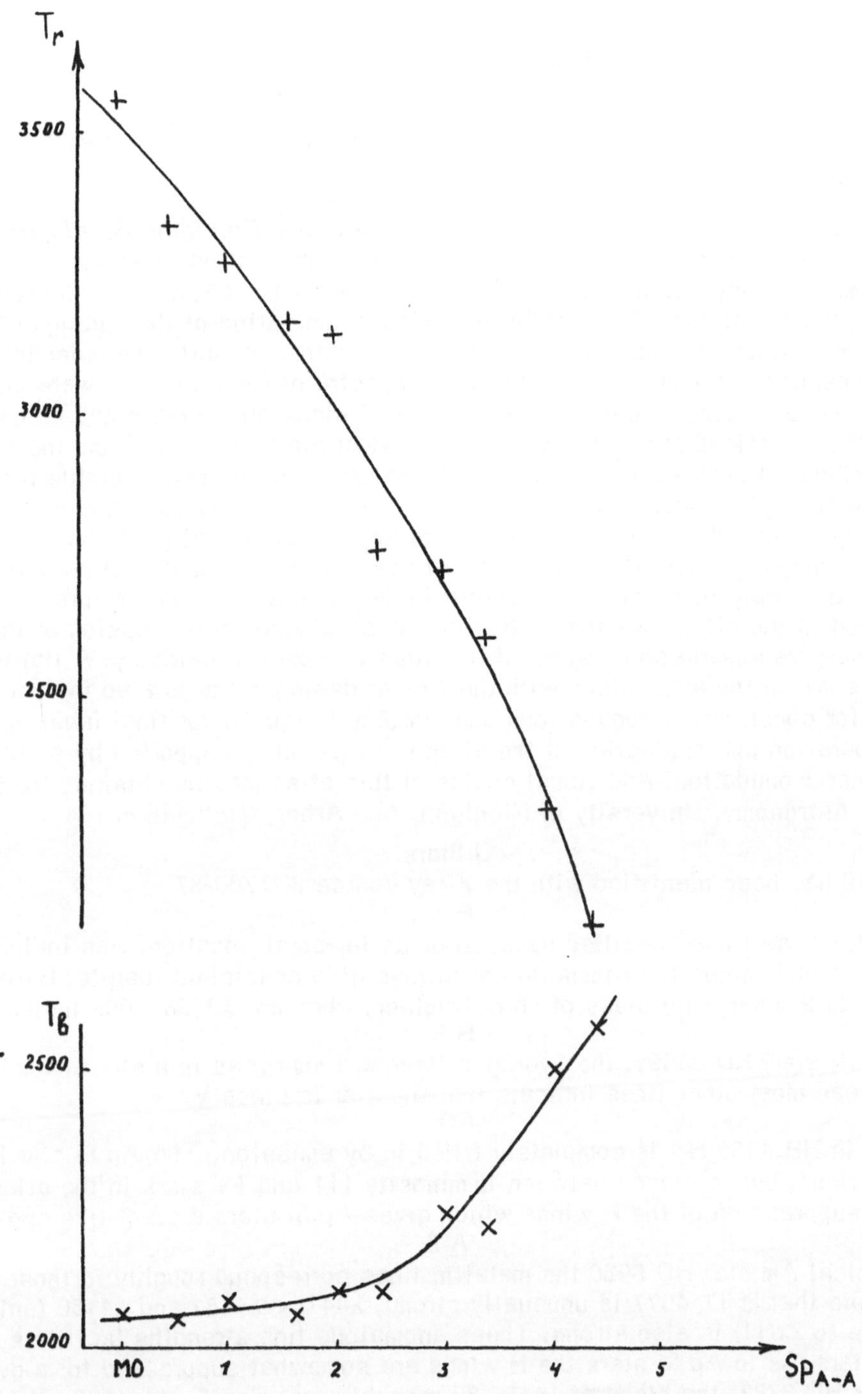

Fig. 4. The variation of the spectrophotometric temperatures with spectral subclass for M-stars.

DISCUSSION

WEST: What are the dispersion and the limiting magnitude of your equipment?
BESHENOVA: The dispersion is 50 A/mm and stars were observed to $6.^m3$.

AN ATLAS OF OBJECTIVE-PRISM SPECTRA

NANCY HOUK, NELSON J. IRVINE AND DAVID ROSENBUSH
THE UNIVERSITY OF MICHIGAN, U.S.A.
PRESENTED BY NANCY HOUK

АТЛАС СПЕКТРОВ, ПОЛУЧЕННЫХ С ПРЕДОБЪЕКТИВНОЙ ПРИЗМОЙ

The atlas of 82 spectra was compiled to accompany the *Michigan Catalogue of Two-Dimensional Spectral Types for the Henry Draper Stars*, and it illustrates the possibilities of approximately MK dispersion. The 108 Å/mm spectra for the catalogue and atlas were taken on IIa-O plates with the 4° + 6° objective-prism combination of the Michigan Curtis Schmidt telescope at Cerro Tololo Inter-American Observatory in Chile. In order to include a full range of spectral and luminosity types, when spectra of MK standards were unavailable or not well-matched in density unknowns were used. Unless otherwise noted below, all of the non-MK classifications are by N. Houk, and most of stars are taken from the first volume of the catalogue. Users are cautioned that these stars are no more suitable for use as standards than any of the other stars given as quality I in the catalogue and were chosen only because they happened to fill gaps and were of appropriate density. Higher dispersion studies might well reveal peculiarities or other reasons why these should not be adopted as standards. In general, only lines which are useful in determining the temperature or luminosity type are labeled in the atlas. See the preface to the catalogue for discussion of the criteria used. Below we give remarks about some of the stars by spectral category. A list by HD number of all the stars in the atlas along with their other designations is also included. We thank Anne Cowley for constructive suggestions and Dorothy Fraquelli for final labeling of the spectra. The preparation and production of the atlas were partially supported by a grant from the National Science Foundation. Additional copies of this atlas may be obtained for $2.50 from Department of Astronomy, University of Michigan, Ann Arbor, Michigan 48104.

O Stars

HD 153919 has been identified with the X-ray source 2U1700-37.

B 1

HD 29138, whose lines are tilted because of its far-south location, was included to illustrate that different luminosity criteria do not always give consistent results. Here the λ4416/4387 ratio yields a luminosity class of Ib or brighter, whereas λ3995/4009 indicates class III.

B 5

In the shell star, HD 33599, the hydrogen lines are sharp, as in a star of luminosity class Ib or II, whereas most other lines indicate that the star is class V.

AO

Note that in HR 4169 Hβ is completely filled in by emission. The Ap Si star HD 103457 is presumably a dwarf, but is placed between luminosity III and IV stars in the atlas to illustrate the typical suppression of the H wings which gives Ap Si stars a giant-like appearance.

A 5

In the typical Am star HD 6930 the metallic lines correspond roughly to those of an early F giant, but note that Sr II 4077 is unusually strong, λ4416<<λ4481 and λ4150 (unlabeled in atlas; mainly due to Zr II) is also strong. These anomalous line strengths facilitate detection of marginal Am stars. As in Ap Si stars the H wings are somewhat suppressed for a dwarf. In the Ap SrEuCr star HD 8783, the blend at λ4128-30 probably includes Eu II 4129, Si II 4128-30 and other lines.

F 2

One characteristic of Fp δDel stars not shown well in the spectrum here of HD 27883 is the typical unusually narrow Ca H and K lines. The same anomalous metal strengths are present as in the Am star discussed above.

F 5

HR 4511, not a known variable, was found to be somewhat variable in spectral type and so is also used as a GO example.

GO

See remark for HR 4511 under F 5.

G 5

This sheet of spectra was completed 3 years ago. At that time, Bidelman's type of G5 Ia was used for HR 6392. More recent comparisons indicate that on our plate a better type would be KO Ia. A possible explanation is that the spectrum is somewhat variable. The area in the vicinity of Fe 1 4045 is continuum, not an emission line.

K O

In the barium star HD 44896 note in addition to the presence of Ba II 4554 that Sr, CH and CN are all very strong. Note that in the no-G-band star HD 18636, whose λ4077/4063 ratio indicates a giant, the CN is as weak as that of a dwarf.

K 3

In the strong and weak CN stars the ratio λ4077/4063 is very similar and indicates that the stars are both giants despite the very different appearance of the CN break.

M 2

The spectrum here of HD 95735 is of poor quality. Unfortunately no photogenic M2 V spectra were available for the atlas, the apparent magnitudes being too faint. A typical long-period variable, an S and a C star also are shown.

The spectra of the Atlas, recently published, are not reproduced here.

Stars in Atlas

HD NO.	HR NO.	NAME	SPEC. TYPE	HD NO.	HR NO.	NAME	SPEC. TYPE
1199			A5 IV	91916	4133	ρ Leo	B1 Ib
3627	165	δ And	K3 III	91572			O7
3651	166	54 Psc	K0 V	92055	4163	U Hya*	C star
6930			Am	92207	4169		A0 Ia
8256			G5 IV	93619			B1 Iab
8783			Ap SrEuCr	93843			O5f
13174	623	14 Ari	F2 III	95129	4278		M2 III
17378	825		A5 Ia	95735			M2 V
18636			no G band	96715			O5
20630	996	κ Cet	G5 V	99148			K0 II
21121			K0 IV	101947	4511		F5 Ia, G0 Ia
27883			Fp δ Del	102567			O/Bne
29138			B1	103457			Ap Si
29488	1479	σ^2 Tau	A5 V	104237			A shell
33599			B shell	111812	4883	31 Com	G0 III
33616			strong CN	113083			weak metal
34503	1735	τ Ori	B5 III	114710	4983	β Com	G0 V
36389	1845	119 Tau	M2 Iab-Ib	121370	5235	η Boo	G0 IV
		CE Tau*		124601	5326	R Cen*	LPV
36988			A5 III	125072			K3 V
37043	1899	ι Ori	O9 III				
				128167	5447	σ_1 Boo	F2 V
44896			Ba star	144470	5993	ω^1 Sco	B1 V
46300	2385	13 Mon	A0 Ib	147084	6081	o Sco	A5 II
47105	2421	γ Gem	A0 IV	147165	6084	σ Sco*	B1 III
47731	2453	25 Gem	G5 Ib	148688	6142		B1 Ia
48737	2484	ξ Gem	F5 IV				
				153919			O6f
50877	2580	o^1 CMa*	K2.5 Iab	155603	6392		G5 Ia
52497	2630	ω Gem	G5 II	163506	6685	89 Her	F2 Ia
58350	2827	η CMa	B5 Ia			V441 Her*	
59612	2874		A5 Ib	164136	6707	ν Her	F2 II
63427			G5 III	164353	6714	67 Oph	B5 Ib
66888	3170	MZ Pup*	M2 II	167884			F5 Iab
71919	3349		A0 V	172052			F5 Ib
74371	3456		B5 Iab	182835	7387	ν Aql	F2 Ib
80057	3688		A0 Iab	186791	7525	γ Aql	K3 II
83754	3849	κ Hya	B5 V	188413			F2 IV
85250	3895		K0 III	195295	7834	41 Cyg	F5 II
86867			composite	196524	7882	β Del	F5 III
87438	3967		K3 Ib	207089	8321	12 Peg	K0 Ib
87887	3981	α Sex	A0 III	209956			F5 V
88092			G0 Ib	212087	8521	π^1 Gru*	S star
				224584			weak CN

* Variable Star

REPORT ON THE 859 ULTRAVIOLET OBSERVATIONS AND ASSOCIATED RESEARCH AT UTRECHT STELLAR SPACE RESEARCH SPECTROSCOPY

K.A. VAN DER HUCHT

SPACE RESEARCH LABORATORY OF THE ASTRONOMICAL INSTITUTE, UTRECHT, THE NETHERLANDS

СООБЩЕНИЕ О 859 УЛЬТРАФИОЛЕТОВЫХ НАБЛЮДЕНИЯХ И ИХ ИССЛЕДОВАНИЯХ, ПРОВОДИМЫХ В УТРЕХТСКОЙ ЛАБОРАТОРИИ СПЕКТРОСКОПИЧЕСКИХ ИССЛЕДОВАНИЙ КОСМИЧЕСКОГО ПРОСТРАНСТВА

(The author has not sent the text of his paper).

THE EFFECTIVE TEMPERATURE OF EARLY TYPE STARS

K.NANDY AND E.KONTIZAS
ROYAL OBSERVATORY, EDINBURGH, U. K.
PRESENTED BY K.NANDY

ЭФФЕКТИВНАЯ ТЕМПЕРАТУРА ЗВЕЗД РАННИХ ТИПОВ

Introduction. In an earlier paper (Nandy and Schmidt, 1975) it has been shown that by the combination of models and the observed spectral distributions from the visible to the far ultraviolet, a temperature scale of B and early A type stars can be derived. During recent years the ultraviolet flux distributions on an absolute scale have been obtained for a considerable number of stars with the sky survey telescope (S2/68) on board the ESRO satellite TDI. The experiment gives spectra at a resolution of 30Å over the wavelength range from 1350Å to 2500Å, and a broad band measurement centred at 2740Å (Boksenberg et al, 1973). We have also started a program of obtaining spectrophotometric observations in the visible wavelength range of a selected number of stars for which ultraviolet data are available. This would give the wavelength coverage of the spectral data from the visible to the far ultraviolet. In this note we shall present the results for 34 stars in the spectral range from BO to A2.

Observations and reductions. Spectrophotometric observations in the visible wavelength range (6000Å - 3400Å) were obtained with a photoelectric scanner at the Cassegrain focus of the 100-cm reflector of Kavalur Observatory, Indian Institute of Astrophysics, India. The observations were made between 3400Å and 5000Å in the second order and between 6000Å and 4500Å in the first order. The spectra were scanned with a slit of 50Å width. Ultraviolet data are taken from the spectra obtained with the sky survey telescope in the TDI satellite. In order to achieve greater photometric accuracy the magnitudes at several wavelengths (2500Å, 2460Å, 2260Å, 2190Å, 2050Å, 1860Å, 1660Å, 1490Å and 1390Å) are derived from consecutive spectral data points. These magnitudes have an effective passband of ~ 100Å. The mean photometric error of the ultraviolet magnitudes obtained in this way is ±0.04 for stars brighter than $V = 5.^m0$ rising to ±0.12 for fainter stars.

The sample of stars studied here contains reddened and unreddened dwarfs and supergiants. The spectral distributions of the reddened stars have been corrected for interstellar reddening, using the mean extinction law derived from the TDI data. The mean value of total extinction per unit visual colour excess, $A(\lambda)\ E_{B-V}$ has been derived by combining all the available ground based and ultraviolet data (Nandy et al, 1975). To illustrate, we have shown in Figure 1 the observed colour $(m_\lambda - V)$ for ζ Per (HD 24398, B1I$_b$, $E_{B-V} = 0.35$) by filled circles and the $(m_\lambda - V)_0$ corrected for reddening by open circles.

Results. For the comparison of the observed spectral distributions of the stars (corrected for reddening) with those from model atmosphere, we have constructed a series of model atmospheres, using the model atmosphere program ATLAS (Kurucz 1970, 1973). The version used allows the inclusion of line opacity through the use of distribution functions based on about 750 000 spectral lines (Kurucz, Paytremann and Avrett, 1972). The effect of line blanketing and the effects of variations in the gravity have been discussed by Nandy and Schmidt (1975). For hotter models the chief effect of line blanketing appears in the form of an increase in the continuum flux from about 5 per cent at 6000Å to about 15 per cent in the ultraviolet around 1500Å. The effect of gravity on the flux distributions is relatively small compared with the effect of temperature changes. Therefore, for B- and early A-type stars the value of log g need not be known with great accuracy. For supergiants we used log g = 2, and 4 for main sequence stars. The temperature is established by fitting the computed $(m_\lambda - V)$ from the models with the observed spectral distributions (as shown in Figure 1 by the solid curve). In Figure 2 we have plotted the effective temperature of the stars studied here, as determined by this method, against MK spectral type and against the parameter $Q = (U - B) - 0.72\ (B - V)$. The temperature scales derived by Johnson (1966) and by Schild, Peterson and Oke (1971) are also shown in Figure 2. It can be seen that the supergiants are separated from the main sequence stars in the log Te vs Q plot. For early B stars the supergiants are cooler than the main sequence stars of the same spectral type, but become hotter near B8.

Details of these results will be presented elsewhere.

Acknowledgement. We should like to thank Prof. M.K.V. Bappu for granting us time on the telescope of Kavalur Observatory, and Mr. Rajmohon and Parth Sarathi for their assistance.

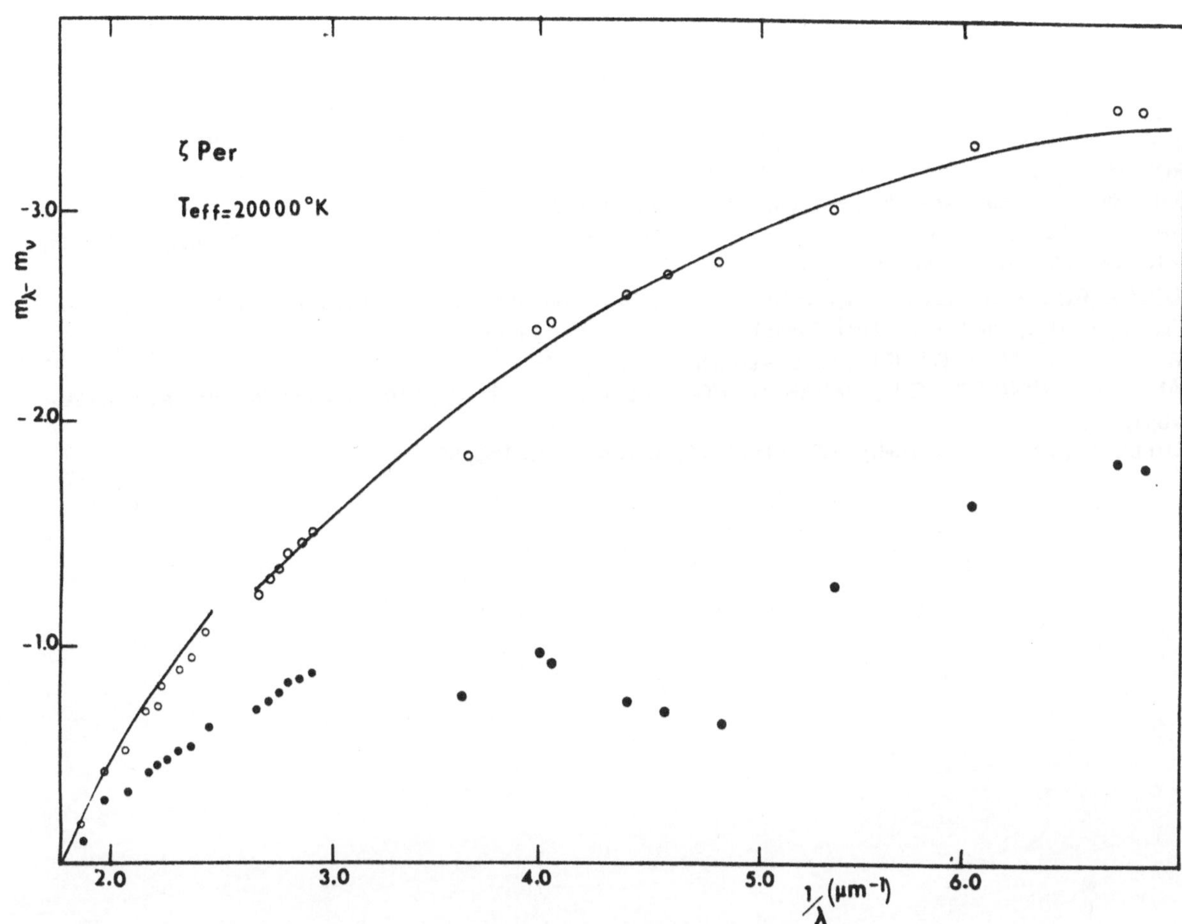

Fig. 1. The observed (m_λ - V) plotted against $\frac{1}{\lambda}$ (μm^{-1}) for Per (HD 24398) is shown by filled circles; the corrected (m_λ - V) is denoted by open circles. The model computations are indicated by a solid line.

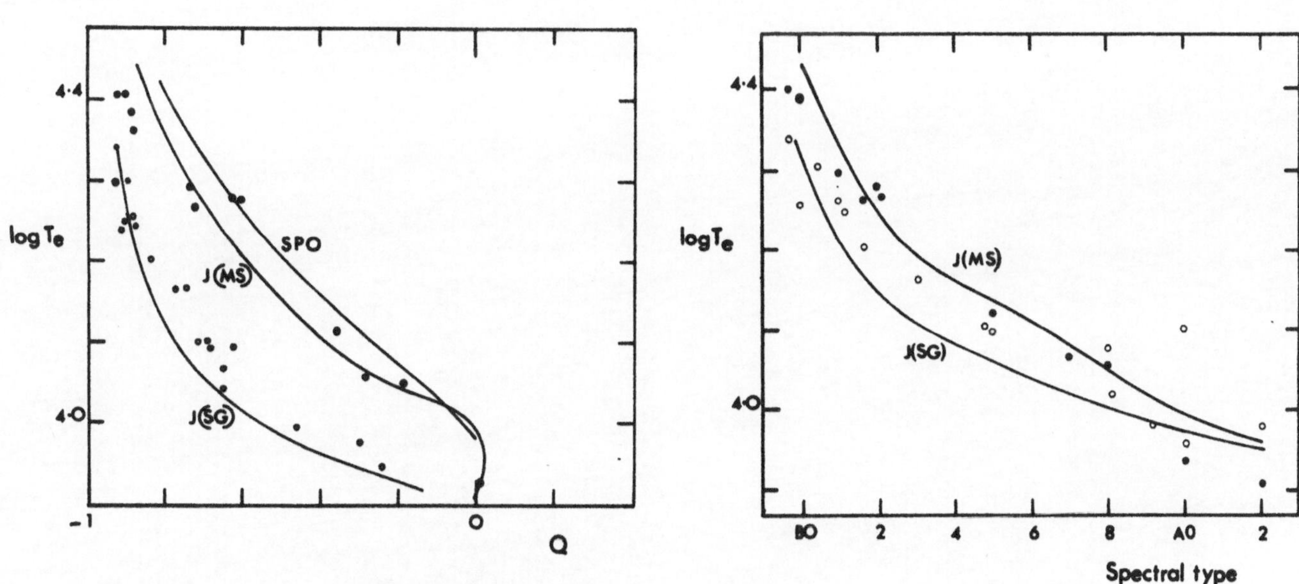

Fig. 2. a) log Te vs Q

b) log Te vs MK spectral type

Open circle are supergiants and filled circles are main sequence stars.

REFERENCES

BOKSENBERG A., EVANS R.G., FOWLER R.G., GARDENER I S K., HOUZIAUX L., HUMPHRIES C., JAMAR C., MACAU D., MALAISE D., MONFILS A., NANDY K., THOMPSON G.I., WILSON R., WROE H., 1973, Mon. Not. R. astr. Soc. 163, 291

JOHNSON H.L., 1966, Ann. Rev. Ast. and Astrophys., 4, 193

KURUCZ R.L., 1970, Smithsonian Astrophys. Obs. Special Report 309 1973a, Centre for Astrophysics Preprint No 15 1973b, Centre for Astrophysics Preprint No 29

KURUCZ R.L., PAYTREMANN E., AVRETT E.H., 1972, Blanketed Model Atmospheres for Early Type Stars (Washington D C, Smithsonian Inst. Press)

NANDY R. and SCHMIDT E.G., 1975, Astrophys. J, 198, 119

NANDY K., THOMPSON G.I., JAMAR C., MONFILS A., WILSON R., 1975 in press in Astronomy and Astrophysics

SCHILD R., PETERSON D.M., OKE J.B., 1971, Astrophys. J., 166, 95

SPECTRAL CLASSIFICATION OF STARS BY THEIR ULTRAVIOLET SPECTROGRAMS

O.V.OHANESYAN
GARNY SPACF ASTRONOMY LABORATORY, EREVAN, U.S.S.R.

СПЕКТРАЛЬНАЯ КЛАССИФИКАЦИЯ ЗВЕЗД ПО УЛЬТРАФИОЛЕТОВЫМ СПЕКТРОГРАММАМ

The possibility of the spectral classification of stars based on the use of mean dispersion spectrograms is known to be confined to stars of magnitude up to 13. This limit is somewhat expanded in particular by way of the use of low-dispersion spectrograms as well as the application of multi-colour photometry.The former technique was utilized by scientists of Abastumani group (Kharadze and Bartaya, 1960), while the latter one by Vilnius group astrophysicists (Straizys, 1973).

Yet another procedure seems plausible and even most perspective, for the spectral classification of faint stars. We mean a possibility of the utilization of the shape and the macrostructures of ultraviolet spectrograms or spectral pictures obtained by means of objective prisms for wave-lengths below 3000Å down to 2000Å. In this case one can classify stars by the lengths of their spectrograms taking no notice of spectral lines, since for these wave-lengths the continuous spectrum proved to be strongly dependent on the spectral class or the efficient temperature of the star. This method may be especially effective for the classification of faint stars (fainter than the 10 stellar magnitude), the spectrograms of which could be obtained by means of the conventional technique- the wide-angle telescopes and objective prisms.

The proposed method of the classification of stars was applied to the group of 2000 non-classified stars in the vicinity of Capella, the spectrograms of which were obtained by means of a wide-angle meniscus telescope and 4^{0} objective prism of the space observatory "Orion-2 " (Gurzadyan et al., 1974; Gurzadyan 1974; Gurzadyan et al., 1975). This sky area was covered by three photographs - F19, F20 and F21 with the exposures of 15 sec., 1.5 min. and 18 min. respectively. The spectrograms were photographed on 100mm wide Kodak 103-0-UV sensitized films. Prelaunching and postlaunching investigations of the photographic films showed the practical absence of photographic fog.

The mentioned property of the ultraviolet spectrograms of stars viz., the strong dependence of their length on the spectral class is clearly seen in fig. 1 - a mosaic of six "Orion" spectrograms pertaining to stars of the same brightness but to different spectral classes. The exposures of these spectrograms made 18 minutes. An analogous mosaic of four other "Orion" spectrograms with the exposure time of 1.5 min. is shown in fig. 2. In this figure one can discern even the difference between the lengths of two spectrograms (two upper pictures) pertaining to the stars differing one from the other by one subclass.

The proposed method of spectral classification of stars is effected in practice not by the comparison of the images of spectrograms, but by the comparison of their microphotometric recordings without reference to the intensity scale. To eliminate the "field effect" we always tried to select the closely situated standard and classified stars.

In fig. 3 we show the superimposition of two microphotometric recordings one of which (solid curve) refers to the known spectral type star, while the other (dashed curve) to the star classified by our technique (in accordance with the numeration of Gurzadyan,1975).

We have used for comparison the spectrograms of stars from the same celestial region yet taken at shorter exposures. In figs. 3, 4 and 5 the examples of classified stars are presented. In all the figures the microphotometric recordings of known spectral type stars are given below those of our classified stars.

Being practised enough one can readily estimate the stars by sight without the comparison with microphotometrical recordings of the spectrograms of known stars.

It is easy to see that the spectral class of the star, $(Sp)_{observed}$, as obtained by this technique always proves slightly later than the true one, $(Sp)_{true}$, depending on its distance from us or, to be correct, on the interstellar absorption;

$$(Sp)_{true} > (Sp)_{observed}$$

Practically for fainter stars we obtain a later type of the spectral class in question i.e. any corrections will render it more early than the directly observed spectral class. The complete list of all 2000stars classified by means of the technique under consideration is given in ref.(Ohanesyan 1975).In table

2

1 we give a fragment of this list. We have also determined the brilliance of the classified stars (V) as well as their colorimetric characteristics (B-V) and (U-B) by measuring their images taken on the Byurakan 40" telescope. It is noteworthy that the colorimetric characteristics of these stars agree with our data on their spectral classes.

TABLE 1

Fragment from Table (Ohanesyan 1975)

No. Garny	SAO	B	B-V	U-B	Sp
1510	040148	12.63	+0.83	+0.22	KO
1511		10.54	0.04	-0.40	B8
1512		12.66	0.58	+0.22	F7
1513		12.61	1.32	0.98	K7
1514		12.87	0.46	0.10	F3
1515		13.73	1.16	0.51	K2
1516		12.70	1.38	0.69	K3
1517		11.45	0.55	0.04	F7
1518		9.55	0.04	-0.40	B8
1519		12.95	0.82	+0.09	KO

The classified stars were statistically analysed with a view of their spectral type distribution. In fig. 7 we give the distribution of the classified group of stars in the spectral types. The earlier classified stars brighter than the 10 stellar magnitude (SAO catalogue) are seen to make in the celestial range considered less than 5% as compared with the "Orion-2" material extending to stars of the 13 magnitude and even fainter. The spectral type distribution of these stars is given in Table 2. The accuracy of the classification was not worse than 2÷3 subclasses over the all spectral range.

TABLE 2

The spectral type distribution of stars according to the SAO catalogue (stars brighter than 10^m) and our classification ($13^m.7$ magnitude) for celestial range in the vicinity of Capella.

Spectral type	B	A	F	G	K	M	Total
SAO stars	9	30	12	8	16	1	76
Our classification	94	411	504	363	455	62	1889

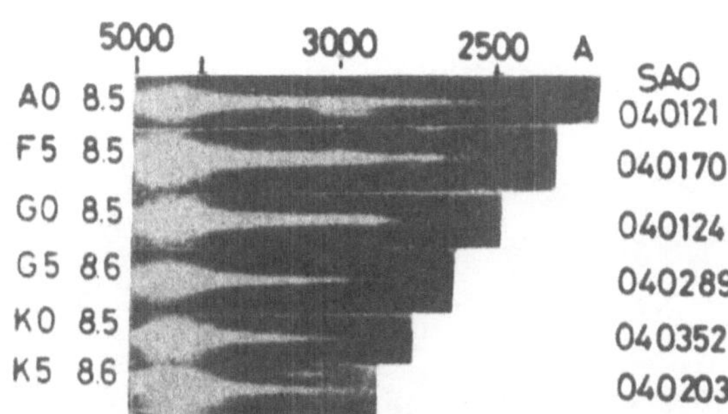

Fig. 1. The mosaic of six "Orion" spectrograms belonging to stars of approximately the same photographic brightness but from different spectral classes. The exposure time was 18min.

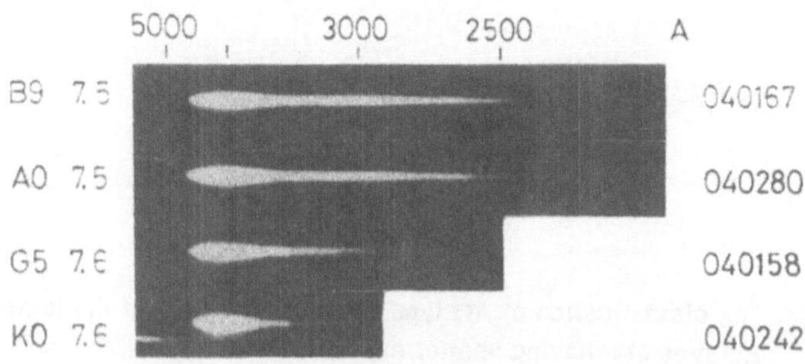

Fig. 2. The mosaic of four "Orion" spectrograms with the exposure of 1.5 min. The difference is seen between the lengths of first two spectrograms, belonging to class B9 and A0 stars.

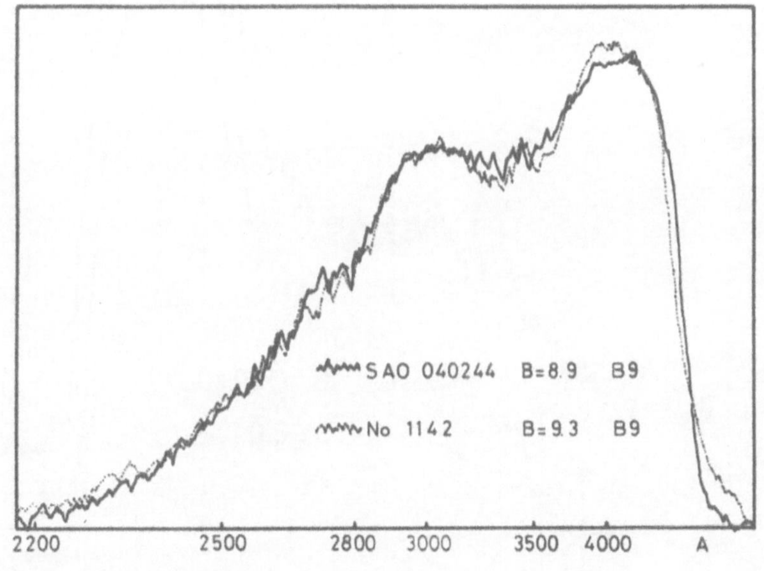

Fig. 3. The superimposition of microphotometric recordings; solid curve pertains to the known spectral type star; dashed curve pertains our classified star.

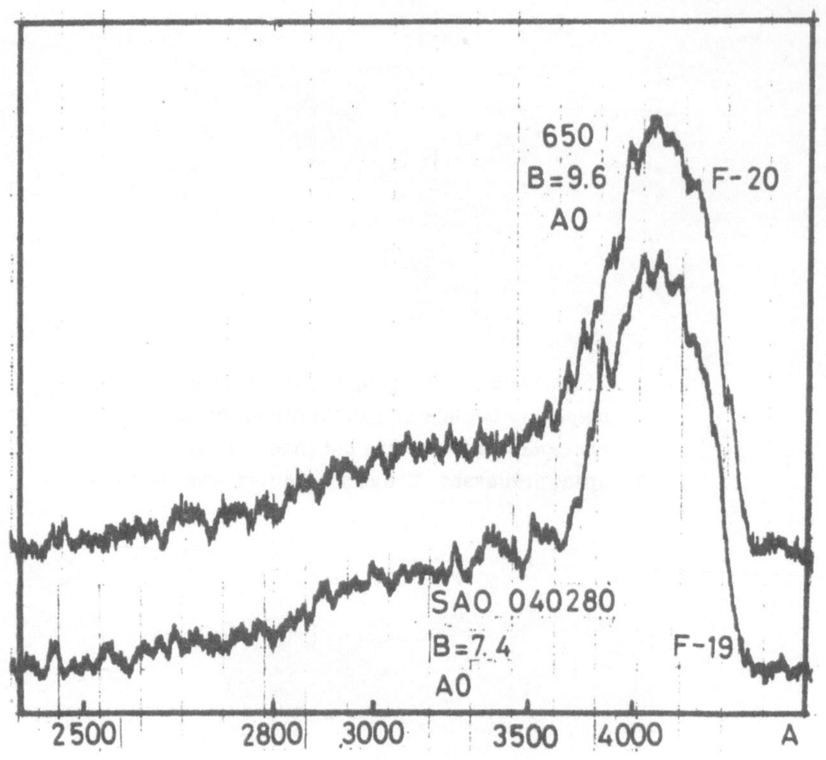

Fig. 4. The classification of AO type star with the help of the known AO type star having shorter exposure.

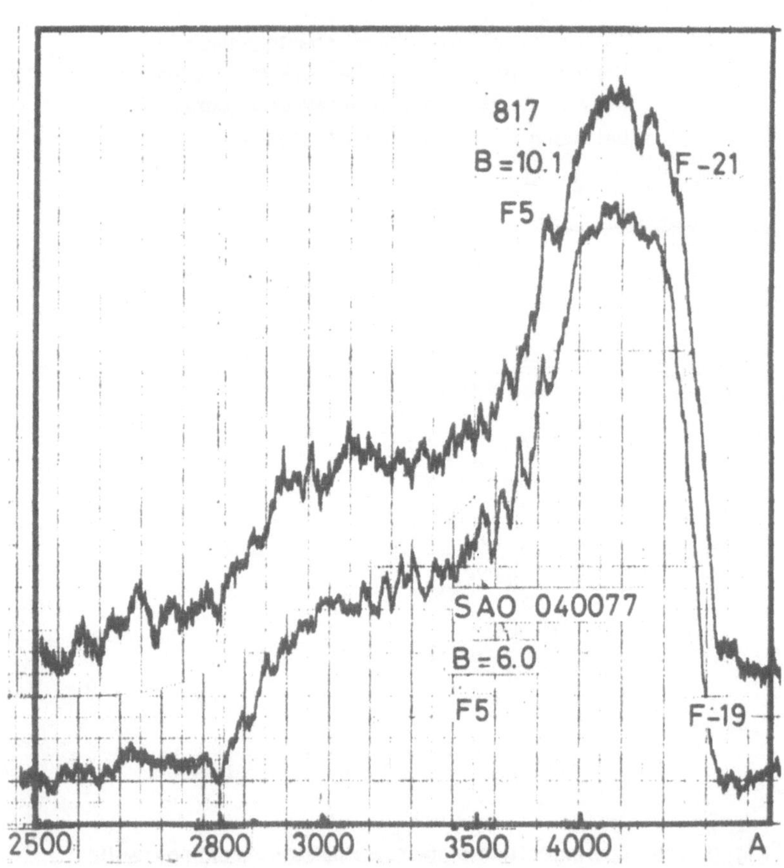

Fig. 5. The classification of F5 type star with the help of known F5 type star having shorter exposure.

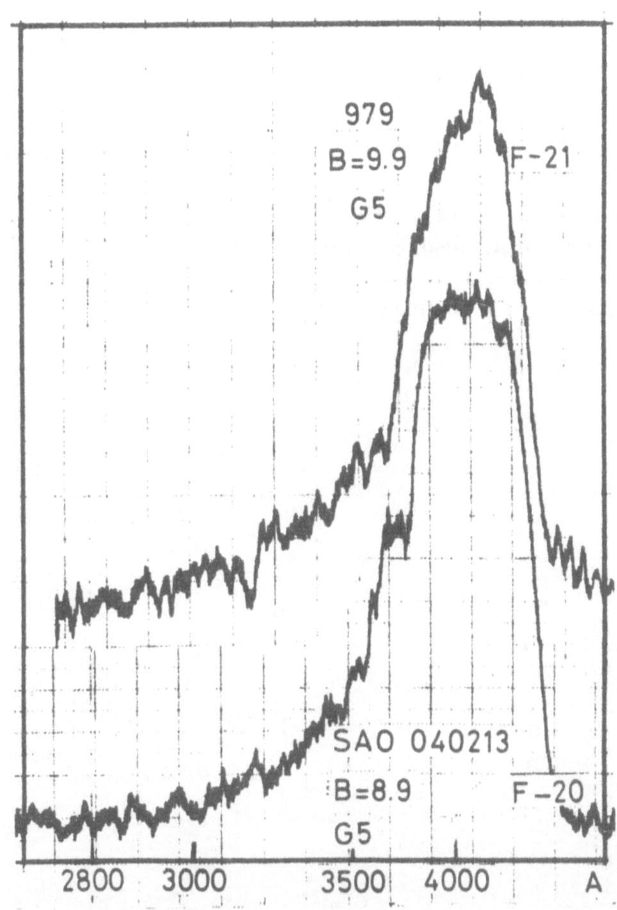

Fig. 6. The classification of G5 type star with the help of known G5 type star having shorter exposure.

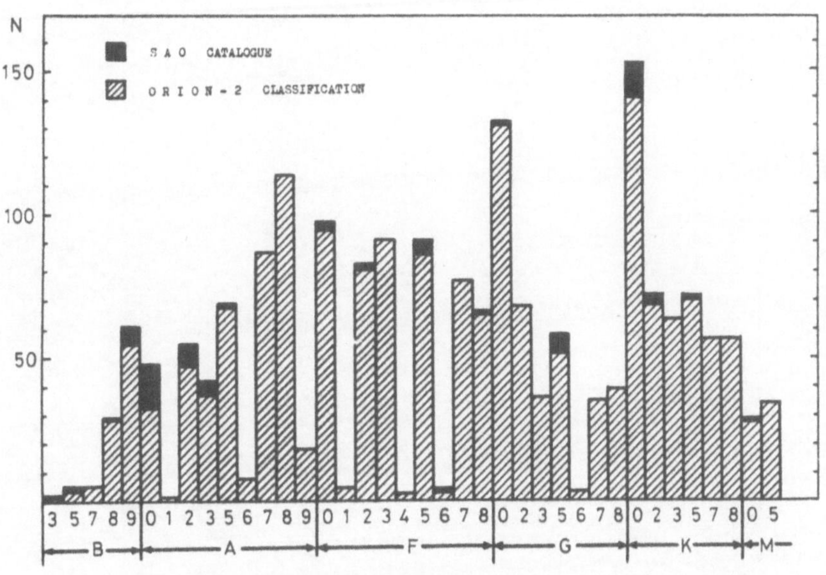

Fig. 7. The spectral type distribution of our classified and SAO catalogue stars from the Capella vicinity.

REFERENCES

GURZADYAN G.A., KASHIN A.L., KRMOYAN M.N., OHANESYAN J.B., 1974, Astrophizika, 10, part 2,

GURZADYAN G.A., 1974, Sky and Telescope, V, 48, 4, 213

GURZADYAN G.A., JARAKYAN A.L., KRMOYAN M.N., KASHIN A.L., LORETZYAN G.M., OHANESYAN J.B., 1975, Astrophys. Space Sci., (in press)

GURZADYAN G.A., 1975, Comm. Byurakan obs., 48, (in press)

KHARADZE E.K. and BARTAYA R.A., 1960, Bull. Abastumani obs., 25, 139

OHANESYAN O.V., 1975, Comm. Byurakan obs., 48, (in press)

STRAIZYS V.L., 1973, "Spectral CL and multicolour photometry", Symp.No. 50, Dordrecht-Boston, p.183

THE ROLE OF THE NUCLEAR ACTIVITY IN THE OVERALL EVOLUTIONARY PROCESSES IN GALAXIES

V.A. AMBARTSUMIAN
BYURAKAN ASTROPHYSICAL OBSERVATORY, U.S.S.R.

РОЛЬ ЯДЕРНОЙ АКТИВНОСТИ В КРУПНОМАСШТАБНЫХ ЭВОЛЮЦИОННЫХ ПРОЦЕССАХ В ГАЛАКТИКАХ

Almost 20 years elapsed since the concept of the activity of nuclei has been introduced. Since then many concrete forms of such activity have been discovered and studied. We now understand that this phenomenon is deeply related to many other physical processes observed in galaxies. Nevertheless many fundamental questions related to the activity of nuclei remain unanswered. Not only we have failed to understand the intrinsic mechanisms of the nuclear activity, but we cannot answer such simple questions as:

1. What kind of a body is an active nucleus? What part of the apparent condensation, that we observe in the central region of galaxies, is the active nucleus?

2. What are the physical state and the form of the masses the nuclei eject during their activity?

3. Which one of the different forms of immense amounts of energy released by an active nucleus is the direct result of the nuclear activity and which part of this energy undergoes transformations in the space surrounding the nucleus?

4. What are the effects of nuclear activity on overall evolutionary processes in the galaxy?

The subject of this report is this last question.

Since we do not know the mechanism generating the activity of nuclei, we shall consider one after another the nuclear processes, which affect the overall properties and structure of galaxies, at purely empirical level.

1. T h e   t r a n s f o r m a t i o n   o f   a   n o r m a l   g a l a x y   i n t o   a   r a d i o g a l a x y. As is well known such a transformation consists in the formation of two clouds of relativistic electrons in two opposite sides of the galaxy. It is widely accepted that these clouds are ejected by the nucleus and thus the whole radiopicture of the galaxy is changed. There remain, however, two different possibilities: a) the particles forming the clouds were ejected immediately by the nucleus or b) the nucleus ejected some compact bodies which are able to emit the high energy particles for a long time. To choose between these two possibilities one must take the duration of the existence of the clouds into consideration.

Now it is estabilished that according to their optical appearance a considerable percentage of radiogalaxies belongs to the morphological class cD. However, the cD radio galaxies are situated almost exclusively in clusters of galaxies which according to Bautz-Morgan classification belong to class I. At the same time according to Tovmassian and others the majority of cD supergiants in clusters are radio-galaxies. This means that the mean life time of radio- emitting clouds is not much shorter than the duration of the phase of the evolution of a supergiant galaxy during which the galaxy remains in the same morphological type cD which in its turn, apparently, cannot be much shorter than 3.10^9 years.

On the other hand, the relativistic electrons in the clouds lose their energy in the timescale of 10^6 years. Therefore the continuous supply of high energy electrons is required. We can imagine two different ways of such a supply: (a) the immediate and more or less continuous supply from the nucleus and (b) the continuous supply from a body which was ejected by the nucleus at the beginning of the process.

At this stage of our knowledge it is difficult to decide which of these alternatives corresponds to reality. But even accepting the case (b) we must not forget the existence of the fairly intense central components in such strong sources as for example Cygnus A and then we must interpret them as the results of recurrent ejections of compact active bodies.

On the other hand the exceptional strength and rarity of such sources as Cygnus A, 3C 295 and 3C 411 apparently are the evidence of their comparative youth. With uncertainty which is typical for the statistics of very small numbers one can estimate from the frequency of such sources that they have passed through 1/10 of their lifetime. This means that their age must be of the order of 3.10^8 years. But it also means recurrent and fairly strong activity at least during such a period of time.

If we accept the hypothesis of continuous supply of energy to outer components of the radio-source then we are forced to conclude that the activity has the duration longer than 10^9 years.

When we add to this that the immense amount of energy emitted by the central part of Cygnus A in optical emission lines (of the order of 10^{43} erg sec^{-1}) is also the result of the continuing activity of the nucleus one can safely accept that *the nuclei of cD galaxies at some stage of their evolution* (perhaps at the initial stage) *remain in active state, at least during several hundred millions of years.*

2. The ejection of gaseous masses from the nucleus. The outflow of large gaseous masses in ordinary spiral galaxies is a widespread form of the nuclear activity. It has been observed in different forms but perhaps the most convincing evidence was found by van der Kruit, Oort and Mathewson from the radio-observations with Westerbork telescope and from the spectral observations in the optical region of the galaxy NGC 4258. According to these observations the galaxy NGC 4258 contains, except the normal spiral arms, an H - arm, which is formed by the gas flowing from the nucleus and interacting with the existing interstellar medium. The rotation speed of the H - arm is considerably lower than that of the normal arms.

In their paper van der Kruit, Oort and Mathewson are going as far as to suppose that in this galaxy we are witnessing an example of the spiral structure formation, as a consequence of the nuclear activity - a process that was generally suggested by the present reporter in his Berkeley report in 1961. But even if we exert caution, we can insist that the observations at least support strongly the view that the gases, ejected by the nucleus, enrich the interstellar material, thus changing its dynamical state.

In another galaxy, NGC 4736 (Sab) van der Kruit has found that an internal ring of 50'' radius consisting of many H II regions is *rotating* around the nucleus and *expanding.* The rotational velocity is about 150 km/sec and the expansion is of the order of 30 km/sec. This expanding ring is apparently independent of the spiral arms which are present in the region outside the ring. One cannot even exclude the possibility that the plane of the ring is somewhat inclined to the plane of the main spiral structure.

Thus there remains almost no doubt that the material of the ring originated in the nucleus about 30.10^6 years ago.

The active state of the nucleus of NGC 4736 is confirmed by the fact that within the galaxy there is a triple system of discrete radio-sources of which one coincides with the nucleus and other two are situated at equal distances in the opposite directions. Their projected distances from the nucleus are almost equal to the radius of the ring.

The fact that within the ring there is no trace of the spiral structure, apparently, is also connected with the activity of the nucleus.

Not less important is the explosive event which happened in M 82 and was discovered by Sandage and Lynds.

Serious objections have been put forward against the first estimates of the ejected mass of gas in M 82. As is known such an estimate must depend on the mechanism of luminescence we observe from the clouds receding from the nucleus in the direction of axis of rotation. It may happen that initially the value of ejected mass was strongly overestimated. Nevertheless, the evidence for an explosive event (perhaps of smaller scale than at first was suggested) about $1.5.10^6$ years ago is very conclusive.

Accepting that the explosion has taken place it is worth to pay attention to another unusual feature of M 82, which is present in the central part of that galaxy. I mean the presence of a number of "superclusters" of very high luminosity (about -16^m in visual light) in the vicinity of the nucleus of M 82. These complexes have been found by van den Bergh from infrared observations.

Each of these clusters must contain a considerable number of O-B stars - much more than usual clusters and associations. It seems that they are complexes of the same nature and of the same order of luminosity as superassociations - a class of systems which is presented in LMC by a single example - 30 Doradus. There is,however,an important difference between superclusters in M 82 and the superassociations.

The superassociations (which are often described as very large H II regions) have diameters between 200 and 1000 parsecs while superclusters,discovered in M 82 by Vandenberg, according to his estimate have diameters of about 60 parsecs.

I am not going to insist that all these superclusters originated just during the explosion mentioned above. But it is very probable that these superclusters are not of very old age. Apparently, they are not older than 10^7 years. Then it is very difficult to imagine that the presence of a considerable number of superclusters can be a perpetual or lasting feature of the central part of the galaxy. Otherwise we should have the problem of unusually high total mass of remnants of many generations of superclusters in the central part of M 82. It is much more probable that the situation we now observe is only phase of a short duration in the life of the central part of M 82.

It is quite natural to suggest that this phase was caused or triggered by some important event during the last 20 or 30 million years in the nuclear region of the galaxy under consideration. At present it is difficult to identify this event with the explosion suggested in the work of Sandage and Lynds. However, it is important that formation in the vicinity of the nucleus of such large complexes as superclusters is to be considered as an important form of the activity of the nucleus.

3. Large complexes of early type stars and gas in the central parts of galaxies. Perhaps the most important fact connected with this kind of nuclear activity is the presence of the *compact* but at the same time massive complex of Sagittarius B2 in the region of the nucleus of our Galaxy. This complex contains dense H II regions, molecular layers and certainly, a number of early type stars ionizing hydrogen. However, owing to the large interstellar absorption in the direction of B2 cloud we cannot observe it in optical frequencies and specify even the approximate number of early type stars. The estimates based on the intensities of molecular radio-lines give for the total mass of the whole complex $M 10^7 M$. *There is no other dense complex of such large mass in the whole Galaxy.* At the same time the Sagittarius B2 complex is a young system. Its age must be of the order of about 10^7 years. In any case it must be less than 10^8 years. The linear diameter of the complex must be of the order of 40 parsecs and, certainly, not larger than 50 parsecs. This is much less than the size of an average stellar association, not speaking about the superassociations. Therefore the complexes comparable in size with Sagittarius B2 are the objects near the centre of M 82. The existence of Sagittarius B2 complex is an evidence of the recent activity of the nucleus of our Galaxy. However, there are galaxies in which this kind of nuclear activity is going on *with much higher intensity.* I would like to mention here as examples the spirals having very intense hot spots which have been found by Sersic and Pastoriza and studied in detail by Osmer, M.G.Smith and D.W.Weedman. Let us take the galaxy NGC 1365. The intensity of emission line H_β of hydrogen radiated from the hot spots in the centre of this galaxy has brought the mentioned authors to the conclusion that there are 870 O-BO.5 stars in these hot spots. Taking for the lifetime of these stars $6 \cdot 10^6$ years and using the Salpeter initial luminosity function, one is forced to conclude that the total mass of stars there must be of the order of a billion solar masses. We cannot suppose that such intense star formation is a lasting feature of the galaxy NGC 1365. Therefore we cannot avoid the conclusion that in the nuclear region we have a relatively short period of star formation in which several billions of solar masses of material are used.

Similar intense star formation processes are going on in the central parts of southern galaxies NGC 1672 and NGC 2992.

Now if we admit that the formation of hot spots is the result of the activity of the nucleus, it is very difficult to suppose that the formation of O-associations and young stellar clusters in spiral arms is connected with some other kind of processes. This means also that the formation of spiral arms and of population I of external regions is also the result of some kind of ejection of material from the nucleus. However, when we try to analyse such a possibility we meet the difficulty of explanation of the large rotation momentum which is connected with the motion of outer parts of galaxies. Even if we suppose that only the stars of population I originate from the material coming from the nucleus the difficulty still remains, since it is the rotational momentum per unit of mass of this population which is difficult to explain under this assumption.

This difficulty of explanation of the rotational momentum is first of all a problem for the theory. It is true that great theoretical difficulties in explanation of some phenomenon must not compel us to disregard almost direct deduction from observations. On the other hand one must state clearly that the theory has failed until now to find *the source and the mechanism of transfer of rotational momentum.*

This problem is of a very general nature, since the rotational momenta of elliptical galaxies, being relatively smaller than the momenta of spirals, are not at all very small. Therefore in this case also an explanation is needed. Among the possible ways of the solution of this problem I can only mention here the recent paper of Dr. Mouradian who has suggested a nontrivial source of rotational momentum of a galaxy in terms of the spin of a supermassive elementary particle, without explaining the mechanism of its transfer to the periphery of the galaxy.

In any case one must not forget that in the objects like NGC 4736, where we observe the expansion of a rotating ring, we have the cases where such transfer of rotational momentum occurs in reality.

4. The Seyfert galaxies and quasars. Now we pass to another kind of nuclear activity represented by violent processes in the nuclei of Seyfert galaxies. The most important characteristic of Seyfert galaxies is the presence of violent motions of gases (or gaseous clouds) in their nuclei. The loss of mass by means of ejection of gaseous clouds speaks in favour of large changes in the mass of the nuclei during the time of the order of 10^8 years. Perhaps this time scale is to be accepted as the duration of Seyfert phase. At the same time the nucleus of a Seyfert galaxy emits a) the thermal radiation of stellar origin, b) non thermal continuous radiation of nonstellar origin which extends far into the ultraviolet, c) the emission line radiation of the gaseous clouds which are in the process of ejection and are in a physical state somewhat similar to H II regions of our Galaxy.

There are some normal spiral galaxies in which the nuclei exhibit with less intensity the spectral features observed in Seyfert nuclei. Therefore it is possible that the Seyfert stage is some phase in the evolution of normal galaxies and the phenomenon itself, being very intense at the beginning of the phase, is dying steadily when the galaxy is transforming itself into a normal one.

Among the Seyfert galaxies of our immediate surroundings we see only objects of the intermediate luminosity or not very luminous giants. However, at larger distances we observe supergiants like famous Markarian 10 with a photographic absolute magnitude of the order of -23. As it was shown by Khachikian and Weedman the object Markarian 102 is a very distant and also exceedingly luminous Seyfert galaxy which reminds some properties of quasars. Thus, the whole class of galaxies with Seyfert spectral features has a sufficiently broad luminosity function which permits us to suppose that perhaps all galaxies including the ellipticals with luminosities between -16^m and -23^m pass through this phase of evolution.

One can estimate that among the galaxies of a given apparent magnitude the Seyferts and galaxies with related emission spectra form about 1% of objects. Since the majority of the galaxies which have a given apparent magnitude belong to objects of luminosities between -18 and -22 in V, we can conclude that the assumption that all galaxies of these luminosities pass through Seyfert phase requires the duration of this phase of the order of 10^8 years in a good agreement with existing estimates.

As it was already mentioned the comparatively high luminosity of nuclei is one of the main characteristics of Seyfert galaxies. But as is known the quasars having many similarities with Seyfert galaxies have much higher luminosity. Therefore following Sandage and his coworkers we can consider quasars and radio quiet quasi stellar objects as galaxies with very luminous nuclei. More exactly the luminosity of nucleus in these objects well exceeds the total luminosity of other parts of the galaxy.

The luminosities of brightest QSO are of the order of $10^{13} L_\odot$. The majority of QSO have the luminosities by one order lower. This means that their power of radiation is of the order of 10^{46} erg/sec. Then during the period of 3.10^8 years the total energy radiated during the lifetime must be of the order of 10^{62}ergs-a quantity which is equivalent roughly to 10^8 solar masses. Even if we suppose that the lifetime of QSO-s is only 10^7 years, we still see that this large amount of radiation, which is, of course, mainly of non-stellar origin speaks on a kind of activity which must play the chief role in the energetics of the QSO-s. And if the QSO is connected with a supergiant galaxy it plays the chief role in the energetics of that galaxy as well.

Let us consider now only the quasars, which are radiosources. We know that their power in radio-frequencies is of the same order as that of strong radio-sources, which are connected with the supergiant cD or elliptical galaxies. Let us introduce the ratio α of duration of the phase of high optical luminosity (t_{opt}) to the duration of strong radio-emission (t_{rad})

$$\alpha = \frac{t_{opt}}{t_{rad}} ,$$

If on the other hand, for the given interval of radiofluxes the ratio of the number of optically observable quasars to the number of radio galaxies in the same volume of space is equal to

$$\beta = \frac{n_{opt}}{n_{rad}} ,$$

then we must have

$$\alpha \geqslant \beta ,$$

since some of radio-galaxies could have another origin. Since the empirical value of β is about 0.1 we can conclude from $t_{rad} = 10^9$ years that the $t_{opt} \sim 10^8$ years.

Thus it is quite probable that the quasi stellar radio sources live as such several hundred millions of years.

Let us now assume only that $\alpha < 1$.

Then it follows immediately that all quasi stellar radio sources are nuclei of the supergiant galaxies (cD or ellipticals). Otherwise we should have observed among the radiogalaxies the galaxies which are not supergiants, which is not the case.

On the other hand the assumption $\alpha > 1$ puts too heavy requirements on the energetics of quasars. Therefore it is almost cetain that *all quasi stellar radio sources are the nuclei* of the supergiant galaxies. Thus we strongly support Sandage's case in this question.

However, the situation is quite different in the case of purely optical QSO-s. It seems certain that *some* purely optical QSO-s are indeed nuclei of supergiant galaxies of elliptical nature. But there is no reason to say the same about *all* optical QSO-s. It seems that all of them are nuclei of some galaxies. But at least some of optical QSO-s are surrounded by less luminous galaxies. Probably their underlying galaxies have a wide range of possible values of luminosities.

5. Some remarks on galaxies of moderate and low luminosities. Thus some optical quasars are nuclei of galaxies of relatively low luminosities (not supergiants). On the other hand many Seyfert galaxies have moderate luminosities. We know also that in many galaxies of moderate luminosities we meet frequently recently formed complexes of young stars - superassociations and associations. Galaxies of irregular type I are good examples of this. Moreover many galaxies of low luminosities appear to us at first sight as "isolated superassociations" (giant H II regions, Sargent). Many examples of such objects have been given recently by Khachikian and Sahakian, then we come to the conclusion that among galaxies of moderate and low luminosity we encounter essentially all types of the activity which occur in supergiant galaxies with the exception of one form - the ability to transform itself into a radiogalaxy (strong radio source). This is especially important when we remember that in many galaxies of moderate and low luminosity we have no immediate optical manifestation of the presence of a nucleus.

We have intentionally avoided the question - whether the totality of different forms of the nuclear activity serves as the basis for *all evolutionary processes* in galaxies. Personally I think that it is so. However, in order to prove or completely reject this opinion we need more complete study of the physical processes in the diversity of types of galaxies.

CERTAIN STATISTICAL PROBLEMS OF SUPERNOVAE

KONRAD RUDNICKI
JAGIELLONIAN UNIVERSITY OBSERVATORY, KRACOW, POLAND

НЕКОТОРЫЕ СТАТИСТИЧЕСКИЕ ПРОБЛЕМЫ СВЕРХНОВЫХ

РЕЗЮМЕ

Довольно "плоское" пространственное распределение Сверхновых в спиральных галактиках создает некоторые трудности в интерпретации, если исходить из гипотезы что Сверхновые являются основными"производителями" тяжелых элементов. Частота Сверхновых в галактиках различных типов ведет к противоречиям, которые нельзя разрешать без уточнения понятия "частоты".
Перечислены различные подходы к Сверхновым, как к индикаторам космических расстояний.Подход Раста и деление им 1 типа Сверхновых на 2 класса рассматриваются как наиболее обещающие. Показано, что некоторые результаты Раста могут рассматриваться как эффекты эволюционные или связанные с звездным населением. Решение вопроса требует статистических данных о весьма далеких Сверхновых. Предлагается программа наблюдений для далеких Сверхновых.

ABSTRACT

The rather flat space distribution of supernovae in spiral galaxies causes some difficulties when interpreted within the hypothesis that supernovae are the main producers of heavy elements. Also the frequency of supernovae in various types of galaxies leads to antynomies which cannot be solved unless the notion of "frequency" becomes more precise.
Various approaches to supernovae as indicators of cosmic distances are listed. The approach of Rust and his division of type I supernovae into two classes is considered as the most promising one. It is shown, however, that some results of Rust can be discussed also as evolutionary or population - effects. Testing of those possibilities is impossible without statistical data on very distant supernovae. An observational program of such distant supernovae is proposed.

Noticeable progress in our knowledge of supernovae has been observed during the last few years. Today we know more than 400 such objects, for several tens of which well - measured spectra exist. We know about 100 supernova remnants in our Galaxy and some tens in other nearby galaxies.Spectra of type I have been finally interpreted by Pskovskij (1968) and Mustel (1971a, 1971b, 1972). The spectral similarities between type I and other types (Kirshner et al., 1973), recent research on a type V supernova 1961v* and other results testify that all five types are of similar physical nature and probably of similar origin. Therefore some phantastic hypotheses of type I explosions are today happily removed. On the other hand some new puzzling facts appeared.

One of these is that type I supernovae explode between old population objects, but in spiral galaxies exhibit a very flat distribution similar to objects of population I. For example, in our Galaxy only the 1604a remnant is placed about 1 kpc from the galactic plane, all the others being located as flat as the interstellar gas. It seems improbable to conceive all the other remnants to be produced by supernovae of types II, III, IV and V only.

The known inconsistencies between the space distributions of supernovae in other late type spiral galaxies and the distribution of supernova remnants in our Galaxy have been successfully removed by Kodaira (1974) and Iye and Kodaira (1975). There remains a small discrepancy between both distributions in the nuclear areas of our and other galaxies. However, keeping in mind that data concerning the distribution of supernovae themselves in other spiral galaxies are averaged data from many galaxies, whereas the distribution of supernova remnants refers to a single object

---

* Designation of the supernovae and suspected supernovae is according to the Catalogue of Karpowicz and Rudnicki (1968), the newer ones - according to the Sargent-Searle-Kowal list (1974).

(our Galaxy) this fact is not surprising. It would be rather astonishing if the characteristics of any individual galaxy were strictly equal to characteristics averaged over all galaxies. It is precisely the nuclei of galaxies which seem to be very individualized. Besides the statistics either of supernovae themselves or of their remnants is less accurate in the vicinity of nuclei.

It was supposed by many authors that supernovae of type I are chronologically the first objects during the evolution of galaxies which produce heavy elements for the formation of population I objects. If this is indeed so, pre-supernovae of type I must have existed earlier than any population I objects, and their expected distribution, and consequently the distribution of type I supernovae should be spherical or intermediate and not a flat one. For this reason the discovered flat distribution seemed to be strange at the first sight. However an opposite line of reasoning may be followed. If population I objects origin out of matter furnished with heavy elements by type I supernovae, their spatial distribution must be the same as that of their ancestors - type I supernovae themselves. Thus we may expect a priori the distribution of supernovae to be flat. These two approaches seem at first glance to be incompatible.

We may pose the problem in a more general way. No matter how we imagine the evolution of stars, if we assume that heavy elements are produced not in nuclei of galaxies but inside old population objects, we meet the same basic difficulty: why do old and young objects have different distributions when the first are parents of the second. One possible solution is to assume that type I pre-supernovae form a spherical subsystem but can erupt only when interacting with something which exists only (with some exceptions) in the galactic planes. Is it the interstellar gas or pre-stellar matter? And if this is so, interaction with what is necessary for the eruption of the same type of supernovae in elliptical galaxies? Unfortunately our knowledge of the composition of elliptical galaxies is so restricted ...

Investigation of the spatial distribution of supernovae in elliptical galaxies is much more difficult than that for spiral ones. The angle between the line of sight and the symmetry plane of an ellipsoid can be established only in a statistical way. An additional stochastic variable has to be introduced to any analysis of this kind, thus requiring much more observational material in order to obtain reasonably reliable statistics. Since there exist at present only about 30 elliptical galaxies in which supernovae were discovered, such an investigation is impossible at present. We have to wait until the number increases at least two or threefold which may, perhaps, happen in the next decades. Those who believe that spiral galaxies are late stages of evolution of all, or of some, ellipticals, and who assume that heavy elements are produced mainly during supernova explosions, should expect the distribution of supernovae in elliptical galaxies also to appear flat.

According to some authors, a distinction should be made between two populations of type I supernovae. The population II members explode in elliptical galaxies and also in spiral and irregular ones and have a spherical distribution (here belongs for example 1604a in our Galaxy). The population I members appear in spiral and irregular galaxies only and form a flat subsystem. This point of view is however not confirmed in distribution investigations. I will return to this problem later and show another argument against such a point of view.

Of course not only the space distribution of supernovae but even the simple investigation of frequency of different kinds of supernovae in various types of galaxies and also clusters of galaxies can be of great importance for solving this and related problems.

Flin (1974a) showed that there is no significant difference between the frequencies of various kinds of supernovae in various types of clusters of galaxies, but more supernovae in general were discovered in compact and medium compact clusters (Zwicky's classification) than in open ones. On the other hand the Sc I galaxies appear more frequently in open clusters, and it is exactly this type of galaxies which is supposed to produce more supernovae than others. This is another puzzle. The special Palomar Sc I supernova search did not, so far, confirm this supposition, but among the 18 galaxies in which more than one supernova explosion were observed, the majority are of type Sc I.

The problem of frequency of supernova explosions in galaxies is in general not an easy one. The mean time T_0 between two supernova explosions in one galaxy is usually calculated from the formula

$$T_0 = \frac{N_g \cdot T}{N_s} , \qquad \text{(i)}$$

where N_g is the number of observed galaxies, N_S the number of supernovae discovered, T - the total time of patrolling. Of course some corrections are introduced here to take into account the frequency of observations, time of visibility of a supernova etc., which makes the problem rather complicated. Anyway, using such a formula, Sargent, Searl and Kowal confirmed recently (1974) the 1970 - Tamman's estimate which is one supernova every 20 or 30 years in one galaxy. For this result, these authors had taken N_g to be the number of bright galaxies, $15.^m7$ and brighter, in the search fields.

Zwicky (1974) as it is well known, has pointed out that the luminosity function of galaxies is an exponential one, thus most of the fainter galaxies visible in search fields are not distant luminous objects, but nearby and intrinsically faint * . Using for N_g the number of all visible galaxies Zwicky confirmed his old value which is one supernova every 300 years in one galaxy, the freguency of Sargent remaining correct for more luminous galaxies. In fact, we can today be sure that the more luminous are the galaxies, the more supernovae they produce in average. (N.B. our Galaxy belongs certainly to those more luminous but the last supernova observed in it was the 1843a; is this merely a fluctuation or an observational selection effect? (On the other hand we cannot even say whether the frequency for an average galaxy lies anywhere between the limits $1/15$ $year^{-1}$ (hypothesis about Sc I galaxies) and $1/300$ $year^{-1}$ (Zwicky's value).

To obtain the lower limit we had to put for N_g the total number of all the galaxies in the surveyed volume of space down to the very end of the luminosity function of galaxies. However this cannot be established, neither theoretically nor observationally. It is rather a matter of definition which smallest aggregates of stars should still be considered to be galaxies. It is well known that supernova 1966O and a probable supernova s1956α both appeared in points of the sky where no galaxies were visible. We call them usually "intergalactic supernovae", but this conveys rather the meaning that these objects exploded in intrinsically faint, invisible galaxies. Thus, we cannot neglect in our calculations the actually invisible but existing galaxies.

Difficulties are even greater with establishing the upper limit of frequency. As it is well known, one cannot discover a supernova in a bright galaxy nucleus. The only supernova with x and y coordinates both equal to zero (1963t) and a couple of others with x, y relatively small were discovered within galaxies possessing rather faint nuclei. All the existing statistics of supernovae are valid for outer regions of galaxies. For solving problems of structure and evolution of galaxies and stars these regions are by no means more important than the inner ones. On the contrary, the supposition that the variability of light of some galaxy nuclei is not only connected but precisely caused by many supernova explosions, was forwarded from the time of the discovery of this variability. Zwicky (1966), introducing his theory of "star studded gaseous ball" supposed even that the supernova outbursts can lose their individuality in high density nuclei. Thus, also for the upper limit problem, two questions appear. One is of observational nature, namely how to observe supernova outbursts in a high-surface-luminosity nuclei. The second is the formal question - what do we call an individual explosion, in areas where individual stars can interact during outbursts.

To solve the problem of supernova frequency in galaxies we have not only to improve materially our observational methods, but first of all we have to formulate more clearly what frequency are we talking about. In my opinion the mean time T_0 between two consecutive supernova outbursts in a galaxy or the frequency f_0 of these outbursts have to be considered as functions of many variables. These variables are, at least, the morphological type X of the parent galaxy, its absolute magnitude M and the distance r of the outburst from the centre of galaxy.

Certainly, to establish a function $f_0/X,M,r/dX \cdot dM \cdot dr$ having at disposal data on 400 supernovae only, many of them being not precise and not complete, is not an easy task. At any rate I would like to emphasize that the "general" frequency of supernovae in galaxies has a very little scientific significance even if it could be established univocally. On the contrary, the frequency interpreted as a function of many variables can lead us to a solution of several interesting problems. Especially the dispersion calculated from the differences between the frequencies $f(X,M,r)$ for individual galaxies and the standard frequency $f_0(X,M,r)$ defined above can be of great informative importance.

---

* Sargent et al. (1974) have tried to contradict this statement using a plot of redshifts versus apparent magnitude of galaxies in which supernovae were discovered. But, of course, to show that Zwicky was not right they should have used rather a plot for all the galaxies, including also those where supernovae were not discovered.

To demonstrate this on an example, let us turn back to the Sc I galaxies problem. The following are the basic facts. 1) The more luminous galaxies produce more supernovae; Sc I belong to the luminous ones. 2) The majority of galaxies with observed repeated supernova outbursts are of Sc I type. 3) During many years of special Palomar patrolling of 40 Sc I galaxies not a single supernova has been discovered. To join these facts together there exists a choice between the following explanations.

1° At least one of the above statements is not true.

2° Because of the small number of galaxies and supernovae involved, no significant statistical results related to the problem can be obtained at present.

3° The Sc I galaxies do not form a uniform class of objects. The special Palomar search galaxies are by chance of different type from those which produce supernovae frequently.

4° Galaxies produce supernovae not at a constant rate, but frequency varies with time.

Both in cases 3° and 4° the dispersion of frequency function would be high, but in case 3° it could be lowered by introducing additional variables (parameters) describing the various subtypes of Sc I galaxies.

Of course, the determination of the standard frequency function f_0 and moreover the frequency f for individual galaxies (even for those producing a great number of supernovae) needs not only more systematic supernova discoveries, but also exact data concerning the population of galaxies confined in search fields. Accidental discoveries, which can be of great importance for other purposes, are of little value for this aim. However, the behaviour of the f_0 function may certainly be established more readily in some intervals of its variables than in others. I think that precisely the intervals of the late spiral galaxies and of low absolute magnitudes should be the first ones to be investigated. It will be important to carefully check again the morphological and spectral peculiarities of galaxies which produce more supernovae and compare them with those of the Palomar Sc I search. It was already noticed by Arp (1974) that some peculiarities of galaxies may be connected with the quality of supernovae produced. It is not impossible that the same may be valid as regards quantity.

To confine myself to within the 20 minutes of time allotted to me, I would like to mention only one class of problems more, related to the book of Rust (1974) on supernovae. The book is devoted rather to cosmological problems which remain outside the topic of our meeting, but many of these results are strictly connected also with evolution or/and population problems. Since the book is not distributed very widely, let us repeat some of its content.

Rust strived to use the type I supernovae as indicators of cosmic distances. The old idea of Zwicky to use for this purpose distance moduli m - M cannot be applied because of the large dispersion of M. The Zwicky-Kwast method (compare Kwast,1970 and Flin,1974b) which uses supernovae as indicators of linear diameters of galaxies, gives cosmological distance estimates independent from the redshift method, but its accuracy cannot compete with the latter method. The Kirshner-Kwan (1974) method gives fairly accurate distances (also independent from redshift interpretation) but can be applied to nearby supernovae only. These were the reasons why Rust used the Doppler effect, not however to electromagnetic wave frequencies, but to some very low frequencies which occur on the light curves. If a galaxy with its supernova has a velocity of recession, all the time intervals on its light curve become larger. Rust used the value Δt_c as a representative time interval. Δt_c is the number of days required for the apparent magnitude to change from $m_0 + 0.^m5$ to $m_0 + 2.^m5$, where m_0 is its magnitude in maximum. The dispersion of Δt_c values for nearby supernovae is rather low.

If we assume that measured redshifts are due to the Doppler effect (or also to some other laws) the correlation between parent galaxy redshifts and Δt_c parameters of their supernovae should be a linear one. The slope of regression line can be calculated a priori; it depends only on the units used, Rust noticed that on the plot Δt_c versus redshift, the regression curve differs significantly from a straight line. Besides the best fitting straight line has a slope ten times larger than the calculated one. When, in spite of all this, Rust applied the Δt_c parameter instead of V_r to establish the Lundmark-Hubble constant, he obtained a value twice as large as the Sandage-Tamman value. The author tried to explain all these unexpected results in cosmological terms. On the other hand, the problem arises whether these results could not be interpreted rather as some intrinsic changes of type I supernova curves with the distance.

Some further investigation of these problems were made by Klimek (1975). He was generous enough to let me present here some of his unpublished results. He considered two portions of type I supernova curves: the fast and the slow ones which he assumed to be rectilinear in the first approximation. Let us denote by ω_f the slope of the fast part of the curve, and by ω_s the slope of the slow one. If we assume we are dealing with a Doppler effect then the difference between them should be

$$\text{(ii)} \quad \Delta\omega = \omega_f - \omega_s = (\Delta\omega)_0 - (\Delta\omega)_0 \frac{V_r}{c}$$

where V_r is the radial velocity of the parent galaxy, c - the velocity of light and $(\Delta\omega)_0$ is the value of $\Delta\omega$ in the frame of reference connected with the galaxy. This last value can be established in practice from data of nearby supernovae. For the ratio of both slopes we get

$$\text{(iii)} \quad \frac{\omega_f}{\omega_s} = \text{const.}$$

In fact the $\Delta\omega$ - V_r relation does not differ significantly from a straight line, but its slope is an order of magnitude larger than that given by formula (ii) (Fig.1). Even more interesting is that the $\omega_f \cdot \omega_s^{-1}$ is not constant but varies with V_r (Fig.2).

Thus it becomes quite clear that the changes of light curve characteristics with the distance cannot be interpreted as a Doppler effect. Apart from some rather eccentric cosmological hypotheses, two simple explanations are possible here. Either type I supernovae reveal a very fast evolutionary effect, or it is a population* effect connected accidentally with distance. To prove the second possibility, Klimek made an additional plot $\Delta\omega$ - versus the distance R from the centre of the Virgo Cluster (Supergalaxy in terminology of de Vaucouleurs). The plot (Fig.3) looks rather strange. One may say that no relation exists between $\Delta\omega$ and R, the plot $\Delta\omega$ versus V_r being more informative. This means that we are observing an evolutionary effect. However if the two last points on the plot can be considered to be reliable, one may also say that, within the limits of our Supergalaxy (R = 25 Mpc on the plot), $\Delta\omega$ stays constant at the level of about 2,4 with a certain dispersion and falls down to about 1,5 outside of the Supergalaxy. This could provide evidence in favour of a population effect. Of course other interpretations are also possible here.

The investigations of Rust who used only the first portion of light curves cover a distance of about 300 Mpc. Klimek, who based on two portions of light curves could include into his considerations supernovae lying not further than 40 Mpc. An evolutionary effect visible along such a small distance would be an enormously fast one.

Rust made a detailed survey of all the possible systematic errors and did not find any which could be responsible for his results. Klimek's investigations are not finished yet, but main sources of errors were investigated and certainly they are not responsible for the general trend of plots, Klimek's results seem rather to back the population effect but the existing observational material is too scanty for any decisive conclusion.

The original results of Rust show that supernovae of type I are rather a mixture of two different subtypes. Allow me to show his known result (Fig. 4). We have here two branches of a light-curve-parameter-Δt_c versus luminosity-in-maximum relation. The gap between two branches is wider than the sum of the dispersions of points around them. If Arp (1974) is right (and I believe he is) that 1971P is a very low luminosity type I supernova, we can eventually expect a third separate branch of this relation. Unfortunately, no characteristic features of light curves or of spectra are known which could be used as indicators of membership of supernova to the higher or lower branch. It is a very important task to search for such indicators.

One may expect that two subtypes of type I supernovae may correspond to the two star populations in a similar way as two branches of period-luminosity curve of cepheids are formed by the young and old population stars, respectively. As early as ten years ago, Bertola and Sussi (1965) suggested that type I supernovae in elliptical galaxies are associated with the population II stars whereas in spiral galaxies - with both populations. Rust's branches have however little to do with star populations. Both luminosity groups occur in elliptical galaxies with roughly the same frequencies. Neither have these groups any relation to the Barbon, Ciatti and Rosino (1973) distinction between fast and slow supernovae. I would not like to suggest that the division of supernovae into two groups obtained by Rust is a final one. Such matters must be checked again and again, but according

---

* The word "population" is used here to denote various kinds of galaxies, not the two star populations.

to my opinion, it is the most reliable division existing at present. It supports the view that type I of supernovae is related to one population stars only. It is the task of further investigations to reveal whether it is an old or a relatively young population. In the second case this would prove that certain population I objects exist already in elliptical galaxies, at least in those of them which produce supernovae.

In conclusion, I have to emphasize that at the present stage of investigation, no unique, consistent, picture can be obtained on the role of supernovae in the evolution of galaxies. As we saw on the basis of investigations of Rust and Klimek many important statistical results can be obtained even using observational material published up to date. Even more facts are still hidden in archives of supernova search centres. Precise measurements of bright supernovae are published promptly, but other, less precise, observations of great value for many statistical investigations are elaborated and published only gradually.

Also, a new observational approach is necessary. The present general Palomar supernova search, which is the most powerful survey supplying us with the most valuable data on these objects, is directed toward detailed research on rather bright objects. The exposure times of search plates are more limited since 1968 but the quality of elaboration improved. That was very expedient for detailed investigation of individual objects and individual phenomena. However some population and evolutionary problems cannot be solved without many data on distant supernovae. To such problems belong those outlined in the work of Rust and Klimek, also problems of supernova membership to various types of clusters (supergalaxies) etc. I am convinced that the Palomar search should be supported by another systematic program devoted to faint and distant supernovae. It would be of great importance if one of the large Ritchey-Chrétien telescopes could be partially devoted to a supernova search down to 20-th magnitude in maximum, for establishing their light curves and colour indices. This could bring us closer to the solution of many interesting problems outlined here.

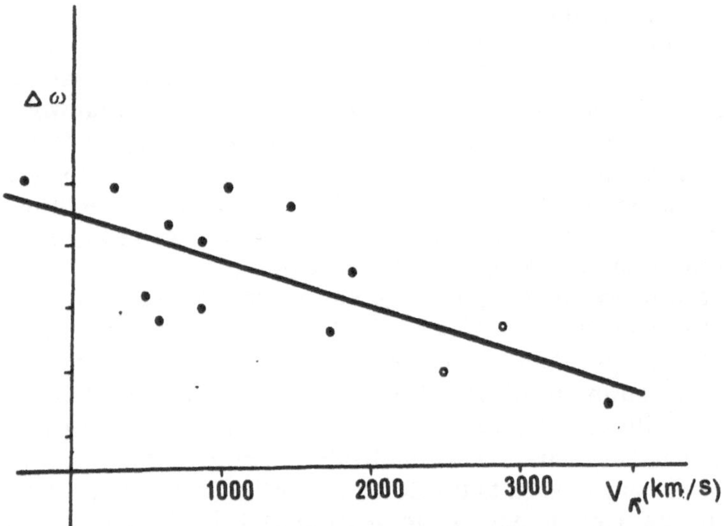

Fig. 1. Relation between $\Delta\omega$ and radial velocity./After Klimek/

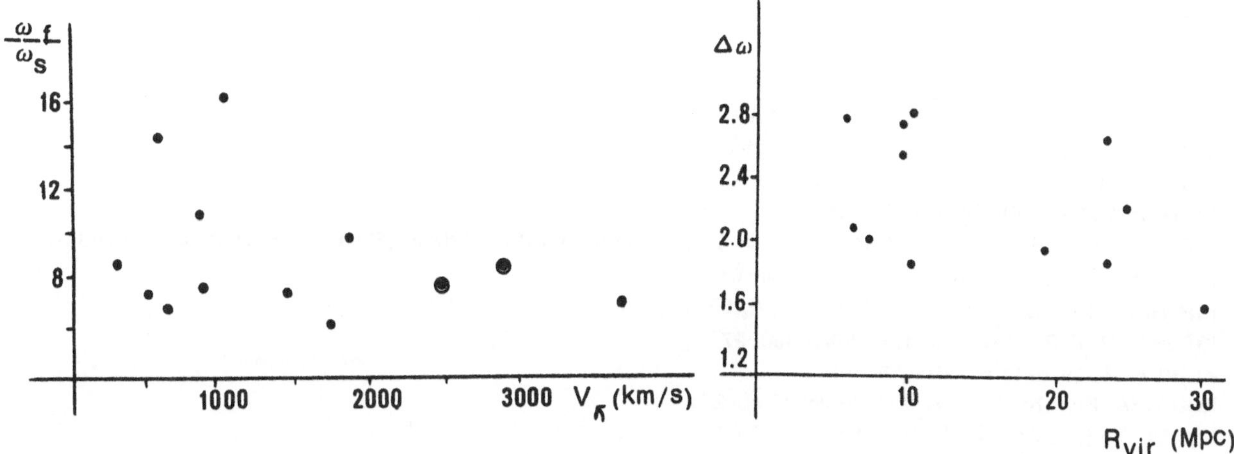

Fig. 2. Relation $\omega_f \cdot \omega_s^{-1}$ versus radial velocity. Circled points correspond to parent galaxies for which an accurate estimate of V_r is lacking./After Klimek/

Fig. 3. Parameter $\Delta\omega$ versus distance from the centre of the Virgo Cluster./After Klimek/

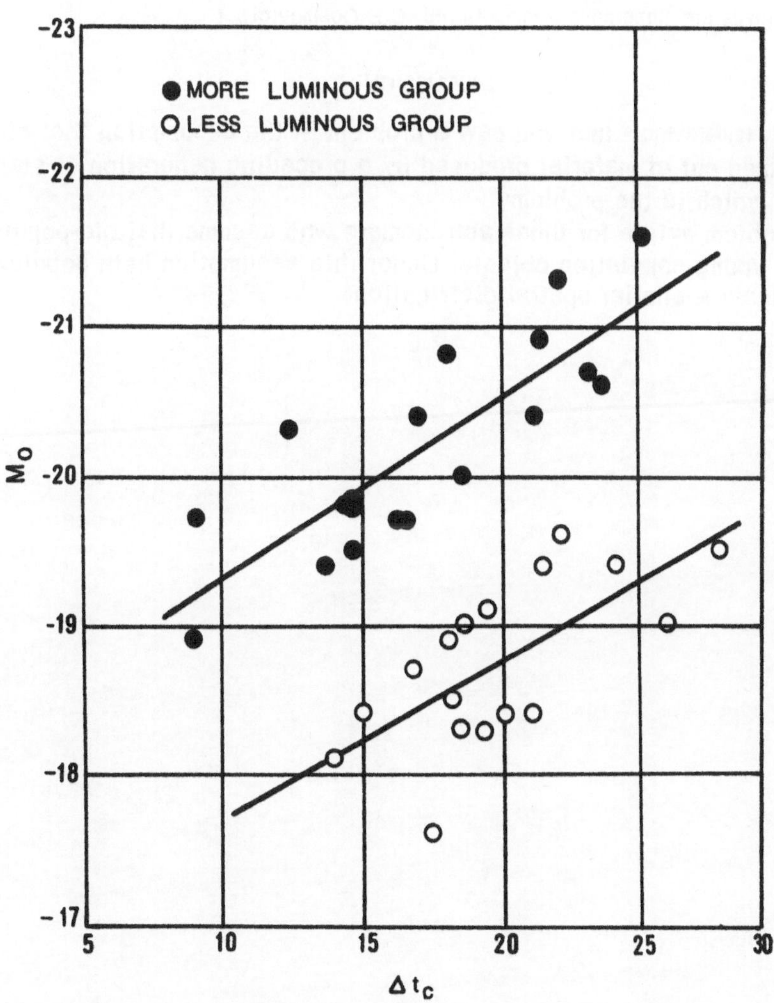

Fig. 4. Relation between the absolute magnitude in maximum and parameter Δt_c for two sub-types of type I supernovae /After Rust/

REFERENCES

ARP H. 1974, Supernovae and Supernova Remnants (ed. C.B.Cosmovici), 89

BARBON R., CIATTI F., ROSINO L. 1973, A.a A. 25, 241

BERTOLA F., SUSSI M.G. 1965, Contr. Asiago 176

FLIN P. 1974a, Acta Cosmologica 2, 21

FLIN P. 1974b, Acta Cosmologica 2, 33

IYE M., KODAIRA K. 1975, "Distribution of Supernovae in Spiral Galaxies" (Seen by the author as a preprint)

KARPOWICZ M., RUDNICKI K. 1968, Publ. Warsaw 15

KIRSHER R.P., OKE J.B., PENSTON M.V., SEARLE L. 1973, Ap. J. 185, 303

KIRSHNER R.P., KWAN J. 1974, Ap.J. 193, 27

KLIMEK Z. 1975, private information

KODAIRA K. 1974, Publ. Astr. Soc. Japan 26, 255

KWAST T. 1970, Astrofizika 6, 404

MUSTEL E.R. 1971a Astronomicheskiy Zhurnal 48, 3

MUSTEL E R. 1971b Astronomicheskiy Zhurnal 48, 665

MUSTEL E.R. 1972, Astronomicheskiy Zhurnal 49, 15

PSKOVSKIJ Yu.P. 1968, Astronomicheskiy Zhurnal 45, 945

RUST B.W. 1974, "The Use of Supernova Light Curves for Testing the Expansion Hypothesis and Other Cosmological Relations" - Oak Ridge National Laboratory Publ.

SARGENT W.L.W., SEARLE L., KOWAL C.T., 1974, Supernovae and Supernova Remnants (ed. C.B.Cosmovici), 33

ZWICKY F. 1965, Atti del Convegno sulla Cosmologia Firenze, 3

ZWICKY F. 1974, Supernovae and Supernova Remnants (ed. C.B.Cosmovici), 1

DISCUSSION

BLAAUW : Did I correctly understand that you saw a problem in the assumption that stars with higher metal content are formed out of material produced by a preceding generation of stars with lower metal content? If so, which is the problem?

RUDNICKI : This problem exists for those astronomers who assume that old-population objects are the "parents" of young-population objects. Under this assumption both populations ought to have the same or at least a similar spatial distribution.

SOME PROBLEMS OF PHYSICS OF SUPERNOVAE

E.R.MUSTEL
ASTRONOMICAL COUNCIL, ACADEMY OF SCIENCES, MOSCOW, U.S.S.R.

О НЕКОТОРЫХ ПРОБЛЕМАХ ФИЗИКИ СВЕРХНОВЫХ ЗВЕЗД

РЕЗЮМЕ

Статья содержит обзор современного состояния вопроса о тех процессах, которые характеризуют сверхновые звезды около максимума блеска (в основном рассматриваются сверхновые 1 типа). В это время сверхновая 1 типа (CHI) аналогична новой звезде, обладая отделившейся оболочкой и "центральным остатком" (Ц.О.), создающим непрерывный спектр звезды и определяющим ее светимость. Отделившаяся оболочка CHI характеризуется большой средней скоростью расширения, в пределах от 6500 до 20000 км/сек у различных CHI , и большой дисперсией скоростей вдоль радиуса, проведенного из центра звезды см.рис.1 . Это создает очень широкие абсорбции и бленды абсорбций в спектре звезды, см верх рис. 2а. У небольшой группы CHI дисперсия скоростей не столь велика и абсорбции в спектре этих звезд значительно уже, см.рис. 2б. Кроме того, отделившаяся оболочка характеризуется сильной неоднородностью, см. вновь рис. 1.

Наблюдения показывают, что излучение от Ц.О. характеризуется, по крайней мере в области $\lambda > 5000$ Å, планковским законом и это дает возможность оценить изменение радиуса фотосферы Ц.О. со временем, см рис 3. Анализ приводит к тому выводу, что газы, составляющие внешние слои Ц.О., его протяженной фотосферы, находятся в состоянии непрерывного расширения и после t_{max}, даже тогда, когда вычисляемый радиус Rp начинает уменьшаться,что наступает через 30-40 дней после t_{max}. Скорость указанного расширения газов, образующих фотосферу Ц.О. убывает со временем.

Анализ спектров CHI и физических процессов в отделившейся оболочке, позволяет сделать вывод о том, что Ц.О. является источником аномально высокого излучения в коротковолновой области спектра. Учет этой особенности CHI а также других особенностей, позволяет сделать определенные заключения о химическом составе различных оболочек CHI (отделившейся оболочки и протяженной фотосферы)

а) Отделившаяся оболочка. Наиболее характерным ее свойством, особенно в ее более внутренних слоях, является наличие высокого относительного содержания атомов N и He. Содержание атомов C, O, H здесь аномально мало. Все это соответствует тому, что должно ожидаться в результате действия термоядерного CNO - цикла при температурах от 10 до 100 миллионов градусов.

б) Центральный остаток. Естественно считать, что газы, вытекающие из фотосферы Ц.О. испытали перед взрывом CHI термоядерные преобразования при еще более высоких температурах, ведущих к сгоранию гелия и к созданию атомов O и C. В пользу этого вывода свидетельствует, например, наличие узких эмиссионных линий [OI] в спектре CHI в IC4182 (1937 г.). Все эти выводы подтверждают наличие термоядерных реакций не только внутри CHI , но и вообще внутри обычных звезд.

From the very beginning it should be reminded that supernovae do not form a group of homogeneous objects. Observations permit to distinguish two types of supernovae which are most frequently observed: Type I (SNI) and Type II (SNII). They differ considerably one from the other. It should be mentioned that SNI constitute a considerably more homogeneous group than SNII. It is quite possible that the relatively rare supernovae of Types III, IV and V, defined by Zwicky (1965) are simply subtypes of SNII.

In this paper SNI for which recently abundant information was obtained is going to be discussed in more detail.

1. General information on the spectra of Type I supernovae. The interpretation of SNI spectra has a very large and complicated story. The spectra of these stars have wide, sometimes very wide, emission maxima separated by wide minima. Minkowski (1939) put forward a hypothesis according to which the emission maxima are the

most characteristic feature of the considered spectra. However, many attempts to identify these maxima with the emission lines of known elements failed. In this connection McLaughlin (1963) suggested that the main property of the SNI spectra are minima, which he identified with the absorption lines of different elements. However, there were objections to his identifications and only in 1968 Pskowskij(1968) was able to identify several distinct minima in the SNI spectra with the strongest, heavily shifted absorption lines of metals and helium in the spectra of ordinary stars - supergiants of classes B and A. The next analysis of the SNI spectra started by Mustel (1971a, 1971b, 1972b) permitted to confirm the absorption hypothesis of the spectra of these stars and to make a number of preliminary conclusions on the physical properties and chemical composition of the SNI envelopes.

An important event in the further studies of SNI were the spectrophotometric observations of the relatively bright supernova SNI 1972e in NGC 5253 made by Kirshner, Willner et al.(1973), Kirshner (1974). The importance of these observations was that for the first time the absolute magnitudes of energy were obtained for SNI 1972e in a wide range of its spectrum from λ=3200Å to λ= 11000Å for a very long period of time, for about 400 days after light maximum (t_{max}). Similar measurements (though for a considerably shorter period) were also made for other SNI and SNII. A detailed identification of absorptions in the spectrum of SNI 1972e was given by Mustel (1973) and reconsidered by Mustel (1975b) with the taking into account the observations given by Kirshner (1974).

Naturally, the construction of a model of an envelope ejected from SNI must be based first of all on the most typical properties of the SNI spectra. These properties of the SNI spectra are the following: a) A very large width of absorptions. This property of the SNI spectra was the main reason why in the interpretation of the spectra of such objects there was such a long delay; b) The displacements of the centres of absorptions in the SNI spectra correspond to very high velocities of expansion of the envelope, on the average of about 10000 km/s, though different SNI have a considerable dispersion around this mean value from 6500 km/s to 15000 km/s, see Mustel (1972a). Apparently, even velocities of the order of 20000 km/s are possible, that follows from a paper of Bergh (1973), c) the SNI spectra correspond (if we reckon the time from t_{max}) to the spectra of supergiants of the spectral classes B, A, F. Such a sequence of spectral classes corresponds after t_{max} to the decrease of the spectrophotometric temperature T of a supernova, which is deduced from the energy distribution in the continuous spectrum of the star, see Mustel (1972a), Mustel and Chugay (1975a). The presence of only of a small number of absorptions in the spectrum of a typical SNI *near* t_{max} itself (classes B) is explained by the fact that in general there are very few sufficiently strong absorption lines in the spectra of ordinary B-stars and the considerable widening of *all* the spectral lines in SNI spectra makes the weak lines not observable *.

On the basis of available observations and other general considerations we have reasons to state that a temperature drop for a typical SNI occurs long before light maximum (see section 3). At t_{max} the temperature of a typical SNI is about 15000-25000°, then during several dozens of days it decreases rapidly down to the values of the order of 6000°-7000°. Apparently, after the temperature drop it rises again, see Mustel (1975b).

It should be noted that in the case of the ordinary novae the temperature of the star also decreases before light maximum, but then almost immediately after t_{max} it begins to rise again. There are all reasons to think that the continuous apectrum of SNI is created by its "central remnant" shown by letter A in Fig.1. The discussion of the problem of temperature of SNI, given in paper of Mustel and Chugay (1975b) shows that only during the first moments after t_{max} the energy distribution in the observed continuous spectrum of the star (between the absorptions) is approximately a planckian one. Twenty-thirty days after t_{max} when the temperature of a supernova is considerably decreased, there appear many lines of metals (produced by the ejected envelope) in the short wavelength part of the spectrum of the star with $\lambda \lesssim$ 5000Å. Due to their large width and mutual superposition the mentioned short wavelength part of the SNI spectrum considerably weakens. Therefore the colour characteristics of SNI at this time (for instance, the magnitudes U, B, V) cannot be used for a quantitative study of the processes in the envelopes of SNI .

---

* There are no hydrogen absorptions near t_{max} in these spectra; see the details in section 4 of this paper.

For the most reliable determination of a SNI temperature, *spectrophotometric* studies of these stars in the *far infrared* region of the spectrum up to several microns are of great importance. Such measurements made for SNI 1972e up to 2.2μ are given in the paper of Kirshner, Willner et al. (1973).

d) The last important characteristic of the SNI spectra is that the displacement of the centres of the wide absorption lines in these spectra practically do not change with time. Some small changes of the mentioned (Doppler) shifts, observed after several hundreds of days after light maximum, are produced, apparently, by the effects of the stratification of elements and of physical conditions in the SNI envelopes, see Mustel (1975b). It is interesting to note that a considerable decrease of the displacement of absorptions after light maximum is observed in SNII spectra, see Chugay (1975).

e) Now let us enumerate the elements, the absorptions of which are detected in the SNI spectra. Mainly all of them coincide with the strongest absorption lines of the spectra of ordinary stars or with their blends. First we shall point out the absorptions of metals. The most distinct and strongest absorptions of the SNI spectra are the blend of H, K, CaII lines and the absorptions of the infrared lines of CaII: 8498, 8542, 8662Å. The absorptions belonging to several multiplets of SiII, especially to the blend of 6371, 6347Å are very distinct. The blend of Mg II lines 7877, 7896Å is very strong. The SNI spectrum has also many absorptions or their blends belonging to FeII. All the lines of FeII originate from metastable levels. Especially wide are the blends of FeII τ and μ, see Fig.2. The absorptions of FeII 5018 and 4924 are very intensive. There are also lines of other metals: CrII, TiII etc., see the lower part of Fig.2. The numerous lines of N, especially in the infrared region play an important role in the SNI spectra. The lines of HeI 5876 and 7065Å are intensive, especially at the end of the evolution of the spectrum. As to the absorptions of O and C they have not been found yet. The presence of hydrogen absorptions in the SNI spectra is also very disputable. In any case, one of the main observational characteristics of the SNI spectra is the absence of clear absorptions of this element . See the details concerning this problem in section 4.

f) So far we have spoken about *absorptions* in the SNI spectra. As to the presence of *emission bands* in their spectra this question is still under discussion. The analysis of the energy distribution in the SNI spectra shows that the widest and brightest maxima of the SNI spectra are those which correspond to the wide regions of the spectra of ordinary novae where practically there are no strong absorption lines. For instance, the maxima δ and ζ (see Fig.2), the maximum ν (Fig.6) and similar others belong to such maxima. At present there are almost no convincing arguments which might indicate the presence of sufficiently bright usual emission bands in the SNI spectra. It is possible that the only exceptions are the above mentioned three infrared lines of CaII, see Kirshner, Oke et al. (1973).

2. The model of an ejected envelope. The first main observational fact requiring explanation is the presence of very wide absorptions in the SNI spectra, regardless to which element the absorption belongs. In paper of Mustel (1971a) the following possible explanations as to the origin of such wide absorptions in the SNI spectra are given: 1) a great number of absorbing atoms in the envelope ejected from SNI; 2) high turbulent velocities of gases in the envelope; 3) very large velocity dispersion of gases in the envelope. A detailed analysis of this problem made in paper of Mustel and Chugay (1975a) led to the conclusion that the very large widths of absorptions in the SNI spectra are due to the presence of a *noticeable dispersion* in the magnitude of gas velocities in the envelope: the gases composing the external layers of the expanding envelope move with much higher velocities than those composing its internal layers. There are reasons to think that a state of free motion is kept in the envelope and therefore for a given moment of time the distance of a gas condensation from the exploded star is proportional to the outward velocity of condensation. In the same paper the relatively high intensity of absorptions in the SNI spectra is discussed, this intensity is of the order of 0.3. A conclusion is drawn from this fact (see ibid) that the envelopes ejected during the SNI explosions are very inhomogeneous and consist of different relatively dense condensations, see Fig.1. Besides the interpretation of the absorption contours in the SNI spectra leads to the conclusion that their central parts are formed mainly by the inner part of the envelope which has a relatively high mean density. The mean velocity of the envelope $\bar{V}$ (see Fig.1) deduced from the displacements of the centres of absorptions, is determined by the internal parts of the envelope.

The existence of a high velocity dispersion in the SNI envelopes not only considerably widens isolated absorption lines of their spectra to the magnitudes of several hundreds of angströms. This velocity dispersion often creates very wide absorption blends in those parts of the SNI spectra in which there are some relatively close and sufficiently strong absorption lines. Such is the case with the blend τ of FeII lines, the wavelengths from 5169 to 5363 Å, see Fig.2a. This figure shows the identification of absorptions in the spectrum of SNI 1960f in NGC 4496, see paper of Mustel (1972a). In the lower part of Fig.2a the spectrum of DQ Her (1934)* is given for the moments immediately preceding light maximum (t_{max} = 22 XII). In the upper part of Fig.2a the curve of energy distribution in the spectrum of the same SNI on May 4, 1960 is given, it is taken from paper of Bloch et al.(1964). The coincidence of the both spectra is obtained for the expansion velocity of the envelope of SNI 1960f equal to about 6800 km/s. In Fig.2a we see that a "cluster" of neighbouring lines from λ= 51690 to 5363Å produces a very wide structureless blend τ in the SNI spectrum. At the same time a relatively small number of SNI have not such a large velocity dispersion in the envelope. In such cases the SNI spectrum has *narrower* absorptions. SNI 1966j in NGC 3198 is an example of such a supernova, see Fig.2b. The upper spectrum (the energy distribution) was obtained on January 9, 1967 by French investigators Chalonge, Burnichon (1968), the lower, right one by Italian investigators Chincarini, Perinotto (1968) for the same day**. The identification of absorptions in the spectra of this SNI, made by Mustel (1972b, 1974) is given in Fig.2b. Here in blend τ we can distinguish clearly some sufficiently well identified absorptions. The Doppler factor $\varkappa$:

$$\varkappa = \frac{\Delta\lambda}{\lambda} = \frac{\bar{V}}{c} \tag{1}$$

is equal for the given star to -0.0255 ($\bar{V} \simeq$ 7500 km/s).

The quantitative physical analysis of envelopes ejected by SNI has been begun only recently. There are difficulties here due to the unusual character of the spectra of these objects. Some preliminary data for the envelope ejected from SN 1972e are given in paper of Mustel, Chugay(1975a). It was accepted (with the corresponding reasoning) that the mass of this envelope is about 0.2M. From this it was found that at the moment t_{max} the mean electron concentration n_e in the envelope of SNI 1972e was about $10^9 cm^{-3}$. Apparently, inside the condensations forming the envelope the value of n_e is larger. In paper of Kirshner and Oke (1975) it was found for the same supernova that at the moment of light maximum $n_e \simeq 10^{10} cm^{-3}$. Thus both estimates indicate rather small values of n_e on SNI envelopes at the time t_{max}. Nevertheless the calculations show that at this moment the main source of opacity is *Thomson electron scattering.* Calculations made in the same paper show that the optical depth of the envelope of SNI 1972e, determined by Thomson electron scattering, was about 0.5 for SNI 1972e at the time t_{max}.

3. The "central remnant" in Type I supernovae and its extended photosphere. The discussion of the problem on a general model of SNI near light maximum, see Mustel (1971a, 1971b) and especially Mustel, Chugay (1975 a), leads to the conclusion that at this time the supernova is similar to novae near light maximum. At this time we observe a certain central body, apparently, a star***which is called a "central remnant" (C.R.) which is surrounded by an ejected envelope, expanding with mean velocity $\bar{V}$, see Fig.1. This C.R. is the main source of energy emitted by a supernova and correspondingly it is the *main* source of its continuous spectrum. According to the generally accepted terminology we shall call those outer layers of C.R., which determine the luminosity of SNI - "the photosphere" of C.R..

As it was already mentioned (section 2 of the paper),at the time of light maximum and some time later, the energy distribution in the observed SNI spectra is approximately a planckian one. Therefore it is possible to estimate, at least approximately, the effective radius of C.R. at the moment t_{max}. Such calculations were performed for SNI in paper of Mustel (1971b). Taking for this moment the value $M_p = -19^m$, it was found that at T=15000° the radius of the "photosphere"

---

* Since the lines of absorption of Balmer hydrogen series in the SNI spectra are not identified yet, the absorption line of H in the spectrum DQ Her is removed.

** In this case microphotograms are reproduced.

***It is also possible that at this time the central body is a pulsar surrounded by a sufficiently dense and hot extended envelope.

is about $15000R_{\odot}$, and that this value is considerably smaller than the dimensions of the *ejected* envelope of SNI.

A more detailed analysis of the behaviour of C.R. of SNI is given in paper of Mustel and Chugay (1975b). The main attention is paid to SNI 1972e for which the most complete spectroscopic information is available. The above mentioned determination of temperature for this supernova is used (Kirshner, Willner et al., 1973) in which the energy distribution in the continuous spectrum up to 2.2μ is studied. For the earlier moments of time, than those used in this paper of Kirshner, the data on the magnitudes B-V were used. In fact, for such moments of time the temperature of the supernova was still sufficiently high and correspondingly the influence of absorptions for $\lambda \lesssim 5000$ Å was rather small. The accepted temperature variations with time for SNI 1972e are shown in the upper part of Fig.3, the right curve "a". The open squares in Fig.3 are based on spectrophotometric measurements, while the black ones correspond to the colour temperature T_c, obtained from the magnitudes B-V. We can see that on the right side of the curve "a" these temperatures T_c are much lower than the spectrophotometric temperatures T and since the temperatures T are here relatively low, the use of T_c for the determination of R_p introduces very serious errors. Note, that according to this graph the temperature of SNI 1972e is expected to be about 20000°-25000° near t_{max}.

The left curve "b" in Fig.3 is plotted for SNI 1970j in NGC 7619. Since the period close to the time t_{max} is studied here, then the temperatures T_c are used, based on the magnitudes B-V. We see that the temperature of this SNI even 12 days before t_{max} can be very high of the order of 50000°. In this connection there are some reasons to think that the bolometric maximum of luminosity of SNI is observed much earlier than the visual one.

With the use of the planckian law the radius of the "photosphere" R_p for SNI 1972e was calculated for different moments, see the lower part of Fig.3. The most characteristic features in the behaviour of R_p are the following: 1) The presence of a maximum for R_p at some moment $t_1 \simeq$ 30-35 days after t_{max}, when the temperature of the star T reaches its lowest value; 2) Approximately the linear growth of R_p before the moment t_1, the same regularity is found* for SNI 1970j as well. Extrapolation of this linear relation to the moment, when R_p=O permits to fix the moment of SNI "explosion" for the both indicated supernovae. This moment for SNI 1972e corresponds to 25^d before t_{max} and for SNI 1970j to 16^d before t_{max}. 3) The velocity of the photospheric expansion $\bar{V}_p$ before the moment t_1 is close to $\bar{V}_p \approx 5000$ km/s. The mean velocity of expansion of the ejected envelope V for SNI 1972e, determined from all the strong absorptions of its spectrum is $\bar{V} \simeq 10500$ km/s, and for about 400 days after light maximum it remained practically constant. Assuming that it was constant before light maximum and considering that the photosphere and the ejected gas masses (envelope) began to expand *simultaneously*, it is possible to find dimensions of the *envelope*, accepting that $R_s = \bar{V}t$ for different moments, see again the same Fig.3. Note that the value of $\bar{V}$ is about twice as high as $\bar{V}_p$.

A very interesting fact is that the photosphere of SNI continues to expand after t_{max}, a fact not observed for the typical novae. Apparently, the *extended* photosphere surrounding C.R. of SNI is in a state of *continuous expansion* after the moment t_1 too, that is even when the radius R_p calculated from observations begins to decrease. This problem is discussed in more detail in the same paper.

Thus, there are some reasons to consider that for a long time after light maximum C.R. is the source of gases continuously ejected from its external layers**. Therefore for a long time there is no emptiness between the photosphere of C.R. and the ejected envelope. In addition it should be mentioned that the extreme parts of the long wavelength wing of the wide absorptions of the spectra of SNI correspond to the very low velocities of outward gas motion from the exploded star. Therefore a very important problem in the physics of SNI is the study of the interaction between gases of both components – the expanding envelope and the gases outflowing from C.R..

The presence of an extended "photospheric" envelope of C.R. and its possible interaction with the internal layers of the ejected envelope immediately suggests that the extended envelope of C.R. may be the source of the enhanced emission in the short wavelength region. And, indeed,

---

* For later moments of time a jump in values of R_p is observed here which is due to absence of data on spectrophotometric temperature for this star.

** There are all reasons to expect that the average velocity of the expansion decreases with time.

the analysis of certain regularities in the SNI spectra leads to the conclusion that such an ultraviolet excess does exist in the spectrum of C.R., though it is difficult to detect it from the outside, mainly because of the absorbing property of the atoms of metals in the ejected envelope. The interaction between the both components of SNI may lead also to the formation of high energy particles in the external layers of the extended envelope around C.R.. The mentioned ultraviolet emission from C.R. and these high energy particles can affect the ionization and the excitation of atoms in the ejected envelope. These questions will be discussed in more detail in the next section of this paper.

4. On the chemical composition of the SNI envelopes. The importance of investigating the chemical composition of the supernova envelopes near light maximum follows from the fact that an analysis of the chemical composition of the supernova "remnants" may give incorrect information on the relative significance of the elements which are ejected during the explosion of SN. In fact all the known supernova remnants were ejected from the corresponding SN hundreds of years ago. During the process of their expansion they had enough time to carry along the interstellar gas which was on their way. Therefore the emission lines in the spectra of these remnants are produced both by the envelopes themselves and by the interstellar gas contained in them. But the separation of these two components of gas in many cases is extremely difficult. For example, the "remnant" SNI Tycho may be mentioned, the spectrum of which has an emission line H_{α}. However, according to van den Bergh (1971) the expanding envelope of this star carried along a mass of interstellar gases which was much larger than the mass of the envelope itself after its ejection.

It is true that the quantitative chemical analysis of the SNI envelopes near light maximum, and during several hundreds of days after it, is also connected with certain difficulties. However they can be overcome, though many further efforts are required for that. The first difficulty of the chemical analysis of envelopes during the first stages of their expansion is that the absorption lines of the SNI spectra are very wide and we must deal mostly with the strong blending of the adjacent absorptions. Further, due to the large widths of absorptions we cannot study the weak absorption lines, which are practically obliterated. The following difficulty is connected with the fact that the envelopes of SNI consist of gases moving with different expansion velocities, from almost zero velocities to those of the order of 20000-25000 km/s. For such envelopes with a gradient of outward velocities, a special theory of the contours of absorption lines must be worked out. It is expected that the theory of Sobolev (1947) of moving stellar envelopes may be very effective in this case.

In his preliminary report the author (Mustel,1975a) came to the conclusion that there exists a stratification of elements and of physical conditions as well inside the envelopes of SNI along the radius drawn from the centre of the star. Namely, there are reasons to consider that the relative abundance of nitrogen and helium, with respect to metals, is higher in the more inner parts of the envelopes than in the more external ones. Finally, the author's conclusion (see section 3 of the paper) that the radiation from C.R. is characterized by two temperatures T and T_{UV} also introduces an additional and a very serious difficulty in the estimation of the conditions of ionization and excitation of atoms in the envelope.

From the above-said it follows that at present we can make only more or less general conclusions on the chemical composition of the ejected envelopes. In some cases certain numerical estimations can be made, see, for instance, Mustel and Chugay (1975a).

The first and at the same time very complicated problem of the chemical analysis of the SNI envelopes is the identification of absorptions and absorption blends in their spectra. The identification must take into account some most important *characteristics* of the SNI envelopes and of their spectra, as well as the evolution of these spectra in time. In the paper of Mustel(1975b) the author enumerated the following principal characteristics: (α) the approximate constancy of the positions of the absorptions during their evolution;(β) The presence of two temperatures describing the radiation field of C.R.: a) the spectrophotometric temperature – T, determined from the continuous spectrum for $\lambda \simeq 5000$Å, b) an excess of radiation from C.R., in the energy range higher than, approximately, 10eV, with the temperature T_{UV} rising with time independently of T(it may be that the equivalent of this temperature T_{UV} is the relatively high energy electrons emitted by an extended "photosphere" of C.R., see section 3); (γ) the accumulation of atoms on metastable levels; (δ) the existence of stratification of elements and of physical conditions along the radius in the SNI envelopes.

As it has been just mentioned, the studies of the *evolution* of spectra of SN in time are very important in the problem of identification of absorptions. This evolution is determined by many factors: a) by the changes of T and T_{uv} with time, see section 3 of the paper; b) by a decrease of the mean density of gases in the envelope and by a decrease of the dilution factor W; c) by an increase of the role played by the metastability of atomic levels.

A preliminary discussion of the evolution of the SNI spectra in time and a more complete analysis of this problem are given in papers of Mustel (1972a, 1975b). Conclusions on the chemical composition of SNI, based on the first paper are made in paper of Mustel (1973). A more complete analysis, with taking into account all the above-mentioned characteristics (α) - (δ) is made in paper of Mustel (1975b). An example of the identification of absorptions in the SNI 1972e spectrum for the period J.D.2441475 - 2441684, taken from this last paper is given in Fig.4. In the upper and the low parts of Fig.4 the spectral lines of different elements are indicated, which have been used in the identification of the corresponding absorptions and of their blends. Fig.4 permits to examine one of the most interesting periods in the star evolution, namely, the relatively long(connected with interruption in observations)period from 2441529 to 2441653. This period was characterized by a rise of temperatures of SN, both of the temperature T and of the temperature T_{uv}. In fact beginning from the moment 2441653 distinct absorptions of N II, belonging to the relatively strong multiplets M3 and M5, appeared in the spectrum of SN. The multiplet M3, N II was also observed in the spectrum of SN 1972e earlier, during the very first moments after t_{max}, but then due to a drop of T, it disappeared, see Mustel (1973). Simultaneously with the appearance of absorptions NII starting again from the moment 2441653, the absorptions HeI 5876 and 7065Å became much stronger and this may be clearly seen in the next Fig.5, which is the continuation of Fig.4.

Both Fig.4 and Fig.5 show the presence of those elements in the envelope of SNI 1972e about which we spoke at the end of the first section (point e). It should be noted that the sufficiently strong absorptions of NI are concentrated mainly in the infrared region of the star spectra. Only the absorptions of those relatively strong multiplets NI which include the lines with $\lambda\lambda$4110, 6485Å are in the optical part of the spectrum. Multiplet NI with 6645A is weaker.

It is shown in the same paper of Mustel (1975b) that the most distinct and strong absorptions of HeI must be the absorptions $\lambda\lambda$ 5876, 7065Å, and this is confirmed by Figs.4 and 5. The other absorptions of HeI are blended with the wide absorptions of other elements. In addition, it is expected that due to the stratification phenomena in the envelopes of SNI,the atoms of He and N must be located on the average in *more internal* parts of the ejected envelope than the metal atoms. Therefore in some cases the absorptions of metals may screen the absorptions of N and He. In conclusion concerning the problem of He I absorption in the SNI spectra, it is necessary to point to the anomalous behaviour of the absorption due to He 10830Å. This anomaly as well as the fact that the absorptions of HeI 5876 and 7065Å appear very early (soon after t_{max}) in the SNI spectra, everything can be easily explained if we postulate the appearance of an anomalously high ultraviolet radiation from C.R. immediately after t_{max}, see section 3 and once more paper of Mustel(1975b).

Thus, the presence of rather strong HeI, NI, NII absorption lines in the SNI spectra speaks in favour of a relatively high content of both He and N atoms in the envelopes of these stars. Let us consider now the problem of the presence of atoms of O,C and H in the SNI envelopes.

From the very beginning it is necessary to say that the relative content of C, N, O atoms is approximately the same in the atmospheres of ordinary stars (there are even more atoms of O than of N). Now the potentials of ionization are approximately the same for these atoms. The excitation potentials for the absorptions of O, C, N about which we are going to speak are also approximately the same. Let us admit for the moment that the relative content of C, N, O atoms in the SNI envelopes is also approximately the same. In this case calculations show (Mustel, 1973) that many absorptions of CI, OI must be about as intense as those of NI. However, it contradicts to observations, see Fig.6. In this figure, besides the positions of absorptions NI, the strongest absorptions OI and CI are also shown. In addition to the mentioned SNI 1972e, the energy distribution in the spectrum of SNI 1871i in NGC 5055 is also given. Fig.6 shows that contrary to the absorptions of NI such strong spectral lines OI as 6158, 7771, 7947Å and CI 9095, 10690Å do not produce *any* appreciable absorptions. Therefore, there are reasons to consider that the abundance of O and C atoms in the SNI envelopes is much lower than that of N atoms. In this connection it is necessary to remind that the spectrum of SNI 1937 in IC 4182 (Minkowski, 1939) at some moments of the star evolution, had bright, *narrow* forbidden undisplaced lines of 01 6300, 6343Å. However, it is doubtful that these lines belonged to the ejected envelope, since in such a case they would be *very wide* ($\bar{V} \simeq 9000$ km/s). In addition, the spectrum of SNI 1972e which was observed for a much longer period of time than the spectrum of IC 4182 had no emission lines of OI. It is quite possible that OI atoms which created the emission lines in the spectrum of SNI IC 4182 were produced by a *relatively slow continuous outflow* of gases from C.R. of this star (see section 3), which means that we observed here the gases of SNI which did not take part in the formation of the ejected envelope.

The problem of the presence of hydrogen in ejected envelopes of SNI is very complicated and has not been solved yet. For instance, at the first moments of its evolution the spectrum of SNI 1972e did not contain any strong absorption lines of Balmer series, see the dashed line in Fig.6a. For the later moments of time a minimum was observed in the place of the shifted line, but it could be equally attributed to the relatively wide and intense blend of absorptions of NI 6483 and 6645 as well, see again Fig.6. Some authors prefer to consider for later moments maximum ν in this figure as an emission line H_α. However, for many reasons this maximum ν can be considered as a "remnant" of the continuous spectrum, limited by strong absorptions (HeI 7065Å and NI 6483Å) on both sides. Other attempts to identify the lines of Balmer series (H_β, H_γ. . .) in the SNI spectra cannot be also considered as convincing enough and therefore this problem requires further and thorough study.

Once more it must be reminded that one of the main observational features of the SNI spectra, according to which these stars are assigned to Type I supernovae is the absence of hydrogen lines in their spectra. Therefore, undoubtedly, the relative abundance of hydrogen in the SNI envelopes is much lower than that in the atmospheres of ordinary stars. Some arguments show that the ratio H:N in SNI is very low (see paper of Mustel, 1973). However, in the connection with the subsequent suggestions on the stratification of elements in the envelopes of SNI (Mustel, 1975) this conclusion on the low content of H is applied only to the internal layers of the SNI envelopes. Hence, in future studies we must look for the H lines which originate in the *external* layers of envelopes.

Independently of all these uncertainties with hydrogen, we should pay greater attention not to the outer layers of the ejected SNI envelopes, but to the more inner ones. In fact we have all reasons to consider that the atoms of these inner layers of the SN envelopes are influenced by the preceding thermonuclear reactions. Bearing in mind the uncertainties related with various published theories of explosion of supernovae, we must proceed from the *observations* and only the observations must be taken as the basis for the theories on supernova explosions!

In this connection let us summarize briefly the conclusions on the chemical composition of the *inner* layers of the SNI envelopes. These layers contain a large amount of helium and nitrogen, relatively a little oxygen, carbon and, apparently, very little hydrogen. These data must be compared with the thermonuclear reactions which have been worked out by physicists and astrophysicists in application to the evolution of stars. It should be indicated at once that such a peculiar chemical composition of the ejected envelopes of SNI (He, N) may be the result of the C,N,O cycle for temperatures from 10 to 100 million degrees, see, for instance, Truran (1973). Thus, there are reasons to believe that when the SNI explosion occurs, the C, N,O cycle with the mentioned temperatures has been already realized not very deeply under the photosphere of SNI before its explosion. Then it is natural to consider that the layers of the stars which are deeper than ones in question were characterized by a temperature higher than 100 million degrees. Calculations show that the next temperature stage is characterized by helium combustion and this leads to the formation of *carbon and oxygen*. If it is so then under favourable conditions we can observe emission lines from these elements, not in the principal (ejected) envelope, but in the slow outflow of gases from C.R. of SNI , which starts soon after light maximum and which creates an extended photospheric shell around the star. Such an outflow explains quite naturally (see section 3) the appearance of intense forbidden O lines in the spectrum of C4182. Thus, the preliminary semiquantitative results of chemical analysis of the SNI envelopes explain the peculiarities of the SNI spectra and moreover, in general, they confirm the presence of thermonuclear reactions inside stars. At last, these results are the starting points from which the model of the SNI explosion may be examined.

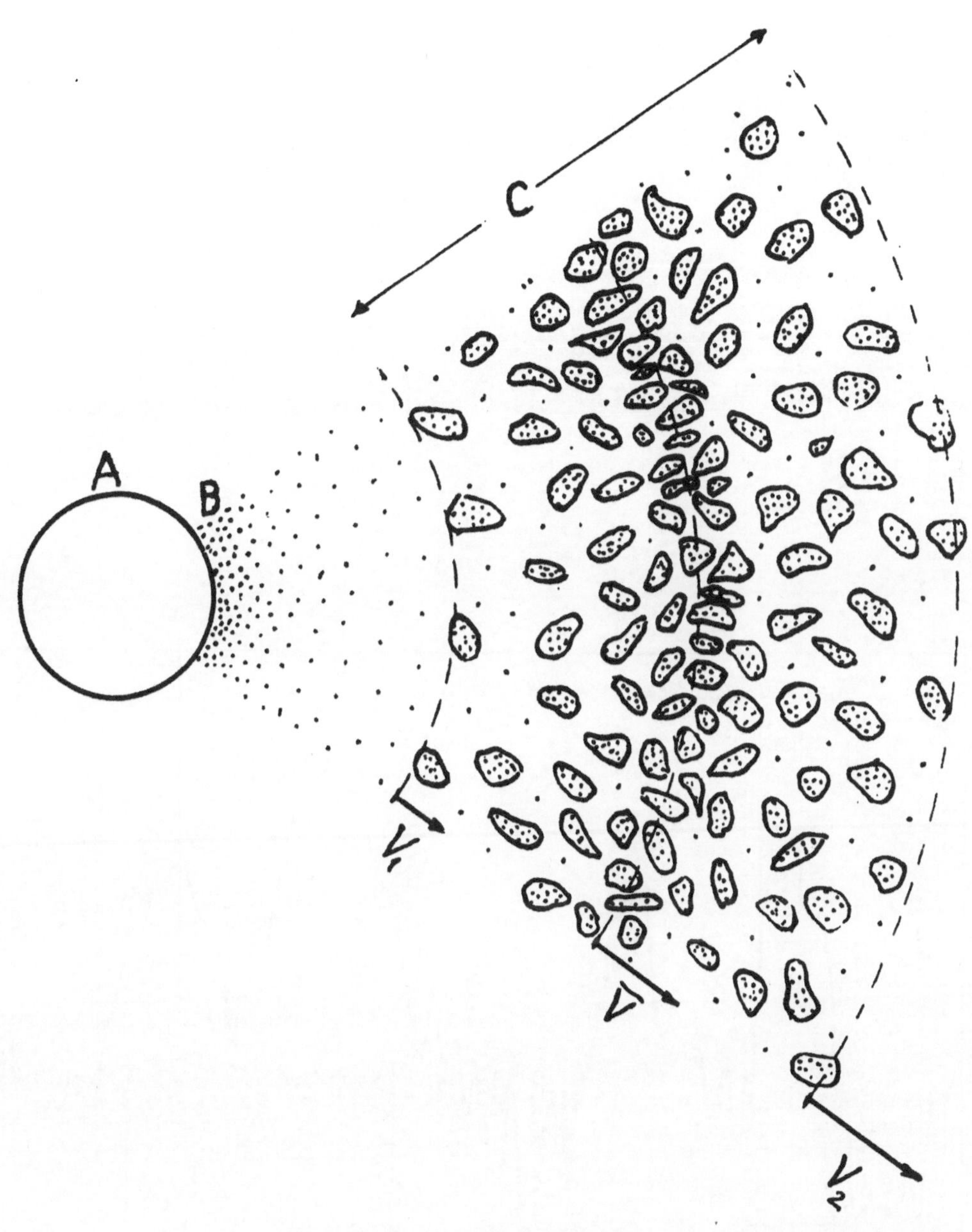

Fig. 1. A model of Type I supernova. A is the "central remnant" of the supernova; B is its extended "photosphere". C is the detached envelope. Velocity V_1 corresponds to the inner boundary of the envelope, V_2 to the outer boundary. Only a relatively narrow section of the model is shown.

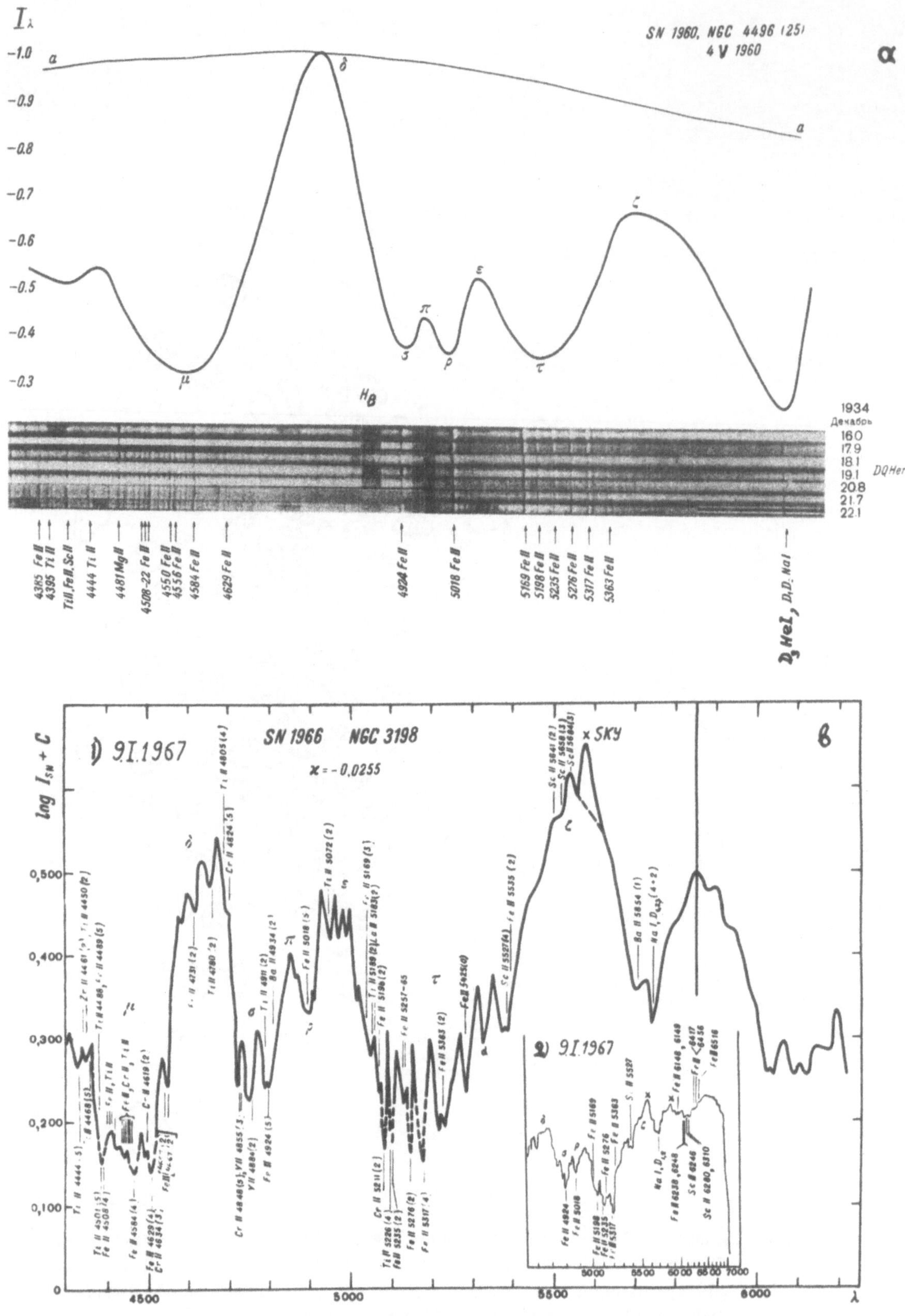

Fig. 2. Identifications of absorptions in the spectra of SNI 1960f in NGC 4496 (Fig.2a) and SNI 1966j in NGC 3198 (Fig. 2b). The second SNI is characterized by anomalously narrow absorptions in its spectrum.

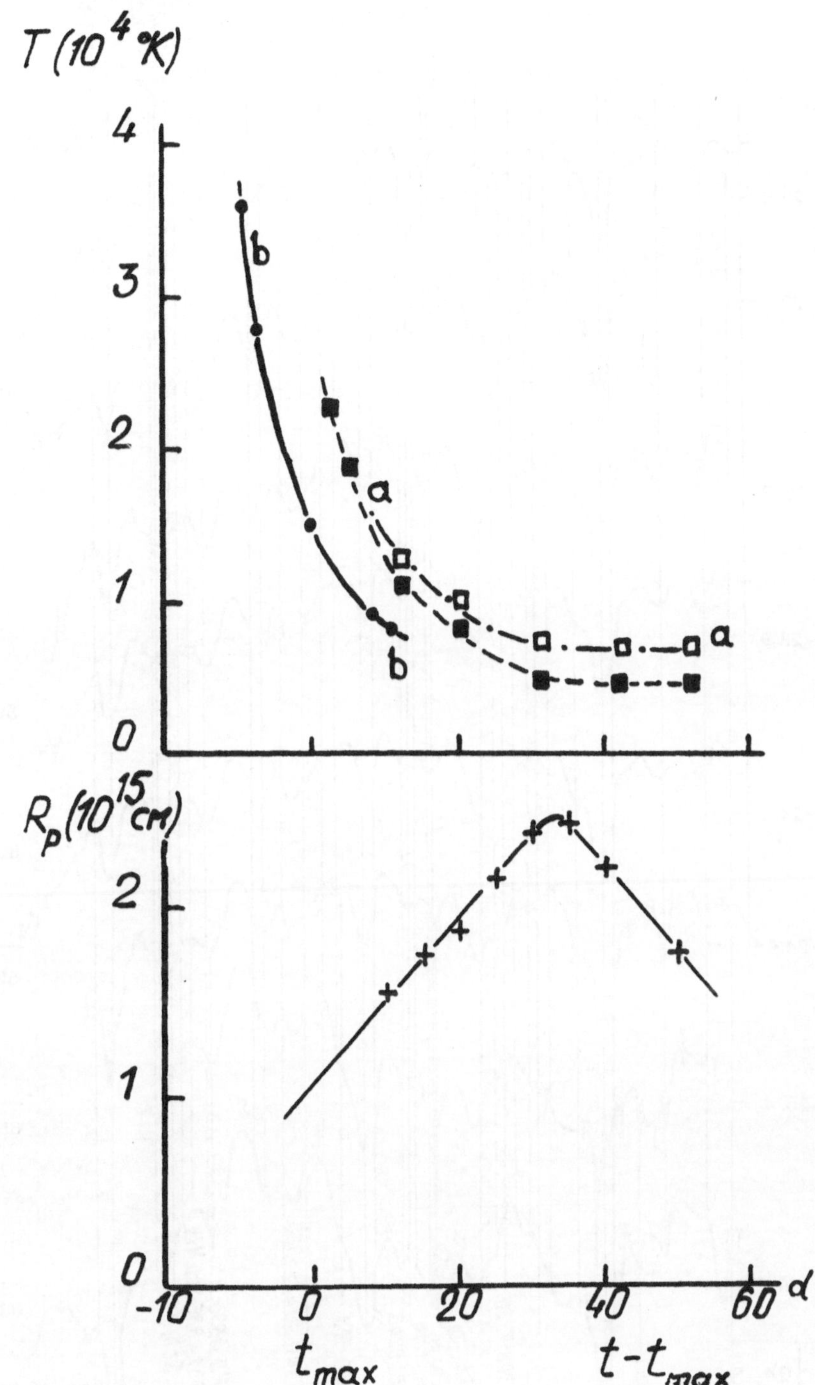

Fig. 3. The change with time (the upper part of the figure) of the spectrophotometric temperature T (open squares) and the colour temperature (black squares) of SNI 1972e. Dots correspond to the change of colour temperature for SNI 1970 in NGC 7619. The changes of the "photospheric" radius of the "central remnant" of SNI 1972e (the low part of the figure).

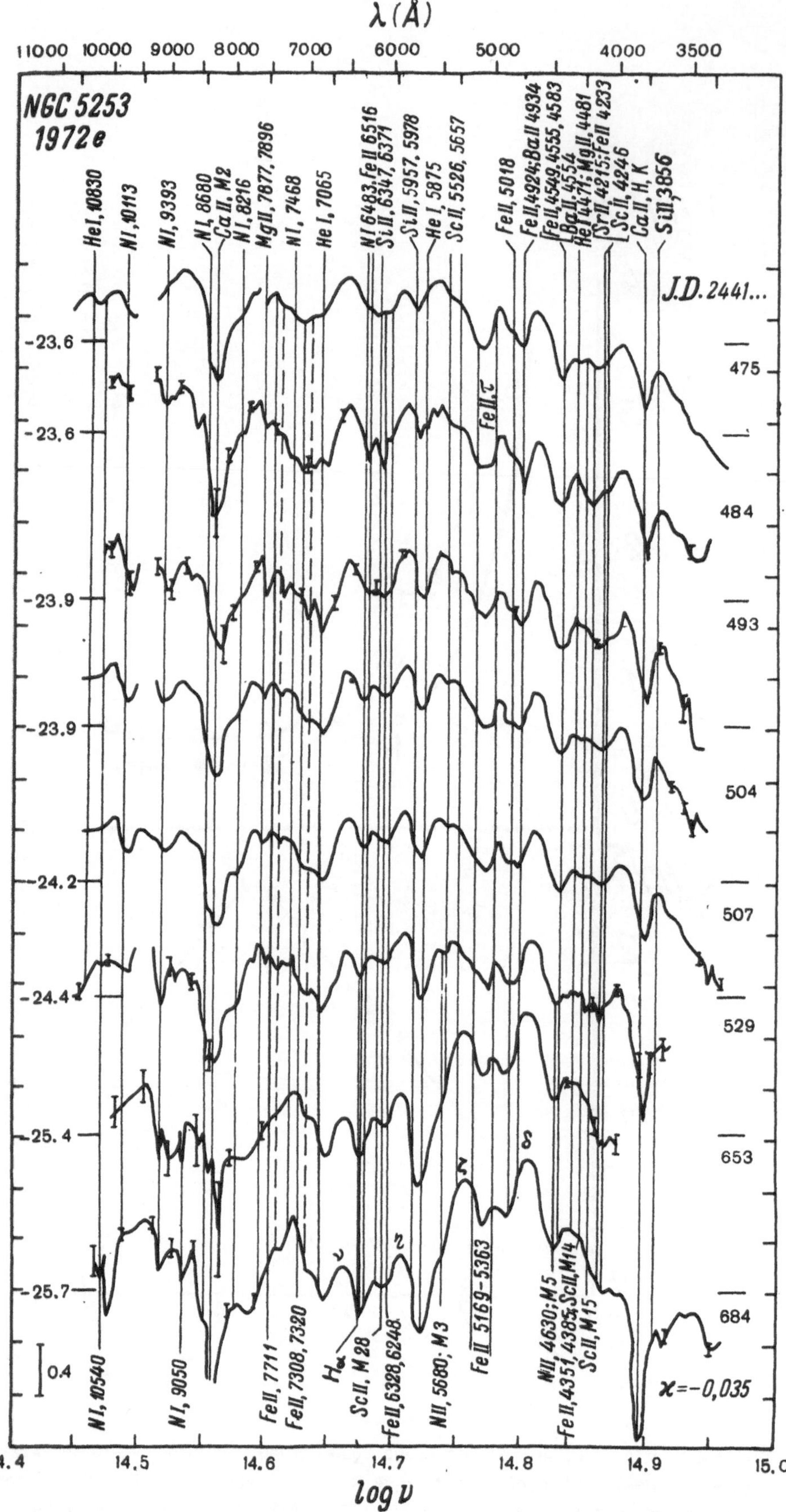

Fig. 4. Identifications of absorptions in the spectrum of SNI 1972e for the period from 2441475 up to 2441684 (J.D.).

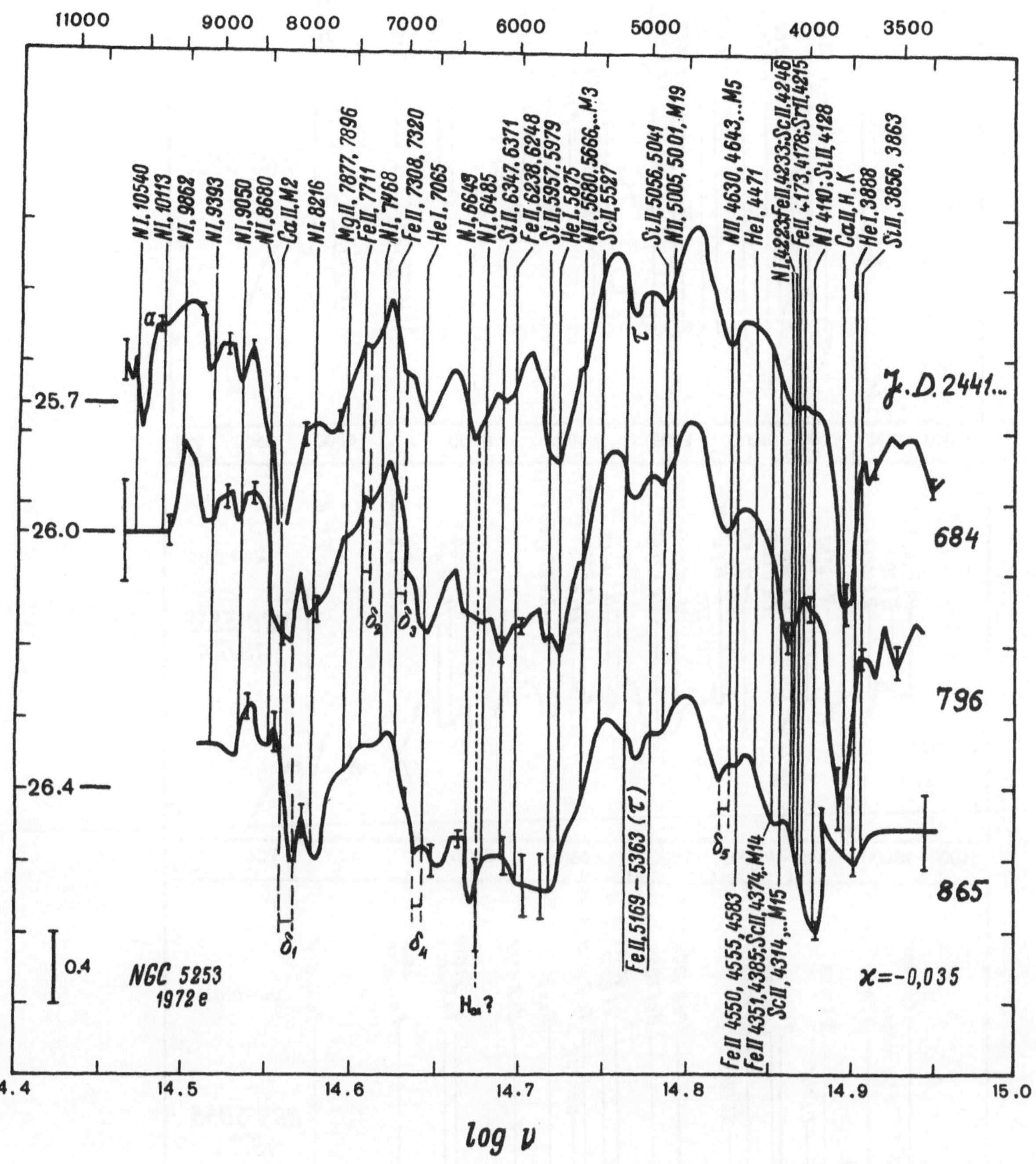

Fig. 5. Identifications of absorptions in the spectrum of SNI 1972e for the period from 2441684 up to 2441865 (J.D.). Greek letters $\delta_1, \delta_2, \delta_3, \ldots$ indicate the stratification effects which are progressing with time.

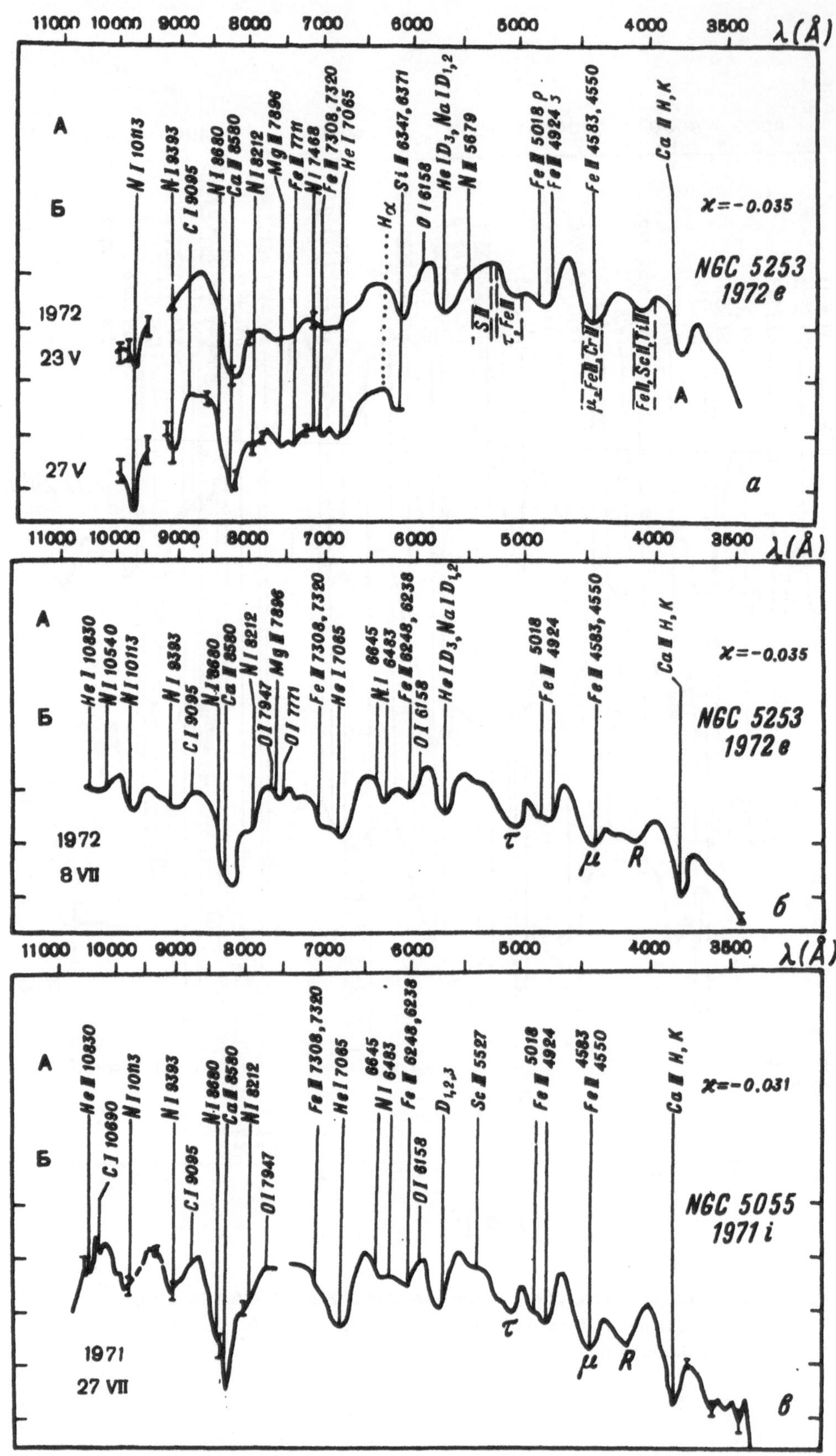

Fig. 6. This figure shows relatively strong absorptions of NI in the spectra of SNI and simultaneously shows the absence of absorptions of CI and OI which also are expected to be strong in case of the relative abundances of C, N, O atoms in the envelopes of SNI are approximately identical.

REFERENCES

BLOCH M., CHALONGE D., DUFEY F., 1964, Annal. Astrophys. 27, 315

CHALONGE D. et BURNICHON M.L., 1968, Journ. Observ. 51, 5

CHINCARINI G., PERINOTTO M., 1968, Asiago Obs. Contrib. N205

CHUGAY N.N., 1975, Astron. Zh. 52, 197

KIRSHNER R.P., WILLNER S.P., BECKLIN E.E., NEUGEBAUER G., OKE J.B., 1973, Astrophys.J. 180, 97

KIRSHNER R.P., OKE J.B., PENSTON M.V., SEARLE L., 1973, Astrophys. J. 185, 303

KIRSHNER R.P., 1974, IAU, Highlights of Astronomy 3, 533

KIRSHNER R.P. and OKE J.B., 1975, Astrophys. J. (In press)

McLAUGHLIN D.B., 1963, Publ. Astron.Soc. Pacific 75, 133

MINKOWSKI R., 1939, Astrophys. J. 89, 156

MUSTEL E.R., 1971a, Astron. Zh. 48, 3; Sov. Astr. 15, 1

MUSTEL E.R., 1971b, Astron. Zh. 48, 665; Sov. Astr. 15, 527

MUSTEL E.R., 1972a, Astron.Zh. 49, 15; Sov. Astr., 16, 10

MUSTEL E.R., 1972b, Astron. Tsirk., N 674

MUSTEL E.R., 1973, Astron. Zh. 50, 1121; Soviet Astr. 17, 711

MUSTEL E.R., 1974, IAU Highlights of Astronomy 3, 545

MUSTEL E.R., 1975a, Astron. Tsirk. N 870

MUSTEL E.R. and CHUGAY N.N., 1975a, Astrophysics and Space Science 32, 39

MUSTEL E.R. and CHUGAY N.N., 1975b, Astron. Zh. 52, 673

MUSTEL E.R., 1975b, Astron. Zh. 52, 1143

PSKOWSKIJ Yu.P., 1968, Astron.Zh. 45, 942; Soviet. Astr. 12, 750, 1969

SOBOLEV V.V., 1947, Moving Stellar Envelopes (in Russian)

TRURAN J.W., 1973, Space Science Reviews 15, 23

VAN DEN BERGH S., 1971, Astrophys.J., 168, 37

VAN DEN BERGH S., 1973, Pub.Astron. Soc. Pacific 85, 335

ZWICKY F., 1965, Stellar Structure 8, Stars and Stellar Structure 367

DISCUSSION

MARTYNOV : Is it true that hydrogen is well abundant in SN II, but not in SN I?

MUSTEL : Yes, there is very little hydrogen in the outer layers of a SN I. The absence of hydrogen lines in the spectra of these stars is the main criterion for classifying them as SN I.

OZERNOY : Do you find individual differences in SN I spectra, which could justify a division of the SN I class into two, physically separate groups, similar to what was mentioned by Prof. Rudnicki?

MUSTEL : There are certainly large differences from object to object, but I think that the spectroscopic material is still too scarce to establish two physically significantly different groups.

OBSERVATIONAL ASPECTS IN STUDYING THE EARLY STAGES OF THE EVOLUTION OF STARS

L.V.MIRZOYAN
BYURAKAN ASTROPHYSICAL OBSERVATORY, U.S.S.R.

НАБЛЮДАТЕЛЬНЫЕ АСПЕКТЫ ИЗУЧЕНИЯ РАННИХ СТАДИЙ ЭВОЛЮЦИИ ЗВЕЗД

РЕЗЮМЕ

На основе наблюдательного подхода к изучению эволюции звезд, разработанного после открытия звездных ассоциаций (1947 г.), критически рассмотрены наблюдательные данные, относящиеся, главным образом, к нестационарным карликовым звездам (звезды типа Т Тельца и примыкающие объекты), представляющим собой ранние стадии эволюции звезд. Отмечены особенности молодых гигантских звезд и случаи их совместного возникновения с молодыми карликовыми звездами. Показано несоответствие наблюдаемых характеристик молодых карликовых звезд с гипотезами о конденсации газа в звезды и о термоядерной природе источников звездной энергии.

I n t r o d u c t i o n. The classical trend in the theory of stellar evolution is based on the hypothesis concerning the condensation of diffuse matter in stars as well as on the theory of internal constitution of stars assuming the thermonuclear nature of stellar energy sources. However, the observational approach to the problems of stellar evolution seems more promising at present, when the evolution regularities of stars are grounded on the generalization and the theoretical analysis of the star characteristics observed that describe their state and eventually the stage of evolution (Ambartsumian, Mirzoyan et al.,1972,1971).

The principal advantage of investigations based on the observational approach lies in practical independence of theoretical assumptions, often not well grounded and essentially predetermining the final results.

The results of investigations have shown that at the early stages of evolution we come across phenomena which not only were unforeseen but, moreover, cannot be interpreted in terms of hitherto worked out theories on structure and evolution of stars.

The possibility of studying the early stages of the evolution of stars based on observations appeared in 1947 after the discovery of stellar associations - systems of recently originated stars (Ambartsumian V.A. 1947). Despite the existing discrepancies in the nature of stellar associations, it is hard to argue against the basic conclusions of the theory of stellar associations about the continuance and group formation of stars in the Galaxy in stellar associations.

It has been stated that the most characteristic feature of stars in their early stages of evolution is their physical instability. These conclusions lie at the bottom of the development of the observational approach in studying stellar evolution.

The observational approach proved especially fruitful in studying the early stages of the stellar evolution before they attain the main sequence when, according to modern concepts (see,for instance, Schwarzschild M. 1971),thermonuclear reactions must come into action.Most interesting results have been derived regarding dwarf stars (with the masses less than 1 $M_{\odot}$).

The present paper will deal mainly with those results.

E a r l y s t a g e s i n t h e e v o l u t i o n o f d w a r f s t a r s. Variable stars of T Tauri type represent one of the earliest stages in the evolution of dwarf stars. Irregular, non-periodic changes of the power and composition of the radiation of those stars characterize the state of the star just after its formation. This is attested first of all by the fact that T Tauri type stars belong to T-associations; therefore they are quite young ($<10^7$ years) (Ambartsumian, Mirzoyan et al.,1972,1971; Ambartsumian,1947).

This early stage of stellar evolution is characterized (see, for instance, Herbig G.H., 1962 ; Kuhi L.V.,and Roy J., 1966), in addition to irregular changes of brightness.by the presence in the spectrum of variable nonthermal continuous emission, of numerous emissions, including forbidden lines and the anomalously rich abundance of lithium in the atmosphere (Bonsack W.K., Greenstein J.L., 1960; Bonsack W.K., 1961) and the genetic connection with diffuse nebulae.

It is also important to note the discovery of powerful excessive infra-red emission in the region 1-3 μ in many stars of this class, (see, for instance, Mendoza 1966,1968; Kuhi L.V., 1974) and radio emission of centimeter waves from T Tauri and $L_{\kappa}H_{\alpha}101$ (Spencer J.H. and Schwarz P.R.,1974).

The changes of the continuum and emission lines in T Tauri type stars correlate badly. According to Walker's observations (Walker M.F. 1963, 1966, 1972) shortwave continuous emission is independent of or depends in small measure on emission lines. Furthermore, as the study of Götz shows (Götz W. 1967), the influence of emission lines on the general radiation of T Tauri type stars is small and decreases with the increase of their brightness . In other words, the changes of the continuum are not directly related to the changes of the emission line spectrum. However, the changes of the line spectrum are sometimes accompanied by qualitative changes of energy distribution in the continuous spectrum of the star. For instance, during the observations by Chalonge et al. (Chalonge D., Divan L., Mirzoyan L., 1971) in the spectrum of RW Aurigae powerful absorption features and positive Balmer discontinuity corresponded to greater photographic brightness of the continuum while at a lower brightness intensive emission lines and negative Balmer discontinuity were present in the spectrum of the star.

It should also be noted that Walker (Walker M.F. 1963, 1966 , 1972) has discovered in the spectra of a small number of T Tauri type stars red displacements of absorption components of emission lines, which he believes to correspond to the gravitational contraction of those stars. This interpretation is hard to agree with the phenomenon of outflow of matter from T Tauri type stars, established for most of them, and this point deserves serious consideration.

The establishment of the phase of flare activity as an evolution phase in the life time of those stars has been of decisive significance in studying the early stages of the evolution of dwarf stars.

The peculiar features of flare stars have been revealed mainly as a result of observing UV Ceti type stars in the neighbourhood of the Sun (Gershberg,1970; Ambartsumian and Mirzoyan, 1971; Kunkel,1973), in particular,through the synchronous spectral and photometric observations of stellar flares made by Kunkel(1973). Gershberg (1970) and Boop and Moffett (1973).

However, the evolutionary significance of flare stars was revealed (Ambartsumian V.A., Mirzoyan L.V.,1971; Haro, 1968; Mirzoyan,1973; Ambartsumian and Mirzoyan, in press), only as a consequence of the statistic study of flares in stellar aggregates - associations and relatively young stellar clusters.

V.A.Ambartsumian (1954) was the first to consider evidence favouring the relation of UV Ceti flare stars with T Tauri type stars which is distinctly manifested at moments of flares. This conclusion has been positively confirmed by Haro's (Haro G.,1957) discovery of flare stars in the Orion association (age $< 10^7$ years (Lesh J.R., 1963)), and then in other stellar aggregates (Haro G., 1962).

It has been determined that the presence of flare stars in not quite old stellar aggregates is a regular feature (Haro G., 1968). The spectral type restricting the lower part of the main sequence of the spectrum - luminosity diagram where flare stars occur, from its upper part where they lack, turned out to shift to later spectral types passing from younger to older aggregates (Haro G.,1968).

The idea that a young star after T Tauri phase passes into that of a flare star began to take shape gradually. This idea, formulated first by Haro (1962) following an analysis of the observational results of flare stars in stellar aggregates of different ages, became incontrovertible after V.A.Ambartsumian (1969) relying on flare statistics in the Pleiades showed that in that relatively young cluster (age of the order of 7.10^7 years (Simpson E., Hills R.E., Hoffman W., Killman S.A. Morton E., Paresce F.Jn., Peterson C., 1970), all dwarf stars (or most of them) fainter than some absolute magnitude must be flare stars.

This conclusion was fully confirmed by subsequent observations (Ambartsumian V.A., Mirzoyan L.V., Parsamian E.S. Chavushian H.S., Erastova L.K., 1970 ; Ambartsumian V.A., Mirzoyan L.V., Parsamian E.S., Chavushian H.S., Erastova L.K., Kazarian E.S., Ohanian G.B., Yankovich I.I., 1973) thanks to which the number of known flare stars in the region of the Pleiades has already exceeded 400. A considerable number of flare stars has also been found in other aggregates

(Orion (Rozino et al., 1969; Kiladze, 1972; Roslund , 1974), Praesepe (Yankovich I., 1975), etc.).

Thus, it has been established that the phase of flare activity is a regular phase in the evolution of stars which all dwarf stars undergo (Ambartsumian V.A.,Mirzoyan L.V., 1971; Mirzoyan L.V., 1973; Ambartsumian V.A., Mirzoyan L., in press.).

V.A.Ambartsumian's (1970) estimation of the relative number of stars with flare activity in a definite sample of T Tauri type stars, in the Orion association has indicated that only one-fourth of T Tauri type stars are capable of flare activity (flares with a photographic amplitude $>0\overset{m}{.}5$), while the flare star phase begins in them shortly before the end of T Tauri type phase (Ambartsumian, 1970). During this period in the life time of dwarf stars T Tauri type and flare star stages coexist.

Assuming that the rate of stellar evolution grows with the increase of their mass, the average luminosity of T Tauri type displaying flare activity should be expected to be above the average luminosity of stars of that type that have not as yet reached the phase of flare activity.

The validity of this assumption in the case of T Tauri type stars of the Orion association is regrettably premature as yet. Flare activity has been revealed only in a small number of stars of this type, presumably on the average in stars with relatively low luminosity, due to the fact that the observed amplitude of flares statistically grows with the magnitude (Ambartsumian V.A., Mirzoyan L., in press). Such selectivity in discovering flares in a small number of stars can seriously affect the expected result.

Assuming the direct dependence of the evolution rates of stars on their mass, the duration of the corresponding phases of the evolution must decrease with the mass. In this case the phase of flare activity can last rather long in stars with small masses. From this viewpoint the existence of flare stars among relatively old dwarf stars can be justified assuming that the above stars possess quite small masses.

Assuming that the flare stars occur overwhelmingly in physical systems it is natural to suppose that UV Ceti type stars in the neighbourhood of the Sun form a physical system.

This possibility was first considered by V.A.Ambartsumian (1957) and Haro (1957). Evidence favouring this concept was further treated by B.A.Arakelian (1969) and A.T.Garibjanian (In press). Indication of low average spatial density of flare stars in the general galactic field much more inferior to the density of stars of the UV Ceti type in the neighbourhood of the Sun is regarded as a telling argument.

However, the existence of flare stars in the general galactic field cannot be denied completely. By Haro's (1969) estimates about 20% of flare stars in the region of the Pleiades are field stars.

Within the framework of the above picture concerning the genesis of flare stars in physical systems, the flare stars occurring in the general galactic field can be regarded either as "runaway stars" flying at great velocity from the existing material systems, or as former members of already disintegrated systems with quite small masses for which the duration of the phase of flare activity must be considerably longer. In the latter case the luminosities of the corresponding flare stars must be quite low.

Clearly for flare stars - "runaways" the criterion of proper motions to determine their belonging to the given aggregate becomes meaningless. For instance, a strong spatial concentration of flare stars in the region of the Pleiades with the motions, different from the mean proper motion of the cluster (Jones B. F., 1973), toward the center of the cluster indicates that most of them belong to the cluster. It is precisely such rapid decrease of spatial density with a removal from generating nuclei that is observed in the case of O-B stars in O-associations (Mirzoyan L.V., 1965).

Herbig-Haro (Ambartsumian V.A., 1954; Herbig G.H., 1974) and FU Orion or fuor (Ambartsumian V.A., 1971) type objects are of major interest in studying the early stages of the evolution of stars.

The similarity of the spectra of Herbig-Haro objects with the spectrum of the nebula associated with T Tauri star and the radical changes found out by Herbig (1969) in one of the objects of this class, testify the states of extreme physical nonstability. Such facts have led V.A.Ambartsumian (1954), before the above observations of Herbig (1969) to assume that the stage of evolution, in which the Herbig-Haro objects are, forms a phase preceding the stage of T Tauri type stars.

Recently interesting optical and infrared observations of Herbig-Haro objects, confirming their physical non-stability, have been made by Strom et al(1974a,b,c).Most interesting are the conclusions concerning the early youth of those objects and the outflow of matter from them even in this early stage of evolution (Strom K.M. et al., 1974c). However, the writers' (K.M.Strom et al., 1974c) interpretation seems controversial: they regard the Herbig-Haro objects as some inclusions in diffuse nebulae illuminated by optically invisible, because of strong absorption ($10\text{-}30^m$), young infrared Herbig-Haro "stars" (like T Tauri type stars). In fact the term "Herbig-Haro objects", implying their extreme youth, concerns in principle (Ambartsumian V.A., 1954) the above-mentioned "stars". But the problem is not only of terminological nature.

The hypothesis of illumination by neighbouring stars can presumably be applied only to diffuse Herbig -Haro-objects. From this standpoint the recent attempt by A.L.Gyulbudagian (1975) to divide the Herbig -Haro objects into two groups differing by their physical nature seems to be more fruitful: objects the luminosity of which can be quantitatively interpreted by the reflection of the light of another source and objects that do not admit such interpretation. As a matter of fact objects belonging to the second group are true Herbig-Haro objects in the initial sense of this term (Ambartsumian V.A., 1954).

Somewhat different is the problem of fuors. The discovery by G.Welin (1971) of the fuor V1057 Cygni owning before "brightening" T Tauri type spectrum (Herbig G.H., 1958) has eventually confirmed the genetic relation assumed earlier between those two classes of stars (Ambartsumian V.A.,1954; Herbig G.H., 1966), and has indicated that the fuor stage can for some time interrupt the T Tauri type stage. This discovery excludes the possibilty of interpreting the fuor phenomenon by fast collapse of an extremely young star originating from diffuse matter suggested by Herbig (1966) in the case of FU Orion.

This stage of evolution is characterized by a similarity of the processes, accompanying the brightening of fuor, with the processes observed in P Cygni as well as with a change of the spectrum of the star in the direction of earlier spectral classes. In consequence of the phenomenon of fuor an object of high luminosity appears in fact.

Regrettably so far we are in possession of more or less complete observational data respecting only two fuors: FU Orion (Herbig G.H., 1966) and V1057 Cygni (Gieseking F., 1974) which relate basically to the period following their "brightening". Prior to "brightening" the brightnesses of both stars and the spectrum of one of them were only known: V1057 Cygnus (Welin G., 1971; Herbig G.H., 1958).

A characteristic feature of the fuor stage is its small duration. If evolutionary stages of T Tauri type and the flare star continue over a million or more years, the fuor stage can last not more than 1000 years, as estimated by Ambartsumian V.A.,(1971). True, it is likely that this phase is recurrent: the young star passes several times through this phase.

To understand better the nature of fuors, their relation to T Tauri type stars, also to determine their place in stellar evolution, the discovery and study of new fuors and an investigation of objects that are likely post-fuors and are still in the maximum of their brightness, are extremely important factors. A probable estimate of the mean time between two consecutive occurrences of fuors within accessible neighbourhood of the Sun (Ambartsumian V.A., 1971) allows to hope for success in discovering the next fuor within the nearest decades.*

The succession stated above respecting the early stages in the evolution of dwarf stars relies completely on observational data and practically does not depend on one assumption or another concerning the nature of protostellar matter.

However, some facts related particularly to differences in the occurrence of flares ("fast" and "slow"), to fuor phenomenon, to the presence or occurrence at times of ultraviolet continuous emission in the spectra of T Tauri type stars and flare stars, admit a simple and natural interpretation based on the hypothesis of superdense protostellar matter and can therefore be taken as indirect evidence in support of this hypothesis (Ambartsumian V.A., 1971).

At the early stages of the evolution of giant stars. The early stages in the evolution of stars of hot giants and supergiants comprising the typical population of O-associations, are investigated much more inadequately.

The strong variability of radiation, characteristic of the early stages of the evolution of dwarf stars, that has contributed in large measure to a study of their evolution, is unfortunately not observed in young giant stars. It can be assumed that this circumstance is accountable in terms of the following fact: the energy of particular flares or rapid changes, at least in the accessible spectral region

---

* It should be noted here that G.Gahm's (In press) conclusion on the inadequate powerful mechanism suggested by V.A.Ambartsumian (1971) for an interpretation of the fuor phenomenon, results from misunderstanding.

is negligible in this case as compared to the quiet radiation of stars. One more fact should be noted in favour of this assumption: the energy of flares grows, for instance, much slower than the luminosity of flare stars (Ambartsumian V.A., Mirzoyan L., In press).

If the expansion of the spectral range of observations does away with that difference,one can expect a more successful evolutionary study of giant stars in stellar associations.

Intensive processes of continuous outflow of matter and the formation of gas envelopes, observed in a number of young giant stars (Wolf-Reyet and P Cygni type stars, Be-stars, etc.), are undoubtedly associated with their evolution and are of considerable interest in studying the early stages in the evolution of giant stars.

From this standpoint observational data on the joint formation of dwarf and giant stars in the association are of key importance.

Strom et al(1972a,b)have derived some facts in support of the proposition that Ae, and Be type stars, associated with nebulae (Herbig G.H., 1960), are presumably extremely young stars, the brightest representatives of quite recently formed stellar groups. These data together with those on the structure of O + T associations (Ambartsumian V.A., 1968) attest that the evolution of giant and dwarf stars proceeds in most cases jointly.

Furthermore, it is likely that the fuor phenomenon illustrates the fact that transitions from dwarf stars to objects of high luminosity are possible in some cases.

That is why the investigation of fuors may throw light on the possible relation between the evolution stages of dwarf and giant stars.

Certain problems of studying young dwarf stars. The investigation of young dwarf stars is of cardinal importance also for the problem of protostellar matter and sources of stellar energy.

For an evolutionary interpretation of the diagram spectrum luminosity of young dwarf stars by means of the theory of initial evolution, based on the hypothesis about the condensation of diffuse matter into stars, the occurrence on this diagram of stars located below the main sequence poses an intricate problem.

This major fact has been authenticated through the investigations of Haro (1966), Herbig (1962), Andrews (1970), and Jones (1972). Poveda's attempt (1965) to account for this fact in terms of the absorption of stellar light by the circumstellar medium of so far unknown nature, seems to be rather artificial, though there is no doubt that circumstellar gas-dust envelopes can exert definite influence on the physical characteristics of young stars (Strom S.E., 1972).

In a recent paper by Grasdalen et al.(1975), dealing with the investigation of one of the nearest T-associations in the Chameleon the very occurrence of T Tauri type stars on the (V, B-V) diagram are regarded as a result of the effect of ultraviolet continuous emission on the colours B-V of those stars. By an empirical consideration of the contribution of continuous emission which, incidentally, spreads also in the long wave region of the spectrum quite far from the Balmer jump (Crasdalen et al.,1975; Mirzoyan,1958) to the colours of B-V, those stars are reduced to the main sequence.

The authors believe that in this way, the discrepancy between the observed location of T Tauri type young stars on the diagram (V, B-V) and their expected situation according to the theory of stellar evolution is eliminated.

It should be noted that such an approach to the problem does not seem quite justified. Virtually it substitutes one quandary, connected with the existence of young stars below the main sequence, for another: the presence of ultraviolet continuous emission of non-thermal nature in the spectra of T Tauri type stars. The point is that the theoretical models suggested so far offer no explanations to the observed peculiarities of the radiation of T Tauri type stars.

Nor can one consider the attempt convincing according to which Ridgren et al.(1975) have conducted extensive studies aimed at interpreting the ultraviolet continuous emission in the spectra of T Tauri type stars in terms of the presence of hot envelopes around them. This idea discussed earlier in literature, is incapable of accounting for a number of peculiar features of continuous emission. In particular, the model "cold star surrounded by a hot envelope" is in contrast with the observed colours of the continuous emission and leaves open the question about the sources of energy of envelope luminescence.

The problem of sources of flare energy as well as fast irregular changes of the brightness of T Tauri type stars, closely connected with the general problem concerning the sources of stellar energy, remains open as before. The interpretations suggested so far of those phenomena within the framework of well-known concepts on the mechanisms of stellar radiation are not in line with observational data.

For most of them the anomalous distribution of energy in the spectrum of continuous emission leading to anomalous colours (Gershberg R.F, 1970; Mirzoyan L.V. 1973; Gorbazki, Mirzoyan, 1969) and the excessively short duration of burning and at times even of the entire flare (Cristaldi, Rodono, 1971), are insurmountable obstacles. Even the most popular of those concepts - nebular hypothesis (see, for instance, Mirzoyan, 1966)) is not deprived of those troubles. It seems therefore well grounded to assume that at least in those cases, the energy sources are not of conventional thermonuclear origin (Ambartsumian, 1954).

It follows from our survey that the study of T Tauri type stars and related objects (Herbig-Haro objects, flare stars and fuors) is of prime importance for establishing and investigating the early stages in the evolution of dwarf stars. There is reason to believe that further work in this direction can yield basically new results on the early stages of the evolution of stars as well as on the so far unknown states of stellar matter and sources of stellar energy.

The author expresses his acknowledgements to V.A.Ambartsumian for his discussion of the present paper.

REFERENCES

AMBARTSUMIAN V.A., MIRZOYAN L.V. et al. Problems of Modern Cosmogony, 2nd Edition, 1972; French Translation: Problemes de'Cosmogonie Contemporaine, 1971

AMBARTSUMIAN V.A., 1947, Stellar Evolution and Astrophysics., in Russian

AMBARTSUMIAN V.A., MIRZOYAN L.V., 1971, IAU Colloquium No. 15, Veröff. Bamberg, 2, 98

AMBARTSUMIAN V.A., MIRZOYAN L.V., IAU Symposium No. 67, In press

AMBARTSUMIAN V.A., 1954, Comm. Byurakan Obs. 13

AMBARTSUMIAN V.A., 1969, Stars, Nebulae, Galaxies; Byurakan Symposium, 283, in Russian

AMBARTSUMIAN V.A., MIRZOYAN L.V., PARSAMIAN E.S., CHAVUSHIAN H.S., ERASTOVA L.K. 1970, Astrofizika, 6, 3

AMBARTSUMIAN V.A.., MIRZOYAN L.V., PARSAMIAN E.S., CHAVUSHIAN H.S., ERASTOVA L.K., KAZARIAN E.S., OHANIAN G.B., YANKOVICH I.I., 1973, Astrofizika, 9, 461

AMBARTSUMIAN V.A., 1970, Astrofizika, 6, 31

AMBARTSUMIAN V.A., 1957, Non-Stable Stars; Byurakan Conference, 9, in Russian

AMBARTSUMIAN V.A. 1971, Astrofizika, 7, 557

AMBARTSUMIAN V.A., 1968, Problems of Evolution of the Universe, in Russian

ANDREWS A.D., 1970, Bol. Obs. Tonantzintla. 5, No. 34, 195

ARAKELIAN M.A., 1969, Non-Periodic Phenomena in Variable Stars, ed. L.Detre. 161

BONSACK W.K., GREENSTEIN J.L. 1960, Ap.J. 131, 83

BONSACK W.K., 1961, Ap.J. 133, 340

BOPP B.M., MOFFETT T.J., 1973, Ap.J. 185, 239

CHALONGE D., DIVAN L., MIRZOYAN L.V., 1971, Astrofizika. 7, 345

CRISTALDI S., RODONO M., 1971, IBVS.Nos. 525, 526
GAHM G., IAU Symposium No.67, Moscow. In press
GARIBJANIAN A.T., Comm. Byurakan Obs. In press
GERSHBERG R.F., 1970, Flares of Red Dwarf Stars, in Russian
GIESEKING F., 1974, Astron. Astrophys. 31, 117,
GORBAZKI V.G., MIRZOYAN L.V., 1969, Stars,Nebulae,Galaxies; Byurakan Symposium. 83
GÖTZ W., 1967, Die Sterne. 43, 16
GRASDALEN G., JOYCE R., KNACKE R.F., STROM S.E., STROM K.M.A.J. 1975, 80, 117
GYUL'BUDAGIAN A.L., A 1975, Astrofizika. 11, In press
HARO G. 1968, Stars and Stellar Systems, eds. Middlehurst B.M., Aller L.H. 7, 141
HARO G. 1957, IAU Symposium No. 3. Non-Stable Stars, ed. Herbig G.H. 26
HARO G. 1962, Symposium on Stellar Evolution, ed. Sahade J. 37
HARO G., CHAVIRA E., 1969, Bol. Obs. Tonantzintla. 5, No. 31,23
HARO G., CHAVIRA E. 1966, Vistas in Astronomy. Vol. 8, eds.A.Beer, K.Aa,Strand. 89
HERBIG G.H., 1962, Adv.Astron.Astrophys. 1, 47
HERBIG G.H. 1974, Lick Obs.Bull.No. 658,
HERBIG G.H. 1969, Non-Periodic Phenomena in Variable Stars, ed. L.Detre, 75
HERBIG G.H. 1958, Ap.J. 128, 259
HERBIG G.H., 1966, Vistas in Astronomy. Vol.8, eds.A.Beer, K.Aa.Strand. 109
HERBIG G.H., 1960, Ap.J.Suppl. 6, 337
HERBIG G.H., 1962, Ap.J. 135, 736
JONES B.F., 1973. Astron.Astrophys.Suppl. 9, 313
JONES B.F., 1972, Ap.J. 172, L57
KILADZE R.I., 1972, IBVS.No.670
KUHI L.V., 1966, J.Roy. Astr.Soc.Canada. 60, 1
KUHI L.V., 1974, Astron. Astrophys. Suppl. 15, 47
KUNKEL W.E., 1973, Ap.J.Suppl. 25, 1
LESH J.R., 1963, A.J. 74, 891
MENDOZA E.E. 1966, Ap.J. 143, 1010
MENDOZA E.E., 1968, Ap.J. 151, 977
MIRZOYAN L.V. 1973, Flare Stars, in Armenian
MIRZOYAN L.V., 1965, Astrofizika. 1, 109
MIRZOYAN L.V., 1958, C.R. Ac.Sci.USSR. 119, 666
MIRZOYAN L.V., 1966, Astrofizika. 2, 121
POVEDA A. 1965, Bol.Obs. Tonantzintla. 4, No.26, 15,
RIDGREN A.E., STROM S.E., STROM K.M., 1975, The Nature of the Objects of Joy: A Study of the T Tauri Phenomenon. Preprint,
ROZINO L., PIGATTO L., 1969, Contr. Obs. Asíago. No.231
ROSLUND C. 1974, Ark.Astr. 5, 381
SCHWARZSCHILD M., 1971, Structure and Evolution of the Stars
SIMPSON E., HILLS R.E., HOFFMAN W., KILLMAN S.A., MORTON E., PARESCE F.Jn., PETERSON C., 1970, Ap.J. 159, 895
SPENCER J.H., SCHWARZ P.R. 1974, Ap.J. 188, L105
STROM S.E., GRASDALEN G.L., STROM K.M. 1974a,Ap.J. 191, 111
STROM K.M., STROM S.E., GRASDALEN G.L.1974b, Ap.J. 187, 83
STROM K.M., STROM S.E., KINMAN T.D.1974c,Ap.J. 191, L93
STROM S.E., STROM K.M., YOST J., CARRASCO L., GRASDALEN G.,1972a,Ap.J. 173, 353
STROM K.M., STROM S.E., BREGER M., BROOKE A.L., YOST J., GRASDALEN G., CARRASCO L.,1972b, Ap.J. 173, 165
STROM S.E. . 1972, P.A.S.P. 84, 745
WALKER M.F. 1963, A.J. 68, 298
WALKER M.F., 1966, Stellar Evolution, eds. Stein R.E., Cameron A,G,W,, 405
WALKER M.F., 1972, Ap.J. 175, 89
WELIN G., 1971, Astron. Astrophys. 12, 312
YANKOVICH I., 1975, Flare Activity of Red Dwarf Stars in the Region of Open Cluster Praesepe, Dissertation

THE CHEMICAL COMPOSITION OF STARS

A.A.BOYARCHUK
CRIMEAN ASTROPHYSICAL OBSERVATORY, U.S.S.R.

ХИМИЧЕСКИЙ СОСТАВ ЗВЕЗД

РЕЗЮМЕ

Обсуждаются химические составы атмосфер различных типов пекулярных звезд. Рассматриваемые группы включают Ap, Am, Fm, δSct, CN, CH, BaII, S, C, R CrB и HdC звезды.

ABSTRACT

A discussion is given of the chemical composition of the atmospheres of various types of peculiar stars. The groups considered include Ap, Am, Fm, δSct, CN, CH, BaII, S, C, R CrB and HdC stars.

In 1929 Russell (1929) determined the chemical composition of the solar atmosphere. Thus Russell founded a new direction (branch) of astronomy - the investigation of the chemical composition of cosmic objects.After that many astronomers investigated chemical compositions of stellar atmospheres. As a result we know the chemical composition of several hundreds of stars.

Although most of stars investigated have the chemical composition of their atmosphere very similar to that of the Sun, there are many stars which show peculiarities of the chemical composition. The observed peculiarities may arise at least due to three reasons.

1) The differences in the chemical composition of pre-stellar matter

ii) "Apparent" anomalies

iii) Changing of the chemical composition of stellar surface during the stellar evolution.

Many astronomers have investigated the chemical composition of stars belonging to the different stellar groups which have different ages. It was established that the older stars have the lower abundances of metals in their atmospheres. It was found from investigations of the clusters that most significant variation of the chemical composition took place during the first 2×10^8 years of the life of Galaxy, when the ratio $[\frac{Fe}{H}]$ increased from - 2 to 0. (Eggen and Sandage, 1969).(Here and later $[\frac{Fe}{H}]$ means the value $\lg \frac{N(Fe)}{N(H)}/\text{star} - \lg \frac{N(Fe)}{N(H)}/\text{Sun}$). After that variations up to factors of ± 2 from the solar metal abundance occured. (See Fig.1). The reason of those variations is still unknown (Arp, 1962). Blaauw (1976) pointed out that the stars which were located near galactic poles had different chemical compositions in comparison with the stars which were located in the galactic plane.

Another important aspect of a problem of chemical compositions of a pre-stellar matter is the relative abundances of individual metals. Pagel (1970) has shown that the s-process elements are less abundant in the stars which have smaller values of $[\frac{Fe}{H}]$. Especially it is true for barium, as we can see from Fig.2,which was taken from Pagel (1970), but three more stars were added.

Thus the general tendency of aging effect of chemical composition is the increasing of the values of $[\frac{Met}{H}]$ and of $[\frac{\text{heavy met}}{Fe}]$ with time.

Lubimkov (1976) has investigated the helium abundance in the atmospheres of early-type stars. He found that the helium abundance in the stellar atmospheres is increased with age. Fig.3 shows the results for different stellar groups.

Now we will consider the stars which have anomalies of chemical composition because of last two reasons mentioned above. Here and later we will understand "the anomalies" as differences between the chemical composition of stellar atmospheres and that of the solar atmosphere. Fig.4 shows schematically the locations of different groups of stars.

We will consider the groups of stars which have "apparent" anomalies. First of all it is necessary to explain how we understand the term "apparent". We will use this term for stars for which we cannot explain their anomalies by the first and third reasons mentioned above. Some astronomers believe that we have here some case of diffusion of the elements (Smith, 1971).

The main sequence of stars has the temperature in the central part high enough only for hydrogen burning. We do not expect any changes of metal abundance in the atmospheres of main sequence stars. Nevertheless there are many stars of spectral type B-F, which have peculiar metal abundance. Those are the Ap, Am, Fm and δSct - type stars.

Fig. 5 shows the relative abundances in the atmospheres of some Ap-stars. Here we used data, which were collected by Hack and Struve (1970). Among the Ap stars we can distinguish several groups having common abundance peculiarities, the Mn group, the Si group, the Sr group the Eu-Cr-Sr group and so on. We consider here only first two groups. Although there is strong scattering we can conclude that both groups have very strong overabundances of heavy metals and moderate overabundances of iron group elements. Helium is underabundant. But there are some differences in details between Mn - and Si - stars. Carbon and oxygen are underabundant in Si - stars while carbon is overabundant and oxygen is normal in Mn- stars.

The metal abundances in the atmospheres of Am and Fm stars are given in Fig.6. We can see that the general picture of abundance peculiarities is similar to that of Ap stars. But overabundances of heavy elements in the atmospheres of Am-Fm stars are much less than that of Ap-stars.

The noticeable features of Am stars are the underabundances of C, Mg, Co, Sc.

Fig. 7 shows metal abundances in the atmospheres of some δ Scuti stars. We can see that the abundance peculiarities are less here than those of Am-stars.

The observed abundance peculiarities in the atmospheres of main sequence stars are explained by a hypothesis of subsurface diffusion of the elements (Watson, 1971, Smith, 1973).

Now we will consider the abundance variations which arise during the stellar evolution. Let us start by the consideration of the lithium problem. Figs.8 and 9 present the lithium abundances in the atmospheres of main sequence stars and giants correspondingly. We can see that the lithium abundances vary in the wide ranges. The explanation of the lithium abundance variations is the following: a star was born with high lithium abundance which is equal to the interstellar value and then the lithium abundance decreased because of the concoction (Boyarchuk, 1976). But very high lithium abundances in the atmospheres of some C and S stars cannot be explained by such hypothesis.

Now we will consider different groups of giant stars. We will start by discussing the normal K-M giants. We called them "normal" for two reasons. First, the number of these stars is above 80% of yellow and red giants. Second, the abundance peculiarities of these stars are smaller than those of other groups.

For M-stars we have now not so well estabilshed data, but there are estimations of abundance of some elements.

The relative abundances of different elements in the atmospheres of normal K-stars do not differ significantly from that in the solar atmospheres (Conti et al., 1967). There is a strong difference of carbon isotopes ratio only. The K stars have the ratio $^{12}C/^{13}C$ less than 10 while the solar value is about 90 (Greene, 1969; Ridgway, 1974).

The chemical compositions of M stars are not known in detail yet. Spinrad and Vardya (1966) have shown that the ratio O/C ≈ 1.05 is less than the solar value 1.7. The nitrogen abundance is not the same for all stars. α Her and o Cet have the normal ratio N/H, but α Ori and R Leo have this ratio ten times more. The ratio $^{12}C/^{13}C$ is less than 8 for M star investigated (Gaball et al., 1972).

Peery (1971) has found technetium lines in the spectra of two M-type miridaes o Cet and R Hya. According to Merchant (1967) M stars have the lithium abundance much less than the Sun.

The next group is the stars with strong cyanogen molecular bands. Greene (1969) has shown that carbon and nitrogen have little enhancement. The abundance of nitrogen for CN stars seems to be very near to that for normal stars (Schmitt, 1969). The ratio $^{12}C/^{13}C$ is equal to 12 for α Ser(Day et al.1973). The metals show a little deficiency about -0.2 dex.

Then we consider a group of Ba II - stars. Here we will follow mainly Warner (1965). In Fig.10 we can see the differences between stellar abundance ratios and that of the Sun. The main features are the following a) large abundance of lithium (see Figure 9), b) large enhancement of elements heavier than Sr, c) small overabundance of Eu and Yb; that is the r - process elements.

In general, metals are overabundant by a factor 1.5 in comparison with hydrogen. The $^{12}C/^{13}C$ ratio in the BaII stars is probably more than 20. No Tc - lines were observed in any Ba II stars. CH-stars are very similar to BaII stars. These stars have strong CH bands; Fig.11 shows us the differential abundance of three CH stars according to the results of the analysis by Wallerstein and Greenstein (1964) and by Lee (1974).

We can see that the peculiarities of abundance ratio $\lg \frac{M}{Fe}$ are very similar to those which we observed in BaII - stars. The heavy elements, which were produced by the s-process are overabundant. The carbon is also overabundant. The metals are deficient with respect to hydrogen by a factor 10. The $^{12}C/^{13}C$ ratio is close to 50. No technetium was observed in CH-star. The strength of CH band in comparison with atomic lines is explained by high abundance of hydrogen. The CH stars have large space velocities. Then we can consider them as BaII-stars of the population II.

Let us consider now S stars. First of all it is necessary to point out that the accuracy of the determination of chemical composition of these stars is much smaller than that of K-type stars or earlier.

But nevertheless we can draw some conclusions on the chemical composition. Fig. 4 shows chemical positions of four S and two C stars according to Tsuji (1962, 1971) Hirai (1969) and Machara (1971). This figure shows that the heavy elements are overabundant here more than in BaII-stars.

Then we can see that C stars contain heavier element than S stars. There are some differences of abundances of r - and s - process elements: the s- process elements as barium are overabundant much more than the r - process elements as europium.The relative abundance of oxygen group elements in the S star, does not differ from that of the Sun more than for a factor 2; $^{12}C/^{13}C = 6$ (Tsuji, 1971). According to Peery (1971) the lines of Tc are present in spectra of variable S-type stars and absent in spectra of nonvariable stars. This is true for all stars which have been checked for Tc.

Davis (1971) has suspected the presence of the lines of prometheum in the spectra of two S-stars V Cnc and T Sgr. The abundance of lithium varies in the wide range.

The main feature of C stars is high abundance of carbon. The ratio O/C is less than unity. The carbon and oxygen form the tightly bound molecule CO. Then the excess of carbon forms such molecules as CH, CN and C_2. But at the same time there is not enough oxygen to form oxides such as TiO and ZrO, which are typical for M - type stars. The ratio O/C in C stars is 20 times smaller than that in the Sun. In most cases the ratio $^{12}C/^{13}C$ is near the equilibrium value in the CH cycle (4.6). But few stars have the ratio $^{12}C/^{13}C$ near 20 (Querci, and Querci, 1970). 6 stars from 8 investigated have Tc lines (Peery, 1971).

Lithium abundances vary in a very wide range, over 6 orders (Torres-Peimbert and Wallerstein, 1966). Some stars have the lithium abundances for 4 orders of magnitude more than of the Sun. Usually these stars have also a high abundance of ^{13}C. The ratio $^{12}C/^{13}C \approx 3$ (Cordon, 1971).

One star has been investigated by Catchpole and Feast (1971) for lithium isotopes. It was found that the whole lithium is ^7Li.

Let us go to the left from the red giants branch and discuss R CrB stars. Only two stars were analysed. They are R CrB itself (Searle, 1961) and RY Sgr (Danziger, 1965). The results are shown in Fig.5. We can see a large deficiency of hydrogen and oxygen. Helium, lithium, carbon and nitrogen are overabundant. Approximately the same position on the HR diagram is occupied by the non-variable stars, which have chemical composition similar to R CrB - type stars. These are the so-called HdC - stars. Fig.6 presents the results of the analysis by Warner (1967). The main features of the abundances are the deficiency of hydrogen and the excess of carbon.

There are a few stars, which are located to the left of the cool HdC-stars and have the same situation with abundances of hydrogen and carbon. Fig.7 shows the chemical compositions of some of these stars (Hill, 1965).

We see, that hydrogen is strongly underabundant. He, C, N, are little overabundant and O little underabundant. We have large scattering for the elements of iron group. This perhaps is a result of weakness of lines of those elements.

In conclusion I would like to pay attention to the unique variable star FG Sagittal. This star was studied by Herbig and Boyarchuk (1968) in1960-1965 and by Langer, Kraft and Anderson (1974) in 1969-1972. The comparison of their results shows that some time between 1966 and 1968, there was a pronounced enhancement of spectral lines of yttrium, zirconium and other heavy elements. The abundances of these heavy elements were increased about 25 times.

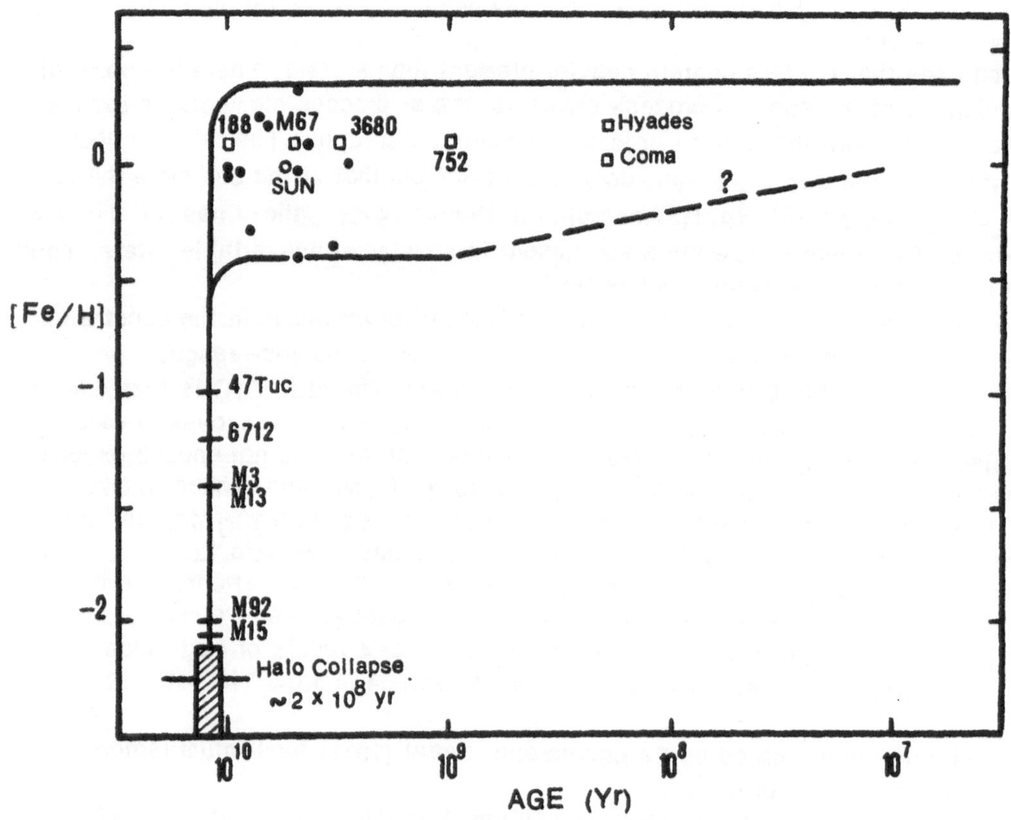

Fig. 1. The variations of the iron abundance of the clusters with their ages.(Eggen and Sandage, 1969).

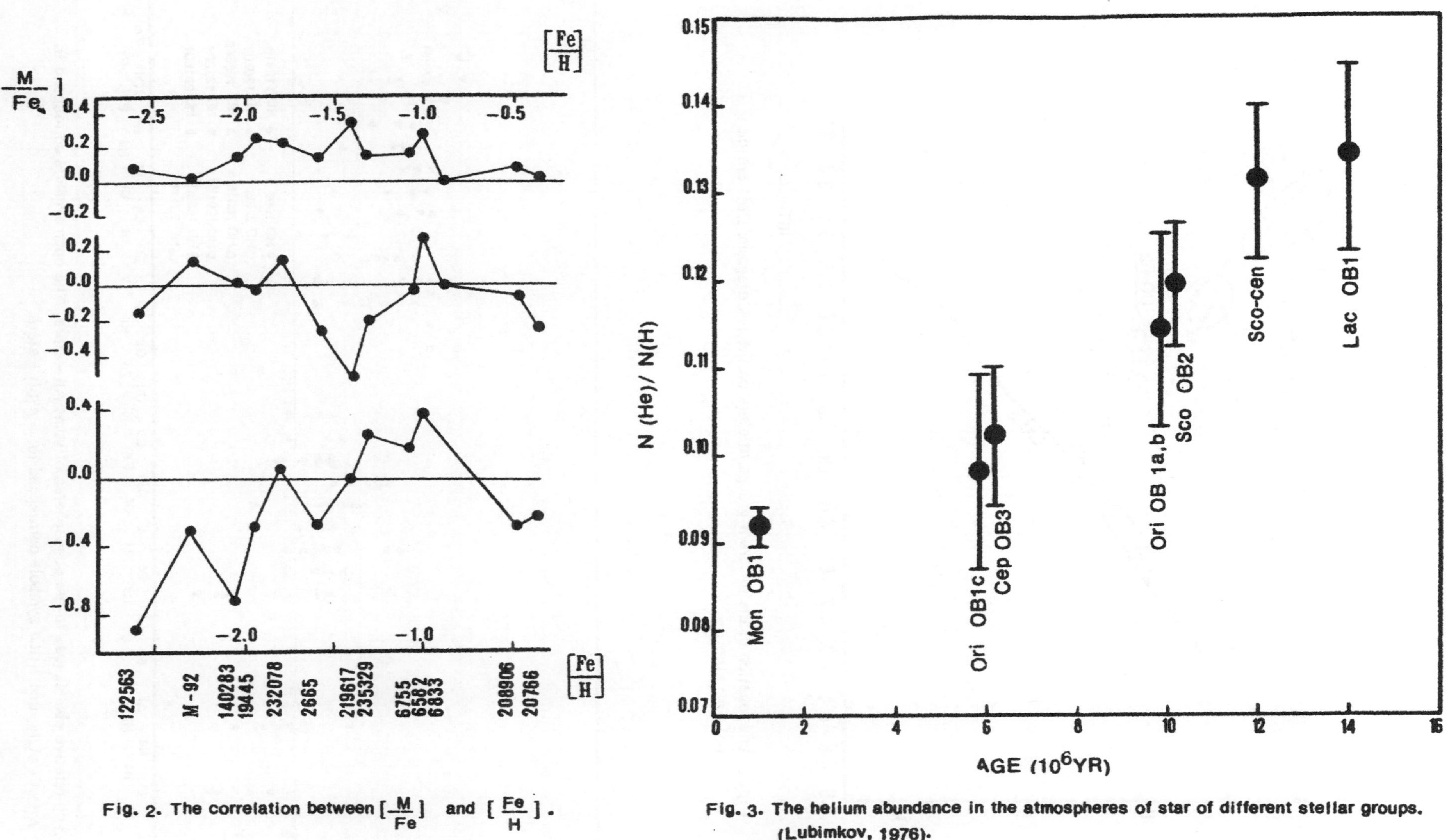

Fig. 2. The correlation between $[\frac{M}{Fe}]$ and $[\frac{Fe}{H}]$.

Fig. 3. The helium abundance in the atmospheres of star of different stellar groups. (Lubimkov, 1976).

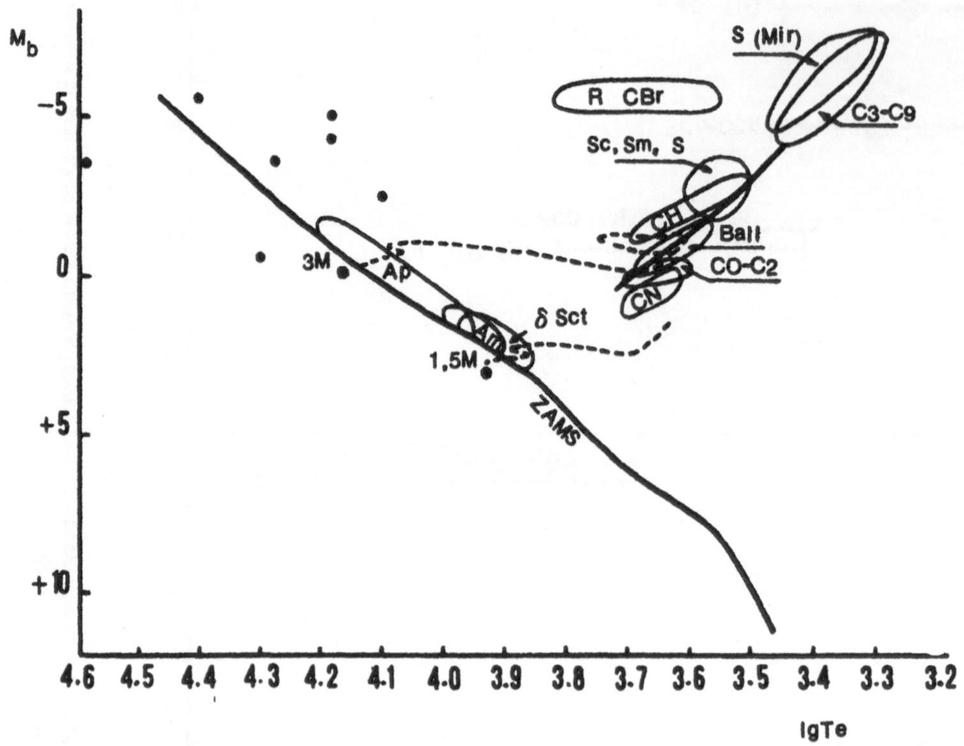

Fig. 4. The location of the different groups of stars on the HR diagram. Dots are the hot HdC stars.

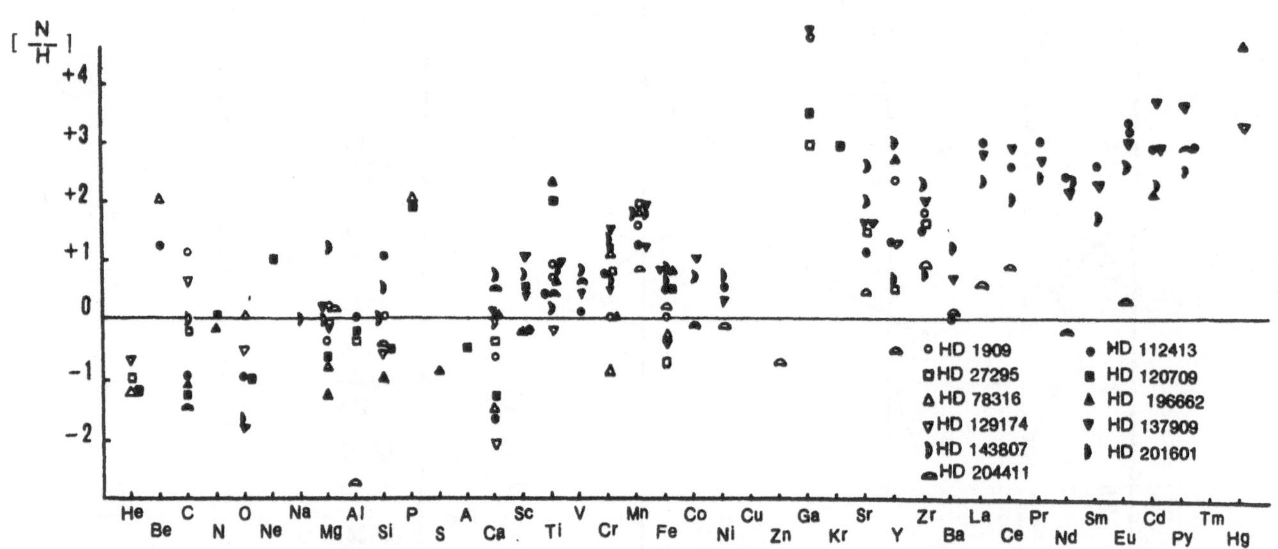

Fig. 5. The relative abundances in the atmospheres of some Ap - stars. The open symbols correspond to A(Mn) - stars, and filled symbols correspond to A(Si) - stars.

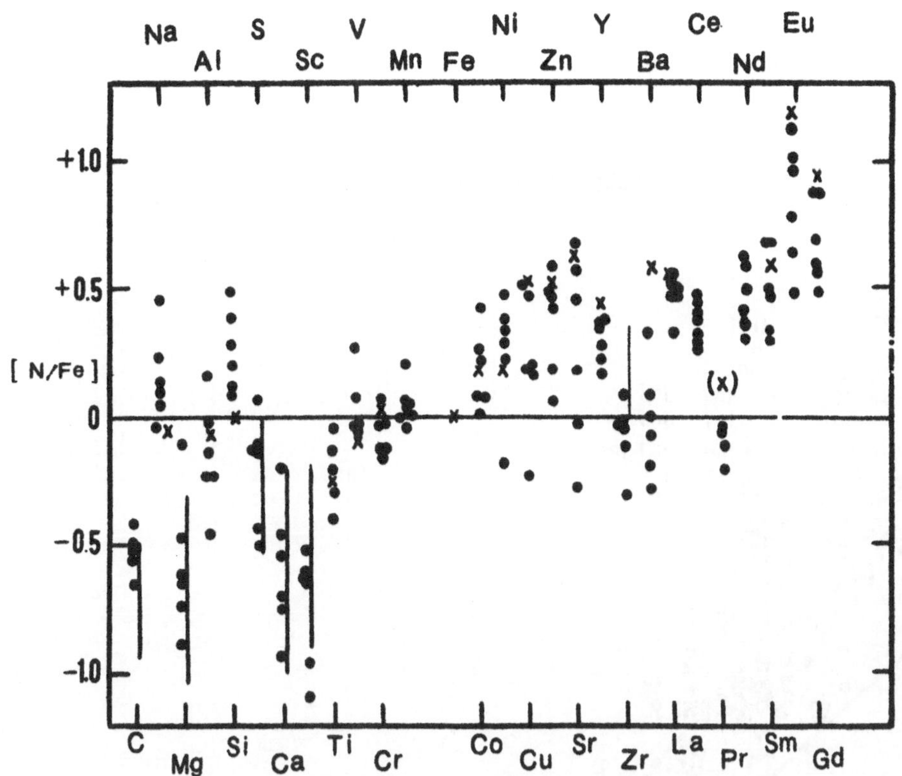

Fig. 6. The metal abundances in the atmospheres of some Fm-stars. Dots, Fm stars; crosses average abundances for Am star (Smith, 1973).

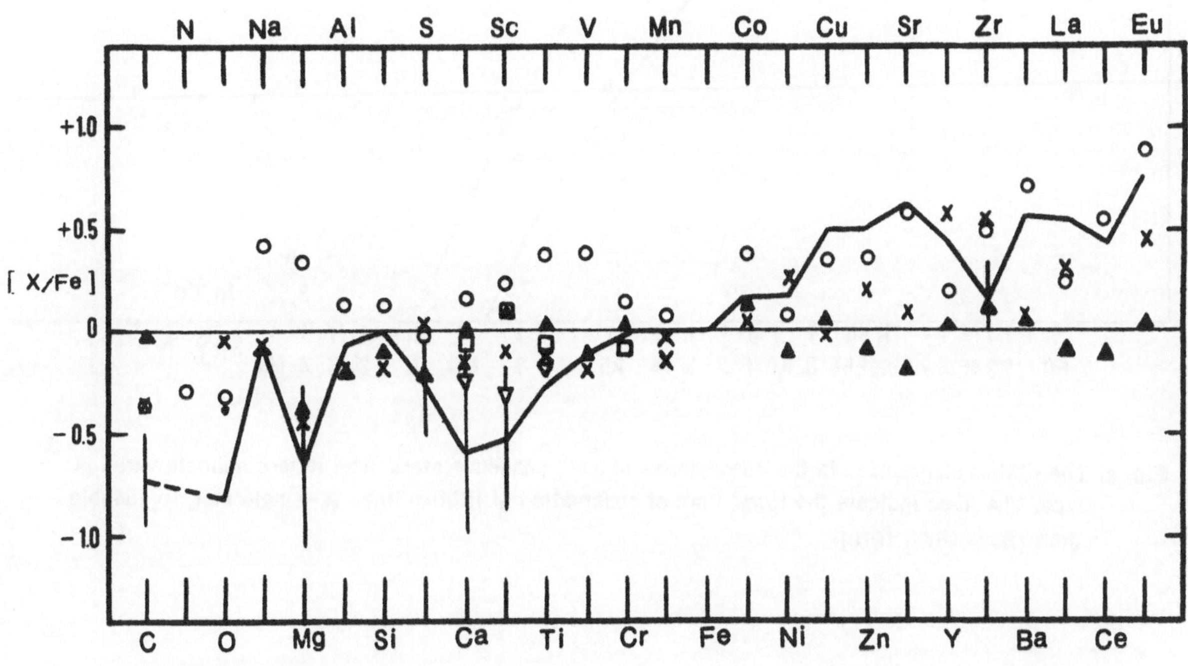

Fig. 7. The metal abundances in the atmospheres of some δ Secunti stars. The solid line represents the metal abundances of Am-stars. (Ishikawa, 1975).

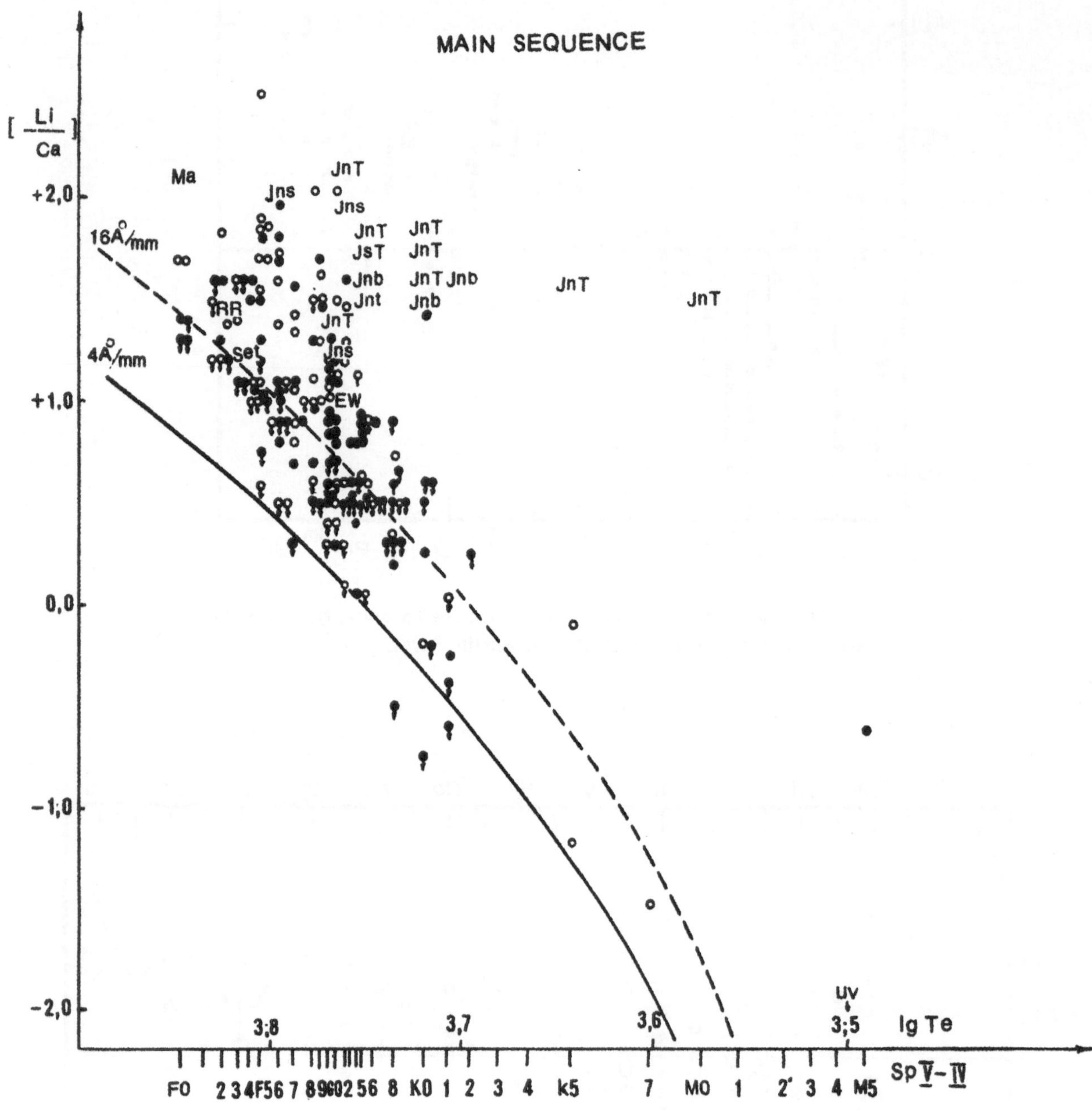

Fig. 8. The lithium abundances in the atmospheres of main sequence stars. The letters indicate variable type. The lines indicate the lower limit of observation of lithium line. ● - single star, o - double Star. (Boyarchuk, 1976).

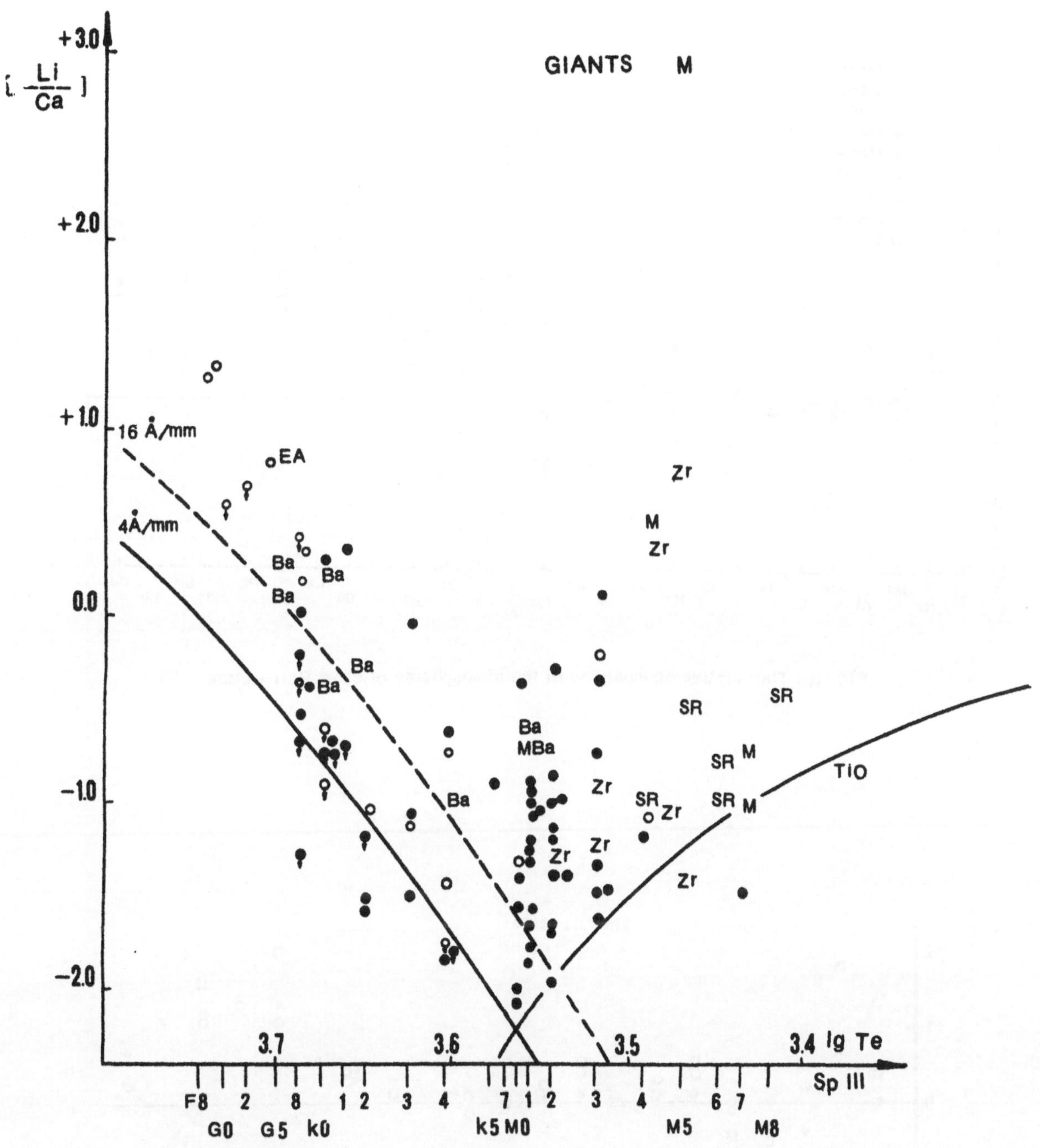

Fig. 9. The lithium abundances in the atmespheres of giants. The symbols are the same as in the figure 8. (Boyarchuk, 1976).

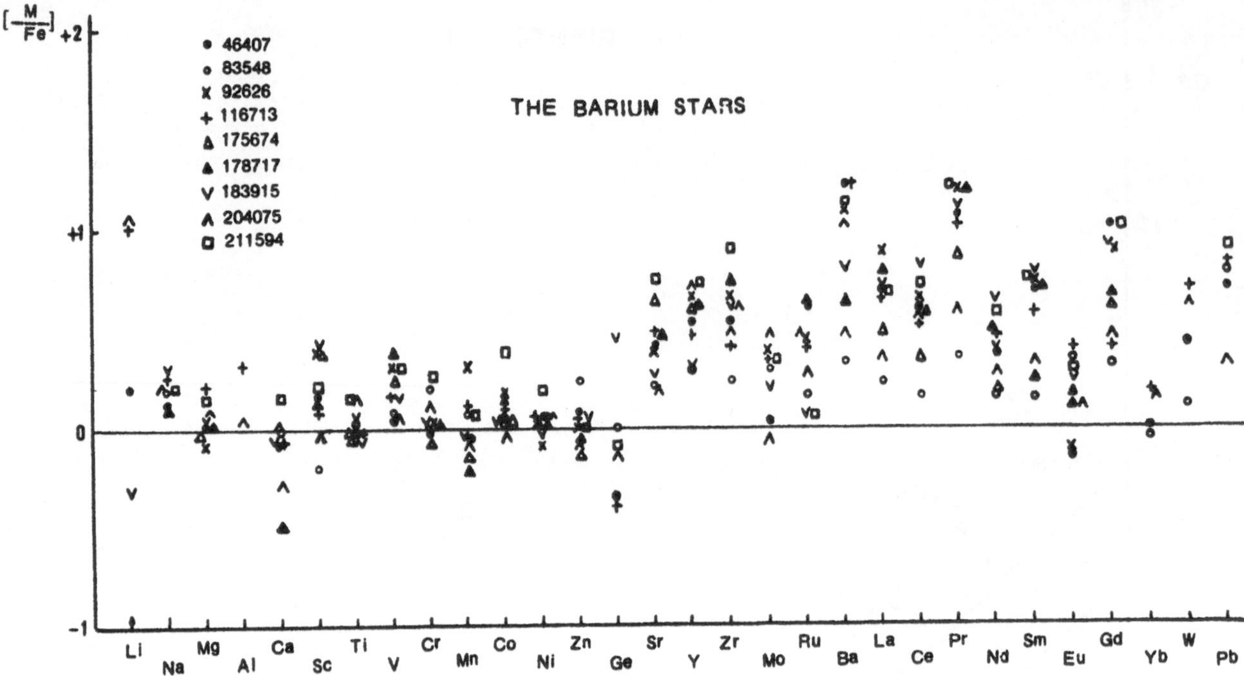

Fig. 10. The relative abundances in the atmospheres of some BaII - stars.

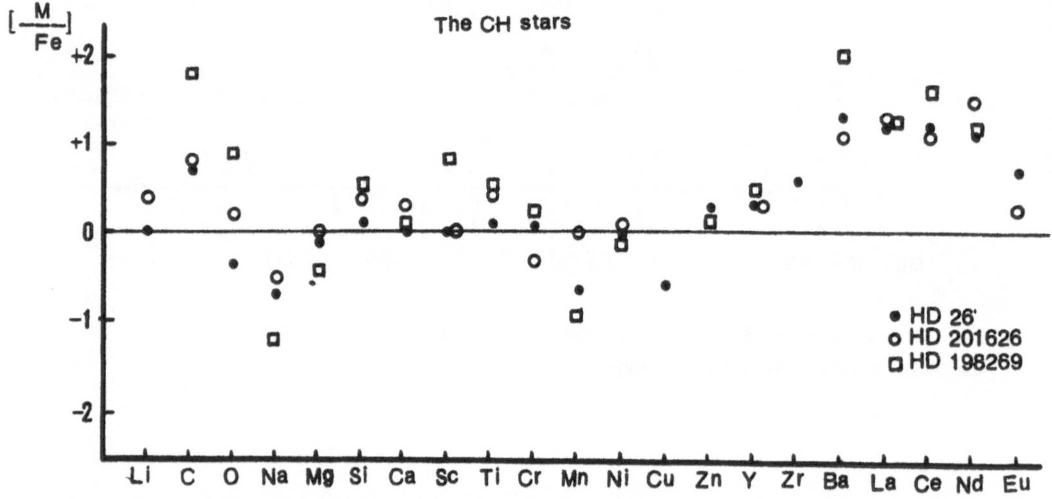

Fig. 11. The relative abundances in the atmospheres of two CH-stars.

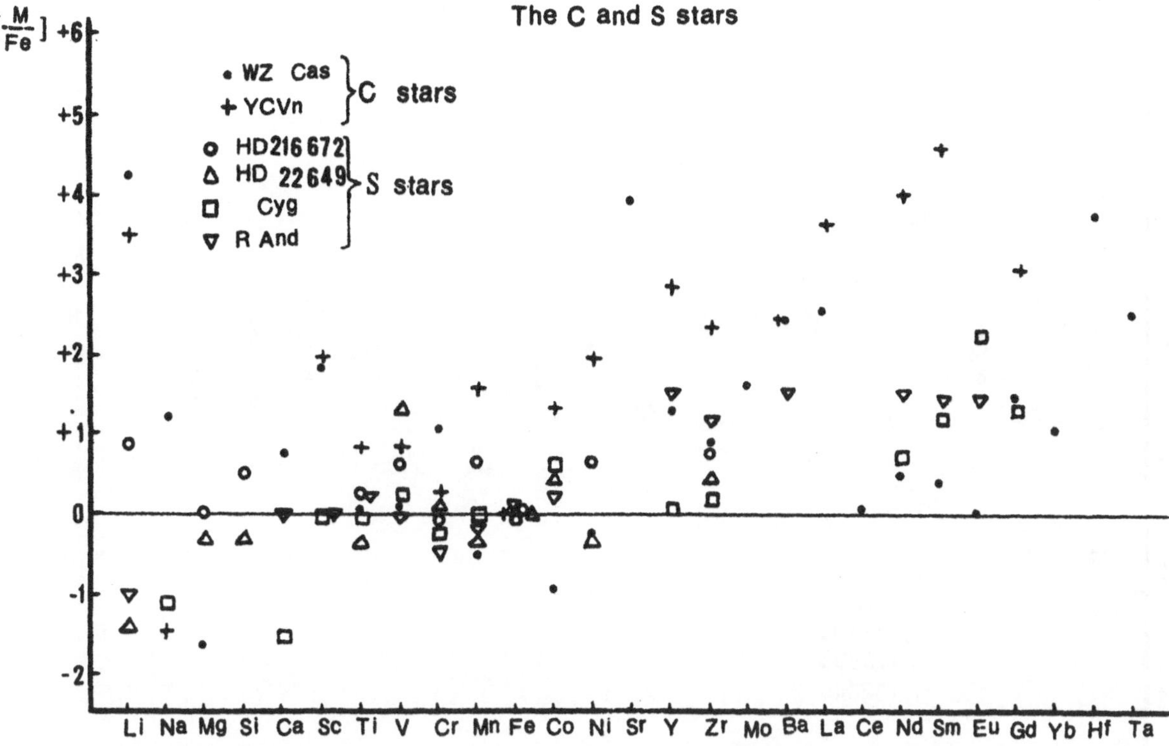

Fig. 12. The relative abundances in the atmospheres of C and S stars.

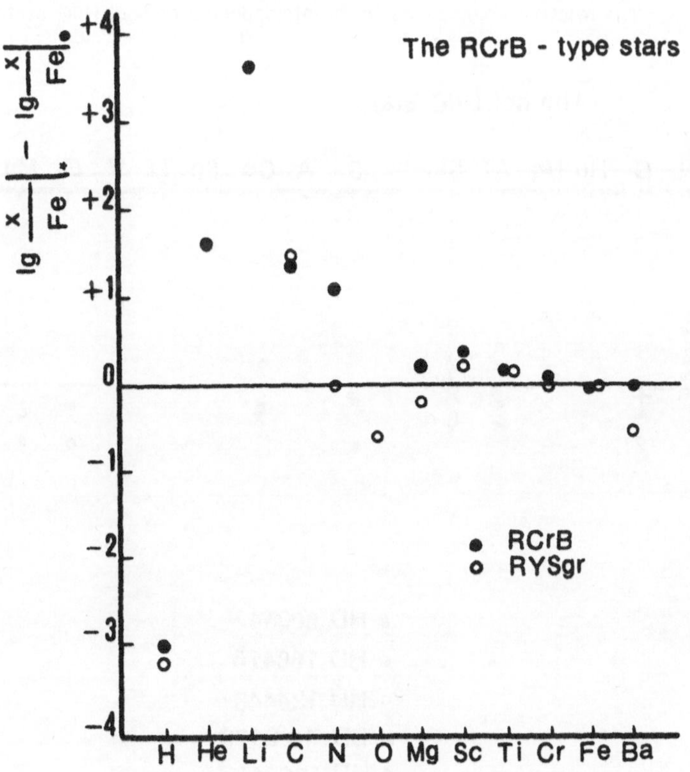

Fig. 13. The relative abundances in the atmospheres of R CrB type stars.

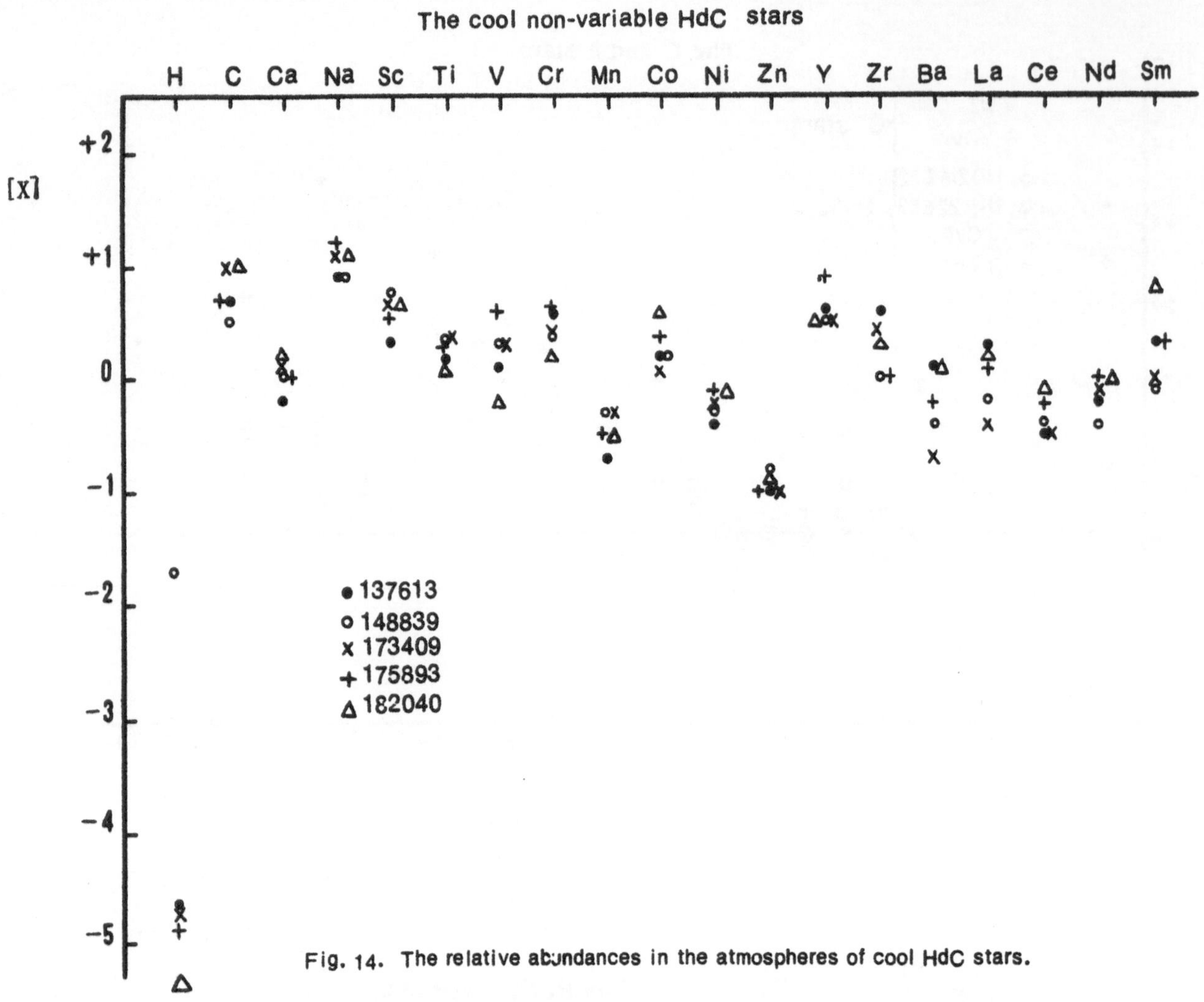

The cool non-variable HdC stars
H C Ca Na Sc Ti V Cr Mn Co Ni Zn Y Zr Ba La Ce Nd Sm
[X]
+2
+1
0
-1
-2
-3
-4
-5
● 137613
o 148839
x 173409
+ 175893
Δ 182040

Fig. 14. The relative abundances in the atmospheres of cool HdC stars.

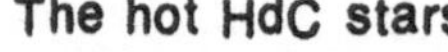

The hot HdC stars
H Ne C N O He Mg Al Si P S A Ca Sc Ti V Cr Mn Fe Ni
[M/Fe]
+1
0
-1
-2
-3
-4
-5
▲ HD 60344
● HD 168476
o HD 124448
x BD+10°2179
+ HD 160641

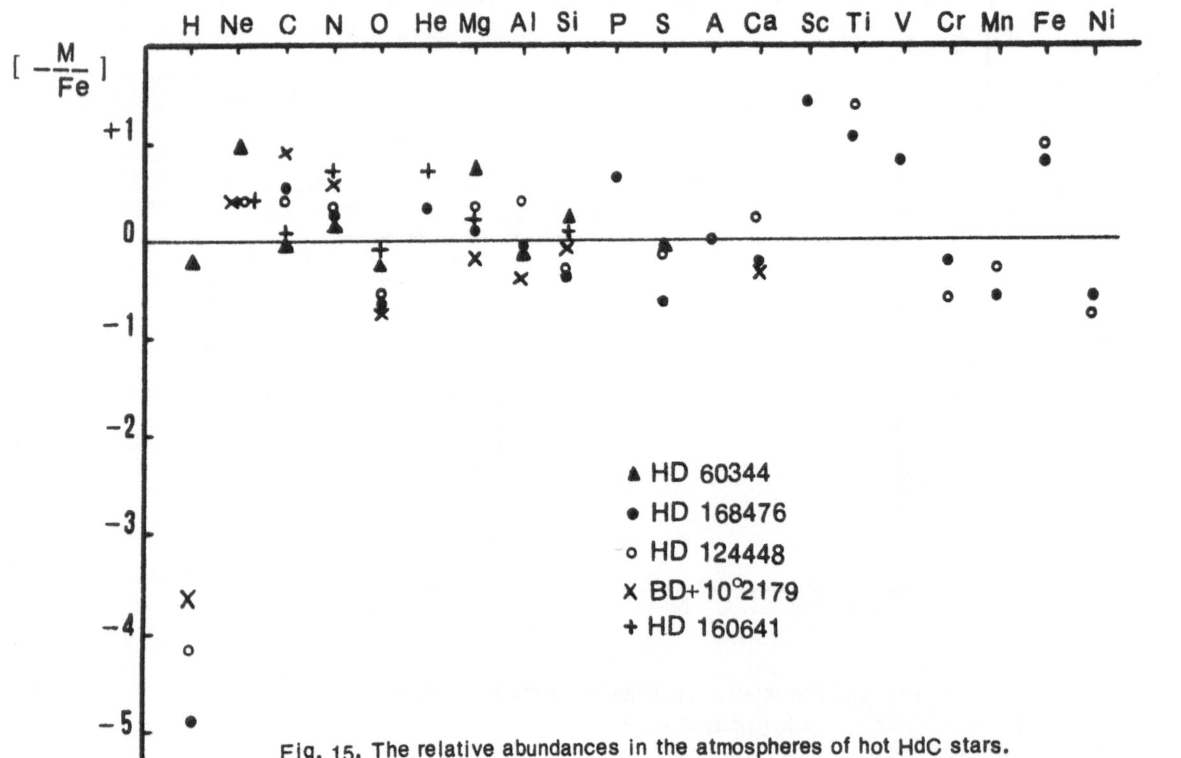

Fig. 15. The relative abundances in the atmospheres of hot HdC stars.

REFERENCES

ARP H.C., 1962, Ap.J., 136, 66
BLAAUW A.H., 1976 this vol
BOYARCHUK M.E., 1967, Izv. Crimean Aph. Obs. 55 (in press)
CATCHPOLE R.M., FEAST M.W., 1971, M.N., 154, 197
CONTI P., GREENSTEIN J.L., SPINRAD H., WALLERSTEIN G., VARDYA M.S., 1967, Ap.J. 148, 105
DANZIGER I.J., 1965, M.N. 130, 199
DAVIS D.N., 1971, Ap.J. 167, 327
DAY R.W., LAMBERT O.L., SNEDEN C., 1973, Ap.J. 185, 213
EGGEN O.J., SANDAGE A., 1969, Ap.J. 158, 669
GABALL T.R., WALLMAN E.R., BANK D.M., 1972, Ap.J., 177, L27
GREENE T.F., 1969, Ap. J. 157, 737
GORDON C.P., 1971, PASP, 83, 667
HACK M., STRUVE 1970, Stellar Spectroscopy, Peculiar stars, p.260
HERBIG G.H., BOYARCHUK A.A., 1968, Ap.J. 153, 397
HILL W.P., 1965, M.N. 129, 135
HIRAI M., 1969, PASJ, 21, 91
ISHIKAWA M., 1975, PASJ, 27, 1
LANGER G.E., KRAFT R.P., ANDERSON K.S.,1974, Ap.J. 189.509
LEE P., 1974, Ap.J., 192, 133
LUBIMKOV L.S., 1976, Astrophysika (in press)
MAEHARA H., 1971, PASJ, 23, 313
MERCHANT A.E., 1967, Ap.J., 147, 587
PAGEL B.E.J., 1970 Q. Jl. RAS 11, 172
PEERY B.F.J., 1971, Ap. J., 163, L1
QUERCI M., QUERCI 1970, A.A. 9, 1
RIDGWAY S.T., 1974, Aph.J., 190, 591
RUSSELL H.N., 1929, Ap.J. 70, 11
SCHMITT J.L., 1969, PASP, 81, 657
SEARLE L., 1961, Ap.J., 133, 531
SMITH M.A., 1971, Astr. Aph. 11, 325
SMITH M.A., 1973, Aph.J.Sup. 25, 277
SPINRAD H., VARDYA M.S., 1966, Ap.J., 146, 399
TORRES-PEIMBERT S., WALLERSTEIN G., 1966, Ap.J. 146, 724
TSUJI T., 1962, PASJ, 14, 222
TSUJI T., 1971, PASJ, 23, 275
WALLERSTEIN G., CREENSTEIN J.L., 1964, Ap.J. 139, 1163
WARNER B., 1965, M.N., 129, 263
WARNER B., 1967, M.N., 137, 119
WATSON W.D., 1971, Astr. Aph. 13, 263

DISCUSSION

KOPYLOV: The determination of the helium to hydrogen ratio is very difficult, especially for hot stars. I got the impression from yourdiagrams that your accuracy is rather high. Which method do you use to determine N(He)/N(H)?

BOYARCHUK: We use model atmospheres. We have made a very careful study of the possible influence of various error sources, and we believe that most of them have been well taken care of, possibly with the exception of those inherent in the determination of equivalent widths.

MASSEVICH: What is the accuracy in the determination of the barium and rare-earth abundances?

BOYARCHUK: Between ±0.2 and 0.5 (logarithmic ratio), depending on the method of analysis.

MARTYNOV: Do the stars without hydrogen lines ("helium" stars) differ from normal stars in other respects than this spectral anomaly?

BOYARCHUK: Yes, for instance they fall above the main sequence like supergiants, and we believe that many of them are binaries.

GASANALIZADE: Which value do you use for the solar iron content in your calculations?

BOYARCHUK: We only determine relative abundances, i.e. $[N/Fe] = [N/Fe]_{star} - [N/Fe]_{sun}$, and the exact value of the zero-point does not enter.

ORLOV: Are the carbon contents of variable and non-variable carbon stars different?

BOYARCHUK: This question is very difficult to answer at the present time, especially since new results obtained by means of synthetic spectra do not at all agree with previous results. The problem with the derivation of abundances from molecular bands is an extremely difficult one and the precision is still very low.

ZHILYAEV: Which were the main changes in FG Sgr during the past 65–70 years?

BOYARCHUK: The luminosity increased by 5^m. From 1955 to 1973 its spectral class changed from B5 to GO.

ŽELWANOVA: Did I correctly understand that Ap stars in your opinion represent a normal stage in stellar evolution?

BOYARCHUK: Yes. This is also supported by their rotational velocities and masses.

MASSEVICH: How can you be sure that your magnetic star is really a member of the Orion association?

BOYARCHUK: This conclusion is based on its position, radial velocity and proper motion.

SUPERNOVAE AND PULSARS AND THEIR EVOLUTION

O.Kh. GUSEINOV., F.K. KASUMOV., I.M. YUSIFOV
SHEMAKHA ASTROPHYSICAL OBSERVATORY, U.S.S.R.
PRESENTED BY F.K.KASUMOV

СВЕРХНОВЫЕ, ПУЛЬСАРЫ И ИХ ЭВОЛЮЦИЯ

The known identification of the pulsar PO531 with the Crab Nebula, great energy release during the Supernova explosions and activity of the supernova remnants (SNR) allow one to connect the pulsar formation with the appearance of SNR. However, in spite of several works concerning the study of the possible connection of pulsars with the SNR (Kiang, 1969., Prentice 1970., Tzarevsky 1972, Guseinov and Kasumov, 1973., Amnuel, Guseinov, 1974) one problem remains undecided: is the formation of a pulsar-neutron star always accompanied by a supernova remnant? At present 100 radio pulsars and 140 pulsars detected are known. In this connection a growing interest has developed in the evolution of these objects and their genetic relation.

1. The relation of pulsars with SNR; pulsar space velocities. Taking into account the above said about pulsar formations and proceeding from the observable distribution of these objects we obtained the average velocity of the space pulsar motion as equal to ~120 km/sec. This value is in a rather good agreement with the results of the independent estimate of the pulsar velocities for the pulsar proper motion in the Crab ~ 100 km/sec (Trimble and Rees, 1971), from the observations of the time of the spectral detail existence for PO328, PO834 etc. equal to 100 km/sec (Ewing et al., 1970)., from the observation of the diffractional picture of 35-120km/sec (Slee et al., 1974) and from the scintillation of radiation of 20-80 km/sec (Lang, 1970). It should be noted that for some pulsars such as PO329, PO829 and P1133 the velocities obtained amount to very great values of order of 200-380 km/sec (Manchester et al., 1974, Galt et al., 1972). On the whole there are 15 pulsars with the space velocities measured. According to the results (Manchester et al. 1974., Galt et al. 1972) three pulsars PO329, PO829 and P1133 have the velocities equal to ~ 200-380 km/sec, while for 13 other pulsars the velocities were found to be equal to 80-120 km/sec.

More recently Hulse and Taylor (1974) have discovered a pulsar P1913 in a binary system with the orbital velocity ~ 200 km/sec. Since this value is identical with the relativistic star velocities which are practically X-ray sources, one can believe that some fraction of pulsars being formed due to the pair break should have velocities more than 200 km/sec. The degree of the supernova explosion asymmetry and the velocity may depend on the closeness of the pair. And as the number of close pairs with the orbital velocities $\lesssim$ 200 km/sec is rather small, the pulsar velocities of 100 km/sec should prevail.

In order to estimate a possible value of the spatial division of pulsars and SNR it is necessary to know the remnant middle age. As far as according to Milne (1970), Downs (1971) and Ilovaisky and Lequeux (1972) at a distance of 3 kpc lie 39,25 and 27 remnants respectively, their total number in the galaxy will be 970,625 and 675 or in the average 800 in all catalogues.

The frequency of the supernova explosions is known to be equal to 1 in 30 yr (Guseinov and Kasumov, 1973., Tammann, 1974) so $\bar{T}_{SNR} = 2.5 \times 10^4$yr. As to the theoretical results concerning the evolution of remnants the upper age limit is $\lesssim 10^5$yr (Shklovsky, 1974), Thus, if a pulsar is the result of the supernova explosion it can move away from the remnants at a distance not exceeding 15 pc. Only two pairs "PO833-VelaX" and "PO611-1C443" in addition to "PO531-Crab" satisfy this limit. The pulsar and remnant ages ~10^4 yr and ~$(0.7\text{-}3) \times 10^4$ yr correspondingly (Tucker, 1971., Ramaty et al., 1971) may also confirm the possibitity of the genetic relations of the first pair. As regards the second pair "PO611-1C443" its genetic relation was assumed for the first time by Davies et al.(1972). New estimates of the remnant age (Winkler et al., 1974, Lozinskaya, 1975) equal to 2.5×10^4 yr are not in accord with that of the pulsar ($T_p \approx 6.5 \times 10^4$yr). This pulsar, however, being rather young after PO531 and PO833 is characterized by a great velocity of the rotational energy loss (by 1-2 orders) relative to the other pulsars with the exception of the two above. If this pair is really related genetically the pulsar age and velocity should be ~2.5×10^4yr and ~350 km/sec respectively and it is quite possible in the case of the close binary system break during a supernova explosion.

The absence of SNR near the other pulsars is explained by the fact that the remnant disappears very rapidly (~ 2.5×10^4yr) and strips the pulsar. It is also proved by the data of the pulsar and remnant spatial densities.

2. Pulsar luminosity and spatial density. As known the determinations of the distance to pulsars are quite uncertain. It is explained by the individual methods used. The methods available are not reliable at all because little is known about the characteristics of both interstellar and peripulsar medium. According to modern concepts the pulsars with minimum dispersion measures DM, maximum galactic widths b and small velocities of the rotational energy loss dE_{rot}/dt should be the nearest ones. Taking into account these facts and the flux in the impluse as well we obtained the pulsar distribution in the volume formed by a rotation of the isosceles triangles with the heights of 0.3,0.45 and 0.65 kpc and base $|2Z| \lesssim 0.3$ kpc. with the rotation centre at the apex of the Sun.

Our pulsar distribution in these volumes is given in Table 1. The ratio of the volumes of the successive rotational bodies is $V_2 V_1 = 2.25$, $V_3 V_1 = 4.7$. The ratio of the number of pulsars will be $n_2/n_1 = 2.3$ and $n_3/n_1 = 5.3$ respectively, i.e. the filling is close to the uniform one. Since the volumes of the rotational bodies considered constitute 2/3 of those of the corresponding cylinders and assuming the filling of the whole volume to be also uniform we obtain $n_1 = 4$, $n_2 = 10$, $n_3 = 24$.

Let us turn now to the recent work dealing with the search for pulsars (Hulse et al.,1974), The region of the sky with galactic longitudes within the interval $35° \leqslant l \leqslant 75°$ and latitudes $/b/ \leqslant 5°$ has been investigated. The sensitivity of the detector was so high that the pulsars with fluxes in the impulse $U_\nu \gtrsim 0.001$ flux units could be subjected to the thorough search. On the whole the search covered 42 square degrees in opposite directions (28 square degrees near 19^h, and 24 square degrees near 6^h). The pulsar survey has resulted in the detection of 11 previously unknown pulsars situated in the galactic widths $/b/ \leqslant 2.°4$ with the mean value $0.°9$ in the 19^h direction. But in the direction toward the galactic anticenter ($\sim 6^h$) no pulsars were discovered.

Earlier 19 pulsars with $b_{max} \sim 6°$ and $/\bar{b}/ \approx 2°.7$ have been discovered. We have chosen only 4 of them (see Table 1). Let us define the average distance to 19 pulsars assuming the applicates of these pulsars and of those which are nearer than 0.65 kpc to be the same. As one can see from Table I the mean value of the applicate for the nearest pulsars is equal to 100 pc.The mean value of the galactic latitude for pulsars situated within 0.45-0.65kpc from the Sun is 7° For 19 pulsars with the fluxes $U_\nu \gtrsim 0.01$ f.u. within 35°-75° the average latitude value is 2°.9. Then the mean distance to these 19 pulsars will be 1.5 kpc. If the pulsars were distributed uniformly at this distance one could discover about 170 pulsars with the same luminosities. But at present, as we know, only ~110 pulsars have been discovered and they constitute of about 2/3 of all pulsars situated at a distance of 1.5 kpc.

As we already noted 11 new pulsars have been detected in this region .Comparing them with the pulsars having mean distances of about 1.5 kpc one can obtain the mean distance to the new pulsars equal to 4.5 kpc. The average dispersion for these pulsars ~1.5 times than that for the pulsars being discovered earlier.

The search (Hulse et al., 1974) did not lead practically to the detection of the nearest but faint pulsars. This confirms the possible presence of some lower limit in the pulsar luminosity. In other words, if while going to fainter objects their number increases, one can expect a considerable number of faint pulsars at an average distance of 1,5 kpc and observe the average latitude increase of newly discovered pulsars. This conclusion is illustrated in Figure 1 where one can see an empirical function of the pulsar luminosities according to Table 1.A dashed line represents the monochromatic luminosity as equal to 2.3×10^{12} W/Hz which may be taken as a mean value of the luminosity. Among all the pulsars known the faintest one has a monochromatic luminosity $\sim 2 \times 10^{11}$ W/Hz or the total luminosity $\sim 3 \times 10^{27}$ erg/sec.

Let us turn now to the estimation of the number of pulsars in the Galaxy. One should note here the case of their non-observability due to the narrowness of the radiation directional diagram. Let some fraction of the visible pulsars in some volume be K. Then knowing that at the distance of $\leqslant 0.65$ kpc /Table 1/ we have 24 pulsars, their whole number in the Galaxy will be

$$N_p = \frac{24}{K} \left(\frac{15}{0.65}\right)^2 = \frac{1}{K} 1.5 . 10^4 .$$

On the other hand, in the region of the thorough search for pulsars with the area within 35° - 75° at a distance of 4 kpc we have 12 pulsars and their whole number may be estimated as follows

$$N_p = \frac{12}{K\eta} \left(\frac{15}{4}\right)^2 \frac{360°}{40°} = \frac{1}{K} 2.5 . 10^4 ,$$

where $\eta = 0.07$ is the region of the thorough search.

Finally, from the existence of two pulsars with ages $\leqslant 10^4$ yr /PO531 and PO833/ situated at a distance $\leqslant 2$ kpc, the number of pulsars with the age less than 5×10^6 yr may be expressed as $1/K \cdot 1.5 \times 10^4$. Thus the number of pulsars in the Galaxy may be adopted as equal to $\frac{1}{K} 2 \cdot 10^4$. At the mean age of $\sim 5 \times 10^6$yr /see below/ the frequency of the pulsar formations will be $0.005 \times 1/K$ per year. Thus, if a pulsar formation is always accompanied by a supernova explosion it is necessery for the orientation coefficient to be 1/6.

3. Orientation effect and frequency of pulsar formations. In works(Davies et al,. 1970, Reifenstein et al.,1969) special pulsar searches in 21 supernova remnants at frequences of 112, 234, 405 and 410 Mhz are described.For mean pulsar velocity ~100 km/sec it would move away from the explosion at a distance not more than 15 pc,i.e. (taking into account the sizes of the remnants investigated ~30 pc) it would be still in the nebula. However, no objects with the pulsation periods from 0.01 to 10 sec have been detected. The searches for the optical pulsars in the locations of radio pulsars and supernova remnants did not yield any results either (Hebo et al., 1974).

So from the 24 remnants investigated only two of them contain pulsars (Crab and Vela X and possibly IC 443). It is possible that in some remnants pulsars were not observed due to their far distances.

About 12 remnants are situated nearer than 2 kpc (Milne 1970., Downs 1971. Ilovaisky and Lequex 1972.,Davieset al., 1970, Reifenstein et al. 1969).

Thus, from 12 remnants only in two or (if one takes into account IC443) three of them pulsars have been detected. If it is due to the pulsar radiation orientation only then one observes 1/6 - 1/4 of the whole number of pulsars. If we confine ourselves to the volume with a radius of 1 kpc we obtain 1/4. According to Henry and Paik (1969) and Zheleznjakov (1971) one can observe 9% and 33% of pulsars respectively.

In order to define the orientation coefficient more exactly as a characteristic of the directional diagram α we used the ratio of the impulse width ΔP to the period. This parameter is known for 92 pulsars and it changes within the range $0.007 \leqslant \Delta P/P \leqslant 0.26$. On the whole, pulsars have $\Delta P/P = 0.02$. The dependence of $\Delta P/P$ on the period or age is not observed. The observed deviation of this quantity is connected with the angle between the magnetic axis and rotational axis θ (in the case of "pencil") and with the rotational axis and line of sight (in the case of "knife", see Fig.2).

Consider now the dependence of $\Delta P/P$ on α and θ . As one can see from Fig. 2 for the central scanning

$$\frac{\Delta P}{P} = \frac{\Delta l}{2\pi r} = \frac{\sin \alpha/2}{\pi \sin \theta} . \tag{1}$$

In the case of non-central scanning of the diagram

$$\frac{\Delta P}{P} \approx \frac{\sin \alpha/2}{4 \sin \theta} . \tag{2}$$

Since the density of the probability of the angle between the line of sight and magnetic axis $P(i) = \text{Sin } i$ the pulsar may be observed from the Earth (see Fig.2) if $\theta - \alpha/2 \leqslant i \leqslant \theta + \alpha/2$. Thus, the probability to observe a pulsar in the case of occasional orientation of the line of sight to the magnetic axis will be

$$K = \int_{\theta-\alpha/2}^{\theta+\alpha/2} P(i)\, di = 2 \,\text{Sin}\, \theta \;\text{Sin}\, \alpha/2 . \tag{3}$$

Combined solution of Eqs.(3) and (2) under assumption of the occasional orientation of the angle of the inclination of the magnetic axis to the rotational one will yield the orientation coefficient of pulsars and angular width of the directional diagram.

The average value of the occasional quantity θ

$$\langle\theta\rangle = \frac{1}{2} \int_0^{\pi} \theta \,\text{Sin}\, \theta d\theta = \pi/2$$

and thus

$$K = 8 \frac{\Delta P}{P} , \qquad \alpha = 2 \text{ arc Sin } \frac{4 \Delta P}{P} . \tag{4}$$

Since for the most pulsars $\Delta P/P$ changes within the range of $0.015 \leqslant \Delta P/P \leqslant 0.035$ one can obtain from (3) $0.09 \leqslant k \leqslant 0.21$ and $7° \leqslant \alpha \leqslant 12°$. In other words, while realizing the pencil diagram for the observable pulsars k changes within the limits of $1/12 \leqslant k \leqslant 1/5$ and it is in accordance with the numbers given above.

Making similar calculations for the knife diagram we obtain $0.99 \leqslant k \leqslant 0.999$ and $5°.5 \leqslant \alpha \leqslant 12°.5$.

The list of remnants in which the search for pulsars should be carried out to define k value more precisely was given by Guseinov and Kasumov (1973).

Thus pulsar searches do not allow to define the directional diagram more certainly though there are rather substantial arguments in favour of the pencil diagram. Let us consider one of them.

Assuming that a supernova explosion always leads to the pulsar formation one can estimate the mean pulsar age. An uncertainty in the definition of the orientation coefficient (due to the deviation of the ratio of the impulse width to the period) amounts to two times. The same uncertainty we note while defining the pulsar age. In order to reduce errors and coordinate different space kinematic pulsar characteristics we make use of the data of mean pulsar distribution.

The ratios

$$|\bar{Z}| = \frac{\bar{T}_p = N_p/\nu_{SN}}{|R \sin b|} = \frac{1}{|\sin b|}\, \bar{T}_p \bar{V}_p = 0,64\, \bar{T}_p \bar{V}_p \qquad (5)$$

yield

$$|\bar{Z}| = \frac{0.64}{K} \left(\frac{R_G}{R'}\right)^2 \frac{\bar{V}_p}{\nu_{SN}} . \qquad (6)$$

From (6) one can find the distance R' for which n pulsars should be detected taking into account the mean space velocity $\bar{V}_p$, mean $|\bar{Z}|$ and the frequency of pulsar formations similar to that of the supernova explosions. A comparison of this value with our pulsar distribution (Table 2) showed the accordance between the mean pulsar characteristics.

In table 2 we give mean pulsar ages for different values of the orientation coefficient. The corresponding distances at which n pulsars should be situated are also given there.

According to this Table one can draw a conclusion that in order to have more or less real notion about the genetic relation of remnants with pulsars one should adopt mean velocity, mean age and orientation coefficient to be equal to 80-100 km/sec, 4×10^6 yr, $1/8 \leqslant k \leqslant 1/5$ respectively.

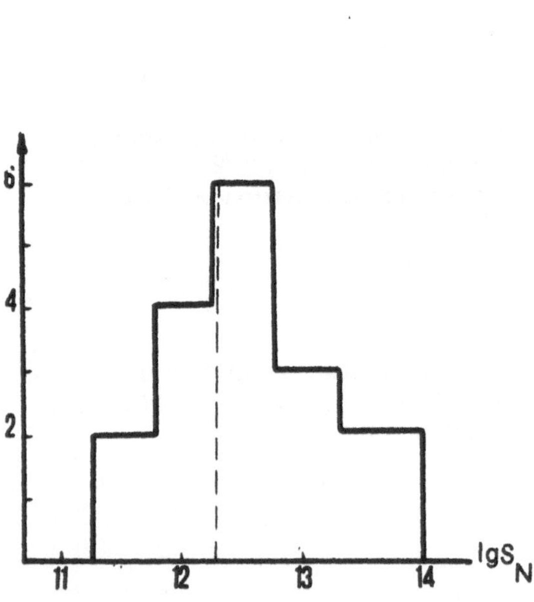

Fig. 1.

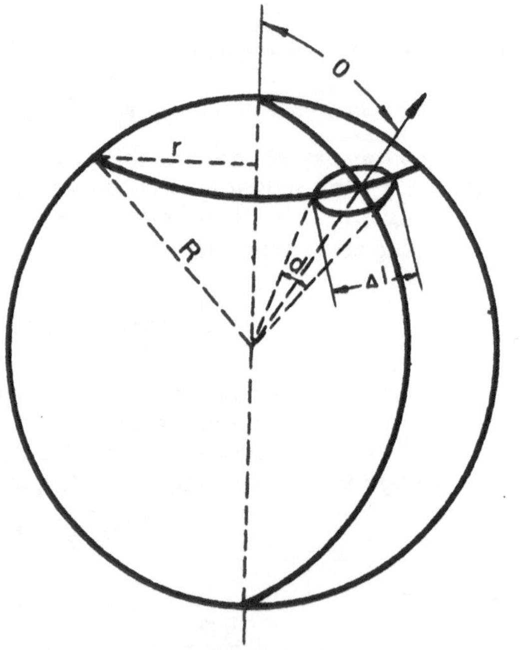

Fig. 2.

Table 1

P (kpc)	Pulsar	P (sec)	R (kpc)	b°	DM (cm^{-3})	/ Z / (pc)	$\frac{dE_{rot}}{dt} \times 10^{-31}$ (erg / sec)	S_ν (10^{-10} w/Hz)
0.3	P1604	0.42	0.3	35.5	11.0	180	1.57	25
	P0809	1.29	0.2	31.6	5.8	104	0.3	41
	P2045	1.96	0.32	33.1	11.5	11.8	5.8	73
	Mean by volume	1.22	0.27	33	9.4	135	2.56	46.5
0.45	P1929	0.23	0.43	-3.9	3.2	27	394.0	338
	P1642	0.39	0.45	26.1	35.7	200	114.0	1290
	P0834	1.27	0.35	26	12.9	154	13.2	80
	P1919	1.34	0.43	3.5	12.4	26	2.16	314
	Mean by volume	0.8	0.35	15	16	101	76.0	50.5
0.65	P0833	0.089	0.5	-2.8	63	1.24	$6.8 \cdot 10^5$	6700
	P1451	0.26	0.48	-8.6	8.6	55	64.0	(214)
	P1953	0.43	0.65	0.7	20	7.8		90
	P2015	0.56	0.54	-4	14.2	130	3.5	6280
	P1706	0.65	0.65	13.7	25.0	156	91.5	157
	P1350	0.69	0.65	11.0	20.0	123		(358)
	P0329	0.72	0.65	-1.2	27.0	19	22.0	8600
	P0628	1.24	0.5	-16.8	34.0	145	5.3	750
	P1530	1.37	0.65	1.9	20.0	19.5		(370)
		0.74	0.52	7.24	21.2	81.0	$5.2 \cdot 10^4$	1492

Table 2

R		0,65 kpc			
		n = 24		|Z̄| = 120 pc	
K		1	1/5	1/8	1/12
T_p	N_p	$2\,10^4$	$1\,10^5$	$1{,}6\,10^5$	$2{,}3\,10^5$
T_p	ν_{SN} = 0,033 yr^{-1}	$6\,10^5$	$3\,10^6$	$5\,10^6$	$7\,10^6$
R^l	$\bar{V}_p$ (km/sec) 80	0,264	0,59	0,75	0,915
R^l	$\bar{V}_p$ (km/sec) 100	0,296	0,66	0,84	1,02
R^l	$\bar{V}_p$ (km/sec) 150	0,361	0,81	1,04	1,25

REFERENCES

AMNUEL P.R., GUSEINOV O.H., 1974, Astron. Astrophys, **31**, 35

BRANDT J., STETCHER T., GROWFORD D., MARAN S., 1971, Astrophys. J. **163**, 99

DAVIES J.G., LARGE M., 1970, Mon. Not. R. Astr. Soc. 149, 301

DAVIES J.G., LYNE A.G., SEIRADAKIS J.H., 1972, Nature, **240**, 229

DAWNES D., 1971, Astron. J. **76**, 305

GALT J.G., LYNE A.G., 1972, Mon. Not. R. Astron. Soc. **158**, 281

GUSEINOV O.H., KASUMOV F.K., LAZAREV V.I., OSIPCHUCK A.V.,1973, Astron. Zh. **50**, 39

GUSEINOV O.H., KASUMOV F.K., 1973, Astron. Zh, **50**, 1166

GUSEINOV O.H., KASUMOV F.K., 1973, Astron. Circ. n. 785

HENRY G., PAIK H., 1969, Nature, **224**, 1188

HULSE R.A., RAMATY J., 1975, Astrophys. J. **195**, 151

HULSE R.A., TAYLOR J.H., 1975, Astrophys. J. **191**, L59

KIANG T., 1969, Nature, **223**, 599

LANG R.R., 1971, Astrophys. J. **164**, 249

LOZINSKAYA T., 1975, Pisma Astron. Zh. n2

MANCHESTER R.H., TAYLOR J.H., VAN Y.Y., 1974, Astrophys. J. **189**, L119

MILNE D.K., 1970, Austr.J. Phys. **23**, 425

PRENTICE A.J., 1970, Nature, **225**, 438

REIFENSTEIN R., BRUNDAGE W., STEALIN D., 1969, Astrophys. J., **156**, 825

RAMATY R., BURGINGON G., GRADER R., STEALIN D., 1971, Astrophys. J. **169**, 87

SLEE O.B., ABLES J.G., BATCHELOR R.A., KRISCNA-MAHAN S., 1974, Mon. Not. R.Astr. Soc. **167**, 31

TAMMANN G.A., 1974, Ap. Sp. Sci. **55**, 155

TRIMBLE V, REES M.J., 1971, The Crab Nebula IAU Sump. n46

TZAREVSKY G., 1972, Astrophys, Lett. **10**, 71

TUCKER W.H., 1971, Astrophys. J. **167**, 85

WILLIS G., 1973, Astron. and Astrophys. **26**, 237

WINKLER P., FRANK J., CLARK G., 1974, Astrophys. J. **191**, L67

EWING M.S., BATCHELOR R.A., FRIEFELD R.D., PRICE R.M., 1970, Astrophys. J. **162**, L172

OPTICAL OBSERVATIONS OF SUPERNOVA REMNANTS

M.A. LOZINSKAYA
STERNBERG ASTRONOMICAL INSTITUTE, MOSCOW, U.S.S.R.

ОПТИЧЕСКИЕ НАБЛЮДЕНИЯ ОСТАТКОВ ВСПЫШЕК СВЕРХНОВЫХ

The latest stage of evolution of supernova phenomenon is the stage of filamentary nebula created as a result of interaction of blast wave with interstellar medium. Observations of this stage of supernovae are important not only for the study of remnant formation processes. They also permit to evaluate the exploded star's character.

In 1966-1974 detailed interferometric observations of faint filamentary nebulae identified with supernova remnants were carried out.

A high contrast Fabry-Pérot etalon and contact image converter mounted in cassegrain focus of the 125-cm or 50-cm reflectors were used. The spectral profiles of Hα and [N II] 6584Å lines at different distance from centre of nebulae were investigated. The parameters of Fabry-Perot interferometer were:

- the free spectral range was 700 km/s in radial velocity scale;
- the linear dispersion in the central interferometric ring was 5-7 Å/mm;
- the actual spectral resolution corresponded to 20 km/s;
- the angular resolution was 10" or 25";
- the exposure time was about 1^h - 2^h for the most faint nebulae;
- about 20-100 films for each object were obtained.

Statistical reduction of the relation between radial velocity or line halfwidth and the distance from the centre of the shell was carried out. An expansion of 10 optical supernova remnants was found and the expansion velocities were evaluated.

The latest, partly unpublished results are:

1. *Five extremely faint filaments identified with supernova remnant VRO 42.05.01.* The mean radial velocity of each filament was measured. Two possible values of the expansion velocity of supernova shell were obtained: 15± 5 km/s or 35±5 km/s (the ambiguity is caused by the unknown localization of the filaments on the two sides of the expanding shell).

An average width of the filaments 0.3-0.6 pc was found from Hα - monochromatic films of nebula. An average Hα line halfwidth over separate filaments was found to be 60-70 km/sec.

2. *The faint complex of diffuse filaments - the supernova remnant HB9.* An expansion of the complex of diffuse filaments was found with the velocity 70 ± 10 km/s. An average Hα line halfwidth was measured to be about 60 km/s.

3. *A thin faint filament located at the edge of the symmetrical radio source CTBl.* Only a mean radial velocity was obtained to be - 33 ± 5 km/s relative to the L.S.R. It corresponds to the kinematic distance 2.6±0.5kpc.(The radio distance to the supernova remnant was not estimated exactly: 0.9 kpc according to Milne (1970); 2.9 kpc according to Willis, Dickel (1971); 3.8 kpc according to Downes (1971); 4kpc according to Willis (1973); 4.7 kpc according to Ilovaisky, Lequeux (1972). The corresponding radius of the nebula is 15 pc. The object is located inside a neutral hydrogen cloud and near a H II region.

4. *IC 443; new interferometric measurements.* Mainly the weak central region was observed. These observations were stimulated by recent theoretical investigations of the shock wave propagation in a cloudy interstellar medium. It was shown (Pickelner and Bytchkov, 1975; McKee, Cowie, 1975) that optical filaments represent shocked interstellar clouds; and consequently the velocity of denser and brighter filaments is lower than of fainter filaments. As a result of our latest observations some filaments were found to have velocity of 160-170 km/s (3 times larger than the mean velocity of the whole shell defined previously by the author). The corresponding age of IC 443 is 25000 years (instead of 60000 years previously estimated).

5. *Simeiz 147 (Shajn 147).* This is the most beautiful optical supernova remnant. Some very faint diffuse filaments were found moving from the center with the velocity of about 100 km/sec. Thus our latest observations are in accordance with the results of Silk, Wallerstein, (1973) concerning the high velocity Ca II line in absorption in the direction of Simeiz 147. This high velocity appears not to be the mean expansion velocity of the system of long filaments S 147, but refers to some faint diffuse filaments.

The most delicate singular filaments were found to be as thin as 0.02 pc and as long as ~ 10 pc.

A mean Hα line halfwidth over separate filaments was estimated to be 40-110 km/s.

Old supernova remnants with the well known velocity of expansion are collected in table 1. The present radius of the shell and the ambient density of interstellar medium are also given. The age and the kinetic energy of the ejected envelope are calculated with the assumption of adiabatic expansion.

Figure I shows "reduced" expansion velocity versus age of supernova. The "reduced" expansion velocity $V(\frac{n_o\,cm^{-3}}{1cm^{-3}})^{1/5}$ km/s is introduced to take into account the difference in the density of ambient interstellar gas. The line corresponds to adiabatic evolution of a supernova remnant with the initial kinetic energy $E_o = 10^{50}$ erg.

What results can one derive from the obtained data?

1. The filamentary nebulae resulting from supernova explosions are objects of uniform class expanding into the medium of different densities.

2. Most objects were created after the explosions with the kinetic energy of about 10^{50} erg. Shajn 59 and Shajn 22 differ from the main group due to lower energy of the explosion. (The kinetic energy of Cass A explosion exceeds 10^{50} erg.).

3. Shklovsky (1975) from the analysis of soft X-ray emission of Cyg Loop and Vela X has shown that both remnants are from supernova explosions of type I with the initial kinetic energy 10^{50} erg. Our results show that this value is typical of the majority of the supernova remnants studied and consequently they all are of the Type I supernovae.

Table 1

supernova remnant	V [km/s]	Observer	r [pc]	n_o [cm^{-3}]	$t \cdot 10^3$ [year]	E_o [erg]
Cyg Loop	116	Minkovsky, 1954				
		Doroshenko, 1970	20	01. - 1.0	70	$0.3 - 3 \cdot 10^{50}$
IC 443	150	Lozinskaya, 1975[b]	10	10 - 20	25	
Monoceros	45	Lozinskaya, 1971	20	4 - 15	200	$0.5 - 1 \cdot 10^{50}$
HB 21	20	Lozinskaya, 1972	13	3 - 5	260	$2 \cdot 10^{49}$
W 28	45	Lozinskaya, 1973	6.5	~100	60	$3 \cdot 10^{50}$
VRO 42.05.01	35	Lozinskaya, 1975a	28	~1	400	10^{50}
HB 9	70	Lozinskaya, 1975a	21	~1	120	10^{50}
Shajn 22	35	Lozinskaya, 1969	4	~1	46	$5 \cdot 10^{47}$
Shajn 59	12	Lozinskaya, Esipov, 1973	3	~10	100	10^{47}
Vela	400 - 500	Wallerstein, Silk, 1971	12-15	0.2 - 4	22	$1 - 10 \cdot 10^{50}$
Puppis	85	from Baade, Minkovsky, 1954	11	~1	50	$3 \cdot 10^{49}$
Shajn 147	90	Silk, Wallerstein, 1973	26	~1	120	$4 \cdot 10^{49}$
	100	Lozinskaya, (1975)				

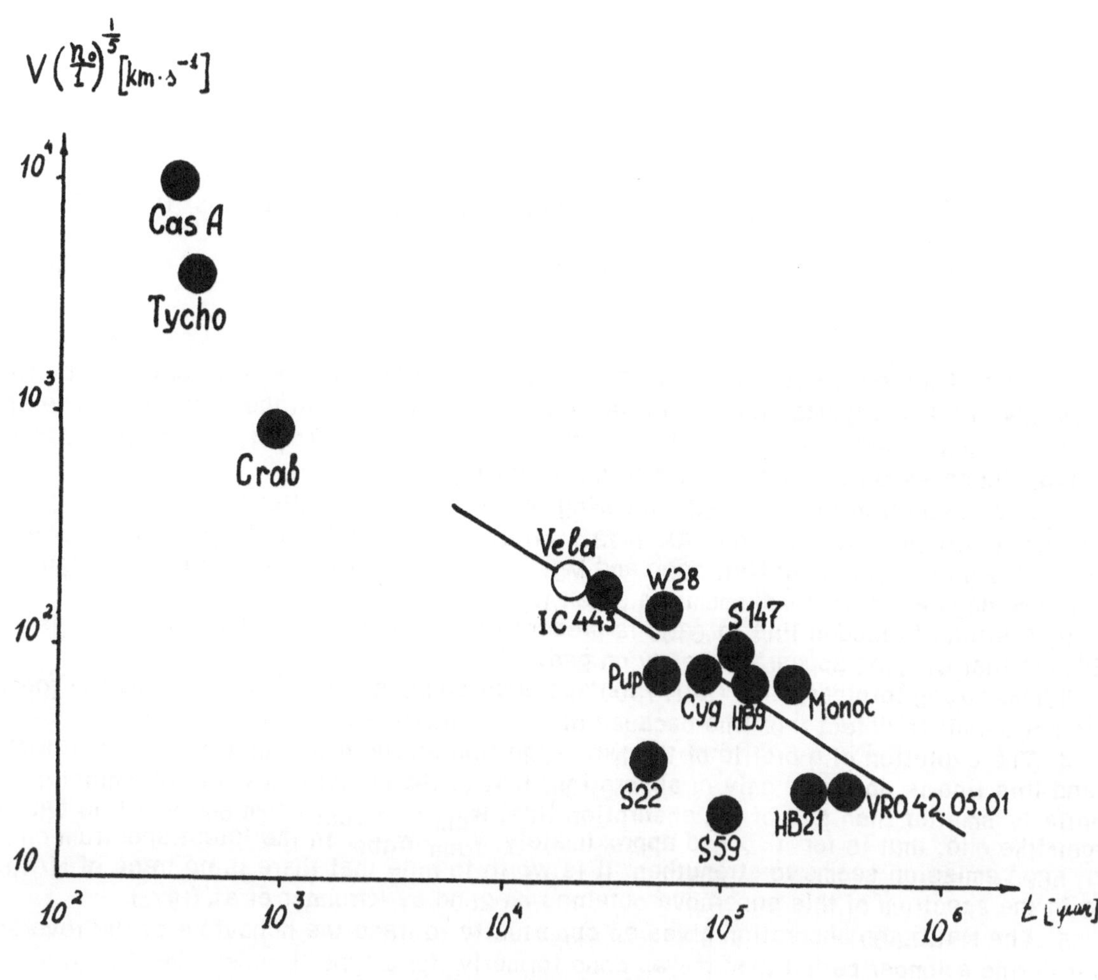

Fig. 1.

REFERENCES

BYCHKOV K.V., PIKELNER S.B., 1975, Sov. Astron. J., Letters, 1, 29
DOROSHENKO V.T., 1970, Sov. Astron. J., 47, 292
DOWNES D., 1971, Astron. J., 76, 305
ILOVAISKY S.A., LEQUEUX J, 1970, Astron. and Astroph., 18, 169
LOZINSKAYA T.A., 1969, Sov. Astron. J., 46, 730
LOZINSKAYA T.A., 1971, Sov. Astron, J., 48, 1145
LOZINSKAYA T.A., 1972, Sov. Astron. J., 49, 265
LOZINSKAYA T.A., 1973, Sov. Astron. J., 50, 950
LOZINSKAYA T.A., 1975a, Sov. Astron. J., 52, 39
LOZINSKAYA T.A., 1975b, Sov. Astron. J., Letters, 1 № 2, 25, 1975
LOZINSKAYA T.A., ESIPOV V.F., 1973, Sov. Astron. J., 50, 702
MCKEE C.F., COWIE L., 1975, Astroph. J., 195, 715
MILNE D.K., 1970, Austr. J. Phys., 23, 425
MINKOWSKI R., 1958, Rev. Mod. Phys., 30, 1048
SILK J., WALLERSTEIN G., 1973, Astroph. J., 181, 799
SHKLOVSKY I.S., 1974, Sov. Astron. J., 51, 3
WILLIS A.G., 1973, Astron. and Astroph., 26, 237
WILLIS A.G., DICKEL J.R., 1971, Astroph., Letters, 8, 203
WALLERSTEIN G., SILK J., 1971, Astroph. J., 170, 289

PRELIMINARY RESULTS OF THE STUDY OF THE SPECTRA OF SN 1970 g.

N.N.CHUGAJ
ASTRONOMICAL COUNCIL, ACADEMY OF SCIENCES, U.S.S.R.

ПРЕДВАРИТЕЛЬНЫЕ РЕЗУЛЬТАТЫ ИССЛЕДОВАНИЯ СПЕКТРОВ SN 1970 g

Spectra of a type II supernova SN1970-g (M101) discussed below were obtained by Drs. V.I.Pronik and K.K.Chuvaev on 2.6-m telescope with image-tube in the Crimea (These spectra will be published elsewhere). The spectra cover the period from 4/8/70 to 22/4/71 that is 262d. These spectra permit to come to the following conclusions.

1. The absorption lines strengthen during the first 3 months after light maximum (which occurredon July 25 (±3), Barbon et al. 1973). Afterwards absorptions decrease but slowly. Rather strong absorptions of NaI, 5890 and MgI, 5180 are seen in the spectrum on 238^d (here and below days after light maximum are given).

2. A strong forbidden line 0I,6300 is present in the spectrum on 189^d. But there is some evidence that 0I,6300 appeared already on 96^d.

3. The strong forbidden line CaII, 7300 appeared on 238^d. The previous spectrum (on 189^d does not permit to detect this line because of an underexposure in the red region.

4. The evolution of a profile of the NaI, 5890 line shows some interesting peculiarities. Up to 40^d this line is observed only in absorption, that is the equivalent width of emission is substantially smaller than that of the absorption line, $W_{em} \ll W_{abs}$. From 65^d NaI-line became P Cygni-like one, that is for $t > 65^d$ approximately $W_{em} \sim W_{abs}$. In the latest spectrum on 270^d NaI, 5890 emission seems to strengthen. It is worth to note that there is no trace of NAI,5890 line in the spectrum of this supernova obtained on 260^d by Kirshner et al. (1973).

5. The NaI, 5890 absorption gives an opportunity to trace the behaviour of the reverse layer during a longer period than it was done formerly for a type II supernova. The radial velocity of the absorption minimum of NaI line decreases from 7500 km/s to 3500 km/s during half a year (or some less) after light maximum. On 270^d radial velocity of NaI absorption is of the order of 4500 km/s. The increase of the radial velocity may be a result of a mentioned strengthening of the emission in this line.

REFERENCES

BARBON R., CIATTI F. and ROSINO L., 1973, Astron. Astrophys., **29**, 57

KIRSHNER R.P., OKE J.B., PENSTON M.V. and SEARLE L., 1973, Astrophys. J. **185**, 303

PRESUPERNOVAE I - COMPONENTS OF BINARIES

E.V.ERGMA, A.V.TUTUKOV
ASTRONOMICAL COUNCIL, ACADEMY OF SCIENCES, U.S.S.R.
PRESENTED BY A.V.TUTUKOV

ПРЕДСВЕРХНОВАЯ ПЕРВОГО ТИПА - КОМПОНЕНТ ДВОЙНОЙ

In the frame of the current theory of evolution of single stars it is impossible to produce the explosions of supernovae in elliptical galaxies, as the mass of their stars does not exceed one solar mass. According to Whelan and Iben (1973) it is possible to explain the explosions of supernovae in the frame of the theory of evolution of binaries. They suggested that the progenitor of type I Supernova is a binary consisting of carbon-oxygen white dwarf accreting the matter lost by its component - red giant with mass about one solar mass.

The evolution of the degenerate carbon-oxygen core with accretion has not yet been studied numerically, but it is clear that if the core mass reaches the Chandrasekhar limit carbon detonation may occur. From the observational point of view there are two kinds of appropriate binaries: nova and novalike stars and symbiotic stars.

We have studied the evolution of central values of temperature and density of carbon-oxygen core by the method proposed by Barkat et al. (1972) and Paczyński (1973). The rate of growth of core mass is taken as constant value and varied from one variant to another from 10^{-9} $M_\odot$/year to 10^{-6} $M_\odot$/year. The core is heated by contraction due to accretion and growth of its mass. The main reason for decreasing core temperature is energy losses due to UFI neutrino. The results of computations show that decreasing of the rate of accretion from $10^{-6} M_\odot$/year to 10^{-9} $M_\odot$/year increases the central density at the moment of the explosive burning of carbon from 3.10^9 g/cm^3 to $\sim 10^{10}$ g/cm^3. High values for central densities is a consequence of decreasing the rate of heating of the core owing to lowering the rate of accretion.

Ivanova et al. (1975) have shown that a neutron star - pulsar may be formed by explosion of the carbon-oxygen stellar core only if the central density exceeds $\sim 6.10^9$ g/cm^3. Thus, it seems rather probable to produce a gravitationally bound body - a neutron star in the process of the explosive burning of carbon of a carbon-oxygen star - component of a binary.

Full paper will be published in "Acta Astronomica"

REFERENCES

BARKAT Z., WHEELER J.C. and BUCHLER J.R., 1972, Astroph. J., 171, 651
IVANOVA L.N., IMSHENNIK V.S. and CHECHOTKIN V.M., 1974, Astroph. Sp. Sc. 31, 477
PACZYŃSKI B., 1973, Acta Astr., 23, 1
WHELAN J., IBEN I.Jr., 1973, Astroph. J., 186, 1007

A SURVEY OF MASS-LOSS EFFECTS IN EARLY-TYPE STARS

THEODORE P. SNOW, JR.
PRINCETON UNIVERSITY OBSERVATORY, U.S.A

ОБЗОР ЭФФЕКТОВ ПОТЕРИ МАСС ЗВЕЗДАМИ РАННИХ ТИПОВ

I. Introduction. Since the discovery by Morton (1967a) that ultraviolet spectra of hot supergiant stars show evidence of mass-loss in strong lines, several observations of these effects have been made by rocket-borne spectrograph (Morton 1967b, 1969, Carruthers 1968; Stecher 1970; Smith 1970, 1972; Morton, Jenkins, and Brooks 1969; Morton *et al.* 1972); and more recently by ultraviolet satellite (Underhill 1974; McCluskey, Kondo, and Morton 1975; Lamers 1975; Stalio and Selvelli 1975; Morton 1967). In general these earlier studies have been concerned with at most a few stars, usually selected for their location in the sky or for their particularly spectacular effects, and data have not been available for a large enough collection of objects to allow a systematic survey to be made. Very recently Henize *et al.* (1975) have used low-dispersion objective-prism plates obtained by the NASA-Skylab missions to study P-Cygni profiles in CIV and SiIV for a large number of stars. In the present study, higher-resolution data obtained with the satellite *Copernicus* (described by Rogerson, Spitzer et al. 1973) are used to define the region in the Hertzsprung - Russell diagram where mass loss occurs. The resolution obtained with *Copernicus* is sufficient not only for detection of the P-Cygni profiles characteristic of large mass-loss rates, but also in many cases allows the detection and analysis of the asymmetric absorption lines which are symptomatic of lower rates of mass ejection

II. The data. a. Properties of the stars. 40 stars have been included in this survey providing as wide a coverage of the H-R diagram as practical. Supergiants are well-covered from O4 to A2, as is the main sequence from O4 to B7, while the gaps exist among the giants in the late O stars and intermediate- to late-B stars.

In figure 1 is shown an H-R diagram containing all the stars used in the survey. The effective temperatures were generally taken from Conti (1973) and Conti and Burnichon (1975) for the O-stars and from the scale of Code *et al.* (1975) for the B-stars. The absolute bolometric magnitudes M_{BOL} for the O-stars are taken from Conti and Burnichon (1975) and for the B-stars from the bolometric corrections of Code *et al.* (1975) with the absolute visual magnitudes of Lesh (1968, 1972). The zero-age main sequence shown in figure 1 was taken from Stothers'(1972) models above $M_{BOL} = -6.50$ and for the fainter stars from Blaauw's (1963) data on young clusters along with the T_{eff} and bolometric correction data of Code *et al.* (1975).

The spectral types shown in labels for other figures are those of Walborn (1972, 1973)(excluding the parenthetical *n* and *f* notation) for the O-stars, while Lesh's (1968, 1972) types were adopted for all the B-stars.

b. Spectral features. The transitions studied for possible mass-loss effects include primarily resonance lines of ions of abundant elements, although some lines arising from excited states such as CIII λ1175.7, SiIV λλ1122.5 and 1128.3 and SIV λ1073.0 are also included. A detailed compilation of spectroscopic data can be found in Snow and Morton (1976) or in Morton (1976)for the principal lines considered in this survey, which are CIII (λλ977, 1175), NIII (λ990), NV (λλ1238, 1242), OVI (λλ1031, 1037), SiIII (λ1206), SiIV (λλ1393, 1402 and λλ1122, 1128), PV (λλ1117, 1128), and SIV (λλ1062, 1073). In addition, information was available for some stars on CIV (λλ1548, 1550), CIII (λ2296), NIV (λ1817), and HeII (λ1640). For the fainter stars, only the region between roughly 1100 and 1300 Å had a strong enough signal to be useful; therefore the data are most complete for the CIII λ1175 and NV λ1238, 1242 features.

c. Ultraviolet scan data. Scans at intermediate resolution (nominally 0.2 Å) covering the region from roughly 1000 Å to 1450 Å have been obtained for a large number of O- and B-stars, over 40 of which were used in the present survey. The spectrometer and the procedures for correcting the data for backgrounds due to charged particles in the upper atmosphere and for stray and scattered light in the spectrometer are described by Rogerson, Spitzer *et al.* (1973); Snow *et al.*(1976); and Bohlin (1975). The overall photometric accuracy of the corrected data is estimated to be ±10% of the continuum level, the largest uncertainty being attributed to fluctuations in the spacecraft guidance which cause slight changes in the position of the stellar image on the spectrometer entrance slit, thereby producing variations in the intensity of light reaching the photocathode.

III. Results: The observed profiles. P Cygni profiles were seen for a wide variety of lines most commonly CIII (λ1175) and NV (λ1240), but also SiIV (λ1400) in several cases, and Si IV ($\lambda\lambda$1122, 1128), C IV (λ1550), S IV (λ1073), and OVI ($\lambda\lambda$1031, 1037) in isolated instances. For the most part, these profiles consisting of longward-displaced emission accompanied by absorption which is shifted to shorter wavelengths, occur in stars of luminosity classes I, II, and III, although NV (λ1240) shows this effect in the dwarf stars 9 Sgr (O4:V), 15 Mon (O7V) and ζ Oph (O9.5V).

Many instances were found of lines showing shifts or extended short-wavelength wings indicating outflowing material, presumably at rates insufficient to produce the red-shifted emission peaks characteristic of P Cygni profiles.

Figures 2, 3, and 4 show a variety of profiles for NV (figure 2), CIII λ1175 (figure 3), and Si IV λ1400 (figure 4). The CIII features were taken from a series of supergiants, and are arranged in order of decreasing temperature. The NV and Si IV features which are shown also are arranged by decreasing temperature, but represent a variety of luminosity classes. In these figures are shown several P Cygni profiles and some cases of line asymmetries (e.g. NV in τ Sco, B0 V; and Si IV in β Cen, B1 III) or shifts (CIII in β Ori A, B8 Ia).

A large degree of variety is evident among the observed profiles, even within small ranges of temperature or luminosity. In some cases widely-spaced doublets such as NV λ1240 or Si IV λ1400 are smeared into a single broad absorption feature, occasionally with narrower lines superimposed such as the NV profile in 15 Mon; but often with no clearly-defined individual components, e.g. the NV profile for ζ Pup.

In many instances the absorption feature shows a sharp edge at the extreme short wavelength side, probably indicative of a real cut-off in the velocity distribution of the outflowing material. In general this terminal velocity exceeds or at least matches the escape velocity of the star, indicating that substantial amounts of material are being lost.

IV. Discussion: distribution in the H-R Diagram. The *Copernicus* data indicate that mass-loss effects occur over a wide portion of the H-R diagram, as shown in figure 5. Among the O-stars, the evidence for mass flow for objects on the main sequence usually consists only of shifts and asymmetries, ζ Oph (O9.5 V) being the only late O-dwarf to show emission components. O-giants and supergiants generally display at least some P-Cygni profiles.

Among the B-stars most mass-loss effects which are seen consist of asymmetries or shifts. On the main sequence, θ Car (B0.5 Vp) is the latest object to show these effects, although in another study of Be stars (Snow and Marlborough 1975) it is seen that emission-line dwarfs as late as B1.5 and possibly even cooler can undergo mass ejection if the rotational velocity is large. No data were available on giants in classes B2 or later, so it is not known to what extent mass-loss effects persist through the B-giants. β Cen (B1 III) has shifts and asymmetries in a few lines. The supergiants show shifts throughout class B to β Ori A (B8 Ia), and indeed the A2Ia supergiant α Cyg shows displaced MgII lines (Lamers 1975).

It appears from figure 5 that mass ejection generally occurs above M_{BOL}= -6.0, although one exception to this is δ Sco (B0.5 IV), implying that the presence of mass flow is not dependent entirely on position in the H-R diagram. Another parameter which might be important for stars near M_{BOL}= - 6.0 could be rotational velocity, as suggested by Snow and Marlborough (1975) for Be stars. Henize *et al.* (1975), using much lower resolution Skylab data, found that marked P Cygni profiles for Si IV λ1400 and C IV λ1550 occur only for M_{BOL} = -8.4 or brighter, whereas at the 0.2 Å resolution of the *Copernicus* data, lines with emission components are seen in stars as faint as M_{BOL}= - 6.7. In the future, higher-resolution (0.05 Å) *Copernicus* data will be obtained to search for line asymmetries in stars fainter than M_{BOL}=-6.0, to see whether very low-volume ejection of material may occur in this region, extending the trend shown in figure 5.

Figures 6 and 7 show the distribution in the H-R diagram of mass-loss profiles in the CIII λ1175 and NV λ1240 features, respectively. For CIII it is seen that emission is generally present in the upper left-hand portion of the diagram, while displaced or asymmetric absorption occurs for a number of stars below this region but brighter than M_{BOL} = -6.0. For NV, on the other hand, there are relatively few cases where displaced or asymmetric absorption is seen without being accompanied by emission (figure 7). The presence of P-Cygni profiles for NV extends to cooler and less luminous objects than for CIII λ1175, possibly because CIII λ1175 arises from a metastable excited level, whereas NV λ1240 is a resonance doublet.

It appears from figure 7 that when NV is present, the mass-ejection rate is usually sufficiently large to produce a P Cygni profile. A similar trend appears when the velocity of the principal absorption is considered; i.e. for CIII λ1175, there is smooth progression of increasing velocities going from lower right to upper left in the H-R diagram, whereas in most cases when NV absorption is seen to be shifted, the velocity is large. This may imply that the presence of substantial quantities of NV above the photosphere helps to drive the mass loss through absorption of radiation in the resonance doublet, in analogy to the role of OVI suggested by Rogerson and Lamers (1975) for τ Sco. Presumably the CIII λ1175 line does not behave the same way because it arises from an excited level which is not highly populated.

Rates of mass loss have not yet been determined for most stars in this survey. Morton (1967b) found from crude estimates based on rocket spectra that δ, ϵ and ζ Ori are ejecting about 10^{-6} $M_\odot$ yr^{-1}, and McCluskey, Kondo, and Morton (1975) found 3×10^{-6} $M_\odot$ yr^{-1} for 29 CMa, based on *Copernicus* data. More refined model-fitting procedures were used by Lamers (1975) for α Cyg (A2 Ia), yielding $dM/dt \geq 3 \times 10^{-10}$ $M_\odot$ yr^{-1}, and by Lamers and Morton (1976) for ζ Pup (O4 I f), yielding a rate of mass loss at least of 10^{-6} $M_\odot yr^{-1}$.

Rogerson and Lamers (1975) have determined that τ Sco (B0 V) is ejecting matter at the rate of $4 \times 10^{-9} M_\odot yr^{-1}$; similar or somewhat smaller rates must prevail in general for the cases where asymmetries without emission are seen.

This research has been supported by National Aeronautics and Space Administration contract NAS5-1810 with Princeton University.

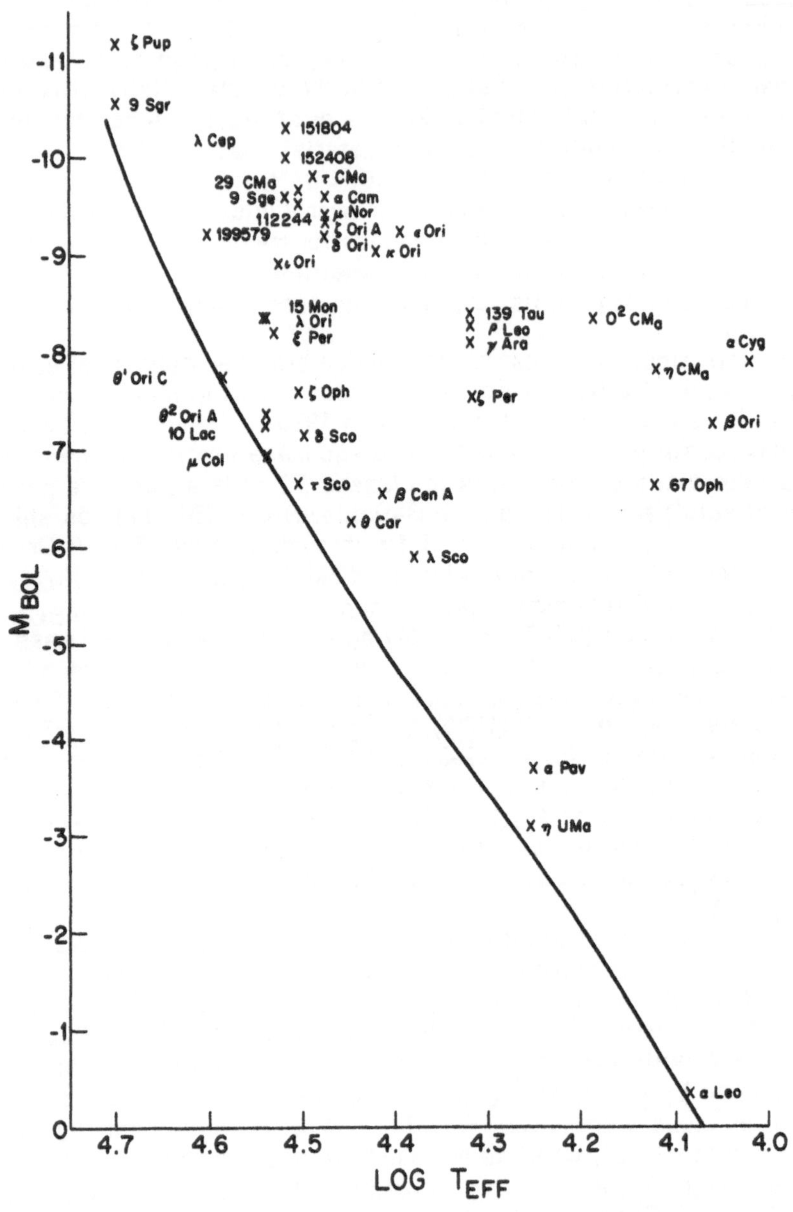

Fig. 1.

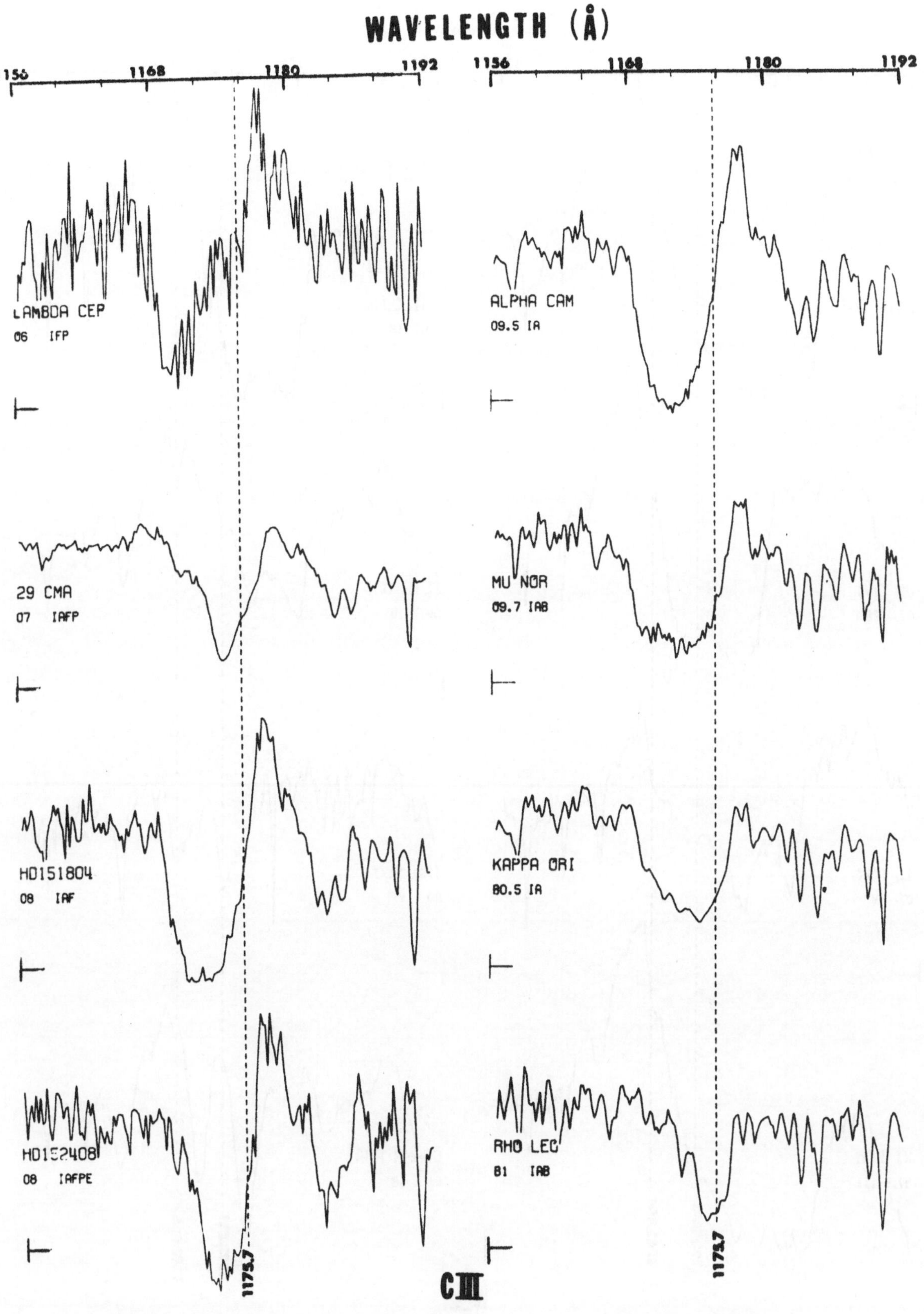

WAVELENGTH (Å)
1156
1168
1180
1192
1156
1168
1180
1192
LAMBDA CEP
06 IFP
ALPHA CAM
09.5 IA
29 CMA
07 IAFP
MU NOR
09.7 IAB
HD151804
08 IAF
KAPPA ORI
B0.5 IA
HD152408
08 IAFPE
RHO LEO
B1 IAB
1173.7
1173.7
CIII

Fig. 2.

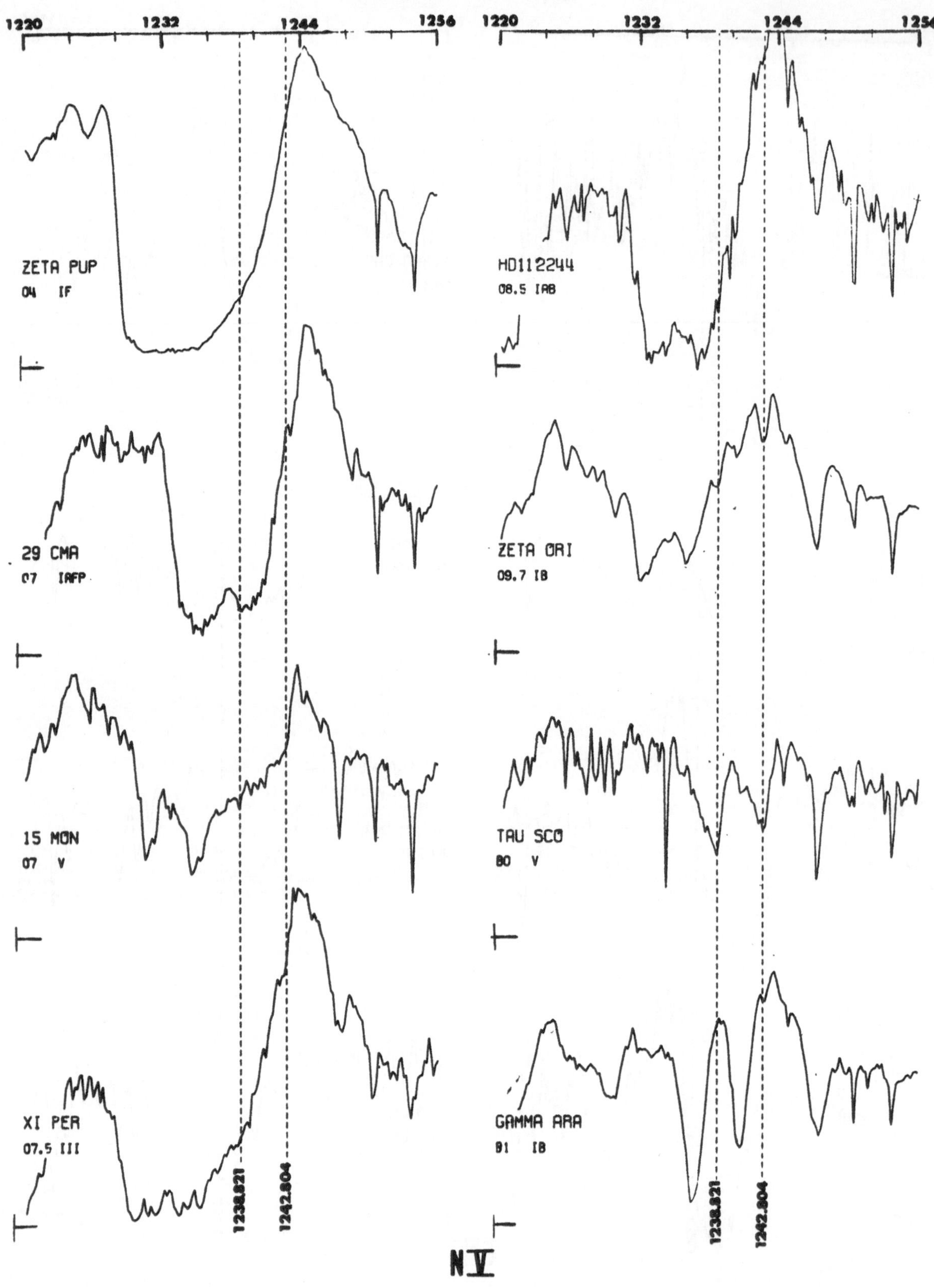

WAVELENGTH (Å)
1220
1232
1244
1256
ZETA PUP
O4 IF
29 CMA
O7 IAFP
15 MON
O7 V
XI PER
O7.5 III
HD112244
O8.5 IAB
ZETA ORI
O9.7 IB
TAU SCO
B0 V
GAMMA ARA
B1 IB
1238.821
1242.804
N V

Fig. 3.

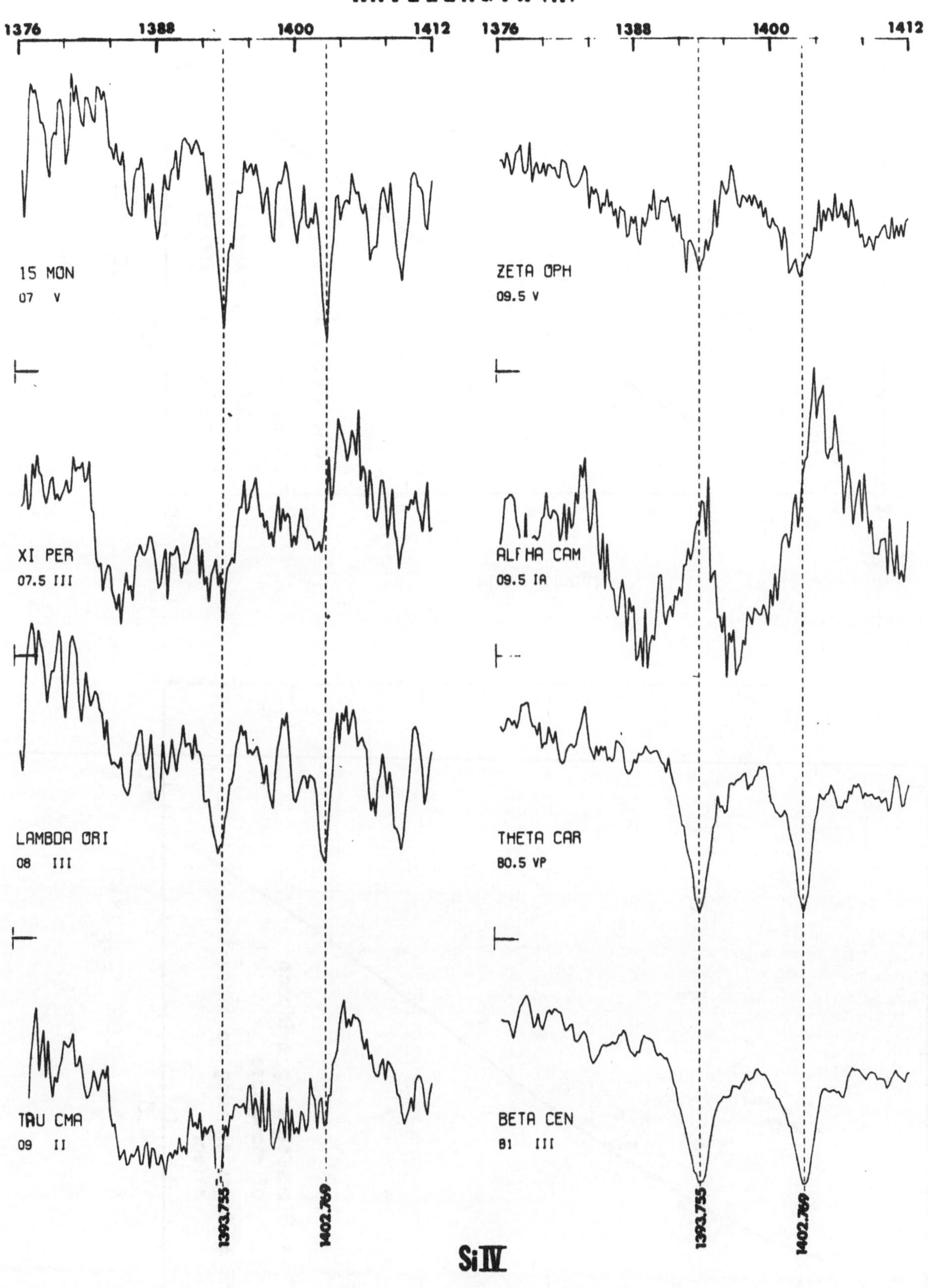
WAVELENGTH(Å)
1376
1388
1400
1412
1376
1388
1400
1412
15 MON
07 V
ZETA OPH
09.5 V
XI PER
07.5 III
ALFHA CAM
09.5 IA
LAMBDA ORI
08 III
THETA CAR
B0.5 VP
TAU CMA
09 II
BETA CEN
B1 III
1393.755
1402.769
1393.755
1402.769
SiIV

Fig. 4.

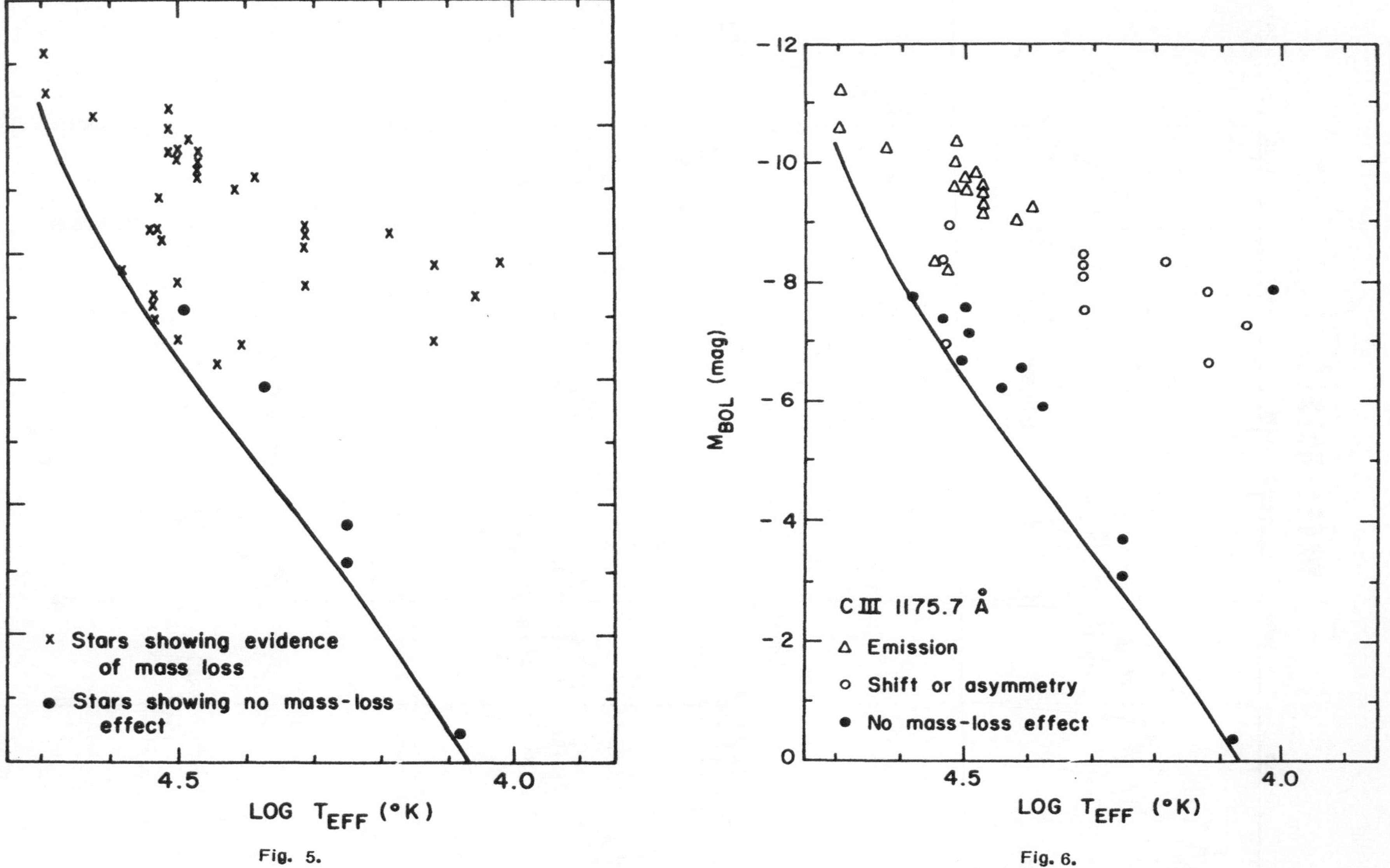
-12
-10
-8
-6
-4
-2
0
4.5
4.0
LOG T_{EFF} (°K)
Stars showing evidence of mass loss
Stars showing no mass-loss effect

Fig. 5.

M_{BOL} (mag)
-12
-10
-8
-6
-4
-2
0
4.5
4.0
LOG T_{EFF} (°K)
C III 1175.7 Å
Emission
Shift or asymmetry
No mass-loss effect

Fig. 6.

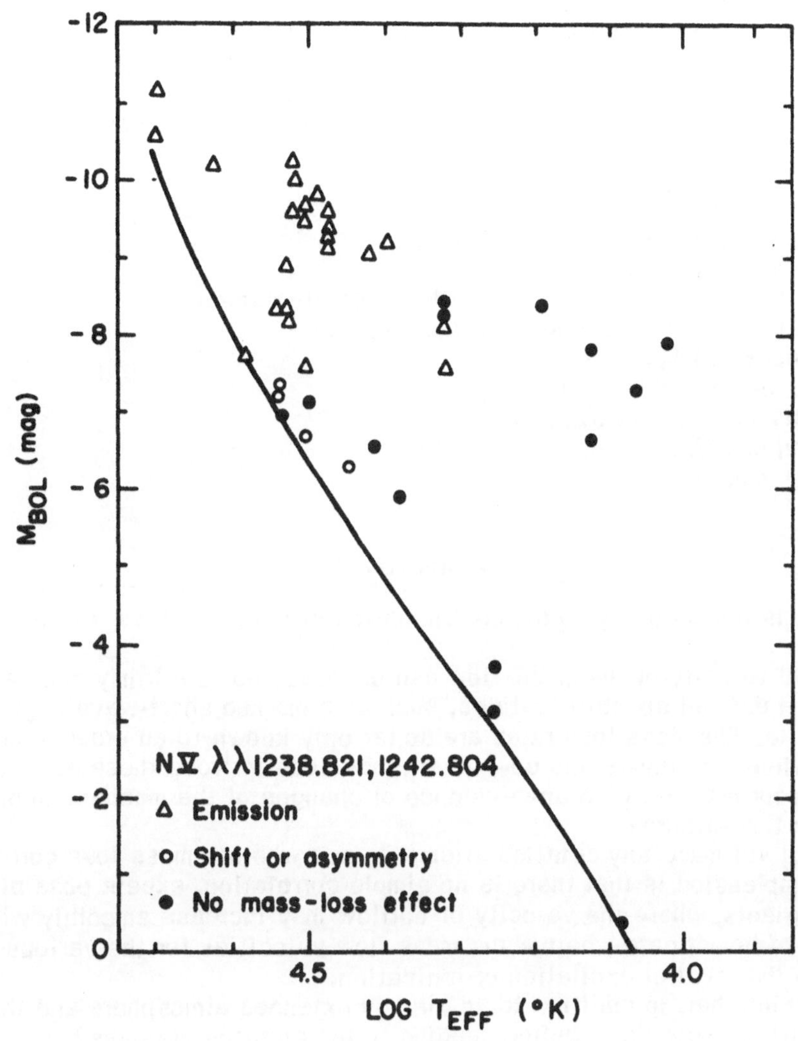

Fig. 7.

REFERENCES

BLAAUW A. 1963, Basic Astronomical Data, ed. K.A. Strand, p.383

BOHLIN R.C. 1975, Ap. J. (In press)

CARRUTHERS G.R. 1968, Ap. J. 151, 269

CODE A.D., DAVIS J., BLESS R.C. AND HANBURY BROWN R. 1975. (Preprint)

CONTI P.S. 1973, Ap.J. 179, 181

CONTI P.S. AND BURNICHON M.L. 1975, Astr. and Ap. (In press)

HENIZE K.G., WRAY J.D., PARSONS S.B., BENEDICT G.F., BRUHWEILER F.W., RYBSKI P.M. AND O'CALLAGHAN F.G., 1975, Ap. J. (In press)

LAMERS H.J.G.L.M. 1975, Phil. Trans. Roy. Soc. London A. (In press)

LAMERS H.J.G.L.M. AND MORTON D.C. 1976. (In preparation)

LESH J.R. 1968, Ap. J. (Suppl.). 17, 371

LESH J.R. 1972, Astr, and Ap. (Suppl.). 5, 129.

McCLUSKEY G.F., KONDO Y. AND MORTON D.C. 1975, Ap. J. (In press)

MORTON D.C. 1967a, Ap. J. 147, 1017

MORTON D.C. 1967b, Ap.J. 150, 535

MORTON D.C. 1969, Ap.J. 158, 629

MORTON D.C. 1976, Ap. J. (In press)

MORTON D.C., JENKINS E.B. AND BROOKS N.H. 1969, Ap. J. 155, 875
MORTON D.C., JENKINS E.B., MATILSKY T.A. AND YORK D.G. 1972, Ap. J. 177, 219
ROGERSON J.B. AND LAMERS H.J.G.L.M. 1975, Nature. (In press)
ROGERSON J.B., SPITZER L., DRAKE J.F., DRESSLER K., JENKINS E.B., MORTON D.C. AND YORK D.G. 1973, Ap. J. (Letters), 181, L97
SMITH A.M. 1970, Ap. J. 160, 595
SMITH A.M. 1972, Ap. J. 172, 129
SNOW T.P. AND MARLBOROUGH J.M. 1976. (Preprint)
SNOW T.P. AND MORTON D.C. 1967. (In preparation)
SNOW T.P., YORK D.G., WELTY D. AND HORNACK P. 1976. (In preparation)
STALIO R. AND SELVELLI P.L. 1975, Astr. and Ap. (In press)
STECHER T.P. 1970, Ap. J. 159, 543
STOTHERS R. 1972, Ap.J. 175, 431
UNDERHILL A.B. 1974, Ap. J. (Suppl.) 27, 359
WALBORN N.R. 1972, A.J. 77, 312
WALBORN N.R. 1973, A.J. 78, 1067

DISCUSSION

MASSEVICH : What is the accuracy in the determination of the mass loss rate for hot supergiants?
SNOW : The outward velocity of the mass flow can be measured to a fairly good accuracy when there are well defined spectral features, such as a marked short-wavelength edge of an absorption profile. The mass loss rates are so far only known to an order of magnitude, but model calculations by Lamers and co-workers may soon improve these estimates.
CHENTSOV : Does your material give any evidence of changes of the mechanism of atmospheric expansion with temperature?
SNOW : We have not yet made any detailed attempt to see whether mass loss correlates with temperature. Our impression is that there is no simple correlation, except possibly among the O and B supergiants, where the velocity of outflow may increase smoothly with the temperature. We hope to investigate whether the mass flow velocities for the various ions in each star correlate with the level of excitation or ionization.
MUSTEL : Do you think that, in addition to an already extended atmosphere and the light pressure, there may still be some other factor, leading to the ejection of gases?
SNOW : Yes, recent work indicates that if material exists above the photosphere which is hot enough to contain abundant ions like O VI and NV, absorption of photons in the resonance transitions of these ions can drive the outward high-velocity mass flow. The heating of such layers may be mechanical of some kind.
VALTIER: Have you studied any β CMa stars?
SNOW : Yes, β Cen and α Vir, both of which have some asymmetric profiles indicative of mass loss. In this context, β CMA stars do not seem to differ noticeably from other stars nearby in the HR diagram.

THE VISIBLE AND U.V. SPECTRUM OF β ORI

L.CRIVELLARI, P. SELVELLI
ASTRONOMICAL OBSERVATORY, TRIESTE; ASTRONOMICAL OBSERVATORY, PADOVA, ITALY

ВИДИМЫЙ И УФ СПЕКТР β ОРИОНА

(The authors have not sent the text of their paper).

OBSERVATIONAL DATA INDICATING EXISTENCE OF MAGNETIC FIELD IN T TAU-TYPE STARS

P.PETROV and A.SCHERBAKOV
CRIMEAN ASTROPHYSICAL OBSERVATORY, U.S.S.R.
PRESENTED BY P.PETROV

НАБЛЮДАТЕЛЬНЫЕ ДАННЫЕ, УКАЗЫВАЮЩИЕ НА СУЩЕСТВОВАНИЕ МАГНИТНОГО ПОЛЯ ЗВЕЗД ТИПА Т ТЕЛЬЦА

For the last thirty years T Tau-type stars are being regarded as the objects that can give us a solution of the stellar evolution problem or at least the key for this solution. However up to date the numerous and partly contradictory data prevent to construct the physical model of these stars.

We suppose that hypothesis of a strong magnetic field in T Tau-type stars can explain a set of not understood observational data.

1. a. One of the brightest and well studied T Tau-type stars, RW Aur, shows increase of the emission spectrum at *minimum* brightness, whereas the absorption features become stronger at maximum brightness (Salmanov, 1972).

b. The same *reversed correlation* between changes of brightness and emission features show T Tau-type stars with intensive emission spectra and low bolometric luminosities, whereas T Tau-type stars of higher luminosities with weak emissions show the *right correlation* (Petrov, 1975).

c. *Relative intensities* of the infrared Ca II emission lines cannot be explained by emission of either optically thin or optically thick homogeneous gaseous shell (Shanin et al., 1975).

d. Increase in the *polarization* of visual light of T Tau itself is accompanied by increase in the infrared *Ca II emissions,* when the brightness being nearly constant (Shanin et al , 1975).

2. a. It is possible to explain all of these phenomena in the framework of one idea, based on the analogy with the solar *surface heterogeneity:* the temperature and the brightness of the solar photosphere in active region are *reduced* due to presence of the sunspots, whereas the chromospheric emission spectrum is *intensified* in comparison with the quiet region ("reversed correlation")

b. *Relative intensities* of the calcium lines are produced by *superposition of the* radiation from the active and the quiet regions, i.e. they cannot be represented by any homogeneous model at all.

c. The cause of appearance of the solar active regions is the emergence of *magnetic field* on the solar surface. We suppose that in a T Tau-star (Fig.1) the Ca II lines are emitted within the active regions (plages), the polarization occurs in the dust shell. The increase in the magnetic field *on the line of sight* leads to intensification of the calcium lines at the chromospheric level and to the increase of the polarization due to more effective orientation of the dust particles within the dust shell.

3. a. Our idea on the T Tau-type stars can be illustrated by Table 1. The causal relationship between the solar characteristics are shown in the left part of Table 1. Magnetic fields and active regions are not observed directly in T Tau-stars, but the causal relationship shown in the right part of Table I indirectly makes up deficiency of observable data on magnetic fields and active regions of T Tau-stars.

b. In the framework of our idea we can understand the difference between photometric behaviour of T Tau-stars of different luminosities. The stars of low luminosities, i.e. of low masses, must have deeper convective zones (Hayashi, 1966) and, consequently, larger sizes of the active regions (Mullan, 1974). This leads to greater range of variability of these stars in comparison with the stars of higher luminosities. The best examples are RW Aur (4L) and T Tau itself (26L).

c. On the other hand, this scheme can explain why the T Tau-stars of low luminosities show stronger pronounced chromospheric activity, whereas right correlation between brightness and emission features is inherent in the T Tau-stars of higher luminosities.

4. The conclusion on existence of strong magnetic field in T-Tau-stars can elucidate the nature of the *FU Ori-phenomenon.*

Let us make a crazy assumption that magnetic field of a pre-main-sequence star can effectively depress the convection throughout the whole star. At the Hayashi track the rate of gravitational contraction depends on the effectivity of the convective transfer of energy. Therefore, taking into account our assumption, magnetic field must reduce the rate of evolution along the Hayashi track, i.e. it reduces the luminosity of the star. So, on the theoretical diagram (Fig. 2) the star with strong magnetic field should occupy the region somewhere down from its "normal" convective track, in the *region of T Tau-stars.*

As far as evolution proceeds, the convective zone eventually becomes so thin that the magnetic field can no longer hold the energy in the star. The star must undergo the transition from the region of T Tau-stars to its "normal" *radiative* track. This transition we can identify with the FU Ori-phenomenon.

Table 1.

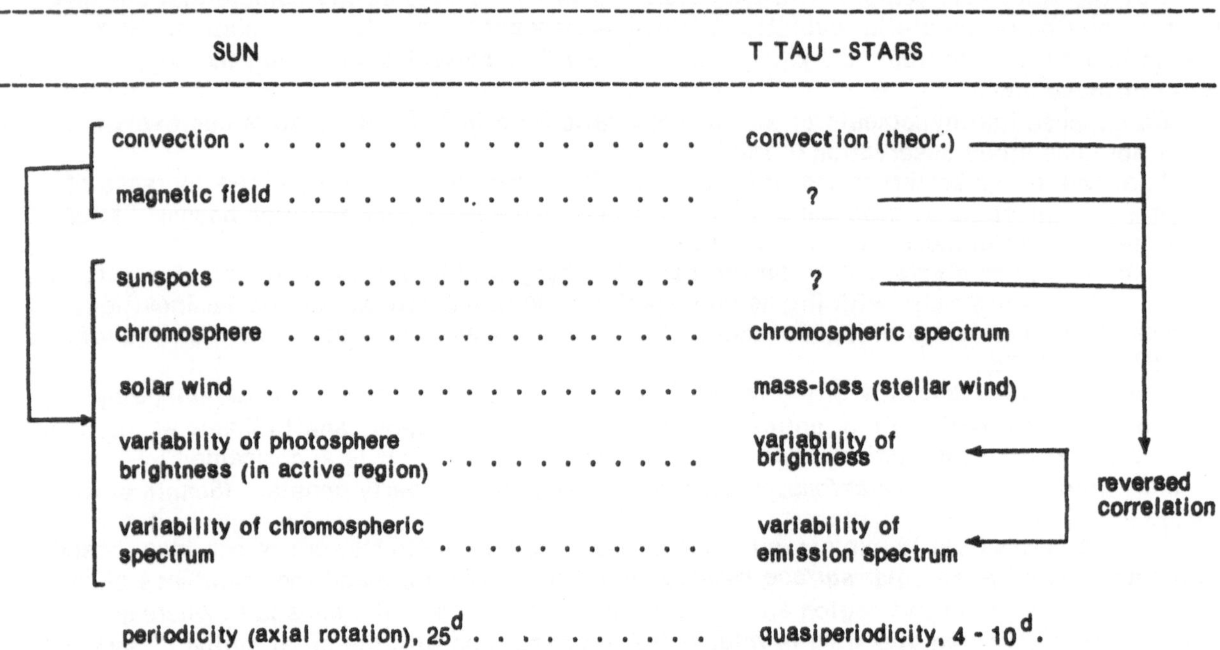

SUN	T TAU - STARS
convection	convection (theor.)
magnetic field	?
sunspots	?
chromosphere	chromospheric spectrum
solar wind	mass-loss (stellar wind)
variability of photosphere brightness (in active region)	variability of brightness
variability of chromospheric spectrum	variability of emission spectrum
periodicity (axial rotation), 25^d	quasiperiodicity, 4 - 10^d.

reversed correlation

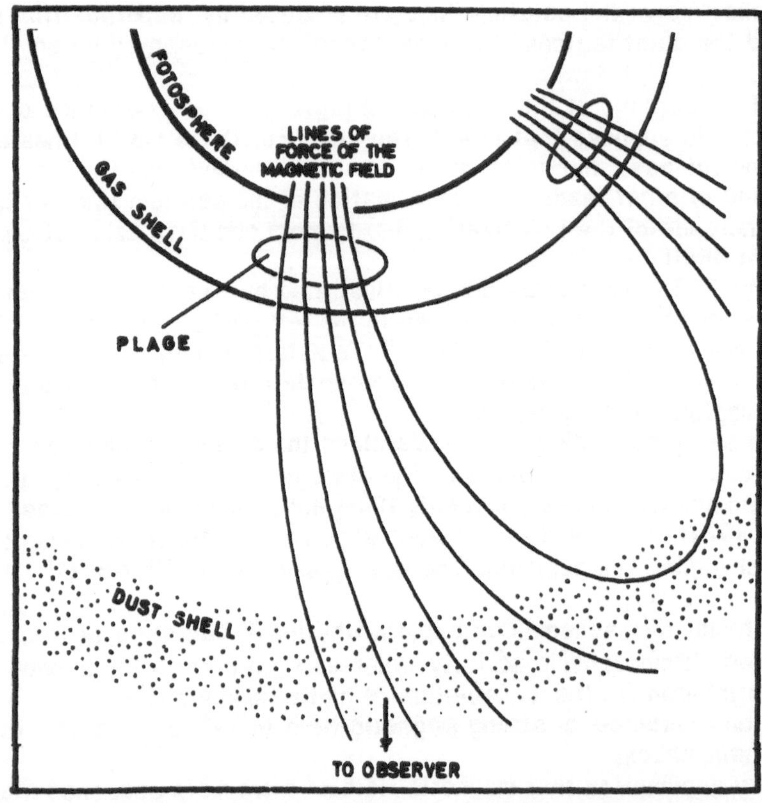

Fig. 1. Magnetic field of a T Tau-star as a cause of variability of the Ca II emission lines and polarisation of visual light.

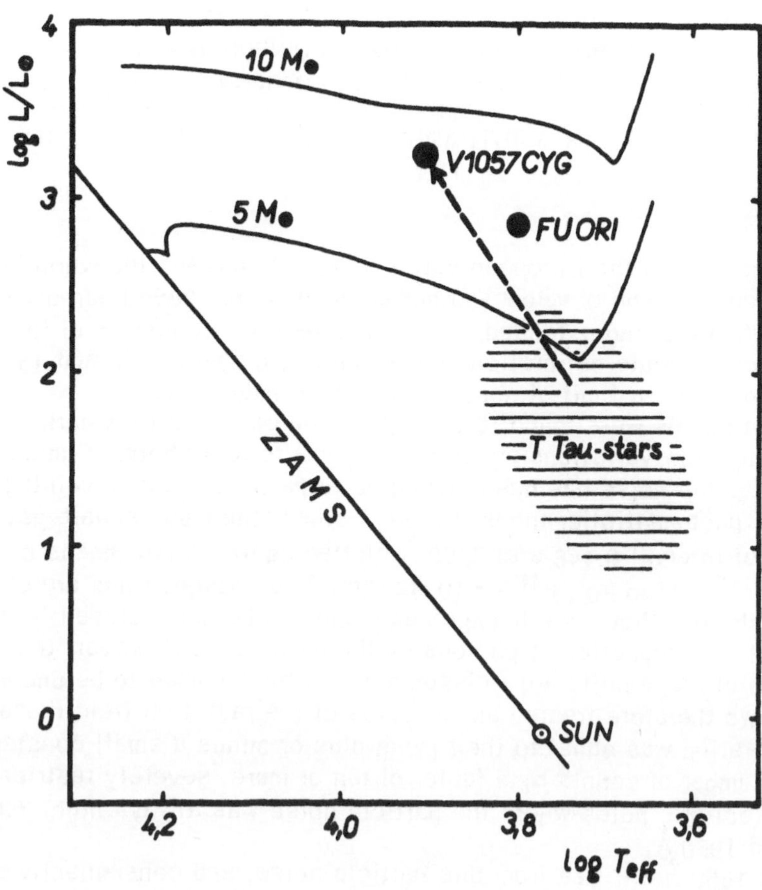

Fig. 2. Location of FU Ori and V1057 Cyg on the evolution diagram (Grasdalen, 1973). Dashed line schematically shows the transition of V1057 Cyg from the region of T Tau-stars to its radiative track.

REFERENCES

EZER D., CAMERON A.G.W., 1967, Can.J.Phys. 45, 3429

GRASDALEN G.L., 1973, Astrophys. J., 182, 781

HAYASHI Ch., 1966, Ann. Rev. Astr. Astrophys. 4, 171

MULLAN D.G., 1974, Astrophys. J. 192, 149

PETROV P.P., 1975, Izv. Krymsk. Astrophys. Obs. 54 (in press)

SALMANOV I.R., 1972, Tzirk. Shemakhinsk. Obs., No 1, 3

SHANIN G.L., SCHERBAKOV A.G., KOVALENKO V.M. et al, 1975, Peremennye Zvezdy (in press)

THE BACKGROUND RADIATION FROM THE S2/68 ULTRAVIOLET TELESCOPE IN THE TD1 SATELLITE *

D.H.MORGAN. K.NANDY AND G.I.THOMPSON
ROYAL OBSERVATORY, EDINBURGH, U.K.
PRESENTED BY D.H.MORGAN

ФОНОВОЕ ИЗЛУЧЕНИЕ ОТ S2/68 УЛЬТРАФИОЛЕТОВОГО ТЕЛЕСКОПА НА СПУТНИКЕ TD1

The S2/68 telescope on the European satellite TD-1A surveys the whole sky in a period of six months from a circular orbit of height 550km,details of which have been given by Boksenberg et al. (1973). When viewing extended objects it effectively consists of four broadband photometers of known absolute calibration centred on 2740, 2350, 1950 and 1550 Å. These can detect B-stars down to a limit of about ninth visual magnitude.

The background counts were found to be higher than expected from dark current and photon estimates, and to show variations sufficient to obscure real photon signals such as that from the Milky Way. This noise has been attributed to particle induced scintillations in the LiF windows of the photomultiplier tubes. Figure 1 shows the mean square deviations of 32 successive counts of interval 0.148 s at 2350 Å plotted against their means on logarithmic scales, for the area $l^{11} = 0°$ to 30°, $b^{11} = -10°$ to -20°. The straight lines are of slope unity and 1.75. The distribution shows the background counts to be non-Poisson: this is attributed to the mechanism of the production of photons by the particles . However, those points that lie close to the line of slope unity are Poisson and can be assumed to be uncontaminated by the particle noise. We therefore treated as measures of the radiation field those counts whose mean square deviation was equal to their mean plus or minus a small constant. This caused a reduction in the number of counts by a factor of ten or more, severely restricting their spatial resolution near the ecliptic poles where the particle noise was always high. The same results occured at 2740 and 1950 Å.

The counts for 1550 Å are free from this particle noise, and consequently show a Poisson distribution. However, on plotting the counts averaged over 100 successive orbits for a region near the north ecliptic pole which is the area of the sky seen on all orbits, a sinusoidal variation was found, whose period was one year and whose maximum, occurring at the summer solstice, was twice its minimum. During the calibration of the photometers, Ly-α radiation was found to scatter into the 1550 Å photometer and, to a lesser extent, the 1950 Å one. The annual variation of the count at the north ecliptic pole can be attributed to the change in the intensity of the Ly-α radiation scattered in the geocorona, which accompanies a change in the solar zenith angle. This has been measured by Meier and Mange (1970) from OGO-IV satellite data. The characteristics of the satellite's orbit cause the solar zenith angle to increase from about 66° in June to about 113° in December. Similar variations in count throughout each orbit have been observed, and these must be separated from the galactic radiation field before the results at 1950 and 1550 Å can be interpreted further.

For 2740 Å the photon counts from the first 2000 orbits have been averaged over intervals of 30° galactic longitude and 5° galactic latitude, and are shown in Figure 2 as a function of galactic latitude for each galactic longitude interval. Each horizontal bar marks the range of galactic latitude in which the ecliptic plane is located for each range of galactic longitude. The zodiacal light is at a maximum on the ecliptic plane and this can be seen here. The Milky Way can be seen clearly at all longitudes where sufficient data exist, and shows a strong variation wtth galactic longitude. For example, it is considerably fainter between $l^{II} = 30°$ and 60° than elsewhere. Other internal structure can be seen: a map of the Orion region at greater resolution shows the Orion nebula area as a bright feature distinct from the Milky Way.

Figure 3 shows results for 2350 Å similar to those for 2740 Å, with the exception that the presence of the zodiacal light is indistinct.

The zodiacal light at 2740 and 2360 Å has been measured by averaging the counts over all galactic co-ordinates with $|b|^{II} \geqslant 30°$ excluding $b^{II} \leqslant -30°$, $l^{II} \geqslant 240°$, for certain ranges of ecliptic latitude. The elongation of the Sun is 90° for all observations. A count difference of 2 was detected between $\beta = 0°$ and $\beta = 45°$ at 2740 Å, but at 2350 Å the difference could only be shown

* The detailed results of this work will be published in Monthly Notices.

to be 0 4. Figure 4 shows the counts at 2740 and 2350 Å averaged over all galactic longitudes after the subtraction of the appropriate zodiacal light plus dark current. The distribution across the Milky Way is the same in each case, and shows the southern galactic hemisphere to be brighter in general than the northern.

The addition of further data will increase the accuracy of the results obtained so far, and, together with the separation of the variable Ly-α airglow component in the short wavelength bands, will permit interpretation in terms of the diffuse galactic light and the scattering properties of the interstellar dust particles which can be compared with Witt and Lillie's(1973) values obtained from OAO-2 data.

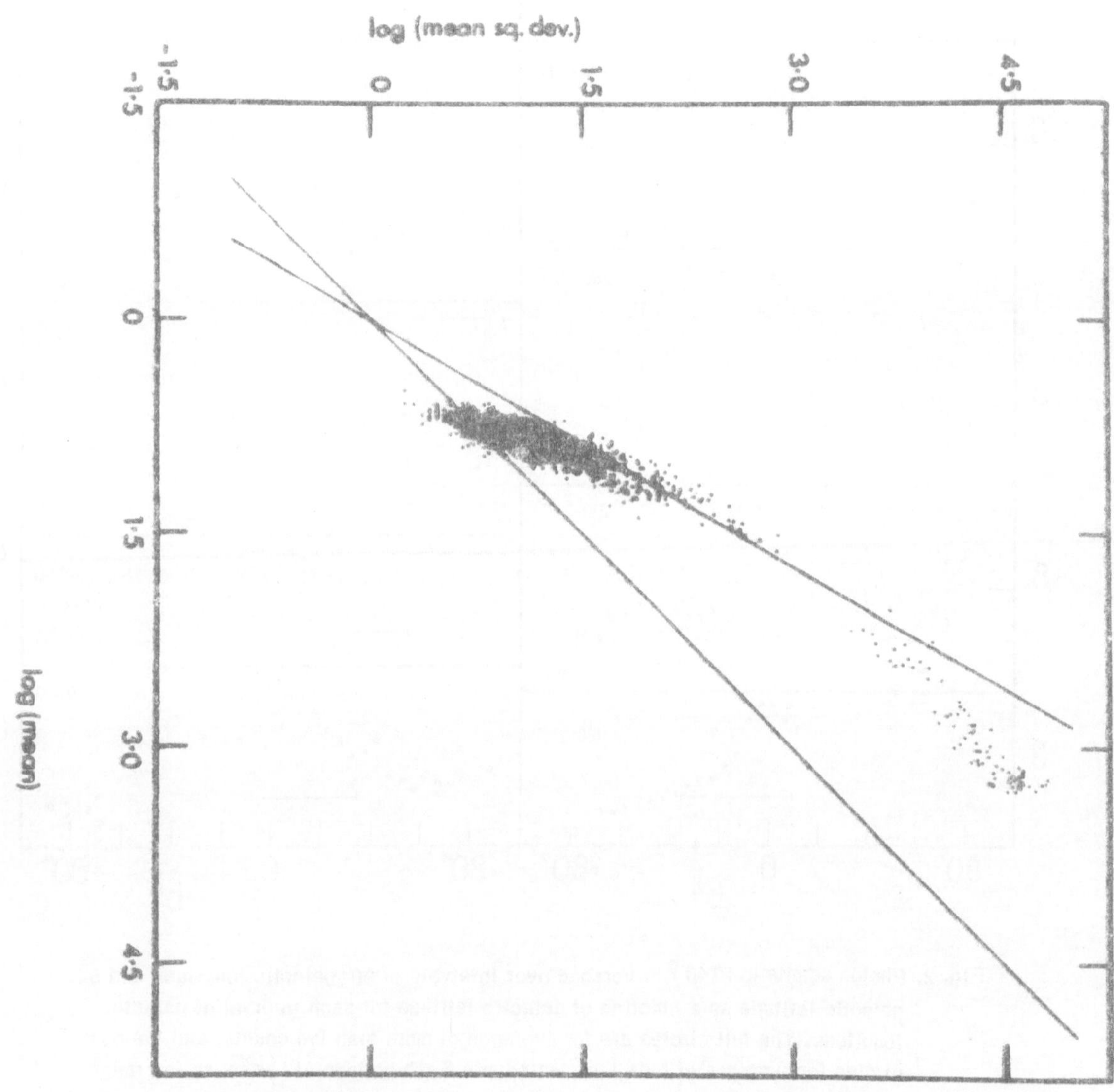

Fig. 1. The mean square deviations of 32 successive counts of interval 0.148 s at 2350 Å plotted against their means on logarithmic scales for the area l^{II}= 0° to 30°, b^{II} =-10 to -20°. The straight lines are of slope unity and 1.75.

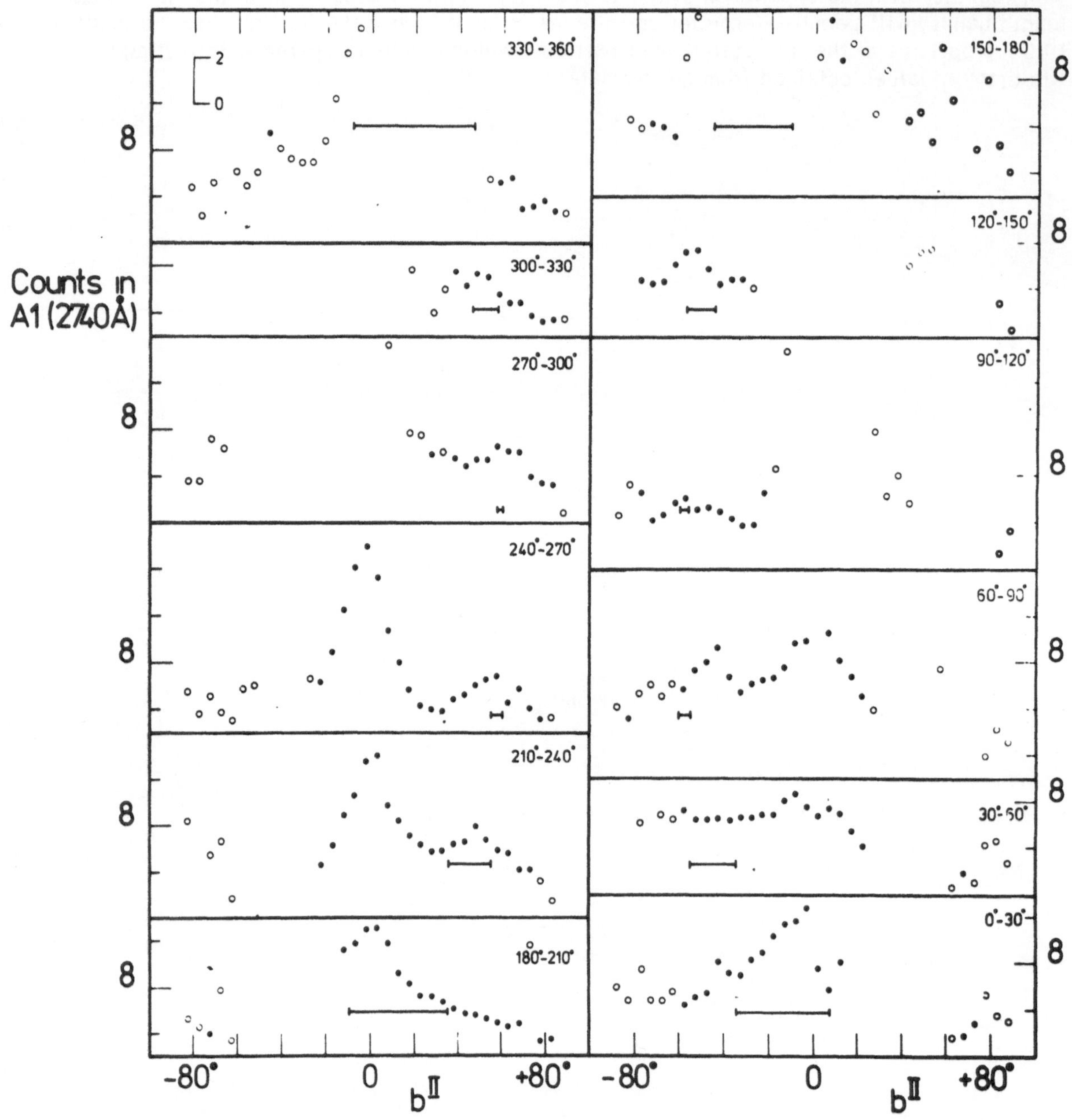

Fig. 2. Photon counts at 2740 Å averaged over intervals of 30° galactic longitude and 5° galactic latitude as a function of galactic latitude for each interval of galactic longitude. The full circles are for averages of more than ten counts, and the open circles for averages of less than ten counts. Each horizontal bar marks the range of galactic latitude in which the ecliptic plane is located for each range of galactic longitude.

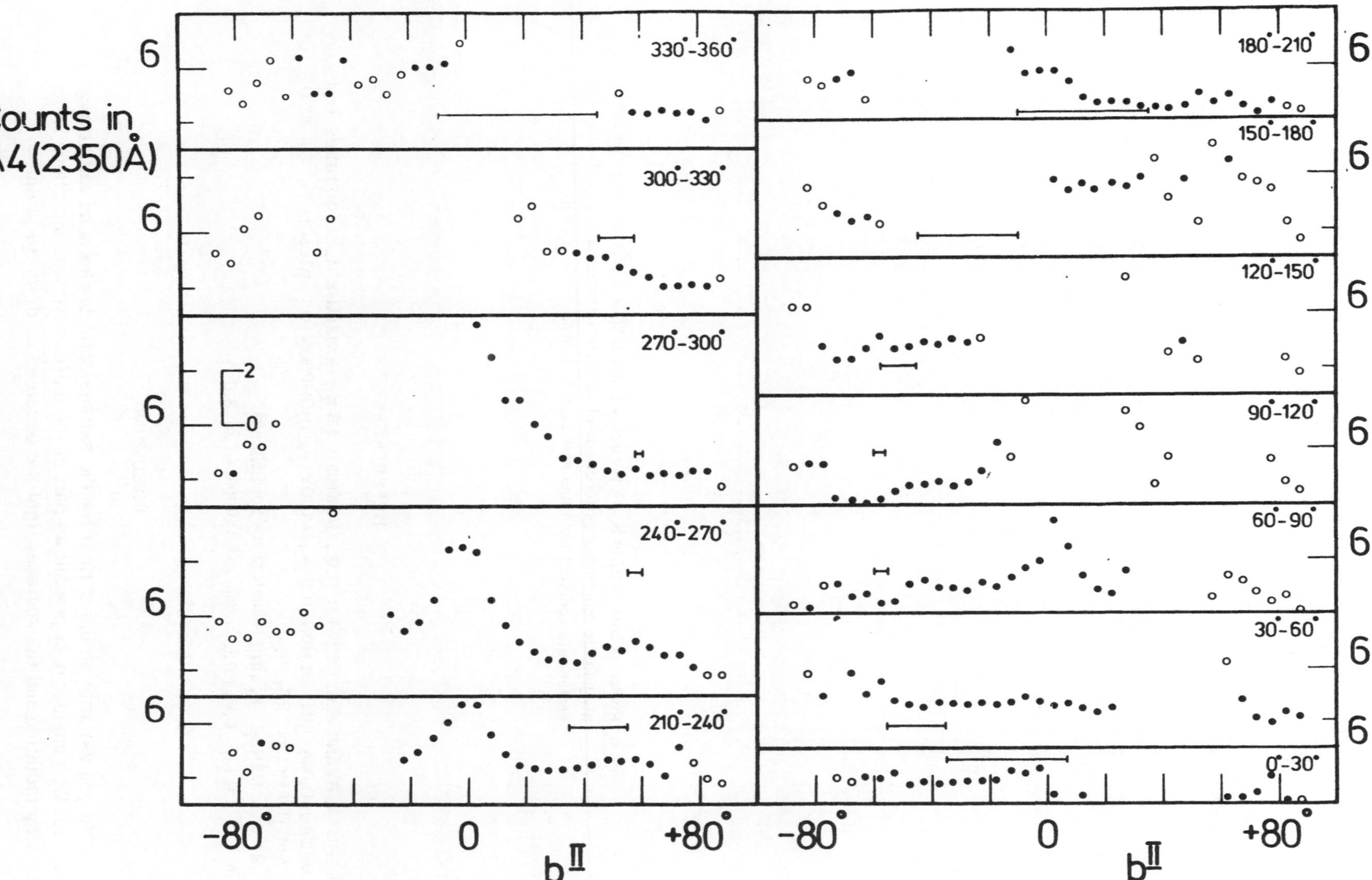

Fig. 3. As Figure 2 for 2350 Å.

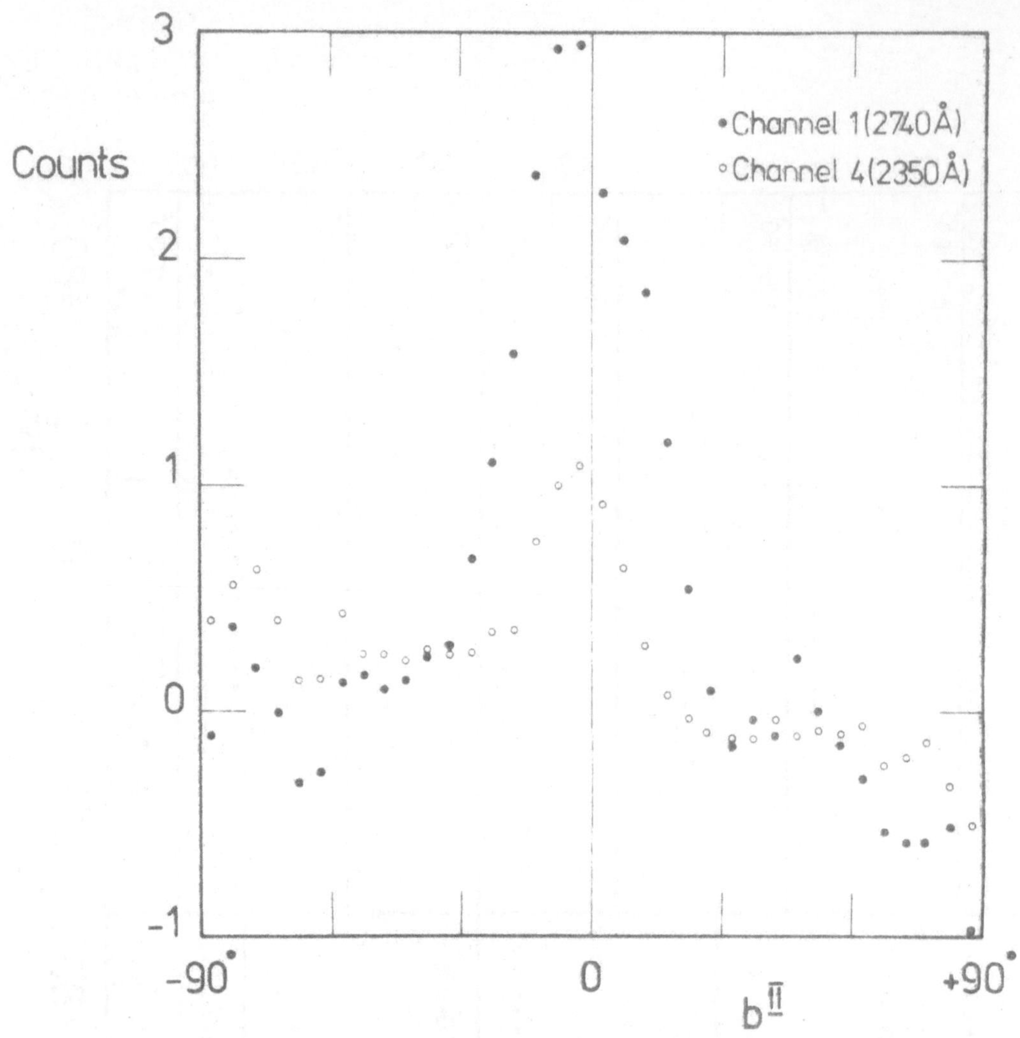

Fig. 4. Photon counts at 2740 Å (full circles) and 2350 Å (open circles) averaged over all galactic longitudes as functions of galactic latitude having subtracted the appropriate zodiacal light and dark current terms.

REFERENCES

BOKSENBERG A, EVANS R.G. FOWLER R.G., GARDNER I.S.K., HOUZIAUX L., HUMPHRIES C.M., JAMAR C, MACAU D, MALAISE D, MONFILS A, NANDY K, THOMPSON G.I., WILSON R. and WROE H. 1973, Mon. Not. R. Astr. Soc. **163**, 291

MEIER R.R. and MANGE P. 1970, Planet. Space Sci., **18**, 803

WITT A.N. and LILLIE C.F. 1973, Astron. and Astrophys., **25**, 397

DISCUSSION

OZERNOY : Did you estimate an upper limit for the extragalactic background radiation?

MORGAN : No. We consider it at present as part of the uniform background above which the Milky Way radiation and the zodiacal light are observed as discrete features.

SPECTROSCOPIC BINARIES WITH Ap COMPONENTS

E. ŽELWANOWA
ZENTRALINSTITUT FÜR ASTROPHYSIK, POTSDAM, D.D.R.

СПЕКТРОСКОПИЧЕСКИЕ ДВОЙНЫЕ С Ap-КОМПОНЕНТАМИ

Introduction. A deficiency or an excess of Ap stars among spectroscopic binaries might be used as an important observational criterion of the plausibility of various hypotheses about the origin of magnetic stars and of their magnetic fields. In the same sense the comparison of the properties of orbits of binary systems containing normal or peculiar components can be used.

1. Statistics of the frequency of spectroscopic binaries with Ap components on the basis of Osawa's catalogue (Osawa, 1959). It was noted several times in literature that there exists a deficiency in the frequency of spectroscopic binaries among the Ap stars (s.f. ex. Abt and Snowden, 1973). In the first part of this paper the results are given of the statistics of spectroscopic Ap binaries, in which we have used, in contrast to earlier studies, a sample of stars, containing Ap stars as well as comparable normal stars. For the stars of this sample we have collected all the necessary data. For details we refer to Želwanowa, Schöneich, Nikolova (1974).

The statistics of the stars of our sample, of the Osawa's catalogue, is presented in Table 1. In the upper part of the table we give the quantitative composition and the percentage proportion of normal, Ap and Am stars in our sample. The obtained frequencies agree well with the results of other investigations.

The next three parts of the table contain the statistics of SB with normal and Ap components of Mn-Hg type and with Ap components of other types. The frequency of normal SB stars is 40%. This value comprises more than 200 cases for which we have insufficient data. Thus we have practically determined only the lower limit of the actual frequency of SB normal stars. The frequency of SB with Ap components of Mn-Hg type is 47%. The number of stars entering this column is, however, rather low, but we have no reason for believing that the frequency of these stars differs considerably from the frequency of binaries with normal components.

On the contrary, for Ap stars of other types there is a distinct deficiency of spectroscopic binary systems. The frequency of SB among these stars is only 26%. Because the radial velocities of these stars are well investigated, this value is near the upper limit of the actual frequency of SB non Mn Ap stars.

II. Comparison of some parameters of the orbits of the spectroscopic binaries with normal and peculiar components. Showing in this way a real deficiency of SB systems among Ap stars in comparison with normal stars, we collected all data of Ap SB systems. From the published material available we found orbital parameters for 21 SB stars with Ap characteristics (without Mn-Hg), for 10 SB stars with Mn-Hg features (Tab. 2), and for 10 SB stars for which it is necessary to verify the peculiarity (Tab. 3).

The orbital parameters for the Ap stars were compared with those for normal stars from The Sixth Catalogue of the Orbital Elements of Spectroscopic Binary Systems (Batten, 1967). We have used 65 B5-B9 (V-III) and 124 AO-FO (V-III) stars. Am types and other abnormal stars have been excluded. We have compared the orbital periods, excentricities and mass functions so far as they are published.

Fig. 1 shows the results. It shows the raction of stars of the corresponding type in intervals of the various parameters in per cents. The rough division of the intervals is due to the comparatively low number of Ap stars with known orbits.

The upper part contains the statistics of the periods. The systems with normal components have 50% of the objects with a period of less than 5 days. Moreover, the data for A and B stars are in good agreement. For stars with peculiar components only 14% of the objects fall in this interval. More than 50% of these systems are concentrated in the last interval of the periods over 20 days. Evidently, there is a lack of Ap SB systems with short periods.

The middle part contains the statistics of the excentricities. For A and B stars the data are in agreement while for the systems with Ap components the distribution of excentricities is distinctly different. In the interval with smallest excentricity, where we have more than half of normal stars, the e is not a single Ap component system.

The lower part contains the statistics of mass functions. There are no substantial differences in the distribution of mass functions for normal and Ap component SB systems.

If we take into consideration that for higher rotation velocities the fine Mn, Hg lines become invisible, the discrepancy for shorter periods between normal and Mn-Hg stars disappears.

We have no reason for believing that the distribution of periods, excentricities and mass functions for Mn-Hg stars differs considerably from those for normal stars.

III. Conclusions. 1. The deficiency of the SB system with Ap components of other than Mn-Hg type is confirmed. For Ap Mn-Hg components the deficiency apparently does not exist.

2. The distribution of orbital periods and excentricities for systems with normal and Ap components of other than the Mn-Hg type is not in agreement. Moreover, in systems with Ap components we observed a lack of shortest periods, as was already noted by Abt and Snowden (1973), and an absence of circular orbits.

3. The distribution of mass functions does not give any support to the probability that secondary components of the normal and Ap SB system differ from each other.

4. As the number of the known Mn-Hg systems is very small, it is not justified to speak of the difference on the characteristics of the orbits between normal and Mn-Hg stars.

IV. Speculations. What might be the origin of the differences between systems with normal and peculiar components?

There are in principle three possibilities:

1. When magnetic fields are present, close binaries cannot be formed.

2. In close binaries magnetic fields and consequently the other characteristics of Ap stars cannot be generated.

3. When magnetic stars are formed, binary systems undergo changes.

There are some arguments against the first and the third explanation. A more detailed discussion of this matter is given in a paper by Želwanowa, Schöneich, Nikolova (1974).

We think that the second possibility, which is indirectly connected with the hypothesis of Dolginov (1973) about the generation of magnetic fields in close binary systems, is a serious alternative explanation.

The generation of the magnetic field, according to Dolginov's hypothesis is strongest at an angle of 45^0 between the rotational axis and the orbital plane. For other values the effect is less pronounced. If the axis is perpendicular to the orbital plane, can the Dolginov - effect interfere with the normal dynamo action that is inhibit the generation of the magnetic field and of the other Ap characteristics?

In this case the detailed investigation of Ap SB's would give much information about the physical conditions for the generation of stellar magnetic fields and last but not least about the structure of those fields.

Tab. 1 STATISTICS OF THE STARS OF OSAWA's CATALOGUE (1959)

a) Frequency of Ap and Am stars
Altogether 500 stars, B5 - A8 (V - III),
brighter than mag. 6.50,
with declination between +10° and +40°.

among them	390 normal stars	78% - 84%
	70 Ap and Ap :	10% - 14%
	40 Am and Am:	6% - 8%

b) Frequency of spectroscopic binaries among normal stars

	390 normal stars	100%
among them	30 stars with orbit	
	69 spectr. bin. without orbit	40%
	56 stars with variable RV	
	235 no binaries or without information	

c) Frequency of spectroscopic binaries among Mn, Hg stars

	17 Mn - Hg stars	100%
among them	5 stars with orbit	
	3 stars with variable RV	47%
	3 stars with const. RV	
	6 stars without information	

d) Frequency of spectroscopic binaries among the other types of Ap stars

	34 Ap stars	100%
among them	5 stars with variable RV	26%
	4 stars with variable RV	
	15 stars with const. RV	
	10 stars without information	

Tab. 2 Ap STARS WITH KNOWN ORBITS

HD	pec	Porb (d)	e	f (m)	References
		Si, Cr, Eu, Sr stars			
8441	Sr, Gr, Eu	106,3			Renson 1965, IAU IBVS 108
11529	Sr	69,92	0,30	0,164	Batten 1967, Publ. Dom.obs. XIII No. 8
15144	Sr, Cr	2,99781	0,22	0,0016	Abt, Snowden 1973, ApJ Suppl. 125
25267	Si	5,9537	0.100	0.032	Batten 1967
25823	Si	7.22743	0.18	0.0033	Abt, Snowden 1973
68351	Si, Cr	585,4154	0,24	0,0985	Abt, Snowden 1973
77350	Si	1401,4	0,35	0,0538	Abt, Snowden 1973
89822	Si (Sr, Hg)	11,57907	0,26	0,0637	Pedoussaut, Ginested 1973, Astr. Astrph. Suppl. 10
90569	Cr, Sr	12658,4	0,75	0,506	Abt, Snowden 1973
98088	Sr, Cr	5,905	0,176	0,229	Batten 1967
112185	Sr, Cr	1515,	0,31	0,0058	Ludendorff 1913, A.N. 195
123515	Si	26,005	0,202	0,0158	Batten 1967
125248	Cr, Eu	1618,0	0,266	0,0661	Abt, Snowden 1973
137909	Sr, Cr, Eu	3833,7	0,406	0,24	Batten 1967
147869	Sr	4,951	0,511	0,0014	Batten 1967
170000	Si	26,768	0,39	0.0408	Abt, Snowden 1973
173524	Si (Hg)	9,810	0,222	0,0210	Batten 1967
183056	Si	35,0225	0,45	0,00048	Abt, Snowden 1973
187474	Cr, Eu	560,0			Eggen 1967, Magn. a. Rel. St., 141
208095	Si, Sr	17,3263	0,224	2,126	Batten 1967
224801	Si, Eu	4,88643			Weiss 1973, Mitt. AG. 25
		Mn, Hg - stars			
358	Hg, Mn	96,699	0,538	0,166	Batten 1967
27295	Mn	4,452064	0,055	0,0004	Dworetsky 1972, PASP 84, No. 501
27376	Mn+Mn	5,0105	0.014	0,135	Batten 1967
78316	Mn, Hg	6,39316	0,319	0.192	Batten 1967
141556	Hg+Am	15,2565	0,00026	0,279	Dworetsky 1972, PASP 84, No. 498
161701	Mn, Hg	12,4519	0,062	0.17	Pedoussaut, Ginestet 1971, Astr. Astroph. Suppl.4
172044	Hg, Mn	1675,	0.16	0.0052	Abt, Snowden 1973
174933	Hg.	6,3624	0.116	0,0036	Batten 1967
191110	Hg+Hg	9,346	0.0120	0.1327	Dworetsky 1974, PASP 86, No. 510
216494	Mn, Hg+(Hg)	3.4298	0.050	0.3099	Wolff 1974, PASP 86, No. 510

Tab. 3 SPECTROSCOPIC BINARIES FOR WHICH IT IS NECESSARY TO VERIFY THE PECULIARITY

HD	pec	$P_{orb.}$(d)	e	f(m)	References
26961	Si:	1,52738	0,047	0,0127	Batten 1967
28319	Sr	140,728	0,750	0,126	Batten 1967
94334	Si:	15,8307	0.305	0.015	Batten 1967
123299	Si:	51,420	0.38	0.436	Batten 1967
132742	Ap	2,3273	0,048	0,115	Batten 1967
162588	AOp	6,1411	0,55	0,0014	Pedoussaut Ginestet 1973
183007	A3p	164,62	0,118	0,0273	Batten 1967
188164	A3p	14,9859	0,563	0,0704:	Batten 1967
196133	Si, Sr	87,687	0,7605	0,0854	Batten 1967
205073	Ap	5,4730	0,29	0.0032	Abt, Sanders 1973, ApJ 186, 177

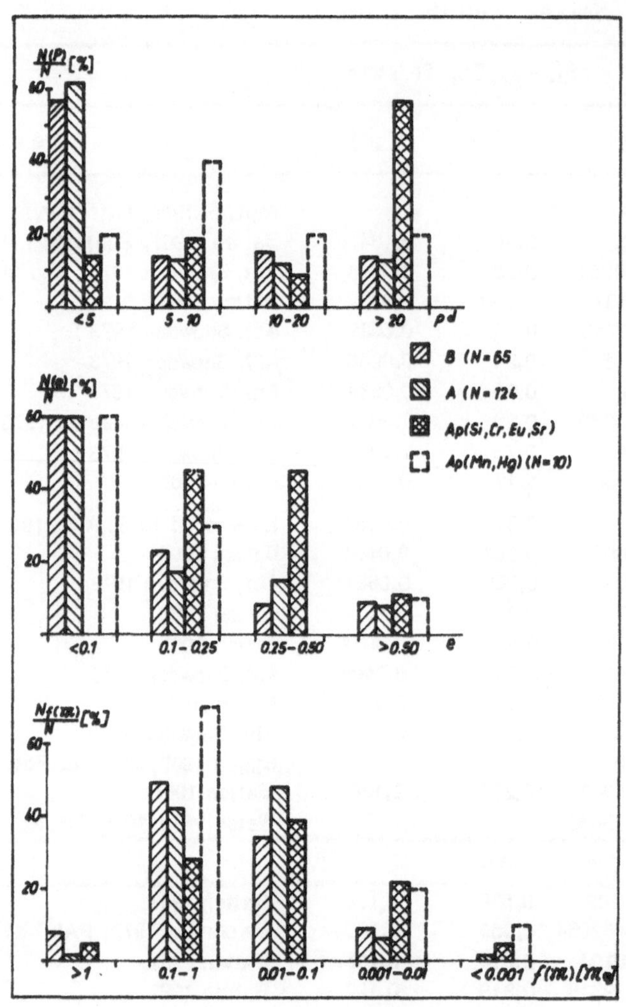

Fig. 1. Comparison of orbital periods (upper part), excentricities (middle part) and mass functions (lower part) for A, B and Ap stars.

PEFERENCES

ABT H.A. AND SNOWDEN M.S. 1973, ApJ Suppl. 25, 137

BATTEN A.H. 1967, Pub. Dom. Ap. Obd. XIII, Nr. 8

DOLGINOV A.Z. 1973, Konf. on Mag. St. in Shemakha Obs. (In print)

OSAWA K. 1959, ApJ. 130, 159

ŽELWANOWA E., SCHÖNEICH W., NIKOLOVA S. 1974, Konf. on Mag. St. in Sonneberg

ON THE ATMOSPHERIC STRUCTURE IN SOME MAGNETIC Ap-STARS

I.A.ASLANOV, Yu.S.RUSTAMOV, V.M.KHALILOV,
A.A.SHAKIR-ZADE
SHEMAKHA ASTROPHYSICAL OBSERVATORY, U.S.S.R.
PRESENTED BY Yu.S.RUSTAMOV

МАГНИТНЫЕ ПОЛЯ В АТМОСФЕРЕ Ap - ЗВЕЗД

One of the problems of particular significance is the problem of the stellar magnetism. Its solution is carried out by the investigation of Ap-stars.

The existence of the strong magnetic fields and amomalous chemical composition are considered to be the characteristic features of these stars. The photometric investigations show the periodical light variation with a little amplitude, not more than 0^m15. A general accepted hypothesis which explains the most observational data is the hypothesis of the inclined rotator according to which the axis of the dipole magnetic field is inclined to that of the stellar rotation. There are some regions in the star with the anomalous chemical composition-"spots". The observed changes of the magnetic field magnitude and sign, spectral characteristics and light in this model are due to the stellar rotation.

The investigators of the Shemakha Astrophysical Observatory, Central Astrophysical Institute of the German Academy of Sciences and Astronomical Council of the Academy of Sciences of the USSR carry out the team-work on the complex investigations of Ap-stars. Both observational and theoretical aspects of the problem under investigation are considered. The spectroscopic observations are carried out with the 2-m telescope at Shemakha Astrophysical Observatory; the photometric observations giving the information about the periods and energy distribution in the Ap-stars spectra are conducted with the aid of the telescopes at the observational station of the Central Astrophysical Institute of GDR and in particular with the original double telescope. The magnetic fields of stars were measured by the spectrograms obtained with the Zeeman analyzer (of Babcock-type) available. The great accuracy of the magnetic field measurements is reached by the coordinate-measuring instrument "Ascorecord" and the unit of Golnov-type made in ShAO which permits one to see simultaneously the line contours in two polarizations on the screen of the doubleray oscillograph.

For the short-time changes the spectrokon was used which gives the gain in time without any resolution loss.

The aim of the present paper is to reveal the relation of the atmospheric structure with the magnetic field. For this purpose we have investigated several Ap-stars with different angles between the axis of rotation and line of sight (the angle of inclination varies from 90° to 11°). (Table 1). The coordinates of the "spots" were determined and, as-one can see from Table 1, the regions of the "spot" formations are situated on the latitudes below 60° with respect to the rotation equator.

But the presence of "spots" is unable to explain the periodic hydrogen line variations obtained for all the stars under consideration.

The comparison with the light variation curve shows that for the minimum light the hydrogen line profiles are wider and deeper than for the maximum one. It should be noted, however, that the light variations have not more than two maxima/period whereas the changes of the hydrogen lines are often of more complex character. The electron concentration changes are also complex and have no simple relation with the light variations. As seen from Fig.2, this concentration has its minimum in those places where the "spots" with the rich abundance of one or another element are situated.

The estimate of the homogenous atmosphere height made by us for some stars in different phases showed the "ragged" structure of the Ap-star atmospheres and the coincidence of the anomalous element spots with the minimum atmosphere heights. Thus one can conclude that the location of "spots" is in close relations with the peculiarities of the atmospheric structure.

Comparing the "spot" coordinates with the magnetic field curves of 17 Com A one can note (see Fig.3) that the field magnitude and the "spot" concentration correlate. While passing through the central meridian of the strongest "spots" of Sr one observes a magnetic field extremum. The analogous regularity is also noted for the star HD 133029: when the "spots" situated on the high latitudes are passing through the central meridian one can observe a maximum value of the field and in the case of the low-latitude "spots" this value decreases (see Fig.4).

Thus the magnetic field is closely connected with the "spotty" structure of the Ap-star atmospheres and has a character more complex than dipole.

To investigate the magnetic field structure we have chosen the star CrB with a great number of narrow, unblended lines. These features allowed us to measure its magnetic field very accurately.

In Figure 5 we illustrate the change of the magnetic field magnitude with the optical depth by neutral Fe lines. To calculate the optical depth of the layer where the given line is effectively formed it is necessary to take into account the anomalies of the chemical composition in "spots" and also the effect of the magnetic field on the layer optical depth. We restricted ourselves by the mean optical depth since this approximation does not influence the qualitative results. It is seen that the field magnitude changes greatly with the depth.

In the literature one usually meets indications of the existence of the short-time light variations which are not connected with the rotation. The nature of these changes is not yet known well though some assumptions concerning their pulsation character were made.

We have found the short-time variations of the angular velocities by H_γ, H_δ, H_ε lines for 17 Com A and HD 224801 with periods of 71^m and 6^h respectively (Fig. 6) These changes being observed for a long time are probably connected with the star pulsation. The observations of H_α - line in 17 Com A made by us this year with the spectrokon showed the presence of the variability in this line with the same period - 71 minutes (Figure 7).

The character of the changes of the central part of H_α - line is likely to be similar with that in stars of β Cep-type which are close to the Ap-stars in H-R diagram.

It should be noted in conclusion that now the cooperative work together with the Central Astrophysical Institute (GDR) and Astronomical Council is being continued.

Table 1

spot number	I	II	III	IV	V	VI	VII	VIII	IX	X
HD 19832 $i=90°$										
ℓ	54°	106°	187°	235°	342°					
φ	12°	13°	12°	12°	15°					
HD 108662 $i=90°$										
ℓ	29°	79°	119°	144°	173°	198°	216°	252°	306°	342°
φ	45°	50°	30°	30°	30°	30°	30°	40°	40°	45°
HD 108945 $i=29°$										
ℓ	32°	86°	140°	200°	230°	300°	346°			
φ	33°	33°	33°	29°	29°	33°	33°			
HD 133029 $i=11°$										
ℓ	0°	47°	79°	131°	166°	203°	238°	267°	313°	
φ	35°	37°	37°	37°	35°	47°	47°	47°	42°	
HD 140160 $i=55°$										
ℓ	0°	80°	146°	234°	310°					
φ	40°	40°	40°	40°	40°					
HD 140728 $i=46°$										
ℓ	65°	144°	200°	288°	340°					
φ	53°	51°	55°	51°	46°					
HD 193722 $i=22°$										
ℓ	25°	81°	116°	225°	290°					
φ	36°	36°	36°	45°	53°					

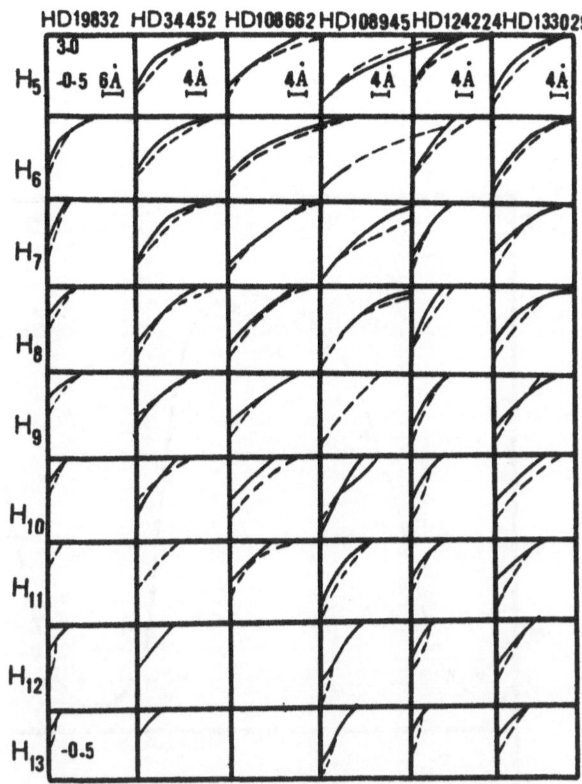

Fig. 1. Hydrogen line profiles. The solid and dotted lines represent the maximum and minimum light, respectively.

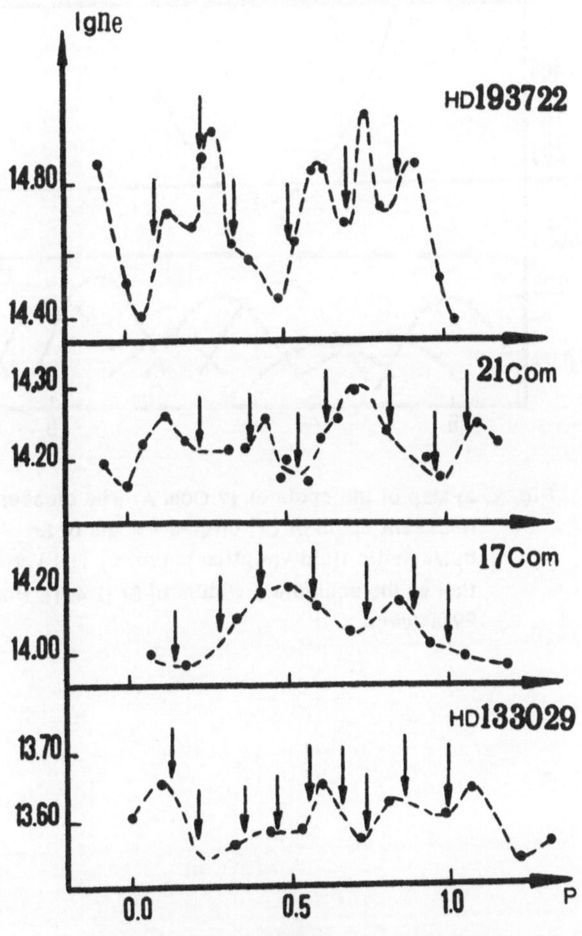

Fig. 2. lg n_e variations for 4 stars. The arrows represent the "spot" positions.

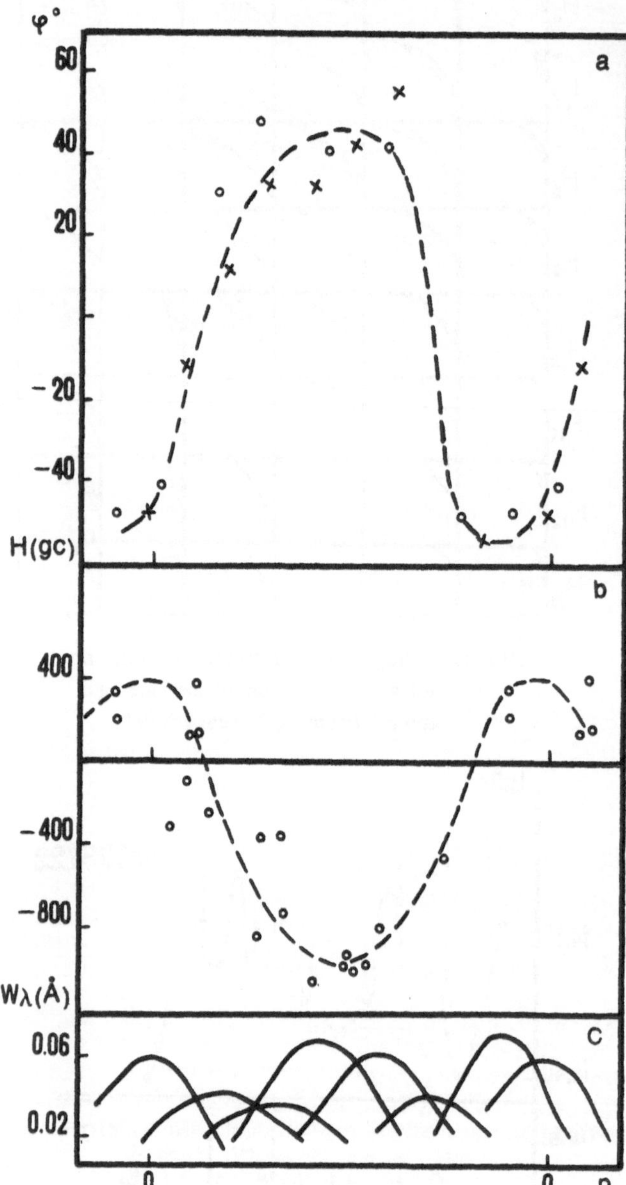

Fig. 3. a) Map of the spots of 17 Com A. The crosses represent spots of Cr, circles - spots of Sr. b) Magnetic field variation curve. c) The variation of the equivalent widths of Sr II 4215 line component.

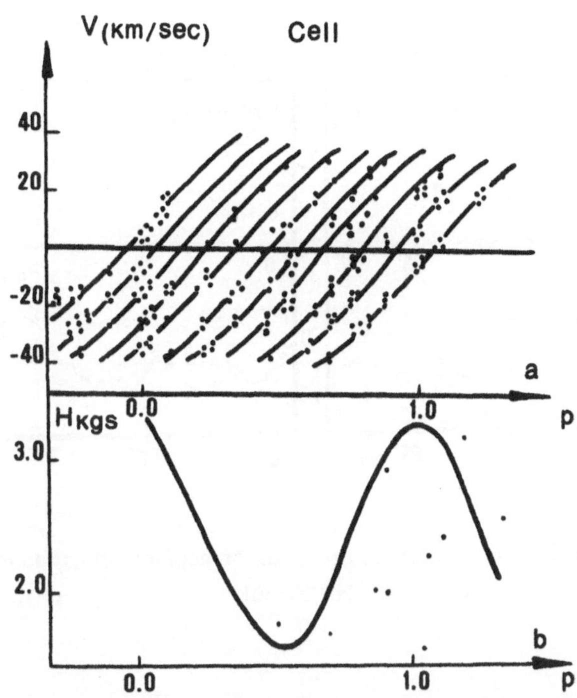

Fig. 4. Comparison of the "spot" coordinates (a) with the magnetic field variation curve (b) for the star HD 133029

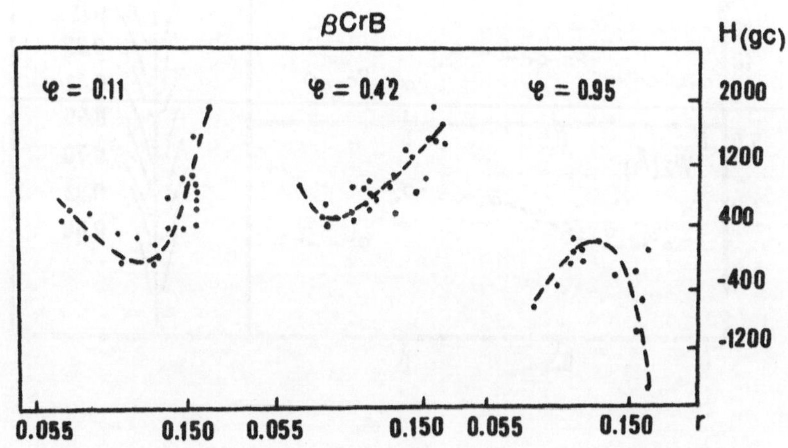

Fig. 5. The variation of the magnetic field magnitude with the optical depth for β CrB.

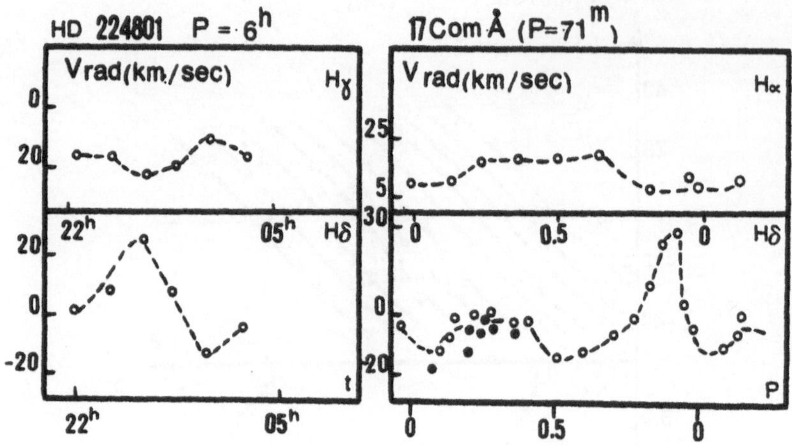

Fig. 6. Short-time variations of the angular velocities in 17 Com A and HD 224801.

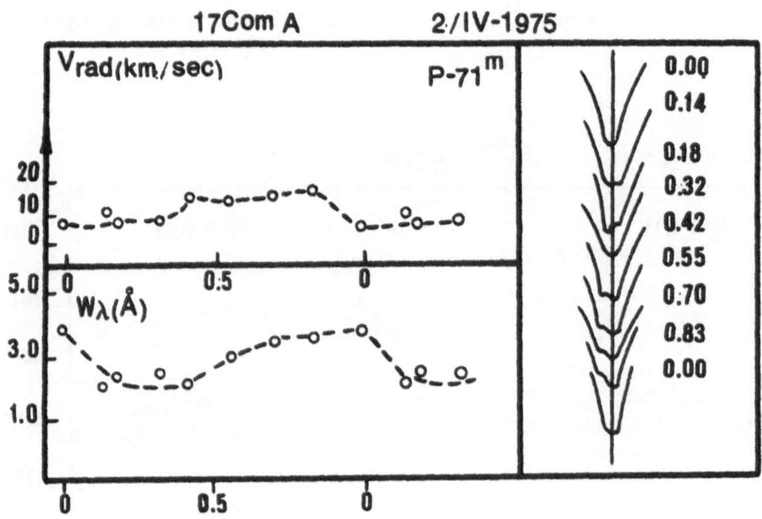

Fig. 7. The observations of the short-time variations of 17 Com A with the spectrokon. On the right there are some contours of Hα line nucleus for different phases.

FRAGMENTATION OF GALAXIES AND THE BIRTH OF DWARF SATELLITES

B.VORONTSOV - VEL'YAMINOV
STERNBERG ASTRONOMICAL INSTITUTE, MOSCOW, U.S.S.R.
READ BY A.R. AZO

ФРАГМЕНТАЦИЯ ГАЛАКТИК И ОБРАЗОВАНИЕ КАРЛИКОВЫХ СПУТНИКОВ

РЕЗЮМЕ

II часть Атласа взаимодействующих галактик содержит 700 фотографий. В ней изложено представление о спокойной фрагментации галактик даже в наше время. Чаще всего отпочковываются небольшие внутренние подсистемы близкие к периферии. Так образуются спутники. Особенно убедительно сопоставление обнаруженных нами прямолинейных формаций со всеми стадиями распада их на минипепочки, состоящие из 3-4 членов в контакте, позднее соединенных перемычками.

From the morphological study of interacting galaxies it is possible to obtain an inkling in the later stages of evolution of galaxies in general.

The II Part of the "Atlas of Interacting Galaxies" is now under compilation. It is based mainly on the MCG with 2000 such objects. But this part is devoted mainly to the M 51-type systems and to the nests of galaxies.

160 systems of the M 51-type were found and studied statistically, M 51 itself in detail. We have found the straight streams of stars and of dark matter from NGC 5194 directed toward its companion NGC 5195, besides the spiral arm of the former which seemingly connects both galaxies. Such direct streams complementing or replacing the spiral filament were not demonstrated by the tidal theory of disturbances. The same phenomenon we have also found in the other interconnected pairs, in VV 19 in particular. It appears also that it is rather NGC 5195 that sereens the northern part of the spiral which runs past it , - but it is not eclipsed by the end of the spiral. So possibly the companion is actually situated at the tip of the spiral arm and it cannot lie lower than 11 kpc under the plane of the spiral, only by chance projecting to us under this arm.

The majority of M 51-type systems have companions at the ends of spiral arms of so insignificant masses, that they by no means could have produced the tidal spiral arms. Sometimes both spiral arms are terminated by very small companions. It appears probable that the companions formed inside a larger stellar system,when at the end of a spiral arm,are liable to separate and become detached satellites. The theoretical possibility of such events, apparent from the comparison of specimens of M 51-type systems, was forseen by Starr and Newall and by Hoyle and Ireland from the study of the mechanical and magnetical torques, respectively. We may add the perturbing forces from the third galaxy and the other modes of transformation of energy.

Apparently, the formation of spirals, predominantly of the outer, additional spirals by tidal ineraction does exist. But this requires massive companions or strangers which are rare phenomena. It is not possible at present to draw the borderline between the companions connected by spiral filaments and companions originated at the ends of the spiral ends of the primary as a result of normal development of some condensations in an isolated spiral galaxy.

The quiet gemmation of dwarf companions from other kind of flattened galaxies seems probable as well. The close dwarf companions as a rule are blue, that is they are composed of hot blue stars and gases with /O II / emission and so are the young formations. In this conception it is important that the companions are formed inside the larger systems and gradually gemmate. No extraordinary energies are required in the process of formation of satellites by gemmation. The drastic expulsion of satellites apparently is a very rare event.

The production of satellites by large galaxies explains why we see them surrounded by dwarf galaxies and why there are but few dwarf galaxies in regions void of the large ones.

We find that the nests and the chains of galaxies are composed mostly of three objects, at most of five. The genuine chains contain close galaxies connected by blue filaments and are enveloped by blue haze. Other chains are probably chance projections of approximately flat groups seen nearly edge-on.

Faint interaction galaxies of the utmost importance lack measured red shifts and none of the tight nests of galaxies were probed for the velocity field.

PHYSICAL CONDITIONS AND CHEMICAL COMPOSITION IN THE NUCLEI OF SPIRAL GALAXIES

D.ALLOIN, P.BENVENUTI AND S.D'ODORICO
OBSERVATOIRE DE NICE,FRANCE;ASTROPHYSICAL OBSERVATORY OF THE UNIVERSITY OF PADOVA, ITALY
PRESENTED BY S.D'ODORICO

ФИЗИЧЕСКИЕ УСЛОВИЯ И ХИМИЧЕСКИЙ СОСТАВ В ЯДРАХ СПИРАЛЬНЫХ ГАЛАКТИК

I n t r o d u c t i o n. We have started at the Asiago Observatory a program of optical observations of spiral galaxies based on spectroscopy and medium to narrow band photography with two purposes :
a) to provide new data such as the structure in the H_α line for the central region, dynamical parameters, non-rotational motions, for individual spiral galaxies which are of interest from the point of view of their radio activity.
b) to compare, on a statistical basis, the radio properties in the central regions of spiral galaxies (total power, dimension of the nuclear source, spectral index) with the physical conditions and abundances which can be inferred from the optical emission line intensities.

Such a comparison between optical and radio properties in spiral galaxies was already published by Alloin (1973): in particular the existence of radio emission in the central region of a galaxy was correlated with the presence of emission lines in the optical range.A preliminary attempt to correlate the nuclear radio power with the H_α/[N II] intensity ratio was also made by Benvenuti and D'Odorico (1974). The new, high spatial resolution radio observations of galactic nuclei, mainly from the Westerbork radiotelescope,now provide a homogeneous set of more accurate radio data. A discussion of the nuclear radio emission in spiral galaxies has been given by van der Kruit (1973) and Ekers (1974). They correlated the existence and power of the nuclear radio source with the absolute optical magnitude and the presence of optical emission lines, but the emission line data usually come from Humason et al. (1956) redshift catalogue, that is low dispersion blue spectra. A positive detection of optical nuclear emission depends strongly on the strength of the underlying continuum and on the dispersion (a narrow emission line is more easily detected when the continuum is spread out at high dispersion and high spatial resolution), on the inclination with respect to the line of sight (the nuclear region is likely to be hidden by absorbing material in edge-on galaxies) and on the distance (in far away galaxies emission from the spiral arms can be merged with the nuclear emission). These effects have to be considered before a significant comparison with the radio data can be made.

In this paper, we report the results of optical observations of the following spiral galaxies IC 342, NGC 2146 and NGC 3079,which do have non-thermal nuclear radio emission and we briefly compare the parameters of the nuclear radio sources with the emission line ratios in a small sample of nuclei of spiral galaxies for which, both radio and optical data are available.

S p e c t r o s c o p i c a n d p h o t o g r a p h i c r e s u l t s.The observations were carried on with the 122 cm and 182 cm telescopes of the Asiago Observatory. The larger telescope was used for direct photography (scale 12 arcsec/mm),while the spectroscopic observations were obtained with the Asiago nebular spectrograph attached to the newtonian focus of the 122 cm telescope. The spectrograph is described in detail by Benvenuti et al. (1975): it employs a WL-30677 image tube and, in the used spectral range ($\lambda\lambda$ 4500 - 7000 Å), the dispersion is 125 Å/mm and the scale perpendicular to the dispersion is 127 arcsec/mm. The width of the slit we used, for these observations, is 4 arcsec on the sky. Spectra and calibration plates were reduced either at the "Centre de Depouillement de Cliches Astronomiques" (INAG, Observatoire de Nice) with a PDS microphotometer and a PDP 11-40 computer, or with the Asiago digitalized microphotometer together with a HP 2100 computer.

A. IC 342

This galaxy has been studied in some detail, at 1415 MHz by Van der Kruit (1973 b). The dimension of the nuclear radio source, according to Ekers (1974), is less than 4 arcsec (100 pc) in diameter, which is smaller than the size (10 x 8 arcs)of the bright central region seen in the optical range. Narrow interference filter photographs, in good seeing conditions, are then needed to investigate a possible detailed optical structure. In addition spectra at high angular resolution and high dispersion would be of some help in understanding the dynamics of the central region and the amount of energy which is associated with possible non circular motions. A photometric study has already been published by Ables (1972). In the optical range the galaxy is of very low surface brightness, except for the bright central bulge, 10 x 8 seconds of arc in size. Two plates of the nucleus, 103a-0 + GG 13 and 103a-E + RG 1, respectively, are shown in fig. 1 a, b. Filamentary emission is appearing, both on the red and blue plates, close to the high surface brightness central region. We have also obtained 5 spectra of the nuclear region, which exhibit a very strong continuum together with the emission lines of H_β, H_α, [N II] 6548-84 Å and [S II] 6717-31 Å. The line ratios are given in table I and will be discussed in section III. For the $I(H_\alpha)/I(H_\beta)$ ratio we are able to derive a lower limit of 8: even after correction for the galactic absorption, this value still indicates a large amount of absorption within the nucleus itself.

B. NGC 2146

This galaxy has one of the strongest non-thermal central radio source, among "normal" quiet spirals. An optical study has been already published by Benvenuti et al. (1975): the nuclear region is partially hidden by a strong dust lane. We have recently obtained new photographs of the galaxy, through a H_α filter of 100 Å half-bandwidth (see fig. 2 a and b), which reveal a strong arm extending to the SE of the nucleus: a possible counterpart of this arm would be hidden by the dust lane to the north of the nucleus. The structure resembles that observed, in H_α light, in the galaxy NGC 4258. It would be of great interest to see whether the radio spiral arms detected in that galaxy are present also in NGC 2146. A blue spectrum of the galaxy has been used to derive a value I ([O III] 4959+5006)/I (H_β) = 1.55 in the region of the H_α arm, while I (H_α)/I([N II] 6548 + 6584) ≃ 1.7 from Benvenuti et al. (1975).

C. NGC 3079

This galaxy shows relatively strong, non-thermal, nuclear radio emission, with a relatively complicated structure (de Bruyn, private communication). An optical mapping of the central region is difficult because the galaxy is seen edge on. Our spectroscopic material consists of five spectra in the red wavelength region and two in the blue one, at different position angles (355°, 345°, 338° and 75°). Photographs in different colours were also obtained. A blue, short exposure photograph is shown in Fig. 3. We were able to measure the line intensity ratios I(H_α)/I ([N II] λ 6548 + 6584) and I (H_α)/ I([S II] λ 6717 n 6731) for 12 emission regions located on the major axis or close to it. The results are given in Fig.4 and 5. Figure 4 is a plot in a logarithmic scale of I (λ6584)/(H_α) versus I ([S II])/(H_α). The bars indicate the estimated errors. The two ratios appear proportional one to the other, the slope being equal to the one observed in M 33 and M 101 (Benvenuti and D'Odorico, 1974). Only the point relative to the nucleus departs from this relation. The electron density measured in the nucleus from the I (λ6717) / I (λ6731) of [S II] (Saraph and Seaton, 1970) is ~ 2 x 10^3 cm^{-3}. With this value of the density, the correction of the S II lines due to collisional deactivation of the 2D level of S^+ is small and the high value of I(λ6584)/I(H_α) is likely to be an indication of N overabundance in the nucleus with respect to the H II regions in the disc (see for a more detailed discussion of this point Benvenuti et al., 1973).

The scatter of the points in Fig.4 is correlated with the distance from the center of the galaxy. This is shown in Fig.5, where the I ([N II])/I ([S II]) intensity ratios are plotted as a function of the parameter ρ/ρ_5 where ρ_5 is the optical radius of the galaxy as given by de Vaucouleurs and de Vaucouleurs (1965) and ρ is the distance from the center in the same units. The observations in M 101 by Comte (1975) have also been included in Fig.5. This author shows how the ratio I ([N II])/I ([S II]) is proportional to the relative abundances in moderate excitation H II regions. The observed trend is the same in the two galaxies and points to an increasing nitrogen abundance toward the center. In the nucleus however the value measured in NGC 3079 is twice that of M 101. From the radio observations we know that the central emission region of M 101 is thermal, while in NGC 3079 the emission is non-thermal. In the next paragraph we will investigate in more detail this correlation between optical and radio properties.

Comparison between radio properties and optical emission line data. Table 1 lists radio and optical data for the nuclei of 9 nearby spirals, including IC 342, NGC 2146 and NGC 3079. With the exception of the latter two, all galaxies are seen close to the face-on position and obscuration by dust does not affect the optical image of the nuclei. They have been considered because they have been studied in some detail both in the optical and radio range and may therefore be used to investigate possible correlations between optical and radio properties. References to the optical data are given in the table. We estimate that the errors in our line ratios are lower than 30%. The radio data are mainly from Ekers(1974) and with a few exceptions have not been published in final form. The morphology of the central radio source is multiform and in many cases complex. It is outside the purpose of this paper to discuss whether the differences are evolutionary stages of the same phenomena or represent unrelated physical conditions and it is unlikely that this problem will be solved until more data are accumulated. When in the central region more than one component is present, we give the properties of the radio component with dimensions comparable to the area of optical measurements.

It is known (Smith, 1975, and references herein) that the nitrogen abundance in the interstellar gas increases toward the center of the galaxies. This effect is enhanced in some nuclei where a sharp rise in the [N II] line strength with respect to H_α is present. Burbidge and Burbidge (1965) first called attention to this point. In the sample of Table 1 it can be seen that the only two galaxies which have just thermal radio emission near the nucleus, M 33 and M 101, do show $I(H_\alpha)/I([N\ II])$ and $I([O\ III])/I(H_\alpha)$ with values appropriate to low-excitation H II regions near the center of galaxies (see Searle, 1971, fig.3). The same could be true for IC 342 and NGC 6946 (where the green spectral region was not observed). The sulphur line ratios however indicate in these nuclei electron densities much higher than in normal H II regions observed in the disc of galaxies. It would be of great interest to obtain emission line ratios for classical H II regions associated with the spiral arms in the two galaxies. In the other nuclei the non-thermal radio emission is coupled with the presence of relatively strong [N II] lines. Optical studies based on spectroscopic data have interpreted the strength of the nitrogen emission as due to nitrogen overabundance in the nuclei (Peimbert, 1968; Alloin, 1973; Warner, 1973). It seems now unlikely that nitrogen overabundance and source of non-thermal radiation occuring in coincident or interacting regions are unrelated phenomena. The mechanisms of nitrogen enrichment which do work in the disc of galaxies (see Comte, 1975) may also work at a faster rate in the nuclear region. In this case the presence of the non-thermal source suggests a high rate of supernova explosions as the possible means of enrichment because the shock waves propagating in the interstellar medium could explain the radio emission. This hypothesis however has to be checked quantitatively. The possibility that the line intensities originate from non-radiative ionization mechanisms can be tested from an observational point of view. Osterbrock and Dufour (1974) compared the spectral characteristic of supernova remnants and galaxy nuclei with Cox (1971) shock wave model and line intensity predictions. They computed also approximate strengths for low excitation lines such as $\lambda\lambda$6717-6731 Å of [S II]. As a result of this comparison, they suggest that in some nuclei much of the energy input to the ionized gas is in the form of kinetic energy that is collisionally converted into heat as in SN remnants. The effect of abundance variations was not considered. We can now extend the optical line intensity comparison between SNR and nuclei of galaxies by use of a sequence which shows for SNR the variation of $I(H_\alpha)/I([N\ II])$, $I(H_\alpha)/I([S\ II])$ and $I(6717)/I(6731)$ as a function of expansion velocity as found by Daltabuit et al. (1975). Firstly, we note that the $I(H_\alpha)/I([S\ II])$ intensity ratio is slightly higher in nuclei than in SNR (in old, decelerated remnants with $v_{exp} \simeq 50$ km sec^{-1} the ratio is still smaller than 2). More detailed theoretical computations are needed to understand whether this difference between galactic nuclei and SNR can be explained with different initial conditions such as electron density, magnetic field and initial energy of the explosion. Figure 6 does show the position of the nuclei in the $I(H_\alpha)/I([N\ II])$ versus $I(6717)/(6731)$ diagram for the galactic SNR. At constant, pre-shock density, the ratio of the two sulphur lines is velocity dependent.

Preliminary computations by Daltabuit et al. (1975) derive from a comparison with the observed data an interstellar density of $\simeq 5$ e.cm^{-3}. If the shock takes place in a denser medium for a given velocity and hence $I(H_\alpha)/I([N\ II])$ ratio we expect a higher density, that is a smaller ratio of the two sulphur lines. The galactic sequence of SNR covers an interval of expansion velocities from ≈ 1000 km sec^{-1} to ≈ 30 km sec^{-1}. Motion of the nuclear gas is known only for a few of the objects listed in Table 1, but from the width and shift of the emission lines we can safely assume an upper limit of 300 km sec^{-1} to the non circular velocity: this corresponds to the central part of the galactic SNR sequence. From the diagram it can be concluded that M 81, M 51 and NGC 3079 agree with the sequence only if in the nuclei an overabundance of nitrogen by at least a factor of two with respect to the average galactic value is assumed. The observed line intensities in NGC 6946, M 64 and NGC 2146 could be understood with shock velocities of the order of 100 km sec^{-1} and densities of the surrounding medium between 5 and 10 cm^{-3}. The interpretation of the nucleus of IC 342 is difficult: it may indicate an underabundance of nitrogen or quite different physical conditions.

In conclusion, an interpretation of the line intensities in nuclei of spiral galaxies has been proposed in terms of collisional excitation and N abundance. A correlation has been observed between the presence of a non-thermal radio source and a nitrogen overabundance in the nuclear gas with respect to the H II regions in the spiral arms.

Two of us (P.B. and S.D.) acknowledge partial support from the Italian National Research Council (C.N.R.). We thank A.G. de Bruyn for communicating his radio results in advance of publication.

Table 1

Radio sources and emission line ratios in nuclei of spirals

Galaxy (NGC)	Hubble type	Distance Mpc	Log P_{1415} $WHz^{-1}ster^{-1}$	Spectral index	Optical size	$\frac{H_\alpha}{[NII]}$	$\frac{H_\alpha}{[S\ II]}$	$\frac{[OIII]}{H_\beta}$	$\frac{\lambda 6717}{\lambda 6731}$	Source
598 (M 33)	Sc	0.72	16.3	thermal	60	2.1	2.35	0.46	1.44	1
2146	S pec	14.5$^+$	21.4	-0.5	40	1.12	-	1.55	1.25	2.3
3031 (M 81)	Sab	3.5	19	+0.3°	2.5	0.32	3.85	0.90	0.87	4
					6	0.43	1.23	1.70	1.15	5
3079	SBc	16.5	-	non thermal	6	0.45	1.50	-	1.15	3
4826 (M 64)	Sab	7	19.4	non-thermal +thermal	6	1.08	0.98	-	1.12	4
5194 (M 51)	Sc	9.5	19.9	-1.0	2.5	0.18	2.27	1 :	1.20	4
					7	0.26	1.20	4	1.00	5,1
5457 (M 101)	Sc	7	18.3	thermal	10	2.35	2.24	-	0.88	1
6946	Sc	10	19.9	-0.9	8	1.21	2.33	-	1.05	3
IC 342	Sc	6	19.6	-1.25	10x8	1.62	3.50	0.15	0.84	3

+) Assuming $H = 75\ Km\ s^{-1}Mpc^{-1}$.

Optical sizes are in seconds of arc

Sources: 1) Benvenuti and D'Odorico, 1974; 2) Benvenuti et al, 1975; 3) This paper; 4) Warner, 1973; 5) Peimbert, 1968.

o) The spectrum indicates internal absorption.

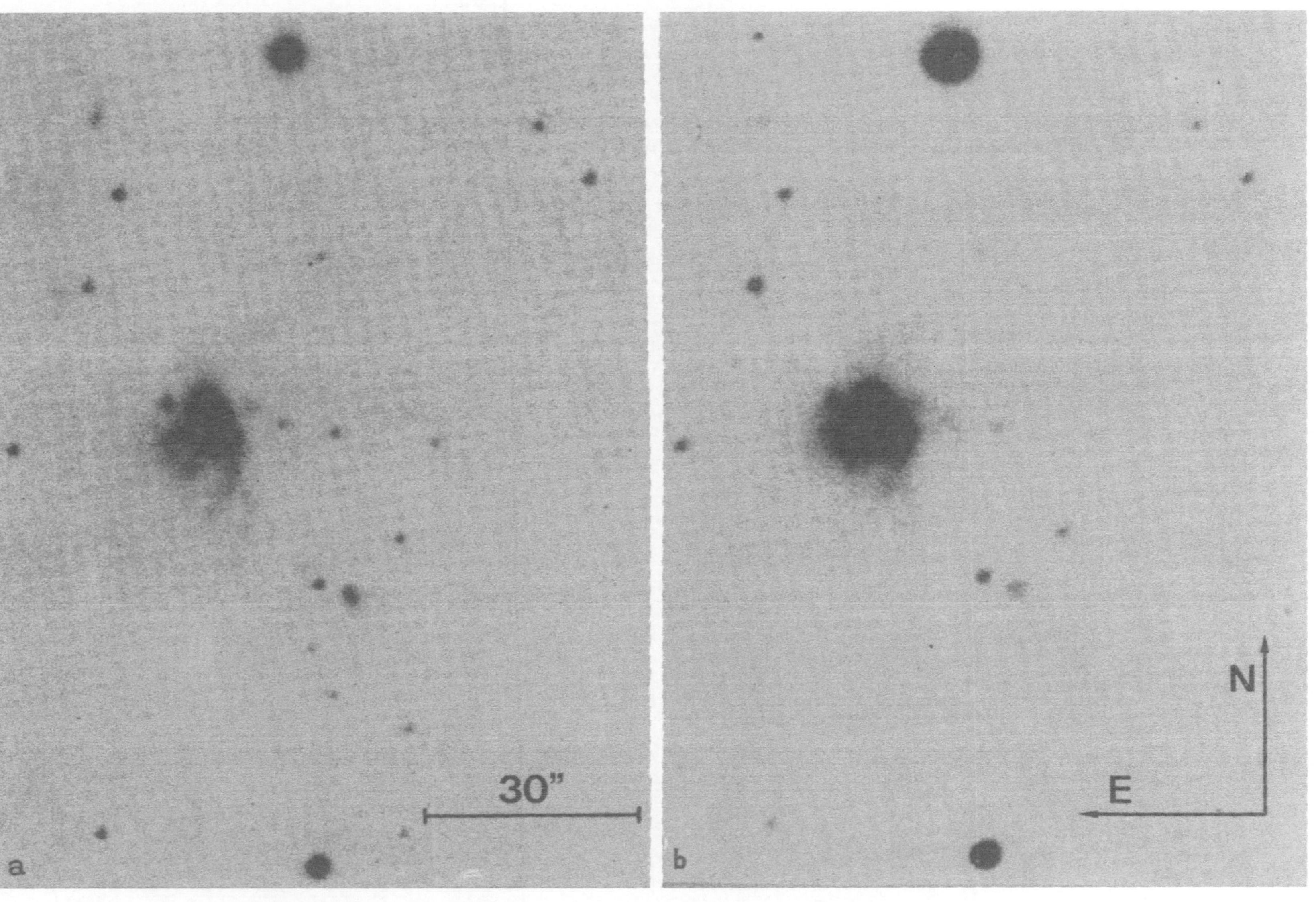

Fig. 1. a) Blue photograph (103a-0) of the nucleus of IC 342. Mount Ekar 182 cm reflector, 37^m exposure.

b) Red photograph (103a-E + RGI) of the nucleus of IC 342. 50^m exposure.

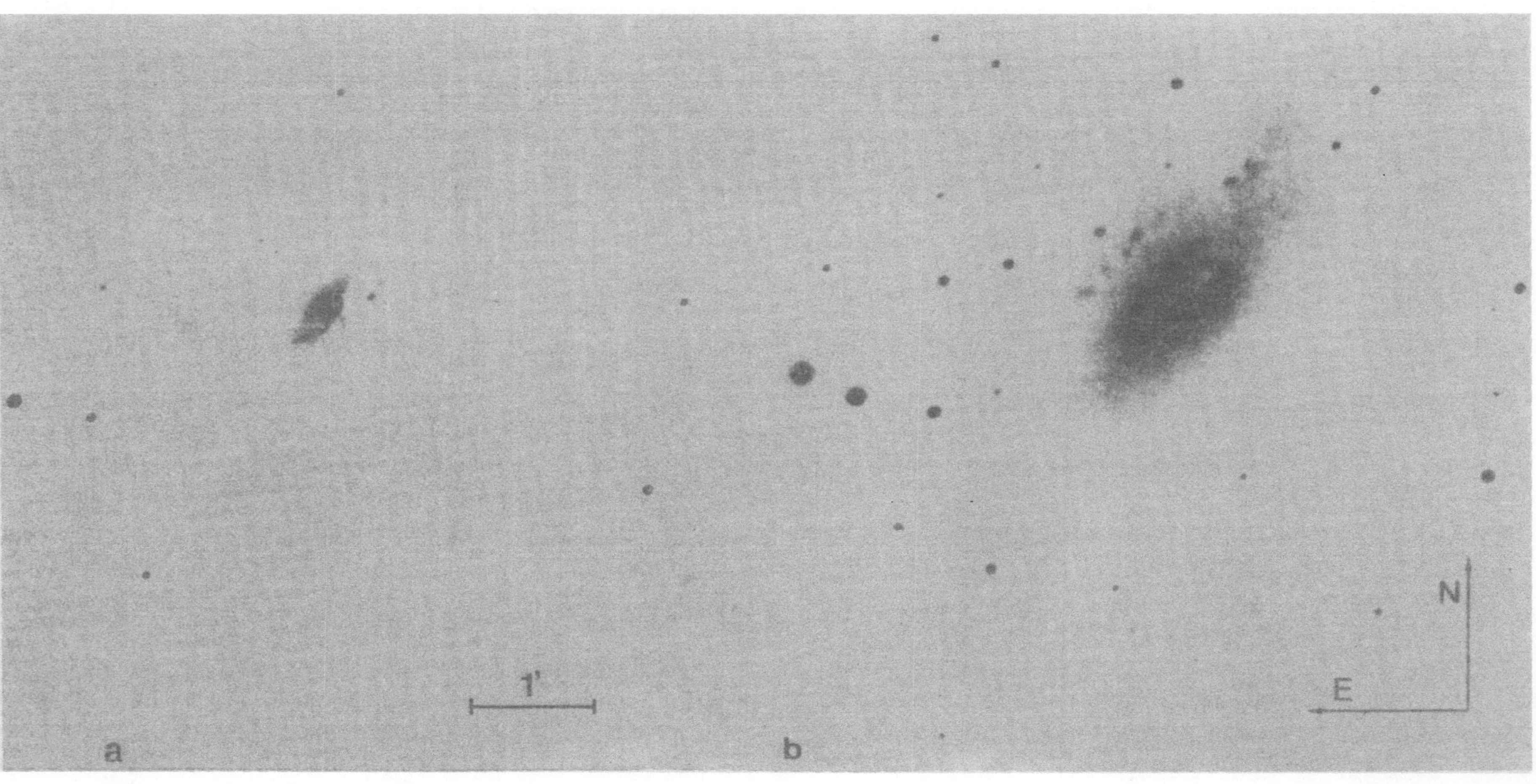

Fig. 2. Photographs of NGC 2146 through a 220 Å half intensity band width interference filter centered at 6590 Å and a WL 30677 image tube. 182 cm Mount Ekar reflector: a) 5^m exposure, b) 13^m exposure.

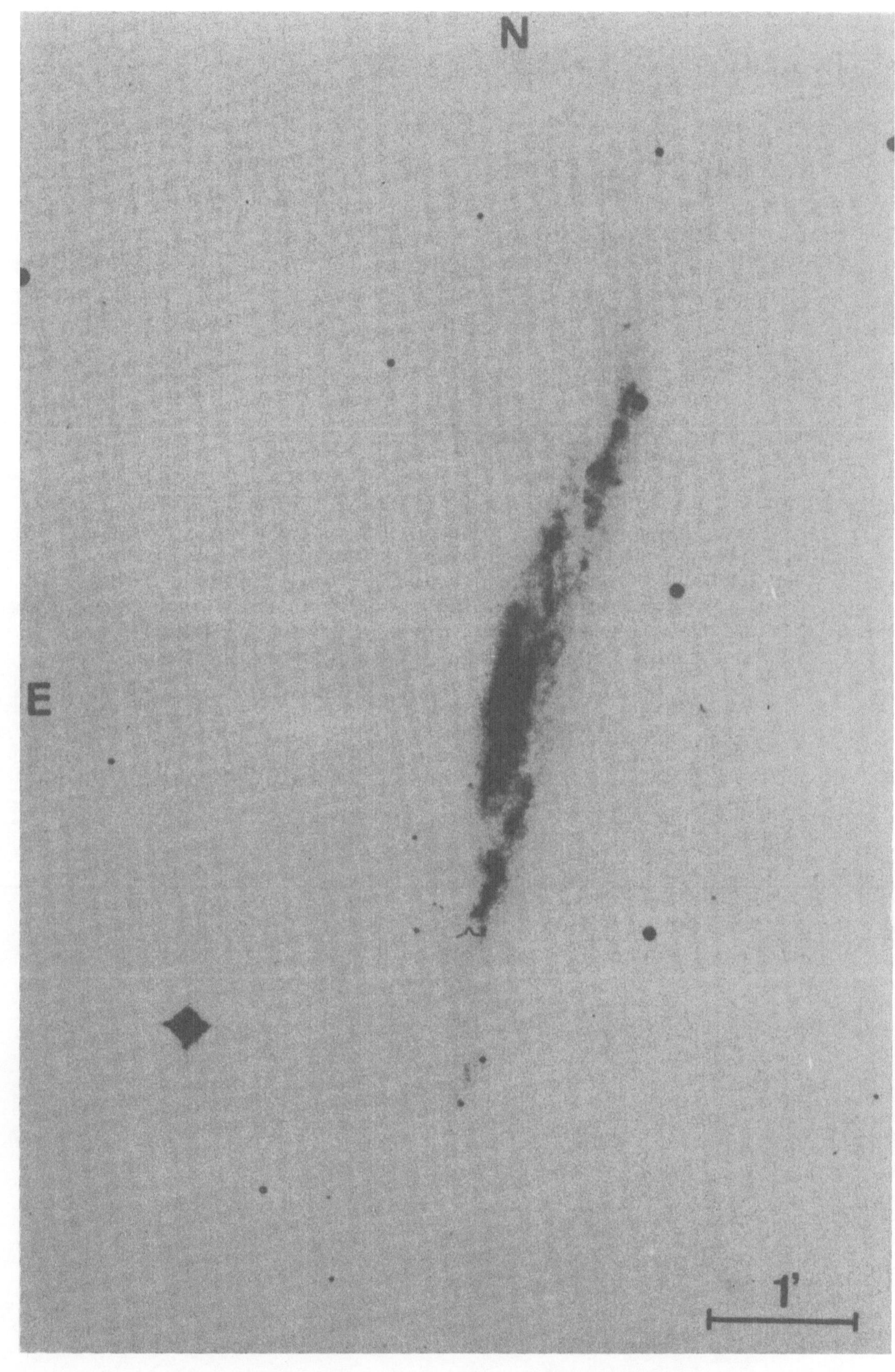

Fig. 3. Blue photograph (103a-0) of NGC 3079. Mount Ekar 182 cm reflector, 15^m exposure.

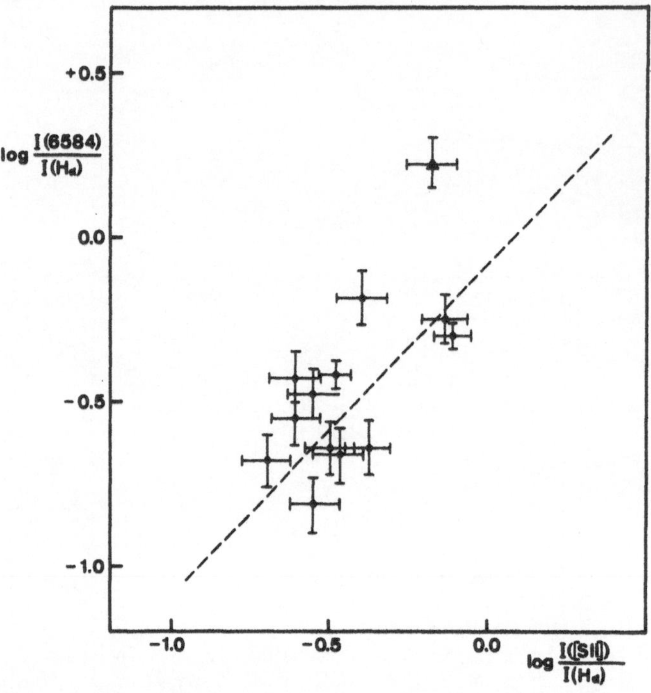

Fig. 4. A plot of the values of the intensity ratio I (λ 6584)/ I (H_α) versus I([SII])/I(H_α) for 12 emission regions in NGC 3079 (filled circles). The triangle indicates the measurement of the nuclear emission, the bars indicate the estimated error. The broken line is the mean relation for H II regions in M 33 and M 101 (Benvenuti et al., 1974).

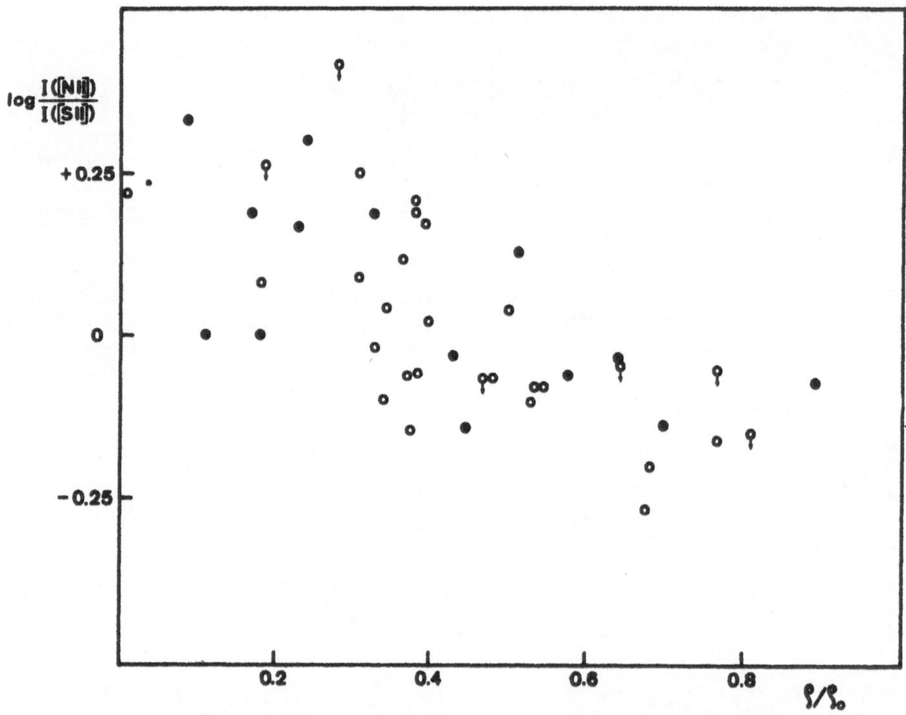

Fig. 5. A plot of the I([NII])/ I([SII]) ratio for H II regions in NGC 3079 (filled circles) and M 101 (open circles) versus the parameter p/p_o where p_o is the diameter of the galaxy and p is the distance from the center in the same units. Data for M101 are after Comte (1974).

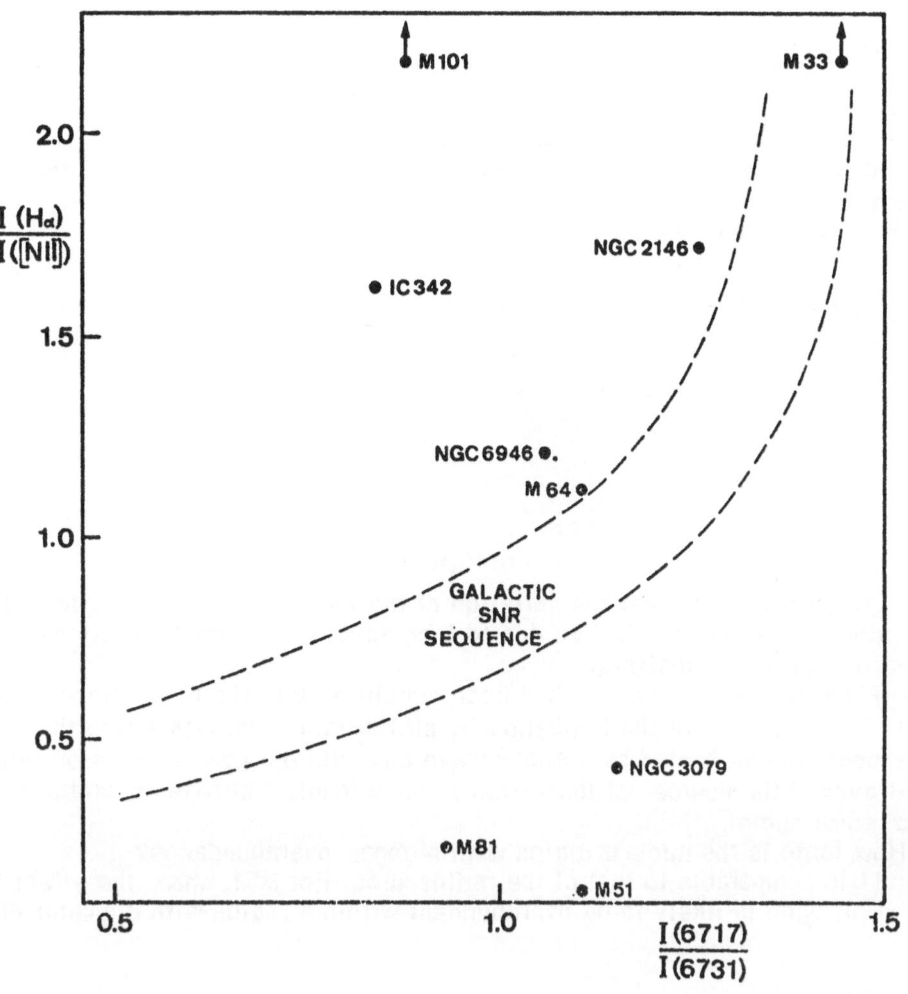

Fig. 6. Line emission intensity ratios for the nuclear region of spiral galaxies. The broken lines delimitate the sequence of evolution of galactic SNR as found by Daltabuit et al. (1974).

REFERENCES

ABLES H.D. 1972, Publ.U.S.Naval Obs. II Ser., Vol. XX, Part IV

ALLOIN D. 1973, Astron. Astrophys., 27, 433

BENVENUTI P., CAPACCIOLI M., D'ODORICO S. 1975, Mem. Soc. Astron. Ital. (in press)

BENVENUTI P., D'ODORICO S. 1974, Proceedings of the 8th ESLAB Symposium A.F.M.Moorwood Ed., ESRO SP-105, p.145

BURBIDGE E.M., BURBIDGE G.R. 1965, Astrophys. J., 142, 634

COMTE G. 1975, Astron. Astrophys., 39, 197

COX D.P. 1972, Astrophys. J., 178, 143

DALTABUIT E., D'ODORICO S., SABBADIN F. 1975, submitted to Astron. Astrophys.

EKERS R.D. 1974, IAU Symposium № 58, Ed. J.R. Shakeshaft, D. Reidel Publ., p.257

HUMASON M.L., MAYALL N.U., SANDAGE A.R. 1956, Astron. J., 61, 97

OSTERBROCK D.E., DUFOUR R.J., 1973, Astrophys. J., 185, 450

PEIMBERT M. 1968, Astrophys. J., 154, 33
SEARLE L. 1971, Astrophys. J., 168, 327
SMITH H.E. 1975, Astrophys. J. (in press)
VAN DER KRUIT P.C. 1973a, Astron. Astrophys., 29, 263
VAN DER KRUIT P.C. 1973b, Astron. Astrophys., 29, 249
VAUCOULEURS G. de VAUCOULEURS A. de 1964, Reference Catalogue of Bright Galaxies, Austin, University of Texas Press
WARNER J.W. 1973, Astrophys. J., 186, 21

DISCUSSION

ARAKELIAN: Why do you interpret the variation of the intensity of the N II lines as an increase in the nitrogen abundance, in turn due to increasing non-thermal radiation? It could also be due to variations in the ionization

D'ODORICO: From detailed studies it has been concluded that the N II strength cannot be explained only by a variation of the ionization by stellar radiation. Comparing the nuclear emission with a model of gas heated by a shock wave and with our observations of galactic SNR, we find that even if the source of ionization is collisional, a nitrogen overabundance must be assumed for some nuclei.

OZERNOY: How large is the nuclear region with nitrogen overabundance?

D'ODORICO: It is comparable to that of the radiosource. For M51, where the effect is most pronounced, the nitrogen is likely to be overabundant within a region with diameter of about 500 pc.

LARGE-SCALE STRUCTURE AND RECENT STAR FORMATION IN THE LARGE MAGELLANIC CLOUD

A. ARDEBERG
LUND OBSERVATORY, SWEDEN

КРУПНОМАСШТАБНАЯ СТРУКТУРА И НЕДАВНИЕ ОБРАЗОВАНИЯ ЗВЕЗД В БОЛЬШОМ МАГЕЛЛАНОВОМ ОБЛАКЕ

Recently the amount of observational data for the Magellanic Clouds has increased quite considerably. Still however, the question of the structure of the Large Magellanic Cloud remains one of very different opinions.

The structure of present-day star formation is conveniently studied by means of the stars of highest luminosity. For such stars observational data are available from investigations covering the main part of the Large Magellanic Cloud.

A catalogue based on data from European Southern Observatory (ESO) in Chile lists more than 400 super-giant stars belonging to the Large Magellanic Cloud (Ardeberg et al. (1972) and a supplement in Brunet et al. (1973)). Sanduleak (1970) has given a finding list of nearly 1300 stars classified as proven or probably members of the Large Magellanic Cloud.

The catalogue based on ESO observations gives MK classes, radial velocities and photometric data. Thus, over 400 super-giant stars concerned can be regarded as definite members of the Large Magellanic Cloud. The stars included were selected from the surveys of Fehrenbach and Duflot (1970, 1973). This means that there is no spectral-type bias (as Fehrenbach and Duflot used radial velocity as a membership criterion) in the material and that the limiting magnitude is practically constant over the field studied.

The field covered by the catalogue based on ESO observations is given in Fig.1. The indication of the field is only approximate. The corresponding surface distribution of super-giant stars is given in Fig.2. The indication of the field is here more detailed than in Fig.1. The coordinates are on the Wesselink (1959) system.

From Fig.2 no significant spectral-type separation can be detected.

Sanduleak made his selection from classifications on low-dispersion (580 Å/mm) objective-prism plates. Therefore, Sanduleak's catalogue is bound to contain a significant spectral-type bias. Fig.3 displays the stars classified by Sanduleak as definite OB stars and B-G super-giant stars. The large-scale agreement between Figs.2 and 3 is good, especially if only the OB stars in Fig.3 are taken into account.

The resulting picture of young structure in the Large Magellanic Cloud is one of mainly four heavy concentrations and two fainter ones.

It seems quite safe to conclude that the Bar is not a place of present-day star formation. On the other hand there is little doubt that the 30 Doradus complex is a centre of intense star formation.

On a direct photograph (such as that in Fig.1) the Bar and the 30 Doradus complex are of highest intensity. The somewhat fainter North-East feature is however matched by a large concentration of luminous stars.

Quite faint on the photographic plate are the features North and South of the North-West part of the Bar. Anyhow, they correspond to high amounts of super-giant stars.

The photographically hardly visible feature down to the South-East is matched by a quite significant concentration of luminous stars.

Finally, there is an arc-like structure of super-giant stars extending southwards approximately from the end points of the Bar.

Watts (1972) recently published a far-ultra-violet photograph of the Large Magellanic Cloud taken from the Moon. It is given in Fig.4 together with a blue plate for the sake of orientation.The agreement between the distributions outlined by Watts' far-ultra-violet photograph and the concentrations of supergiant stars is excellent.

The distributions of stellar associations (Lucke and Hodge, 1970), H II regions (Henize, 1956) and large H I complexes (McGee and Milton, 1966) are all in good agreement with that defined by the super-giant stars.

The resulting over-all picture of young structure in the Large Magellanic Cloud is one of well-defined concentrations. This is in very good agreement with the ideas of super-associations (Westerlund, 1974).

From direct photographs, star counts and surface photometry de Vaucouleurs and Freeman (1972) made a study of the structure of the Large Magellanic Cloud. They summarized their results as in Fig. 5. It is pointed out that the orientation of Fig.5 is reversed with respect to that of the previous figures.

De Vaucouleurs and Freeman gave a detailed account of what they defined as "an extensive spiral structure".

It is noted that, whereas de Vaucouleurs and Freeman characterized the South-East feature as "one of those marked by the richest concentrations of super-giant stars", they did not mention the arc-like structure South of the Bar. It is shown here that the arc structure is richer of the two in luminous stars.

Except for this discrepancy the concentrations of super-giant stars confirm the major features given by de Vaucouleurs and Freeman.

However, from the distributions of luminous stars indication of spiral structure is not found to be extremely evident. Especially it seems hard to recognize the connections between spiral arms and central region as outlined by de Vaucouleurs and Freeman. A striking example is the rich South-West concentration.

It is emphasized that the difficulties to recognize the spiral structure defined by de Vaucouleurs and Freeman are equally manifested if one regards the far-ultra-violet photograph published by Watts.

The stars taken from the catalogue based on ESO observations all have photometric data measured. Further, the distance modulus of the Large Magellanic Cloud is quite well-known,and the colour excesses are small and fairly easy to determine. Thus, with very reasonable accuracy values of absolute luminosity and effective temperature may be attributed to the stars.

Using data for the initial main sequence by Sandage (1958) and model calculations by Stothers (1963, 1964, 1966a, 1966b) and Iben (1965, 1966a, 1966b, 1967) the stellar data may be calibrated in terms of age. This should give quite reasonable ages, especially relative ages, considering the fact that all the stars included have absolute visual magnitudes brighter than-5.

Fig.6 gives resulting "apparent birth function" versus age. It covers 2 x 10^7 years and it is corrected for the influence of bolometric correction but not for the effect of high-luminosity depletion.

The corresponding depletion function is shown in Fig.7. The quantity given is the ratio of all stars formed to those still being around. This ratio is displayed versus age. Constant starbirth rate has been assumed. It seems evident that the influence of depletion on the birth function is of a minor order only.

It is therefore indicated that the formation of stars in the Large Magellanic Cloud during the last 2 x 10^7 years has been mainly a one-event feature. Systematic errors in the evolutionary models used would tend to push the time of this "birth flash" forwards or backwards but hardly smooth it out.

Hodge (1973) recently studied the evolutionary history of the Large Magellanic Cloud. He used age determinations for clusters and covered the last 14 x 10^6 years.

Hodge concluded that clusters are formed in several bursts, well distributed in time and space.

However, if Hodge's clusters are plotted in an "apparent birth function" diagram the result is that given in Fig.8. In this figure the "apparent birth functions"for super-giant stars and for clusters are displayed together. None of the two functions in Fig. 8 has here been corrected for the influence of bolometric correction. It seems safe to conclude that the two functions agree quite well.

If for the concentrations outlined, individual functions are constructed, one-peak distributions are formed. The peak values of these distributions occur for ages very much the same as that for the peak value of the total birth function.

It was concluded above that the structure of the extreme Population I in the Large Magellanic Cloud hardly indicated spiral structure. Evidently, the one-event star formation even less favours spiral-structure type generation of stars.

The parent-mass concentrations could well have been ejected from a common point of origin some time outside the range of the diagrams given above. Then, about 7-8 x 10^6 years ago star formation started simultaneously in these aggregates or super-associations.

This would be in good agreement with the suggestions made by Ambartsumian (1975)

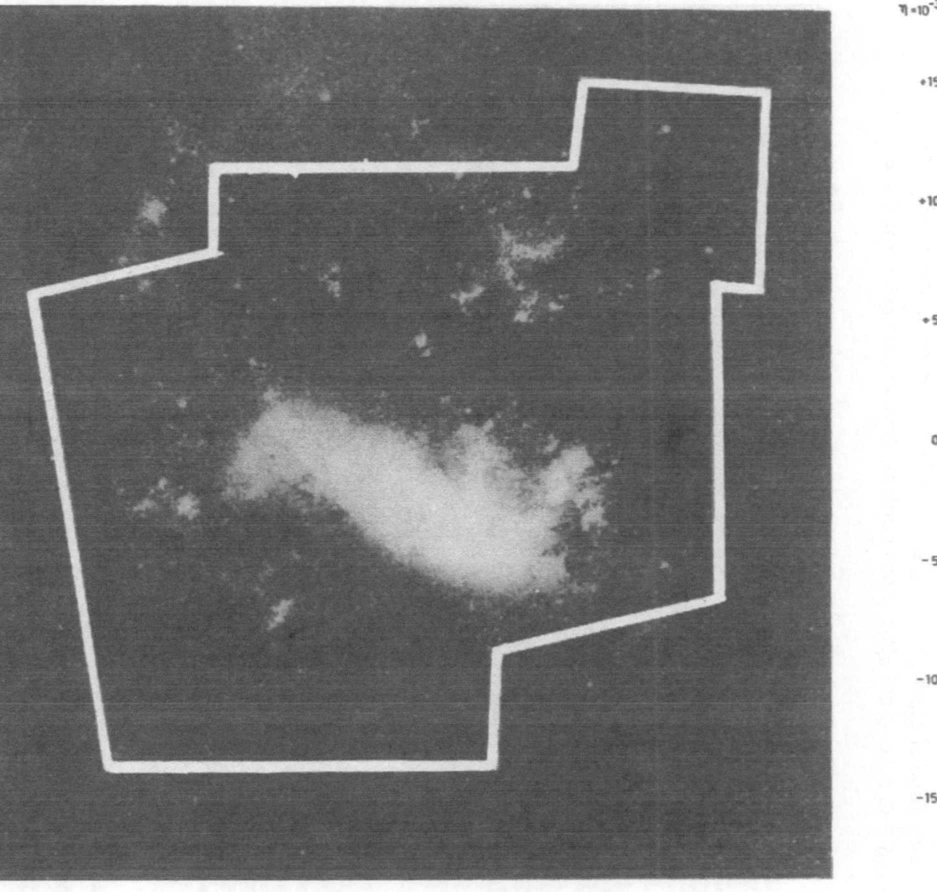

Fig. 1. The Large Magellanic Cloud from a blue-range (photographic) Harvard plate. The indications give the approximate limits of the field studied in the work by Ardeberg et al.(1972). North is up and West is to the left.

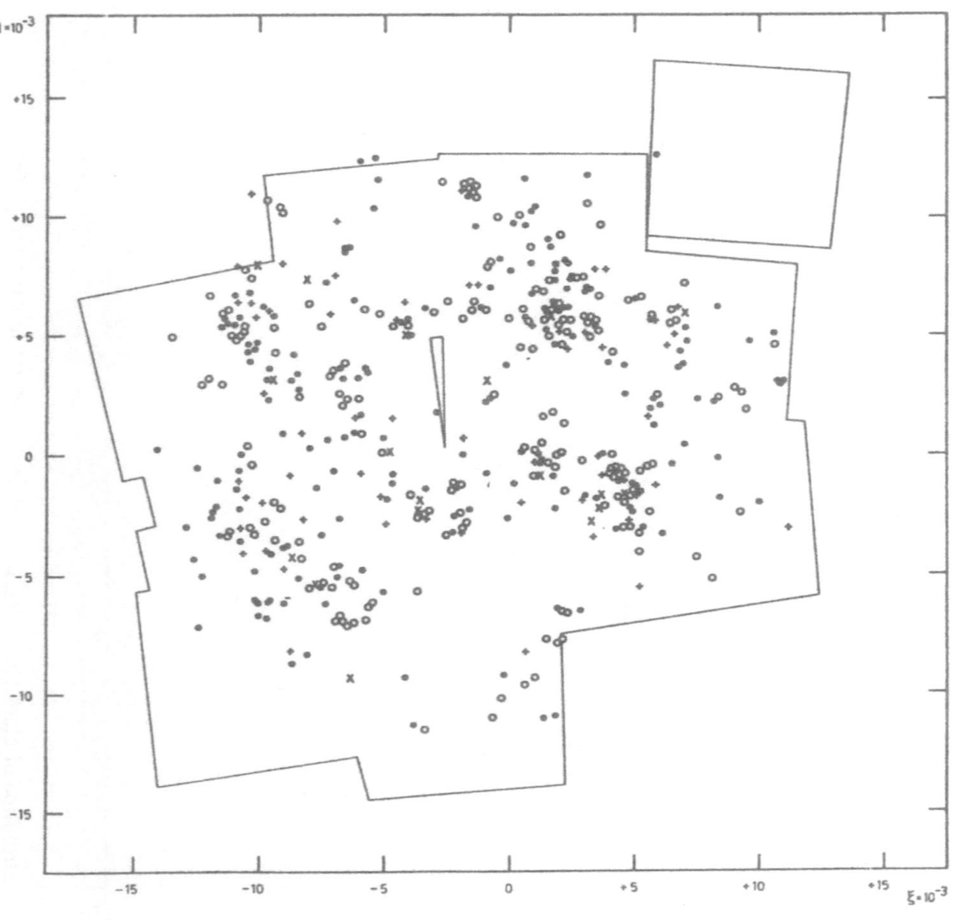

Fig. 2. Surface distribution of the super-giant stars taken from the catalogue based on ESO observations. The field indication is the same as in Fig. 1 but more detailed. Stars of spectral types O and B are denoted by open circles, stars of spectral type A by filled circles, and stars of spectra types F and G by horizontal-vertical crosses. Stars classified pec. are denoted by diagonal crosses. Coordinates are given on the Wesselink (1959) system.

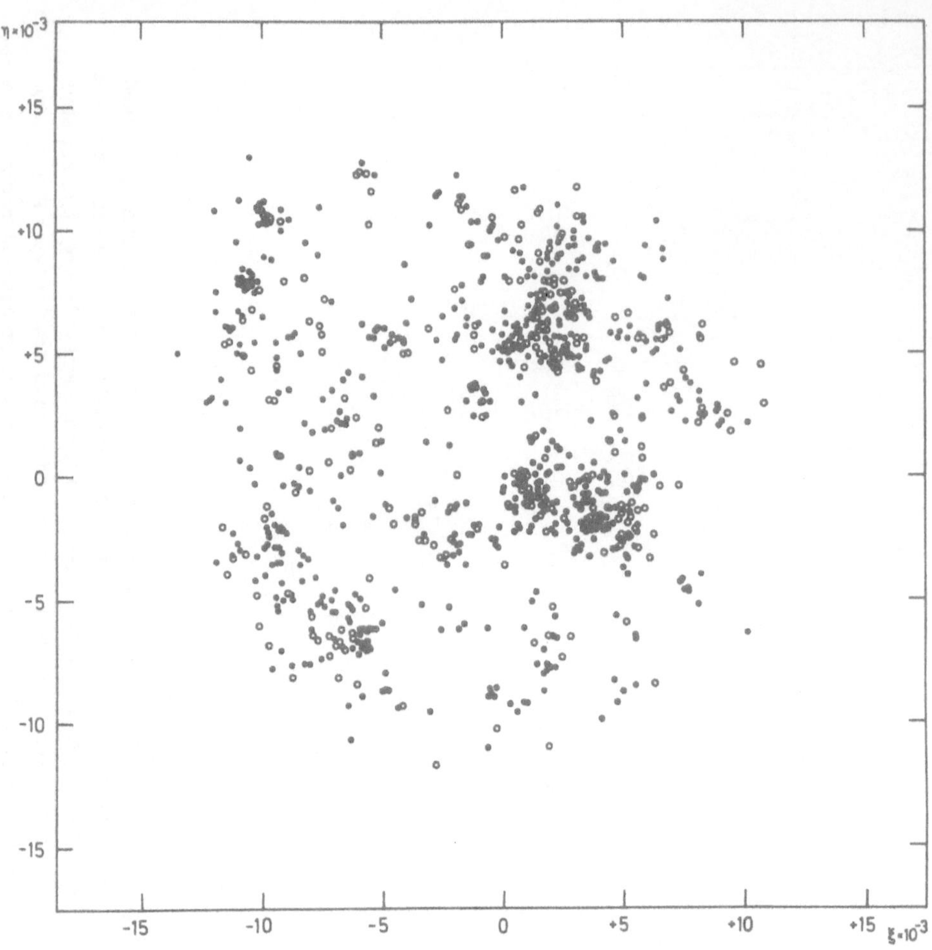

Fig. 3. Surface distribution of stars classified by Sanduleak as definite OB stars (filled circles) and B-G super-giant stars (open circles). Coordinates are the same as in Fig. 2.

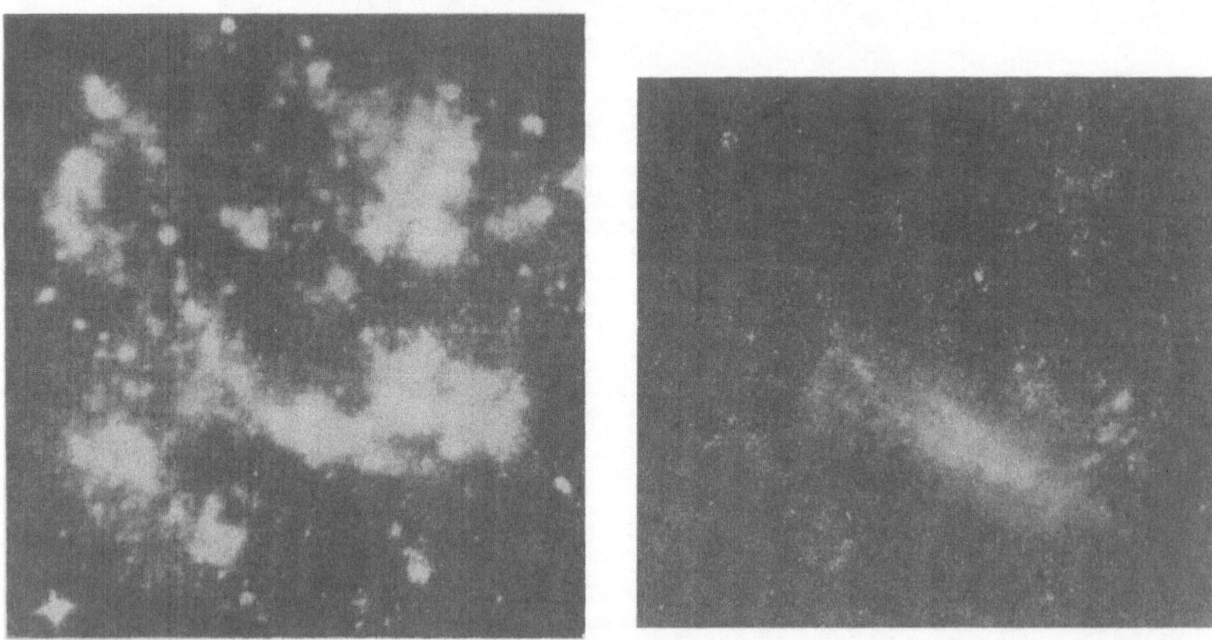

Fig. 4. Far-ultra-violet photograph (left frame) of the Large Magellanic Cloud published by Watts (1972). The right frame gives a blue-light Harvard photograph on the same scale and orientation as the left frame. North is up and West is to the left.

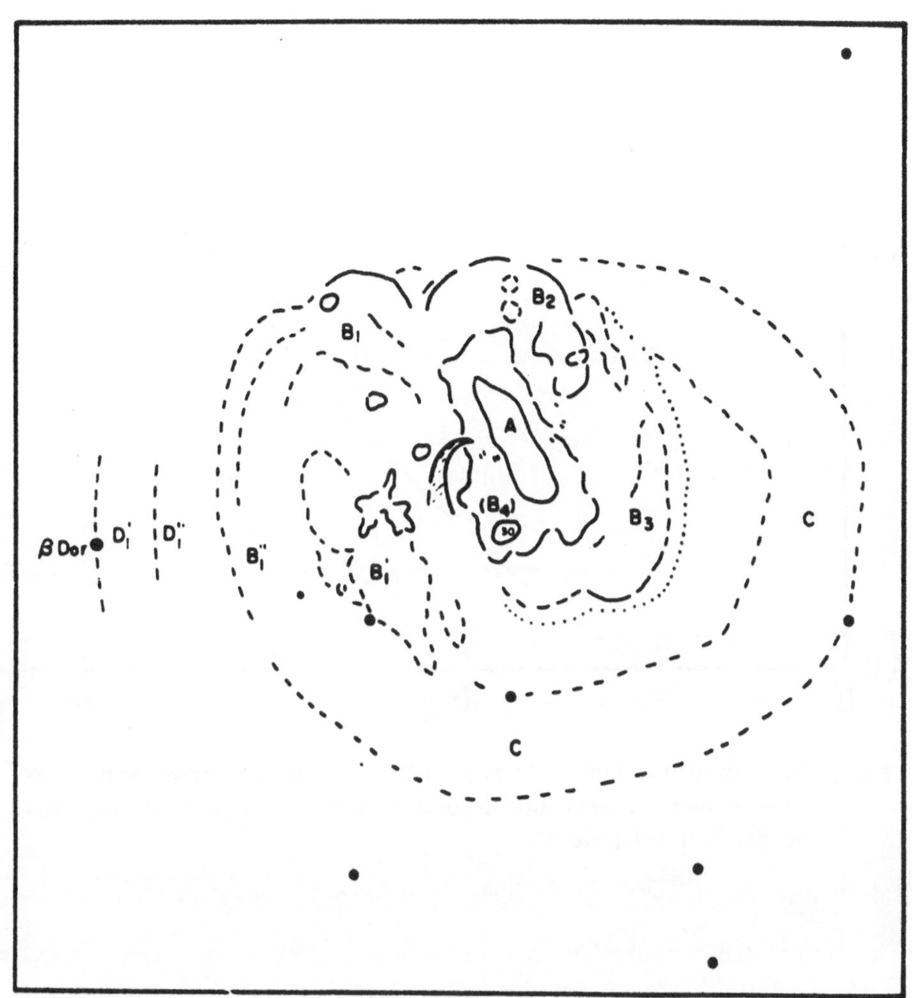

Fig. 5. The structure of the Large Magellanic Cloud according to de Vaucouleurs and Freeman (1972). North is to the left and West is up. Note that the field is reversed and turned 90° with respect to those of Figs. 1, 2, 3 and 4.

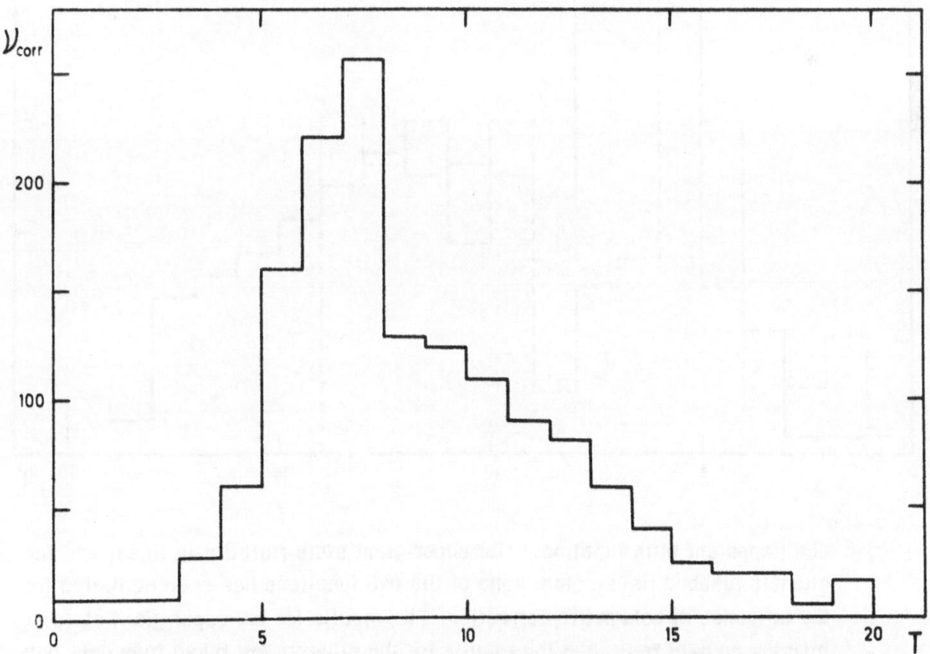

Fig. 6. "Apparent birth function" for the Large Magellanic Cloud over the last 2×10^7 years. Correction is made for the influence of bolometric correction but not for the effect of high-luminosity depletion. The apparent birth frequency in births per 10^6 years is given versus age, expressed with 10^6 years as unit.

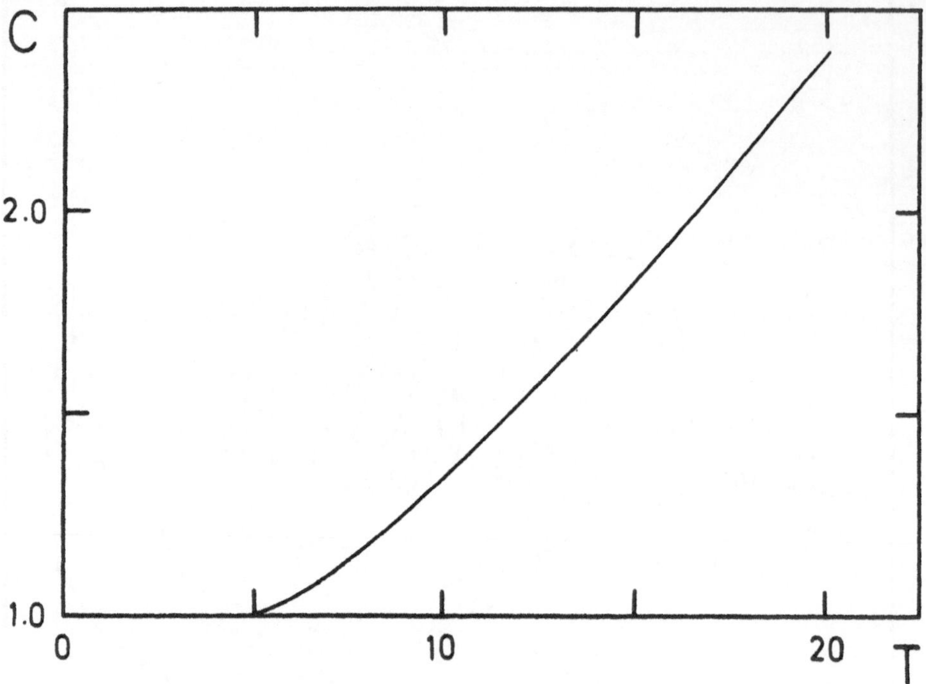

Fig. 7. The correction for high-luminosity depletion versus age, expressed with 10^6 years as unit. The correction factor displayed is the ratio of all stars formed to those still being around.

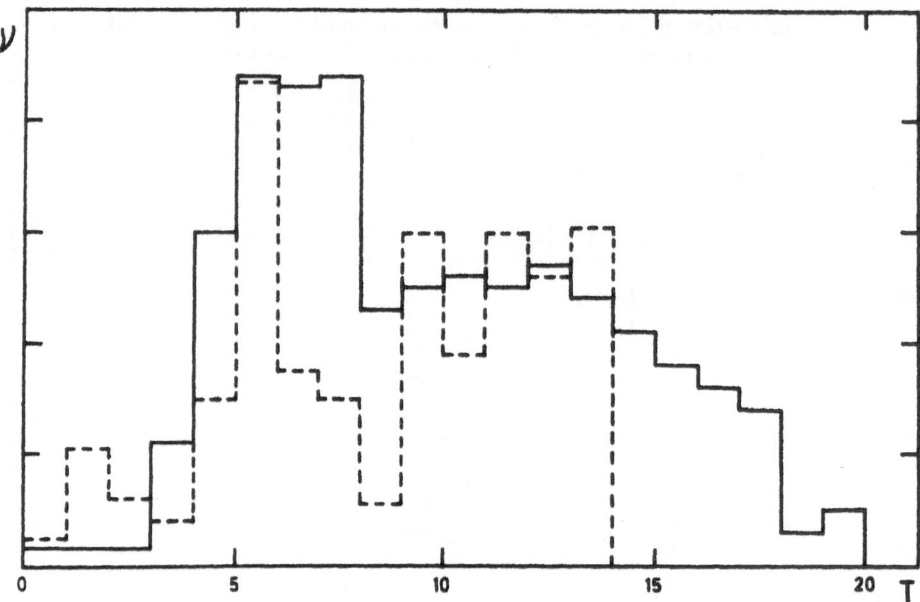

Fig. 8. The "apparent birth functions" for super-giant stars (full-drawn lines) and for clusters (dashed lines). Here none of the two functions has been corrected for the influence of bolometric correction. The results for the super-giant stars are from the present work, and the results for the clusters are taken from data published by Hodge (1973).

REFERENCES

AMBARTSUMIAN V.A.1975. The Role of the Nuclear Activity in the Overall Evolutionary Processes in Galaxies, in Proc. Third European Astronomical Meeting.

ARDEBERG A., BRUNET J.P., MAURICE E., PRÉVOT L. 1972, Astron.and Astrophys. Suppl. Ser. 6, 249

BRUNET J.P., PRÉVOT L., MAURICE E., MURATORIO G.,1973.Astron.and Astrophys. Suppl. Ser. 9, 447

FEHRENBACH Ch., DUFLOT M. 1970,Astron.and Astrophys. Special Suppl. Ser. 1

FEHRENBACH Ch., DUFLOT M. 1973. Astron. and Astrophys. Suppl. Ser. 10, 231

HENIZE K.G. 1956, Astrophys. J. Suppl. Ser. 2, 315

HODGE P.W. 1973, Astron. J. 78, 807

IBEN Jr., I. 1965, Astrophys. J. 142, 1447

IBEN Jr., I. 1966a, Astrophys. J. 143, 483

IBEN Jr., I. 1966b, Astrophys. J. 143, 516

IBEN Jr., I. 1967, Astrophys. J. 147, 650

LUCKE P.B., HODGE P.W. 1970, Astron. J. 75, 171

McGEE R.X., MILTON J.A. 1966, Australian J. Phys. 19, 343

SANDAGE A. 1958, Luminosity Functions of Galactic Clusters, Globular Clusters, and Elliptical Galaxies, in Stellar Populations, Ed. D.J.K. O'Connell, S.J., North Holland Publishing Company, Amsterdam, Interscience Publishers, Inc., New York, p.75

SANDULEAK N. 1970, Contr. Cerro Tololo Int. - Amer. Obs. 89

STOTHERS R. 1963, Astrophys. J. 138, 1074

STOTHERS R. 1964, Astrophys. J. 140, 510

STOTHERS R. 1966a, Astrophys. J. 143, 91

STOTHERS R. 1966b, Astrophys. J. 144, 959

VAUCOULEURS G., de, FREEMAN K.C. 1972, Structure and Dynamics of Barred Spiral Galaxies, in particular of the Magellanic Type, in Vistas in Astronomy, Ed. A. Beer, Pergamon Press, Oxford-New-York-Toronto-Sydney-Braunschweig, p. 163.

WATTS Jr., R.N. 1972, Sky Tel. 44, 6

WESSELINK A.J. 1959, Monthly Notices Roy. Astron. Soc. 119, 576

WESTERLUND B.E. 1974, The Magellanic Clouds, in Proc. First European Astron. Meeting, Vol. 3; Galaxies and Relativistic Astrophysics, Eds. B. Barbanis and J.D. Hadjidemetriou, Springer.Verlag, Berlin-Heidelberg-New York, p.39.

IS THE CHAIN OF GALAXIES NEAR NGC 247 ANOMALOUS ?

C.BALKOWSKI, P.CHAMARAUX
OBSERVATOIRE DE MEUDON, FRANCE
PRESENTED BY C.BALKOWSKI

АНОМАЛЬНА ЛИ ЦЕПЬ ГАЛАКТИК ВБЛИЗИ NGC 247 ?

I n t r o d u c t i o n. The Burbidge's chain is located at 20' NE of NGC 247; it consists of 5 late-type galaxies A, B, C, D, E lying about 8' from north to south (Fig.1). Two components are predominant (86% of the total luminosity); their spectra were taken by Burbidge *et al.*(1963). They derived recession velocities of 6164 km s^{-1} and 6308 km s^{-1} respectively for B and E. In view of the emission features observed in these spectra and of the remarkable alignment of the members of the chain, these authors concluded at a possible recent formation of this object.

On the other hand, Arp (1973), by statistical considerations of the neighbourhood of nearby spiral galaxies, proposed that there could be a physical association between NGC 247 and the chain, and therefore anomalous redshifts in the group. An answer to Burbidge's and Arp's proposals can be given by the use of 21-cm line and optical observations.

21 - cm l i n e o b s e r v a t i o n s. The 21-cm line observations were carried out with the Nançay radiotelescope, with 15 adjacent channels spaced by 63.3 km s^{-1}; the beam of the instrument is 4'x 24' at this declination, so that the different members of the chain are not resolved. Figures 2 and 3 present the profiles obtained at the optical positions of B and E. As B and E are predominant in luminosity the main contribution to these profiles is provided by these galaxies, even if A, C and / or D have velocities in the range displayed in the observed line. Note the dissymmetry of the two lines with a reinforcement at the optical velocity of E, and which is more pronounced on the line obtained at the position of E; note also that the centre of the profile obtained on B is nearly at the optical velocity of B. These peculiarities can be accounted by the fact that B is nearly edge-on, whereas E is face-on.

D i s t a n c e d e t e r m i n a t i o n s. The derivation of the distance of a galaxy is based on relations between its total parameters (Balkowski et al.,1973). The distance criteria are applied to the galaxy B which is presumably predominant in the total profile. For that purpose the most likely values of the width and of the area of the 21-cm line of B must be drawn from the total profile; three cases have been considered according to the contribution of the other galaxies to this profile.

The distance values with their errors are given in Table 1. A special care was taken for the error calculations, in particular the true dispersions were used in the relations involved.

In the three considered cases, the calculated values d are in complete agreement with the cosmological distance $d_C = 89^{+16}_{-13}$ Mpc using a Hubble constant H = 70 ± 13 km $s^{-1}Mpc^{-1}$ (Durand,1975) obtained from the same distance criteria as ours. On the contrary d is not compatible with $d' = 1.8^{+0.7}_{-0.5}$ Mpc, the distance of NGC 247; the discrepancy being more 3.9 times the mean uncertainty.

C o n c l u s i o n. Therefore a physical association is ruled out between B and NGC 247. This conclusion is probably true for the whole chain.

On the other hand no evidence of youth of the chain is brought to light, neither from the dynamical state (probable stability of the chain), nor from the neutral hydrogen content which is fairly normal for the types involved.

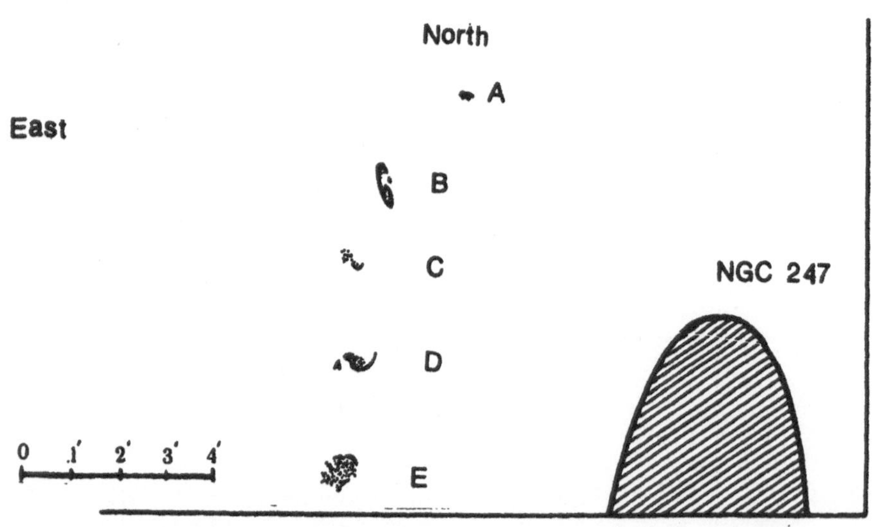

Fig. 1.

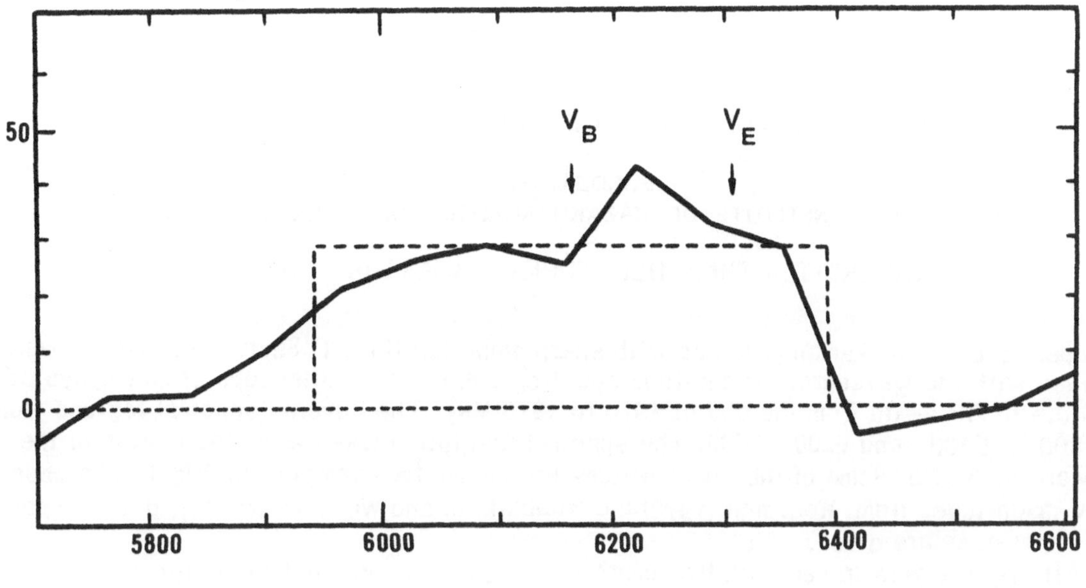

Fig. 2.

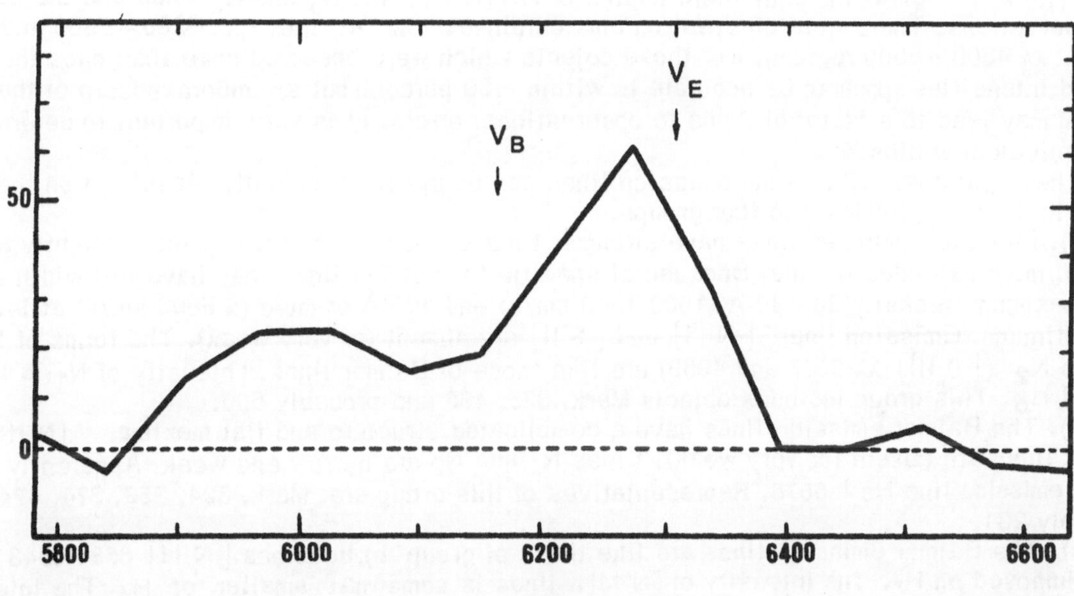

Fig. 3.

REFERENCES

ARP H. 1973, Astrophys. J. 185, 797

BALKOWSKI C., BOTTINELLI L., CHAMARAUX P., GOUGUENHEIM L., HEIDMANN J. 1973, Astron. and Astrophys. 25, 319

BALKOWSKI C., CHAMARAUX P. 1975, this work, Astron. and Astrophys. in press.

BURBIDGE E.M., BURBIDGE G.R., HOYLE F. 1963, Astrophys. J. 138, 873.

DURAND N. 1975, Thèse d'Université, Paris VII.

THE CHARACTERISTICS OF SOME SEYFERT GALAXIES

E.K.DENISYUK
ASTROPHYSICAL INSTITUTE OF KAZAKH ACADEMY OF SCIENCES, U.S.S.R.

ХАРАКТЕРИСТИКА НЕКОТОРЫХ ГАЛАКТИК СЕЙФЕРТА

Spectra of 22 Markarian galaxies with sharp emission lines ("Seyfert galaxies") were obtained with the Cassegrain image-tube spectrograph on 70cm telescope of Astrophysical Institute in Alma-Ata from the end of 1970 to 1975 May. The covered spectral regions were λλ 4800 - 5500 and 6200 - 7500. The spectral resolution was 4-8 Å, the widths of the slit were 5."5-7."5. Some of the spectrograms are shown for example in Fig.1. The observed V-magnitudes from Weedman,(1973) and Khachikian and Weedman,(1974), dates,exposures and dispersions are given.

All spectra were traced with two-canal microphotometer of right intensity. Second canal was used for subtraction of the night sky spectrum. The results of the observations are given in Fig.2 A, B and C in form of the profiles of the emission lines, normalized to continuum. It should be noted that galaxies in Fig.2 C were observed only in red light. Some emission lines in red are distorted with absorption night sky lines (λλ ~6840 and 6880).These absorption lines are shown by vertical broken lines.

Table I gives the equivalent widths of H_α+[N II], H_β, N_1 and N_2 lines and the dates of observations. There were 60 spectrograms obtained: 42 at the λλ 6200 - 7500 and 18- at the λλ 4800 - 5500 regions. For those objects which were observed more than once the observed intensities appear to be accurate to within ±30 percent but an underexposure of the continuum may lead to a factor of 2 due to observational errors. It is very important to determine the equivalent widths.

These galaxies with strong emission lines can be spectroscopically classified on the basis of their line profiles into four groups.

a). Emission hydrogen lines have strong but narrow nuclei and broad wings, slightly asymmetric, more extended to blue. Because of specific form of the lines they have full width at half-maximum intensity 20 - 30 Å (1000-1500 km/s) and 100 Å or more (> 5000 km/s) at level of continuum. Emission lines [N II] and [S II] are absent (or very weak). The forms of lines N_1 and N_2 ([O III] λλ 5007 and 4959) are like those of Balmer lines. Intensity of N_1 is like that of H_β. This group includes objects Mark. 335, 486 and probably 506.

b). The Balmer emission lines have a complicated structure and flat maxima. [N II] and [S II] lines are absent (or very weak). Lines N_1 and N_2 are narrow and weak. Apparently there is the emission line He I 6678. Representatives of this group are: Mark. 304, 352, 374, 474 and probably 231.

c). The Balmer emission lines are like those of group b), but lines [N II] 6584-6548 are superimposed on H_α. The intensity of [N II].- lines is somewhat smaller of H_α. The intensity of line 6584 is stronger than that of lines [S II] 6517-6731 is stronger too. The intensity of line N_1 is like that of H_β. Lines N_1 and N_2 have full width at half-maximum intensity ~20 Å. Group c) includes objects Mark. 10, 78, 273, 372, 376, 504 and probably 9, 133, 382, 507.

d). Emission hydrogen lines have an average width and smooth forms. There are intensity lines of [N II] and [S II]. N_1 and N_2 - lines are extremely strong, brighter than all other lines of optical spectra. Mark.3, 34 and probably 1 and 348 are representatives of this group.

Groups a) and b) consist of objects of SyI -class (see Weedman, 1973, Markarian, 1973) group d) includes objects of Sy2-class and galaxies with weak Seyfert features. These are SyI and Sy2-class objects in group c).

Table 2 contains average U - B and B - V colors (from Weedman, 1973 and Khachikian and Weedman, 1974) for our groups of galaxies. It can be seen from Table 2, that our Seyfert galaxy classification by the shapes of their emission lines agrees quite well with division into SyI and Sy2 types according to the two-color diagram. That indicates the close correlation between the processes of emission line radiation and integral characteristics of continuum. As the spectra are more informative than the UBV-photometry, for example, the spectra study permits us to obtain more detailed differences and so to ascertain much more physical various groups among our objects. It is evident that the increase of the number of Seyfert galaxies with detail studied spectra, the improvement of the accuracy of these investigations and the extension of the spectral range will tend us to more complete understanding of the nature of these interesting objects.

Table 1.

Mark No.	Equivalent width				Date of observ. *	
	Hα+[NII] . 6548 6584	[SII] . 6717 6731	[OIII] . 5007 4959	Hβ	Hα	Hβ
1	210	85	not observed		10.71	
3	265	75	800	40	5.75	4.75
9	(>)** 150	-	45	55	5.75	5.75
10	260	20	55	55	5.75	4.75
34	100	32	480	75	5.75	12.70
78	(>) 200	85	175	38	5.75	4.75
133	80	-	not observed		5.75	
231	(>) 230	-	" "		5.75	
273	(>) 140	30	50	25	5.75	4.75
304	400	-	24	100	10.73	10.73
335	300	-	75	150	10.73	10.73
348	100	55	not observed		11.73	
352	250	-	10	70	10.73	10.73
372	(≥) 100	10	15	?	10.73	10.73
374	(>) 120	-	40	45	10.73	4.75
376	(>) 380	20	42	140	11.72	4.75
382	100	30	32	40	5.75	4.75
474	(>) 600	-	80	105	3.73	3.73
486	(>) 325	-	not observed		10.73	
504	(>) 150	-	" "		9.73	
506	(>) 240	-	" "		10.73	
507	70	-	" "		10.73	

* Month and year for observation of the best spectrogram, which was used for obtaining equivalent width.

** These lines are distorted with absorption night sky lines.

Table 2.

Groups	U - B	B - V	Quantity
a	- 0.64	+0.53	3
without a?	- 0.69	0.42	2
b	- 0.39	0.66	5
without b?	- 0.52	0.62	4
c	- 0.10	0.80	10
without c?	- 0.31	0.81	6
d	+0.12	1.01	4
without d?	+0.11	1.10	2

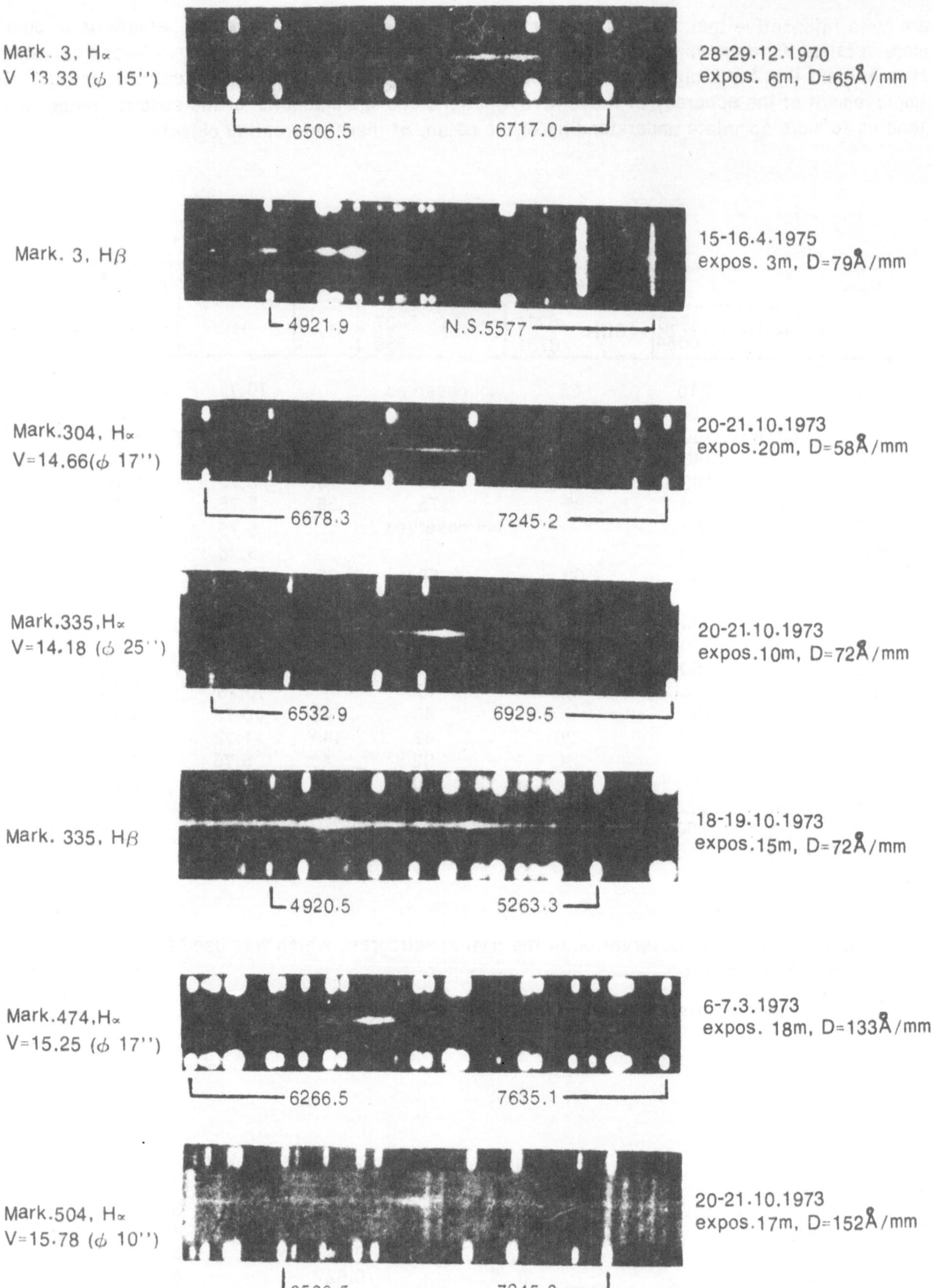
Mark. 3, Hα
V 13.33 (ϕ 15'')
28-29.12.1970
expos. 6m, D=65Å/mm
6506.5
6717.0
Mark. 3, Hβ
15-16.4.1975
expos. 3m, D=79Å/mm
4921.9
N.S.5577
Mark.304, Hα
V=14.66(ϕ 17'')
20-21.10.1973
expos.20m, D=58Å/mm
6678.3
7245.2
Mark.335,Hα
V=14.18 (ϕ 25'')
20-21.10.1973
expos.10m, D=72Å/mm
6532.9
6929.5
Mark. 335, Hβ
18-19.10.1973
expos.15m, D=72Å/mm
4920.5
5263.3
Mark.474,Hα
V=15.25 (ϕ 17'')
6-7.3.1973
expos. 18m, D=133Å/mm
6266.5
7635.1
Mark.504, Hα
V=15.78 (ϕ 10'')
20-21.10.1973
expos.17m, D=152Å/mm
6506.5
7245.2

Fig. 1.

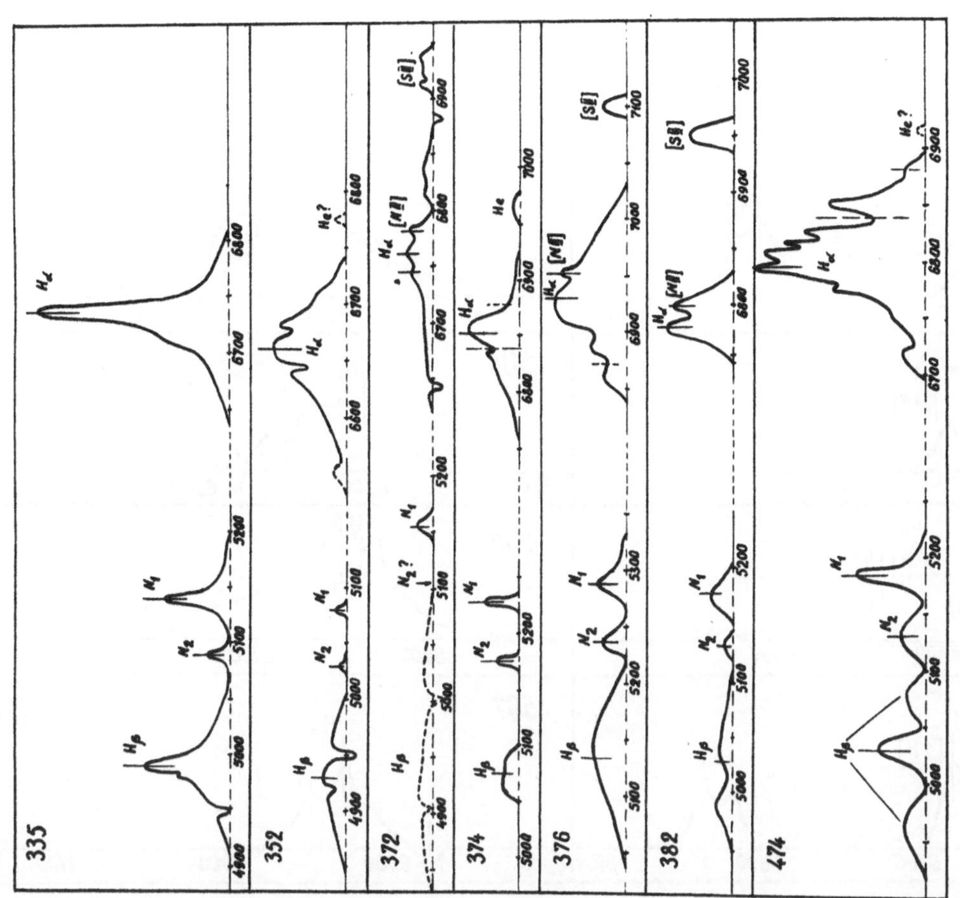

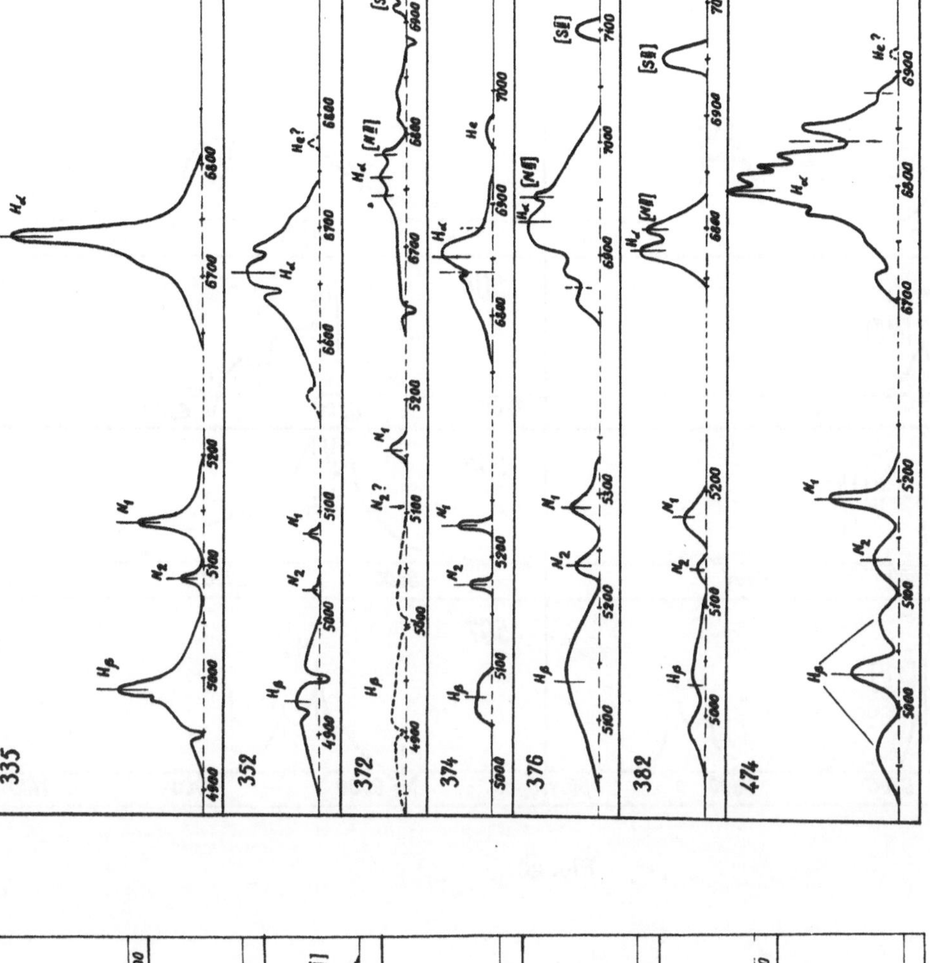

Fig. 2B

Fig. 2A

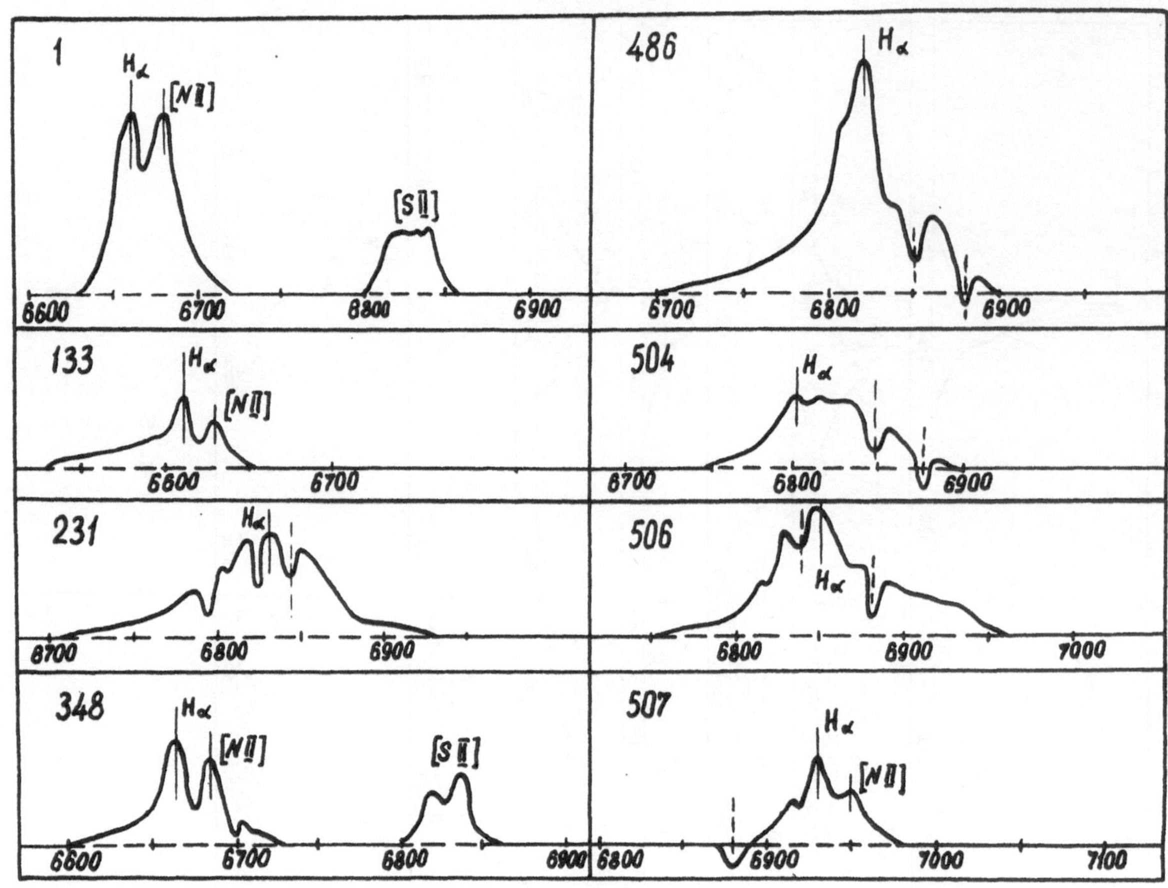

Fig. 2C

REFERENCES

KHACHIKIAN E.Ye., WEEDMAN D.W., Ap.J. 192, 581, 1974

MARKARIAN B.E., Astrofizika, 9, 6, 1973

WEEDMAN D.W., Ap. J. 183, 29, 1973

DISCUSSION

TOVMASSIAN: How many galaxies are used for the classification and how many of them are there in each of your groups?

DENISYUK: We have used 22 galaxies for the classification. The four groups contain: a) 14%, b) 22%, c) 37 %, and d) 27%, respectively. The statistics may be somewhat weak, also because different authors may use different emission line strengths in order to classify a galaxy as of Seyfert type.

COMPACT AND INTERACTING GALAXIES AMONG ISOLATED GALAXIES

V.Ye.KARACHENTSEVA
SPECIAL ASTROPHYSICAL OBSERVATORY, ZELENCHUK, U.S.S.R.

КОМПАКТНЫЕ И ВЗАИМОДЕЙСТВУЮЩИЕ ГАЛАКТИКИ СРЕДИ ИЗОЛИРОВАННЫХ ГАЛАКТИК

On the basis of Palomar Sky Survey the catalogue of isolated galaxies (field galaxies) containing 1051 objects with apparent magnitudes $m_p \leq 15.7$ and with declination $\delta > -3°$ has been made up (Karachentseva, 1973).

Field galaxies do not show clumping effect, and do not concentrate to the equator of the Local Supercluster (fig.1). This conclusion drawn from the analysis of their distribution in the supergalactic coordinates is valid for any samples of our catalogue - by the apparent magnitudes, the types, and the radial velocities.

We have then a uniformly distributed sample, i.e. some "etalon" for comparison of characteristics of galaxies included in systems.

M o r p h o l o g i c a l t y p e s. The spiral galaxies, mainly of the later types, with true axis ratio about 0.1-0.2, contain more than 80 per cent of field galaxies, Markarian galaxies make about 1 per cent. The frequency of occurence of Markarian galaxies among the isolated galaxies is three times as less as that among the pairs components (Karachentsev , Karachentseva, 1974). The compact galaxies make about 7 per cent; the galaxies with peculiar features make about 10 percent (including peculiar galaxies according to Arp, and interacting to Vorontsov-Vel'yaminov).

C o m p a c t g a l a x i e s. Compact galaxies among the isolated galaxies are represented by all subtypes according to Zwicky; compact, very compact, extremely compact. Their mean surface photographic brightness calculated taking into account the inclination of galaxy and the Galaxy absorption of light is $m/\square'' = 21.2 \pm 0.08$ (mean error of the average). The mean apparent axis ratio is 0.70 ± 0.03 (mean error of the average).

The distribution of compact isolated galaxies over the sky is uniform (fig.2).

P e c u l i a r g a l a x i e s. Peculiar isolated galaxies may be divided into:

a) galaxies with a distorted spiral structure (for example, NN 130, 152, 250, 251, 583, 1006*).

b) galaxies with several condensations in central part, with blue condensations, and faint compact companions (NN 1, 33, 45, 72, 77, 85, 147, 235, 267, 494, 508, 625, 972, 1030).

c) galaxies having features of interaction, i.e. "jets", ejections, deformed envelopes(NN 76, 92, 194, 293, 341, 349, 369,787, 819, 870, 940, 946).

A combination of several features of peculiarity occurs frequently.

The existence of isolated "interacting" galaxies (not being members of systems, and having no massive enough companions) is especially interesting, since in this case one has to suggest the presence of active processes inside the galaxies.

---

* All numbers are given from (Karachentseva, 1973). The identification and morphological description see in (Karachentseva, 1973).

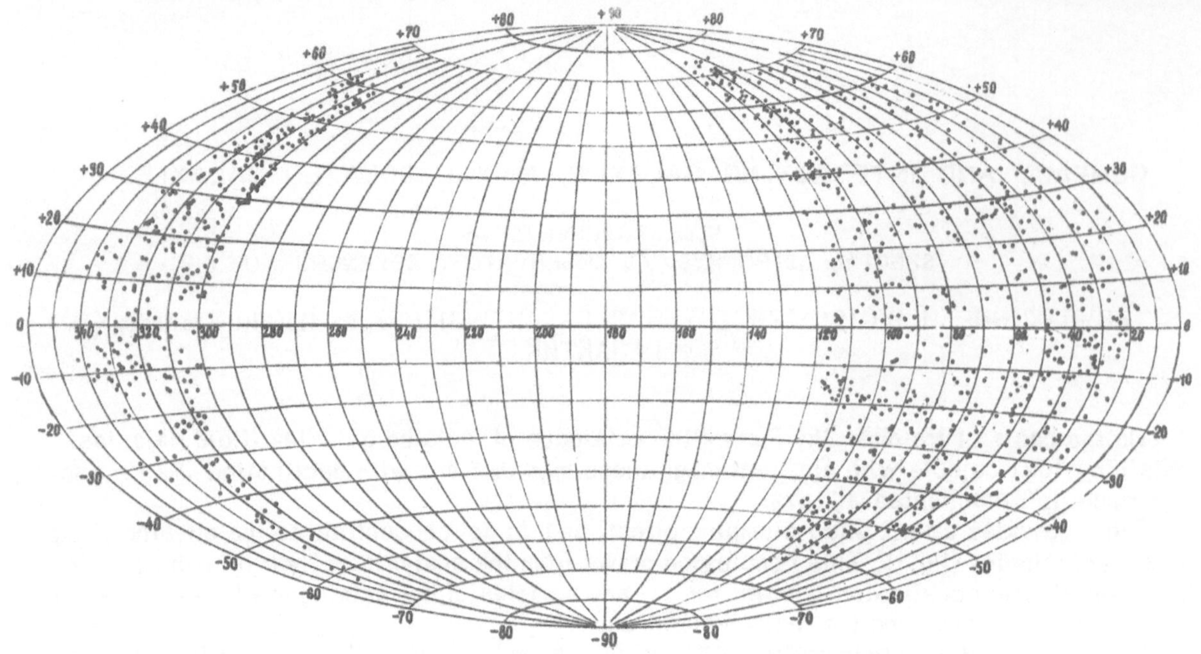

Fig. 1. The distribution of 1051 isolated galaxies with m ≤ 15.7 in supergalactic coordinates.

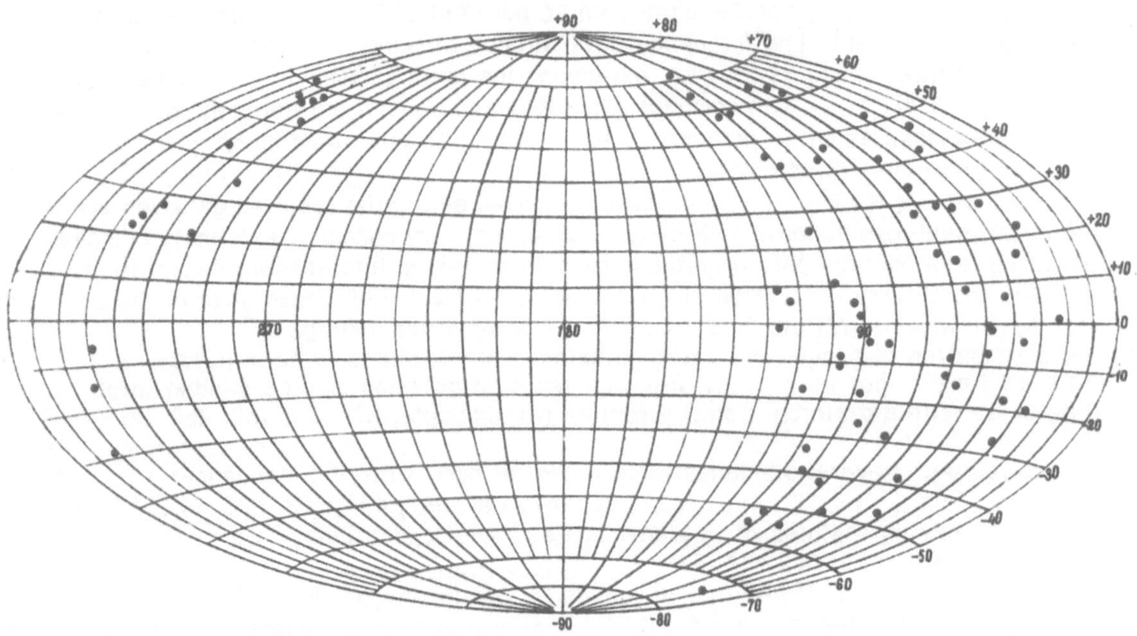

Fig. 2. The distribution of compact isolated galaxies in supergalactic coordinates.

REFERENCES

KARACHENTSEV I.D., KARACHENTSEVA V.Ye., 1974, Astron. Zhu. 51, 724

KARACHENTSEVA V.Ye. 1973, Soobsch. SAO USSR Acad. Sci., 8, 3

DISCUSSION

CHERNIN: How many of spiral galaxies are interconnected?

KARACHENTSEVA: At the most a few percent.

KOMBERG: What is the velocity dispersion for field galaxies?

KARACHENTSEVA : Radial velocities have so far only been measured for 65 isolated field galaxies. This material is much too small for any quantitative estimate of the dispersion.

DEUTSCH: How did you get 15^m7 as limiting apparent magnitude?

KARACHENTSEVA: From the Zwicky catalogue.

THE BAR-LIKE FORMATION IN THE GALAXY CENTRE AS A POSSIBLE ENERGY SOURCE FOR THE MAINTENANCE OF THE SPIRAL STRUCTURE

V.I.KORCHAGIN and L.S.MAROCHNIK
ROSTOV STATE UNIVERSITY, U.S.S.R.
PRESENTED BY L.S.MAROCHNIK

БАРОПОДОБНОЕ ОБРАЗОВАНИЕ В ЦЕНТРЕ ГАЛАКТИКИ КАК ВОЗМОЖНЫЙ ИСТОЧНИК ЭНЕРГИИ ДЛЯ ПОДДЕРЖАНИЯ СПИРАЛЬНОЙ СТРУКТУРЫ

It is now a most attractive point of view that the spiral arms in galaxies are the density waves (Lin et al. 1964, 1969). The density-wave hypothesis allows to solve in principle the persistence problem of the spiral patterns in the differentially rotating galaxies. However, Toomre (1969) and Mark (1971) have shown that in Lin et al. theory the inward propagating waves are absorbed in the location of the inner Lindblad resonance. So the persistense problem exists in the wave theory as well.

On the other hand Marochnik et al. (1972) have shown that only flat subsystem contributes to the density wave dynamics. This result leads to a conclusion that only outward propagating trailing wave fit the observations. It is then natural to assume that there is some "generator" in the centre of a galaxy which determines the spiral wave frequency and feeds them by the energy. Such a generator may be bar - like formations observed in many spiral galaxies (Holmberg 1958, Lindblad 1958).

In this paper we adduce shortly results of calculations of the flat subsystem response on the rigidly rotating bar in the centre of a galaxy. The actual wave pattern is determined by the physical properties of the generator and the system. The central highly asymmetric formation ("narrow" bar) generates in differentially rotating disk, excluding the neighbourhood of corotation, the wave pattern given by the formula

$$\tilde{\sigma}(r,\theta,t) = \frac{A}{\sqrt{r}} \left[F\left(\theta - \Omega_b t - \frac{\Omega_d - \Omega_b}{c} r\right) + F\left(\theta - \Omega_b t - \frac{\Omega_d - \Omega_b}{c} r + \pi\right) \right]. \qquad (1)$$

Here A is a constant determined by the parameters of the system and the generator, $\Omega_d(r)$ and C are the angular speed and the velocity dispersion of the flat subsystem, respectively.

The actual shape of the function F is determined by the distribution of the matter in the bar along the r and θ - coordinates.

From (1) it follows that the two-armed pattern is generated far from the "narrow" bar. The winding of spiral arms is determined by the local relation between the angular velocities of the bar and the disk. In regions where $\Omega_b > \Omega_d(r)$ the trailing spiral pattern is generated. The leading spiral pattern is generated if the opposite inequality takes place. The angular velocity of the spiral wave packet is equal to that of the bar. The size of the packet (the thickness of the spiral arm) is determined by the density distribution in the bar and by its length. The smaller is the bar thickness, the smaller is the thickness of the spiral arm, the larger is its amplitude for the fixed value of the bar mass M_b, Ω_d, Ω_b and the bar length 2L. The arm thickness is $\to$ 1kpc. for the values $C\ 7\frac{km}{sec}$, $|\Omega_d - \Omega_b| \sim 10\ \frac{km}{sec.\ kpc}$, $L \sim 2$kpc.

Replacing the "narrow" bar by the arbitrary asymmetric formation and then by the axisymmetrical generator the wave pattern transforms from the two-armed into the composite many-armed one and then disappears. The detailed consideration also shows that the wave pattern is not destroyed at the locations of the Lindblad resonances in case of generation by a "narrow" bar. So for the generation of the two-armed pattern it is necessary for a "narrow" bar to exist in the centre of the galaxy.

On approaching the corotation point r_k the spirals become tightly wound. The response is nonspiral in region $(r-r_k)^2\Omega_d^2(r_k) \ll C^2$ where the relative velocity of the flat subsystem and generator is small.

The question about the bar influence on the flat subsystem in the neighbourhood of corotation has been considered by Feldman and Lin (1973).

Let us estimate the value of the bar mass which is required for the maintenance of the spiral pattern in the galactic disk. Linearizing the expression for the density of surface energy and taking into account the first three powers of the expansion we obtain for the total energy change of the system in the $V_0 \gg C$ approximation

$$\Delta W \lesssim \int \frac{V_0^2}{\sigma_0} \tilde{\sigma}^2 dS \lesssim \alpha^2 \Omega_d^2 M_{fl}$$

Where σ_0, V_0 are the unperturbed values, σ is the perturbation, $\alpha = |\frac{\tilde{\sigma}}{\sigma_0}|$, $V_0 \sim \Omega_d R$, R is the radius of the system, M_{fl} is the mass of that part of the flat subsystem where the wave pattern is generated. Equating the wave energy to the bar energy $E_b \sim M_b \Omega_b^2 L^2$ we get

$$M_b \lesssim \frac{\alpha^2 \Omega_d^2 R^2}{\Omega_b^2 L^2} M_{fl}.$$

we find that $M_b \lesssim 0.25 M_{fl}$ for $d \sim 0.1$, $\Omega_d \sim \Omega_b$, $\frac{R}{L} \sim 5$.

Consequently a bar with a mass $M_b \sim M_{fl}$ could maintain the spiral pattern during ten or more galactic revolutions.

Thus, under the reasonable assumptions on the character of the generator and its physical parameters the latter may ensure the generation of large variety of spiral patterns and the maintenance of density-waves during a long time.

The results mentioned in the report were published partly in Korchagin, Marochnik(1973, 1975), the rest of them are to be published.

REFERENCES

FELDMAN S.I., LIN C.C., 1973, Stud. Appl. Math., 52, 1

HOLMBERG E., 1958, Meddel, Lund. Obs., Ser, 1, N 136

KORCHAGIN V.I., MAROCHNIK L.S., 1973, Astron. Tsirk. N 800, 1975, Astron. Zh., 52, 15

LIN C.C.,SHU F., 1964, Ap. J., 140, 646

LIN C.C.,YUAN C.,SHU F. 1969, Ap.J., 155, 721

LINDBLAD B., 1958, "Scripta varia", N 16, 33

MARK J., 1971, Proc. Nat. Acad. Sci, 68, 2095

MAROCHNIK L.S., MISHUROV Yu.N., SUCHKOV A.A. 1972. Astrophys.,Space Sci., 19, 285

TOOMRE A., 1969, Ap.J., 158, 899

DISCUSSION

DAVIES: There is evidence for a neutral hydrogen bar in the galactic centre with a mass of $\sim 10^6$ $M_\odot$. Is this sufficient to generate spiral structure?

MAROCHNIK: Probably not. Calculations show that the mass of the bar must be at least 10% of the mass of the surrounding flat system in which the spiral structure is to be formed.

SHULMAN: Numerical experiments normally show first a spiral structure from which a bar later emerges. How does this fit with your model?

MAROCHNIK: The bars which we observe in galaxies often consist of old stars, whereas spiral arms consist of young stars. We have therefore adopted a model with a stiffly rotating bar at $t = 0$, and this bar then generates the spiral structure.

ZHILYAEV: What do you do with ring galaxies which do show some spiral structure, but no bar?

MAROCHNIK: It is not obvious that the same mechanism should apply to all galaxies, but we sometimes observe in our models a tendency to approach ring-like structures.

CHERNIN: Would the introduction of a quadrupole momentum for the bar facilitate the construction of galaxies with a specified number of arms, e.g. $n = 2$?

MAROCHNIK: We have not yet carried through such an analysis, but we hope to do so soon.

KHACHIKIAN: Where does the bar come from in the very beginning?

MAROCHNIK: We simply start out with the bar present; i.e. our model treats only galaxies from this evolutionary stage and onwards.

THE EXTRAGALACTIC BACKGROUND BRIGHTNESS AT 4000 Å

KALEVI MATTILA
OBSERVATORY AND ASTROPHYSICS LABORATORY, UNIVERSITY OF HELSINKI, FINLAND

ЯРКОСТЬ ВНЕГАЛАКТИЧЕСКОГО ФОНА ПРИ 4000 Å

1. Observations. The extragalactic background brightness in the optical part of spectrum (the abbreviation EBL, extragalactic background light, will be used in the sequel) is an observational quantity of fundamental interest in several fields of cosmology. Questions involved are the decision between different cosmological models, the existence of luminous stellar matter between the galaxies, the emission by intergalactic gas, and evolutional effects in the luminosity and the number of galaxies. Early theoretical studies of the problem by Chéseaux (1744) and Olbers (1823) led to the result known as Olbers' Paradox: in a static, homogeneous and infinite universe the sky background would be as bright as the Sun's surface.

The measurement of the EBL is very difficult because of its weakness and the great complexity of the composite light of the night-sky. So far only upper limits have been measured for the EBL. Roach and Smith (1968) using ground based observations have placed an upper limit of 6.10^{-9} erg $cm^{-2}s^{-1}sr^{-1}Å^{-1}$ ($= 5\ S_{10}/□°$) at 5300 Å, while Lillie (1968) using rocket measurements has given an upper limit of $5 \cdot 10^{-9}$ erg $cm^{-2}s^{-1}sr^{-1}\ Å^{-1}$ ($\approx 2\ S_{10}/□\ °$) at 4100 Å.

In the present study a new method is presented for the measurement of the EBL at 4000Å. It utilizes the screening effect of a dark nebula, situated at a high galactic latitude, on the EBL. The difference in surface brightness between the dark nebula and its surroundings is due to two components only: (1) the EBL, and (2) the diffusely scattered starlight from interstellar dust. In order to separate the extragalactic and the galactic components measurements are made at two wavelengths, just shortward and longward of the 4000 Å discontinuity. (For practical reasons the two wavelengths could not be selected very close to 4000 Å in the present study, but were centered at 3800 Å and 4200 Å.)

A major feature in the spectrum of the integrated starlight is the discontinuity at 4000Å. This same feature is present in the integrated spectra of all older stellar systems (Schild 1972). Thus, for the galactic light component we can write

$$I_{gal}\,(3800\ \text{Å}) = 10^{-0.4\ \Delta m}\ I_{gal}\,(4200\ \text{Å}), \qquad (1)$$

where $\Delta m = 0.49$ is the size of the 4000 Å discontinuity in magnitudes. The spectrum of the extragalactic background light, on the other hand, has no such discontinuity since radiation from galaxies and other luminous matter over a vast range of distances, from $z = 0$ up to $z \approx 2$ or 3 at least, contributes to it. Thus we can assume that for the EBL component I_o :

$$I_o(3800) = I_o\,(4200). \qquad (2)$$

From the observed values of the surface brightnesses $I_{obs}(3800)$ and I_{obs} (4200) we form the quantity

$$J = \frac{I_{obs}(3800) - 10^{-0.4\Delta m}\ I_{obs}\ (4200)}{1 - 10^{-0.4\Delta m}}\ .$$

Because of Eq.(1), J is independent of the galactic light component. On the other hand, the extragalactic background light, fullfilling the condition (2), is included with full weight in J. Consider now the change of J across a high latitude dark nebula In the direction of the nebula the EBL is blocked out, i.e. $I_o=0$, while in the surroundings of the nebula, I_o remains undiminished. Thus, J should show a minimum in the direction of the dark nebula if the EBL component is present in measurable amount.

Photoelectric observations in the area of the high-latitude ($b \approx 36°$) dark nebula L134 have been carried out using the 91-cm telescopes of the Kitt Peak National Observatory and the Steward Observatory. In Fig. 1 the observed values of J across the nebula are shown. They are given relative to a zero point (indicated by a cross) in the most opaque portion of L134. The large foreground components, the zodiacal light, the airglow, and the atmospheric scattered light do not influence these relative values of J, because they are in front of the dark nebula, and thus the same both in the direction of the nebula and in its surroundings. The starlight down to a limiting magnitude of $\sim 20^m$ could be avoided by selecting the measured 2' areas using the PSS prints.

The observations indicate that J has a larger value in the transparent surroundings of the dark nebula than in the direction of the nebula. Furthermore, as shown by Fig. 1, the values of J have a good correlation with the number of galaxies, N, counted across the same region (Shane and Wirtanen 1967). Thus it is indicated that the excess observed in the surroundings of L134 for the quantity J is due to the extragalactic light. Taking the mean of the five regions I, II, III, X, and XI of good transparency, we find for the EBL the value

$$\Delta J = I_0 e^{-\tau_0} h = 20 \cdot 3 \pm 3 \cdot 3 \cdot 10^{-9} \text{erg cm}^{-2} \text{s}^{-1} \text{sr}^{-1} \text{Å}^{-1}.$$

Applying the correction h for the scattered extragalactic light in the dark nebula, and omitting the small correction $e^{-\tau_0}$ for the general galactic extinction, the following value is found for the EBL at 4000 Å:

$$I_0 = 23 \pm 8 \cdot 10^{-9} \text{erg cm}^{-2} \text{s}^{-1} \text{sr}^{-1} \text{Å}^{-1}.$$

2. Possible consequences for galaxy evolution and cosmology. The value of the EBL found in the present study is several times higher than the integrated light of galaxies as predicted by most of the model calculations (see e.g. Sandage and Tammann 1965, Peebles and Partridge 1967, Tinsley 1973). One of the following three factors might offer an explanation:

(i) *Local luminosity density.* Several recent observations support the idea that the presently accepted value of the local luminosity density should be considerably increased (Arp 1965, Arp and Bertola 1971). This would lead to a corresponding increase in the EBL. Then, also the mass density of the universe should be substantially increased, and its value would approach the critical value needed to close the universe.

(ii) *Formation and evolution of galaxies.* Tinsley (1973) has studied the influence of different galaxy evolution models and formation epochs on the EBL. Her results indicate that for models including galaxy evolution (Curve 1 in Fig. 2) the background is brighter by about a factor of 3 - 10 as compared to the models with no evolution (Curve C1). But also the evolving models produce a background intensity of only $\sim 1 \cdot 10^{-9}$ erg cm^{-2}s^{-1}sr^{-1}Å^{-1} at 4000 Å when the galaxies are assumed to form at large redshifts, $z>8$, as was the case for models C1 and 1.

A very different result was obtained by Tinsley when galaxies were assumed to have formed quite recently, at a redshift $z = 3$ (her models 12 and Y12). In this case the strong far-UV radiation of young galaxies at $\lambda \approx 1000$ Å is shifted into the optical wavelengths, and a very high and relatively narrow peak in the EBL spectrum is obtained. The measured value in Fig.2 is seen to fall between Tinsley's models 12 and Y12, which differ in their assumed far-UV evolution for galaxies, and between which the "best far-UV evolution model" should lie.

(iii) *Galaxy density evolution.* A third possibility, which has been suggested by Rowan-Robinson (1974), is that there is a strong density evolution for normal galaxies in the same sense as found for quasars and strong radio sources. However, such a rapid density evolution of normal galaxies would be physically very difficult to understand.

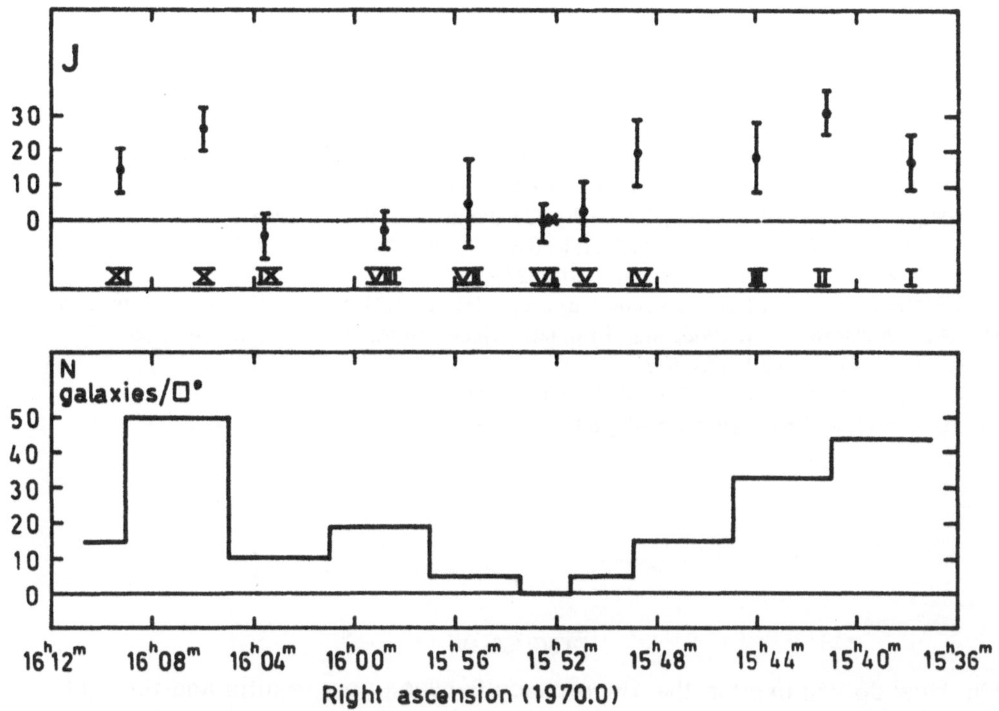

Fig. 1. *Upper part* : The observed values of the quantity J (as defined in the text) across the dark nebula L134. The values are given relative to a zero point (indicated by a cross) in the most opaque part of L134. Unit is 1.10^{-9} erg $cm^{-2}s^{-1}sr^{-1}Å^{-1}$.
Lower part : The number of galaxies per □° accross the area of L134. (According to Shane and Wirtanen 1967).

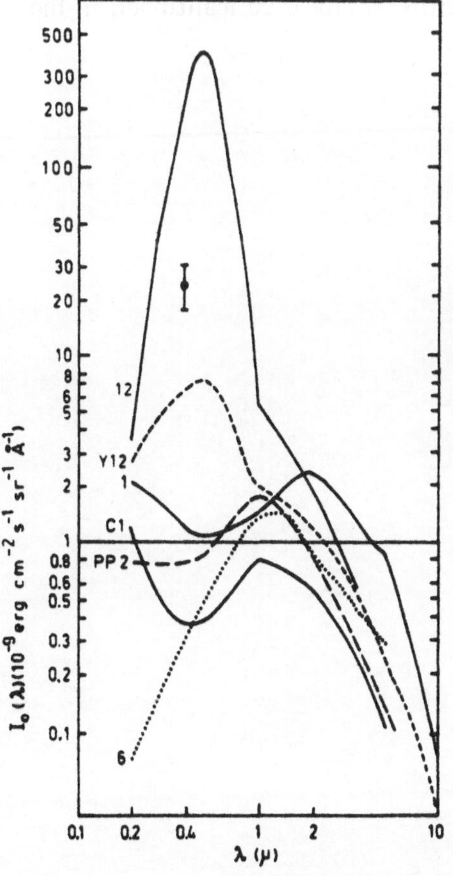

Fig. 2. Calculated spectral energy distributions of the integrated light of galaxies. Curve PP2 is model No.2 of Peebles and Partridge (1967), while all the other curves have been adopted from Tinsley (1973). The observed value from the present study is indicated with its error bar.

REFERENCES

ARP H. 1965, Astrophys. J. 142, 402

ARP H., BERTOLA F. 1971, Astrophys. J. 163, 195

CHESEAUX L. de 1744, Traite de la cométe qui a paru en décembre 1743

LILLIE C.F. 1968, Doctoral thesis, University of Wisconsin (unpublished)

OLBERS H.W.M. 1923, Astronomisches Jahrbuch 1826 (herausgeg. von J.E. Bode), 110

PEEBLES P.J.E., PARTRIDGE R.B. 1967, Astrophys. J. 148, 713.

ROACH F.E., SMITH L.L. 1968, Geophys. J. 15, 227.

ROWAN-ROBINSON M. 1974, Paper presented at a specialist meeting of the R.A.S. on October 11th, 1974

SANDAGE A., TAMMANN F.A. 1965, Ann. Rep. Mt. Wilson and Palomar Obs. 1964-65, 35

SCHILD R. 1972, Astrophys. J. 178, 617

SHANE C.D., WIRTANEN C.A. 1967, Publ. Lick Obs. 22, Part 1

TINSLEY B.M. 1973, Astron. Astrophys. 24, 89

DISCUSSION

LONGAIR: How do you explain the discrepancy between your results and those of other authors?

MATTILA: The upper limits by Roach and Smith, and Lillie, were deduced from an absolute surface photometry of the night sky and large corrections were made for zodiacal light, starlight, diffuse galactic light and in some cases for the airglow. The discrepancy can only be explained on the basis of observational errors, either I am wrong or they are.

SLYSH : Do you suppose that the dark nebula absorbs all the energy at 4000 Å?

MATTILA : Monte Carlo computations were made for the scattering of light in a spherical model dust cloud in an isotropically incident radiation field. The efficiency of the dark nebula as a shield against the extragalactic background light depends on the albedo of the dust grains and the optical thickness. The correction factor 0.90 mentioned in the text was determined from the MC computations.

THE EXPLOSIONS OF YOUNG GALAXIES AND THE MISSING MASS PROBLEM

A.A.SUCHKOV, Yu.A.SHEKINOV
ROSTOV STATE UNIVERSITY, U.S.S.R.
PRESENTED BY A.A.SUCHKOV

ВЗРЫВЫ МОЛОДЫХ ГАЛАКТИК И ПРОБЛЕМА СКРЫТОЙ МАССЫ

Many authors conclude that the early evolution of galaxies is connected with a violent star formation period (see the extensive discussion of this problem in Partridge and Peebles, 1967 a,b). A young galaxy at this stage is very bright, its luminosity exceeds by a few orders of magnitude the ordinary luminosity of the galaxy. Taking the model of a young galaxy proposed by Partridge and Peebles (1967a,b) we have considered the effect of radiation in a young galaxy on its internal dynamics.

As has been argued by Partridge and Peebles (1967 a,b), the first stars in a galaxy like our own release in a short time - of about $\Delta t \approx 3 \cdot 10^7$ years - the energy $E^G \approx 3 \cdot 10^{61}$ erg Perhaps this value is to be increased by an order of magnitude. The greater part of this energy is due to OB-type stars, the 22% of it being radiated in Lyman continuum. The ionizing quanta are reradiated by interstellar gas, yielding up to 10% of the radiation energy in the form of L_α - quanta.

Now, a number of arguments can be brought which support the view that a region of star formation is surrounded by HI gas, which is the main contributor to the total mass of a young galaxy (Suchkov, Shekinov, 1975). The L_α-quanta diffuse through this gas. If the diffusion time turns out to be larger than the time of the bright stage and the dynamical time-scale of young galaxy (the expansion time) then the increasing pressure of L_α-quanta will expell the gas out of the galaxy when their energy exceeds the gravitational energy of the system, that is when $E_{L_\alpha} \geq U$. Combining these requirements together, we find the mass of the gas which can be expelled out of a galaxy at its bright stage:

$$M \leqslant 4 \cdot 10^{-25} E_{L_\alpha}^{3/5} n^{-1/5} ; \qquad (1)$$

where M is the mass in units $M_\odot = 2 \cdot 10^{33}$g, n is the mean particles number density. For a system with the total stellar mass M_{st} we obtain

$$M/M_{st} \leqslant 5 \cdot 10^5 M_{st}^{-2/5} n^{-1/5} \qquad (2)$$

So we see that at the bright stage of young galaxies there is released the energy sufficient to expell a mass of the gas exceeding by a few orders of magnitude the total stellar mass at this stage. For our Galaxy at the bright stage ($M_{st} \approx 2 \cdot 10^{10}$) the ratio (2) is equal to ≈ 40, and for a globular cluster ($M_{st} \approx 10^6$) it is equal to $\approx 10^3$.

Now, the loss of the half mass by a system makes its total energy positive. Those galaxies which lose their mass at the bright stage in a manner described above will decay, their stars scattering in the intergalactic space. Perhaps these stars are main contributors to the missing mass in the clusters of galaxies. A number of factors can inhibit this process. For example, if the greater part of the gas is ionized by the star radiation then the L_α-quanta could escape from the gas cloud in time smaller than the expansion time of the cloud. The other, more plausible possibility is that the mass of a galaxy exceeds the limiting value given by(2).

After the bright stage of our Galaxy, its disk population stars, which contain up to 77 % of the total stellar mass are still to be born. So at the bright stage this mass was in the gaseous form and could be thrown out of the Galaxy by L_α-radiation pressure. The Galaxy could have survived if its mass is larger than $10^{12} M_\odot$. The corresponding mass excess, if it really exists, should be attributed perhaps to the coronas of Galaxy. The coronas also could be the contributors to the missing mass.

In summary, we conclude that 1) many galaxies could be decayed at their bright stage of evolution, and their nowadays old stars fill the intergalactic space, perhaps accounting for the missing mass; 2) those galaxies which managed to survive possess probably the massive coronas, which are perhaps the main contributors to the mass of galaxies.

REFERENCES

PARTRIDGE R.B., PEEBLES P.J., 1967a, Ap.J., 147, 868
PARTRIDGE R.B., PEEBLES P.J., 1967b, Ap.J., 148, 377
SUCHKOV A.A., SHEKINOV Yu.A., 1975, Astrofizika, 1975, in press

THE FINE STRUCTURE OF THE GALACTIC SPIRAL ARMS

Th. SCHMIDT-KALER
ASTRONOMISCHES INSTITUT DER RUHR-UNIVERSITÄT, BOCHUM, F. R. G.

ТОНКАЯ СТРУКТУРА ГАЛАКТИЧЕСКИХ СПИРАЛЬНЫХ РУКАВОВ

There is increasing optical and radioastronomical evidence (summarized by Schmidt-Kaler, 1975, Hill, Kilkenny, Schmidt-Kaler, 1975) that the spiral arms of our Galaxy are not just a phenomenon of the galactic plane. Because of the inaccuracy of distance determinations any fine structure of the spiral arms is best investigated by analysing the positions at the sphere (or restriction to the immediate neighbourhood of the Sun).

Photographs by means of a wide-angle camera (field diameter 140°) in the ultraviolet revealed a characteristic wave-like fine structure of the Sagittarius arm. Between $l = 270°... 20°$ the next-inner spiral arm -I appears composed of three very narrow, partially overlapping filaments ("shingles"), about 1.3 kpc long and 70 pc thick (Schmidt-Kaler, Schlosser, 1973). The phenomenon is also evident from photoelectric ultraviolet surface photometry (Pfleiderer and Mayer 1971). It can be shown that this structure is not due to the configuration of the local dark clouds but is genuine to arm -I. The most pronounced spiral tracers, e.g. early-type open clusters and OB-associations are concentrated in these shingles. It is remarkable that also the distribution of supernova remnants and X-ray sources in $l = 260°... 30°$ appears related to these filaments. Thus the correlation to the shingles may be used to find age and distance parameters of population I objects.

From an analysis of recent 21 cm-line surveys Quiroga (1974) found the same wave-like pattern with an average length of 1.7 kpc and an average thickness of 90 pc in the Sagittarius and Scutum-Norma arms. These investigations have recently been extended to the tangential points up to $l = 90°$ with similar results. Also, in the vicinity of the Sun the local hydrogen follows the trend of the local shingle closely. The distribution of the stars of Gould's Belt has been described by a disk of roughly 1 kpc diameter with a thickness of about 100 pc, inclined by 17° to the galactic plane (Stothers and Frogl, 1974). Correspondingly, the local neutral hydrogen has been described by an expanding disk (Lindblad, 1973). Close inspection, however, reveals that a shingle of 1.3 kpc length, 0.5 kpc width and 70 pc half-intensity thickness is a more accurate description. Heavy dark clouds in a typical cellular network are also arranged in the local shingle. Gould's Belt is just one of a sequence of shingles along the local arm.

The shingle phenomenon can be explained as a density wave normal to the galactic plane (Schmidt-Kaler and House, 1975). In analogy to the density-wave theory of spiral arms in the R,θ-plane we solved the Boltzmann and Poisson equations neglecting the R-coordinate, i.e. assuming essentially circular arms and decoupling the motions perpendicular to the plane. This is permissible since the length of a shingle is small compared to the circumference $2\pi R_0 \doteq 50$ kpc. First-order perturbations are superposed on a flat subsystem with approximately constant density within $|z| < 0.2$ kpc. A self-consistent solution results for wave-numbers $k \doteq 40$, i.e. $\lambda = 2\pi R_0/k \doteq 1.3$ kpc. These waves can be excited and maintained by the Jeans instability modified by the two-stream situation due to the relative flow of disk population and extreme population I.

REFERENCES

HILL P.W., KILKENNY D., SCHMIDT-KALER Th. 1975, Monthly Not.Roy. Astr.Soc. 171, 353
LINDBLAD P.O. 1973 In: Proceed. First Europ.Astronom. Meeting, Vol.2, 65, 5 Springer. Berlin
PFLEIDERER J. and MAYER U. 1971., Astron.J. 76, 691
QUIROGA R.G. 1974, Astrophys. Space Sci. 27, 323
SCHMIDT-KALER Th. 1975, Vistas in Astronomy 19, 69
SCHMIDT-KALER Th. and HOUSE F. 1975, Astron. Nachr. in press
SCHMIDT-KALER Th. and SCHLOSSER W. 1973, Astron. Astrophys. 25, 191
STOTHERS R. and FROGL J.A. 1974, Astron. J. 79, 456

DISCUSSION

SANDQVIST: In your example of an external galaxy with the "shingle" phenomenon, I was not convinced that the apparent inclination of the "shingles" was due to an inclination with respect to z. Could they possibly be due to inclinations in the plane with respect to x and y?

SCHMIDT-KALER : There are dozens of galaxies, especially of type Sb, showing the phenomenon. On the minor axis the aspect effect is at minimum. Schlosser and I have calculated the shingle inclination for about a dozen cases and we found around 10°.

GOSACHINSKIJ: I should like to note that such an inclination of the hydrogen is not only present in the inner part of the Galaxy, but is also seen in the outer spiral arms, as we have found at Pulkovo.

SCHMIDT-KALER : I am afraid that I am not aware of this work. But very recently Quiroga and Schlosser, and Rolfs, have found the same phenomenon in the local arm and in the Perseus arm, respectively.

MAGNITUDE - REDSHIFT RELATION IN SYSTEMS OF GALAXIES

T. JAAKKOLA
OBSERVATORY AND ASTROPHYSICS LABORATORY HELSINKI, FINLAND

ОТНОШЕНИЕ ВЕЛИЧИНЫ К КРАСНОМУ СМЕЩЕНИЮ В СИСТЕМАХ ГАЛАКТИК

1. Pairs, Groups and Clusters. The magnitude-redshift relation within the systems of galaxies is a direct kinematic test of the stability of the systems, if inclusion of projected galaxies can be avoided and if redshifts are Dopplerian. The present data consist of 1043 galaxies in 16 clusters, 64 groups, 121 pairs and 14 other systems. The results of calculation of the (m, z) - relation are given in Table I. Column 1 gives the category of the systems, 2 the number of systems, 3 the number of galaxies, 4 the ratio of the frequency of systems with positive (m, z) - correlation to that of systems with negative correlation, 5 the combined regression coefficient (b_m) between the magnitude-difference $\Delta m = m - \bar{m}$ and the normalized residual velocity $u = (V_o - \bar{V}_o)/\sigma_V$ (the mean values $\bar{m}$ and $\bar{V}_o$ and the velocity dispersion σ_V refer to individual systems), 6 the standard deviation of b_m, 7 the chance probability of b_m.

"C I" and "C II" mean clusters collected using a stricter and a looser selection of members, respectively.The value of a parameter for "adopted clusters" is the mean of the values for C I and C II. "Other systems" are for cases where only two members have measured redshifts, the system being not identified as a pair in the literature. The group and pair data are based mainly on the lists of Karachentsev (1970a, 1970b).

One can see that adopted clusters, groups and pairs have significant positive (m,z) - correlations. For a sample where all separate systems have been put together, $b_m = +0.09 \pm 0.04$, with the corresponding probability P = 0.007. Hence, if the (m,z) - relation is used as a kinematic test, the systems seem to be in general expanding. The characteristic expansion time-scales are 6.9×10^8, 2.5×10^9, 1.1×10^{10} years for pairs, groups and clusters, respectively.

This relation was studied also as a function of the size of the systems,and the results are shown in Fig. 1 in which h is in km $s^{-1}Mpc^{-1}$, meaning the rate of change of radial velocity within the system, calculated separately for different intervals of radius from the formula $h = b_m \sigma_m \sigma_V / R$ (σ_m is the dispersion of Δm and R is the radius in Mpc). Curve 1 is for pairs, 2 isolated groups, 3 sub-groups, 4 CI-clusters, 5 CII-clusters. It is found that the "instability"-parameter, h, is increasing towards small R-values. This appears both within and between the various categories and is opposite to the result obtained using the virial theorem (Karachentsev,1966, Rood et al., 1970), according to which the mass-discrepancy increases when the size increases. The trend found in the study here reported that both b_m and σ_V in general increase towards decreasing R, similarly as h, supports the present result, while the opposite result from the virial theorem may follow directly from factor R which is in the nominator in the virial mass formula.

Karachentsev (1966) and Rood et al. (1970) concluded that the contamination by field galaxies is insignificant in their data. Also increase of σ_V and b_m towards decreasing R supports the view that the present results are not mostly caused by optical members. The short time scales given above also make the kinematic explanation of the (m,z) — relations doubtful.

2. Local Supergalaxy. The (m, z) - relation within the Local Supergalaxy(LSG) is naturally of interest in the present context and because it is closely adjacent with the universal (m, z) - relation and gives information on the deceleration parameter q_o (Sandage et al ,1972), it is also important cosmologically. This relation is still rather poorly known, as shown by the discrepant results obtained on it (de Vaucouleurs, 1972; Sandage and Tammann, 1975a,b). I shall pay attention to two kinds of evidence which I consider relevant in discussion of the supergalactic redshift field.

First, larger values of the Hubble constant have been obtained for the nearby objects, most of which belong to LSG, than for distant objects. From several studies, H 100 km $s^{-1}Mpc^{-1}$ for the former (de Vaucouleurs, 1972; Durand, 1975; Heidmann, 1970; Roberts,1972: note also the case of the Virgo cluster in Abell, 1972 and in Gudehus, 1973) and H 50 km $s^{-1}Mpc^{-1}$ for the latter (Abell, 1972; Durand, 1975; Gudehus, 1973; Sandage and Tammann, 1975b). In all these studies the commonly accepted estimates of distances to the Local Group members are used in calibration of the distance scale. However, there are differences in other steps of calibration which partly explain the difference in H-values, reducing it to ΔH 30 km $s^{-1}Mpc^{-1}$. If this is taken at face value, this implies a steeper (m, z) - relation within LSG than in the Metagalaxy in general. The data given by Sandage and Tammann(1975 a, b) do not contradict this result. The effect of luminosity selection on the quoted results of the value of H (Teerikorpi, 1975) should be studied further.

Secondly, considering the redshift anisotropy discovered by Rubin et al. (1973), it is interesting that the region of high redshifts (region II) includes central parts of LSG. Fig. 2 illustrates the anisotropy of the (m, z) - relations of six kinds of objects in the distance interval corresponding to the symbolic velocity V from about 3000 km/s to 10000-15000 km/s, and summarizes the results of Rubin et al. (1973) Jaakkola et al. (1975a,b,c,) and Karoji and Nottale (1975). Considering the "Hubble modulus" HM = log V - 0.2 m, the weighted mean difference in HM between the two regions of sky, calculated for the six samples, equals 0.097±0.016. This is statistically significant at 6σ level. The corresponding average redshift anisotropy is 1300±210 km/s. At distances with V larger than 10000 - 15000 km/s the anisotropy vanishes. This is against the interpretation of the effect through galactic absorption.So is also the fact that small HM are not associated with regions of exceptionally high absorption (denoted HA in Fig.2) and large HM with regions of low absorption (LA), these regions being obtained on the basis of counts of clusters of galaxies (Holmberg, 1974). Also the other conventional ways of interpretation appear unsatisfactory with respect to this anisotropy of the (m, z) - relation (Jaakkola et al.,1975c).

In the latter paper it has been suggested that the supergalactic concentration of galaxies would be associated with a concentration of redshifting medium which would bring about the discussed high H-value in LSG. Such an effect can be predicted if one accepts the interaction theory of redshifts by Pecker and Vigier (Merat, 1974). The supergalactic medium would also redshift the light from objects in the background, explaining the Rubin-Ford anisotropy. But this excess redshift, $\Delta z \sim 0.004$-0.005 would not be observable in the (m, z) - relation for z larger than about 0.04, and there the anisotropy does, indeed, vanish.

3. Concluding Remarks. As briefly discussed in Section 1, the (m, z)- relations in pairs, groups and clusters are hardly explicable in commonly accepted terms.These could be interpreted through intrinsic redshifts in galaxies of faint absolute luminosity and/or, similarly as assumed above for LSG, through an intergalactic non-Dopplerian redshift within the systems.By these means a variety of peculiarities of redshifts, occurring in systems of largely different scales, can be explained in a coherent way. Alternatively, if the present results are expressed kinematically, matter would be expanding most rapidly in places where it is most densely packed. Taking into account gravitation, and that the systems are still in the sky, the latter interpretation does not seem plausible.

Acknowledgements. Part of the study reported was supported by a scholarship from the French government. I am grateful to Institut d'Astrophysique, Institut Henri Poincare and Observatoire de Meudon for hospitality and several researchers of these institutes for their cooperation. In particutar, I am indebted to J.C.Pecker, J.P.Vigier and S.Depaquit.

Table 1

Category	N_s	N_G	N_S^+/N_S^-	b_m	D_m	P_m
1	2	3	4	5	6	7
Clusters C I	17	541	0.55	-0.01	0.05	0.42
Clusters C II	16	576	1.00	+0.18	0.04	$<10^{-3}$
Adopted clusters	16	559	0.78	+0.08	0.05	0.04
Isolated groups	57	306	1.38	+0.10	0.07	0.06
Sub-groups	11	53	4.50	+0.22	0.12	0.04
All groups	64	345	1.46	+0.11	0.06	0.03
Pairs in the field	63	126	1.18	+0.12	0.11	0.17
Pairs in groups	30	60	1.50	+0.07	0.24	0.39
Pairs in clusters	28	56	2.25	+0.83	0.51	0.03
Karachentsev's pairs	96	192	1.40	+0.02	0.11	0.42
Other pairs	25	50	1.63	+1.18	0.52	0.01
All pairs	121	242	1.44	+0.20	0.13	0.06
Other systems	14	28	1.00	+0.05	0.21	0.40
All separate systems	162	1043	1.27	+0.09	0.04	7.10^{-3}

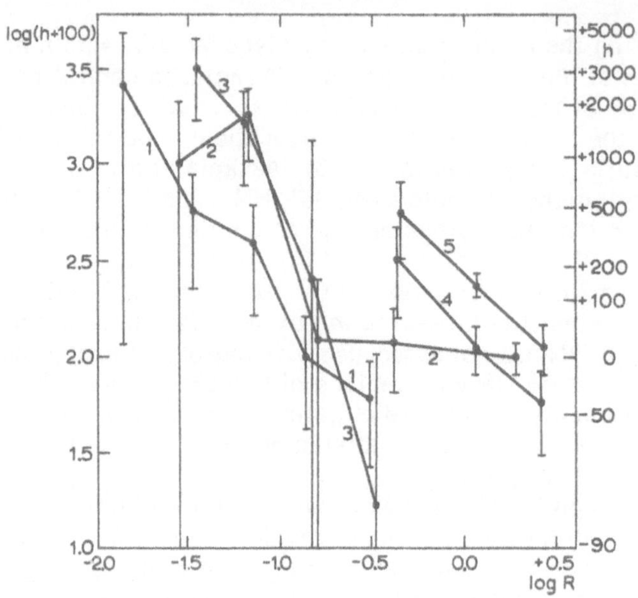

Fig. 1. Correlations between the "instability" - parameter h and radius R, in logarithmic scales, for different categories. R is in Mpc and h in km s^{-1} Mpc^{-1}.

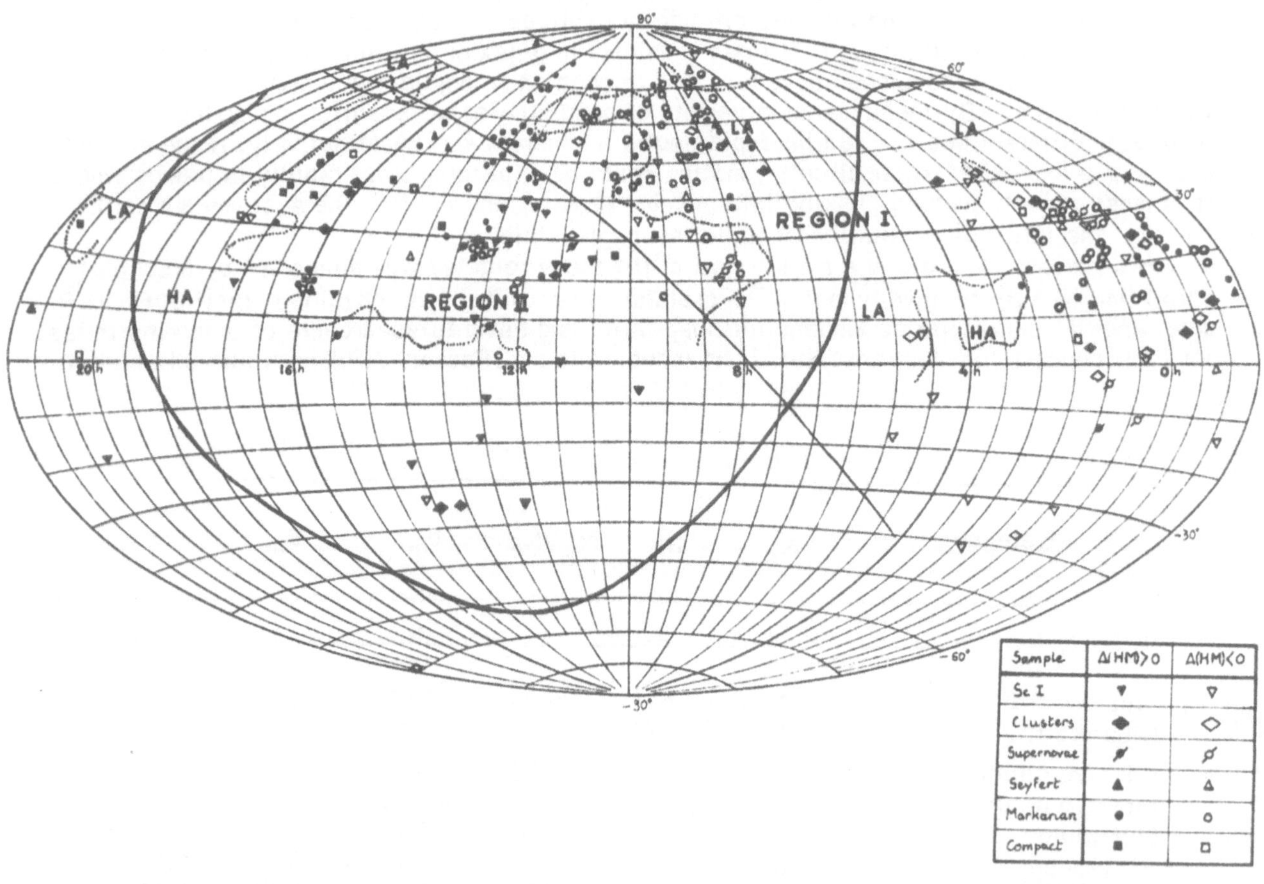

Fig. 2. Distribution in the sky of objects belonging to six samples, given separately for objects with lower-than-average value of HM = log V - 0.2 m (open symbols) and those with higher-than-average HM (closed symbols). The full curve from upper left to lower right shows the approximate borderline between the regions of high and low average HM. The other full curve shows the galactic equator. The areas between the latter and the dashed curves are those of exceptionally high, or exceptionally low absorption, denoted by HA and LA, respectively. These have been transformed from Holmberg (1974), and show areas where absorption deviates from the law $A = 0.25 \operatorname{cosec} |b|$ by $\pm$ 0.15 mag or more.

REFERENCES

ABELL G.O., 1972, in D.S. EVANS (ed.), 'External Galaxies and Quasi-Stellar Objects', IAU Symp. 44, 341

VAUCOULEURS G., 1972, in D.S.EVANS (ed.), 'External Galaxies and Quasi-Stellar Objects', IAU Symp.44,353

DURAND N., 1975, thesis, Obs. de Meudon.

GUDEHUS D.H., 1973, Astron. J.78, 583

HEIDMANN J., 1970, Compt. Rend. Acad. Sci. Paris 271B, 658

HOLMBERG E., 1974, Astron. Astrophys. 35, 121

JAAKKOLA T. and le DENMAT G., 1975 to be published.

JAAKKOLA T., KAROJI H., MOLES M., and VIGIER J.P.:1975a, Nature 256, 24

JAAKKOLA T., le DENMAT G., MOLES M., PECKER J.C., and VIGIER J.P., 1975b, to be published.

KARACHENTSEV I.D., 1966, Astrofizika 2, 81

KARACHENTSEV I.D., 1970a, Problemy kosmicheskoi fiziki 5, 201, Izd-vo Kievskogo universiteta.

KARACHENTSEV I.D., 1970b, Vestnik Kievskogo universiteta, ser. astron. 12, 103

KAROJI H. and NOTTALE L., 1975, to be published.

MÉRAT P., PECKER J.C., and VIGIER J.P., 1974, Astron. Astrophys. 30, 167

ROBETRS M.S., 1972, in D.S. EVANS (ed.) 'External Galaxies and Quasi-Stellar Objects', IAU Symp. 44, 12

ROOD H.J., ROTHMAN V.C.A., and TURNROSE B.E., 1970, Astrophys. J. 162, 411

RUBIN V.C., FORD Jr., W.K., and RUBIN J.S., 1973, Astrophys. J. Lett.,183, L111

SANDAGE A. and TAMMANN G.A., 1975a, Astrophys, J. 196, 313

SANDAGE A. and TAMMANN G.A., 1975b, Astrophys. J. 197 , 265

SANDAGE A., TAMMANN G.A., and HARDY E., 1972, Astrophys. J. 172, 643

TEERIKORPI P., 1975, Observatory 95, 105

THE MARKARIAN OBJECTS AS OVERLUMINOUS GALAXIES

L.BOTTINELLI, R.DUFLOT, L.GOUGUENHEIM AND J.HEIDMANN
OBSERVATOIRE DE MEUDON AND OBSERVATOIRE DE MARSEILLE, FRANCE
PRESENTED BY L.BOTTINELLI

ОБЪЕКТЫ МАРКАРЯНА КАК ГАЛАКТИКИ ОБЛАДАЮЩИЕ СВЕРХСВЕТИМОСТЬЮ

I n t r o d u c t i o n. The Markarian galaxies exhibit a large range of properties. Even among the class of galaxies with narrow emission lines, at least two main subclasses arise (Sargent, 1972):

1) luminous galaxies with 'hot spots'
2) dwarf objects resembling the compact galaxy II Zw 40.

A previous 21-cm line study of eleven Markarian non-Seyfert galaxies (Bottinelli *et al.*, 1973) has shown that the spread of the integral parameters among this sample is the same as among the classical galaxies of the Hubble sequence. However, no conclusion was reached as to whether a precise type could be assigned to each Markarian galaxy, from its integral characteristics.

In order to investigate further this problem, a program of joint optical and 21-cm line spectroscopic observations has been undertaken at the Haute-Provence Observatory and at the Nançay Observatory (Bottinelli *et al.*, 1975). Because previous studies of Seyfert galaxies had shown that the integral parameters are not affected by the anomalous characteristics of the nucleus, we selected just non-Seyfert Markarian galaxies; the dwarfs have been also excluded, owing to a previous 21-cm line study (Balkowski *et al.*, 1974).

D e t e r m i n a t i o n o f m o r p h o l o g i c a l t y p e s. New results have been obtained for nine galaxies (Ma 7, 8,12,314,325,326,370,391 and 401).

In order to determine whether all the integral parameters of a given galaxy have the values expected for a single morphological type, this type was determined from the measured values of the five following parameters:

- neutral hydrogen projected density σ_H
- neutral hydrogen mass-to-luminosity ratio M_H/L_0
- neutral hydrogen mass-to-indicative total mass ratio M_H/M_i
- indicative total mass-to-luminosity ratio M_i/L_0
- internal velocity spread W_t
- neutral hydrogen quasi-volumic density σ_H/a_0

An r.m.s. error was also estimated using the method described by Bottinelli and Gouguenheim (1975).

The results (Fig.1) show some agreement for 7 galaxies but not for Ma 370 and 401. In the case of classical galaxies, agreement is always expected. *This suggests that at least some of the Markarian galaxies are peculiar from the point of view of their integral features.*

However, the luminosity effect which modifies slightly the expected values of the parameters, for each type, must be also taken into account. The results obtained (Table 1) show that *the seven galaxies for which a type can be assigned, appear to be overluminous for their type. In particular, Ma 7 and Ma 8 appear to belong to a new class of giant irregulars, with absolute magnitudes - 19.8 and -20.7.*

The results found here for Ma 7,8,12 and 326 are in general agreement with those obtained from optical considerations by Börngen and Kalloghlian (1974) and by Kalloghlian (1974).

P a i r s o f G a l a x i e s. Among the galaxies studied, the two pairs Ma 7-8 and Ma 325-326 have been investigated from the point of view of their stability. The difference of velocities between the two components has been obtained from our velocity measurements. From it and from the linear separation of the two galaxies, the minimum mean mass that each component should have for a relative circular motion has been compared to our total mass determination (Table 2).

It comes that the pair Ma 7-8 is not bound and that the pair Ma 325-326 may be bound.

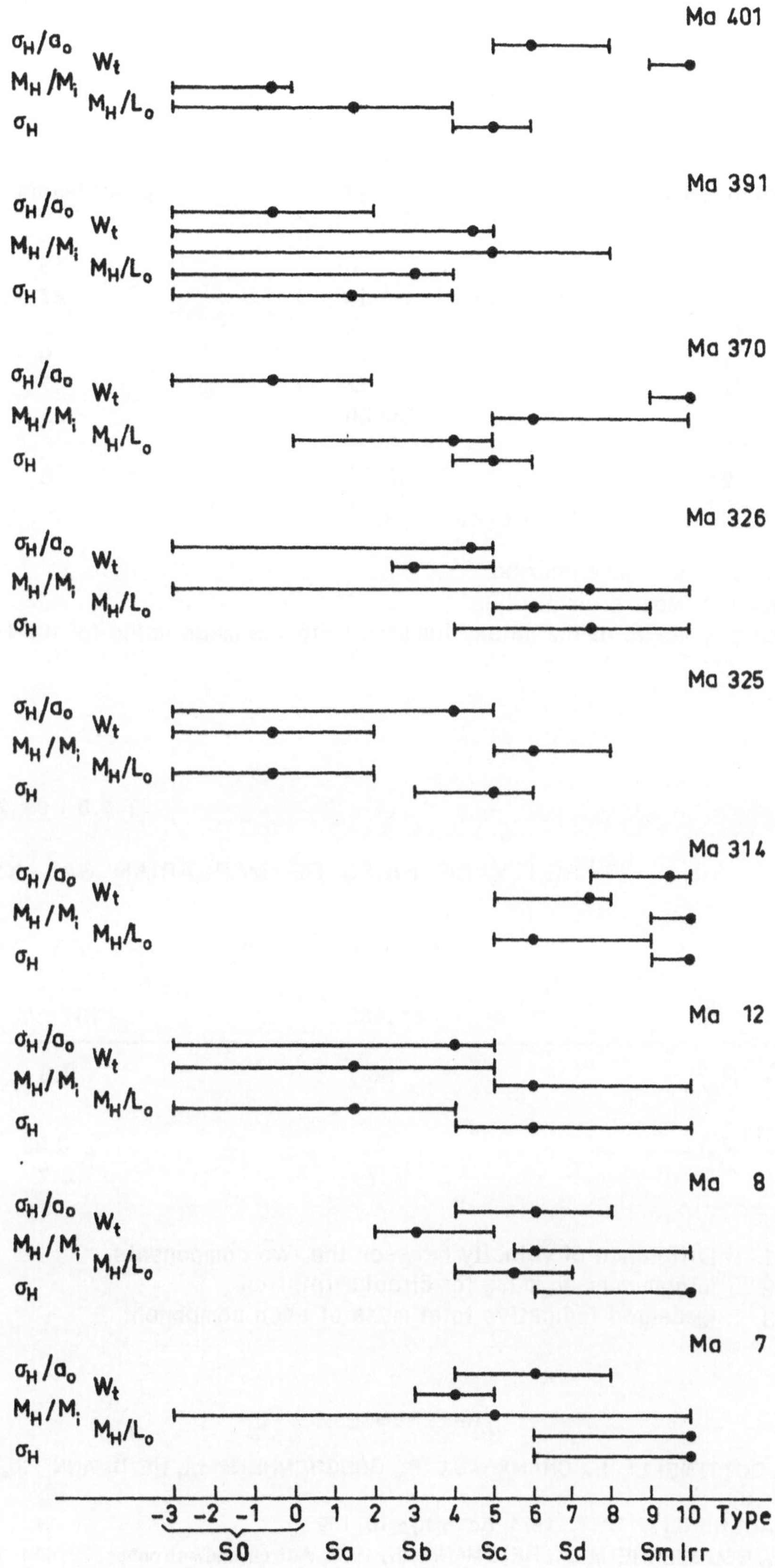

Fig. 1. The morphological type corresponding to the integral properties for the observed Markarian galaxies. The integral parameters are indicated on the left; the dots correspond to the most probable type and the bars to types such that the measured integral properties are less than one standard deviation from the statistical value.

Table 1

MORPHOLOGICAL TYPE OF MARKARIAN GALAXIES

Markarian Number	Type	Overluminosity
7	Irr	10
8	Irr	27
12	Sc	6
314	Irr	9
325	Sc	7
326	Sc-Sd	5
370	-	-
391	Sa-Sb	3
401	-	-

Column 1 : Markarian number
Column 2 : Most probable type
Column 3 : Ratio of the galaxy luminosity to the mean value for the type.

Table 2

STABILITY OF PAIRS OF MARKARIAN GALAXIES

Pair	Ma 7 - 8	Ma 325 - 326
ΔV (km s^{-1})	475±85	157±35
M_c ($10^{11} M_\odot$)	98	2.8
M_i ($10^{11} M_\odot$)	{ 4 2.7	{ 0.88 3.7

Line 1 : Difference of velocity between the two components
Line 2 : Minimum mean mass for circular rotation
Line 3 : Measured indicative total mass of each component

REFERENCES

BALKOWSKI C., BOTTINELLI L., CHAMARAUX P., GOUGUENHEIM L., HEIDMANN, J., 1974, Astron. and Astrophys. 34, 43
BÖRNGEN F., KALLOGHLIAN A.T., 1974, Astrophys 10, 159
BOTTINELLI L., GOUGUENHEIM L., HEIDMANN J., 1973, Astron.and Astrophys. 22, 281
BOTTINELLI L., DU=LOT R., GOUGUENHEIM L., HEIDMANN J. 1975, Astron.and Astrophys. 41, 61
BOTTINELLI L., GOUGUENHEIM L., 1975, Astron.and Astrophys. 39, 341
HEIDMANN J., KALLOGHLIAN A.T. 1973, Astrofizica, 9, 71
KALLOGHLIAN A.T. 1974, Astrofizica (in press)
SARGENT W.L.W. 1972, Astrophys. J. 173, 7

MORPHOLOGICAL STUDY OF MARKARIAN GALAXIES IN PAIRS *

C. CASINI and J. HEIDMANN
INSTITUTO DI ASTRONOMIA DELL'UNIVERSITA DI MILANO, ITALY,

OBSERVATOIRE DE MEUDON, FRANCE
PRESENTED BY J. HEIDMANN

МОРФОЛОГИЧЕСКОЕ ИССЛЕДОВАНИЕ ПАР ГАЛАКТИК МАРКАРЯНА

We present a study of the morphology of 8 pairs of galaxies containing Markarian galaxies based on large scale electronographs or photographs. We compare their morphology to the one of isolated Markarian galaxies obtained by Kalloghlian and Börngen and bring to light a new class of irregular galaxies.

DISCUSSION

KOMBERG: How do you select the class of irregular Markarian galaxies?
HEIDMANN: The clumpy irregulars we isolated are characterized by UV radiation, clumpy structure, large diameters and luminosities, and large internal velocities. By clumpy structure, we mean half a dozen or so large comparable condensations more or less evenly distributed in a common envelope.

* (published in Astron. Astrophys. and Astron. Astrophys. Supp., 1975)

ELECTRONOGRAPHIC STUDY OF THE NUCLEAR REGION IN M 31

D. ALLOIN, D. PELAT, A. BIJAOUI
OBSERVATOIRE DE NICE ; OBSERVATOIRE DE MEUDON, FRANCE
PRESENTED BY D. PELAT

ЭЛЕКТРОНОГРАФИЧЕСКОЕ ИССЛЕДОВАНИЕ ЯДЕРНОЙ ОБЛАСТИ В М 31

We present some preliminary photometric results on the nuclear region of M 31, using electronographic plates obtained through three narrow interference filters, respectively centered at 4670, 5550, and 6130 Å. According to the linear response of the electronic camera, we have obtained *intensity* isophotos of the nuclear region : in the 4''-60'' radius region, they appear to be asymmetric, elongated in the S-W direction, following the same trend as within a 2'' radius region (Light *et al*, 1974). The intensity profiles, we get over a 30'' diameter, are nearly identical in the three colors and show a slope discontinuity occuring at 2''-3'' from the center, in agreement with the existence of the 1''x1.6'' nucleus. We bring further evidence to the presence of dense absorbing matter close to the center (10-20 pc) in the northern part of the nucleus. This dense material could be linked to some higher rate of star formation on the northern side of the inner region (2'' in radius), and therefore explain the displacement to the N-E of the intensity peak in the very nucleus, as observed by Light *et al* (1974). Six among the studied objects surrounding the nucleus in M 31 (within 140'' from the center) are found to be globular clusters, on the basis of colors and width-arguments.

REFERENCE

LIGHT E.S., DANIELSON R.E., SCHWARZSCHILD M. 1974. Astrophys. J., 194. 257

WEDNESDAY, JULY, 2 FIFTH SCIENTIFIC SESSION. "STARS"
CHAIRMAN: PROF. R.E.GERSHBERG

EMISSION LINES IN STELLAR SPECTRA: OBSERVATIONS AND INTERPRETATION

R.E.GERSHBERG AND L.S.LUUD
CRIMEAN ASTROPHYSICAL OBSERVATORY, TARTU INSTITUTE OF ASTROPHYSICS AND ATMOSPHERE PHYSICS, U.S.S.R.
PRESENTED BY L.S.LUUD

ЭМИССИОННЫЕ ЛИНИИ В ЗВЕЗДНЫХ СПЕКТРАХ : НАБЛЮДЕНИЯ И ИНТЕРПРЕТАЦИЯ

(The paper was published in 1975, preprint 7, W.Struve Tartu Astrophysical observatory, Estonian Academy of Sciences).

MODELS OF STELLAR ATMOSPHERES IN COMPARISON WITH OBSERVATIONS

A.A. SAPAR
TARTU INSTITUTE OF ASTROPHYSICS AND ATMOSPHERE PHYSICS, U.S.S.R.

МОДЕЛИ ЗВЕЗДНЫХ АТМОСФЕР СОПОСТАВЛЕНИИ С ДАННЫМИ НАБЛЮДЕНИЙ

(The author has not sent the text of his paper).

BALMER AND PASCHEN DECREMENTS IN BE STARS

D. BRIOT
OBSERVATOIRE DE PARIS, FRANCE

ДЕКРЕМЕНТЫ БАЛЬМЕРА И ПАШЕНА В Be ЗВЕЗДАХ

The aim of this study is the determination of physical conditions in the envelopes of Be stars, from the relative intensities of emission hydrogen lines.

The first study was made on the Balmer line series of 55 stars whose spectral types range from BO to B9 (Briot 1971). This sample was large enough to obtain statistical conclusions, that permitted us to check various theories of the envelopes of the Be stars. We cannot find any agreement between the observational values and the theories based on the hypothesis of a stationary envelope. On the contrary, the observational values agree rather well with Sobolev's theoretical decrements (1974). Following the Sobolev theory, the star envelope is moving with a constant velocity gradient. In a next step we used more sophisticated calculations from the Sobolev's theory and we compared to theoretical values both Balmer and Paschen decrements.

We determined the Paschen decrements of 12 stars with emission Paschen lines whose spectral types range from BO to B5. We studied spectra taken on Kodak IN plates with a dispersion of $19mm^{-1}$ and $39mm^{-1}$. So we can study the Pashen series from PII, or P12, to the end of the series. The absorption lines of the underlying star were taken from standard stars of the same spectral type. We took the infrared excess of Be stars into account for the determination of the continuum radiation.

We studied the behaviour of Paschen decrement versus Balmer decrement using observational values and theoretical values obtained from various approximations of the Sobolev's theory. The various calculations are made by Doazan (1965), Hirata and Uesugi (1967), Boyarchuk (1966), Luud and Ilmas (1971) and Ilmas (1974). Only when the collisional terms are considered in calculations (Luud and Ilmas's calculations) there is a good agreement between theoretical and observational values obtained for the decrements of the hottest stars. It is not possible to find an agreement between theoretical and observational values for the latest type stars of our sample. It will be necessary to find what is the physical phenomenon that is important in the envelopes of these stars and that was neglected.

REFERENCES

BOYARCHUK A.A. 1966, Izv. Krym. astrofiz. Obs. **35**, 45
BRIOT D. 1971, Astron. and Astrophys. **11**, 57
DOAZAN V. 1965, Ann. Astrophys. **28**, 1
HIRATA R., UESUGI A. 1967, Mem. Coll. Sci. Kyoto Univ. Ser. A, **31**, 199
ILMAS M. 1974, Hydrogen Emission lines in the spectra of Early type stars, Tartu
LUUD L., ILMAS M. 1971, Emission lines in stellar spectra, Tartu
SOBOLEV V.V. 1947, Moving envelopes of stars

ON THE INHOMOGENEITY OF CHEMICAL COMPOSITION OVER THE SURFACE 21 PER

Ju.V.GLAGOLEVSKIJ, K.I.KOZLOVA, V.S.LEBEDEV, N.S.POLOSUKHINA
SPECIAL ASTROPHYSICAL OBSERVATORY, ZELENCHUK, U.S.S.R.
PRESENTED BY Yu.V.GLAGOLEVSKIJ

О НЕОДНОРОДНОСТИ ХИМИЧЕСКОГО СОСТАВА НА ПОВЕРХНОСТИ 21 PER

The magnetic variable star has been studied from 4 and 8 Å/mm spectra obtained with the 2.6 - meter reflector of the Crimean Astrophysical Observatory. Spectral line intensities (W_λ) and radial velocities (V_r) have been measured.

It turned out that Eu, Gd, Ti and Mn concentrate in two regions. The second group of elements is formed out of Fe, Ca, Si, Sr, Cr, Mg and others, concentrated in four regions. The longitudes of the centres of Fe concentration make 357°, 101°, 180°, 270° from the meridian passing across the center of the stellar disk at 0.0 phase, and the latitudes are -20°, +10°, -20°, -10°, respectively.

The centers of Eu, Gd, Ti and Mn concentration regions coincide with the first and third iron spots. Heavy elements, Hg 1, Am II, Ag, Os I, W I, are suspected to be present in the atmosphere of 21 Per.

21 Per (HD 18296) is a magnetic variable A_p-star. Its spectrum contains lines of anomalous intensity belonging to Si, Sr, Eu, Cr. The magnetic field intensity value He is determined with uncertainty because of the complex line profiles. H.Babcock has determined that polarity varies, and the extreme values of He equal -1270 - +1350 gs. Preston's results proved to be less confident relative to the variability of He, namely, the extreme values were slightly larger than the measurement errors and turned out to be -370 - +790 gs. The period of photometric and spectral variations is $2^d.88422$ (Preston, 1969).

The purpose of the present investigation is to study the distribution of elements over the surface of 21 Per. The observational material has been obtained with the 2.6 - meter reflector of Crimean Astrophysical Observatory using a Coude spectrograph with dispersions 4 and 8 Å/mm. 23 spectra have been obtained in the region 3900 - 4600 Å and several spectra up to 6600 Å. From the character of radial velocity variations with a period of (Vr, P) and equivalent width variations (Wλ, P) with period it has been concluded that in 21 Per all the elements can be divided into two groups (Fig.1 and 2):

Group I Ti, Mn, Eu, Gd

Group II Fe, Ca, Si, Sr, Cr, Mg etc.

The analysis of the curves (V_r,P) and (W_λ,P) has shown (Glagolevskij, et al. 1974) that the elements of group I concentrate in two opposite areas, "spots", and the elements of group II are distributed in a more complex manner. A more careful analysis (Glagolevskij, et al. 1975) has shown that Fe, Sr, Si and apparently, all the rest of the elements of group II are concentrated in four regions. Two of them pass across the visible central meridian in phases 0.0 and 0.5 and coincide with the areas occupied by the elements of group I, two others occupy the regions between them. The curves of (V_r, P) and (W_λ, P) for the elements of group I, those for Fe II λ4351.76 Å of group two are in Fig.2. The intersection of the curves of (Vr, P) with the line V_r = 0 and the position of the maximum of the curve (W_λ, P) give the longitude of the center of the spot relative to the original meridian. Radii and latitudes of the spots are obtained on the basis of a theoretical model in which it is supposed that the spots are round and the chemical elements are distributed uniformly. The darkening of the limb is not taken into account. A preliminary value of the inclination angle i is determined on the basis of the known empirical relations ($R_*/R_\odot$, T_e) using the values of P and V Sin i = 24 km/s. If one assumes two values of luminosity classes - V and IV, then i=35° and 21°, respectively.

For Eu (and also on the average for other elements of group 1) and for Fe (for the whole group II on the average) the following parameters of the spots are obtained (Preston, 1969; glagolevskij, 1975) (see Fig.3):

	Longitude			
Eu	0°	–	180°	–
Fe	357°	101°	180°	270°
	Latitude			
Eu	-20°	–	-10°	–
Fe	-20°	+10°	-20°	-10°

	Radius			
Eu	75°	–	75	–
Fe	55°	20°	50°	30°

The regions between the spots do not contribute markedly to intensities of Eu lines, for Fe II, however, they contribute a share which is about 0.1 of the maximum. Relative intensities of lines formed in the spots are equal to:

Eu	1.0	–	0.71	–
Fe	1.0	1.30	0.75	1.50

The best fit of the theoretical and observed curves (Vr, P) and (Wλ, P) occurred at 50° for Eu and i= 40° for Fe.

Hydrogen lines in 21 Per are variable. H_δ line showed the most considerable variation of the parameters with phase. Comparison of the observed contours with theoretical (calculated by D.Mihalas) showed that the temperature at the surface of the star varies within the errors (200° - 300°). Brightness variation, if ascribed to temperature variations, correspondsto $\Delta Te = 200°$. The character of variation of contours is such that one may suppose variability of effective surface gravity of the star (Fig.4). This result confirms K.D.Rakosch's et al(1974)suggestion that under the influence of the magnetic field the effective surface gravity may considerably change. As it is well known, however, inhomogeneities of chemical composition lead to violations of the atmospheric structure, therefore the variability of hydrogen lines is apparently due to both factors.

A portion of contribution of each of them may be determined through calculation. For the last years the problem of existence in the atmospheres of magnetic stars of a considerable number of various heavy elements, particularly short-lived, which are formed as a result of the supposed nuclear reactions in the active regions of the surface has been often discussed. We attempted to find in the spectrum of 21 Per those heavy elements which had been detected by other authors in a number of magnetic stars. It turned out that individual lines may be attributed with a great degree of probability to the following heavy elements:

Hg I λ 3984.00 Å
Ag I λ 4210.94 Å
Os I λλ 4400.58; 4420.47; 4213.86 Å
W I λλ 4044,29; 4306.88 Å
Am II λ 4575.59 Å

The intensities of these lines lie within 20-80 mÅ with a limit of detectability of 5-7mÅ The averaged curves of line equivalent width variations for these elements are presented in Fig.5. The presence of variability is of no doubt; however, the absence of the maximum at phase 0.5 is unexpected. This result may be attributed to the uncertainty of the considered part of the curve due to the insufficient number of points. On the basis of the above - stated we draw the following principal conclusions. The elements are distributed in "spots" over the surface of 21 Per. The character of distribution of the spots and the composition of the element groups is, apparently, unique. Spectral variability can be readily explained in the framework of the oblique rotator hypothesis. The uniquity of the distribution of the chemical elements poses considerable difficulties in interpreting the phenomena considered. All existing hypotheses suggest the presence of quite definite regularities in the composition of groups and in the character of distribution of elements over the surface.

The variability of hydrogen lines may evidence for the presence of a dipole magnetic field at the surface of 21 Per, if Rakosch's et al. supposition of the effect of He on g were true. In this case the poles of the magnetic field must coincide with the regions occupied by rare earth elements and I, III spots of Fe. The presence of heavy elements suspected in 21 Per and other stars makes one pay attention to the hypothesis on the existence of nuclear reactions at the surface of magnetic stars.

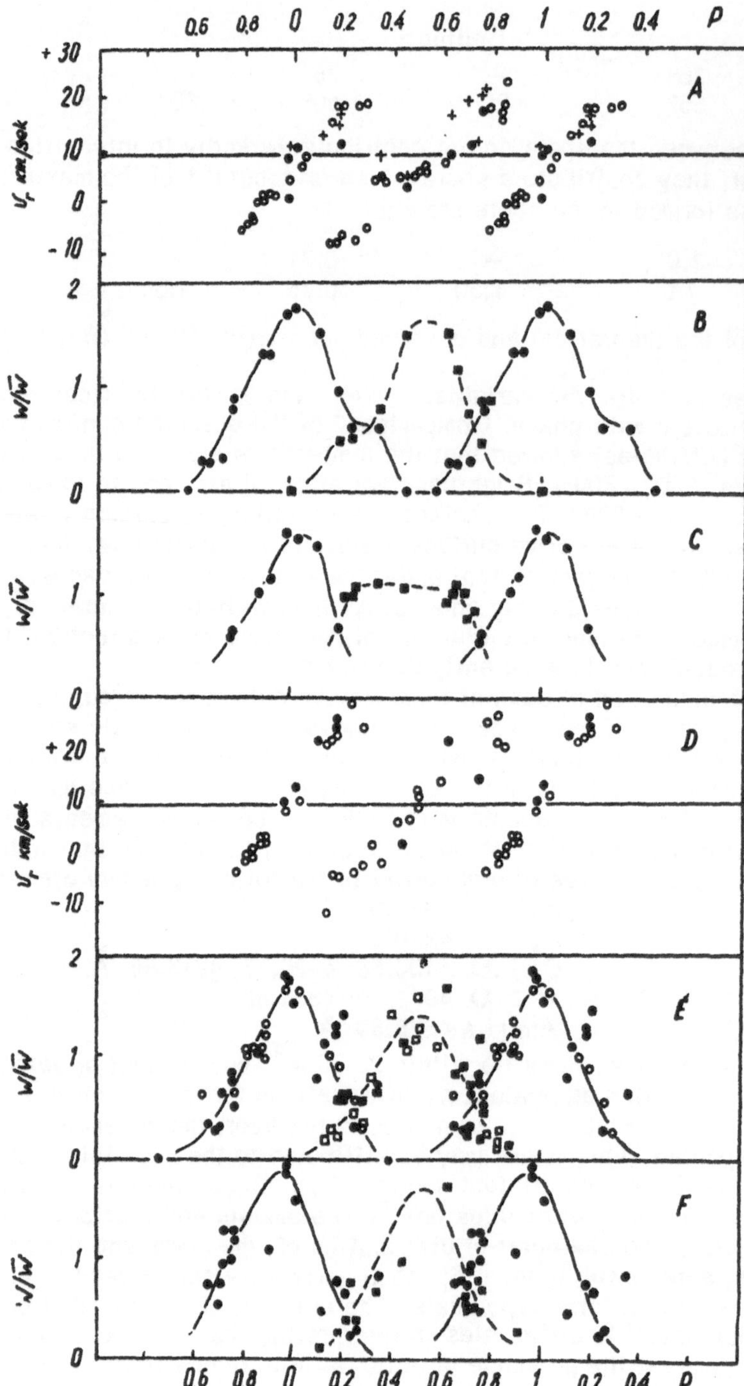

Fig. 1. Radial velocity (Vr) and relative equivalent line widths (W_λ / $\bar{W}_\lambda$) variations during the period for the elements of group 1 (Ti, Mn, Eu, Gd) :

A) Ti+Mn (our measurements); o(Ti,Mn) Preston [1]

B) Ti• primary spot; ▪ secondary spot (our measurements)

C) Mn• primary spot; ▪ secondary spot (our measurements)

D) Eu•(our measurements); o (Eu, Gd) Preston [1]

E) Eu• primary spot; ▪ secondary spot (our measurements)
Eu o; □ Preston [1]

F) Gd• ▪ (our measurements)

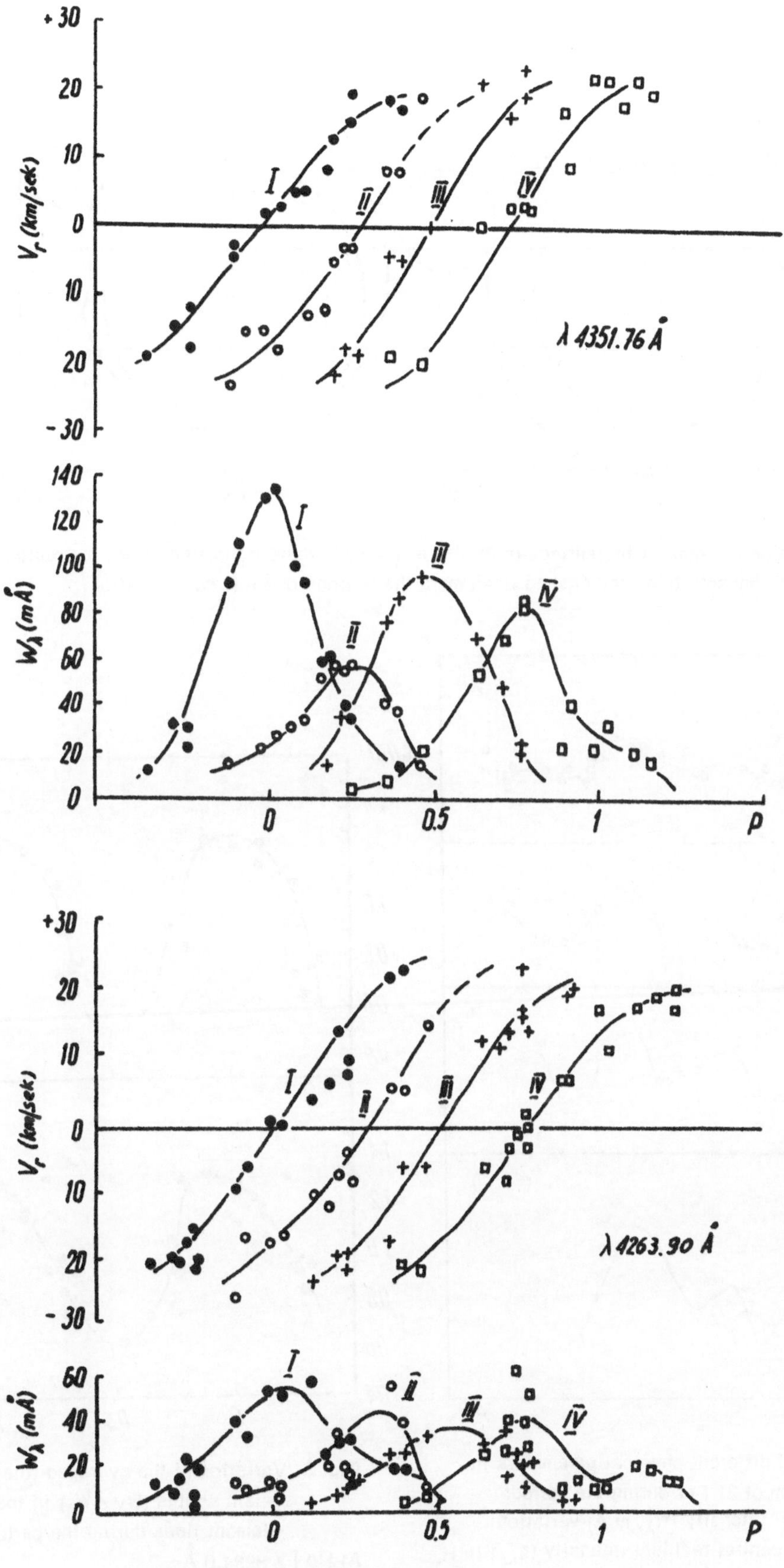

Fig. 2. Radial velocity (Vr) and equivalent line width (W_λ, m Å) variations during the period for the elements of group II (Fe II λλ4351.76, 4263.90 Å).

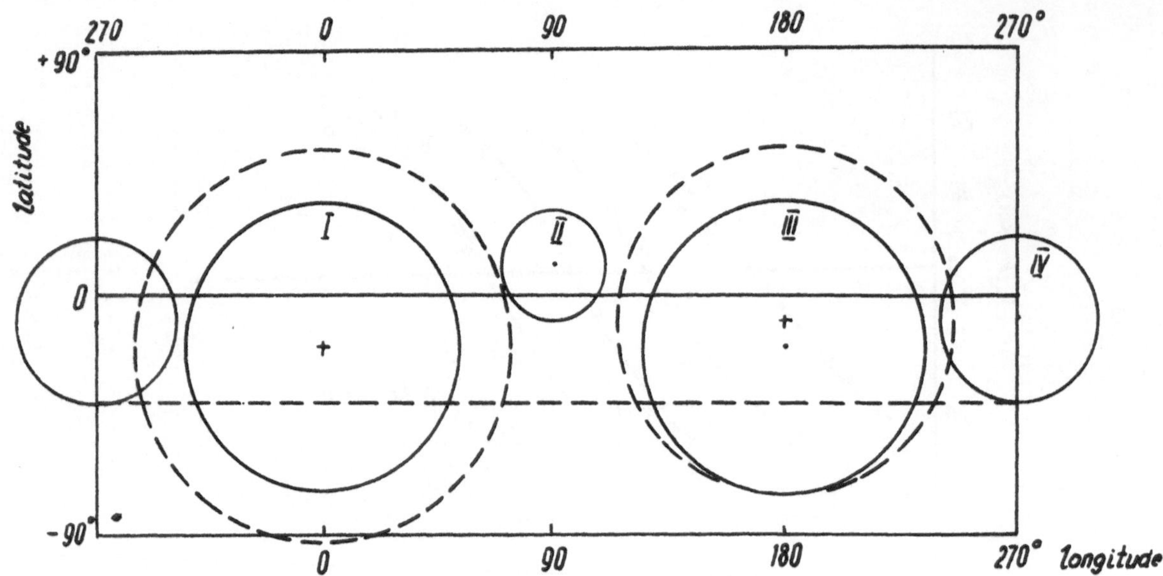

Fig. 3. Position of spots at the surface of 21 Per at i = 40°, where solid lines mark the regions of Fe II concentration, and dashed lines mark the regions of Eu II concentration.

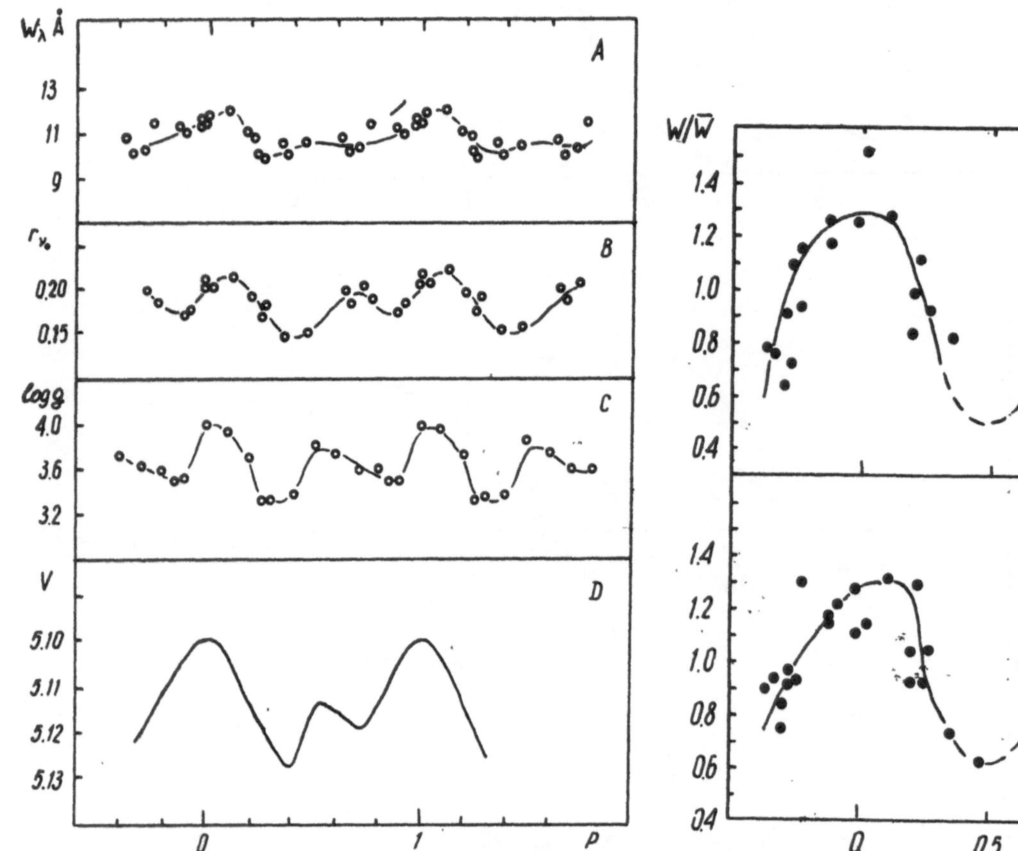

Fig. 4. Variation of different physical parameters in the spectrum of 21 Per during the period:

A) H_γ hydrogen line intensity (W_λ, m Å) variation;

B) variation of the central residual intensity ($r_{\nu o}$) in H_γ hydrogen line;

C) variation effective gravity (logg) acceleration obtained from H_γ line with the help of D Mihala's model at $\Theta_e = 0.425$ (11700°).

D) 21 Per brightness curve from the data of [1].

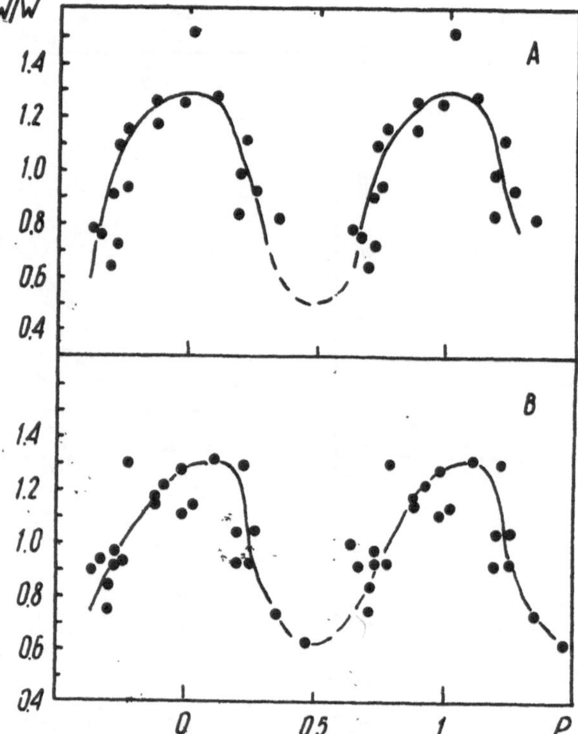

Fig. 5. Variation of the averaged relative equivalent widths ($W_\lambda / \overline{W}_\lambda$) in the heavy element lines during the period (P) :

A) Hg I λ 3984.0 Å

B) W I λλ 4044.29, 4306.88 Å
Os I λλ 4213.86, 4400.58, 4420.47 Å
Ag I λ 4210.94; Am II λ 4575.59 Å

REFERENCES

GLAGOLEVSKIJ Ju.V., KOZLOVA K.I., POLOSUKHINA N.S. 1974, Astrophysics, 10, Iss.4, 517

GLAGOLEVSKIJ Ju.V., KOZLOVA K.I., LEBEDEV V.S., POLOSUKHINA N.S. 1975, Astrophysics. (In print)

PRESTON C.W. 1969, Ap. J., 158, 251

RAKOSCH K.D., SEXL R., WEISS W.W. 1974, Astron. and Astrophys., 31, N. 4, 441

SIGNIFICANCE OF ASYMMETRIC EMISSION LINE PROFILES ON MASS-FLUX AND CHROMOSPHERE FROM T TAURI STARS

S.DUMONT, N.HEIDMANN and R.N.THOMAS

INSTITUT D'ASTROPHYSIQUE, PARIS, FRANCE

PRESENTED BY N. HEIDMANN

О ЗНАЧЕНИИ АСИММЕТРИЧНЫХ ПРОФИЛЕЙ ЭМИССИОННЫХ ЛИНИИ ДЛЯ ПОТОКА МАССЫ И ХРОМОСФЕРЫ В ЗВЕЗДАХ ТИПА T TAURI

We make a tentative interpretation of the spectra of T Tauri stars different from the usual interpretations by an extended expanding atmosphere with radiative energy sources only or by some non-thermal emission processes. We test a third alternative, the existence of a deep lying chromosphere, in agreement with the idea that the chromosphere is a normal region of stellar atmosphere regarded as a transition zone between a quasi-equilibrium stellar interior and a non-equilibrium interstellar medium. The chromosphere would differ from star to star by the depth of its beginning and would be heated from velocity fields (convection, pulsation, for instance) in the sub-atmosphere. The outward component of the velocity field which would produce the chromospheric temperature rise is also responsible for the mass-flux from the star and may be estimated from asymmetry and displacement of the lines.

We have continuous spectral distributions of eighteen T Tauri stars, all in the Taurus-Aurigae dark cloud, taken from Kuhi (Astron. and Astrophys, Suppl. *15*, 47, 1974). They were obtained by scanner measurements at Palomar and Lick Observatories, from λ3300 to λ11000 Å. We have also profiles of H_α and Ca II H and K lines from two stars, T Tau and RY Tau, obtained at Lick and Mount Wilson Observatories with a dispersion of 16 and 10Å/mm, respectively (see L.V.Kuhi, Ap.J. *140*, 1409, 1964).

Fig. 1 displays the variable continuous spectra of the two stars, at two epochs, which show the Balmer continuum in excess and H_α line in emission. Fig.2 displays H_α line profiles which are in emission, asymmetric, displaced and variable with time.

We make the hypothesis that the Balmer continuum is formed in the chromosphere and compute H_α as produced by photo-excitation followed by recombination (see Dumont et al., Astron, and Astrophys. 29, 199, 1973).

From the observed continuous flux, compared to blackbody flux, we deduce radiation temperatures T_P and T_B^* in Paschen and Balmer continua and the difference $\Delta T^* = T_B^* - T_P$. By interpolating the continuum at H_α wavelength we obtain, by difference, the observed integrated flux F_{H_α} in the line. Fig.3 shows a correlation between $\bar{F}_{H_\alpha}$ and ΔT^*.

We try to predict theoretically this correlation by taking a two-level hydrogen atom in a rough atmosphere made of a photosphere over- spread with a chromosphere having a mean temperature T_B and by using the formalism of a non-LTE atmosphere established by Jefferies and Thomas (Ap. J. *129*, 401, 1959). Comparison with the precedent correlation gives a modified mean temperature T_B of the chromosphere with its corresponding mean, finite, optical depth τ.

The computed H_α line profiles from T Tau and RY Tau are shown in Fig.4. The integrated flux in these lines is equal to the observed one but we see that the shapes of the computed profiles, even with the best damping parameter, do not resemble the observed ones. They have always a central reversal and they are symmetric, unshifted and not so broad as the observed ones.

Indeed, in agreement with the logic of the existence of the chromosphere, a velocity field is needed. Fig.5 shows the effects of introducing a linear velocity field which increases towards the surface. Profiles are plotted for several values of the velocity-gradient. These gradients modify the shape of the static profiles by acting on the optical depth of the line, but they are low enough to leave the source-function practically unchanged. This fact simplifies the computations (see E.Simonneau, Astron. and Astrophys., *29*, 357, 1973).

On the other hand, Thomas (Astron. and Astrophys. *29*, 297, 1973) has shown that such a systematic velocity field is forced to be slightly less or equal to the thermal velocity. So the velocity gradient gives the temperature variation.

The best fit for T Tau H_α-line profile is obtained,as shown in Fig.6. with a velocity field having three different gradients increasing towards the surface, through the chromosphere. The corresponding temperature increases from 4000 K at the depth of the transition with the photosphere to about 10000 K at the depth where H_α is formed in the chromosphere (see Table 1).

We are now in the process of computing Ca II H and K lines in order to test the above atmosphere. We intend to study the problem of the variability of the spectrum for which simultaneous observations of lines and continua are necessary.

Table 1

τ	T	$\sqrt{\pi}\,\dfrac{\nu_0}{c\,\Delta\nu_{\text{Doppler}}}\,\dfrac{dv}{d\tau}$
60 000	4 000 K	
		10^{-6}
4 000	4 130	
		10^{-4}
400	5 060	
		3×10^{-3}
0	9 950	

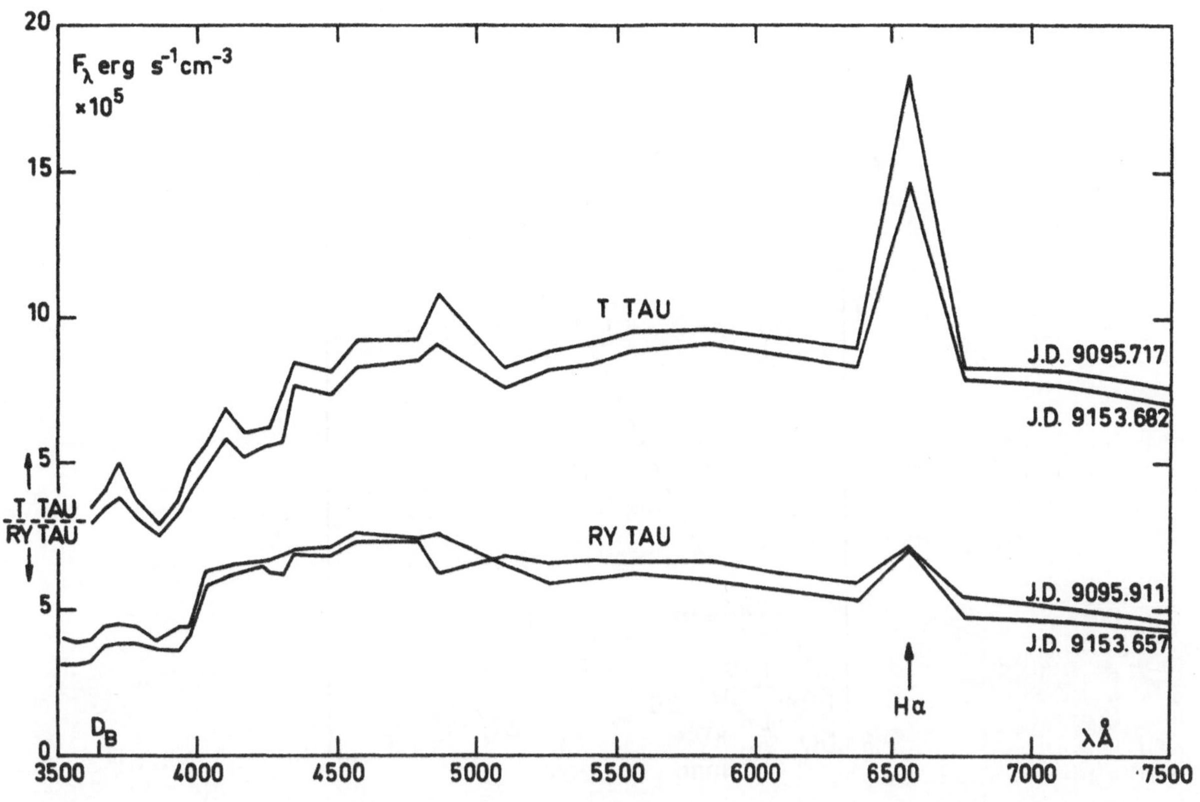

F_λ erg $s^{-1}cm^{-3}$
×10^5
T TAU
RY TAU
J.D. 9095.717
J.D. 9153.682
J.D. 9095.911
J.D. 9153.657
Hα
D_B
λÅ

Fig. 1.

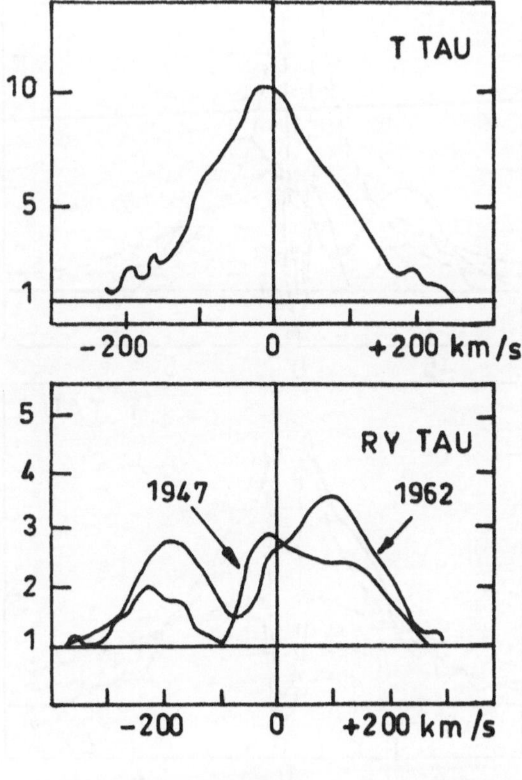

T TAU
-200
0
+200 km/s
RY TAU
1947
1962

Fig. 2.

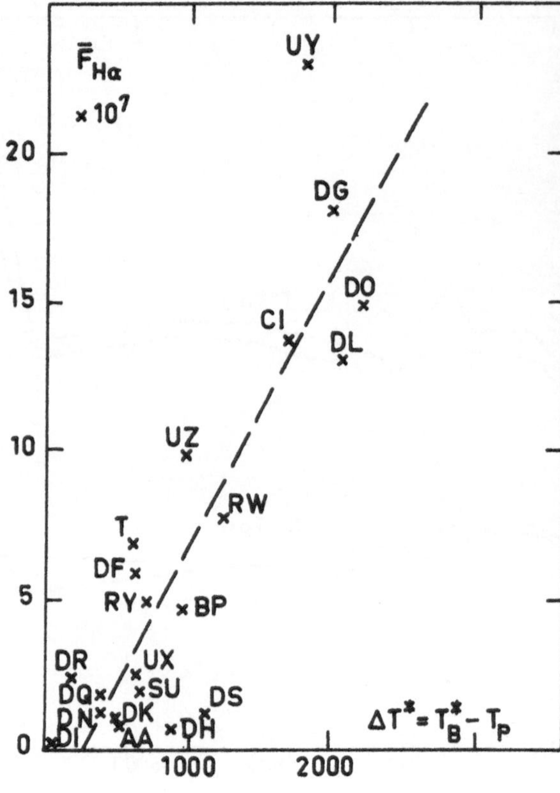
$\bar{F}_{H\alpha}$
$\times 10^7$
UY
DG
DO
CI
DL
UZ
RW
T
DF
RY
BP
DR
UX
DQ
SU
DS
DN
DK
DI
AA
DH
$\Delta T^* = T_B^* - T_P$
0
5
10
15
20
1000
2000

Fig. 3.

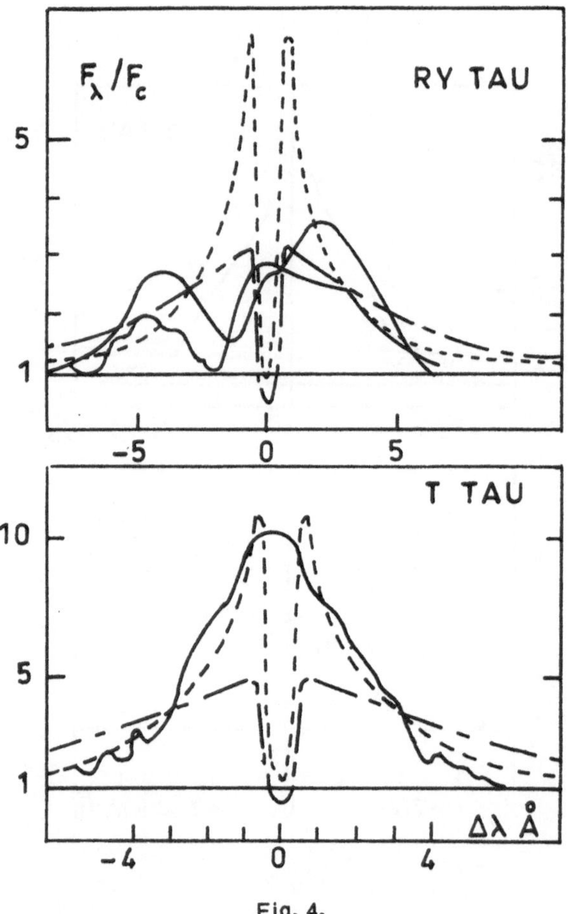
F_λ / F_c
RY TAU
5
1
-5
0
5
T TAU
10
5
1
$\Delta\lambda$ Å
-4
0
4

Fig. 4.

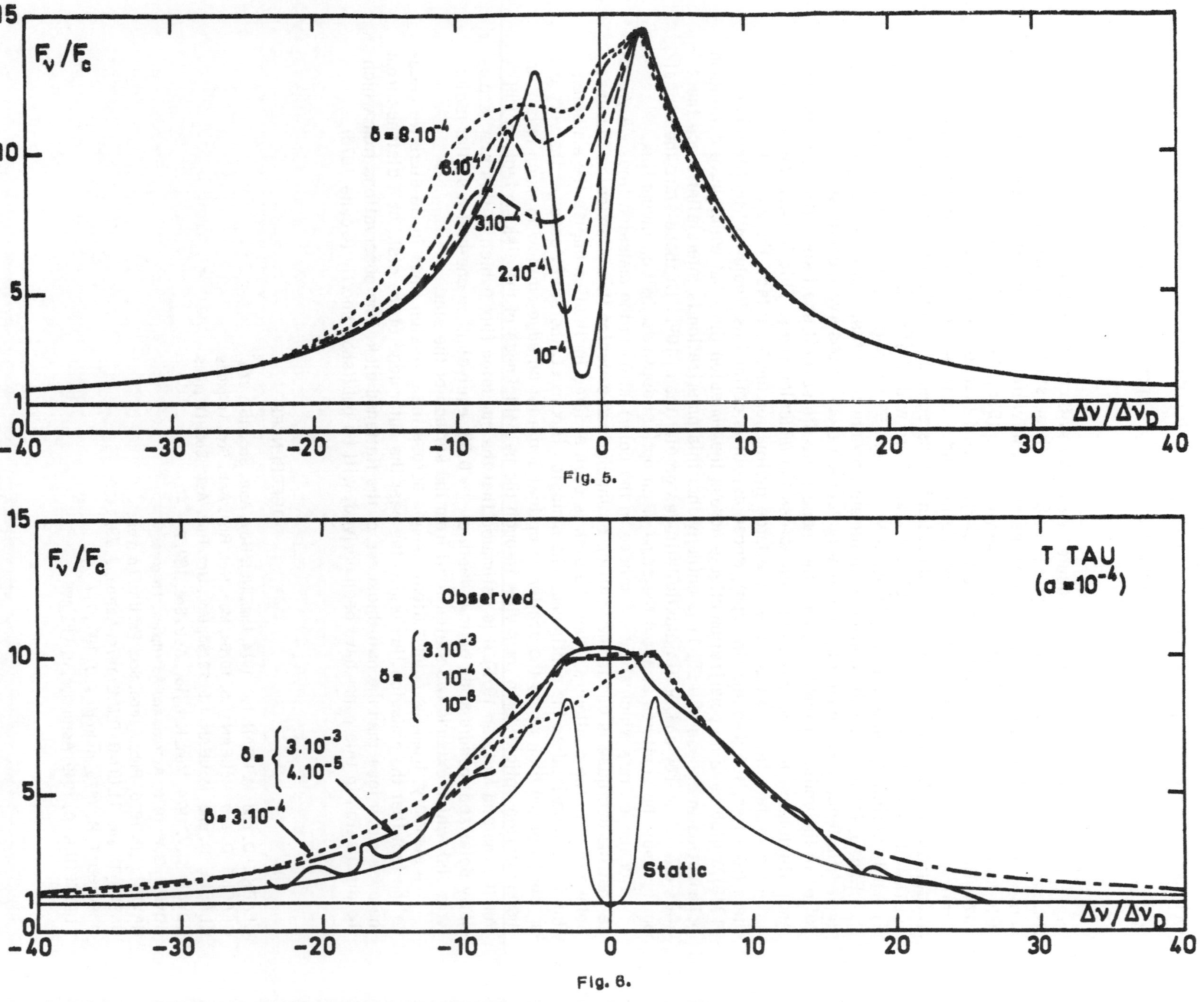

15
10
5
1
0
F_v/F_c
δ = 8.10^-4
6.10^-4
3.10^-4
2.10^-4
10^-4
Δν/Δν_D
−40
−30
−20
−10
0
10
20
30
40

Fig. 5.

15
10
5
1
0
F_v/F_c
T TAU
(a = 10^-4)
Observed
δ = 3.10^-3 10^-4 10^-6
δ = 3.10^-3 4.10^-5
δ = 3.10^-4
Static
Δν/Δν_D
−40
−30
−20
−10
0
10
20
30
40

Fig. 6.

POLARIZATION MEASUREMENTS IN THE H-ALPHA AND H-BETA LINES OF EARLY-TYPE EMISSION-LINE STARS

GEORGE V. COYNE, S.J.
VATICAN OBSERVATORY, VATICAN CITY STATE AND UNIVERSITY OF ARIZONA U.S.A.

ИЗМЕРЕНИЯ ПОЛЯРИЗАЦИИ В ЛИНИЯХ Н-АЛЬФА И Н-БЕТА РАННИХ ЗВЕЗД С ЭМИССИОННЫМИ ЛИНИЯМИ

Linear polarization has been measured with narrow-band filters across the H-alpha and H-beta emission lines of 13 early-type stars. Six of these show a decrease in polarization across the emission lines. We combine these new observations with those already published (Clarke and McLean 1974, 1975a, 1975b, Coyne 1974, 1975, Hayes 1975, Hayes and Illing 1974, Coyne and McLean 1975, Poeckert 1975). We find that for 27 Be stars for which polarization measures have been made in the H-alpha and H-beta emission lines 10 show a definite decrease in the polarization of the emission flux as compared to the continuum flux. All but two of these (Pi Aqr and HD 20862) are shell stars with V sin i in excess of 300 km/sec. Pi Aqr is definitely an exception. It has a V sin i of 153 km/sec and it is not a shell star

Except for this star the general requirement appears to be that one must be seeing an extended flattened disk about a Be star nearly edge-on order to observe a decrease in the polarization in the emission lines. On the other hand there are several shell stars (Phi And and Omi And) which show no polarization effects across the emission lines. This is undoubtedly because in these stars the disks are seen at a large inclination angle. In fact, there are three Be stars (R48 Per, 105 Tau and Chi Oph), whose aspect is defined as "pole-on" by Slettebak (1949), which have no polarization effects across the emission lines but which have continuum polarization in excess of 0.6%. It is unlikely that this polarization is interstellar since these stars are all nearby and/or at galactic latitudes greater than $|10^o|$. In these stars the emission flux must be scattered at about the same mean optical depth as the continuum flux.

For those stars which show a decrease in the polarization in the emission lines we propose that the decrease is due to the addition of unpolarizad emission flux or emission flux less polarized than the continuum. When the disk is seen at large inclination angles the emission flux is as polarized as the continuum. The emission flux must, therefore, occur at the very outer edges of the disk where the scattering optical depth for an edge-on view is much less than the scattering optical depth for a view through the half-thickness of the disk. In fact, in Phi Per (Coyne and McLean 1975) it is estimated that the emission flux (which in this star is partially polarized) occurs at an optical depth of $\tau = 0.03$, equivalent to one-half a stellar radius for a disk which extends two stellar radii from the surface of the star.

In summary, from the polarization changes across the emission lines in Be stars we deduce the fact that the emission flux originates near the outer edge of the disk, at a distance from the edge much less than the half-thickness of the flattened disk. The observations from which the discussion in this paper have been derived, will be published shortly (Coyne 1975).

REFERENCES

CLARKE D. and McLEAN I.S. 1974, Mon. Not. Roy. Astr. Soc. 167, 27p
CLARKE D. and McLEAN I.S. 1975a, Mon. Not. Roy. Astr. Soc. In press
CLARKE D. and McLEAN I.S. 1975b, Mon. Not. Roy. Astr. Soc. In press
COYNE G.V. 1974, Mon. Not. Roy. Astr. Soc. 169, 7
COYNE G.V. 1975, Astron. and Astrophys. In press
HAYES D.P. 1975, Pub. Astron. Soc. Pacific. In press
HAYES D. and ILLING R.M.E. 1974, Astron. J. 79, 1430
POECKERT R. 1975, Astrophys. J. 196, 777
STETTEBAK A. 1949, Asrtophys. J. 110, 498

ON THE INHOMOGENEITY OF ATMOSPHERIC PARAMETERS OF MAGNETIC STARS

I.M.KOPYLOV, R.N.KUMAJGORODSKAYA, L.I.SNESHKO, V.V.SOKOLOV, N.M.CHUNAKOVA
SPECIAL ASTROPHYSICAL OBSERVATORY, ZELENCHUK, U.S.S.R.
PRESENTED BY I.M.KOPYLOV

О НЕОДНОРОДНОСТИ ПАРАМЕТРОВ АТМОСФЕРЫ МАГНИТНЫХ ЗВЕЗД

Investigation of the nature of stellar magnetism and chemical composition anomalies in the atmospheres of magnetic and peculiar stars is sufficiently hampered by the complex picture of distribution of chemical elements and by the differences in the atmospheric structure over different parts of the star surface. These features of the outer layers of magnetic stars show spectral variability, light and color variability with a period equal to that of the star's rotation (oblique rotator hypothesis).

In particular, in a number of studies it has been shown that light variations of Ap-stars in a wide spectral range can be explained in the first approximation by variations of the effective acceleration of gravity, g_{eff}, and the effective temperature, T_{eff} over the surface or in the atmosphere of a star (Sneshko, 1972; Rakosch et al., 1974; Sneshko, 1975; Khokhlova, 1971.)

Since the amplitudes of light and color variations of magnetic stars with period are comparatively small ($0.^m05$ - $0.^m20$), one should also expect on the face of it relatively small variations of g_{eff} and T_{eff} (or θ_{eff}).

As modern spectrophotometric techniques of determination of g_{eff} and T_{eff} are burdened with great errors, to investigate temperature and acceleration of gravity variations over the surface of magnetic and peculiar stars it is important to choose such a method which would allow to obtain the two parameters simultaneously, using homogeneous spectral material.

The fact that it suffices to observe only relative variations of these characteristics with period makes the task easier.

It is known that for many peculiar stars considerable variations of equivalent widths and other hydrogen line parameters during the period are discovered (Kumajgorodskaya and Chunakova, 1973; Ryabchikova, 1972.) Since the values of these parameters are functions of temperature and acceleration of gravity in the atmosphere, a method of determination of g_{eff} and T_{eff} from the contours of hydrogen lines may be then (the unknown) one sought for.

We applied this method to two peculiar A-stars, α^2CVn and HD 184205, for which a detailed investigation of variations of different parameters of Balmer series hydrogen lines during the photometric period (Kumajgorodskaya and Chunakova, 1973, 1975) had been already carried out using homogeneous spectral material (dispersion 15 Å mm^{-1}).

Temperatures and accelerations of gravity have been determined by comparison between the measured on spectrograms and theoretically calculated by Mihalas models (1965) $H\gamma$ and $H\delta$ line contours. The contours are calculated (Sneshko, 1975) using a more precise theory of hydrogen line broadening.

Comparisons have been made for each line separately from several (up to 9) pairs of halfwidths $\Delta\lambda$ in the range of 3 to 12 Å since the innermost ($\Delta\lambda \leq 2$ Å) and the outermost ($\Delta\lambda \geq 13$ Å) parts of line contours are constructed less reliably than the middle parts both theoretically and from spectrograms. θ_{eff} and lg g_{eff} found for each pair have been averaged for each line and for each phase of observation.

Variation of the values of θ_{eff} and lg g_{eff} for α^2CVn and HD 184905, respectively, is shown in Figs.1 and 2. For α^2CVn there are given the mean values of θ_{eff} and lg g_{eff} found from $H\gamma$ and $H\delta$, for HD 184905 - those from $H\delta$ alone (a portion of spectrograms for this star in $H\gamma$ region is underexposed). The dashes indicate the mean values of each determination of θ_{eff} and lg g_{eff}.

From Fig.1 it can be seen that variations of θ_{eff} over the surface of α^2CVn are relatively small: the amplitude of the effective temperature variation is +500° when its mean value is 9500°. At the same time, variation of lg g_{eff} with phase is considerable: its total amplitude is ~0.80, i.e. the effective radius of the atmosphere over different points of the star's photosphere varies by more than a factor of 2.5.

From Fig.1 it also follows that smaller values of lg g_{eff} correspond, in general, to the low-temperature areas of the star's surface and vice versa.

The extreme Θ_{eff} and lg g_{eff} do not coincide with the magnetic poles and are not located on the equator of the star. It is interesting to note that one of the maximum Θ_{eff} and g_{eff} (P = 0.35-0.4) areas of the star's atmosphere corresponds to the region of the highest concentration (maximum line intensity) of Mg II and Si II and is away from the positive magnetic pole (p=0.4) by 35°-50°. A second region with the maximum values of Θ_{eff} and lg g_{eff} (P=0.80-0.85) does not coincide with the negative magnetic pole (P=0.0) and is away from it by about 65°-70°.

Therefore, the effective extension and the mean temperature of the star's atmosphere, where hydrogen lines are formed, vary considerably over the surface, neither denser and hotter zones nor less dense and hot ones being coincident with the magnetic poles and magnetic equator.

Similar results are obtained for the peculiar star HD 184905, but somewhat less confident due to a little number of spectrograms.

However, if in α^2 CVn the effective temperature varies definitely during the period of rotation of the star, in HD 184205 it is almost constant (its amplitude of variation ±350° (with an error determination of 150-200°). At the same time, lg g_{eff} shows considerable variations whose amplitude is ~0.6. The magnetic field for HD 184205 has not yet been determined, comparison of g_{eff} variation with magnetic elements of the star does not seem possible. As it is shown, for instance, (Rakosch et al., 1974) the effect of a powerful magnetic field on the acceleration of gravity may be large. However, the lack of correspondance of the regions of the extreme g_{eff} values on the surface of the star α^2CVn shown by Fig.1 either with the magnetic poles or with the magnetic equator makes us assume that a real picture of the atmospheric structure and magnetic field configuration in peculiar stars is much more complicated than it has been suggested (Rakosch et.al., 1974).

Further examination of the results obtained by us may consist of:

1) setting a definite law of variation of Θ_{eff} and lg g_{eff} over the surface of the star; 2) calculation of the integral (over the star's surface) brightness at different wavelengths or the continuous spectrum for different phases; 3) comparison of the calculated variations of brightness and color index or energy distribution with the observed ones (obtained in the narrow-band multicolor system, in ten-color, for example or by spectrophotometric methods).

Now, however, it seems to us quite evident that the techniques of determination of chemical composition of the atmospheres of magnetic and peculiar A - stars applied until recently and based on the assumption that the atmospheric structure of magnetic and normal stars is slightly different should not be considered sufficiently reliable.

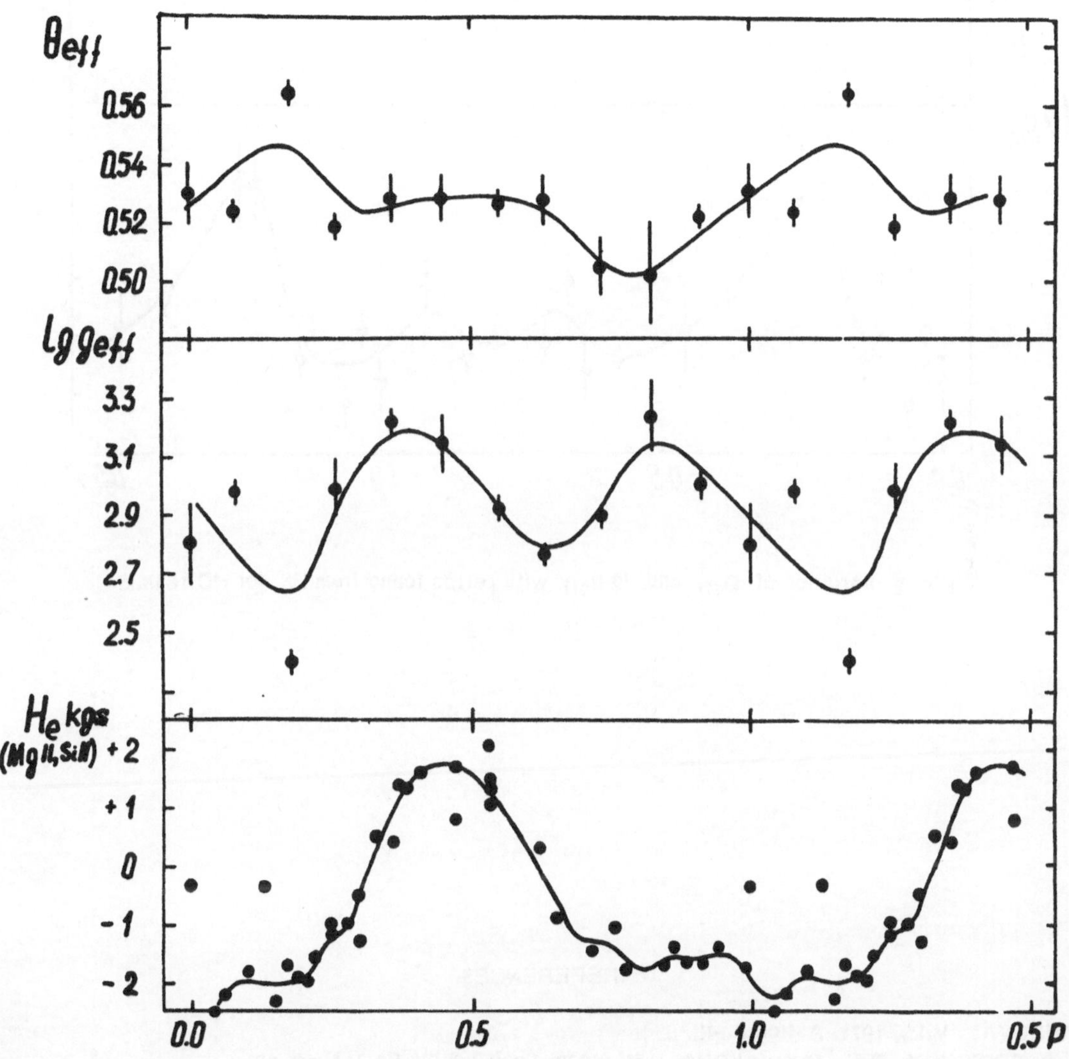

Fig. 1. Variation of the mean values of Θ_{eff} and lg g_{eff} with period found from H_{γ} and $H\delta$ and H_e (Mg II, Si II) for α^2 CVn.

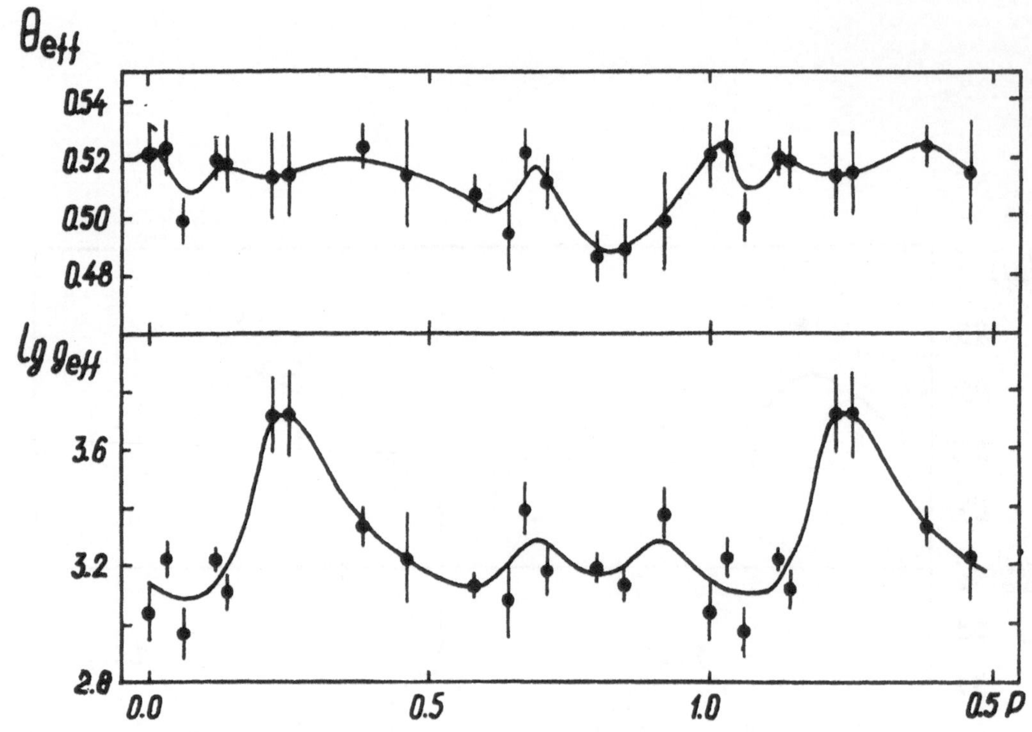

Fig. 2. Variation of Θ_{eff} and lg g_{eff} with period found from Hδ for HD 184905.

REFERENCES

KHOKHLOVA V.L., 1971, Astron.J.,48, 534
KUMAJGORODSKAYA R.N., CHUNAKOVA N.M., 1973, Soobsheniya SAO, N 10, 12
KUMAJGORODSKAYA R.N., CHUNAKOVA N.M., 1975, Astrophys. Research (Izvestiya SAO), 7, 3
KUMAJGORODSKAYA R.N., KOPYLOV I.M., 1972, Astrophys. Research (Izvestiya SAO), 4, 50
MIHALAS D., 1965, Astrophys.J. Suppl. Ser., 9, 321
RAKOSCH K.D., SEXL R., WEISS W.W., 1974, Astron.Astrophys., 31, 441
RYABCHIKOVA T.A., 1972, Izvestiya of Crimean Observatory, 45, 146
SNESHKO L.I., 1972, Astron. Circ., N 741, 3
SNESHKO L.I., 1975, Astrophys. Research (Izvestiya SAO), 8, (in print)

THE CHEMICAL COMPOSITION OF τUMa - AN A_m STAR

DANUTA KUROCZKIN

ASTRONOMICAL INSTITUTE OF NICOLAUS COPERNICUS UNIVERSITY
TORUŃ, POLAND

ХИМИЧЕСКИЙ СОСТАВ τ UMa - A_m ЗВЕЗДЫ

I n t r o d u c t i o n. τUMa is an A_m star. The first coarse analysis of its chemical composition was undertaken by Greenstein in 1948. This analysis has been carried out in the blue spectral region using relative to the Sun curve of growth technique. Therefore it was decided to repeat his analysis in the yellow and red regions,where the lines are not so seriously blended. The gf-values especially for FeI have been considerably improved since 1948 year.

The spectra were taken by Dr.Greenstein with the 200'' telescope. The observational data are presented in Table 1.

Table 1. Observational material UMa = HD 78362

Plate	Emulsion	Dispersion (Å/mm)	Wave-Length Region
7300[a]	11a-0	4.5	3830 - 4170
7660[a]	11a-D	6.4	4700 - 5300
7661[a]	11a-D	6.4	5125 - 6050
7662[a]	11a-D	6.4	4700 - 6170
7660[b]	11a-F	6.4	6200 - 6700
7661[b]	11a-F	6.4	6122 - 6543
7662[b]	11a-F	6.4	6200 - 6680
7303[b]	11a-F	6.4	6130 - 6670

The mean values of equivalent widths were calculated with the mean error $\Delta \log W/\lambda = 0.10$ for weak lines and 0.05 for medium and strong lines.

M e t h o d o f A n a l y s i s. The curve of growth method has been improved by Cayrel and Jugaku (1963) in that way, that they include the contribution of different atmospheric layers to the line formation. The abscissa log X on the curve of growth can be considered as the sum of three terms:

$$\log X = \log gf + \log \Gamma + \log N_{el}/N_H \qquad (1)$$

which equals $\log W/\lambda$ for the weak-line approximation. This form shows the spectroscopic term of the line, log gf; the thermodynamic term $\log\Gamma$; and the term proportional to the abundance of the element considered.

The new gf-values for FeI were reduced to one common system, which is almost identical with that used by Foy (1972). For other elements the gf-values were also brought to one scale. The sources of gf-values used in the analysis are given in Table 4.

The Γ-values are very sensitive to the distribution of the temperature with depth $T = T(\tau)$. For example even for the Sun, where the temperature distribution is well known the Γ-values for FeI lines differ by an amount in $\Delta \log \Gamma = 0.17$ if the slightly different gradients are adopted (Foy, 1972). This of course will introduce a difference in the abundance determination of $\Delta \text{Log } N = 0.17$.

The Γ-values for τUMa with $\theta_{exc} = 0.67$, $\log g = 4.0$ were calculated using Cayrel - Jugaku's tables. This fact implies of course a serious assumption that the temperature distribution $T = T(\tau)$ for τUMa is identical with that of the Sun.

Besides Wrubel's theoretical curve of growth for $B_0/B_1 = 2/3$ and $\log a = -2.6$ was also applied.

R e s u l t s. The final results are given in Table 2. The abundances of elements are given in the scale log H = 12.00. Generally, the abundance increases with the increasing atomic number. Carbon, magnesium, silicon, sulphur, calcium and scandium are underabundant, whereas the iron group, barium and rare earth elements are overabundant. The most abundant elements are cobalt, nickel, strontium and rare earth elements.

Table 2. Final abundance of UMa (log H = 12.00)

Element	Number of lines	Present analysis Γ-method	Wrubel's c. of g.	Sun	$[El/H]_\odot$	Other determinations Greenstein	Miczaika	Smith (log N)	Source of gf - values
C I	3	-	7.73	8.72[a]	-0.99	-	-	8.61	1a
Na I	9	6.48	6.38	6.30[a]	0.18	-	0.88	6.54	1
Mg I	4	7.20	6.94	7.36[b]	-0.16	-	0.2	6.86	1,10
Si I	10							7.30	1,2
Si II	3	7.31	7.23	7.79[b]	-0.48	-	-	8.27	1,3
S I	3	-	6.49	7.30[a]	-0.81	-	-	7.24	1
Ca I	17	5.64	5.57	6.04[b]	-0.40	-1.05	-0.62	5.69	1
Sc I	3								
Sc II	6	2.74	2.77	3.85[b]	-0.11	-	-	2.55	4,5
Ti I	5	5.20	5.00	4.63[b]	0.37	-0.45	-0.20		4,11,12
Ti II	21							4.67	8,13,14
V I	3								4,5
V II	10	5.12	4.62	4.17[b]	0.95	-0.62	-0.40	4.15	8
Cr I	51							5.89	4,5
Cr II	16	6.63	6.11	5.85[c]	0.78	0.13	0.34	6.28	18,20
Mn I	22	6.16	6.00	4.80[b]	1.36	-0.05	0.31	5.48	4,5,6
Fe I	221							6.93	21,22,23 24,25,26
Fe II	40	7.76	7.56	7.60[d]	0.06	0.00	0.16	6.86	3,16,17 18,19,20
Co I	8	5.46	5.91	4.70[b]	1.24	-0.25	-	5.74	4,5
Ni I	48	7.48	7.28	5.66[b]	1.82	0.61	0.98	6.21	4,5
Zn I	3	5.26	5.20	4.40[a]	0.86	0.58	0.74	3.99	9
Sr II	3	4.08	4.13	2.60[b]	1.48	-0.26	0.01	3.50	4,5
Y II	14	3.04	2.82	3.20[b]	-0.16	0.06	0.11	3.09	7
Ba II	4	-	3.13	2.50[b]	0.63	0.2	0.12	2.35	15
R.E.II	27	-	3.11	1.53[e]	1.58	0.0	-	2.64	4

Remarks:

a) Aller, L.H. 1961, "The abundances of elements", Interscience Publishers, Inc., New York.

b) Muller E.A. 1968, "Origin and distribution of the elements", L.H. Ahrens (ed), Pergamon Press.

c) Cocke C.L., Curnutte B., Brand J.H. 1971, Astr. Astrophys. 5, 244

d) Garz, T., Holweger H., Kock M., Richter J. 1969, Astr. Astrophys. 2, 446

e) Grevesse N., Blanquet L. 1969, Solar Phys. 8,5.

The results are not different when Wrubel's curve of growth is applied. This suggests that the one layer approximation is not so bad, especially when keeping in mind, that the Γ- method implies the solar temperature distribution for an A_m star.

The electron pressure log p_e and the ionization temperature Θ_{ion} were calculated from the log N_I/N_{II} values for iron and silicon. The results are given in Table 3.

Table 3. Physical parameters of τ UMa

Present analysis		Greenstein (1948)	Miczaika et al. (1956)	Smith (1973)
Θ eff	0.67		-	0.67
log g	4.0	4.0	4.0	4.0
Θ_{exc}	0.72	0.91	0.89	-
Θ_{ion}	0.75	0.86	0.86	-
log p_e	1.41	0.0	0.12	-
$\xi t \frac{km}{s}$	4.8	3.8	3.8	6.5

The so called microturbulence parameter was also determined and is equal to 4.8km.s^{-1} This result confirms the conclusion given by Andersen (1973), that the new gf-values decrease the microturbulence and our conviction about high value of that parameter in the A and A_m stars must be revised.

Table 4. Sources of gf - values

1a. Wiese, W.L., Smith M.W., Glennon, B.M. 1966, *NSRDS-Natl. Bur. Std. U.S. 4,* Vol.1
1. Wiese W.L., Smith M.W., Miles B.M. 1969, *NSRDS-Natl. Bur. Std. U.S. 22,* Vol. 2
2. Lambert D.L., Warner B. 1968, *Mon. Not.Roy.Astr.Soc., 138,* 181
3. Groth H.G. 1961, *Z. Astrophys., 51,* 231
4. Corliss C.H., Bozman W.R. 1962, *Natl. Bur. Std. U.S. Monograph 53*
5. Takens R.J.1970, *Astron. Astrophys., 5,* 244
6. Woodgate B. 1966, *Mon.Not.Roy.Astron. Soc., 134,* 287
7. Krueger T.K., Aller L.H., Ross J., Czyzak, S.J. 1968, *Astrophys. 152,* 765
8. Roberts J.R., Andersen T., Sørensen G., 1973, *Astrophys. J., 181,* 567
9. Lambert D.L., Malia E.A., Warner B. 1969, *Mon. Not. Roy. Astr. Soc., 142,* 71
10. Lambert D.L., Warner B. 1968, *Mon.Not.Roy.Astr.Soc., 140,* 197
11. Tatum J.B. 1961, *Mon.Not.Roy. Astr.Soc., 122,* 311
12. Klempt M. 1973, *Astron. Astrophys., 29,* 419
13. Mendlowitz H. 1968, *Astrophys. J.. 154,* 1099.
14. Wolnik S.J., Berthel R.O. 1973, *Astrophys. J., 179,* 665
15. Miles B.M., Wiese, W.L. 1969, *Natl. Bur. Std., Tech. Note No. 474.*
16. Roder O. 1962, *Z.Astrophys., 55,* 38.
17. Baschek B., Garz T., Holweger H., Richter J. 1970, *Astron. Astrophys., 4,* 229
18. Warner B. 1967, *Mem. Roy. Astron. Soc., 70,* 165
19. Wolnik S.J., Berthel R.O., Wares, G.W. 1971, *Astrophys. J. Letters, 166,* L 31.
20. Warner B. 1968, *Mon. Not. Roy. Astr. Soc., 138,* 229
21. Foy R. 1972, *Astrophys. Astron., 18,* 26
22. Bridges J.M., Kornblith R.L., 1974, Astrophys. J., *192,* 793
23. Kurucz R.L. 1974, *Smithonian Ap. Obs. Spec. Rept. No.* 359
24. Huber M.G., Parkinson W.H. 1972, *Astrophys. J.,* 172, 229
25. Morton D.C., Smith W.H. 1973, *Astrophys. J. Suppl., 26,* 333
26. Wolnik S.J., Berthel R.O., Wares G.W. 1970, *Astrophys. J., 162,* 1037

REFERENCES

ANDERSEN P.H. 1973, Pub. Astron. Soc. Pacific, 85, 666
CAYREL R., JUGAKU J. 1963, Ann. Astrophys., 26, 495
FOY R. 1972, Astron. Astrophys., 18, 26
GREENSTEIN J.L. 1948, Astrophys. J. 107, 151
MICZAIKA G.R., FRANKLIN F.A., DEUTSCJ A.J., GREENSTEIN J.L. 1965, Astrophys. J., 124, 134
SMITH M.A. 1973, Astrophys. J. Suppl., 25, 277
WRUBEL M.H. 1949, Astrophys. J., 109, 66

WAVELENGTH DEPENDENCE OF POLARIZATION IN THE PLEIADES*

T.MARKKANEN
OBSERVATORY AND ASTROPHYSICS LABORATORY
UNIVERSITY OF HELSINKI, FINLAND

ЗАВИСИМОСТЬ ПОЛЯРИЗАЦИИ ОТ ДЛИНЫ ВОЛНЫ В ПЛЕЯДАХ

The linear polarization in UBV has been measured for ten bright and two fainter members of the Pleiades.

The directions of the polarizations coincide well with filament directions in the nebulae. The wavelength dependence of five stars is found to be similar to that of general interstellar polarization. The ratios of total to selective extinction and lower limits of colour excess are estimated for them. The colour excess estimates agree with those derived earlier from photometry, except for one star. Its photometric extinction is very low.

Five stars show highest amount of polarization in the ultraviolet with a minimum in the blue. An explanation would be dust particles smaller than generally in the interstellar dust clouds. It seems puzzling, however, now such small particles could exist around the brightest members of the Pleiades, which are not very young any more. No correlation between infrared excesses and anomalous wavelength dependence is found.

Two stars show polarization typical to early type shell stars. One of them is Pleione which is known to be a shell star. The other one, 19 Tau, is a normal main sequence star. It might be of interest to look for indications of emission in a high dispersion spectrum of 19 Tau. The directions of polarization are almost parallel to the nebular filaments, i.e. the magnetic field. If there is a disk around the star, the direction would tend to be perpendicular to the disk. This would mean that the rotational axis of the star would be parallel to the magnetic field.

* The full paper will be published elsewhere.

THE HELIUM ABUNDANCE OF STARS AS DERIVED FROM PHOTOELECTRIC OBSERVATIONS OF THE HeI λ 4026 LINE

P.E.NISSEN
INSTITUTE OF ASTRONOMY, UNIVERSITY OF AARHUS, DENMARK

СОДЕРЖАНИЕ ГЕЛИЯ В ЗВЕЗДАХ ПО ФОТОЭЛЕКТРИЧЕСКИМ НАБЛЮДЕНИЯМ ЛИНИИ HeI λ 4026

(The author has not sent the text of his paper).

ATMOSPHERE STRUCTURE OF LATE-TYPE STARS

J.I.K.STRAUME
BALDONE RADIOASTROPHYSICAL OBSERVATORY, U.S.S.R.

СТРУКТУРА АТМОСФЕР ЗВЕЗД ПОЗДНИХ ТИПОВ

In the Radioastrophysical Observatory of the Academy of Sciences of the Latvian SSR model atmospheres of late-type stars have been constructed. A computer program is written based on LTE and a plane-parallel atmosphere in hydrostatic and radiative equlibrium. The Runge-Kutta method and straight mean absorption coefficient have been used for solution of the equation of hydrostatic equilibrium. Opacity sources include absorption by negative hydrogen H^-, hydrogen and negative molecule H_2^-. Equation of state includes:H, H^-, H^+, H_2, H_2^- and 10 metals: Na, K, Ca, Mg, Al, Si, C, O, S and Fe. Equation of state has been solved according to Mihalas (1967). The "grey" temperature distribution has been used. Model atmospheres for effective temperature T_e = 2500, 3000, 4500°K and surface gravity log g = 0, 2, 4 and for T_e = 3500 and 4000°K and log g = 1, 3, 5 and model atmosphere for Sun (T_e = 5785°K, log g = 4.44) were calculated. These model atmospheres have been calculated for solar abundances. Besides these, model atmospheres for T_e = 2500 and 4500°K and log g = 2 and 4 with solar metal abundances reduced by factors 0.1,10 and 100 have been constructed. Our model atmospheres have been compared with Pierce and Waddell (1961) model atmospheres for Sun and Auman (1969) model atmospheres for late-type stars. From the comparison it is seen that our results are in satisfactory agreement for $T_e > 3000$°K. The model atmospheres for $T_e<3000$°K differ greatly, probably because scattering on H and H_2 and molecular line blanketing were not taken into account. From analysis of the models for non solar abundances we see that changes in metal abundances have great influence on the structure of atmospheres. A decrease of metal abundances causes an increase of the gas pressure and decrease of pressure gradient. This influence is stronger for dwarfs (log g = 4) than for giants (lig g = 2).

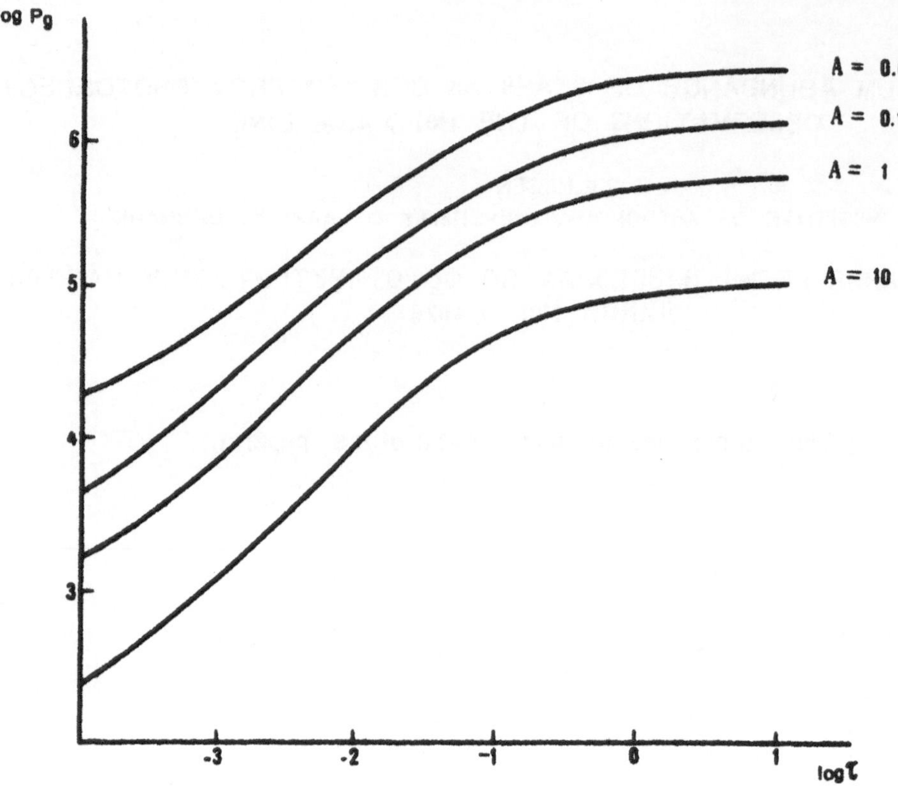

Fig. 1. Dependence of log p_g from log τ for $T_e = 2500°K$, log g=2. "A" denotes the factor by which solar metal abundances are multiplied.

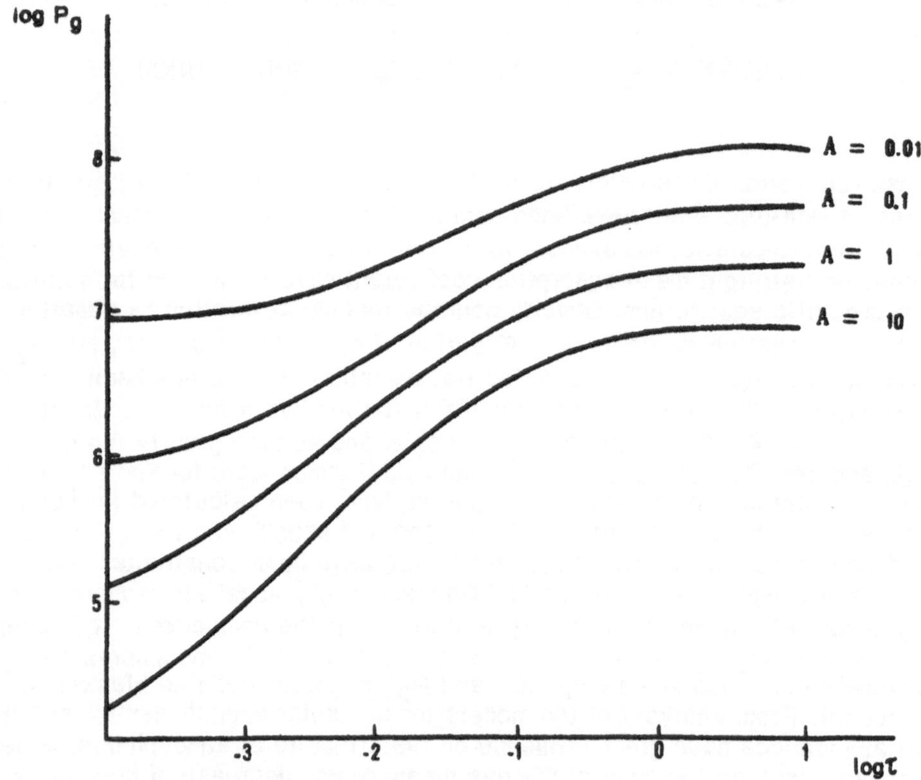

Fig. 2. Dependence of log p_g from log τ for $T_e = 2500°K$, log g=4.

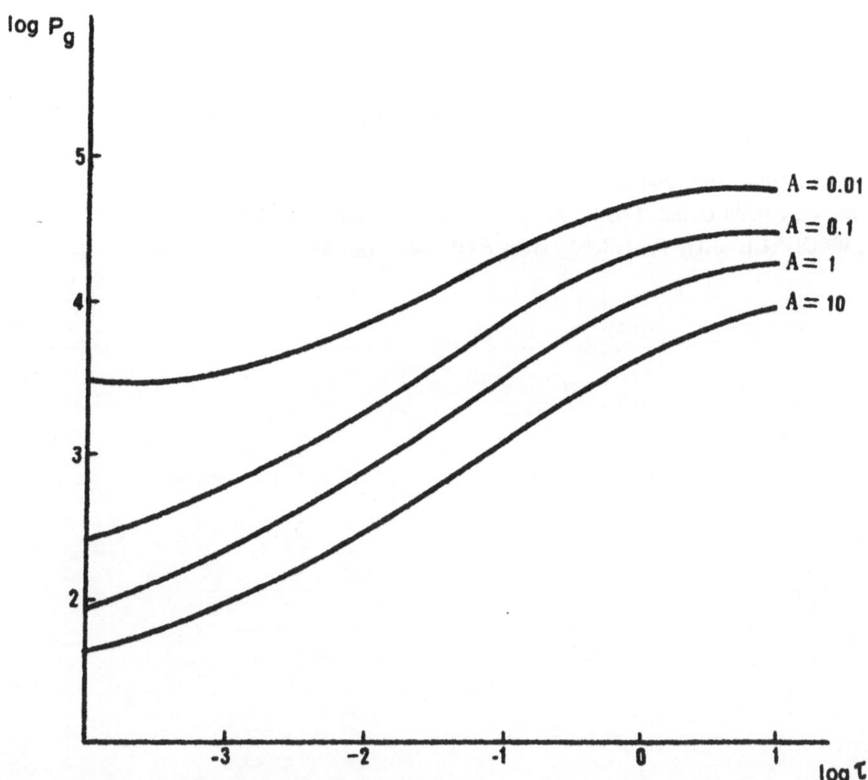

Fig.3. Dependence of log p_g from log τ for T_e =4500°K, log g=2.

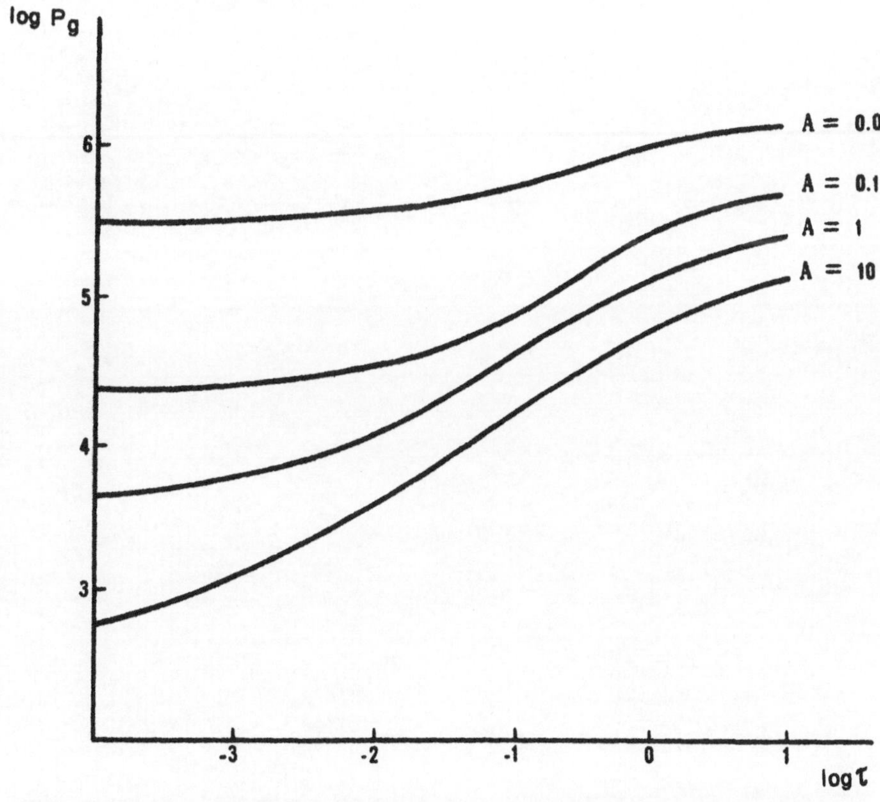

Fig. 4. Dependence of log p_g from log τ for T_e =4500°K, log g=4.

REFERENCES

AUMAN J.R., Ap.J., 1969, 157, 799

MIHALAS D. 1967. Meth. in Comp. Phys., 7, 1, New York-London, Acad. Press

PIERCE A.K., WADDELL J.H. 1961, Mem. Roy. Astr. Soc., 68, 89

COMPACT GROUPS OF COMPACT GALAXIES

V.A.AMBARTSUMIAN, L.V.MIRZOYAN, M.B.PETROSIAN, R.K.SHAHBAZIAN
BYURAKAN ASTROPHYSICAL OBSERVATORY, U.S.S.R.
PRESENTED BY L.V.MIRZOYAN

КОМПАКТНЫЕ ГРУППЫ КОМПАКТНЫХ ГАЛАКТИК

РЕЗЮМЕ

Отмечена пионерская работа Ф.Цвикки по изучению компактных галактик, указавшего на большую частоту встречаемости компактных галактик в кратных системах и существование их скоплений. Уточнено определение компактности галактик (средняя поверхностная яркость больше 20^m с квадратной секунды, в красных лучах). На картах Паломарского атласа выявлено существование более 250 систем, компактных и состоящих из около десятка галактик, в своем большинстве компактных, часто эллиптических или сферических. Многие из этих систем имеют необычные конфигурации (цепочки, криволинейные структуры с пустотой в центре и т.д.). На примере систем Шахбазян 1 и 123, обладающих очень небольшой дисперсией радиальных скоростей, что указывает на крайне низкое значение отношения масса-светимость, показано, что, по крайней мере, часть членов компактных групп различается, по природе, от обычных галактик тех же светимостей.

The study of compact galaxies and their systems indicates the unusual physical nature of these objects, their major significance to physics and the evolution of galaxies.

Zwicky (1971) was the first to appreciate the importance of compact galaxies and conducted their systematic study. He found out the following regular features:

1. Both blue and quite red objects occur among compact galaxies;
2. Frequently compact galaxies are encountered in pairs, triplets, etc.;
3. Clusters of compact galaxies exist; they consist of at least several dozen objects most of which are described as quite red. Blue objects in those clusters occur very seldom. The dimensions of clusters of compact galaxies are comparable to those of usual clusters of galaxies.

The significant investigation of Robinson and Wampler (1973), gave a new impetus to the research of compact galaxies; they showed that cluster Shahbazian 1 discovered in Byurakan (Shahbazian, 1957) is a distant compact cluster of compact galaxies possessing quite unusual properties.

This has led Arp, Burbidge and Jones (1973) to consider it as a unique in characteristic features of clusters of galaxies. In fact, the Palomar Atlas contains no other group of galaxies that might equal or excel the group Shahbazian 1 simultaneously in number of members, in its compactness and the compactness of the members of the group.

However, the group Shahbazian 1 can also be regarded as the extreme representative of a broader class of groups of compact galaxies, groups of compact galaxies which are poorer as to the number of their members or groups richer yet less compact. Such an approach to the problem of the existence of compact groups of compact galaxies proved to be quite fruitful.

Compactness is the principal characteristic of galaxies - members of compact groups under review. Therefore, to distinguish compact galaxies and their groups from the totality of all the galaxies one must be equipped, foremost of all, with definite practical criteria in order to assign the galaxies to the compact type. In other words, one must have a definite criterion of compactness.

Zwicky was quite right in suggesting to consider the high average surface brightness as a criterion of compactness of galaxies. The great value of this quantity in fact characterizes somewhat the "excited" state of the galaxies producing an unusually high surface brightness of compact galaxies.

The experience of studying compact galaxies has led to the conclusion (Ambartsumian et al., 1975), that as a correction for this definition of compactness offered by Zwicky, it is convenient to consider compact those galaxies the average surface brightness of which in the red is higher than 20^m from the square second of arc.

An examination of the maps of the Palomar Sky Survey has shown that a considerable number of compact groups of compact galaxies exist, made up of a small number (of the order of one dozen) of members. Most of the members of those groups satisfy the specified condition of compactness.

On the other hand, in accordance with the data of de Vaucouleurs' Catalogue (1964) only 1/25 part of all the galaxies possesses the average surface brightness in the red above $20^m/_\square$ i.e. are compact (Ambartsumian et al., 1975). Therefore there are no doubts that the great majority of the observed compact groups of compact galaxies are real physical systems and not the random result of projection. Most galaxies of these groups possess red magnitudes in the interval $17^m_.5$ - $18^m_.5$. Saturation of the images on the maps of the Palomar survey is for them a sufficient condition to be compact, as has been shown (Ambartsumian et al., 1975).

Spectroscopic observations of the group Shahbazian 78 (Ambartsumian et al., 1975;Mirzoyan et al., 1975) have shown that its brightest members are stars, while on the photograph of the group Shahbazian 129 obtained with the 5-m telescope (Ambartsumian et al., 1975) of the Palomar Observatory, all the objects have star-like images. Only the spectroscopic observations of this group enable us to determine its true nature. Those data attest that an insignificant number of groups, consisting of stars or casual groupings of galactic stars and galaxies may occur in the published lists of compact galaxies (Shahbazian, 1975; Shahbazian and Petrosian, 1974; Petrosian, 1974, Baier et al., 1974; Baier and Tiersch, 1975). However, most of those groups are undoubtedly real physical systems, consisting of compact and some non-compact galaxies. Naturally, objects projected on those groups are likely to be included in the lists of the members of compact groups.,(Shahbazian, 1975; Shahbazian and Petrosian, 1974; Petrosian, 1974; Baier et al., 1974; Baier and Tiersch, 1975).

Over 250 systems (Shahbazian, 1975; Shahbazian and Petrosian, 1974; Petrosian, 1974; Baier et al., 1974; Baier and Tiersch, 1975), have so far been included in the lists of compact groups of compact galaxies, compiled by the astronomers of the Byurakan Astrophysical Observatory of the Academy of Sciences and the Central Institute of Astrophysics of the Academy of Sciences of the GDR. Clearly, the total number of compact groups of compact galaxies up to the limit of the Palomar Atlas all over the sky must be considerably larger: of the order of one thousand.

The group Shahbazian 1 (Robinson and Wampler, 1973; Shahbazian, 1957) containing at least 20 galaxies, is a relatively rich group of compact galaxies. The linear diameter of this group is of the order of 200-300 thousand parsecs (Robinson and Wampler, 1973; Ambartsumian et al.,1975). Some of the members of the group are quite compact.

Croups more compact than Shahbazian 1 are met among compact groups of compact galaxies included in the lists, however, most of them are less compact.

Spectroscopic observations of the compact group Shahbazian 1 (Robinson and Wampler,1973) as well as of the brightest galaxies of the compact group Shahbazian 123 (Mirzoyan et al., 1975) testify to a very small dispersion of radial velocities in those systems as compared to other groups of non-compact objects. Although, as the observations of Lynds and Khachikian of the group Shahbazian 4 (Lynds and Khachikian,private) have shown, the small dispersion of radial velocities is not a property characteristic of all compact groups of compact galaxies, yet the presence between those groups of systems with small dispersion of galaxies is a highly important property for this class.

As distinct from the case of great dispersion of radial velocities, when a strong violation of the virial theorem cannot be excluded, such great violation is ruled out in this case. Worded differently, even in case of instability of the system the velocity dispersion cannot be many times less than the virial one., if we have not caught the system right in time of the beginning of the collapse. The latter is, of course, unlikely. Therefore the value of the mass of the system derived from the virial theorem should be taken as close to the reality. This leads to a very low value (of the order of unity and less) of the ratio mass to luminosity - M/L. In this way it turns out that at least several members of the compact groups of compact galaxies differ by their nature from ordinary galaxies of the same luminosities.

Supposing that red compact galaxies consist of a physically homogeneous class of objects, it should be assumed that even in cases when the dispersion of radial velocities in the group is large, for instance in the group Shahbazian 4 (Lynds and Khachikian, private), the ratio M/L is of equally small value. The large dispersion of radial velocities can in this case be accounted for by the violation of the stability of the system.

It is to be noted in connection with the problem under consideration that the data on the rotation of compact galaxies 1 Zw 129 and 11 Zw 70 confirm the small value of the ratio M/L (O'Connel and Kraft, 1972). In particular, for the compact galaxy I Zw 129 this ratio is estimated equal to 0.18 (O'Connel and Kraft, 1972). On the other hand, it is well known that for normal galaxies this ratio is equal to several unities or considerably more. According to the investigation by O'Connell and Kraft (1972), the small value of the ratio M/L and the observed energy distribution in the spectrum of galaxy 1 Zw 129 attest that the initial luminosity function of stars formed in it differed substantially from the corresponding function for the vicinity of the Sun. However, as the observations show (O'Connel and Kraft, 1972), kinematically compact galaxy 1 Zw 129 differs but little from normal galaxies.

It should be noted that compact galaxies included in the lists of compact groups, are often very red objects. For instance, according to Börngen and Kalloglian's determination (1974),most of the compact galaxies in the compact groups Shahbazian 17, 18, 41 and 42 have colour indices B-V well surpassing 1. This is not surprising as during the search for compact groups (Shahbazian,1975;Shahbazian and Petrosian,1974;Petrosian,1974;Baier et.al.,1974;Baier and Tiersch, 1975) particular attention was paid to colour and groups were chosen that consisted mainly of objects red in colour.

However, the existence of blue compact galaxies admits no doubt . From this more general point of view Sargent (1970) believes that compact galaxies, form an extremely unhomogeneous class of objects.

Among the blue compact galaxies are known, for instance, (1 Zw 17 and 11 Zw 40) for which the ratio mass-luminosity is essentially larger than unity (Sargent and Searle, 1970).

Compact galaxies differ sharply as to the relative content of gas. For instance, the mass of ionized hydrogen in the galaxy 1 Zw 129 by an approximate estimation comes to 0.07 of its total mass (O'Connell, Kraft , 1972) while its portion in the galaxies I Zw 17 and II Zw 40 is at least of one order less (Sargent and Searle, 1970).

It cannot be excluded that there is some heterogeneity also among red compact galaxies. The unusually high surface brightness of compact galaxies can be due to both uncommonly low values of the ratio mass-luminosity, M/L. and the unusually high concentration of stars in them. Which of these possibilities is realized in each particular case is a matter for special investigation.

Most compact groups of compact galaxies contain non-compact galaxies in their composition as well. However, it is of interest to note that most of the galaxies, at a definite interval of absolute stellar values, are compact in the clusters of compact galaxies especially in Zwicky's clusters of compact galaxies (Zw Cl 0152 + 3337, Zw Cl 1700.5 + 3322, Zw Cl 0054.6-127, Zw Cl 1710.4 + 6401 (Zwicky F. and Zwicky M., 1971). This means that in the above clusters a considerable number of compact galaxies have luminosities close to each other. As a result of which there is a certain maximum in the luminosity function of these galaxies corresponding to the above interval of absolute magnitudes.

Such a distribution of compact galaxies according to their luminosities indicates that the observed state of a compact galaxy can last for an essential portion of the age of the cluster.

Compact galaxies in compact groups are mostly nearly elliptical or spherical in form. Besides, spiral galaxies also occur, though much less in number. However, bright irregular galaxies are presumably lacking in compact groups (Ambartsumian et al., 1975).

The geometrical configurations of the compact groups themselves are of great interest. As distinct from usual groups and clusters of galaxies, the compact groups do not show, for the most part, noticeable concentration toward the center of the system. Less than 10% of all the known groups show signs of concentration. Moreover, low density of galaxies is observed in the central part of some groups. A considerable part of the compact groups are anomalous in form. Chains or systems of chains are frequently met among compact groups of compact galaxies. Of special interest are groups that are nearly closed curves in shape (with a cavity in the middle). We can describe them as groups of peripherical structure.

Those structures are sometimes extremely unusual. For instance, the structure of the group Shahbazian 65 is so much unusual that it calls for some special interpretation. This is not a group but rather a cluster of compact galaxies in the form of the Greek letter "Ω" with a cavity in the center. The cluster of galaxies Shahbazian 65 is certainly at a distance greater than the group Shahbazian 1 and considerably richer than the latter. If we assume, judging by the visible brightnesses of galaxies, that the radial velocity of the group Shahbazian 65 is close to 50,000 km/sec, we shall obtain about 800 Kpc for its diameter. This is appreciably larger than the diameters of the group Shahbazian I. The group Shahbazian 65 excels the group Shahbazian I in other characteristics too (the number of members, mass, etc.).

Most of the compact groups of compact galaxies seem to be well isolated, however we come across cases one can suspect that the group is a kind of nucleus of a larger system,the peripherical part of which consists of considerably weaker galaxies. Calculations of the number of galaxies in consecutive rings surrounding the systems Shahbazian 31, 41 and 84 on plates obtained in the primary focus of the 4-m telescope of the Kitt Peak National Observatory (Ambartsumian et al., 1975), showed a marked decrease in their density, as the distance from their nuclei increased, which definitely indicates the presence of objects connected with the given compact group, in addition to background galaxies. In the case of the groups Shahbazian 34, 35 and 43 this decrease is less pronounced. Thus one can suspect that, at least in some cases, the compact group of compact galaxies is quite a dense nucleus of a larger extended cluster. This question deserves serious attention.

In this connection it is necessary to mention the interpretation of compact groups of compact galaxies given by J.E.Einasto et al., (1974). They consider them as nuclei of what they call giant hypergalaxies - a class of systems which are quite different from clusters of galaxies.

It should be added that the group Anon (Ambartsumian et al., 1975) seems to be the richest among the studied groups of compact galaxies. Group Anon consists of at least two dozen compact galaxies; it has not as yet been entered in the published lists of compact groups of compact galaxies (Shahbazian,1975; Shahbazian and Petrosian, 1974; Petrosian, 1974; Baier et al., 1974; Baier and Tiersch, 1975).

We should like to note in conclusion that the compact groups of compact galaxies form an interesting class of systems of galaxies, and their detailed study can throw light on the physics and evolution of galaxies.

REFERENCES

AMBARTSUMIAN V.A., ARP H.C., HOAG A.A., MIRZOYAN L.V. 1975, Astrofizika, 11, 193

ARP H.C., BURBIDGE G.R., JONES T.W. P.A.S.P. 1973, 85, 423

BAIER F.W., PETROSIAN M.B., TIERSCH H., SHAHBAZIAN R.K., 1974, Astrofizika, 10, 327

BAIER F.W., TIERSCH H. 1975, Astrofizika, 11, 217

BÖRNGEN F., KALLOGLIAN A.T., 1974, Astrofizika, 10, 21

EINASTO J., JAANISTE J., JÔEVEER M., KAASIK A., KALAMEES P., SAAR E., TAGO E., TRAAT P. VENNIK J., CHERNIN A.D. 1974, Tartu Astr.Obs.Preprint No. 48

LYNDS C.R., KHATCHIKIAN E.YE. Private communication

MIRZOYAN L.V., MILLEN J.S., OSTERBROCK D.E., 1975, Ap.J. 196, 687

O'CONNELL R., KRAFT R.P. 1972, Ap. J. 175, 335

PETROSIAN M.B., 1974, Astrofizika, 10, 471

ROBINSON L.B., WAMPLER E.J., 1973, Ap. J. 179, 135

SARGENT W.L.W. 1970, Ap.J. 160, 405

SARGENT W.L.W., SEARLE L. Ap. J. 1970, 162, L155

SHAHBAZIAN R.K., 1975, Astron. Circ. USSR. No. 177, 11

SHAHBAZIAN R.K. 1973, Astrofizika, 9, 495

SHAHBAZIAN R.K., PETROSIAN M.B. 1974, Astrofizika. 10, 13

VAUCOULEURS G. de VAUCOULEURS A. de 1964, Reférence Catalogue of Bright Galaxies

ZWICKY F., ZWICKY M. 1971, Catalogue of Selected Compact Galaxies and Post-Eruptive Galaxies

DISCUSSION

DEUTSCH: Is it not difficult to define the mean surface brightness of compact objects, since the diffusion of light in the photographical emulsion always enlarges the images?

MIRZOYAN: No, the criterion of the mean surface brightness is indeed very useful for the definition of compactness of galaxies. It is, however, quite another question how to define this parameter on photos taken with different telescopes.

GOUGUENHEIM: You mentioned that the red compact galaxies look like elliptical galaxies; but we find from 21-cm line observations at Nançay that their neutral hydrogen content is quite large. How did you determine the masses?

MIRZOYAN: The main difference we have noticed is that the red compact galaxies in compact groups look more spherical than the classical elliptical galaxies. We estimate the masses by the virial theorem, using the practical absence of dispersion in the radial velocities of the galaxies in the compact groups Shahbazian 1 and 123. We also used the data of O'Connel and Kraft from their study of the rotation of two compact galaxies 1Zw 129 and 11Zw 70.

SOME RESULTS ON COMPUTER INVESTIGATION OF THE CATALOGUE OF BRIGHT GALAXIES COMPILED ON MAGNETIC TAPE

1. ON PROBABLE BRIGHT COMPACT GALAXIES

N.G.KOGOSHVILI
ABASTUMANI ASTROPHYSICAL OBSERVATORY, U.S.S.R.

НЕКОТОРЫЕ РЕЗУЛЬТАТЫ КОМПЬЮТЕРНОГО ИССЛЕДОВАНИЯ КАТАЛОГА ЯРКИХ ГАЛАКТИК НА МАГНИТНОЙ ЛЕНТЕ

1. О ЯРКИХ ГАЛАКТИКАХ ПОДОЗРЕВАЕМЫХ В КОМПАКТНОСТИ

РЕЗЮМЕ

Приводится описание Абастуманского Каталога Ярких Галактик, АбКЯГ, (30000 галактик) на магнитной ленте ЭВМ, основанного на Морфологическом Каталоге Галактик Б.А.Воронцова-Вельяминова и др. и использующего данные других каталогов и списков галактик. Каталог, в основном, предназначается для статистического исследования распределения галактик, а также изучения их различных морфологических характеристик.

С помощью АбКЯГ были исследованы яркие компактные галактики. Была подсчитана поверхностная яркость для всех галактик каталога и были отобраны 157 компактных галактик, севернее - 33°, для которых поверхностная яркость оказалась ярче $20^m.0/\square''$, а также 829 галактик с поверхностной яркостью $20^m.0/\square''$ - $21^m.0/\square''$.

Была подсчитана также средняя поверхностная яркость для 271 яркой галактики, найденных в Каталоге Компактных Галактик Цвикки и перенесенных на магнитную ленту АбКЯГ. Она оказалась равной $22^m.0/\square''$.

ABSTRACT

The magnetic tape version of the Abastumani Catalogue of Bright Galaxies, AbCBG, (30000 galaxies) based on the Morphological Catalogue of Galaxies by Vorontsov-Vel'yaminov et al. and other major catalogues and lists of galaxies, is described. The AbCBG Catalogue is especially suited for computerized statistical investigation of galaxy distributions and studies of their morphological characteristics.

Bright, compact galaxies were searched for in the AbCBG Catalogue. The surface brightness values were computed for all the galaxies in the Catalogue and we found 157 compact galaxies north of declination -33° with surface brightness brighter than $20^m.0/\square''$, and 829 galaxies with surface brightness between $20^m.0/\square''$ and $21^m.0/\square''$, which were not before recognized as compact galaxies

The mean surface brightness of 271 bright galaxies, listed in the Catalogue of Selected Compact and Post-Eruptive Galaxies by Zwicky as well as in the AbCBG Catalogue, was found to be $22^m.0/\square''$.

1. Introduction: The Abastumani catalogue of bright galaxies. The compilation of a machine-readable catalogue of bright galaxies was undertaken at the Abastumani Astrophysical Observatory in 1968 in order to facilitate the statistical analysis of the presently available, large observational material on galaxies from the Palomar Sky Survey. The data were mainly taken from the Morphological Catalogue of Galaxies by Vorontsov-Vel'yaminov et al. (1962-1968), completed with photographic magnitudes from the Catalogue of Galaxies and Clusters of Galaxies by Zwicky et al. (1961-1968), and morphological classification types, integrated colour-indices and radial velocities from the Reference Catalogue of Bright Galaxies by de Vaucouleurs (1964). Further data were added from several other catalogues and lists (about 100 references altogether). Preliminary information on our Catalogue, in what follows referred to as the Abastumani Catalogue of Bright Galaxies, AbCBG, has been given by Kogoshvili (1972). A full description is being published (Kogoshvili, 1975).

The AbCBG Catalogue comprises about 30000 galaxies, mostly brighter than $15^m.1$ (pg). It is available on magnetic tape and the appropriate software for its efficient utilization has been written on the M-220 computer at the Institute for Applied Mathematics of Tbilisi University. The structure of the magnetic tape data has been specifically planned for easy extension as more data on galaxies already in the Catalogue become available.

The AbCBG Catalogue lists for each galaxy, as far as data have been found in the literature published before the end of 1972, the following parameters:

1. MCG, NGC or IC identification nos,
2. (α, δ) at epoch 1950.0,
3. m (pg), mostly from CGCG, some from other accurate sources (found in the MCG), the remaining are the rough values given by the authors of the MCG,
4. diameters for inner and outer parts from MCG,
5. inclination estimate from MCG,
6. surface brightness estimate from MCG,
7. detailed morphological description from MCG,
8. U-B, B-V and (U-B)(O), (B-V) (O) corrected for inclination from BGC, the first two also from some other sources,
9. morphological classification types by de Vaucouleurs, van den Bergh and Morgan from BGC,
10. radial velocity,
11. number of observed supernovae,
12. radio-flux at five frequency wavelenghts,
13. spectral class,
14. presence of optical emission lines,
15. rotational velocity,
16. membership in galaxy pairs and multiple groups, interaction from MCG,
17. whether or not of Markarian type

A broad range of statistical investigations are being undertaken by means of the AbCBG Catalogue. The present paper discusses some of the first results concerning the compact galaxies in the Catalogue. Quite surprisingly, a rather large number of galaxies were found, which were hitherto not recognized as compact, but for which the surface brightness clearly indicates that they belong to this interesting group of galaxies.

II. The surface brightness of bright compact galaxies in the CCG catalogue. The surface brightness of galaxies has been studied in detail by several authors. Heidmann et al. (1972) investigated the influence of various factors on galaxy surface brightness for different morphological types, using the BGC Catalogue and Holmberg's Survey (1958), which are believed to be the most uniform catalogues with optical data. Arakelian (1974, in press) studied the surface brightness of some interesting galaxy groups, including Markarian galaxies as well as those, referred to as the galaxies with emission lines, and concluded that their surface brightness is 1^{m} higher than that of normal galaxies.

The present investigation first concerned the study of bright compact galaxies. The CCG Catalogue (Zwicky, 1971) contains 507 compact galaxies brighter than $15\overset{m}{.}1$ (pg) and a few fainter, of which 383 are included in the AbCBG Catalogue. Of these, 271 have known diameters and photographic magnitudes and have individual numbers in the MCG Catalogue. The diameters as well as the magnitudes were all taken from the MCG, but the latter came from three sources: the CGCG (230 galaxies), the MCG (36) and other accurate sources (5). The corresponding mean surface brightness and the dispersion for the three samples are shown in Table 1. Within the accuracy, there is a very good agreement between the samples, which proves that the apparent magnitudes of the MCG Catalogue do not deviate significantly from those of the CGCG, at least for Zwicky's compact galaxies.

Zwicky's criterion (1964, 1971) for a compact galaxy states that its surface brightness (for the whole surface or some part thereof) must be brighter than $20\overset{m}{.}0/\square''$ for any particular wavelength. For Zwicky's galaxies we found $m_{sb} = 21\overset{m}{.}9$ and a very broad distribution ranging from $19\overset{m}{.}5$ to $24\overset{m}{.}0$ (Fig. 1). This discrepancy is readily explained by the fact that Zwicky (1971) included in his Catalogue many bright peculiar galaxies (e.g. eruptive and post-eruptive galaxies) for which only part of the surface fulfills his criterion. This has been earlier noted by Sargent (1970).

III. New bright compact galaxies. In order to investigate whether there are other bright compact galaxies than those 271 mentioned above, the surface brightness was computed for all galaxies in the AbCBG Catalogue, for which diameters and photographic magnitudes were available. The galaxies with the surface brightness brighter than $20\overset{m}{.}0/\square''$ and those brighter than $21^{m}.0/\square''$ were picked out. A correction was applied for galactic absorption ($-0^{m}.25$ cosec b^{II}), but not for inclination. However, taking into account the considerable dependence of the galaxy surface brightness upon the inclination, as revealed in the mentioned papers by Heidmann and Arakelian, we confined ourselves to the galaxies with an observed ratio of the axes smaller than 2. The galaxies situated at low galactic latitudes were also excluded.

We found 157 compact galaxies with $m_{sb} \leqslant 20^{m}.0/\square''$ and 986 with $m_{sb} \leqslant 21^{m}.0/\square''$. The mean surface brightnesses are shown in Table 1.

Since the mean surface brightness of Zwicky's compact galaxies in the AbCBG Catalogue is only $21^{m}.9/\square''$, there can be no doubt that these galaxies do indeed belong to the compact class.

IV. Discussion. The question naturally arises why these 986 bright galaxies are not included in the CCG Catalogue. First of all, about half of them lie south of declination 0°, the practical limit of the CCG Catalogue, The surface distribution of the new galaxies is more uniform than those from the CCG Catalogue, and it is quite probable that several fields were not thoroughly studied by Zwicky. The surface distribution of Zwicky's compact galaxies and those of the new compact ones are shown in Figs 2 and 3.

A comparsion of the morphological classes of the compact galaxies from these different lists shows a significant difference (Table 2). Whereas the CCG sample mainly consists of types S and N; H, i.e. galaxies with well defined structure, the new sample has a considerable majority of galaxies, showing little or no structure (types N and (N) according to morphological description of galaxies from MCG Catalogue).

The mean surface brightness for each morphological type is shown in Table 3. Quite a good agreement is found between the surface brightness of various morphological types with that of the newly found galaxies. However, in the case of Zwicky's galaxies there is an appreciable difference clearly showing that morphologically different objects with a marked deviation in surface brightness were classified into one group.

The computed values of m_{sb} depend on two parameters only, the angular diameters and the photographic magnitude. The diameters were measured by Vorontsov-Vel'yaminov on the Palomar Sky Atlas prints which show the faintest details visible on the original plates. It is therefore unlikely that any significant errors in m_{sb} enter through the diameter values. On the other hand, the prints are more contrasty than the original plates and bright galaxies do not show any details in the center. This may partly explain the difference in morphological types and, to some extent, lead to too high estimates of the photographic magnitude. However, as shown above, the effect cannot be very significant.

The mean values of the observed axis ratios and the diameters for the above groups of galaxies are given in Tables 4 and 5. Whereas the mean magnitude, $15^{m}.0$, is approximately the same for all groups of galaxies, the mean diameter of the new compact galaxies is markedly smaller than for Zwicky's galaxies. Moreover, since all high-inclination galaxies were eliminated in the present study, we are confident that apart from accidental errors, only compact galaxies are included in the new list.

The number of pairs and multiple groups in the above lists with compacts only and with compacts and normal galaxies (having individual numbers in MCG Catalogue), was estimated. These close pairs and multiple groups were already pointed out by the authors of MCG Catalogue,on the basis of apparent closeness of the components, small difference in magnitudes, as well as the presence of interaction.

It appears that there is an excessive number of close, double and multiple groups among the new compact galaxies. Statistical analysis shows (Table 6) that the number of pairs is much higher than should be expected from a random sky distribution. This tendency towards grouping among compact galaxies was noted by Zwicky (1971) and other authors, especially at the Byurakan Observatory in the papers by Shahbazian (1973) and Shahbazian and Petrosian (1974).

The analysis in Table 6 shows Zwicky's tendency to include in his list some bright peculiar galaxies found in pairs and groups, although they frequently did not fulfill the main compactness criterion.

A new list of compact galaxies will be published in current issue of the Abastumani Astrophysical Observatory Bulletin.

A c k n o w l e d g e m e n t. It is a great pleasure to thank Professor V.A.Ambartsumian, Byurakan Observatory, for valuable advice and the Director of the Abastumani Astrophysical Observatory, Professor E.K.Kharadze, for his very positive encouragement during the various phases of the compilation of the Abastumani Catalogue of Bright Galaxies.

Table 1. Comparison of mean surface brightness m_{sb} for different lists of compact galaxies with different sources of magnitudes.

The groups of galaxies investigated	m_{sb}	σ	m_{sb}(Zw)	σ	m_{sb}(V-V)	σ	m_{sb}(Oth.)	σ
Compact galaxies with surf. bright. up to $21^m.0/\square''$	20.4 n=986	±0.39	20.5 n=507	±0.36	20.4 n=435	±0.41	20.3 n=44	±0.49
Compact galaxies with surf. bright. up to $20^m.0/\square''$	19.7 n=157	±0.24	19.8 n=61	±0.21	19.7 n=81	±0.25	19.7 n=15	±0.15
Zwicky's compact galaxies from CCG	21.9 n=271	±1.01	22.0 n=230	±1.03	21.7 n=36	±0.76	21.1 n=5	±1.10

Table 2. Comparison of the morphological classes for different lists of compact galaxies

Morphological types of galaxies	Numbers of galaxies with surf. bright. up to $20^{m}.0/\square''$	Number of galaxies with surf. bright. up to $21^{m}.0/\square''$	Zwicky's compact galaxies from CCG
S (spirals)	5	72	78
N; H (bright nucleus with considerable envelope)	11	108	98
N or (N) (very concentrated image)	130	677	46
F (edge-on galaxies without visible structure)	10	126	42
Others	1	3	7
Total number of compact galaxies	157	986	271

Table 3. Comparison of mean surface brightness for different lists of compact galaxies according to morphological classes

The groups of galaxies investigated	m_{sb} (S)	σ	m_{sb}(N;H)	σ	m_{sb}(F)	σ	m_{sb}(N)	σ
Compact galaxies with surf. bright. up to $21^{m}.0/\square''$	20.6 n=72	±0.30	20.6 n=108	±0.39	20.5 n=126	±0.32	20.4 n=677	±0.40
Compact galaxies with surf. bright. up to $20^{m}.0/\square''$	19.8 n=5	±0.12	19.7 n=11	±0.23	19.8 n=10	±0.20	19.7 n=130	±0.24
Zwicky's compact galaxies from CCG	22.3 n=78	±0.93	22.0 n=98	±0.87	21.7 n=42	±0.87	21.0 n=46	±0.77

Table 4. Comparison of mean values of diameters for different lists of compact galaxies

The groups of galaxies investigated	d in 0.1'	σ	d (S) in 0.1'	σ	d(N;H) in 0.1'	σ	d (F) in 0.1'	σ	d(N) in 0.1'	σ
Compact galaxies with surf. bright. up to 21^m.0/□''	3.3	±1.46	4.7	±2.40	3.6	±1.38	4.0	±1.51	3.0	±1.16
Compact galaxies with surf, bright. up to 20^m.0/□'	2.4	±1.14	2.8	±1.63	2.8	±1.48	3.1	±1.58	2.3	±1.01
Zwicky's compact galaxies from CCG	9.2	±7.83	12.1	±7.40	7.8	±4.18	9.9	±7.41	4.2	±2.41

Table 5. Comparison of mean values of observed galaxy inclination for different lists of compact galaxies

The groups of galaxies investigated	i	σ	i(S)	σ	i(N;H)	σ	i(F)	σ	i(N)	σ
Compact galaxies with surf. brigh. up to 21^m.0/□'	1.3	±0.33	1.5	±0.34	1.2	±0.29	1.6	±0.36	1.3	±0.29
Compact galaxies with surf. brigh. up to 20^m.0/□'	1.3	±0.32	1.3	±0.31	1.2	±0.24	1.6	±0.37	1.2	±0.29
Zwicky's compact galaxies from CCG	1.7	±1.05	1.8	±1.04	1.3	±0.47	2.8	±1.55	1.4	±0.55

Table 6. A number of pairs and groups among different compact galaxy lists

The groups of galaxies investigated	Total number of gala-xies	Number of compact galaxy pairs	Number of compact galaxy tri-plets	Number of 4 and more-memb. groups of compact galaxies	Number of compact– normal galaxy pairs	Number of compact-normal galaxy triplets	Number of 4 and more-memb. mixed groups of galaxies	Number of expected compact galaxy pairs clo-ser than 4'	Number of expected mixed ga-laxy pairs closer than 4'	Rel. numb. of galax. in groups of com-pact gala-xies	Rel.num. of galax. in mixed groups of gal.
Compact galaxies with surf.bright. up to $20^m.0/\square'$	157	20^m- 20^m -	- -	- -	29(29)	9(10)	10(12)	0.01	0.9	0.11	0.44
		20^m- 21^m 13(13)	2(2)	2(3)		3(3)	7(21)				
Compact galaxies with surf.bright. up to $21^m.0/\square'$	986	21^m- 21^m 12(24)	- -	- -	128(128)	24(28)	24(41)	0.03	6.3	0.06	0.32
		20^m- 21^m 25(50)	2(6)	2(8)	157(157)	33(38)	34(53)				
Zwicky's comp. galax.from CCG with comput. surf.bright.	271	26(52)	-	-	10(10)	3(4)	6(12)	-	-	0.19	0.29
Total number of Zwicky's comp.galaxies in AbCBG	383	44(88)	-	4(17)	12(12)	3(4)	6(12)	0.1	4.8	0.27	0.35

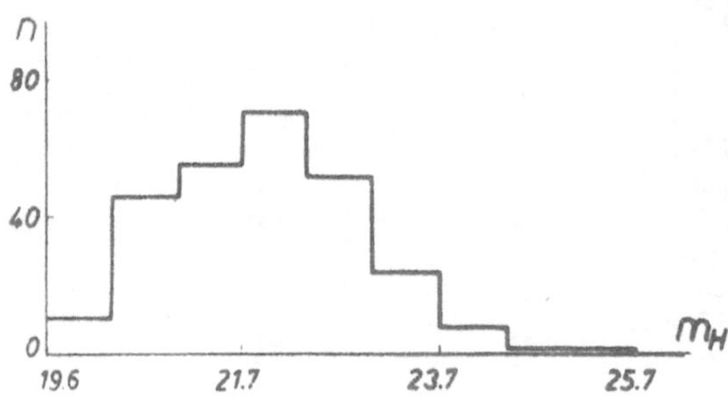
n
80
40
0
m_H
19.6
21.7
23.7
25.7

Fig. 1.

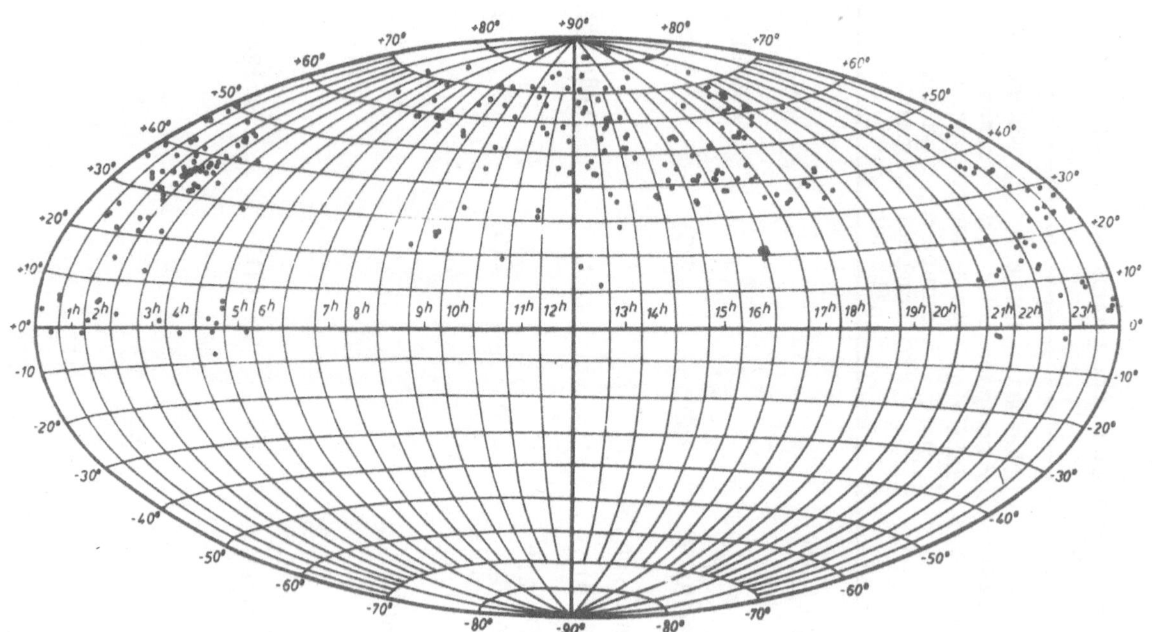
+90°
+80°
+70°
+60°
+50°
+40°
+30°
+20°
+10°
0°
-10°
-20°
-30°
-40°
-50°
-60°
-70°
-80°
-90°
1h
2h
3h
4h
5h
6h
7h
8h
9h
10h
11h
12h
13h
14h
15h
16h
17h
18h
19h
20h
21h
22h
23h

Fig. 2.

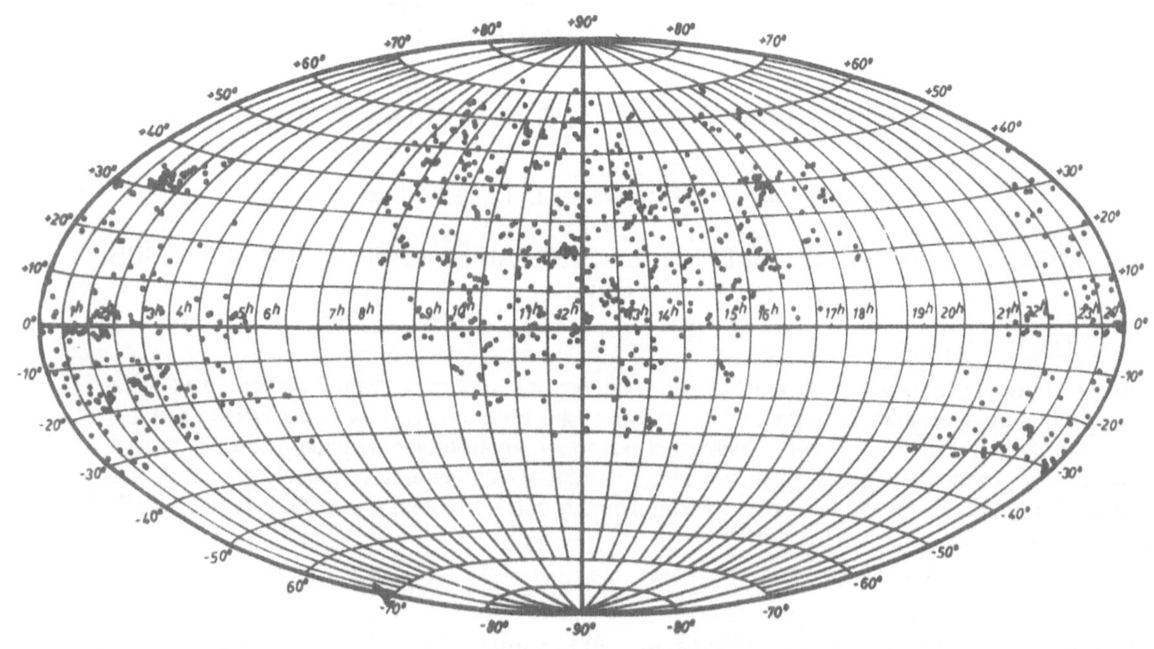

Fig. 3.

REFERENCES

ARAKELIAN M.A. 1974, Astrofizika, 10, 4,507
ARAKELIAN M.A. Astron. Astrophys. (In press)
HEIDMANN J., HEIDMANN N., VAUCOULEURS G. de 1972, Mem. R.A.S. 75, 1; 76, 11, 111
HOLMBERG E.B. 1958, Medd. Linds. Obs. 11, 136
KOGOSHVILI N.G. 1972, Astron. Circ. Ac. Sci. USSR. No 706
KOGOSHVILI N.G. 1975, Bull. Abast. Astrophys. Obs. No 46
SARGENT W.L.W. 1970, Ap.J. 160, 405
SHAHBAZIAN R.K. 1971, Astrophzika. 9, 4, 495
SHAHBAZIAN R.K., PETROSIAN M.V. 1974, Astrofizika. 10, 1, 13
VAUCOULEURS G. de, VAUCOULEURS A. de. 1964, Reference Catalogue of Bright Galaxies. BGC
VORONTSOV-VEL'YAMINOV B.A., KRASNOGORSKAYA A.A. 1962, Morphological Catalogue of Galaxies, MCG
VORONTSOV-VEL'YAMINOV B.A., ARKHIPOVA V.P. 1964, II; 1963, III; 1968, IV, MCG
ZWICKY F., 1964, Ap.J. 140, 1467
ZWICKY F., HERZOG E., WILD P. 1961, 1; 1963, 2, Catalogue of Galaxies and of Clusters of Galaxies, CGCG
ZWICKY F., HERZOG E. 1966, 3; 1968, 4, CGCG
ZWICKY F., KARPOWICZ M., KOWAL C. 1965, 5, CGCG
ZWICKY F., KOWAL C. 1968, 6, CGCG
ZWICKY F. 1971, Catalogue of Selected Compact and Post-Eruptive Galaxies, CCG

DISCUSSION

JAAKKOLA : Did you include Zwicky's galaxies which were designated as galaxies with compact parts, when calculating the average mean surface brightness?

KOGOSHVILI : The Zwicky galaxies with compact parts were included in the investigation, but we found that among the Zwicky galaxies which are brighter than $15^m_.1$, most were galaxies with compact parts.

SOME RESULTS OF COMPUTER INVESTIGATIONS OF THE CATALOGUE OF BRIGHT GALAXIES COMPILED ON MAGNETIC TAPE

2. STUDIES OF APPARENT DIRECTIONS OF GALAXY SPIRAL ARMS' COILING

T.M.BORCHKHADZE
ABASTUMANI ASTROPHYSICAL OBSERVATORY, U.S.S.R.

2. ИССЛЕДОВАНИЕ ВИДИМЫХ НАПРАВЛЕНИЙ ЗАКРУЧИВАНИЙ СПИРАЛЬНЫХ РУКАВОВ ГАЛАКТИК

РЕЗЮМЕ

Рассматриваются результаты исследования право и левосторонних ориентаций спиральных рукавов галактик на основе машинного анализа Абастуманского Каталога ярких галактик на магнитной ленте. Показано, что число галактик, имеющих правостороннюю ориентацию рукавов (вида "S") составляет 53% от общего числа видимых на небе плашмя спиральных галактик, содержащихся в Каталоге (Табл.1). Вероятность того, что подобное преобладание числа галактик вида "S" обусловлено случаем составляет $\sim 5.1 \times 10^{-9}$

The Abastumani Catalogue of Bright Galaxies, abbreviated AbCBG, comprises among the other parameters the morphological descriptions of the Morphological Catalogue of Galaxies (Vorontsov-Vel'yaminov, 1962-1968).

One of the important parameters for spiral galaxies is the orientation of the spiral arms, as seen projected in the sky. B.A.Vorontsov-Vel'yaminov used the descriptions "S" and "reverse S" for the two orientations.

Without regard to the question of whether spiral arms are leading or trailing, the relative occurence of "S" and "reverse S" systems is of cosmological importance.

With the aid of the M-220 computer we have carried out an investigation of this question on the basis of the AbCBG Catalogue. The results appear in Table 1.

Table 1

m(pg)	No. of gal. in AbCBG	No of spirals	% of spirals	No. of "S"/"rev S"	Ratio "S"/"rev S"
$<11^m$	301	165	55	78/69	1.13
11^m-12^m	607	328	54	148/139	1.06
12^m-13^m	1651	753	46	340/293	1.16
13^m-14^m	6600	2369	36	1077/950	1.13
14^m-15^m	14576	4286	29	1908/1661	1.15
15^m-16^m	3724	892	24	376/324	1.16
16^m-19^m	1546	265	17	109/91	1.20
Total	29005	9058	31	4036/3527	1.14

Of the 29005 galaxies in the Catalogue, 9058 (31%) have spiral features. The dependence on the photographic magnitude is shown in the first four columns of Table 1. Column 5 gives the actual numbers of spiral galaxies with inclination estimates 1, II and III (according to Vorontsov-Vel'yaminov) and with orientation "S" and "reverse S", respectively. The ratio is given in column 6.

As it can be seen from the table the ratio "S" to "reverse S" spiral galaxies for the sky north of -33° in all cases is well above unity. A detailed study of the individual Palomar regions revealed that although the ratio "S"/"rev S" varies much from region to region., from 8/1 to 1/12, there are no obvious regional deviations from the mean ratio 1.2 when larger sky areas are considered. No evident dependence was found on right ascension, declination, galactic longitude, galactic latitude and known galaxy clusters.

Fig.1 shows the distribution in supergalactic coordinates (de Vaucouleurs, 1971) the centres of those Palomar Sky Survey zones for which the ratio "S"/"rev S" equals 4:1 and more (open circles) or "S"/"rev S" 1:4 and less (filled circles).

The dotted line shows the region of the sky south of declination -33°.

If the distribution of spiral galaxies and their spatial orientations are random, then we should observe nearly equal numbers of "S" and "reverse S" systems.

If the chances "S" and "reverse S" are equally probable and mutually independent for different galaxies.then the number of "S" chances, while observing 7563 galaxies, has a binominal distribution with the parameters: p=0.5 and N=7563. With so large N this distribution can be approximated with high accuracy by means of a normal distribution with standard deviation $\sigma = \sqrt{Np(1-p)} = 43.48$.

The observable number of "S" chances equals 4036 and its deviation from the mean amounts to 254.5, i.e. 5.85 times higher than the standard deviation.

The probabillity of this or a larger devitation from the mean is about 5.1×10^{-9} (Owen. 1973). This means that the observed predominance of "S" spirals over "reverse S" spirals is very unlikely to be accidental.

In order to verify the descriptions of Vorontsov-Vel'yaminov we have spot-checked his designations for more than 100 spiral, mainly faint, galaxies on the different Palomar Sky Atlas prints, and no errors were found. Furthermore, the observed ratio "S"/"reverse S" is independent of magnitude and inclination (Table 1, column 1 and 6). Observational errors may therefore be excluded in from context.

The presently demonstrated anisotropy of spiral galaxy orientations is difficult to explain. Several hypotheses may be considered, but I prefer not to put forward any far reaching conclusions, except that a large-scale asymmetry of the local region of the Universe is strongly indicated.

It would be interesting to pursue this type of investigations to a fainter limiting magnitude in a limited number of fields. Likewise, observations of galaxies south of declination - 33° are highly desirable to have a complete sky coverage and thus to be able to check whether the ratio "S"/"rev S" is indeed, within the statistical uncertainty, the same in all directions.

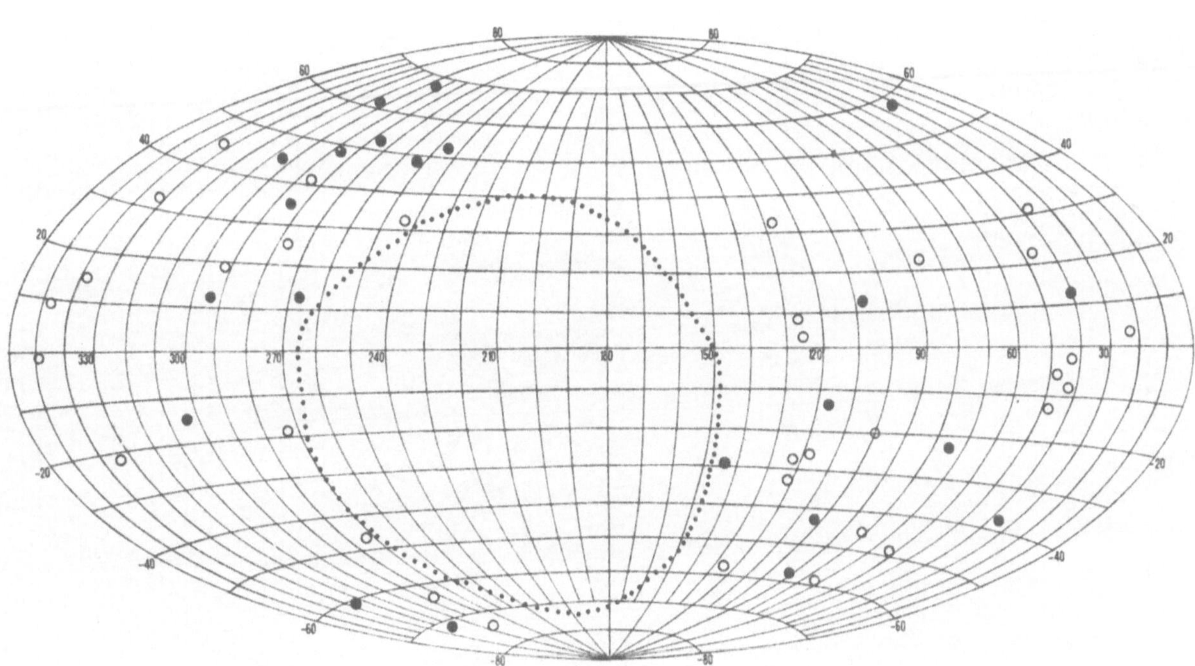

Fig. 1.

REFERENCES

OWEN D.B. 1973, Handbook of statistical tables.

VAUCOULEURS G. DE 1971, Publ. Astron. Soc. Pac. 83, 113.

VORONTSOV-VEL'YAMINOV B.A. et al 1962-1968, Morfologicheskij Katalog Galactic. Vol. 1, II, III and IV.

DISCUSSION

J.HEIDMANN: Maybe one should look for an explanation of the difference between the "S" and "rev-S" numbers in some psychological effect? Did you make tests on artificial samples of "S" and "rev-S"?

AMBARTSUMIAN: Would it not be useful to test this result for personal errors by having several persons assigning "S" or "rev-S" type to the same galaxies, if possible on different plates, in different colours?

BORCHKHADZE: No, we did not test artificial samples, but, as I said, we did check the classification of Vorontsov-Vel'yaminov without finding a single error. I could hardly see how psychological effects could be attributed to Vorontsov-Velyaminov and his collaborators en-block.

OZERNOY: Do you find any correlation between the preferred orientation of spiral galaxies in some regions with the orientation of the system to which these galaxies belong?

BORCHKHADZE: No, but we did not study this in detail.

DODD: How large is the region of asymmetry?

BORCHKHADZE: Well, if it is real, then it must be of the order of at least 100 Mpc.

MORPHOLOGY IN TIGHT EXTRAGALACTIC SYSTEMS

J. HEIDMANN
OBSERVATOIRE DE MEUDON, FRANCE

МОРФОЛОГИЯ В ТЕСНЫХ ВНЕГАЛАКТИЧЕСКИХ СИСТЕМАХ

РЕЗЮМЕ

Обзор выполненных за последние годы исследований морфологии как отдельных галактик, расположенных в тесных системах (двойных и кратных), так и самих систем (кроме двойных), включая компактные группы компактных галактик. Рассматриваются сложившиеся на основе оптических и, частично, радионаблюдений представления, теоретические морфологические модели, построенные с применением компьютеров, и соответствующие интерпретации.

I n t r o d u c t i o n. We define a tight extragalactic system as one containing at least two extragalactic objects with separations between one and several times their diameters. The number of objects may range from 2, up to a few thousand for rich Abell clusters of galaxies. However these will not be considered here, since they are not tight enough. In practice we shall stop at a dozen or so objects, such as the Shahbazian compact groups of compact galaxies.

We shall of course cover the optical domain; the radio domain will also be studied for double radiogalaxies. The objects will be not only straightforward galaxies, but also quasistellar objects (QSO), compact objects and mixtures such as quasar-galaxy, radiosource-galaxy etc...

For pairs, we cannot define a morphology for the system (there are only 2 points!) but we do have the internal morphology of each component, and their relational morphology (i.e. the morphology of one with respect to the other).

When the number of objects is larger, we have in addition the morphology of the system itself.

There are two further aspects, over and above the purely observational identification of morphology:

- *experimental morphology :* this involves numerical simulations of systems using computers. For 2 objects, we can again obtain internal and relational morphologies through the gravitational interaction of 2 systems of points (and gas); for more than 2 objects, we can obtain the morphology of the system, with eventually special processes; this has been pushed up to 100 or more objects, resulting in the production of subunits which are tight systems.
- *theoretical interpretations* of morphology using various hypotheses: tidal interactions in encounters, ejections from supermassive bodies in galaxies, pure pair formation from unknown bodies, gemmation, fragmentation of elliptical galaxies.

This review covers largely recent work, going back at most a few years. However, I list below a number of important compendia on morphology in extragalactic systems, or on subjects related to it :

- Handbuch der Physik, *53,* 1959: Stellar systems,
- Conference on the Instability of Systems of Galaxies, Santa Barbara, Astron.J. *66,* 533, 1961,
- Problems of Extragalactic Research; I.A.U. Symp. *15,* Santa Barbara, 1962,
- External Galaxies and Quasistellar Objects, I.A.U. Symp. *44* , Uppsala, 1970,
- Nuclei of Galaxies, Pontif. Ac.Sc., 1971,
- The Formation and Dynamics of Galaxies, I.A.U. Symp. *58* , Camberra, 1973,
- Stars and Stellar Systems, *9* , : Galaxies and the Universe, plus more personal ones such as
- Zwicky, Morphological Astronomy, 1957,
- Zwicky, Compact and Dispersed Cosmic Matter, Advances in Astronomy and Astrophysics, 1967 and 1970,
- Abell, Clustering of Galaxies, Annual Review of Astronomy and Astrophysics, 1965,

and Atlases which will be quoted later.

We shall not deal with kinematic and dynamical information; this is a subject for other review papers. Consequently, our interpretations will be somewhat limited, since they do not go beyond the morphological information limit.

O b s e r v a t i o n s. R a d i o m o r p h o l o g y. Up to 80% of extragalactic radiosources are double. Their total dimensions are typically of the order of 100 kpc, the size: separation ratio ranging from unity to a few percent. The largest is *Centaurus A* (NGC 5128), nearly 1 Mpc in size.

The best radio maps, with a 2" resolution have been made at a wavelength of 6 cm using the 5 km Cambridge radiotelescope; this has been done for 50 radiosources, mostly from the 3 CR (Pooley and Penbest 1974).

The two components are often linked by a faint bridge, sometimes with fainter radio condensations, or just have tails; they have sharper leading edges, as if propagating in some braking medium from an initial central event. Some are quadruple and resemble a repetition of the double pattern at two different stages (e.g. *Cen A*) with different position angles.

They are generally associated with optical galaxies or quasars, which fall in between the components. These are often an E galaxy with some peculiarity (most frequently a large heavy dust lane); this object may be a radiosource itself. Possible relations between the line of the radio components and the polar axis of the E galaxy have been investigated: polar ejection is ruled out, but isotopic or equatorial is not (Gibson 1975).

The hardest attempt to obtain optical counterparts has been made using the 5 m Hale telescope on 46 radiosources of the 3 CR catalogue; the lower magnitude limit was 23.5 (Longair and Gunn 1975). 78% of the objects were identified: 35 were galaxies and 8 were QS. 20% were not identified, because the radio size was too large, the radiosource was just too asymmetrical , the optical object was misaligned, or simply because the object is too distant.

The radio structure of 40 quasars has been investigated by Wardle and Miley (1974) with a 1" resolution; 22 were resolved - they showed generally a double structure similar to radiogalaxies. Some have a third component which coincides with the optical object and a few have 4 aligned components.

These objects are interpreted as exhibiting some kind of ejection from a galaxy or a quasar and in particular from the nucleus of a galaxy, as in the case of *Virgo A* (NGC 4486, M 87) where a jet of synchrotron radiation and a counter jet originate at the nucleus.

There is no completely satisfactory model. The most recent is due to Blandford and Rees (1974). In their model, relativistic plasma escapes along two opposite nozzles bored through the static thermal gas: this gives two beams which impringe on the extragalactic gas and produce bright radio ends.

The radiotails trailing behind moving radiogalaxies (Miley *et al* . 1972) can also be interpreted in terms of just such an extragalactic gas.

P a i r s o f g a l a x i e s. The physical pairing of galaxies is suggested from evidence of mutual interaction, and also from statistical comparisons of their actual distribution on the celestial sphere with a random distribution of points.

Pairs which exibit interaction. The interaction signatures are :

bridges or connecting filaments; common envelopes (either in the optical range or in the neutral hydrogen line);

radio continuum emission enhancement due possibly to mutual compression of the gaseous component or each galaxy;

tidal effects (distortions such as warping of galactic disks, tidal arm and counter-arm);

structure reminiscent of ejection (e.g. luminous bullets at the ends of faint bridges).

The most significant evidence comes from three important surveys:

i) Vorontsov-Vel'yaminov's *Atlas and Catalogue of Interacting Galaxies* (1959), which contains 355 examples taken from the *Palomar Sky Survey* prints; a second atlas with 700 examples is in preparation for the *Astronomy and Astrophysics Supplements.* One should also consult Vorontsov-Vel'yaminov's paper on gemmation (1974).

Different more or less well differenciated classes may be found, however, to start with, these should be considered as being morphological descriptions only, although the terms in which they are described would often suggest a real physical process such as an encounter or a separation. The classes are: with a bridge of the Messier 51 type, with a bridge, with a long or bright bridge, fusionning, in a common envelope, elliptical with bridge, in contact with a strong perturbation, distant interaction.

ii) Arp's *Atlas of Peculiar Galaxies* (1966) reproducing 338 5 m Hale telescope plates; they were selected on the basis of Vorontsov-Vel'yaminov's Atlas and Holmberg's (1937) study of multiple galaxies. Arp gives a rough preliminary classification. The largest class contains 102 peculiar spirals, not necessarily double; its largest subclass comprises 66 spirals with companions attached to spiral arms - these companions may be small or large, they may be elliptical, they may have low or high surface brightness.

The second class contains 44 E-like galaxies actually connected to spirals or repelling spiral arms, or perturbing spirals or with material emanating from them.

The third class contains the objects which do not clearly belong to either of the first two: there may be neighbouring rings, filaments, counter tails (diffuse or narrow), or ejected material. One very important subclass contains 21 objects which appear to be in a state of fission, and as well as 10 objects in irregular clumps.

The fourth class contains about 20 proper double objects: spirals with connecting arms, interacting galaxies, some exhibiting a wind effect or, more specifically, the infall of matter from one component to the other as if sucked in by the main component's nucleus.

According to Arp and Vorontsov-Vel'yaminov most of these peculiar pair morphologies suggest that there exists, in addition to a normal gravitational interaction, some kind of viscous interaction, perhaps arising from the presence of magnetic fields.

iii) *Zwicky's survey* of the sky is a third source of interacting pairs: the data exist in the form of notes in the *Catalogue of Galaxies and of Clusters of Galaxies* which he assembled with a number of collaborators (1961-8), and in the *Catalogue of Selected Compact Galaxies and of Post-Eruptive Galaxies* (1971) set up by himself and his wife and supplemented by a posthume list (Zwicky *et al.*, 1975).

Some of the nicest examples suggestive of pairs produced as a consequence of ejection are provided by IC 1182, NGC 3561 and I ZW 96 (Arp 1973 a) : small compact galaxies appear to be ejected from larger galaxies, and are still connected by nearly straight filaments suggesting a rapid process. In the first two, the compacts and jets have emission lines, thereby indicating the presence of gas; in the third, the compact shows K, H and G absorption lines, implying the presence of stars. NGC 3561 even has a neighbouring quasar with redshift 2.2.

It would appear that certain galaxies can eject plasma, ionized gas, stars and compact objects of high redshifts. These are to be distinguished from pairs of interacting galaxies which at first sight (see later) would seem to result from the classical gravitational interaction of two galaxies which pass each other. A good example of this latter is given by NGC 4038-4039, the "antennae galaxies", whose shape has been numerically simulated by the Toomres. Recent 21-cm line observations at Bonn (Huchtmeyer and Bohnenstengel, 1975) have detected 10^9 solar masses of neutral hydrogen along the antennae, rather a very large amount considering that the main bodies have $2 \times 10^9 M_\odot$; there is also some continuum emission from the interacting interface of the two components.

Pairs deduced by statistical arguments. Pairs of normal galaxies have been for a long time known. Karachentsev (1971) has exhibited their existence by a systematic statistical analysis of galaxies brighter than m=14 contained in the catalogue of Zwicky *et al.* The relative number of normal galaxies in physical pairs is estimated to be 12-24%, the mean spatial separation of the components being 110 kpc.

From their *Catalogue of Isolated Pairs of Galaxies* (Karachentsev, 1972) and *Catalogue of Isolated Galaxies* (Karachentseva 1973) constructed from the *Palomar Sky Survey* red prints, Karachentsev and Karachentseva (1975) were able to compare the morphological types of normal galaxies in pairs against those which are isolated; their main result is that the relative proportion of ellipticals is much higher in pairs than in singles and that "flat" (Population 1) systems are relatively more frequent in singles. These two facts reflect a division by population: Population I is favoured in isolated galaxies, Population II being found preferentially in pairs.

Furthermore, doubles whose components have the same type (E or S) occur more often than random pairing would give indicating a common origin; and barred spirals are more frequent in pairs.

A statistical study of the distribution of Markarian galaxies has been made by Heidmann and Kalloghlian (1973); they also found an excess of physical pairs, with mean projected separation 55 kpc, an excess which is four times larger than one might expect on the basis of the relative frequency of Markarians among normal galaxies (Heidmann, 1974). Their morphology has recently been investigated by Casini *et al.* (1974) and Casini and Heidmann (1975 b): compared to the isolated Markarians studied by Kalloghlian (1968) and Börngen and Kolloghlian (1975), these are more often spirals, freguently peculiar, and plain spirals are apparently favoured over barred spirals.

A 21-cm study showed that they are overluminous for their type (Bottinelli *et al.*,1975).

A class of giant *clumpy* irregulars has emerged from this work: the class is characterized by a clumpy structure, large diameters, high luminosity and large internal velocities.

These pairings have been extended to pairs made up of one Markarian and one normal galaxy by Casini and Heidmann (1975 a), and to one Markarian and one Zwicky compact by Heidmann and Kalloghlian (1975); their morphology is being investigated by Casini and Heidmann with large telescopes. One of the most famous and most debated pairs is NGC 4319-Ma 205 (Arp,1973 b; Walker et *al.*, 1974).

In a morphological study of the Zwicky compact galaxies from the *Palomar Sky Survey*, Bertola *et al.* (1971) noticed that compact galaxies seemed to occur frequently in pairs.

A very curious fact has been found by Turner and Gott (1975) for the celestial distribution of pairs of galaxies. They divided galaxies brighter than m= 14 in two groups: *associated* galaxies when a neighbour was closer than 45', and *isolated* galaxies otherwise. The celestial distribution of isolated turns out to be uniform, while that of *associated* is very clumpy. The simplest interpretation is that the universe is filled with two populations: a uniform field containing 40% of all galaxies, on which are superposed groups and clusters containing the rest. How this arose is not yet understood, but should be relevant to theories concerning the growth of structures in the universe.

An extreme case of pairing may be furnished by fissioning or parting of dividing galaxies (for example Ma 212) or even by galaxies with double nuclei such as those noted by Karachentsev and Karachentseva in their Catalogue.

Very close pairs may in some cases be simply different parts of a single galaxy: this is so, for example, for the "pair" Ma 94 - III Zw 0834 +51 where the former is a large H II region of the latter (Arp and Khachikian,1974).

Two optical objects. We are now led naturally to consider pairs consisting of one galaxy and one quasistellar object. Recent reviews have been given by Arp (1973 a) and by Sargent (1973 a).

The best case where components are linked is provided by the distorted spiral NGC 7603 from which two curved bridges reach a compact object (Arp,1971; Walker *et al.*, 1974). Because of the curvature of the links, Arp proposed that the compact body had been ejected slowly by the spiral; however the redshift of the compact is 8000 km s^{-1} higher.

An other curious case is provided by the galaxy CGCG 1124 +54 (Arp and O'Connel,1975): this is a galaxy of the blue dwarf compact type, probably young. It has a main body consisting of central core with two fainter lobes on either sides, all in a faint halo. At the edge of the halo is a sharp edge condensation containing some internal structure suggestive of a dust lane. From spectrographic information, this may be the wake of a massive body ejected from the nucleus.

Three cases of statistical close associations where a quasar was within 1' of a galaxy were collected at Camberra.Two more have recently been given by Arp *et.al.*(1975)One is 1'6 from the NGC 5862 - Ma 474 pair: it has a redshift of 1.94, while the velocities of the components are 2200 and 12300 km s^{-1}. The other is 1'3 from IC 1417 and has a redshift of 0.73. If the density of QSOs per unit area of sky has not been underestimated by several orders of magnitude, these pairs are statistically highly significant. No specific peculiarity has been noted for the galaxies.

We consider now pairs containing two compact objects: pairs of blue compact objects are rare (Zwicky, private communication, 1973). One was found by Heidmann and Kalloghlian(1973) 2' from the Ma 261 - 262 pair. Further work (Arp *et al.*,1974, Casini *et al.*,1974) has shown that the components have strong emission line spectra and absolute magnitudes - 17; they are separated by 6 kpc and their diameters are 2kpc, Apart from the large diameters the objects are similar to dwarf blue compacts. It is interesting to note that another , fainter , double system can be seen in between the components. Together with the pair Ma 261-262, this double double makes a very extraordinary system , whose radius is only 35 kpc.

Another double object was found between the components of the Ma 38-39 pair (Casini *et al.*,1974, Casini and Heidmann, 1975 b), one component being blue and the other red. Six other small compact pairs have been found by Iskudarian in the vicinity of galaxies (private communication).

Two bona fide QSO pairs have been reported by Stockton (1972) and by Wampler *et al.*(1973). One has a separation of 35" and the components have redshifts 0.55 and 1.17, while the components of the other, are separated by 4.5", have redshifts 0.44 and 1.90. In the latter case,there is a double radiosource, one component of which coincides with one of the QSOs. Bolton (1973 b) and Peterson (1973 b)claim to have found five pairs of QSOs whose separations are less than 30". Sargent (1973 b) estimated that, if there are 10^6 QSOs up to m = 19 in the sky, one should find 100 pairs whose components are separated by less than 5".

Multiplets. Examples of multiplets or small compact groups are labelled in Vorontsov-Vel'yaminov's and Arp's Atlases as nests or chains. Two of the best known are Stephan's quintet and Seyfert's sextet. Many of these systems show signs of interaction, which would favour their being physical groups; however, in a few cases, the components seem to have inconsistent redshifts, and this has led to the so-called redshift controversy.

Other evidence is of a statistical nature. Holmberg (1969) searched the *Palomar KSky Survey* for companions to 174 large spirals. Each spiral has one to five physical companions or satellites out to a distance of 50 kpc; his search was limited to companions larger than 600 pc and brighter than absolute magnitude - 10.6. It is very important to note that their distribution avoids the equatorial plane of each spiral in contrast to the distribution shown by radiogalaxies. Holmberg suggested that the satellites were formed from matter ejected by the nuclei of the spirals: if the material is ejected close to the equatorial plane, it is stopped by the galactic gas.

In the context of this ejection scheme for multiple satellites, one should mention the case of 3C 371, a radio N-galaxy around which six compact companions are distributed in two opposite sectors, some being linked to the central galaxy by luminous bridges (Arp and Visvanathan, 1970).

In addition to the usual interaction signatures already described for pairs, a widely discussed morphology is the chain; the brightest galaxy may be in the middle, and the chain is sometimes wary. It has often been suggested that these systems are young, since such a configuration is gravitationally unstable. However, Turner and Sargent (1973) have recently numerically simulated the evolution of (stable) groups of 6 galaxies under the mutual interaction of their gravitational forces, and they have shown that for 8% of the time their configuration will project into the sky as some kind of chain-like structure. We note, however, that the radial velocities in the observed chain-like groups often span a 1000 or 2000 km s^{-1} range even omitting groups exhibiting anomalous redshifts; this gives rise to a crossing time shorter than 10^9 years (Burbidge and Sargent, 1971) so that these systems would in fact appear to be unstable and young.

In some cases, faint outer envelopes have been found by Kormendy and Bahcall (1974); these have been interpreted in terms of the individual large halos found by Arp, Bertola and de Vaucouleurs for the ellipticals, and there is no evidence for supplementary intergalactic matter (except in the case of the *Coma* cluster).

In view of the red-shift issue, it is worth-while noting some recent morphological data which has been obtained for two multiplets. The low redshift galaxy NGC 7320 in Stephan's quintet shows a tidal-like extension, suggesting that it belongs with the four other high redshift galaxies (Arp, private communication). The high redshift "d" galaxy in Seyfert's sextet shows no sign of interaction (Walker *et al.*, 1974), according to Chincarini and Martins (1975) the sextet may be a condensation in a more extended group of galaxies.

Arp proposed that certain high redshift multiplets were physically associated with nearby spirals. One such example is the chain close to NGC 247; however, using distance criteria derived from 21-cm line observations, Balkowski and Chamaraux (1975) have shown that this chain is at its cosmological distance (60 Mpc), and so has no relation to NGC 247 which is at 2 Mpc.

In the case of Stephan's quintet, however, a radio bridge has been found by Kaftan-Kassim and Sulentic (1974) linking it to the nearby spiral NGC 7331.

Finally, we note that there are 8 spirals in the tight octet VV 282 (Burbidge and Sargent, 1971), and the very tight triplet of blue objects 0901 +:77 (Kalloghlian and Börngen 1974).

Groups. We end this review with a very strange class of groups, somewhat more populated than multiplets : the compact groups of compact galaxies. The first example was given long ago by Zwicky, and the most extensive search is due to Shahbazian;175 such structures are now catalogued (Petrosian,1974). They contain 5 to 15 objects and sometimes more than 20. They may be in chain (no. 98), they may form an open cluster (no. 83) or even a bubble (no. 106); one even has an Ω shape (no. 65).

It would appear that their component galaxies have a non-statistical grouping into pairs (nos 1, 3, 26). The Shahbazian groups may be mixed with normal galaxies (no. 245) or with a tight subgroup of galaxies (or a clumpy irregular ?) (no. 26).

Four-color photometry has been published for four groups (Börngen and Kalloghlian, 1974).

Radial velocities have been obtained for two (nos 1, 123) (Robinson and Wampler, 1973 ; Mirzoyan *et al.* 1975) ; their radii are 60 and 120 kpc and the mean separation of their components is 16 and 50 kpc. The diameters of the compacts are about 3 kpc and the brightest of them reaches absolute magnitude - 22, which is very bright indeed. Their most intriguing feature is that, unlike other tight systems, they seem to be "too" stable, the radial velocities having a dispersion ≤ 60 km s^{-1}. This indicates a low mass to luminosity ratio, similar to the one obtained by Karachentsev *et al.* (1974) for a pair of compact galaxies.

E x p e r i m e n t a l   m o r p h o l o g y. Computers have been used to simulate numerically the evolution of a number of mass points under their gravitational interaction. There are several kinds of simulations:

i) *a single system of points:* most work has been in the field of star clusters, but some results may be applied to groups of galaxies. These latter lend themselves particularly well to computer analysis because a) the number of points need not be large (~ 10), b) there is no need to follow the evolution for a long time, since the maximum age of groups is only of the order of several crossing times, and c) a cut-off may be introduced on the inverse square law at short distances, since galaxies have a finite size, and this leads to a reduction of computer time.

A number of interesting results have emerged, such as the formation of tight subclusters at the expense of a number of runaway objects. Van Albada (1968) obtained a surprising effect: a group of 10 points stayed tight for about 30 crossing times; then, suddenly, it literally exploded, dispersing nearly all of its components into space.

More sophisticated models have been made by Karachentsev and Terebizh (1969) - galactic fission processes were made to take place during the evolution - while Haggerty and Janin(1974) introduced an expansion bias.

ii) a *single system of points plus a perturbing point which passes by :* Eneev *et al.*(1973) studied the effect produced by a massive body passing by a disk of up to 2000 gravitating points (a galaxy) in order to investigate how spiral structure might be formed. They did in fact obtain spiral arms: there was generally a diffuse arm towards the perturbing body and a thinner arm away from it, a configuration sometimes observed for spirals in pairs of Markarian galaxies (Casini and Heidmann 1975 b). Important distortions of the galactic plane and ejection of matter, which sometimes fell back in, were observed.

iii) *two systems of points :* Toomre and Toomre (1972) investigated the passage of two flat disks of points each gravitating around its own nucleus. They obtained a bridge and a far side counter-arm in the case of a small companion, and a long and curving tail of escaping debris from the far side for the passage of companion having the same size. They also gave examples which reproduced in a striking way four remarkable pairs of interacting galaxies: Arp 295 with a long thin bridge and counter-tail, Messier 51, NGC 4676 ("the mice") and NGC 4038-4039 ("the antennae").

iv) a *system of points ejecting one or two bodies :* Clairemidi and Hayli (1973) have started to investigate what happens when the nucleus of a disk of points gravitating around it ejects, in the plane of the disc, one massive body: they find that a diffuse arm is formed, which at first is leading and then becomes trailing. In the case of a symmetrical ejection of two bodies, two leading arms are formed, which then disappear, leaving a kind of bar.

A fifth domain should be investigated with computers in view of possible applications to pair production of galaxies (cf. infra) : the sudden separation of two blobs of matter in the form of two compact spheroidal systems of mass points. Here also spiral structures should be expected; such an event can be thought of as the second half of the scenario emerging from the passage of two systems and so spiral structure is to be expected; moreover, Toomre and Toomre's simulations show that important morphological changes occur mainly after the closest approach.

T h e o r e t i c a l   i n t e r p r e t a t i o n s. We have seen that the morphology of certain tight pairs can be interpreted in terms of the gravitational interaction of two gravitating systems which pass each other. However, this is not a complete interpretation, since it concerns only the individual morphology of each component ; the fundamental problem is : how is it that two such systems just happen to pass each other? It might perhaps be more correct to ask how it is that two or more such systems happen to be close neighbours.

It is rather unreasonable to consider that the components originating in widely separated regions of space, just happened to meet.

Some of the groups could be stable multiplets produced by density fluctuations of the primeval gaseous substratum of the universe. This cannot be generally true, since the ratio of the virial mass to the "luminous" mass of the component galaxies is usually much larger than one (and even 10), sometimes reaching 1000.

The groups could be loose aggregates of neighbouring field galaxies which are expanding apart in the Hubble flow, like unbound density enhancements. Turner and Sargent(1974) consider that such is the case for 75% of the small nearby groups of galaxies for which crossing times are of the order of 10^{10} years. However, 25% have much shorter crossing times and have positive energies.

The groups could have condensed recently from large clouds of (neutral) hydrogen; however although neutral hydrogen intergalactic clouds are now being discovered (e.g. Mathewson *et al.* 1975), their masses are too small.

Much of this line of interpretation depends on the outcome of the "missing mass" problem which will be covered by Bertola and Karachentsev at this Meeting. I shall consequently consider other lines.

Ambartsumian (1971) has proposed that the galaxies originate from supermassive objects ejecting "pellets" of matter in the form of gas or plasma, 10^8 solar masses at a time; they do this at intervals, and also continuously. This process is imagined to last a sufficient time to generate a galaxy, whose nucleus contains the supermassive object; companion galaxies could be formed when the "pellets" move out faster than the escape velocity, and even groups and clusters of galaxies by successive fragmentations.

This scheme has mainly been taken up by Arp, who gave it morphological support, some of which we have already seen. He proposed a *sequence:* in the first frame (Arp no. 49), a compact object travels through a galactic disk, leaving behind it a wake; in the second (Arp no.58), the pellet begins to *unfold*, attracting behind it a spiral arm-bridge; In the last frame (Messier 51), it is a full sized companion galaxy (Arp,1969).

A variant of Ambartsumian's idea has been proposed by Casini *et al.* (1974) as a method for generating pairs of galaxies. They suggest that an unknown parent body ejects two lumps of matter in opposite directions; then, each of these as they separate evolves along the sequence *compact object → Markarian galaxy → normal or compact galaxy*. The evolution is mediated mainly by their mutual gravitational interaction.

Another model has recently been proposed by Vorontsov - Vel'yaminov (1974): rather than being ejected from nuclei, secondary galaxies would be formed peacefully inside a spiral arm of a mother-galaxy, and would then part in a smooth, slow way which he calls gemmation.

In 1968, Sersic suggested that the fragmentation of giant elliptical galaxies triggered by an implosion of their cores gives rise to dwarf ellipticals, irregulars and spirals.

I shall end with a completely different process: it was invoked by Freeman and de Vaucouleurs (1974) to explain two very peculiar classes of pairs consisting of a spheroidal galaxy and either a ring galaxy (Arp. no. 147) or a chaotic object containing several nuclei (Arp no. 143). These would result from the encounter of normal spirals with intergalactic neutral hydrogen clouds. In such a process, the stellar bulge would carry on essentially undisturbed, while the gaseous ring - shaped disk component would be slowed down and separated from the bulge. This ring would subsequently collapse into a chaotic object.

C o n c l u s i o n s . The morphology of tight extragalactic systems poses a number of very interesting questions concerning the, perhaps recent, origin and evolution of galaxies, not mentioning the anomalous redshift issue. High resolution photographs obtained with large telescopes are much needed. Together with numerical experiments, as well as information concerning the kinematic and physical state of the systems obtained from high dispersion and high angular resolution spectroscopy, they should help us to understand some of the mysteries of the universe and perhaps uncover some as yet unknown fundamental physical processes.

REFERENCES

AMBARTSUMIAN V.A. 1971, *Nuclei of Galaxies*, 9, Acad. Vatican

ARP H. 1966, *Atlas of Peculiar Galaxies*, Astrophys.J. Supp. XIV, 123

ARP H. 1969, Astron. Astrophys. 3, 420

ARP H. 1971, Astrophys. Let. 7, 221

ARP H. 1973 a, I.A.U. Symp. 58, 380

ARP H. 1973 b, *The Redshift Controversy*, 48

ARP H., BALDWIN J.A., WAMPLER J. 1975, Lick Obs. Bull. 694

ARP H., HEIDMANN J., KHACHIKIAN E.Ye., 1974, Astrofizica, 10, 7

ARP H., KHACHIKIAN E.Ye. 1974, Astrofizica 10, 173

ARP H., O'CONNELL R.W. 1975, Astrophys. J. 197, 291

ARP H., VISVANATHAN N. 1970, Astrophys. Let. 5, 73.

BALKOWSKI C., CHAMARAUX P. 1975, Third European Astron. Meeting, Tbilisi and Astron.Astrophys.

BERTOLA F., LUCCHIN F., NASI E. 1971, Mem. Soc. Astron. Ital. 42, 517.

BLANDFORD R.D., REES M.J. 1974, Mon. Not. Roy. Astron. Soc. 169, 395

BOLTON J.G. 1973, I.A.U. Symp. 58, 197

BÖRNGEN F., KALLOGHLIAN A.T. 1974, Astrofizica, 10, 21

BÖRNGEN F., KALLOGHLIAN A.T. 1975, Astrofizica, 11,

BOTTINELLI L., DUFLOT R., GOUGUENHEIM L., HEIDMANN J. 1975, Astron. Astrophys. 41, 61

BURBIDGE E.M., SARGENT W.L.W. 1971, *Nuclei of Galaxies*, 351, Acad. Vatican

CASINI C., HEIDMANN J. 1975a, Astron. Astrophys. 39, 127

CASINI C., HEIDMANN J. 1975 b, Third European Astron. Meeting, Tbilisi, and Astron. Astrophys.

CASINI C., HEIDMANN J., LELIÈVRE G. 1974, Second European Astron. Meeting, Trieste

CHINCARINI G., MARTINS D. 1975, Astrophys. J. 1969 196, 335

CLAIREMIDI S., HAYLI A., First European Astron. Meeting, Athens, 405

ENEEV T.M., KOZLOV N.N., SUNYEV R.A. 1973, Astrophys. 22, 41

FREEMAN K.C., DE VAUCOULEURS G. 1974, Astrophys. J. 194, 569

GIBSON D.M. 1975, Astron. Astrophys. 39, 377

HAGGERTY M.J., JANIN G. 1974, Astron Astrophys. 36, 415

HEIDMANN J. 1974, Second European Astron. Meeting, Trieste,

HEIDMANN J., KALLOGHLIAN A.T. 1973, Astrofizica, 9, 71

HEIDMANN J., KALLOGHLIAN A.T. 1975, Astrofizica, 11, 229

HOLMBERG E. 1973, Lund Obs. Annals, 6

HOLMBERG E. 1969, Ark. Astron. 5, 305

HUCHTMEYER W.K., BOHNEN-STENGEL H.D. 1975, Astron. Astrophys. 41, 477.

KAFTAN-KASSIM M.A., SULENTIC J.W. 1974, Astron. Astrophys. 33, 343

KALLOGHLIAN A.T. 1968, Astrofizica 4, 475

KALLOGHLIAN A.T., BÖRNGEN F. 1974, Astrofizica. 10, 295

KARACHENTSEV I.D. 1971, Acta Astron. 21, 237

KARACHENTSEV I.D. 1972, Soob. Special Astrophys. Obs. 7

KARACHENTSEV I.D., PRONIK V.I., TSUVAIEV K.K. 1974, Astrofizica 10, 441

KARACHENTSEV I.D., TEREBIZH V.Yu. 1969, Soob. Byurakan Obs. 41, 99

KARACHENTSEV I.D., KARACHENTSEVA V.E. 1975, Soviet Astron. 18, 428

KARACHENTSEVA V.E. 1973, Soob. Special Astrophys. Obs. 8

KORMENDY J., BAHCALL J.N. 1974, Astron. J. 79, 671

LONGAIR M.S., GUNN J.E. 1975, Mon. Not. Roy. Astron. Soc. 170, 121

MATHEWSON D.S., CLEARY M.N., MURRAY J.D. 1975, Astrophys. J. 195, L 97

MILEY G.K., PEROLA G.C., van der KRUIT P.C., van der LAAN H. 1972, Nature 237, 269

MIRZOYAN L.V., MILLER J.S., OSTERBROCK E. 1975, Astrophys. J. 196, 687

PETROSIAN M.B. 1974, Astrofizica, 10, 471

POOLEY G.C., HENBEST S.N. 1974, Mon. Not. Roy. Astron. Soc. 169, 477

ROBINSON L.B.., WAMPLER E.J. 1973, Astrophys. J. 179, L 135

SARGENT W.L.W. 1973 a, I.A.U. Symp. 58, 195

SARGENT W.L.W. 1973 b, I.A.U. Symp. 58, 197

SERSIC J.L. 1968, Astrofizica 4, 105

STOCKTON A.N. 1972, Nat. Phys. Sc. 238, 37

TOOMRE A., TOOMRE J. 1972, Astrophys. J. 178, 623

TURNER E.L., GOTT III J.R. 1975, Astrophys. J. 197, L 89

TURNER E.L., SARGENT W.L.W. 1973, Pub. Astron. Soc. Pacific 85, 538

TURNER E.L., SARGENT W.L.W. 1974, Astrophys. J. 194, 587

VAN ALBADA T. 1968, Bull. Astron. Netherlands, 19, 479
VORONTSOV - VEL'YAMINOV B.A. 1959, *Atlas and Catalogue of Interacting Galaxies,* Sternberg Institute
VORONTSOV - VEL'YAMINOV B.A. 1974, Astron. Astrophys. *37*, 425
WALKER M.F., PIKE C.D., Mc GEE J.D. 1974, Astrophys. J. *194*, L 125
WAMPLER E.J., BALDWIN J.A., BURKE W.L., ROBINSON L.B. 1973, Nature *246*, 203
WARDLE J.F.C., MILEY G.K. 1974, Astron. Astrophys. *30*, 305
ZWICKY F., HERZOG E., WILD P., 1961-8, *Catalogue of Galaxies and of Clusters of Galaxies,* Cal. Inst. Tech.
ZWICKY F., SARGENT W.L.W., KOWAL C.T. 1975, Astrophys. J.
ZWICKY F., ZWICKY M.A. 1971, *Catalogue of Selected Compact Galaxies and of Post-eruptive Galaxies, Cal. Inst. Tech.*

DISCUSSION

AMBARTSUMIAN: What are the largest distances to which isolated HI clouds in the intergalactic space have been observed?
HEIDMANN Not more than corresponding to a recession velocity of about 10,000 km s^{-1}, and on the average 1000 to 2000 km s^{-1}, since these clouds are rather small.
JAAKKOLA: While the absolute luminosities of field compact galaxies are often very high, compact members of systems which are mainly made up of normal galaxies are always fainter, on the average about 2^m, than the brightest normal galaxies. Would you care to comment on this?
HEIDMANN: I am sorry, we did not look at normal-compact pairs.
KHACHIKIAN: I have recently obtained photographs of Markarian 305 and 306 with the Kitt Peak 4 m telescope. They show that 306 has spiral structure and it appears that 305 and 306 are surrounded by a common envelope, z - 0.019 for both, and the distance between them is about 12 kpc. They could possibly represent an early "M 51 - type" pair.
ARAKELIAN: I have studied the tendencies of normal vrs. peculiar galaxies to appear in groups and pairs and I found very little difference. Thus it seems to me that the conclusion about the youth of Markarian and other peculiar galaxies previously suggested from the statistics of pairs cannot be considered to be definitely established. We can only conclude that the galaxies in pairs have a common origin.
HEIDMANN: From the number of Markarian-normal pairs and from the percentage of Markarian galaxies one finds that the expected number of Markarian-Markarian pairs is four times smaller than what is observed. Recent origin is deduced from the low kinetic age of two positive energy Markarian pairs (Ma 7-8 and 56-57); pairing only points to common origin.
OZERNOY: The colour meassurements of some Markarian galaxies indicate that they contain very red and therefore - as generally believed - very old stars. Is this compatible with a very recent common origin of some of the Markarian pairs?
HEIDMANN: Of course not, in case such old stars are indeed found in pairs thought to be young. But Markarian galaxies are very diverse; indeed some are old galaxies owing the Markarian status to a Seyfert activity or to the presence of the large superassociation.

RADIO AND X-RAY OBSERVATIONS OF CLUSTERS OF GALAXIES

M.S. LONGAIR
MULLARD RADIO ASTRONOMY OBSERVATORY, CAVENDISH LABORATORY, U.K.

РАДИО И РЕНТГЕНОВСКИЕ НАБЛЮДЕНИЯ СКОПЛЕНИЙ ГАЛАКТИК

РЕЗЮМЕ

Изложены результаты наблюдательных и теоретических исследований радио и рентгеновских свойств богатых звездами скоплений галактик. Радиоисточники в скоплениях отличаются от радиоисточников общего поля наличием протяженных радиоисточников и источников с крутыми спектрами. Это свидетельствует о высокой плотности тепловой материи в скоплениях. Рентгеновская эмиссия скоплений Эбеля, относящихся к I I классн по числу звезд, является тормозным излучением горячего газа внутри скоплений. Описаны модели распределения газа в скоплениях, считая что газ образует адиабатическую атмосферу. Она образуется вследствие коллапса газа в скоплениях и может объяснить наблюдаемые свойства рентгеновского излучения. Подчеркивается важность поисков рентгеновской эмиссии из скоплений с большим красным смещением.

1. I n t r o d u c t i o n. The study of clusters of galaxies at radio, optical and X-ray wavelengths has proved to be one of the important growth points in extragalactic astronomy in recent years. The observations made in these wavebands complement each other and have led to a much improved understanding of the structure and evolution of clusters as a whole. In this brief review I will not discuss the wealth of new optical observations which are now available but will concentrate upon those aspects related to the radio and X-ray properties of rich clusters. I intend to show how the observational evidence derived from studies at radio and X-ray wavelengths support the view that there is a considerable amount of hot gas in clusters of galaxies. This in turn leads to the construction of improved models for the distribution of the gas in clusters which also involves a discussion of the origin of such gas. These investigations suggest that very soon it will be possible to study clusters at large redshifts through their X-ray emission - during formation and cooling clusters should be very intense X-ray sources.

2. S u m m a r y o f t h e r a d i o p r o p e r t i e s o f c l u s t e r s o f g a l a x i e s. It has been known for a long time that there is a highly significant statistical association between radio sources and rich clusters of galaxies in the sense that within 0.3 R_c an excess of radio sources is found in complete sky surveys; R_c is the radius of a rich Abell cluster according to the definition of Abell (1958). Many authors have studied these correlations but most of them have been based upon radio surveys with low angular resolution and therefore it has not been possible to understand the exact relation between the radio properties of the cluster and the galaxies themselves. High resolution studies of a complete sample of Abell clusters of galaxies have recently been completed at Cambridge in order to find out the nature of the radio sources associated with the richest clusters and to find out they differ from radio sources in the general field. The surveys were restricted to Abell clusters known to contain radio sources in a well defined region of sky since the optical selection criteria are most uniform for the Abell sample.

The surveys were made by Riley and Slingo with slightly different aims in view. Riley (1975a) performed aperture synthesis observations of about 25 Abell clusters coincident with 4C radio sources in the 20° to 40° declination range to discover how many of the radio sources were genuinely associated with Abell clusters and to investigate their morphology and physical properties. Slingo (1974a, b) performed similar observations but was primarily interested in those radio sources which possess very steep low frequency spectra. It had been shown by Baldwin and Scott (1973) that there exists a strong correlation between the presence of a very steep low frequency radio spectrum ($\alpha \geqslant 1.2$ where the spectral index α is defined by $I_\nu \propto \nu^{-\alpha}$) and the association of the source with an Abell cluster. Slingo's surveys were primarily intended to investigate the origin of those radio components with very steep spectra but because of the overlap with Abell cluster programme, the combined results of the surveys of Riley and Slingo can be used to generate complete samples suitable for statistical analyses.

2.1 Statistical Results. Riley (1975b) has summarised the results of statistical studies of Abell clusters of galaxies and their radio properties:

(i) 10 to 25% of all *strong radio sources* are in Abell clusters. By strong radio source one means a radio source with intrinsic radio luminosity at 178 MHz in the range $10^{24} \leqslant P_{178} \leqslant 10^{26.5}$ W $Hz^{-1}sr^{-1}$. This definition is somewhat arbitrary but at least for this luminosity range (a) one can be certain that the statistics are complete and (b) one is dealing with samples of sources which are ''typical'' of powerful radio sources in the sense that most of them have double, complex or radio trail structures. Obviously if one takes a low enough luminosity limit, all galaxies are radio emitters.

(ii) *More than 25% of all Abell clusters* contain strong radio sources according to the above definition.

(iii) There is little or no correlation between the probability of an Abell cluster containing a strong radio source and the richness of the cluster. Richness is a measure of the total number of galaxies in the cluster and therefore this result suggests that the correlation between the presence of radio emission and cluster membership must be with some property of the cluster which is independent of the number of galaxies in the cluster.

(iv) More than 70% of those clusters which possess a dominant cD galaxy (i.e. Bautz-Morgan class I clusters) are strong radio sources. This result has been found by McHardy (1974) from a study of radio sources in Abell clusters in the 3C and 4C catalogues.

2.2 Results of the Aperture Synthesis Surveys. Riley summarises the results of the high resolution studies of the complete Abell sample as follows:

(i) The complete sample consists of only 25 sources in the declination range 20° to 40°. She has shown that 4 of these sources are probably *not* in the clusters with which they are coincident on the sky and one of them could not be detected. These results are not unexpected since (a) the association is only a statistical one and must be contaminated by unrelated background sources; (b) some sources may not be detected either because they have very steep spectra or because they are of very low surface brightness or because they are very close to the flux density limit of the catalogue where confusion may result in the inclusion of faint sources with overestimated flux densities and erroneous positions. These numbers indicate the size of these effects for sources in the 4C catalogue.

(ii) 75% of strong radio sources according to the above definition are associated with the *brightest object* in the cluster. Very often this means a dominant cD galaxy but in many cases there are additional complexities. In 2 cases the brightest object has a multiple nucleus; in 2 cases the systems are dumb-bell galaxies; in 4 cases the brightest object is a very compact group of galaxies. It is not clear whether or not these phenomena are different aspects of the same process which leads to the formation of massive central objects.

(iii) 5 of the 20 clusters contain more than one radio source with $P_{178} \geqslant 4 \times 10^{23}$W $Hz^{-1}sr^{-1}$. This result indicates that some galaxies more than a magnitude fainter than the brightest object in the cluster can become relatively strong radio sources. In one of Slingo's clusters (not included in the statistical sample 4C 64.20.1) there are 4 radio sources in the cluster.

(iv) The spectral index distribution has a long high-spectral index tail; sources with steep spectra tend to lie in clusters with a dominant optical object.

(v) A very wide variety of structures is found including double, complex, radio trail and compact sources. In linear size, the objects range from less than 30 kpc to about 800 kpc. The steep spectrum sources have a very similar distributions of physical size and morphological type to those sources which have normal spectra (i.e. $\alpha \sim 0.75$).

An important question is in what way radio sources in rich clusters differ from those in the general field. The answer appears to be "Not very much at all'' except in two important respects. The ways in which they do *not* differ are in overall physical dimensions (Hooley 1974) and in the relative proportions of compact, double and complex sources. They *do* differ in that

(i) *the radio trail sources* appear to belong exclusively to clusters. This may be partly a selection effect since observers have concentrated upon clusters of galaxies in order to find these sources but even so none of the 14 examples known so far has been found outside a cluster.

(ii) The excess of *sources with steep low frequency spectra* discussed above appears to be a characteristic only of sources in rich clusters. This is because in Baldwin and Scott's complete sample. most of the sources with steep spectra should be associated with clusters of galaxies, leaving no sources with steep spectra to be associated with non-cluster objects. On the other hand, as noted above, other radio properties of these sources are remarkably similar to sources with normal spectra.

3. Summary of the X-ray properties of clusters of galaxies. The strong X-ray emission from clusters of galaxies is one of the most important discoveries made by the UHURU satellite (Kellogg et al 1973). At the present time there are 21 X-ray sources associated with rich clusters of galaxies; this sample is sufficiently large that it is already possible to make a number of important inferences from the data. It is convenient to follow the analysis of N. Bahcall (1974) and divide the sample into strong and weak X-ray emitters according to the luminosity criterion

$$L_X > 2 \times 10^{44} \text{ erg s}^{-1} \equiv \text{strong}$$
$$L_X < 2 \times 10^{44} \text{ erg s}^{-1} \equiv \text{weak}$$

Figure 1 shows the distribution of these sources as a function of cluster richness. It should be remembered that the relative proportions of Abell clusters in richness classes 0, 1 and 2 is roughly 3:3:1 respectively and thus Fugure 1 indicates that there is a strong correlation of X-ray emission with richness class 2 clusters.

Weak X-ray clusters. For three of these sources, those associated with 3C 66, Virgo A and 3C 264, the X-ray position of the source does not coincide with the dynamical centre of the cluster but rather with an active galaxy displaced from the centre. It would be rash to conclude from these poor statistics that all the weak X-ray sources are associated with active galaxies rather than with the cluster itself but these three examples show that some of the cluster sources are probably not associated with diffuse intergalactic material filling the central region of the cluster. They may be associated with the emission of hot gas in the vicinity of the active galaxy or with inverse Compton radiation if there exist large fluxes of relativistic electrons originating in the active galaxy.

Strong X-ray sources. The excess of richness class 2 clusters is the most striking feature of this distribution. In addition, in those cases for which accurate optical and X-ray data exist, these X-ray sources are coincident with the dynamical centres of the associated clusters. Several of the sources have measured angular diameters, generally of the order of 30' arc which are greater than the core diameters of the distribution of galaxies in the cluster centre.

It is of particular interest to note the radio properties of the richness class 2 X-ray clusters since this gives some clues to the origin of the emission.

4. Comparison of the radio and X-ray properties of rich clusters of galaxies. Most of the strong X-ray sources associated with richness class 2 clusters are also associated with well-known radio sources which possess the distinctive features noted in Section, 2 (i.e. radio trails and steep low frequency spectra). Table 1 indicates the radio properties of the sources shown in Figure 1 which satisfy the above X-ray criteria. This evidence can be interpreted as favouring the thermal bremsstrahlung interpretation of the X-ray emission from clusters of galaxies.

(a) All models to explain the radio trail sources require substantial amounts of hot cluster gas to be present in order to confine the particles in the trail and also to sweep the trail backwards by the ram pressure of the intergalactic gas over time-scales of 10^8 years (e.g. Taffe and Perola 1973). This is one aspect of the more general problem of extragalactic radio sources that there must be sufficient hot gas present in the vicinity of intense sources to confine the irregular regions of diffuse radio emission which are frequently observed (see Longair, Ryle and Scheuer (1973)). In the case of the radio trail sources, it is important that they are not located in the central regions of the clusters and therefore the hot cluster gas must extend throughout a large region of the cluster. The gas must be hot or else it will collapse to the centre of the cluster forming a compact X-ray source which will have a short lifetime because the particle density would be high and the cooling rapid. The natural interpretation of the radio observations is to suppose that hot gas fills the cluster with such density that it does not cool within cosmological timescales. The motion of the active galaxy through the intergalactic gas in the cluster can account for the various complexities of the radio structure (see e.g. Taffe and Perola 1973).

(b) The association of rich clusters with sources which have steep low frequency spectra can be explained if the confining mechanisms for the particles are more effective in clusters of galaxies so that there is more time for the particles to lose a substantial fraction by their energy synchrotron or inverse Compton losses. Some of the radio spectra are so steep at high frequencies that it appears that effectively all the high energy relativistic electrons are absent. The natural explanation is that there has been no replenishment of the relativistic electron flux and that synchrotron losses have completely exhausted the energy present in the highest energy electrons (Slingo 1974b). This argument also favours the presence of high densities of hot gas in clusters to provide more effective confinement of the radio emitting regions. This argument is not, however, on as firm a quantitative basis as argument (a).

The more detailed calculations which have been performed suggest that the observed intensity of X-ray emission expected from gas dense enough and hot enough to effect confinement in

Table 1

The properties of radio sources associated with intense X-ray sources in Abell clusters of galaxies of richness class 2.

These sources have X-ray luminosity $\geqslant 2 \times 10^{44}$ erg s^{-1} (see Bahcall 1974).

Abell cluster	Radio source	Presence of a radio trail source	Presence of a component with very steep low frequency spectrum
A 401	4C13.17		
A 426 (Perseus)	3C83.1B/84	(1)	
A 754	3C218	no information	
A 1656 (Coma)	5C4		(2)
A 2199	3C338	-	-
A 2256	NB 78.26	probably (3)	
-	Cygnus A(4)	-	-
-	3C129/129.1(5)		(6)

Notes

(1) There are three radio trail sources in the Perseus cluster.
(2) Steep spectrum radio component associated with intergalactic medium (Wilson 1970).
(3) Although not noted by Slingo, morphologically NB 78.26 resembles a radio trail and has a steep low frequency spectrum.
(4) Cygnus A belongs to a cluster of richness class 2 according to Matthews, Morgan and Schmidt (1964).
(5) 3C129/129.1 is included because it belongs to a cluster of unknown richness - it lies in an obscured region; it is one of the most striking examples of a radio trail.
(6) The steep spectral component occurs towards the end of the radio trail but does not result in an overall steep low frequency spectrum for the source.

the above senses, is close to that observed from those rich clusters which are strong X-ray emitters. At present this argument cannot be considered conclusive but further radio and X-ray studies should clarify the picture and there is a need for improved models of the distribution of gas in clusters necessary to explain both the radio and X-ray observations. One line of approach to this problem is outlined in the next section.

5. Models for hot gas in clusters. These models have been developed by Gull and Northover (1975), the motivation arising from their previous studies of the effects of buoyancy on the dynamics of clouds of relativistic material (Gull and Northover 1973). Many authors have made the simplest assumptions about the distribution of gas in clusters, namely that its distribution is isothermal and follows the overall spatial distribution of galaxies. The observations now merit improved models and Gull and Northover have adopted models in which the equation of state of the gas in the cluster is taken to be *adiabatic*. The justification for this assumption is that, in practice, the mean free path of the particles in the intergalactic gas in the cluster is very small, being determined by the gyroradius of protons in the intergalactic magnetic field. Even if this field is very small indeed, $\sim 10^{-9}$ gauss, the gyroradius of a proton is many orders of magnitude smaller than the scale of the cluster. Therefore the effective conductivity of the gas is very small in contrast to the assumption of those models which adopt an isothermal distribution. Further more, the gas in the cluster is likely to be well mixed as a result of stirring by the motions of galaxies through the intergalactic medium and through high energy events such as those associated with double radio sources of the explosion in NGC 1275. The adiabatic equation of state means that when two elements of volume are interchanged in the mixing process, the properties of the interchanged volumes are related by the adiabatic identity.

The programme of Gull and Northover is then the following: they assume that the gravitational potential distribution in the cluster is entirely determined by the distribution of mass in galaxies - i.e. they assume the cluster is bound and the distribution of mass follows that of the galaxies. They then fill up this potential well with an adiabatic atmosphere. Because an adiabatic equation of state is assumed, the atmosphere is bounded unless the central temperature of the gas is greater than or equal to $T_0 = \frac{2GM_{cl}m_p}{P_{cl}}$,

where m_{cl} and R_{cl} are the mass and radius of the cluster and m_p is the proton mass. These models are unlikely because they lead to the prediction of excessive X-ray background emission. The more likely situation is illustrated in Fig.2 - a schematic diagram showing the gravitational potential distribution in a cluster of galaxies. Hot gas forms an adiabatic atmosphere in the potential well. The height to which the well is filled is determined only by the temperature of the gas T_g. T_0 is the temperature of gas which would just fill the potential well. T_∞ is the effective temperature of the cluster gas as R tends to infinity. Atmospheres which do not fill the potential well have negative values of T_∞. R_{cl} is roughly the radius within which half of the total mass of the cluster lies. This figure illustrates an essential feature of these models. The galaxy distribution follows closely an isothermal distribution which is very reasonable for a gas of particles with very long mean free paths interacting only through gravitation. The density scale high of such a distribution is determined by the *core radius* of the distribution. However, the gas distribution is determined by the gravitational scale height of the potential distribution in the cluster as a whole and by the temperature of the gas as illustrated in Figure 2. In fact, for the models discussed by Gull and Northover, $R_{core} \approx 0.05\ R_{cl}$. The only free parameters left once the potential distribution is defined are the temperature of the gas T_g, defined as shown on figure 2 and the central gas density n_0. For the standard cluster a mass of $5\times10^5\ M_\odot$ is assumed. The space distribution of galaxies is assumed to follow the de Vaucouleurs' law for galaxies in the outer regions and the Hubble law in the core of the cluser. The following results are obtained:

(i) The temperature decreases outwards and therefore the predicted X-ray spectrum is the superposition of a number of thermal bremsstrahlung spectra at varying values of Te Figure 3. shows an unscaled fit of the models to the observations of the Perseus cluster for an assumed central density $n_0 = 10^{-3} cm^{-3}$. It is clear that the shape of the distribution is in good agreement with the observations and can be improved by adjusted n_0. Similar satisfactory results are obtained for the Coma cluster.

(ii) The models are only strictly applicable when the mass of intergalactic gas is much less than the mass in galaxies. However, they note that in principle one can hide a great deal of mass, of the order of the total mass of the cluster, in the outer regions of the cluster where it is cool and of low density. Such mass could bind the outer parts of the cluster but because the gas density does not increase as greatly as the galaxy density towards the central regions, it is not possible to bind the core of the cluster in this way.

(iii) The angular size of the X-ray source depends upon the depth to which the potential well is filled. A satisfactory fit to all the data is obtained if $T_\infty = -0.25\ T_0$. This result contrasts with that of the isothermal models in which the angular size of the X-ray source is of the order of the size of the isothermal core of the galaxy distribution.

(iv) As noted by Syunayev and Zel'dovich (1972), the presence of hot gas in a cluster of galaxies results in a significant optical depth to Compton scattering for background radiation as it passes through the cluster. This effect is due to Compton scattering of the microwave background radiation by the thermal electrons of the intracluster gas and the result is a dip in the microwave background in the direction of the cluster for frequencies in the Rayleigh-Jeans part of the spectrum (i.e. $\lambda > 1$ cm.) In the present case the expected effect is quite large because there is a significant amount of hot gas in the outer parts of the cluster. For a wide range of values of T_∞, the expected dip in the direction of a cluster such as Coma is about 1 mK. Parijskij (1973) claims to have detected such a dip in the direction of the Coma cluster but the result has not been confirmed. This experiment is difficult but feasible and we are at present attempting it using the 26-metre telescope at Chilbolton. This is a crucial test of the hypothesis that there is hot gas in clusters. Note incidentally that this experiment measures directly $n_e kT_e$ which is the pressure of the cluster gas and the important parameter so far as the confinement of the radio components is concerned.

The adiabatic models are thus able to account for many of the observed properties of the X-ray emission of clusters of galaxies.

Dynamical models. The models considered above are stationary models corresponding to stable adiabatic atmospheres in clusters of galaxies. They are only applicable within a certain range of time-scales. Thus the condition that the cluster must have had time to settle down to a stationary adiabatic state sets a lower limit to the appliable range of time-scales: on the other hand the time-scales must not be so long that cooling of the hot gas by bremsstrahlung becomes important. Gull and Northover have run a set of dynamical models for the infall of gas into a pre-existing cluster of galaxies to study the type of atmosphere set up. Infall is an attractive source of the energy for heating the intergalactic gas because its energy requirements are very large indeed. These computations, which are described in detail in their paper, show that the infalling gas eventually settles down to an adiabatic atmosphere. Undoubtedly, this model is a vast oversimplification of the full problem because it is much more likely that the gas falls in at the same time that the galaxies are forming. However, it seems plausible that the mixing processes will be equally effective in such situations and that the adiabatic models will give a reasonable description of the distribution of gas in the cluster.

A second set of models was run in order to investigate the effects of cooling. These show that cooling begins in the centre of the cluster where the density and temperature are highest and that when it does happen it is catastrophic in the sense that the resulting inflow of gas into the central regions takes place on the same time scale as the cooling. Thus once cooling begins all the gas condenses into the central regions.

For the richness class 2 clusters which are observed as X-ray sources, the cooling times are of the order of 10^{11} years and the hydrodynamic time scale to establish the adiabatic atmosphere only about 10^9 years according to the present models. It is interesting that significant variations from these values ($T_0 = 6\times10^8$K; $n_0 = 10^{-3}$cm^{-3}) would either result in excess cooling in timescales less than 10^{10} years or else a much smaller X-ray luminosity since the emission is proportional to $n_e^2 T_0^{-1/2}$ and the cooling time to $n_e^{-2} T_0^{-1/2}$.

The corollory to this result is that there must have been some clusters in which the gas cooled on time scales much less than 10^{10} years and the X-ray luminosities must have been correspondingly higher by factors of 100 or possibly more. Such clusters should be readily observable with X-ray sensitivities about an order of magnitude better than that of the UHURU survey. Indeed we cannot exclude the possibility that some of the unidentified UHURU high latitude sources are clusters at large redshifts which cooled on time-scales of the order of 10^9 years.

On general grounds it is clear that cluster formation must result in vast energy deposition and it is likely that a significant fraction of this energy will be emitted at X-ray wavelengths. This must be one of the most exciting perspectives for the future of X-ray astronomy - that one may be able to study clusters forming and cooling at large redshifts.

As a final note, it is interesting that the spectrum of the X-ray emission of clusters shown in Fig.3 is strikingly similar to that of the X-ray background. Fig.3 illustrates a comparison of the predicted spectrum of cluster X-rays sources with the observed spectrum of the Perseus cluster. The corresponding densities and temperatures are

$$T_\infty = 0 \quad ; \quad n_0 = 5.6 \times 10^{-4} cm^{-3}$$
$$T_\infty = -0,25 \; ; \quad n_0 = 5.2 \times 10^{-3} cm^{-3}$$

At low energies ($\mathcal{E}$<20 keV) the spectral slope is $I(\mathcal{E}) \propto \mathcal{E}^{-\alpha}$, $\alpha \sim 0.5$, which is very close to the observed figure. Notice that the spectrum is not that of thermal bremsstrahlung at a single temperature but is the result of integrating the temperature and density distributions throughout the cluster. There is an abrupt cut-off at about 40 keV which corresponds to the depth of the gravitational potential well of a rich cluster of galaxies. This corresponds closely to the observed break in the spectrum of the X-ray background. If the present value of the space density of richness class 2 Abell clusters is used, the estimates of background intensity fall short by a considerable factor. However, at large redshifts their behaviour may have been quite different in that all Abell clusters may have been more powerful emitters and certainly all clusters must have been able to lose their binding energies. It is the points of principle which are important - clusters must cool at large redshifts and within each cluster a range of temperatures will be found: there must be a break in the X-ray spectrum corresponding to the X-ray energy at which the thermal energy of the electrons is equal to the gravitational potential energy which the matter gains on falling into a rich cluster.

Future X-ray observations will reveal whether any of these exciting possibilities will be realised.

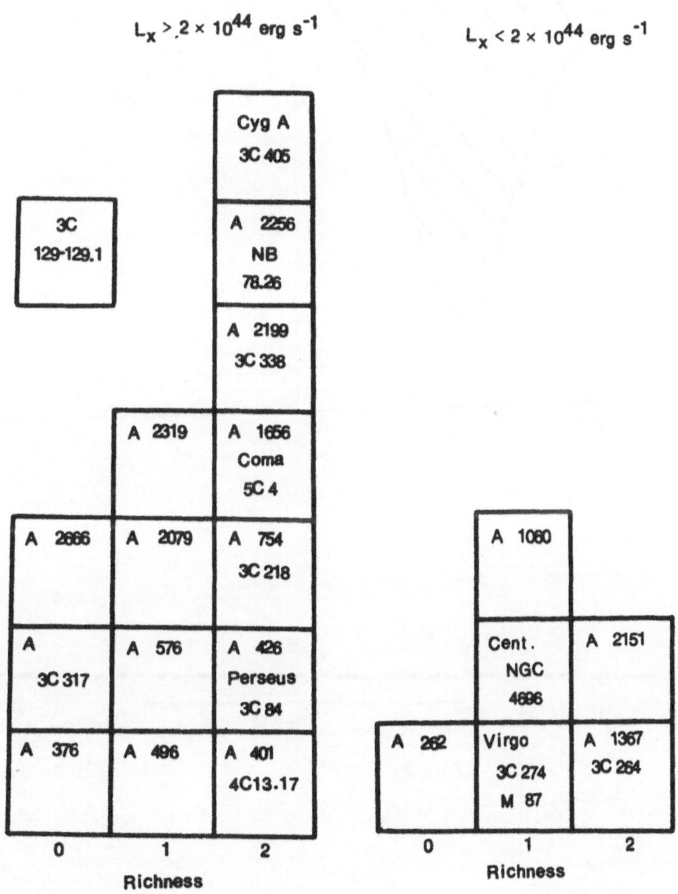

The Distribution of Strong and Weak Cluster X-ray Sources as a function of Richness Class. The richness of the cluster around 3C 129 + 129.1 is not known. "A" stands for Abell.

Fig. 1.

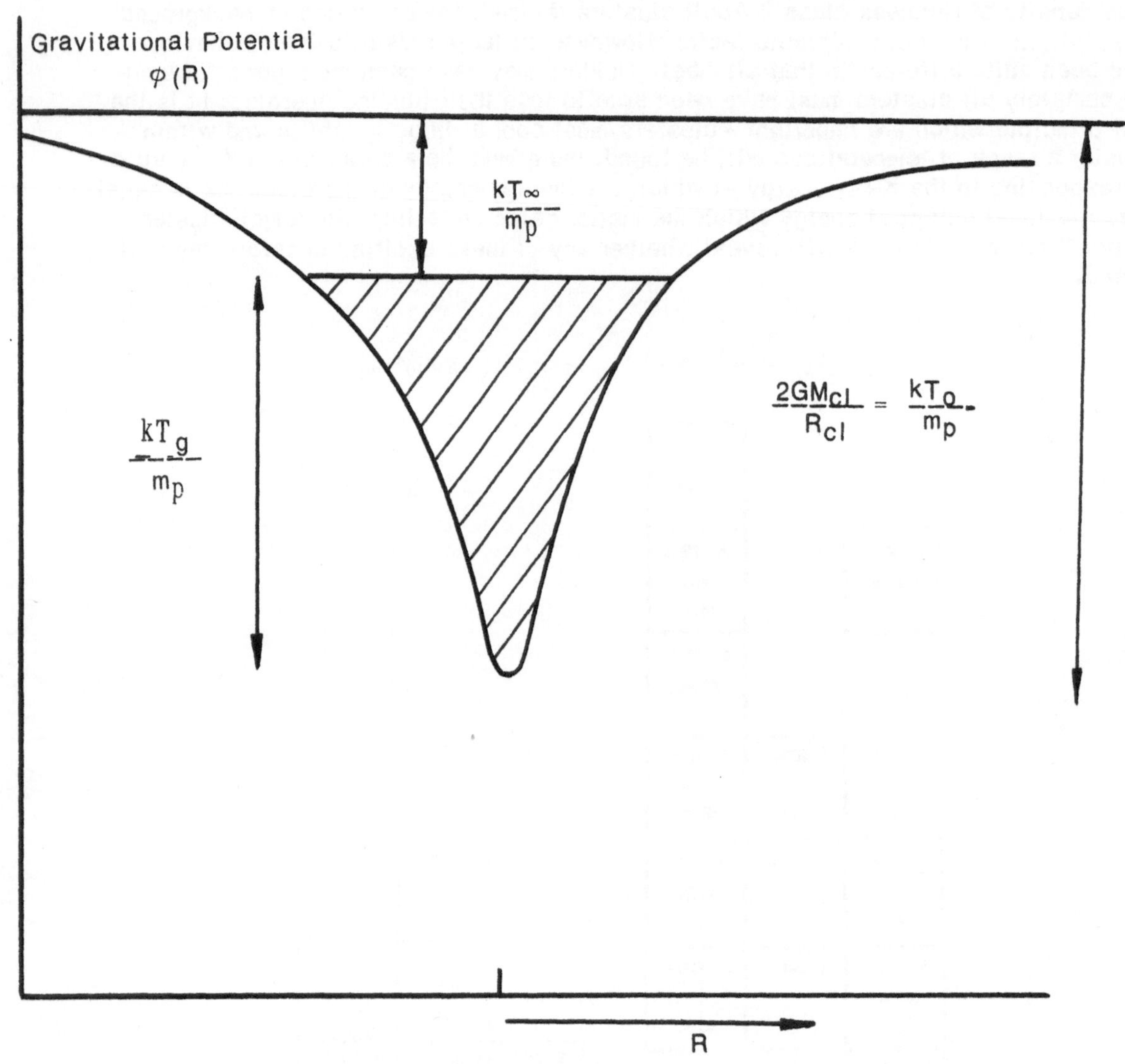
Gravitational Potential
φ (R)
$\frac{kT_\infty}{m_p}$
$\frac{kT_g}{m_p}$
$\frac{2GM_{cl}}{R_{cl}} = \frac{kT_o}{m_p}$
R

Fig. 2.

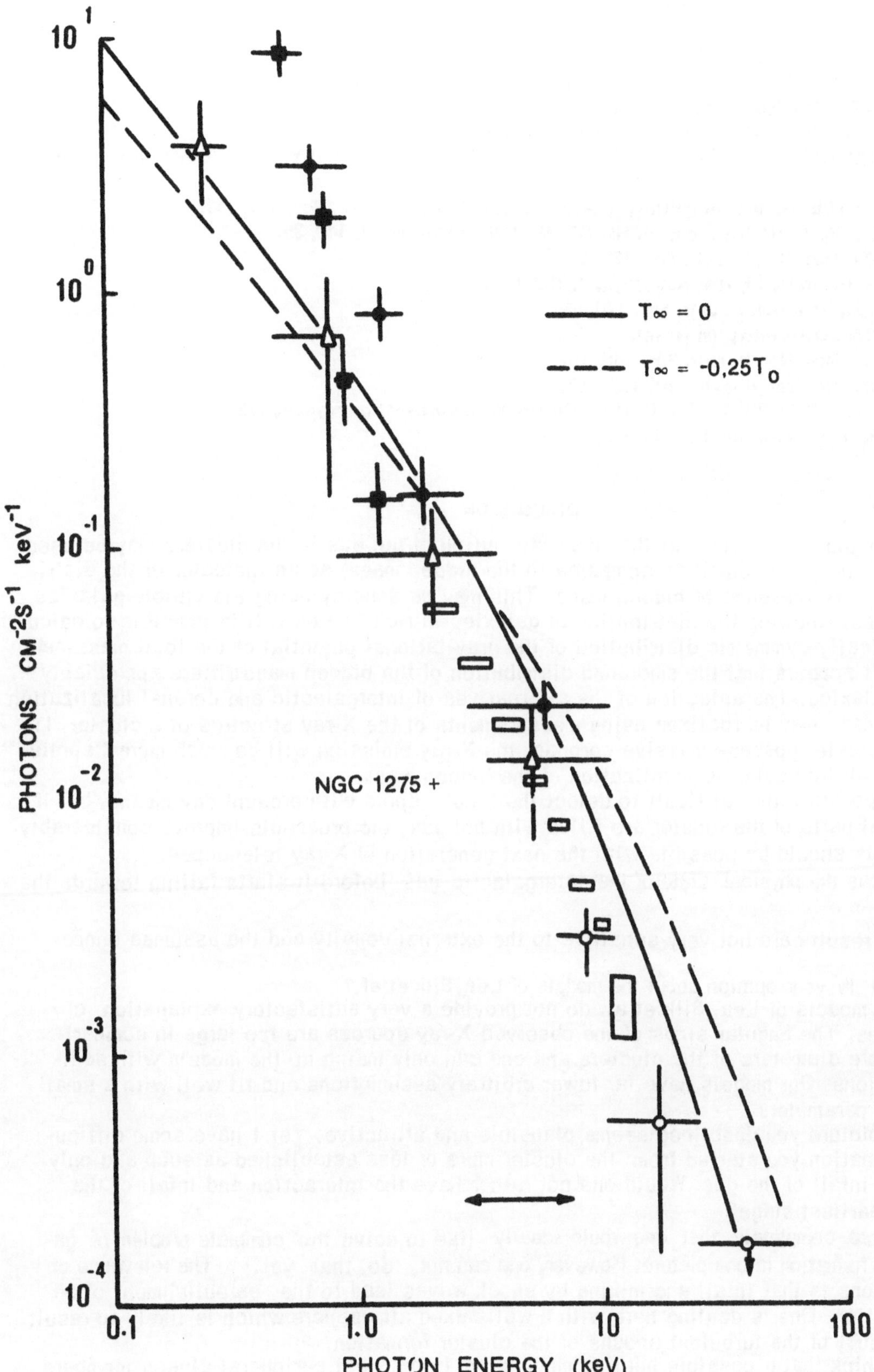

T∞ = 0
T∞ = -0,25T$_0$
NGC 1275 +
PHOTONS CM^{-2}S^{-1} keV^{-1}
PHOTON ENERGY (keV)
10^1
10^0
10^{-1}
10^{-2}
10^{-3}
10^{-4}
0.1
1.0
10
100

Fig. 3.

REFERENCES

ABELL G.O. 1958, Astrophys. J. Suppl. 3, 211.

BAHCALL N.A. 1974, Astrophys. J. 193, 529.

BALDWIN J.E. and SCOTT P.F. 1973, Mon. Not. R. astr. Soc. 165, 259.

GULL S.F. and NORTHOVER K.J.E. 1973, Nature. 244, 80.

GULL S.F. and NORTHOVER K.J.E. 1975, Mon. Not. R. astr. Soc. (in press).

HOOLEY T., 1974, Mon. Not. R. astr. Soc. 166, 259.

JAFFE W.J. and PEROLA G.C. 1973, Astr. Astrophys. 26, 423.

KELLOGG E., MURRAY S., GIACCONI R., TANANBAUM H. and GURSKY H. 1973, Astrophys. J. 185, L 13.

LONGAIR M.S., RYLE M. and SCHEUER P.A.G. 1973, Mon. Not. R. astr. Soc. 164, 243.

MATTHEWS T.A., MORGAN W.W. and SCHMIDT M. 1964, Astrophys. J. 140, 35.

MCHARDY I. 1974, Mon. Not. R. astr. Soc. 169, 527.

PARIJSKIJ 1973, Radiofiz. 16, 784; Astrophys. J. 180, L 47.

RILEY J.M. 1975a, Mon. Not. R. astr. Soc. 170, 53.

RILEY J.M. 1975b, Observatory, (in press).

SLINGO A. 1974a, Mon. Not. R. astr. Soc. 166, 101.

SLINGO A. 1974b, Mon. Not. R. astr. Soc. 168, 307.

SYUNAYEV R.A. and ZEL'DOVICH Ya. B. 1972, Comments Astrophys: Sp. Phys. 4, 173.

WILLSON M.A.G. 1970, Mon. Not. R. astr. Soc. 151, 1.

DISCUSSION

OZERNOY: I should like to suggest that the distribution of hot gas in the clusters may be used (if the mass of the gas is small as compared to the hidden mass) as an indicator of the distribution and the very presence of hidden mass. This may be done by using the visible galaxies as test particles. Knowing the distribution of galaxies in rich clusters, it is possible to calculate the spherically-symmetric distribution of the gravitational potential of the total mass (mainly hidden). It appears that the smoothed distribution of the hidden mass differs appreciably from that of galaxies. The selection of the alternatives of intergalactic and coronal localization of the hidden mass may be realized using measurements of the X-ray structure of a cluster. If galaxies in a cluster possess massive coronae, the X-ray emission will be much more "spotted" as compared with intergalactic localization of the hidden mass.

LONGAIR: It appears rather difficult to detect the "hot" spots with present day means, buf if only the central parts of the cluster are filled with hot gas, the prospects improve considerably. The experiments should be possible with the next generation of X-ray telescopes.

CHERNIN: What is the physical state of the intergalactic gas before it starts falling towards the cluster?

LONGAIR: The results are not very sensitive to the external density and the assumed temperature is zero.

CHERNIN: What is your opinion about the models of Lea, Silk et al.?

LONGAIR: The models of Lea, Silk et al. do not provide a very satisfactory explanation of the observations. The angular sizes of the observed X-ray sources are too large in comparison with the core diameters of the clusters and one can only match up the models with additional assumptions. Our models have far fewer arbitrary assumptions and fit well with a small number of free parameters.

BLAAUW: The picture you described seems plausible and attractive. Yet I have some difficulty with the situation you started from: the cluster more or less established as such and only then starts the infall of the gas. Would one not rather have the interaction and infall of the gas from very earliest stage?

LONGAIR: I agree completely that one would ideally like to solve the complete problem of galaxy and cluster formation in one picture. However, one can not do that yet. The relevance of these calculations is that infall and mixing by shock waves lead to the establishment of an adiabatic situation. One is dealing here with a well-mixed atmosphere which is likely to result as the end product of the turbulent epochs of the cluster formation.

KOMBERG: I think that a possible heating mechanism is that fast peripheral cluster members that are rich in gas pass through the central giant ellipticals.

LONGAIR: I agree that some of the X-ray sources are probably primarily associated with active galaxies, rather than the cluster itself. It must be remembered that the energy requirements are enormous. However, even in that case, it is likely that the gas distribution will be adiabatic, once the gas attains a high temperature.

OBSERVATIONS OF EXTRAGALACTIC RADIO SOURCES WITH THE CAMBRIDGE 5-KM TELESCOPE

P.J.HARGRAVE M.S.LONGAIR
MULLARD RADIO ASTRONOMY OBSERVATORY, CAVENDISH LABORATORY, U.K.
PRESENTED BY P.J.HARGRAVE

НАБЛЮДЕНИЯ ВНЕГАЛАКТИЧЕСКИХ РАДИОИСТОЧНИКОВ С КЭМБРИДЖСКИМ 5-КИЛОМЕТРОВЫМ ТЕЛЕСКОПОМ

РЕЗЮМЕ

Наблюдались более 100 дискретных внегалактических радиоисточников при разрешении 2" x 2" cosec δ. Они представляют собой типичные примеры внегалактических радиоисточников. Для многих из них выполнены поляризационные измерения.
Представленные наблюдения относятся главным образом к расширенным источникам. Рассмотрены статистические свойства, особенно, те, которые более доступны интерпретации. Наблюдения истолкованы в смысле критически оцененной модели непрерывного истечения. Описаны наблюдения, указывающие на корреляцию между центральной нетепловой радио и оптической эмиссиями.

ABSTRACT

More than 100 discrete extragalactic radio sources have been observed with a resolution of 2" x 2" cosecδ by the 5-km Telescope. These sources form a representative sample of the extragalactic radio source population. Polarization measurements have been made for many of these sources. The observations presented are mainly of extended sources. The statistical properties of this sample are surveyed, concentrating upon those parameters of most relevance to the interpretation. The observations are interpreted in terms of the continuous flow model of extragalactic radio sources which is critically appraised. Observations indicating a correlation between central non-thermal radio and optical emission are described.

1. I n t r o d u c t i o n. The Cambridge 5-km Telescope (Ryle 1972) has so far mapped about 120 extragalactic radio sources at 5 GHz with a resolution of 2" x 2" cosecδ. Pooley and Henbest (1974) have presented maps of 48 of these sources, and some others of particular interest such as Cygnus A, M82, M87 and 3C 66 have been published separately (Hargrave and Ryle 1974; Hargrave 1974; Turland 1975a; Northover 1973). For the purpose of this talk we shall concentrate on the most recent observations which are either in press or in preparation (Hargrave and McEllin 1975; Longair 1975; Riley and Pooley 1975; Turland 1975b). The maps reveal many new details of the structure of the sources, both in total intensity and linear polarization, and provide severe tests for some of the more plausible theories that have been proposed to explain double extragalactic sources.

2. T h e o b s e r v a t i o n s. To give some idea of the increased resolution provided by the 5-km Telescope over the previous best synthesis observations we present (Fig.1) the map of 3C 390.3 obtained by Harris (1972) at 5 GHz with the Cambridge One Mile Telescope. The map, which has a resolution of 6".5 arc, reveals a ''classic double radio source'' with two components on opposite sides of, and approximately collinear with, the optical identification. In this case it is an N galaxy, whose nucleus coincides with a compact central radio component. The map also shows evidence for a bridge of emission, and convolution reveals that it extends all the way between the outer components and contributes $\approx$45 per cent of the total 5-GHz flux density.

Fig. 2 is the new map of the Sf component obtained with the 5-km Telescope. It shows, as is also found in many other sources, a characteristic ''head-tail'' structure. of particular interest is the polarization map of this component which is shown in Fig.3. The peak of the polarized emission coincides with the peak of total intensity in the head; here it is 38 per cent polarized, but over much of the tail the polarization exceeds 30 per cent and the directions of the polarization vectors indicate a highly ordered magnetic field. Interpreted as polarized synchrotron radiation, the magnetic field must run tangential to the leading edge of this component. Since the maximum theoretical polarization from the synchrotron process is $\approx$75 per cent, these high percentages indicate that there cannot be much magnetic field structure on a scale smaller than the resolution of the instrument ($\approx$ 3 kpc).

The Np component is shown in Fig. 4; it is seen to consist of two ''hot-spots'' on a much lower surface brightness plateau. This illustrates another fairly common occurrence, the presence of multiple ''heads'' in a source component whose alignment seems to be unrelated to the position angle of the main source axis.

The presence of a central component coincident with the nucleus of the galaxy would, on some theoretical models, be taken as a by-product of an energy source which is continuously ''beaming'' energy to the ''heads''. A striking result from the 5-km observations is the large number of sources for which central components have been detected. They have been observed in 45 per cent of the sources which would be classified as ''doubles'' (Jenkins 1975) (see also Section 4). Also of relevance to those models in which energy is beamed to the heads is the relationship between the radio source axis and the axis of the optical galaxy. In the case of Cygnus A it was argued (Hargrave and Ryle 1974) that the radio source axis is aligned with the rotation axis of the galaxy. This is definitely not the case for 3C33 (Fig. 5). Here there is a weak central component coincident with the nucleus of a DE4 galaxy. Matthews, Morgan and Schmidt (1964) describe the galaxy as having a central nucleus with diameter 2.5 arc, surrounded by an elongated structure 8.4 x 3.3 arc, with minor axis along p.a. 163°. This is in turn surrounded by a faint envelope with major axis along p.a. 163°, which can be seen out to diameters of 22'' x 9".4 arc From the tilt of the emission lines, Matthews *et al.* conclude that the major axis of the outer envelope is the rotation axis of the galaxy, an axis which makes an angle of 36° with the radio source axis on the plane of the sky. It is very important that this type of optical work be extended to other radio galaxies.

Turland (1975b) has noticed the presence of so called ''jets'' near the central components of a few of the sources so far mapped. A good example of such a feature is shown in Fig.6; this is a map of 3C 219 on which is visible a compact source centred on the galaxy and a ''jet'' pointing along the axis of the source. The jet is shown in more detail in Fig. 7. This is the first time that such a jet has been associated with a classical double radio source. It is unresolved perpendicular to its axis, but has at least two maxima along its length. Previously such jets have only been observed in the ''non double'' sources M87, 3C 66B, 3C 273 and 3C 279. From the sources so far mapped with the 5-km Telescope, Turland cites 5 other ''double'' sources which may contain jet-like features: 3C 153, 207, 249.1, 434 and perhaps 3C 346. Such features might represent the usually invisible beam revealed by some mechanism which causes the entrainment of intergalactic matter, and they are therefore of much theoretical interest.

A further feature revealed by the observations is that many sources exhibit interesting asymmetries. Fig. 8 is the map of 3C 154 in which the outer components have a flux density ratio of 17:1. There are also sources which have their optical identification and/or central component significantly off the line joining the heads; an example of this is 3C 430 (Fig. 9). A particularly asymmetrical source is 3C 351 (Fig. 10). The lower map was obtained with the One Mile Telescope at 1.4 GHz (Mackay 1969) and shows two outer components with a flux density ratio of 8:1. The 5-km map reveals that the brighter of these two components is itself double, with a flux density ratio of 1.8 : 1, separated along a line almost orthogonal to the main source axis.

Lest the impression may be given that all sources conform to the picture of ''hot-spots'' on lower surface brightness regions, we present in Fig. 11 the map of 3C 332 in which there is no evidence for compact features in the components.

Finally, Fig. 12 is a map of the ''radiotrail'' galaxy 3C 83.1B. The lower map is with the full resolution of the 5-km, and in the upper map the data have been convolved in right ascension to produce a circular beam. The most significant feature is the lack of any compact structure along the two radio ''trails'' which extend away from the nucleus of the associated galaxy is marked with a cross. The only compact feature is an unresolved component, with size <1'' arc, coincident with this nucleus. The ''trails'' also appear to be well collimated in that there is no evidence for any emission ioining them in the region immediately to the North of the galactic nucleus.

To summarize the observational evidence, the picture of double radio sources which is emerging is one in which the outer components consist of one or more compact high surface brightness ''head'' regions together with lower surface brightness tails. These regions are often found to exhibit strong linear polarization. In a high proportion of cases central components are found coincident with the optical identification. Many sources, however, show interesting asymmetries.

3. Interpretation. There are very large gaps in the theory of extragalactic radio sources and so in interpreting the results we will concentrate upon those aspects of the theory most readily testable by the new observations. These concern the overall dynamical behaviour of sources and the relation between the compact features and the lower brightness tails and bridges observed in many sources.

The observations of Cygnus A (Hargrave and Ryle 1974) suggested that some form of continuous flow model must be adopted. In this model, energy is continually generated in the form of relativistic particles in the compact heads and then escapes into the low brightness tails or bridges. The energy in the tails and bridges eventually comes pressure balance with the external pressure of the intergalactic gas. The energy supply to the heads advances through the intergalactic medium. either by ''burning'' its way through as a relativistic jet or possibly as a discrete object.

In the case of Cygnus A, the following observational pieces of evidence favoured this scenario. (i) the spectral indices of the head components are flatter than those of the tail regions suggesting that the particles remain in the heads for shorter times than the light travel time from the galactic nucleus, which is the presumed ultimate source of energy for the source. (ii) the energy densities in the heads are too large to be contained by the most efficient containment mechanisms. The cold matter densities derived from the polarization information were too small to give effective confinement. Therefore the energy must escape from the heads in times much shorter than the age of the source and hence the need for a continuous replenishment of the particle supply to the heads. (iii) Granted that escape of particles does take place from the heads into the tails the energy requirements of the tails could be supplied by such energy transfer. (iv) X-ray observations of Cygnus A were consistent with the presence of hot intergalactic gas in the vicinity of the radio source of sufficient pressure to confine the tail regions.

One of the major programmes of the 5-km Telescope has been to test this model for other double extragalactic radio sources since there is no guarantee that Cygnus A is indeed the typical extragalactic double radio source. 5 more sources have now been studied in much more depth using a refined version of the continuous-flow model developed by Hargrave and McEllin (1975). These sources are 3C 33, 61.1, 379.1, 390,3 (Hargrave and McEllin 1975) and 3C 219 (Turland 1975b). All of these sources possess compact features in the double source components and consequently high internal energy densities, together with strong tails and bridges associated with the components. Three of them, 3C 33, 219 and 390.3, also possess compact central radio components suggestive of continuous energy supply.

To summarise the results of this investigation, we compare the above results for Cygnus A with those obtained for the 5 sources.

(i) Spectral index information. Only in the case of 3C 61.1 is there any evidence that the spectral indices of the heads may be smaller than those of the tails or bridges. In general, for the other sources the errors on the spectral indices are too large to make definite statements. The major observational problem is the differing resolutions and sensitivities with which the sources are observed at different frequencies and by different observatories.

(ii) *Component confinement.* In all of the compact source components, it is demonstrated that ram pressure and inertial confinement are inadequate to account for the presence of such high energy densities and therefore the particles must escape from the heads in times short compared with the source lifetime.

(iii) *Energy supply to the tails and bridges.* The improved models of energy injection into the tails and bridges take account of the finite velocity of escape of material from the heads to the bridges and give a proper description of the adiabatic losses from heads to bridges (Hargrave and McEllin 1975). These models indicate that there is no problem in accounting for the observed radio emission from tails and bridges as originating in the compact heads with continuous energy supply.

(iv) *X-ray emission from double radio sources.* There is X-ray emission detected only from 3C 390.3 (Charles, Longair and Sanford 1975); the large column densities of neutral hydrogen inferred from the spectrum and the variability of the X-ray emission argue againts the association of this emission with the hot cluster gas. However, the ambient intergalactic particle densities and temperatures needed to confine the low radio brightness regions are such that the lack of thermal X-ray emission from the vicinity of the 5 radio sources poses no problem for the model.

How far do these new observations provide support for the continuous flow model? Hargrave and McEllin have shown that there is no inconsistence with the model but of the above results, only (ii) provides definite evidence for the necessity of continuous replenishment of the energy in the compact features. The most important areas for testing the model further lie in refining (i) by further observations of higher sensitivity using instruments of similar resolution and by making more detailed polarization measurements at a number of frequencies. Both of these programmes are now in hand at MRAO.

In the meantime, the continuous-flow models seem the simplest explanation of the essential features of the observations. Typical parameters for the model on the basis of the 6 sources for which it has been tested in detail are :

Ambient intergalactic gas density and temperature	$n_{1G} \sim 10^{-3}\ cm^{-3}$, $T \sim 10^{8} K$
Cold matter density in compact features	$n_e \sim 10^{-4}$-$10^{-3}\ cm^{-3}$
Magnetic field in head	$B \sim 10^{-5}$-10^{-4} G
Magnetic field in bridges	$B \sim 5 \times 10^{-6}$ G
Velocity of advance of heads through intergalactic gas	$v \sim 0.01 - 0.03c$
Escape velocity of relativistic matter from heads	$v_s \sim 0.1c$
Overall lifetime of radio source	$t \sim 10^{7}$ years
Rate of energy supply to heads	$W \sim 10^{45}$ ergs s^{-1}

The model leaves a number of features unexplained and whilst one can rationalize their existence in terms of the model, these are best regarded as unsolved problems.

a) *Multiple Compact Features in Heads.* These features are now found frequently (e.g. Cygnus A, 3C 390.3 and 3C 351). They may be interpreted in a variety of ways. For example, they may represent separate points at which the beam supplying energy from nucleus encounters the intergalactic gas or they may result from large scale instabilities in the interaction of a single beam with the intergalactic gas.

b) *Gross Asymmetries in Radio Source Structure.* In some sources, the component intensities are very different and they are located at very different distances from the galactic nucleus. Sometimes the component morphology is entirely different, some components containing compact features with diffuse tails, others being amorphous *in the same source* (e.g.3C 300). This may be ascribed to differing conditions in the intergalactic gas in the vicinity of the radio source or possibly to asymmetry in the ejection of materials from the nucleus. It is striking that the jets observed in radio sources are all asymmetric (Turland 1975b). These propositions are no more than guesses and to find if there is any truth in them would require optical and X-ray observations of high sensitivity. It is to be hoped that some of the more grossly asymmetric nearby sources will command the attention of optical and X-ray astronomers.

c) *Non-aligned Radio Structure.* Very often the whole radio structure is displaced with respect to any symmetry axis of the Galaxy and the radio source. In some extreme cases, the radio trail sources (e.g. 3C 83.1B, NGC 1265 Fig.12), this is attributed to motion of the radio galaxy with respect to the intergalactic gas. If this is indeed the case, it would seem natural to account for many of the less extreme non-aligned radio structures by similar relative motions. Small scale irregularities may be due to turbulent velocities in the intergalactic gas, possibly due to the passage of galaxies through the gas; large scale irregularities may be due to large scale motions in the gas, possibly associated with rotation of the gas in a cluster. These are no more than guesses which badly need independent evidence as test of their plausibility.

d) *Jets and the Source of Energy.* Turland (1975b) makes the point that if indeed energy is supplied continuously from a galactic nucleus it is surprising that more jets such as that seen in 3C 219 (Fig.7) have not been seen. This negative result may be due to the stability of beams or it may simply indicate that the whole picture is wrong. More high resolution observations of high sensitivity are needed to clarify this point as well as a better understanding of the stability of relativistic beams.

The ultimate source of energy is inferred to be the galactic nucleus from the observation of compact central radio components. VLBI observations in conjunction with optical studies would seem to give the best chance of studying the nature of these obscure regions.

4. Unidentified radio sources - why are there no central components? A new fact concerning double radio sources has arisen as a by-product of the study of unidentified radio sources (Longair 1975) A complete sample of 46 3CR radio sources had been investigated by Longair and Gunn (1975) A complete faint optical magnitudes using deep plates taken with the 200-inch Hale Telescope. 10 sources remained unidentified at the limit of these plates (m ~ 23.5 - 24). It was noted that the radio structures of these unidentified sources were poorly known and observations of all of them have now been completed with the 5-km Telescope. 9 of the 10 sources have been found to be classic double sources with angular sizes in the range 6'' - 60''. Only one probable new identification was found on the basis of the new structural data 3C 41 A striking feature of this sample of sources is the total *absence of central radio components*. Since the sources are unidentified at m ~ 23.5, a conservative lower limit to their redshifts is 0.4 and therefore we can compare their radio properties with those of sources which must have comparably high radio luminosities. This comparison for all sources observed with the 5-km Telescope is shown in Fig. 13. It is apparent that among identified radio sources, compact central radio components are very frequently observed, especially if the optical counterpart has a strong non-thermal optical component. This contrast with the situation for unidentified sources for which no central components are found. It is natural to associate the unidentified radio sources with radio galaxies and then the present result suggests a correlation between the presence of *non-thermal radio and optical emission from the parent object.* This result is similar to that of Wall (1975) who has shown that it is possible to identify 95 per cent of radio sources observed at 2700 MHz which have flat radio spectra. The present result suggests that this is also true of the parent objects of double radio sources. These results suggest a physical correlation between the intensity of non-thermal radio and optical emission in the central regions of powerful extragalactic radio sources where it is assumed the energy generating events take place.

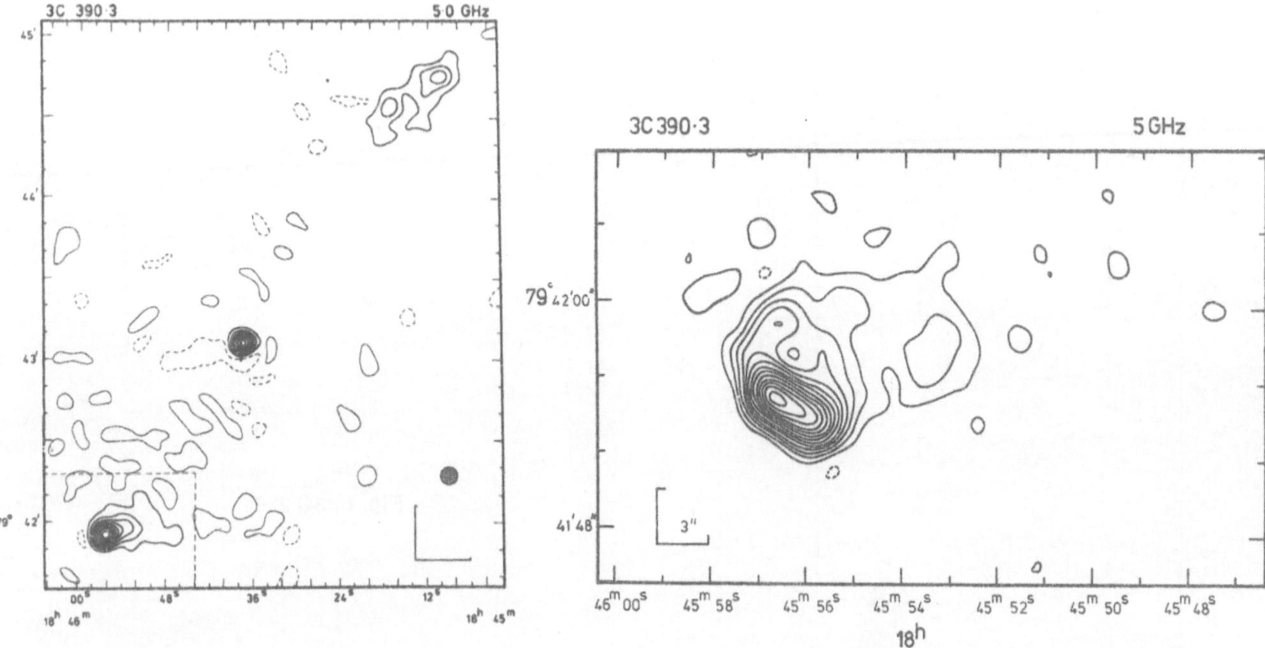

Fig.1. The map of 3C 390.3 obtained by Harris (1972) with the One Mile Telescope. For the sake of clarity every other contour has been omitted within the rectangle marked around the southernly component.

Fig. 2. The Sf component of 3C 390.3 mapped with the 5-km Telescope

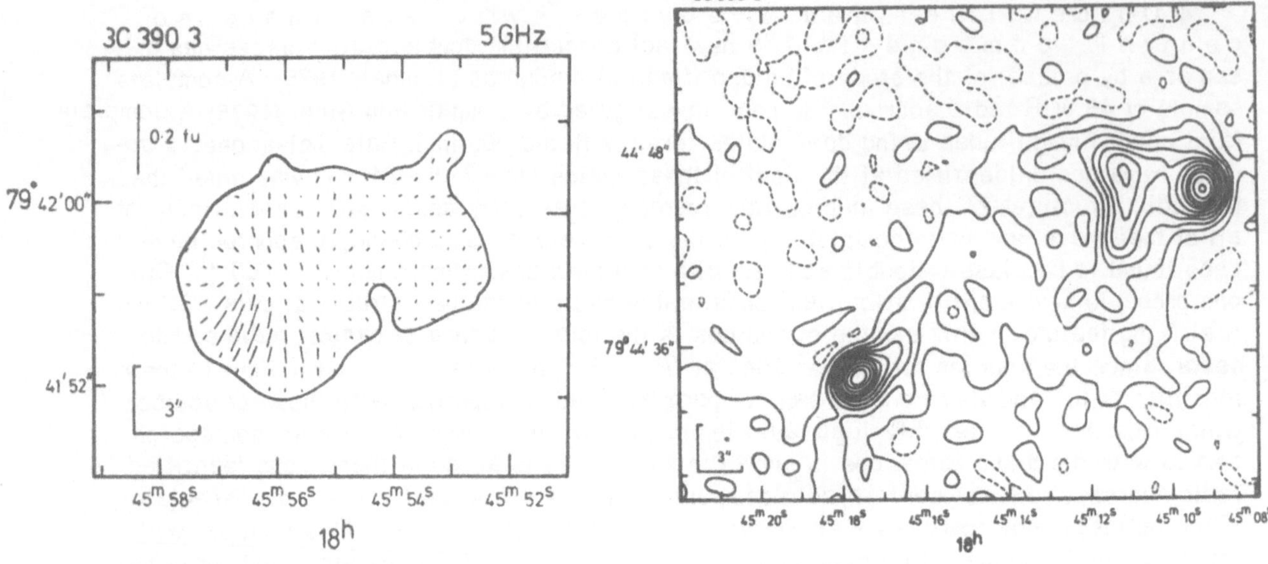

Fig.3. The polarization structure of the Sf component of 3C 390.3

Fig. 4. The Np component of 3C 390.3

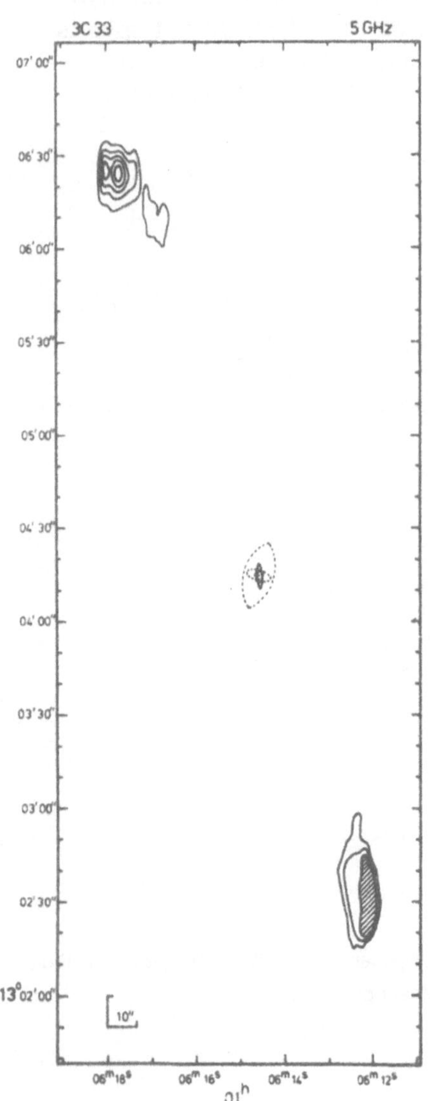

Fig.5. The map of 3C 33. The dotted lines surrounding the central component are a schematic representation of the optical galaxy described by Matthews, Morgan and Schmidt (1964)

3C219 5GHz

Fig. 6. 3C 219

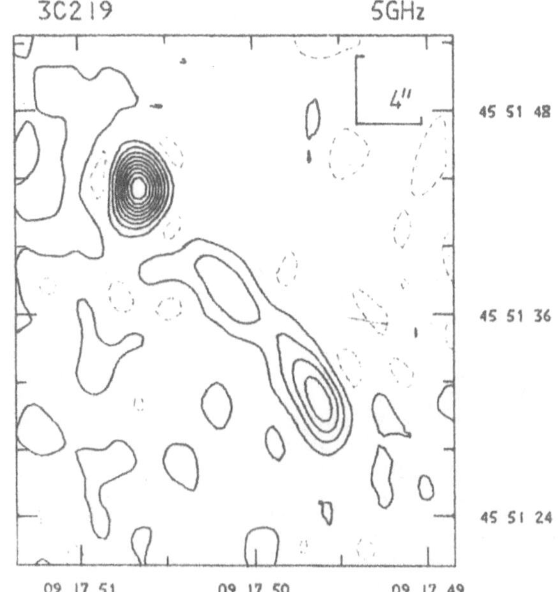

Fig. 7. Detail of the central component and jet in 3C 219

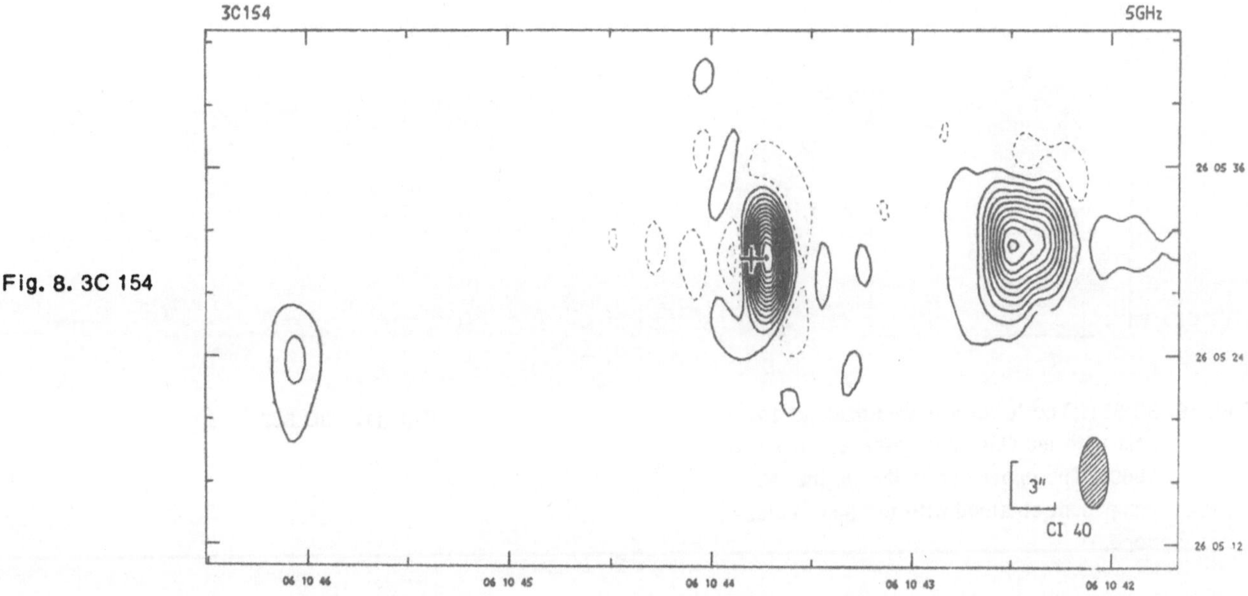

Fig. 8. 3C 154

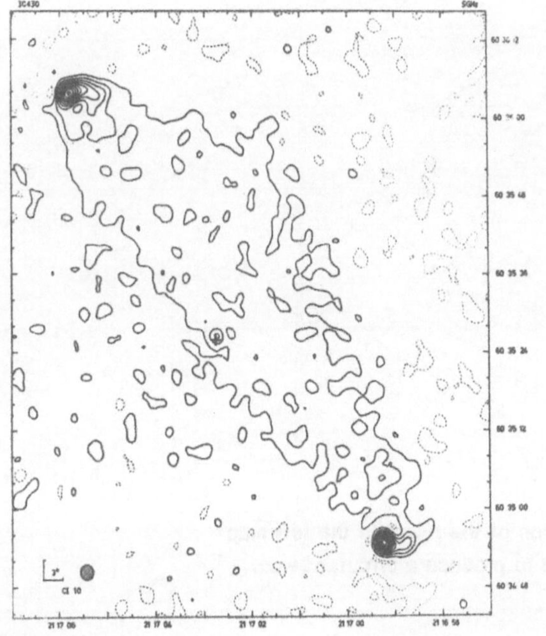

Fig. 9. 3C 430

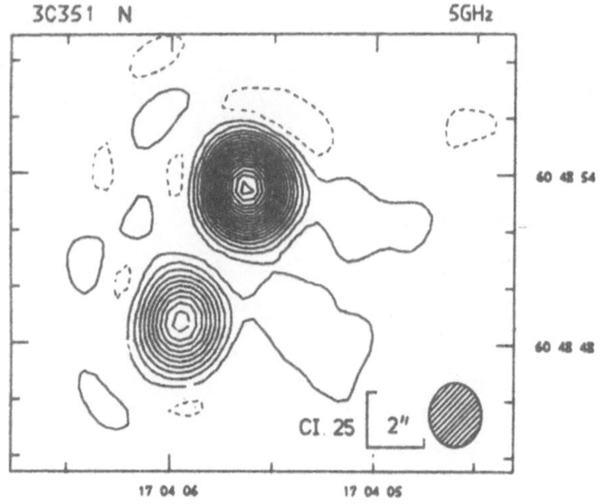

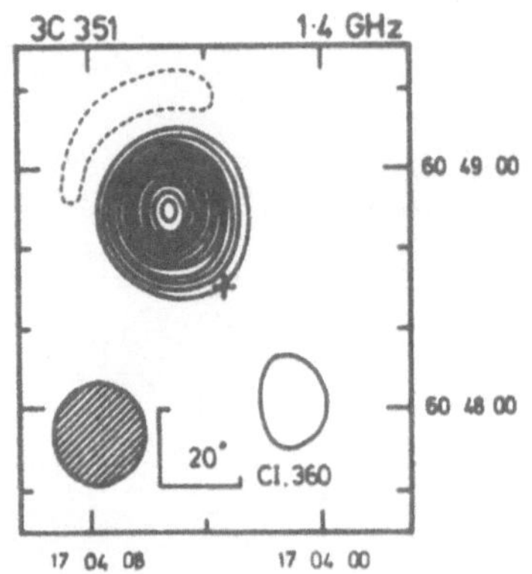

Fig. 10. 3C 351. The lower map was made at 1.4 GHz with the One Mile Telescope (Mackay 1969). The upper map is that of the Np component obtained with the 5-km Telescope.

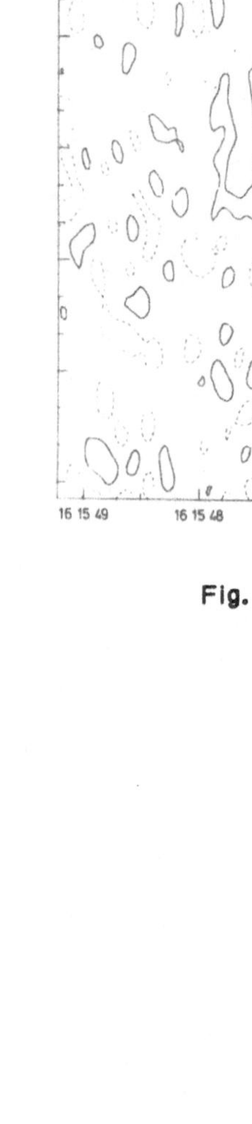

Fig. 11. 3C 332

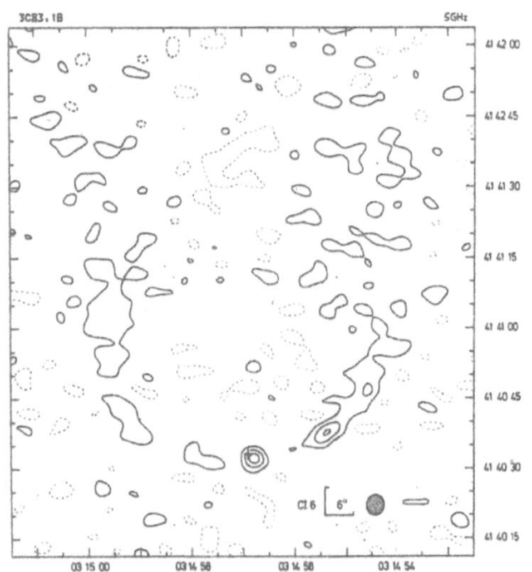

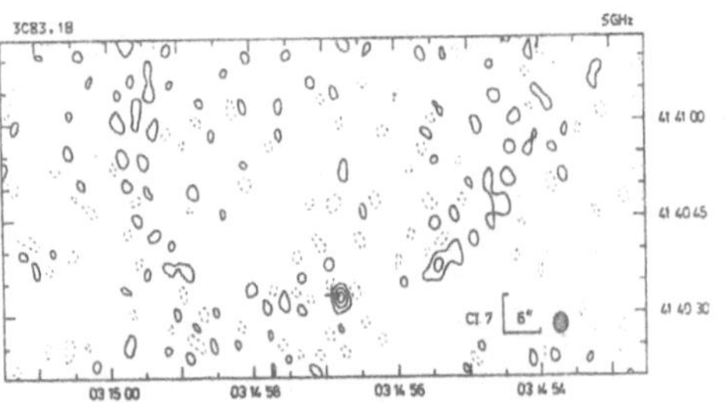

Fig. 12. 3C 83.1B. The right map is with the full resolution of the 5-km. In the left map the data have been convolved in right ascension to produce a circular beam.

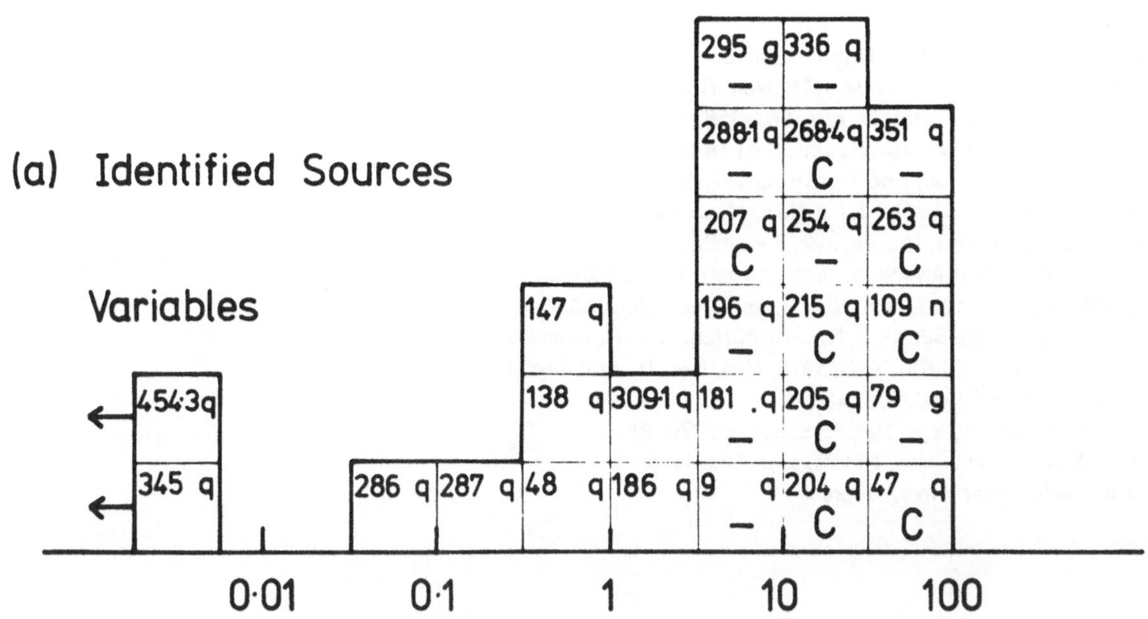

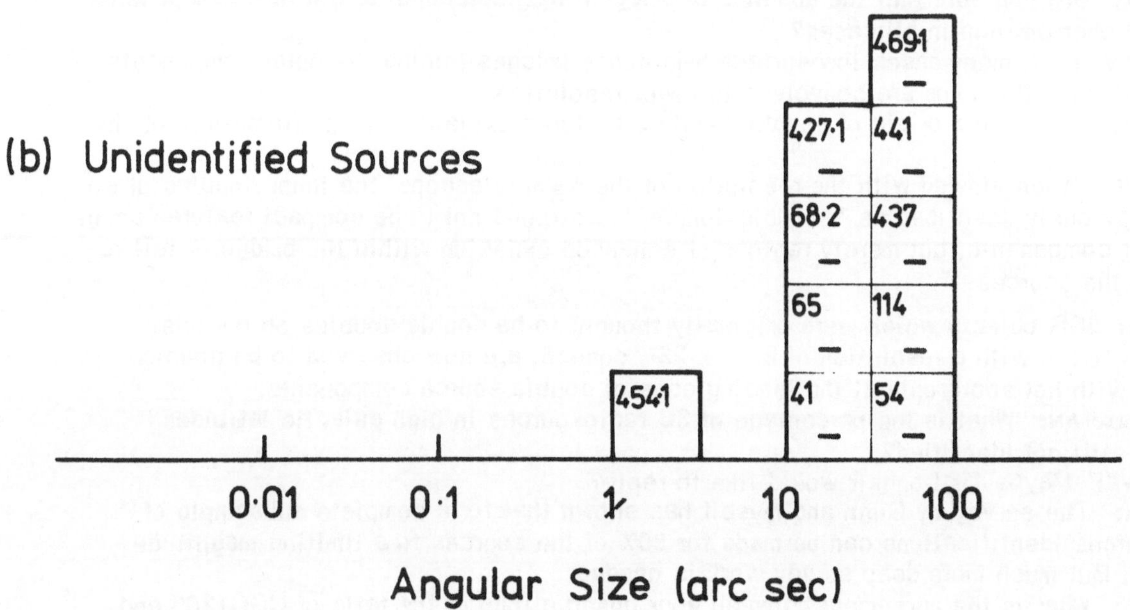

Fig. 13. Histograms of the largest angular size for (a) the sample of identified sources and (b) the sample of 10 unidentified sources. The sample of identified sources consists of all those in the complete 3CR 200 sample which have been observed by the 5-km Telescope and which have measured redshifts with $P_{178} > 5.3 \times 10^{26}$ W Hz^{-1} sr^{-1}. Legend : q = quasar ; n = N galaxy; g = radio galaxy; C signifies the presence of a compact central component. – signifies no central component.

REFERENCES

CHARLES P.A., LONGAIR M.S. and SANFORD P. 1975, Mon. Not. R. astr. Soc. 170, 17
HARGRAVE P.J. and RYLE M. 1974, Mon. Not. R. astr. Soc. 166, 305
HARGRAVE P.J. 1974, Mon. Not. R. astr. Soc. 168, 491
HARGRAVE P.J. and McELLIN M. 1975, Mon. Not. R. astr. Soc. In press
HARRIS A.B 1972, Mon. Not. R. astr. Soc. 158, 1
JENKINS C.J. 1975, Mon. Not. R. astr. Soc. In preparation
LONGAIR M.S. 1975, Mon. Not. R. astr. Soc. In press
LONGAIR M.S. AND GUNN J.E. 1975, Mon. Not. R. astr. Soc. 170, 121
MACKAY C.D. 1969. Mon. Not. R. astr. Soc. 145, 31
MATTHEWS T.A., MORGAN W.W. and SCHMIDT M. 1964, Astrophys J. 140. 35
NORTHOVER K.J.E. 1973, Mon. Not. R. astr. Soc. 165, 369
POOLEY G.G. and HENBEST S.N. 1974, Mon. Not. R. astr. Soc. 169, 477
RILEY J.M. and POOLEY G.G. 1975, Mon. Not. R. astr. Soc. In preparation
RYLE M. 1972, Nature Lond. 239, 435.
TURLAND B.D. 1975a, Non. Not. R. astr. Soc. 170, 281.
TURLAND B.D. 1975b, Mon. Not. R. astr. Soc. In press
WALL J.V. 1975, Observatory. In press

DISCUSSION

OZERNOY: Are you sure that the absence of bridges between small bright radiocomponents is a real phenomenon in all cases?

HARGRAVE: In many cases low surface brightness bridges joining the outer components are found when the maps are convolved to lower resolutions.

OZERNOY: How is the origin of double-double structures explained in the framework of the beam model?

HARGRAVE: When viewed with the resolution of the 5-km telescope, the inner doubles of sources previously described as ''double-double'' are found not to be compact features as in the outer components, but merely regions of enhanced emission within the bridge or tail regions of the sources.

LONGAIR: 3CR objects which were originally thought to be double-doubles on the basis of observations with a resolution of 23'' x 23''cosecδ, are now observed to be double sources with hot spots only at the leading edge of double source components.

AMBARTSUMIAN: What is the percentage of 3C radiosources in high galactic latitudes that are still not identified?

HARGRAVE: Maybe Dr.Longair would like to reply?

LONGAIR: The survey by Gunn and myself has shown that for a complete subsample of 3CR sources identifications can be made for 80% of the sources to a limiting magnitude of $23^{m}_{.}5$. But much more deep survey work is needed.

BERTOLA: What is the agreement between your observations of the tails of NGC1265 and the Westerbork observations?

HARGRAVE: Our new map has a resolution that is about three times better. It shows that the tails do not contain compact ''blobs'', but rather are diffuse features.

SOME RESULTS OF SPECTRAL OBSERVATIONS OF GALAXIES OF HIGH SURFACE BRIGHTNESS

M.A.ARAKELIAN
BYURAKAN ASTROPHYSICAL OBSERVATORY, U.S.S.R.

НЕКОТОРЫЕ РЕЗУЛЬТАТЫ СПЕКТРАЛЬНЫХ НАБЛЮДЕНИЙ ГАЛАКТИК ВЫСОКОЙ ПОВЕРХНОСТНОЙ ЯРКОСТИ

Statistical investigation of some types of extragalactic objects with active nuclei had revealed definite correlation between the nuclear activity of galaxies from the one hand and the gradient of surface brightness, or mean surface brightness itself from the other hand (Zasov and Lyutyj, 1973; Arakelian 1972, 1974a; Arakelian and Balkowski 1975). Proceeding from the existence of such correlation Arakelian (1975) compiled the list of about 600 galaxies with the mean surface brightness which is not less than $22^m.0$ from square second of arc in the system roughly coinciding with the Holmberg's one. The apparent magnitudes M_p by Zwicky et al. (1961, 1963, 1965, 1966, 1968a,b) Catalogue of Galaxies and of Clusters of Galaxies (CGCG) and angular sizes by Vorontsov-Vel'yaminov et al. (1962, 1963, 1964) Morphological Catalogue of Galaxies (MCG) have been used for computation of the mean surface brightness.

600 galaxies with high surface brightness are located in the area of nearly 15000 square degrees (north hemisphere excluding galactic lane with $|b^{II}| < 20^o$) and have been selected from nearly 15000 galaxies of this area for which relevant data are given in both catalogues. Thus, the mentioned galaxies compose nearly 4 per cent of all the galaxies considered.

The high percentage of Markarian, Zwicky and Haro galaxies among the galaxies with high surface brightness must be emphasized. Indeed, 200 galaxies from Markarian lists (Markarian 1967, 1969a,b; Markarian and Lipovetsky 1971, 1972, 1973, 1974), 220 galaxies from Zwicky (1971) Catalogue of Selected Compact Galaxies and Post-Eruptive Galaxies and 24 galaxies from Haro's (1956) list of blue galaxies were among 15000 objects for which the mean surface brightness had been computed. Meanwhile our list of galaxies with high surface brightness contains 34 Markarian galaxies, 34 Zwicky galaxies and 7 Haro galaxies (several cases of coincidence of these galaxies exist). Thus, one can see that the percentage of the mentioned galaxies among galaxies of high surface brightness exceeds significantly the value corresponding to randomly selected objects (~4 per cent). This fact is not surprising, since all the mentioned types of galaxies are known as ones possesing high surface brightness.

The spectral observations of 320 galaxies with high surface brightness have been carried out in 1974, 1975 by the use of 125-cm reflector of Sternberg Institute Crimean Station (Arakelian, Dibaj and Yesipov 1975a,b, 1976; Doroshenko and Terebizh 1976). The emission lines were detected and redshifts were measured in the spectra of nearly 55 per cent of the observed objects. I should like to discuss now two peculiarities connected with the statistics of luminosities and the statistics of Seyfert type objects.

1. T h e s t a t i s t i c s o f l u m i n o s i t i e s. Let us consider the distribution of absolute magnitudes of 120 galaxies of high surface brightness for which the redshifts are measured till now (Figure 1a). For comparison the distribution of absolute magnitudes of 192 galaxies with $M_p \geqslant 13.5$ for which the radial velocities are given in CGCG is presented in Figure 1b. Finally the distribution of absolute magnitudes of all 608 objects of the same catalogue for which the radial velocities are given is presented in Figure 1c. (The members of Local Group were not included into these samples) In all cases the absolute magnitudes have been computed by the use of apparent magnitudes given in CGCG with Hubble's constant $H = 75$ km sec^{-1} Mpc^{-1} and with correction for absorption in Galaxy -0.25 cosec $|b^{II}|$. The galaxies with radial velocities less than 600 km sec^{-1} were considered as being at the distance of 8 Mpc and several double galaxies were considered as two objects with twice less luminosities.

The differences between Figure Ia and two other ones are quite obvious. Though the difference between Figures 1b and 1c is not strong, it seems more reasonable to compare Figures 1a and 1b since the bulk of galaxies with high surface brightness have $m_p \geqslant 13.5$. The mean values of absolute magnitudes and their dispersions for three samples considered are respectively

$$< M_p > = -20.0, \sigma^2 (M_p) = 3.08 \quad \text{for Ia}$$

$$< M_p > = -19.0, \sigma^2 (M_p) = 1.52 \quad \text{for Ib}$$

$$< M_p > = -19.8, \sigma^2 (M_p) = 1.34 \quad \text{for 1c}$$

Though the small differences between the mean values are not statistically significant,the distributions Ia and Ib do differ with high level of significance. The significance of difference between Ia and 1b was established by the use of both Fisher's method of comparison of two dispersions and Pearson's method of comparison of two distribution functions.

Thus,the sample of galaxies with high surface brightness being compared with random sample of galaxies shows an excess both of objects with high and low luminosities. The abundance of objects of high luminosity can be understood since the dispersion of linear sizes of galaxies is much less than the dispersion of luminosities. Therefore the majority of objects of high surface brightness must be of high luminosity. But if so, then the abundance of objects of very low luminosity (or even their normal content) among the galaxies of high surface brightness seems to be rather surprising.

It must be finally emphasized, that the features mentioned are yet established only for objects with emission lines. For investigation of galaxies without emission lines the spectral observations with high resolution are necessary.

2. The Statistics of Seyfert-type objects. It is usually thought that Seyfert galaxies compose several per cent of galaxies of high luminosity. But it was shown by Markarian's survey, which is now the most powerful method of detection of such galaxies, that this value is strongly overestimated. Indeed, the percentage of Seyfert-type objects among Markarian galaxies is about ten, and these galaxies themselves compose not more than ten per cent of all galaxies of high luminosity (Sargent 1973 ; Huchra and Sargent 1973; Arakelian 1974b). Thus, the relative number of Seyfert-type objects in random sample of galaxies must be not more than 0.01. More correct estimates obtained by Huchra and Sargent (1973) and Arakelian (1974) show in mutual agreement that Seyfert-type objects compose 0.005 of all field galaxies with absolute magnitude $M_p < -18.5$.

Meanwhile the spectral observations of about 320 galaxies of high surface brightness have revealed 6 objects having Seyfert-type properties. Thereby the high percentage of Seyfert-type object cannot be entirely explained by the abundance of galaxies of high luminosity. Since one can see in Figure 1 that the percentage of galaxies with absolute magnitudes Mp -18.5 in all three samples considered is close to 0.8. It must be noted that 3 Seyfert-type objects are from 120 galaxies with surface brightness $\tilde{B} \leqslant 21.5$ and remaining 3 - from 200 galaxies with $21.5 < \tilde{B} \leqslant 22.0$. Though the statistics is small it can be stated that even in the considered rather small interval of surface brightness the relative number of Seyfert-type objects increases with growth of surface brightness.

Therefore, the correlation between nuclear activity and surface brightness, suggested in the beginning of this paper can be spread out on the appearance of Seyfert-type spectral features as well.

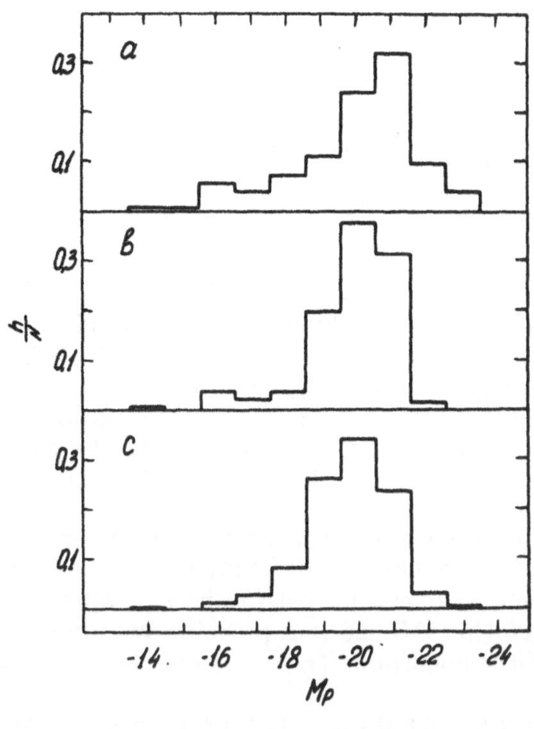

Fig. 1.

REFERENCES

ARAKELIAN M.A. 1972, Astrofizika, 8, 624

ARAKELIAN M.A. 1974a, Astrofizika, 10, 507

ARAKELIAN M.A. 1974b, Astron. Zhurnal. 51, 730

ARAKELIAN M.A. 1975, Soobsch. Byurakan Obs., 47, 3

ARAKELIAN M.A. and BALKOWSKI C. 1975, unpublished

ARAKELIAN M.A., DIBAJ E.A. and YESIPOV V.F. 1975a, Astrofizika, 11, 15

ARAKELIAN M.A., DIBAJ E.A. and YESIPOV V.F. 1975b, Astrofizika, 11, 321

ARAKELIAN M.A., DIBAJ E.A. and YESIPOV V.F. 1976, Astrofizika. In press.

DOROSHENKO V.T. and TEREBIZH V.Yu. 1976, Astrofizika, In press

HARO G. 1956, Bol. Obs. Tonantzintla y Tacubaya 14, 8

HUCHRA J. and SARGENT W.L.W. 1973, Ap. J. 186, 433

MARKARIAN B.E. 1967, Astrofizika. 3, 55

MARKARIAN B.E. 1969a, Astrofizika. 5, 443

MARKARIAN B.E. 1969b, Astrofizika. 5, 581

MARKARIAN B.E. and LIPOVETSKY V.A. 1971, Astrofizika. 7, 571

MARKARIAN B.E. and LIPOVETSKY V.A. 1972, Astrofizika. 8, 155

MARKARIAN B.E. and LIPOVETSKY V.A. 1973, Astrofizika. 9, 487

MARKARIAN B.E. and LIPOVETSKY V.A. 1974, Astrofizika, 10, 307

SARGENT W.L.W. 1973, Ap. J. 173, 7

VORONTSOV-VEL'YAMINOV B.A. and KRASNOGORSKAYA A.A. 1972. Morphological Catalogue of Galaxies, vol.1.

VORONTSOV-VEL'YAMINOV B.A. and ARHIPOVA V.P. 1963, Morphological Catalogue of Galaxies, vol.II.

VORONTSOV-VEL'YAMINOV B.A. and ARHIPOVA V.P. 1964, Morphoolgical Catalogue of Galaxies, Vol.III.

ZASOV A.V. and LYUTYJ V.M. 1973, Astronom. Zhurnal. 50, 253

ZWICKY F. 1971, Catalogue of Selected Compact Galaxies and Post-Eruptive Galaxies.

ZWICKY F., HERZOG E. and WILD P. 1961, Catalogue of Galaxies and of Clusters of Galaxies, vol. I

ZWICKY F., HERZOG E. 1963, Catalogue of Galaxies and of Clusters of Galaxies, vol. II

ZWICKY F., KARPOWICH M. and KOWAL C.T. 1965, Catalogue of Galaxies and of Clusters of Galaxies, Vol.V

ZWICKY F. and HERZOG E. 1966, Catalogue of Galaxies and of Clusters of Calaxies, vol.III

ZWICKY F. and HERZOG E. 1968a, Catalogue of Galaxies and of Clusters of Galaxies, vol. IV

ZWICKY F. and KOWAL C.T. 1968b, Catalogue of Galaxies and of Clusters of Galaxies, vol. VI

WESTERBORK OBSERVATIONS OF LARGE RADIO GALAXIES

R.G.STROM, J.P.HAMAKER, A.G.WILLIS
LEIDEN OBSERVATORY, DWINGELOO OBSERVATORY, LEIDEN OBSERVATORY, THE NETHERLANDS
PRESENTED BY R.G.STROM

НАБЛЮДЕНИЯ БОЛЬШИХ РАДИО ГАЛАКТИК В ВЕСТЕРБОРКЕ

I n t r o d u c t i o n . The suggestion (Bridle et al., 1972) that the radio galaxies 3C 236 and DA 240 might be objects of unexpectedly large size was subsequently confirmed by observations made with the Westerbork telescope (Willis et al., 1974). The original measurements were made at 50 cm, and our investigation has continued with observations of both objects at 21 cm and 6 cm. Such measurements enable us to study spectral index and polarization characteristics over a wide frequency range. The preliminary results discussed here indicate, among other things, high degrees of linear polarization in both objects at all three wavelengths.

We are also extending our investigation to other objects of known or suspected large size, and have begun a survey to search for more candidates. Among the radio galaxies studied are PKS 0634-20 and 3C 326. The former is found to be somewhat larger than previously suggested, stretching for 1.25 Mpc. 3C 326 is a probable very large radio galaxy, though its exact size remains somewhat uncertain. The survey we have begun is briefly discussed.

O b s e r v a t i o n s o f l a r g e r a d i o G a l a x i e s . We arbitrarily define a " large " radio galaxy as a radio source whose maximal projected linear dimension exceeds 1 Mpc. The first such object to be investigated was Centaurus A (Cooper et al.,1965), a double radio source whose total extent is between 1 and 2 Mpc. The uncertainty arises because the optical galaxy, NGC 5128, lies nearby and its distance is consequently uncertain. One of the most important facts about the morphology of Cen A is that in addition to two huge outer components, the object contains an inner double radio source. The inner and outer component pairs are roughly aligned with each other. A similar though much more precise alignment has been found in 3C 236 (Willis et al., 1974).

As far as the morphology of the outer components is concerned, however, Cen A is more like DA 240. In Fig.1 the relative sizes of 3C 236 and DA 240 are compared with several other radio galaxies, and the orientation of the inner and outer components of 3C 236 is illustrated.

At 50 cm, the two large components of DA 240 were found to have a relatively smooth structure. This is also true in our much higher resolution 21 cm observations: there is relatively little fine scale structure. In the western component, only the highest peak observed at 50 cm is fairly compact at 21 cm. Its brightness does not approach that found for the intense peak near the center of the eastern component, however. We have studied the eastern peak with high resolution at 6 cm, and find it to be elliptical in shape with major and minor diameters of about 11 kpc by 7 kpc. The elongation of this hot spot runs transverse to the major axis of the entire radio source. Its outer edge, which faces away from the parent optical galaxy, is slightly flattened. The linear polarization is uniform and everywhere about 25% of the total intensity. Its direction indicates that the magnetic field running through the hot spot is perpendicular to the major axis of the entire radio source.

A 21 cm map of 3C 236 (Fig.2) shows that the structure of most of the western component is smooth. About two-thirds of the way along the component, however, there are one or two bright knots of emission. As with the bright peaks found in DA 240, these are not located on the outer-most edge of the component. This is in contrast to the eastern component, which does have a bright, sharp outer edge. In an effort to study this feature in greater detail, we have observed it at 6 cm. Our map (Fig. 3) shows, firstly, that the two objects which lie near the outer peak of the component, but are off the component axis, almost certainly are unrelated to 3C 236. As for the component itself, only the emission from its outer edge is clearly present. As in other sources, a single hot spot lies well to one side of the way in which the peak merges with extended emission to the north, it is difficult to determine its exact size. A very rough estimate is 20 kpc by 10 kpc. The position angle of its major axis is roughly 20°, which is nearly perpendicular to the major axis of the entire radio source.

The hot spot in 3C 236 is very strongly polarized, with a position angle which indicates that, as in DA 240, the magnetic field direction is transverse to the major axis of the entire radio source. The degree of linear polarization is 65%±20%, suggesting that within the hot spot itself the magnetic field is almost perfectly ordered. When convolved to the resolution we obtain at 21 cm, the degree of polarization drops to about 20%, which is similar to that actually measured at 21 cm. Along the entire frontal part of the component, the magnetic field is parallel to the sharp outer boundary. This is, of course, not surprising, for we would expect the magnetic field to be aligned with the shock front, whose presence the sharp radio boundary presumably indicates.

The radio galaxy PKS 0634-20 has been observed at 21 cm (Fig.4). The map obtained shows that the outer edges of the two components are separated by 14' arc, which corresponds to a total linear extent of 1.25 Mpc. Morphologically, the components have the narrow, confined appearance of those found in 3C 236, and as in that object, the brightest outer edge is the most distant from the optical galaxy. Further study of this object is under way, in which we hope to be able to determine its polarization structure.

3C 326 was considered to be part of the north polar spur by Mackay (1969). Baker (1974) suggested that it is more likely to be a huge radio galaxy. The apparent magnitude of two optical galaxies found near the radio centroid indicates, assuming one of them to be the correct identification, a linear size of 4 or 5 Mpc. The Westerbork 50 cm observations would appear to support the suggestion that 3C 326 is a very large radio galaxy.

C o n c l u s i o n s. Our observational attack on the topic of large radio galaxies has been a three-pronged one: (1) To study known large galaxies, such as DA 240 and 3C 236, in as great a detail and at as many frequencies as possible; (2) To make preliminary observations of candidate objects, such as PKS 0634-20 and 3C 326; and (3) to attempt a systematic search for other candidate galaxies. In connection with(3),we have begun observing all confused sources listed in the BDFL catalogue (Bridle et al., 1972) with Westerbork at 50 cm. Only a few of the nearly 70 maps have as yet been processed. The quality to be expected is illustrated by the detailed map we have obtained of 3C 35 (Fig.5). Although 3C 35 is itself a fairly large double source, it is clearly not connected with the object to its northeast.

Any candidate turned up by our survey would be observed further at 50cm and, if sufficiently interesting, at 21 cm and 6 cm as well. Because of their great angular extent, large radio galaxies offer us an almost unique opportunity to study the double radio source phenomenon in immense detail. If we can observe a few more outer component hot spots with the resolution attained in our 6cm maps of 3C 236 and DA 240, we may be able to impose important boundary conditions on the machanisms occuring in these apparent seats of activity. Studies of spectral variations in the extended regions may, because of the great distances involved, enable us to assign ages to different regions of the component. A better understanding of the outer components may guide us in our attempts to decipher the mysterious events which occur in galactic nuclei. The alignment of the inner and outer components in 3C 236 has already suggested that the radio components may have been expelled along the rotation axis of a massive body. Will observations of other large radio galaxies support this conclusion?

A c k n o w l e d g e m e n t s. The observation of PKS 0634-20 is part of an investigation being carried out by J.L. Casse and K.L.Wellington. Other people who have participated in the study of large radio galaxies include A.S.Wilson and H. van der Laan. We thank many colleagues in Westerbork and Leiden for their invaluable help. The Westerbork Observatory is operated by the Netherlands Foundation for Radio Astronomy with financial support from the Organization for the Advancement of Pure Research (Z.W.O.).

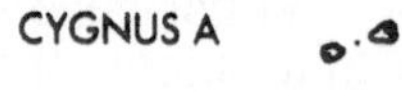

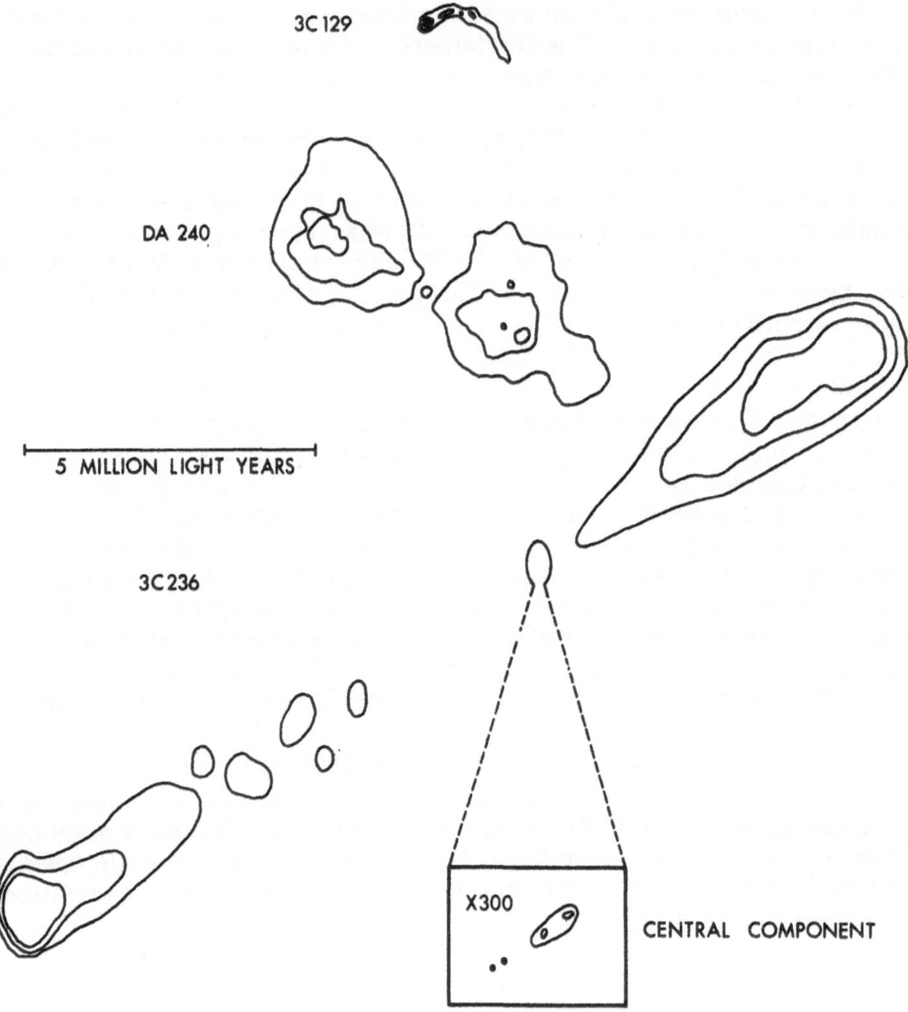

Fig. 1. The relative linear sizes of four radio galaxies. Cygnus A, a classical double radio source, has an extent of 200 kpc. 3C 129, associated with a cluster galaxy, is a typical example of a radio tail, with a total length of about 800 kpc. DA 240 and 3C 236 are both considerably larger than 2 Mpc. A sketch map of the central component of 3C 236, recently made from NRAO observations by E.B. Fomalont and G.K. Miley, is also shown. The similar orientation of inner and outer components is apparent.

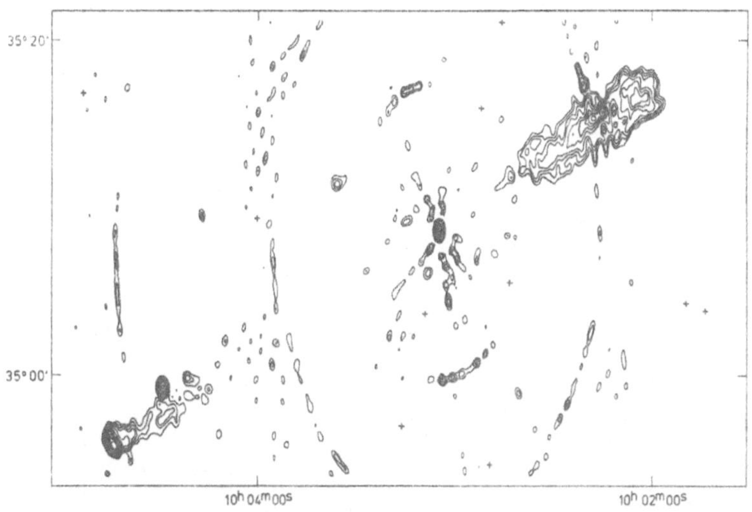

Fig. 2. A 21 cm map of 3C 236 made with the Westerbork telescope. The elliptical rings centered on, and linear features emanating from, the central component are caused by rapid phase fluctuations introduced by the atmosphere. A procedure for removing these unreal effects is under development at the Dwingeloo Observatory. Crosses mark the positions of background sources.

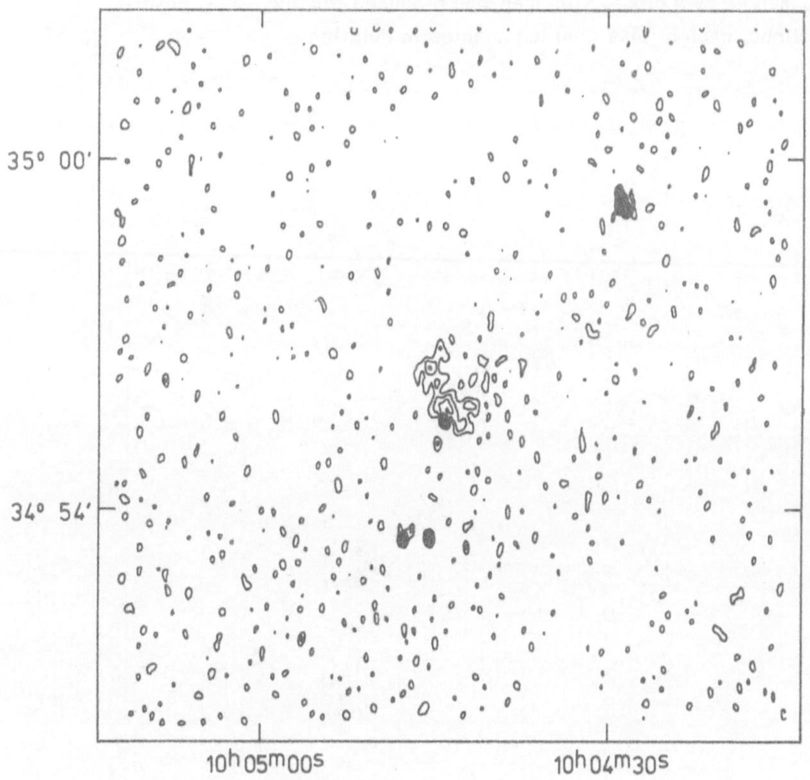

Fig. 3. A 6 cm map of the eastern component of 3C 236, clearly showing the two background sources and the outer edge with its hot spot. The latter has a peak flux density of 7.5 mJy, and is found to be about 65% linearly polarized.

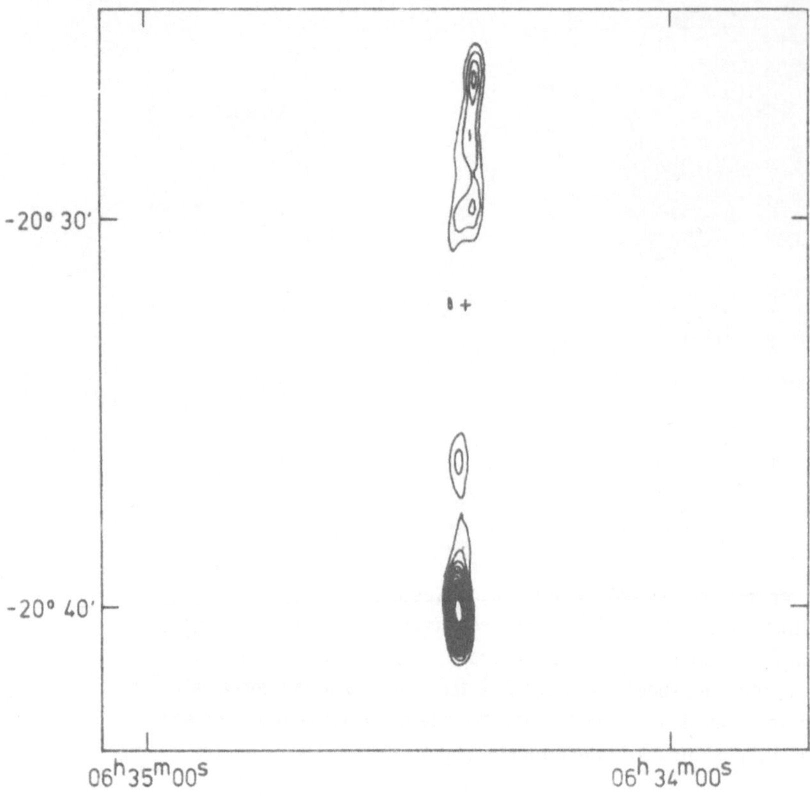

Fig. 4. A 21 cm map of the radio galaxy PKS 0634-20. The optical galaxy is marked by a cross. This map was produced from two short observations, each of less than ten minutes in duration.

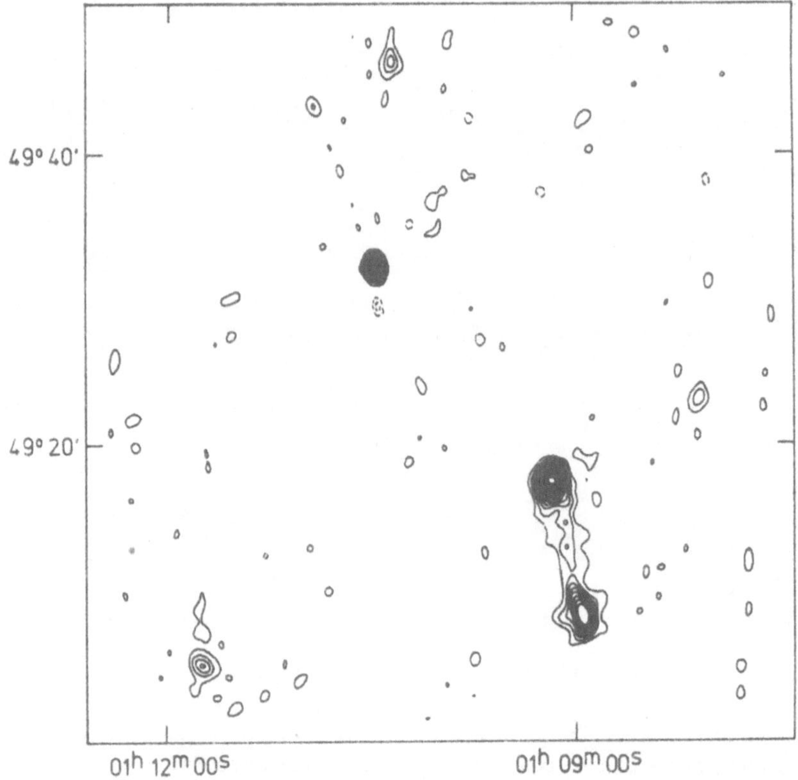

Fig. 5. One of the fields from the 50 cm survey of the BDFL confused regions. The large s.p.double source is 3C 35, while the strong n.f. object is the confusing source. It is clear that the two are unrelated.

REFERENCES

BAKER J.R., 1974, Paper delivered at the II EAM in Trieste, Italy.
BRIDLE A.H., DAVIS M.M., FOMALONT E.B., and LEQUEUX J., 1972 Astron. J., 77, 405
COOPER B.F.C., PRICE R.M., and COLE D.J., 1965. Aust. J. Phys., 18, 589
MACKAY C.D., 1969, Mon. Not. R. A. S., 145, 31
WILLIS A.G., STROM R.G., and WILSON A.S., 1974. Nature, 250, 625

DISCUSSION

LONGAIR: What is the axial ratio of the hot spot in the largest radio source of all?
STROM: The ratio of distance from the optical galaxy to diameter of the hot spot in the eastern component is about 200:1.
LONGAIR: This is certainly the world record. Congratulations.

U B V - PHOTOMETRY OF IRREGULAR GALAXIES TYPE M 82

B.P.ARTAMONOV
STERNBERG ASTRONOMICAL INSTITUTE, MOSCOW, U.S.S.R.

U B V - ФОТОМЕТРИЯ НЕРЕГУЛЯРНЫХ ГАЛАКТИК ТИПА М 82

1 .The irregular galaxies Irr II were marked out by Markarian (1963), at contradiction between the class of integral spectrum and the colour indices. Lynds and Sandage (1963), have discovered in the prototype of these galaxies - M 82 the attributes of the phenomenon similar to explosion in the region of the galactic centre.

Numerous works followed (Lynds and Sandage 1963), and showed that M 82 was a very complicated object by its structure and kinematics. We have investigated Irr II galaxies M 82 and NGC 3077 with the help of the photographic material, recieved by 2,6 m telescope (ZTSH) (Artamonov et al, 1974, Artamonov and Nasarova, 1974) and by Tautenburg-Schmidt telescope. (Artamonov et al. 1975, Antropova et al. 1975, Artamonov , Börngen, Shapovalova, 1975a,1975b).

2. The observational material for the photographic photometry was selected from numerous photographic plates of glass library of Tautenburg's observatory. For the group of galaxies M 81 photographs were chosen: two for each - U, B, R and three - for V, taken in 1968-1969. For the galaxy NGC 520, according to our request, doctor F. Börgnen supplied six photographs in U, B, V - system.

The treatment of the negatives was made in 1973-1974 in SAO AS USSR with microphotometers: "spectr-cod" and "Schnell". The photometric sections have been built mainly in the direction of the major and minor axes of galaxies. The colour indices (U-B), (B-V) were obtained. The equidesitometry in the system UBVR was made for the four galaxies in the photographs of the group M 81. The absolute zero points of the photographic colour indicas have been obtained in comparison with the photoelectric observations of other authors. We have carried out the photoelectric observations of NGC 3077 in the beginning of 1974 by the telescope "Zeiss-600" in SAO AS USSR. The results of colorimetry of galaxies were presented by two colour diagrams (U-B), (B-V). The principal conclusions for the studied galaxies are presented below.

3. M 82. The observational model obtained for the galaxy M 82: the position of axes in space, the distribution of stars and gas-and-dust cloud. The comparison of the colorimetric results of observations with spectroscopic ones pointed out that the region of the galactic centre is surrounded by dust clouds, which have the radial components of velocity. There are many hot stars ionizing gas in central region of M 82. The regions of H II and hot stars illuminate dust clouds which have bluer light than the stars in the centre of M 82. The data on polarization of light are not in contradiction with this picture. We observe the gradual decrease of dust component along the major axis of M 82 and a change of stellar population from early-type to late-type stars (from A to F-G). The possible cause of the radial velocity of dust clouds is the influence of radiation pressure of early-type stars in the central region of M 82.

4. NGC 3077. According to the distribution of colour indices in galaxy there was suggested a model, in which the main radiation in the central region is conditioned by reddened B-stars ($\tau \sim 1$-2)in periphery mixture of reddened A-F stars with scattering light of the stars of the galactic centre on dust cloud.

5. NGC 520. It is shown that by colour and spectral characteristics the galaxy NGC 520 is probably a single physical system. The superassociation in the northern part of the galaxy consists of possibly early-type stars which are inside a dust cloud at the optical depth of $1 \div 1.5$. The superassociation excites the region H II whose size is$\geq$500 pc. and mass is$\sim 10^6\ M_\odot$ For ionization of the region H II$\sim$900 BO-stars or$\sim$30 stars of type 07 are needed.

6. A similarity between the three studied galaxies has been observed: abundance of dust, groups of early-type stars and H II- regions, kinematic characteristics. There are probably similar processes of secondary formation of stars.

REFERENCES

ANTROPOVA L.G., ARTAMONOV B.P., BÖRNGEN F., 1975, Astrophys.issled. (Izw. SAO AS USSR), 7

ARTAMONOV B.P., NASAROVA L.S., Astrophys. issled.., 1974 (Izw. SAO AS USSR), 4, 143

ARTAMONOV B.P., NASAROVA L.S., 1974, Izw. Krims. Obs., 50, 115

ARTAMONOV B.P., BÖRNGEN F., NOTNI P , 1975, Astrophys.issled. (Izw. SAO AN USSR), 7

ARTAMONOV B.P., BÖRNGEN F., SHAPOVALOVA A.I., 1975a, Astrophys. issled. (Izw. SAO AS USSR), 8, in press

ARTAMONOV B.P., BÖRNGEN F., SHAPOVALOVA , 1975b Soob. SAO AS USSR, N13, in press

LYNDS C.R., SANDAGE A., Astrophys J., 137, 1005, 1963

MARKARIAN B.E., Soob. Byurakan. Obs., 34, 19, 1963

SURFACE COLORIMETRY OF GALAXIES WITH ULTRAVIOLET CONTINUUM

F. BÖRNGEN, A.T. KALLOGHLIAN
CENTRAL INSTITUTE OF ASTROPHYSICS, G.D.R.
BYURAKAN ASTROPHYSICAL OBSERVATORY, U.S.S.R.
PRESENTED BY A.T.KALLOGHLIAN

ПОВЕРХНОСТНАЯ КОЛОРИМЕТРИЯ ГАЛАКТИК С УЛЬТРАФИОЛЕТОВЫМ КОНТИНУУМОМ

The main feature of Markarian galaxies is the existence of ultraviolet continuum and emission lines. Many Seyfert type galaxies have been discovered among them. An important question is to reveal the possible influence of active events in nuclei on the physical features and structure of the outer parts of such galaxies. In this aspect a morphological and colorimetrical investigation of galaxies with ultraviolet continuum and normal galaxies has been undertaken. Here we report about colorimetrical results on eight Markarian galaxies.

UBVR surface photometry of galaxies Markarian 7,8,10,11,12 and 13, and UBV surface photometry of Markarian 185 and 190 have been carried out on the Schmidt plates of Tautenburg two-meter telescope.

Some data obtained in the course of investigation are collected in Table 1. One can see that U-B as well as B-V of nuclei or of brightest condensations (in the case of No.7 and 8)usually are smaller than the integral values. The contrary is true for V-R. However the nucleus of Markarian 185 is red enough to make difficult the explanation of the existence of ultraviolet continuum. It is remarkable that the last one is quite faint but extended far to short wavelength region. The emission lines are faint too. Apparently the UV continuum comes from the blue condensation in the spiral arm at 15'' from the nucleus.

The nucleus of Markarian 12 in U-B, B-V, V-R and luminosity is identical with the super-association in the south arm of the galaxy.

For calculation of the weighted surface brightnesses β_o different weights were given to the regions in a galaxy brighter and fainter than B-$21^m.5$ per second of arc. It is seen that the brightest galaxies are Markarian 12 and 190 of Sc and SO type, respectively. The faintest is Markarian 13 of type SBb. At the same time the nucleus of this galaxy is the absolutely faintest (M_B =-15.3) among others. It should be pointed out that in this case the dispersion of β_o is 1.5 times smaller than the dispersion of mean surface brightnesses calculated in usual manner.

Table 2 shows the distribution of relative intensities in the zones parallel to the black-body line on the (U-B, B-V) diagrams. The first four zones are located above the black-body line. It is seen that for various galaxies from 4 to 73 percent of total brightness is due to the regions located above the black-body line. The highest percentage for Markarian 190 apparently is due to the highly peculiar nuclear part consisting of several condensations. The high percentage for Markarian 10 is caused by Seyfert nucleus.

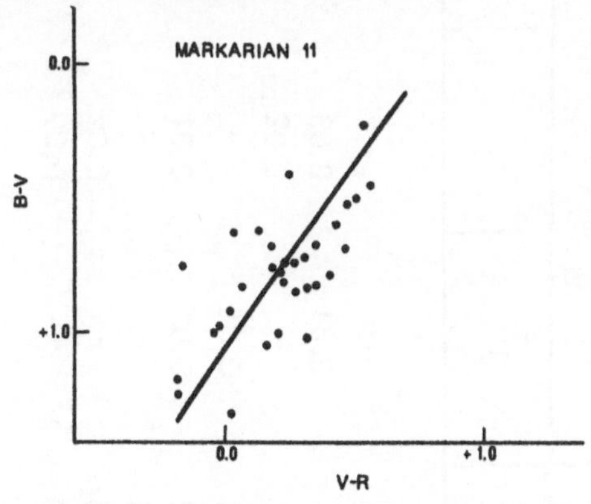

Fig. 1. Relation between B-V and V-R for Markarian II.

A well defined relation between B-V and V-R exists for Markarian 8, 10 and 11. Fig. 1. shows this relation for Markarian 11, It is seen that B-V is decreasing with increasing V-R. Apparently besides UV excess here we have an excess in red.

A similar study of normal galaxies on the same plates is in progress to compare with the results obtained for galaxies with ultraviolet continuum.

Table 1

Photometric characteristics of Markarian Galaxies

	No 7 Irrl	No 7 d 2	No8 Irrl	No8 d1	No10 SBb	No10 s1	No11 SO	No11 s3	No12 Sc	No12 d3	No13 SBb	No13 sd3	No185 SB(r)b	No185 sd3	No190 SO	No190 sd2
	int.	nuc.	int.	nuc.	int.	nuc.	int.	nuc.	int.	nuc.	int.	nuc.	int.	nuc.	int.	nuc.
B	14.68		14.43		14.40		15.12		13.46		15.01		13.70		13.70	
U-B	- 0.35	-0.42	-0.37	-0.84	-0.33	-0.87	+0.12	+0.10	-0.15	-0.39	+0.01	-0.30	+0.43	+0.36	-0.06	-0.23
B-V	0.29	0.26	0.43	0.20	0.70	0.50	0.56	0.22	0.36	0.39	0.44	0.06	1.13	0.95	0.82	0.67
V-R	0.15	0.30	0.13	0.31	-	-	0.36	0.51	0.26	0.28	0.23	0.47	-	-	-	-
B_o	14.59	17.21	14.28	17.17	14.22	15.76	14.96	16.51	13.42	16.33	14.64	16.78	13.36	16.08	13.35	15.24
$-M_{B_o}$	19.0	16.4	19.5	16.6	21.6	20.1	-	-	20.8	17.9	17.5	15.3	19.9	17.2	17.4	15.5
$B/\square''$	22.48	20.75	22.85	20.68	22.80	18.98	22.68	19.85	22.10	19.87	23.10	20.33	22.45	19.62	22.00	18.78
β_o	21.63		21.65		21.50		21.60		21.30		22.18		21.97		21.22	

Table 2

Relative Intensities in the zones parallel to the Black-body line

Cal. No	Zone No 1	2	3	4	5	6	7	Morph. Type	Sp.	$I_1 = \sum_{K=1}^{4} i_K$	$I_2 = \sum_{K=5}^{7} i_K$
7	-	-	0.02	0.11	0.45	0.39	0.03	Irrl	d2	0.13	0.87
8	0.01	0.03	0.06	0.19	0.55	0.15	0.01	Irrl	d1	0.29	0.71
10	-	0.02	0.44	0.14	0.22	0.12	0.06	SBb	sl	0.60	0.40
11	0.01	-	0.01	0.08	0.36	0.10	0.44	SO	s3	0.10	0.90
12	0.01	0.01	0.01	0.01	0.30	0.49	0.17	Sc	d3	0.04	0.96
13	-	-	0.01	0.07	0.16	0.48	0.28	SBb	sd1	0.08	0.92
185	-	0.02	0.12	0.27	0.53	0.05	0.01	SB(r)b	sd3	0.41	0.59
190	-	0.09	0.16	0.48	0.19	0.08	-	SO	sd2	0.73	0.27

CHAIRMAN (after the break) : PROF. J.COURTES

OBSERVATIONAL EVIDENCE OF HIGH - ORDER CLUSTERING OF GALAXIES

M.KALINKOV, K.STAVREV, IL.KANEVA, V.DERMENJIEV
DEPARTMENT OF ASTRONOMY, BULGARIAN ACADEMY OF SCIENCES, BULGARIA
PRESENTED BY M.KALINKOV

НАБЛЮДАТЕЛЬНОЕ ДОКАЗАТЕЛЬСТВО ОБРАЗОВАНИЯ СКОПЛЕНИЙ ГАЛАКТИК ВЫСШЕГО ПОРЯДКА

1. Introduction. There are many arguments at present supporting the idea of the existence of a second-order clustering of galaxies. But is it possible to extrapolate this idea for third-order clustering? One may expect for the characteristic size of third-order clusters a value of $\sim 200\ h^{-1}$ Mpc (where h is the Hubble's constant in units of 100 km sec^{-1} Mpc^{-1}). According to many well established results, the second-order clustering has a scale of $(40\div50)h^{-1}$ Mpc (e.g. Abell 1958, 1963; Karachentsev 1966; De Vaucouleurs 1971; Kalinkov 1973, 1974, 1975). Characteristic sizes of 100, 150, 200 - 250 h^{-1} Mpc (Kalinkov, 1975) have been found quite unexpectedly. It seems that there is a tendency to clustering with a continuous spectrum (Kiang and Saslaw, 1969, De Vaucouleurs, 1971). A very complicated picture may be derived as a result of investigations of Peebles (e.g. Hauser and Peebles, 1973, Peebles, 1974).

To clarify the question of the characteristic size, we treated with the help of several statistical methods two samples drawn from the fundamental catalogues of Abell (1958) and Zwicky et al. (1961-68) around the NGP. An area of about 5000□° centered in the NGP for Abell clusters has been chosen. There are 938 clusters in the area. We used Lambert coordinates (equivalent azimuthal projection) to avoid the projection effect.

The first method used here consists of digital processing of discrete fields, representing the distribution of clusters, with smoothing functions (SF) and filtering functions (FF). SF are denoted as [0.5], [1], [2], [4], and their responses are given by Kalinkov et al. (1976b). These four SF generate six FF whose responses are shown in Fig.1. Note that the FF are normalized - i.e. the amplitudes for a maximum transmission through filters are restored. The data interval for our discrete fields is of 5.25□°. Therefore, in order to have the true scale size, the scale size L in Fig.1, which is in arbitrary units, must be multiplied with a factor of $\sqrt{5.25}$. Using the formulae given by Kalinkov et al. (1976a), we have Table 1, where L_0 is the scale size, corresponding to maximum transmission. The Nyquist scale size here is $4^{\circ}.6$.

II. Results. We present here some maps showing the filtered distribution of Abell clusters around NGP. Everywhere densities are in cl/□°. All maps are uniform. They contain L and B, data interval 5.25□° - black squares, half-beam width (HBW) for those SF, from which FF are constructed and m_{10}. Two smoothed and one filtered (non-normalized) maps are given by Kalinkov (1975) for the same area.

The maps in Figs. 2-3 are related to Abell clusters with $m_{10} \leqslant 17^{m}.0$ - filters [0.5] - [2] and [0.5] -[4]. It is noteworthy that none of the second-order clusters registered by Kalinkov (1975) are visible here. It is easy to explain because the mean distance R to the mentioned second - order clusters is from 440 to $560h^{-1}$ Mpc. The weighted mean distance to clusters for Figs. 2 - 3 is about $296h^{-1}$ Mpc, if we take into account the calibration procedure of Kalinkov et al. (1975).

Figs. 4 - 8 present maps of clusters having $m_{10} \geqslant 17^{m}.1$, with weighted mean distance of $500h^{-1}$Mpc. The known second-order clusters can be recognized here, and also some others may be added (especially for Figs.4, 5 and 7). The most important of these maps is Fig.8, where two well-defined areas of positive density can be seen, for $0° < L < 90°$ and $180° < L < 270°$ respectively. FF [2] -[4] has $L_0 = 30°.0$. So, the maximum transmitted square size is of 900□°.

The first condensation is of $\sim 200\square^\circ$ and it is passed through the filter. If it is not an effect of random clumpiness of five or more second-order clusters (compare with Fig. 4), the alternative possibility must be accepted, namely that there is a third-order cluster. The above consideration remains valid for the second condensation, too. The diameters of aggregates of this kind is (130 ÷ 160) h^{-1} Mpc. Another very interesting feature for the map under study (Fig. 8) is also the presence of two symmetrical areas with negative densities. Actually, their existence can be expected because they are noted in Figs. 4 - 7. They are caused either by a real deficiency of clusters or by absorption.

Some of our considerations can be verified immediately. For instance, additional filtered maps are given (Figs. 9-11) but for clusters with $m_{10} \geqslant 17^m.5$. The statistics is twice smaller but never mind - the characteristic features remain the same.

Finally Figs. 12 - 16 give the filtered maps for all Abell clusters (Σ) in the investigated field. Here we have more than 10 second-order clusters and again 2 third-order clusters and again almost the same size (120 - 180) h^{-1} Mpc.

Some other results we want to add here are related to the application of other statistical methods.

A careful study of a field around NGP (Kalinkov 1975, 1976; Kalinkov and Tomov, 1976) with the generalized $\chi 2$ - test leads to the conclusion that for both Abell and Zwicky clusters the characteristic size of second-order clusters is between 40 and 60h^{-1} Mpc. But there are other sizes, 150 and 200h^{-1} Mpc, which can be connected with the existence of third-order clusters.

According to the nearest neighbours test (two- and three-dimensional), we obtain several characteristic sizes - 50, 80, and 120h^{-1}Mpc, which is difficult to explain.

Unfortunately we have even more characteristic sizes if we apply correlation methods - 50 80, 150, 200, 250, 350h^{-1} Mpc, for various distance classes. That is why we think that there really is a tendency towards clustering with an almost continuous spectrum. Here only one result is given which shows the high internal convergence of the results. In Fig.17 three correlation functions (method B, according to Kalinkov, 1974) for Zwicky ED clusters, located in a square region of $1296\square^\circ$ around NGP are given, together with ±1 s. d. The three functions are related to various breakings up, i.e. various data intervals, namely 4, 9, and $16\square^\circ$. It can be seen that the correlation functions are almost identical. It is indeed very hard to determine the position of the first zero but we roughly accept 10° as the correlation radius. Then the angular characteristic size will be 20° to which a linear characteristic size of 260h^{-1} Mpc will correspond (because the radial velocity for ED clusters, according to Zwicky, is 67.5×10^3 km sec^{-1}).

Finally, we must add that contrary to many other authors we have established a very close resemblence between the global characteristics of the distribution of cluster of galaxies according to both Abell and Zwicky catalogues (which can be seen by comparing maps for Zwicky clusters, e.g. Kalinkov, 1974, with maps for Abell clusters given in this paper.).

Table 1

Scale size L_0 for maximum transmission of the filters

FF	L_0
[0.5] - [1]	7°5
[1] - [2]	15.0
[2] - [4]	30.0
[0.5] - [2]	11.8
[1] - [4]	23.7
[0.5] - [4]	19.8

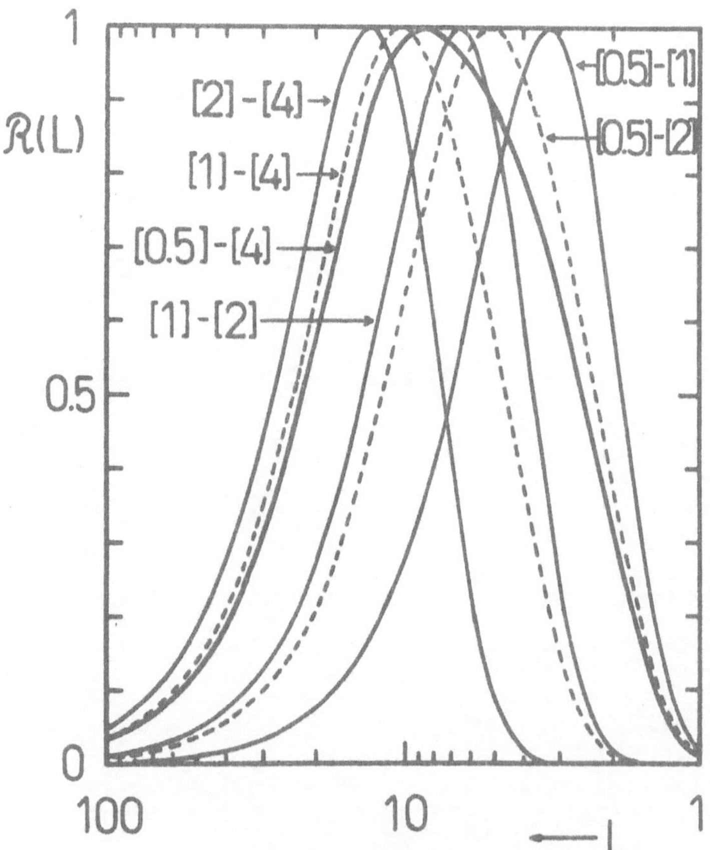

**Fig. 1. Normalized responses of the filters used in this paper.
L is an arbitrary scale size.**

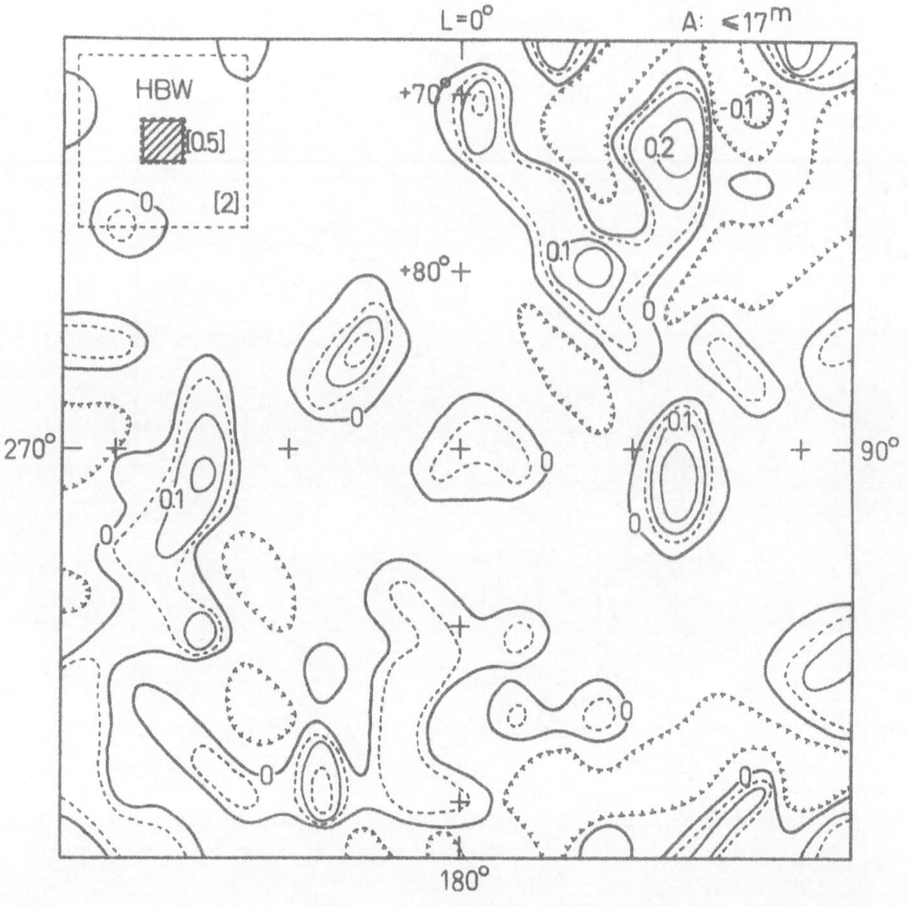

Fig. 2.

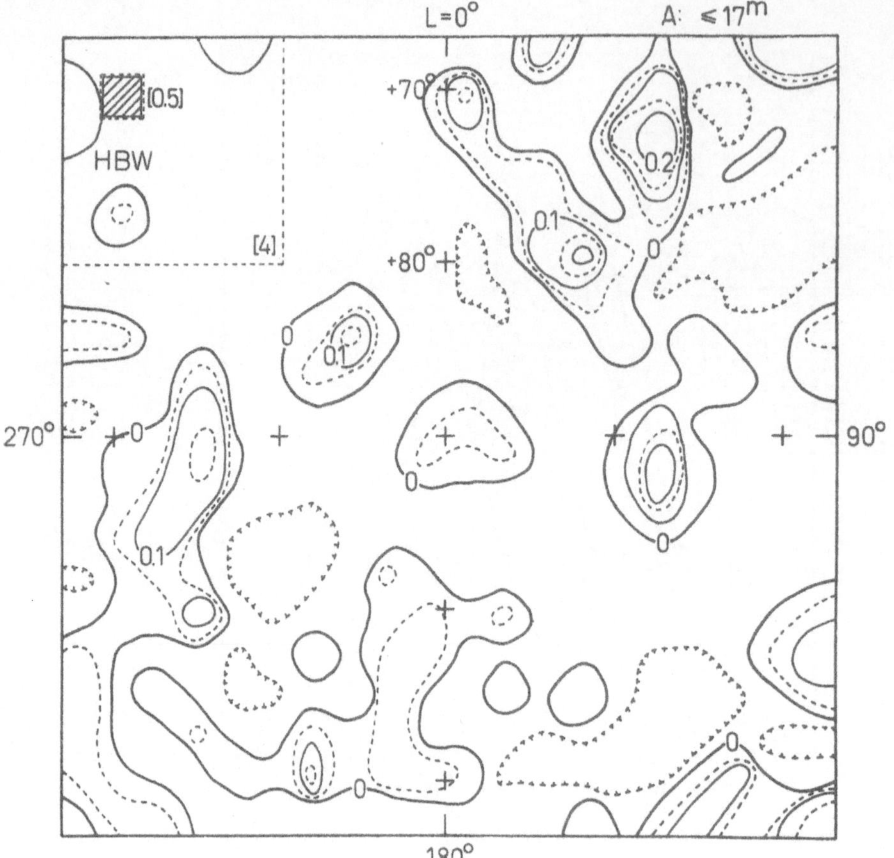
L=0°
A: ≤17m
[0.5]
HBW
[4]
+70°
+80°
0.2
0.1
0
270°
90°
180°

Fig. 3.

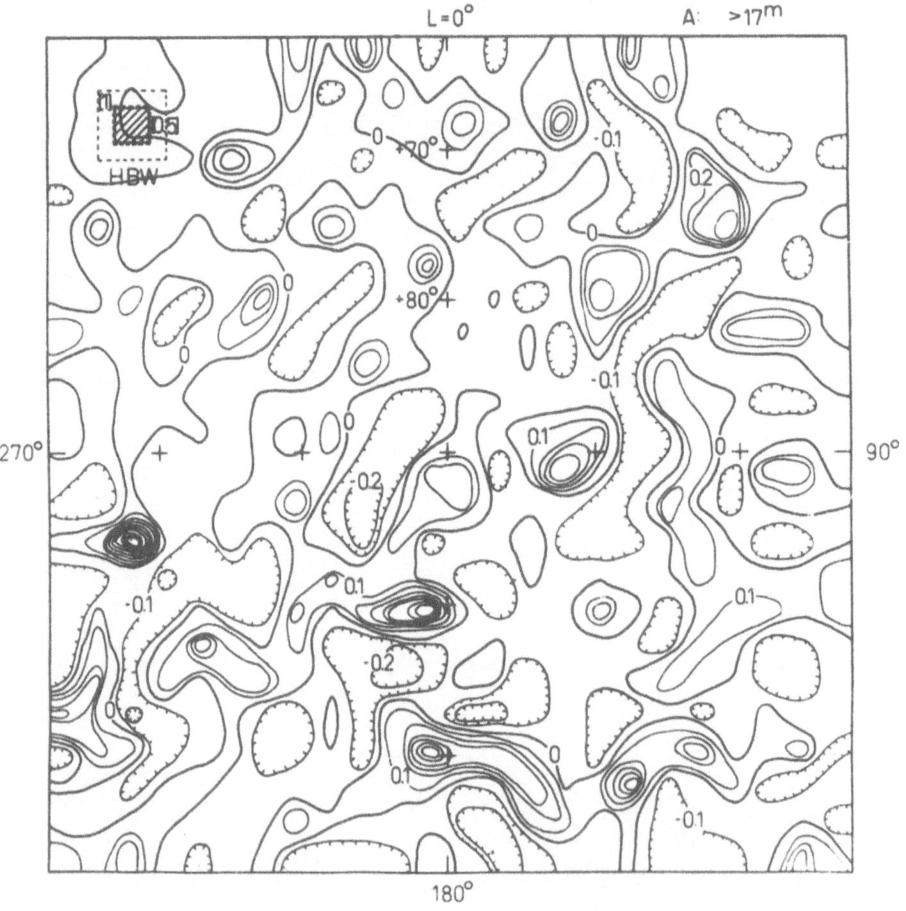
L=0°
A: >17m
[1]
0.5
HBW
+70°
+80°
0.2
-0.1
0.1
-0.2
270°
90°
180°

Fig. 4.

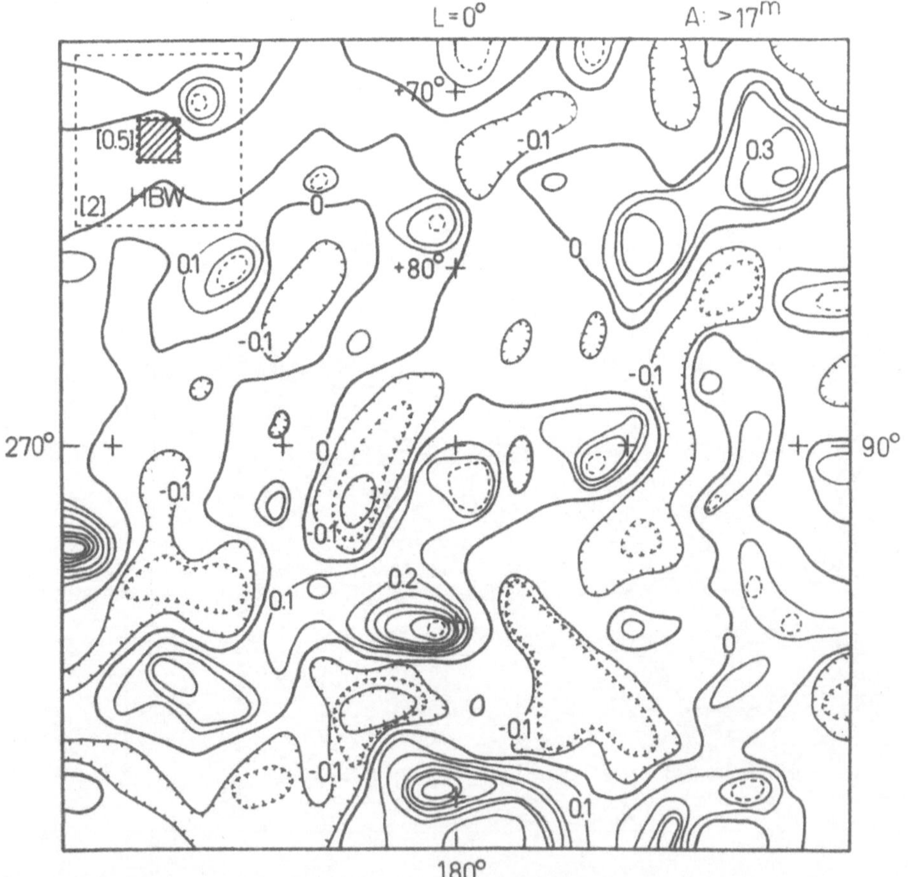
L=0°
A: >17m
+70°
-0.1
0.3
[0.5]
[2] HBW
0
0.1
+80°
-0.1
-0.1
270°
90°
0
-0.1
-0.1
0.2
0.1
0
-0.1
-0.1
0.1
180°

Fig.5.

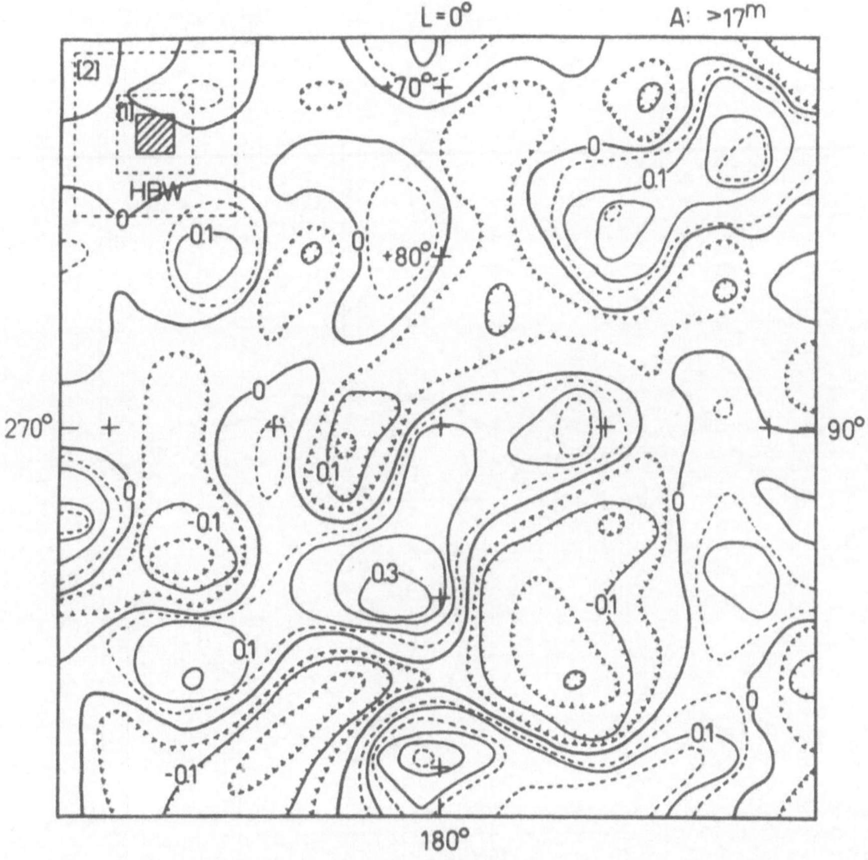
L=0°
A: >17m
[2]
+70°
[1]
HBW
0
0
0.1
0.1
0
+80°
0
270°
90°
0
-0.1
0
-0.1
0.3
-0.1
0.1
0
-0.1
-0.1
180°

Fig. 6.

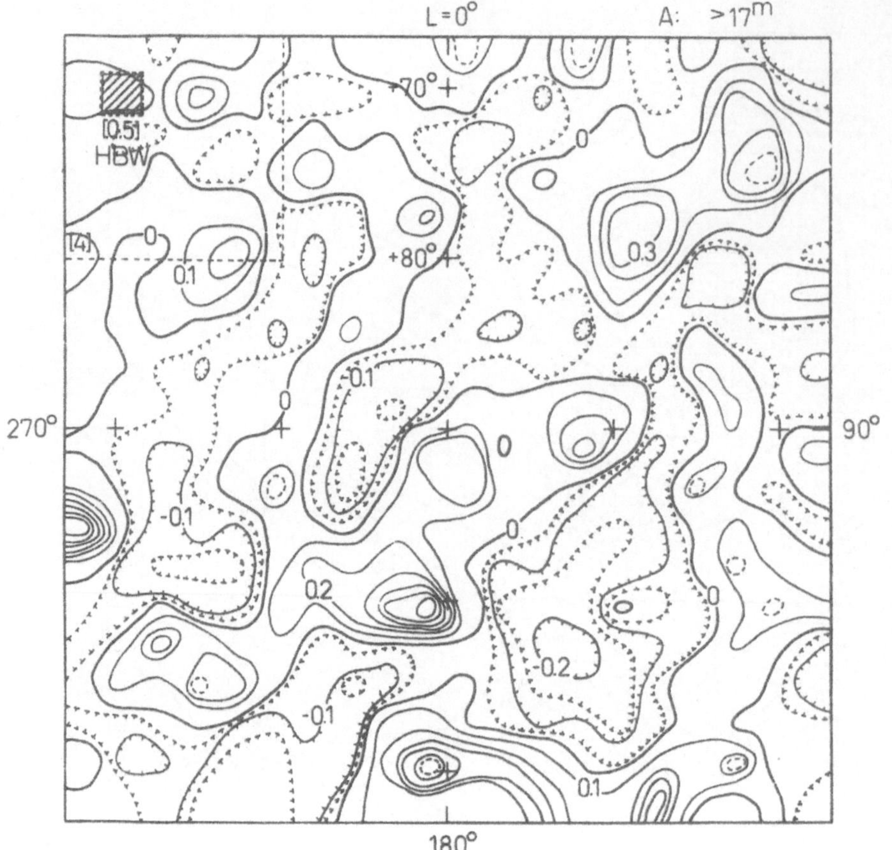
L=0°
A: >17m
[0.5]
HBW
+70°
[4]
0
0.1
+80°
0.3
0.1
0
270°
90°
0
-0.1
0
0.2
0
-0.2
-0.1
0.1
180°

Fig. 7.

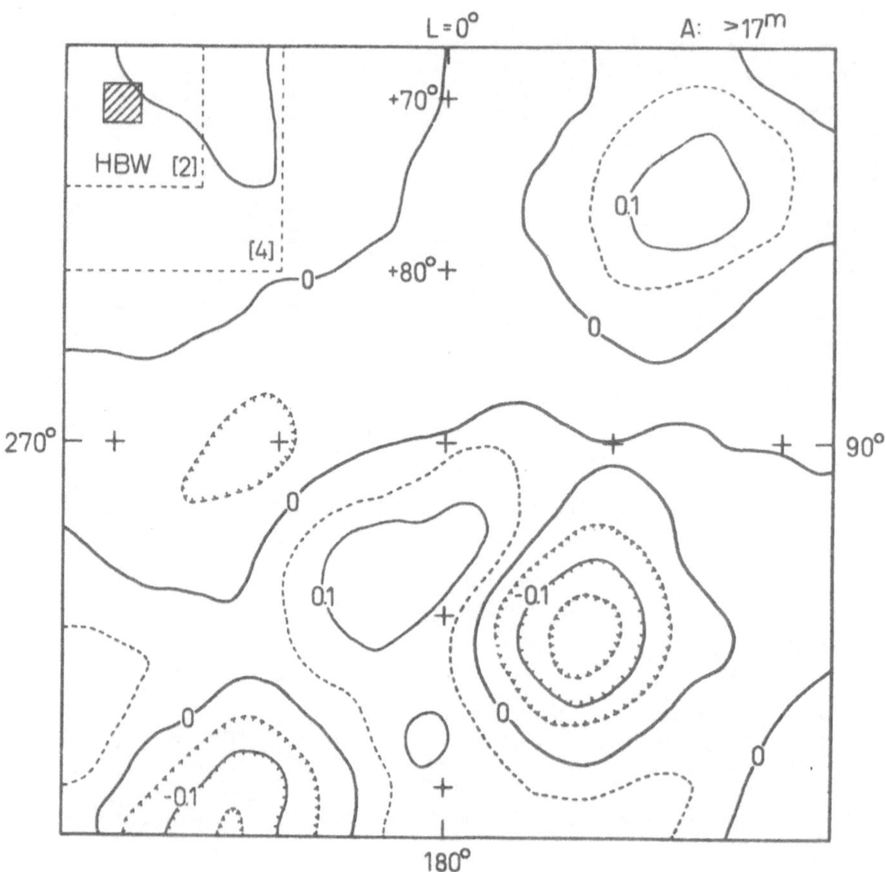
L=0°
A: >17m
+70°
HBW [2]
0.1
[4]
0
+80°
0
270°
90°
0
0.1
-0.1
0
0
0
-0.1
180°

Fig. 8.

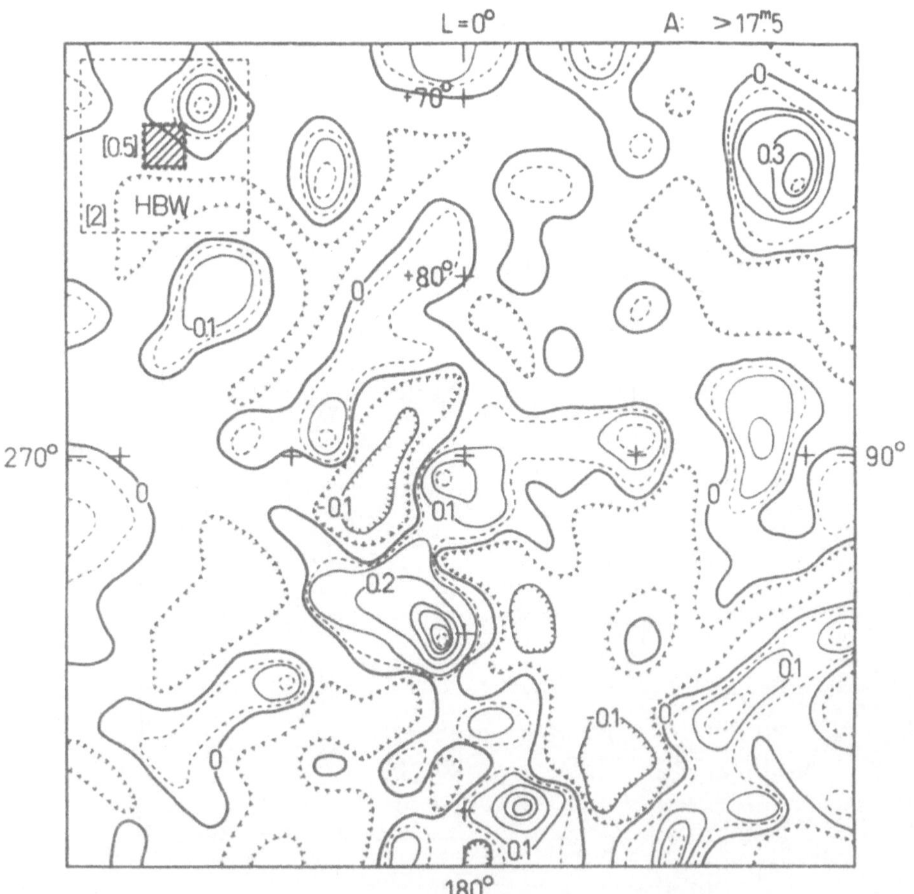
L=0°
A: >17ᵐ5
+70°
[0.5]
[2]
HBW
0.3
+80°
0
0.1
270°
90°
0.1
0.1
0.2
0.1
-0.1
0
0.1
180°

Fig. 9.

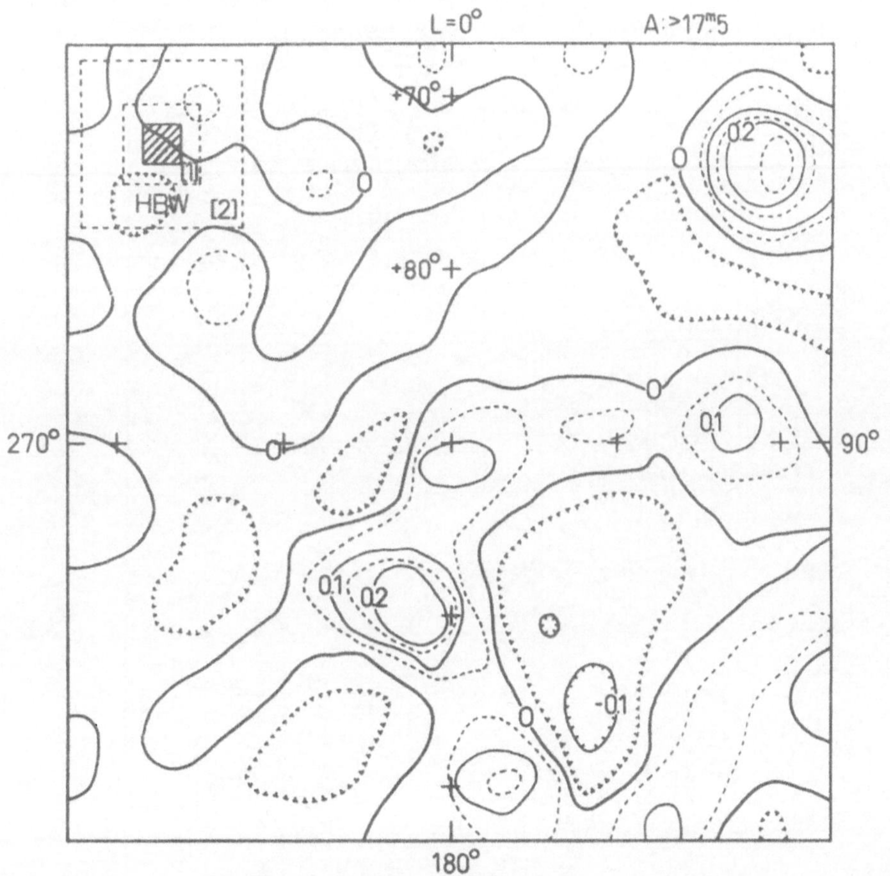
L=0°
A: >17ᵐ5
+70°
[1]
HBW
[2]
0.2
0
+80°
0
0.1
270°
90°
0.1
0.2
-0.1
0
180°

Fig. 10.

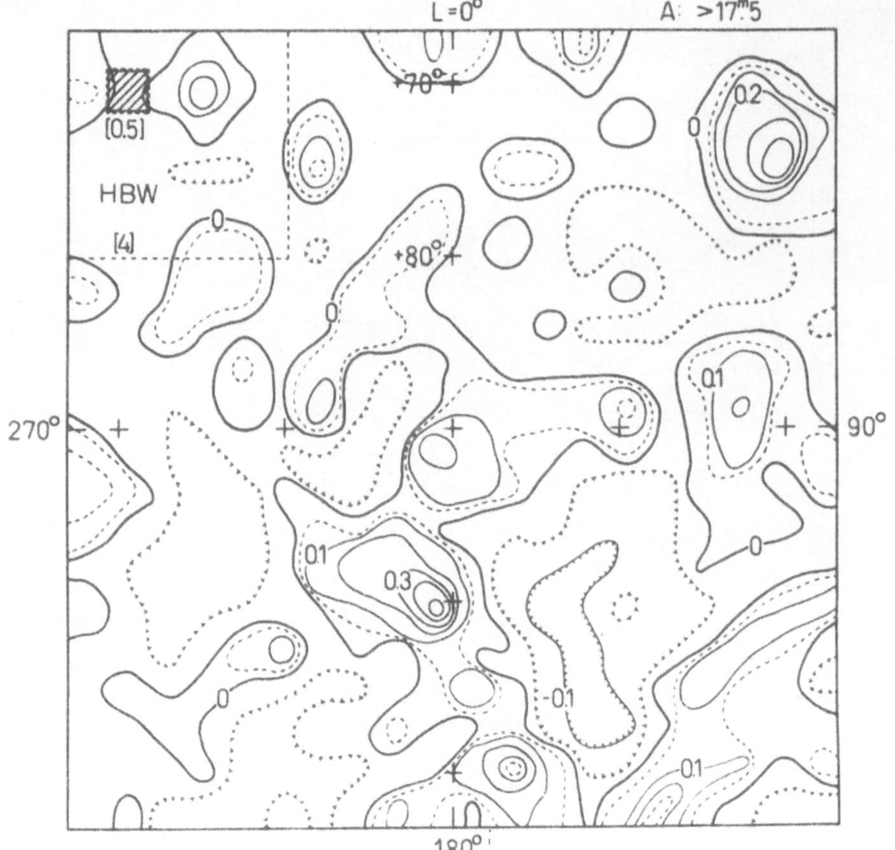
L=0°
A: >17ᵐ5
+70°+
[0.5]
HBW
[4]
0
0.2
+80°+
0
0.1
270°
90°
0.1
0.3
0
-0.1
-0.1
180°

Fig. 11.

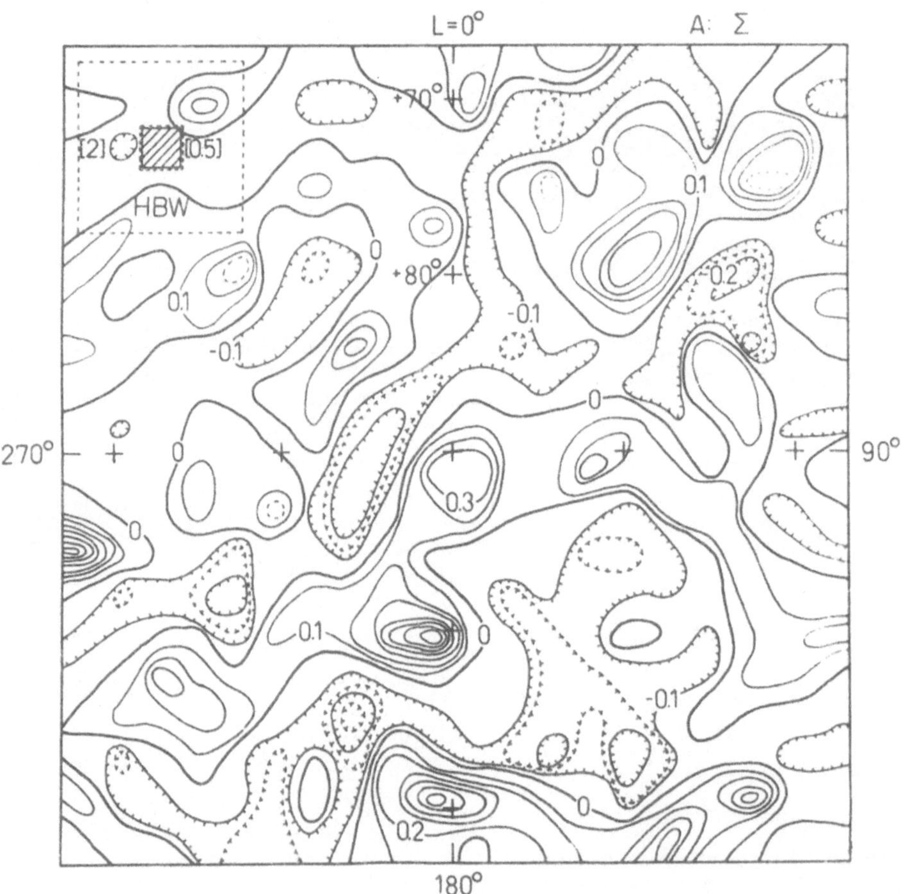
L=0°
A: Σ
+70°+
[2]
[0.5]
HBW
0
0.1
+80°+
0.1
-0.1
-0.1
-0.2
0
270°
90°
0
0.3
0
0.1
0
-0.1
0
0.2
180°

Fig. 12.

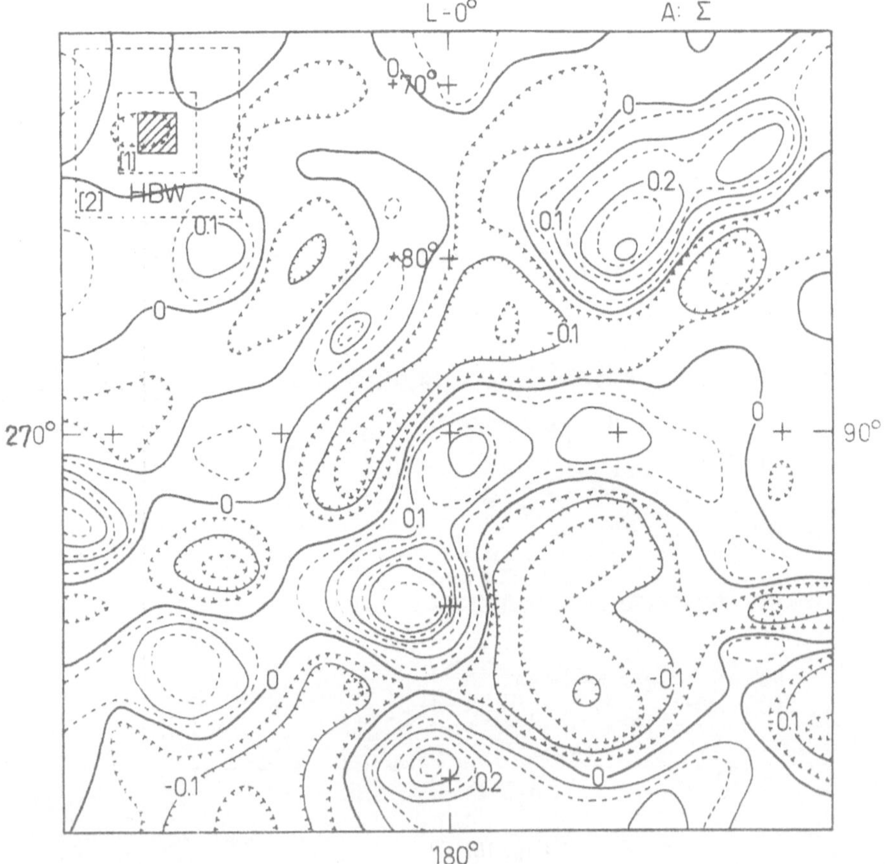
L-0°
A: Σ
+70°
+80°
270°
90°
180°
HBW
[1]
[2]
0
0.1
0.2
-0.1
-0.2

Fig. 13.

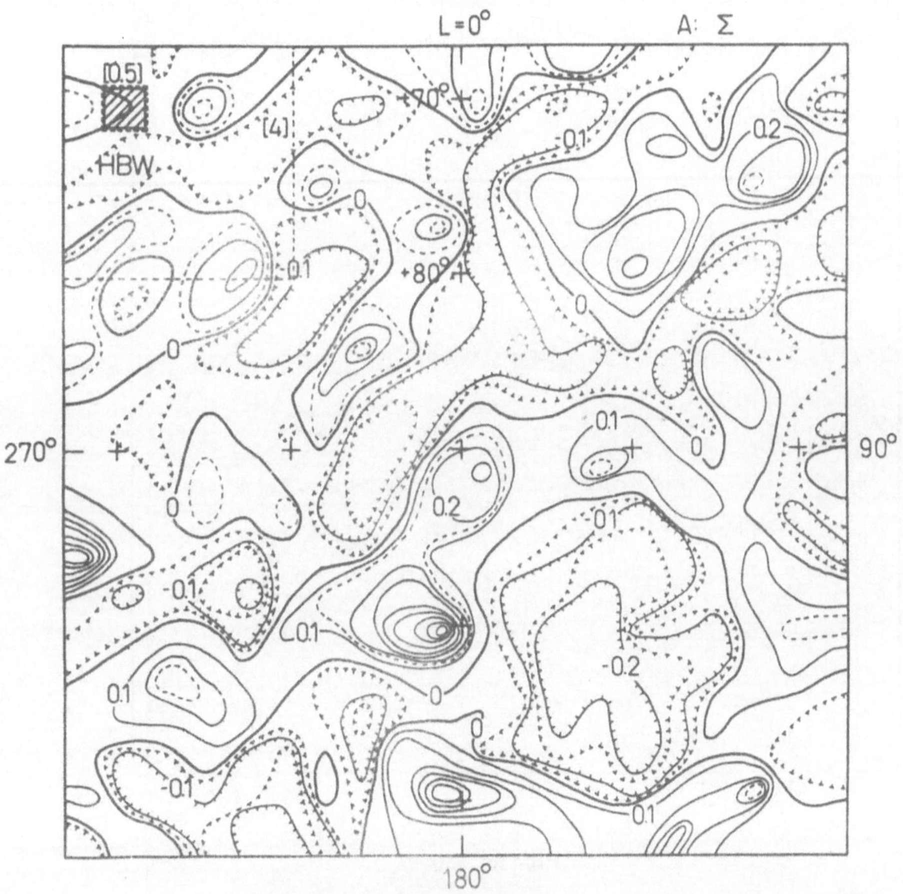
L=0°
A: Σ
[0.5]
HBW
[4]
+70°
+80°
270°
90°
180°
0
0.1
0.2
-0.1
-0.2

Fig. 14.

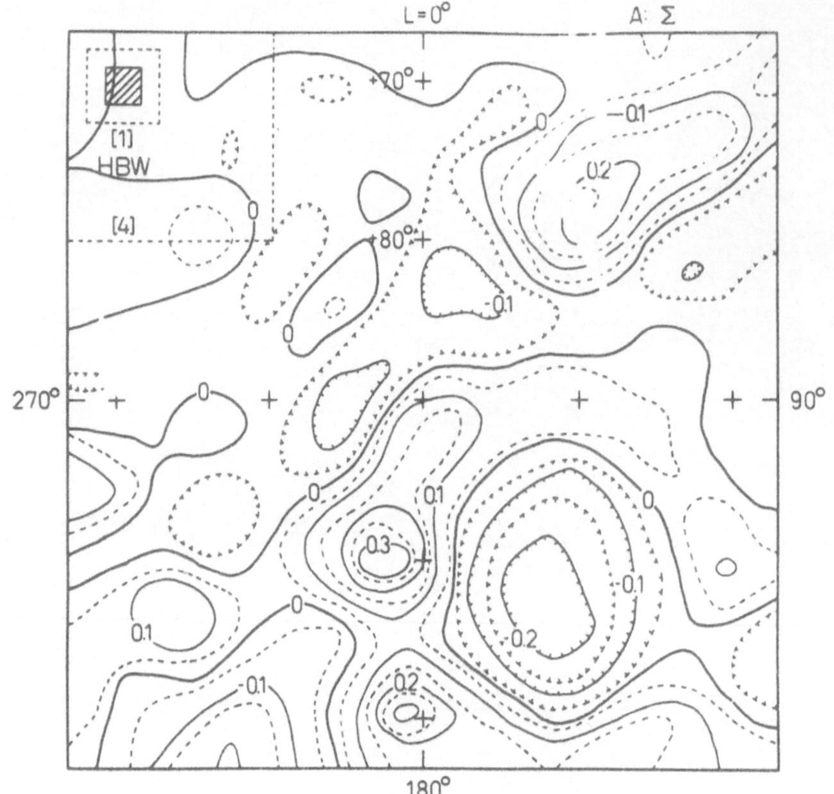
L=0°
A: Σ
[1]
HBW
[4]
+70°
+80°
270°
90°
180°
0
-0.1
-0.2
0.1
0.2
0.3

Fig. 15.

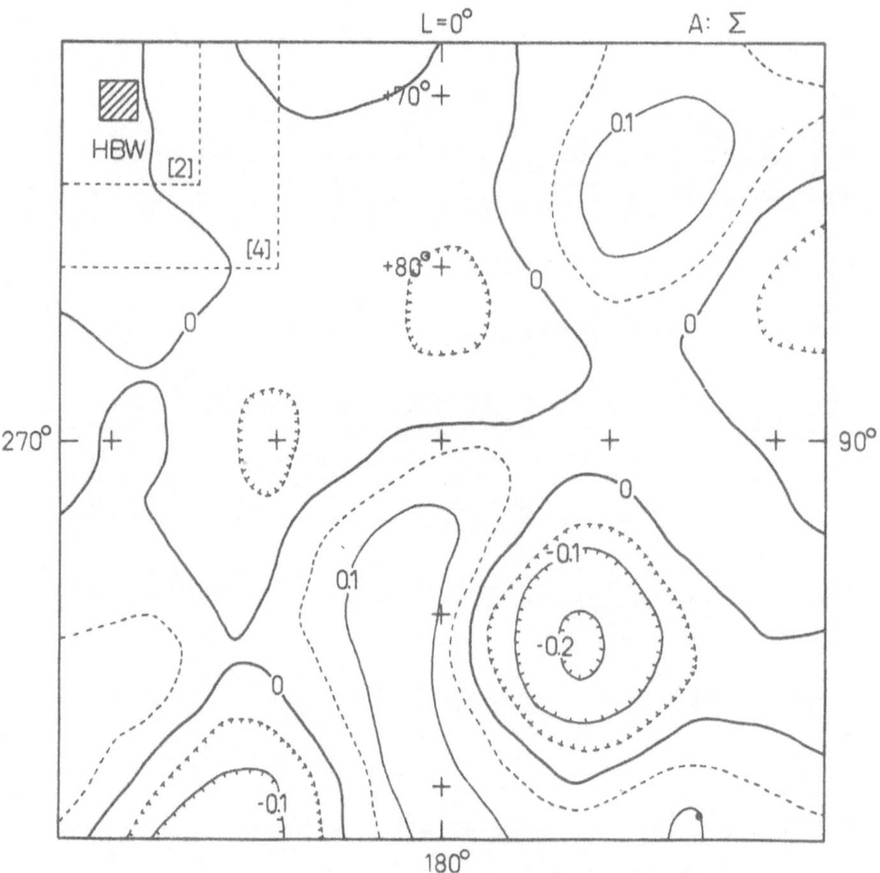
L=0°
A: Σ
HBW
[2]
[4]
+70°
+80°
270°
90°
180°
0
0.1
-0.1
-0.2

Fig. 16.

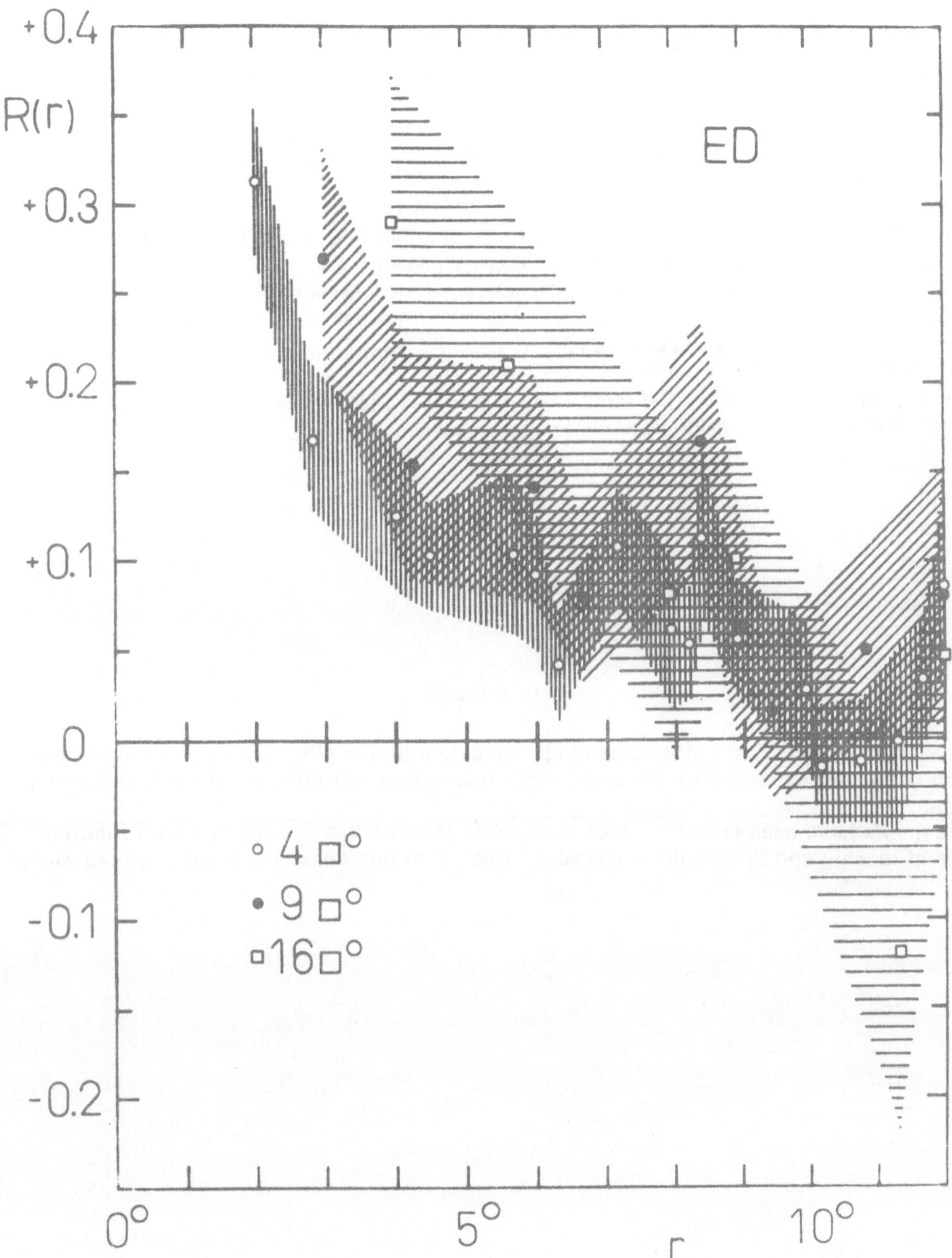

Fig. 17. Two-dimensional normalized correlation functions for ED clusters around NGP. Three breakings up are used. The lag r is in degrees.

REFERENCES

ABELL G.O., 1958, Astrophys. J. Suppl. 3, 211

ABELL G.O., 1965, Ann. Rev. Astron. Astrophys. 3, 1

HAUSER M.G., PEEBLES P.J.E., 1973, Astrophys. J. 185, 757

KALINKOV M., 1973, Compt.Rend.Acad. Bulg.Sci. 26, 1155

KALINKOV M., 1974, Proc. First Europ. Astron. Meet. Springer Verlag, Berlin, Vol. 3, 142

KALINKOV M., 1975, Proc.Second Europ. Astron. Meet. (in press)

KALINKOV M., 1976, Astrophys. Investigations 2 (in press)

KALINKOV M., KANEVA IL., STAVREV K., TOMOV B., VLAHOVA K., JANEV K., 1976 a) Compt. Rend. Acad. Bulg. Sci. (in press)

KALINKOV M., STAVREV K., KANEVA IL., 1975, Lett. Astr. Zu. 1, Nr. 2

KALINKOV M., STAVREV K., KANEVA IL., DERMENJIEV V., RUDNICKI K., 1976b), This volume

KALINKOV M., TOMOV B., 1976, Astrophys. Investigations 2 (in press)

KARACHENTSEV I.D., 1966, Astrofizika 2, 307 (English translation in Astrophysics 2, 159)

KIANG T., SASLAW W.C., 1969, Monthly Notices Roy. Astron. Soc. 143, 129

PEEBLES P.J.E., 1974, Astrophys. J. Suppl. 28, 37

VAUCOULEURS G. de, 1971, Publ. Astron. Soc. Pacific 83, 113

ZWICKY F. HERZOG E., WILD P., KARPOWICZ M., KOWAL C.T., 1961-68, Catalogue of Galaxies and of Clusters of Galaxies, in 6 Vols. California Institute of Technology, Pasadena

DISCUSSION

OZERNOY: Are superclusters of third and higher orders a global effect? If so, I am afraid that they may be in contradiction with the small-scale temperature variations of the 3°K background radiation.

KALINKOV: We have used two catalogues of clusters (Abell's and Zwicky's) with a limiting distance of roughly 600 Mpc (with H= 100 km s^{-1} Mpc^{-1}). In this sense our results are not global

THE STRUCTURE OF CLUSTERS OF GALAXIES FROM OBSERVATIONAL POINT OF VIEW

M.KALINKOV, K.STAVREV, IL.KANEVA, V.DERMENJIEV, K.RUDNICKI
DEPARTMENT OF ASTRONOMY, BULGARIAN ACADEMY OF SCIENCES, BULGARIA;
KRAKOW UNIVERSITY, POLAND
PRESENTED BY K. RUDNICKI

СТРУКТУРА СКОПЛЕНИЙ ГАЛАКТИК С НАБЛЮДАТЕЛЬНОЙ ТОЧКИ ЗРЕНИЯ

1. I n t r o d u c t i o n. It is known that the information which may be obtained from the counting of galaxies (a very laborious procedure) is not used exhaustively. Sometimes only general features of visible galaxy distribution are received. The purpose of the present paper is to pay attention to the fact that with simple digital methods some important phenomenological results from usual galaxy counts may be obtained. We are intending to process in future all accesible to us countings of galaxies, especially those which have been carried out in Warsaw and Krakow Observatories.

A first approach is made for two near clusters in the Zwicky catalogue (Zwicky and Herzog, 1963) Zw 156 - 5 (=Zw 155-19), 1105.3+2835, mc, population of 1090, angular radius r= 1°.96, which is Abell cluster (Abell, 1958) A1185, 1108.2+2857, r=0°.9, m_{10}=14.3 (or the distance is R=106 Mpc, with H=100 km sec^{-1} Mpc^{-1}, Kalinkov et al.,1975) and Zw 156-14 (=Zw 155-4), 1115.2+3013, mc, population of 950, r=1°.44, or A1213, 1113.8+ 2933, r=0°.8, m_{10}= 14.5 (R=117 Mpc). A general view of these clusters is presented in Fig.1 (according to Zwicky catalogue, with usual symbols).

The observational data obtained through the method of Rudnicki (1963) are given by Rudnicki and Baranowska (1966). But here the original galaxy counts on 48" Palomar Schmidt plates are used.

The first step to digital processing consists of a construction of a set of smoothing functions (SF) and filtering functions (FF). A representation of the frequency responses of five gaussian SF, denoted as [0,25], [0.5], [1],[2] and [4] (Kalinkov 1973, 1974) is given in Fig.2. Two-dimensional SF are used in this paper but the frequency responses are one-dimensional, since each central trace of a symmetrical SF gives the frequency response. More precisely in Fig. 2 L is not the frequency but the scale size (i.e. the period for the one-dimensional case), which are reciprocal. The responses for all the ten FF (which are non-normalized) constructed for our five SF are also given in Fig.2.

The data interval in the original counts is 5'x5'. As the scale size L in Fig.2 is in arbitrary units we must specify the characteristics of SF and FF. It may be easily shown (Kalinkov et al. 1976) that SF [0.25] is not applicable to the case under study. Following the formula in the paper cited we have Table 1, where R_{max} and L_0 are respectively the maximum transmitted amplitude and the corresponding scale size (in degrees). Note that the Nyquist scale size is 10'=0°.167. Therefore, in order to be restored, the transmitted amplitudes must be multiplied by a factor of 1/0.472 for cases of FF [0.5] - [1], [1] - [2], and [2] - [4] .

II. S o m e r e s u l t s . Smoothed and filtered fields, containing Zw 156-5,14 are given in Figs. 3 - 16. Everywhere SF or FF, the half-beam width (HBW), the data interval (black squares) and the position of a star having $\alpha = 11^h 11^m.1$, $\delta = +29°08'$ (1950.0) are marked. B and Y stand for blue and yellow plates, and M and L - for medium and long exposures respectively. All isopleths are in gal/□ ° (but for filtered maps they are not restored).

Fig. 3 gives an idea of the complicated picture in the investigated area. Many subclusters may be recognized. A well-defined area between both clusters shows a deficiency of galaxies, probably caused by absorption effects. One may expect a small number of subclusters on the next map (Fig.4), treated with [1]. Some other small areas with a deficiency of galaxies may be seen in Figs. 5-6, SF [1], but for blue, medium and long exposures, and their positions do not coincide with the areas from Fig.1. We wonder why the position of the deficiency area between both clusters is shifted to the north in respect to Fig.7 (yellow) - e.g. in Figs. 8-9 (blue). This effect does not exist for SF [4] - Figs. 10 - 12.

Investigated clusters show many peculiarities some of which are very important - e.g. Fig. 13, presenting a smoothed map for the blue counts *minus* the yellow counts. A glance is enough to establish the unusual appearance of the isopleths and the new position of cluster centers. Here we have a colour segregation.

Three filtered maps are also constructed, Figs. 14 - 16. One might expect the situation in Fig. 14 to be repeated to a certain extent in Fig.15, too. But it is quite surprising that both clusters are divided into two parts (Fig.16). We think that the behaviour of the isopleths in this case is unpredictable.

III. D i s c u s s i o n . Our method of smoothing and filtering discrete fields representing galaxy counts (preferably for different colours) allows to obtain many interesting features. We have to point out that the responses of SF and FF are of great importance. We should also add that the famous Lick counting of galaxies (Shane and Wirtanen, 1967) is not well smoothed (SF, used for their maps "averaging square arrays of four", p.12, has a bad response with a reversal of polarity). An important advantage of the proposed method is that cluster sizes are obtained independently from the number and the distribution of background galaxies (but after many FF have been used). It is curious that a great number of subclusters which cannot be discovered by other methods may be found. Besides, the segregation may be studied more carefully. We have a slight segregation for the brightest galaxies which is nonradially symmetrical. A colour nonradial segregation may also be found. We refer to the above-figures to demonstrate the importance of determination of cluster centers in different colours, because it is very important to see why centers do not coincide.

An attempt may be made on the basis of well smoothed and filtered maps to obtain the space distribution of galaxies in spherical symmetrical clusters.

A paper containing a larger number of isopleth maps and also a correlation analysis of the investigated area will appear in one of the next issues of "Acta Cosmologica".

Table 1

Scale size L_0 for maximum transmitted amplitude R_{max}

FF	L_0	R_{max}
[0.5] - [1]	0.27	
[1] - [2]	0.54	0.472
[2] - [4]	1.09	
[0.5] - [2]	0.43	0.779
[1] - [4]	0.86	
[0.5] - [4]	0.72	0.922

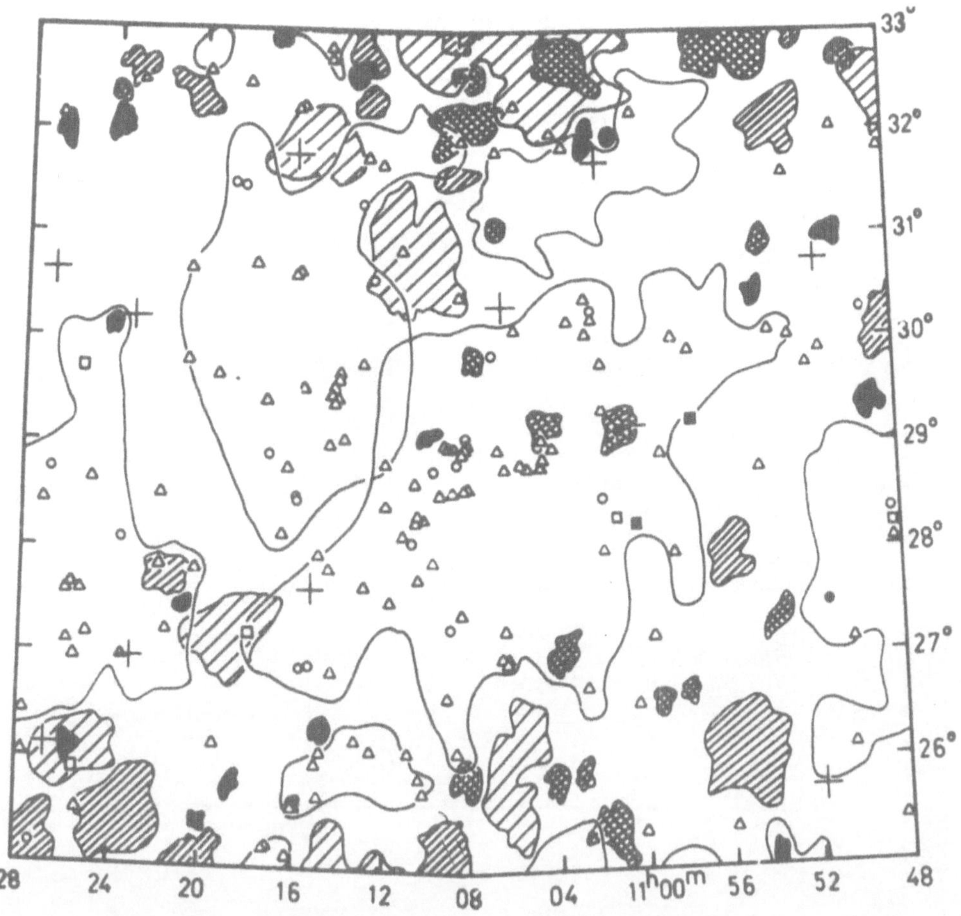

Fig. 1. Clusters Zw 156-5,14 = A1185, A1213. ED clusters are black, and N clusters are marked only with their contour lines. Other distance classes can be easily distinguished. Construction by the Zwicky Catalogue.

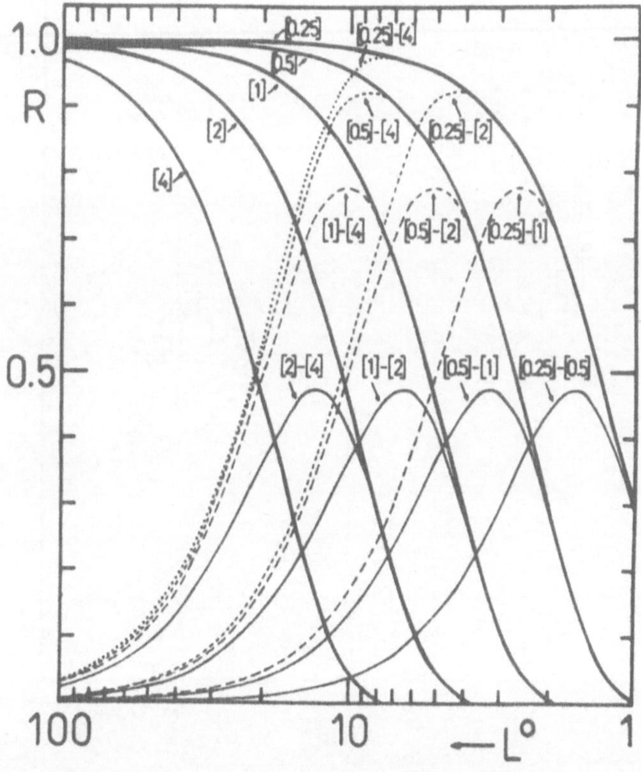

Fig. 2. Responses for a set of SF and FF. L is the scale size in arbitrary units.

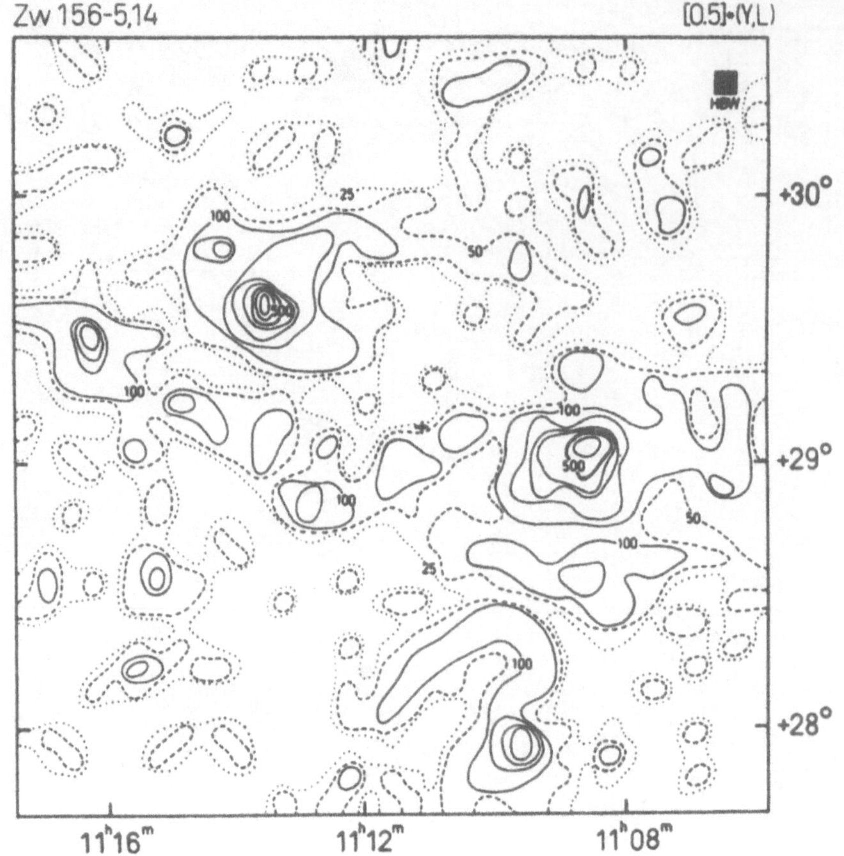
Zw 156-5,14
[0.5]•(Y,L)
HBW
+30°
+29°
+28°
11h16m
11h12m
11h08m

Fig. 3.

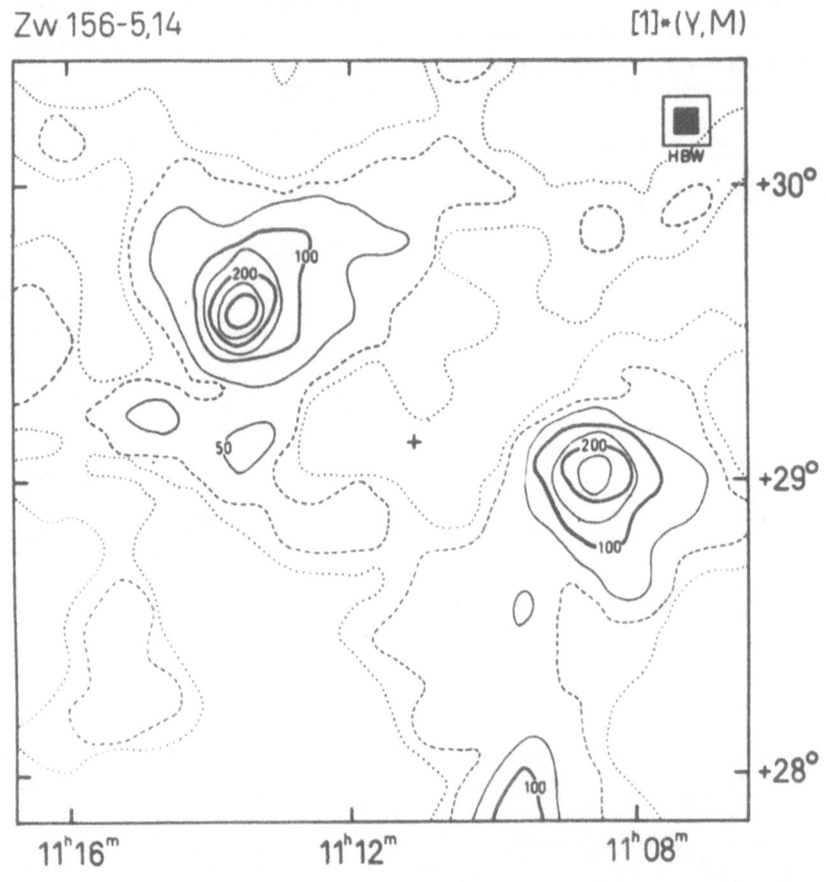
Zw 156-5,14
[1]•(Y,M)
HBW
+30°
+29°
+28°
11h16m
11h12m
11h08m

Fig. 4.

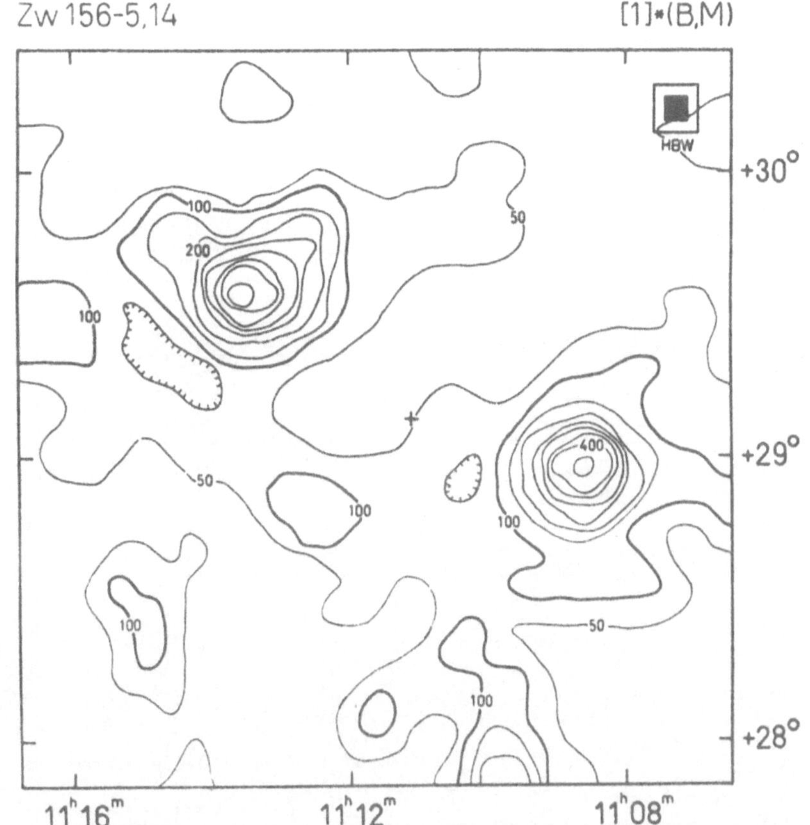
Zw 156-5,14
[1]*(B,M)
HBW
+30°
+29°
+28°
100
200
100
50
400
50
100
100
100
50
100
11h16m
11h12m
11h08m

Fig. 5.

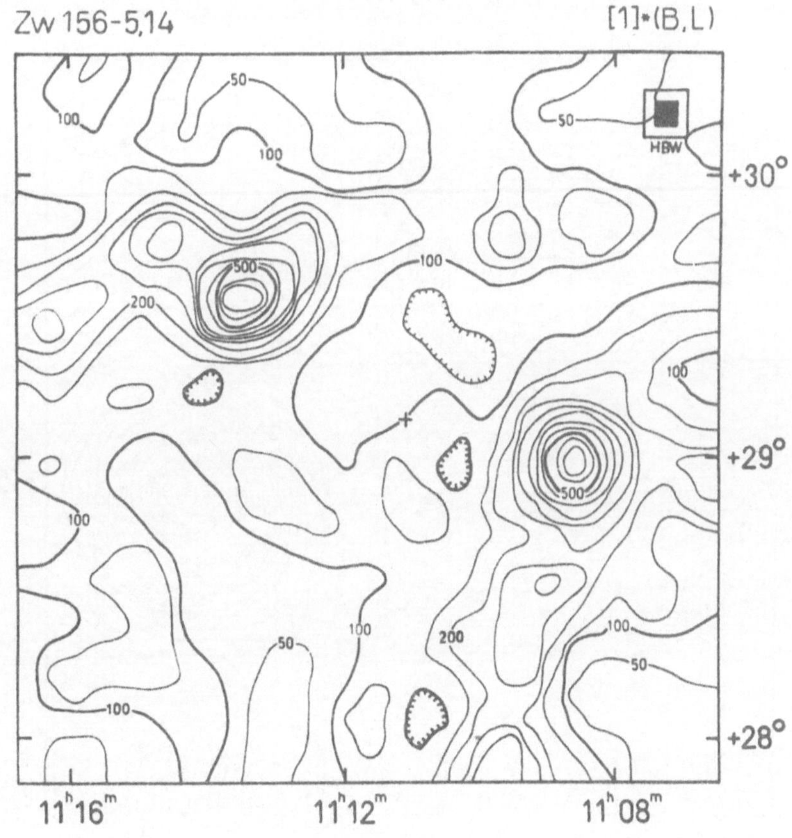
Zw 156-5,14
[1]*(B,L)
HBW
+30°
+29°
+28°
50
100
100
50
100
500
200
100
500
100
100
50
100
200
100
50
100
11h16m
11h12m
11h08m

Fig. 6.

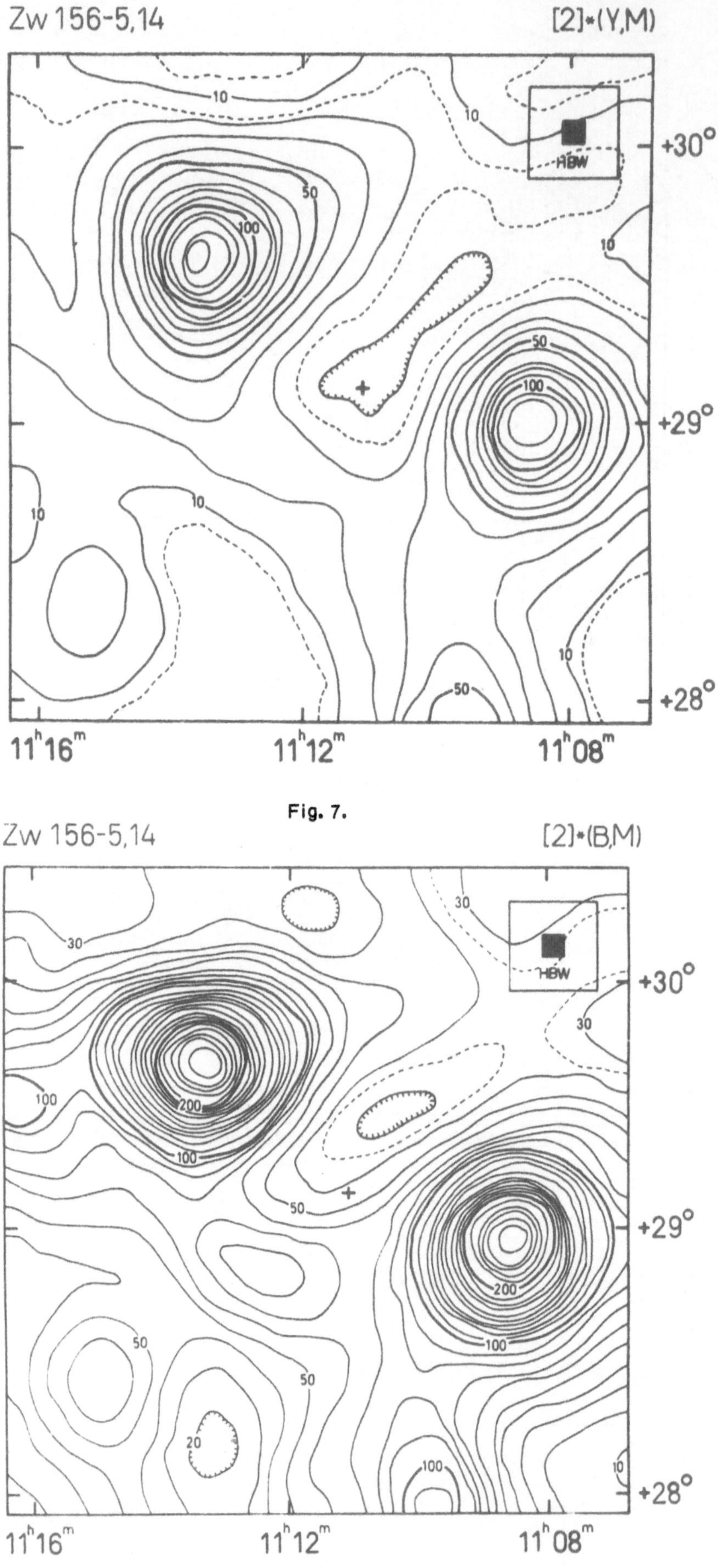
Zw 156-5,14
[2]*(Y,M)
10
50
100
HBW
+30°
+29°
+28°
11h16m
11h12m
11h08m
Zw 156-5,14
[2]*(B,M)
30
100
200
50
20
HBW
+30°
+29°
+28°
11h16m
11h12m
11h08m

Fig. 7.

Fig. 8.

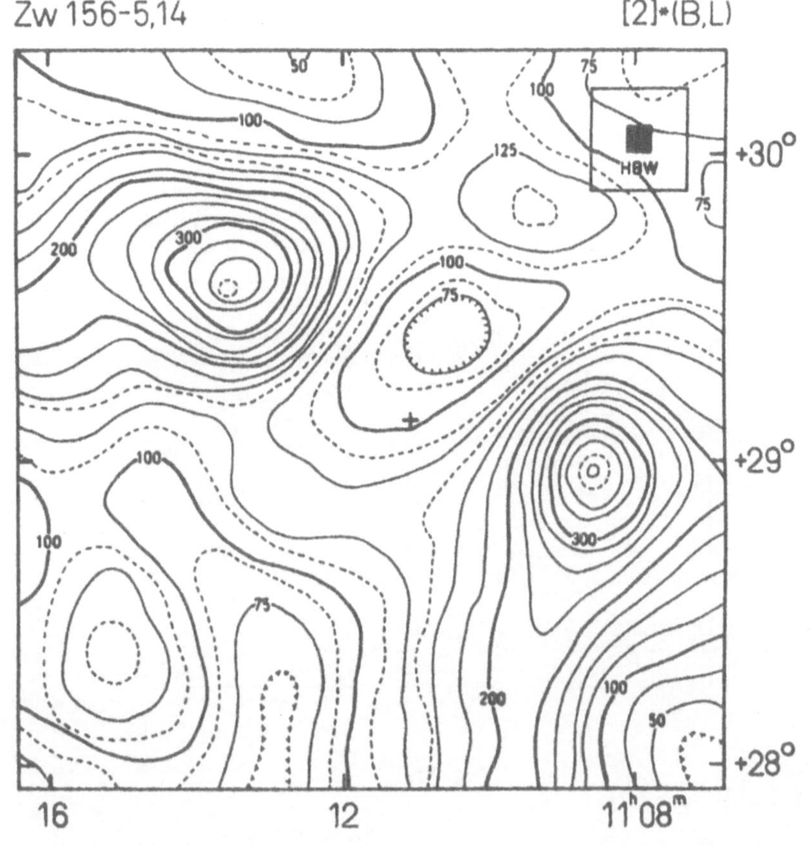

Zw 156-5,14
[2]*(B,L)
50
75
100
100
125
HBW
+30°
75
300
200
100
75
+29°
100
100
300
75
200
100
50
+28°
16
12
11h08m

Fig. 9.

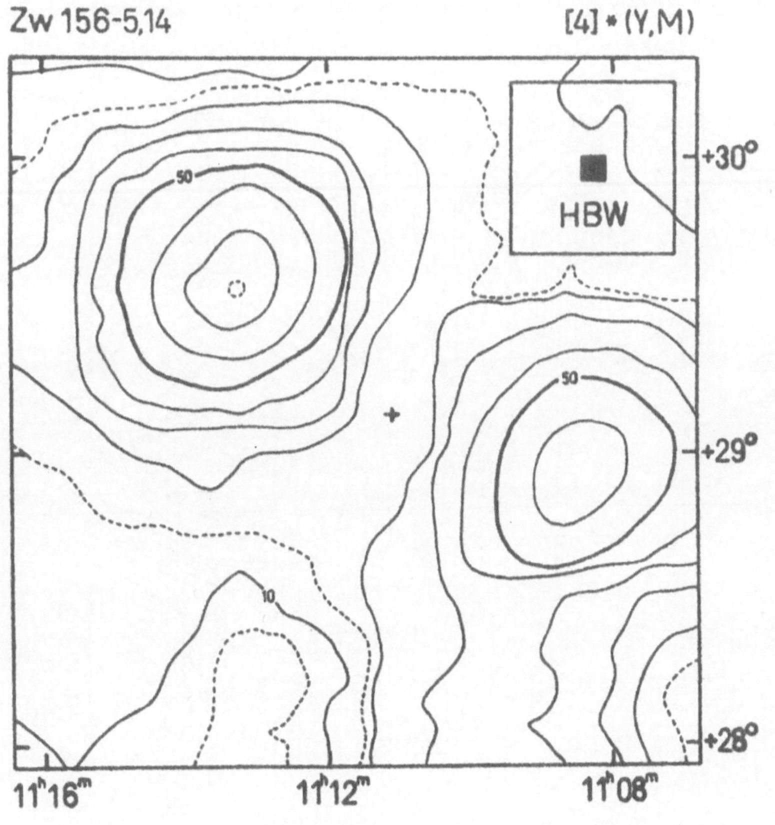

Zw 156-5,14
[4] * (Y,M)
+30°
50
HBW
50
+29°
10
+28°
11h16m
11h12m
11h08m

Fig. 10.

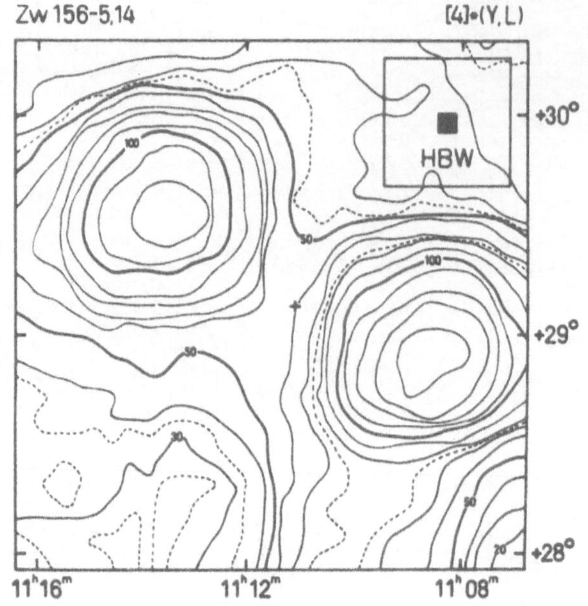
Zw 156-5,14
[4]•(Y,L)
HBW
100
50
30
20
+30°
+29°
+28°
11h16m
11h12m
11h08m

Fig. 11.

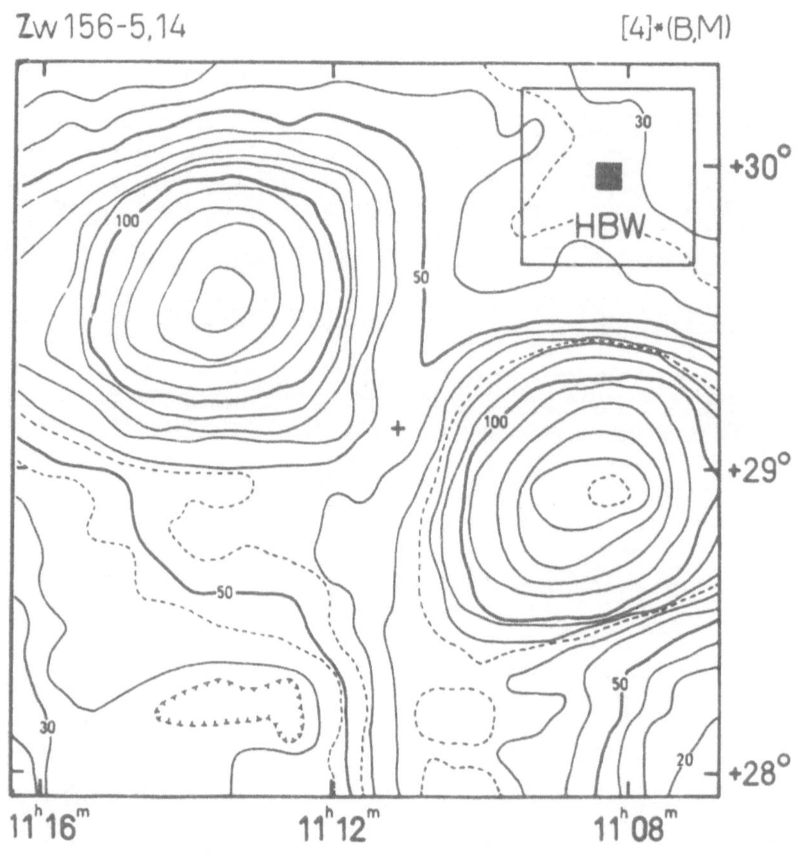
Zw 156-5,14
[4]•(B,M)
HBW
30
100
50
20
+30°
+29°
+28°
11h16m
11h12m
11h08m

Fig. 12.

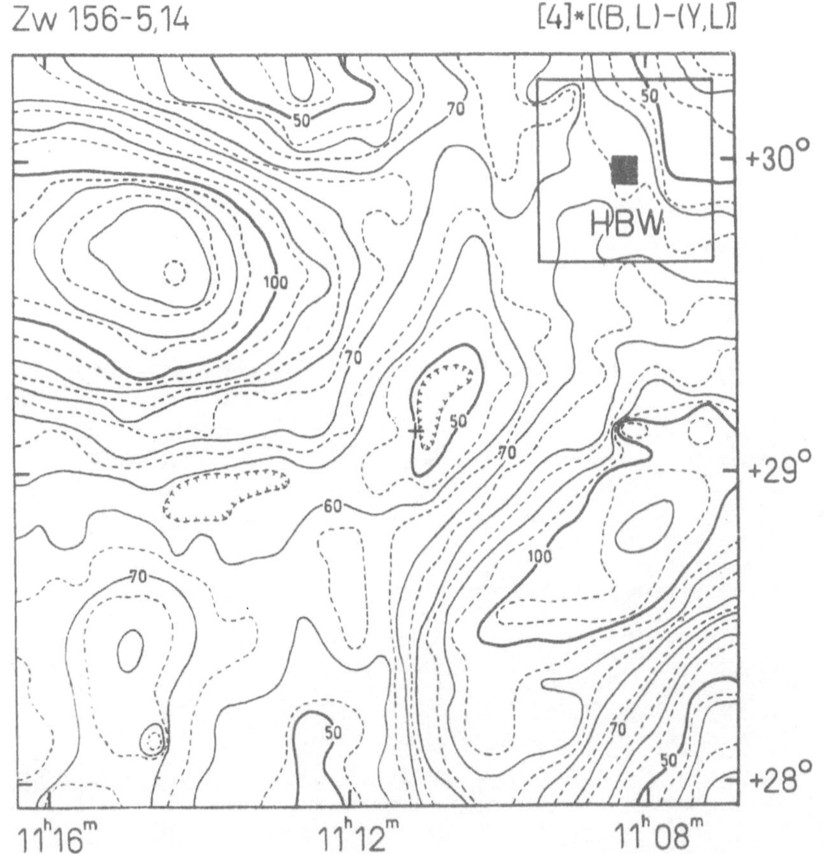
Zw 156-5,14
[4]*[(B,L)-(Y,L)]
50
70
HBW
+30°
100
70
50
70
60
+29°
100
70
50
70
50
+28°
11h16m
11h12m
11h08m

Fig. 13.

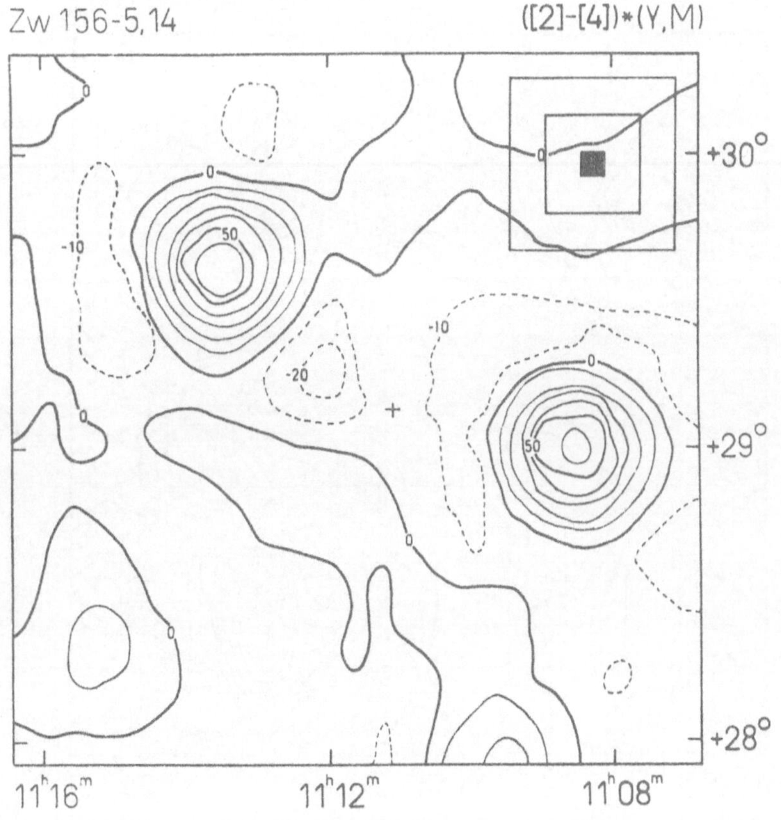
Zw 156-5,14
([2]-[4])*(Y,M)
0
0
+30°
0
-10
50
-10
-20
0
0
50
+29°
0
0
+28°
11h16m
11h12m
11h08m

Fig. 14.

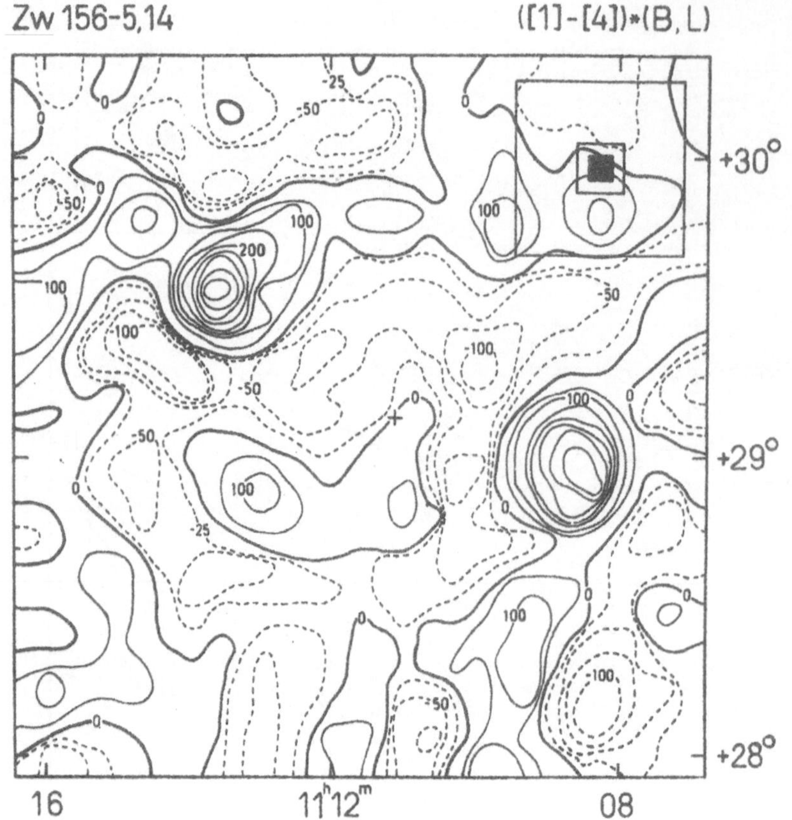
Zw 156-5,14
([1]-[4])*(B,L)
+30°
+29°
+28°
16
11h12m
08

Fig. 15.

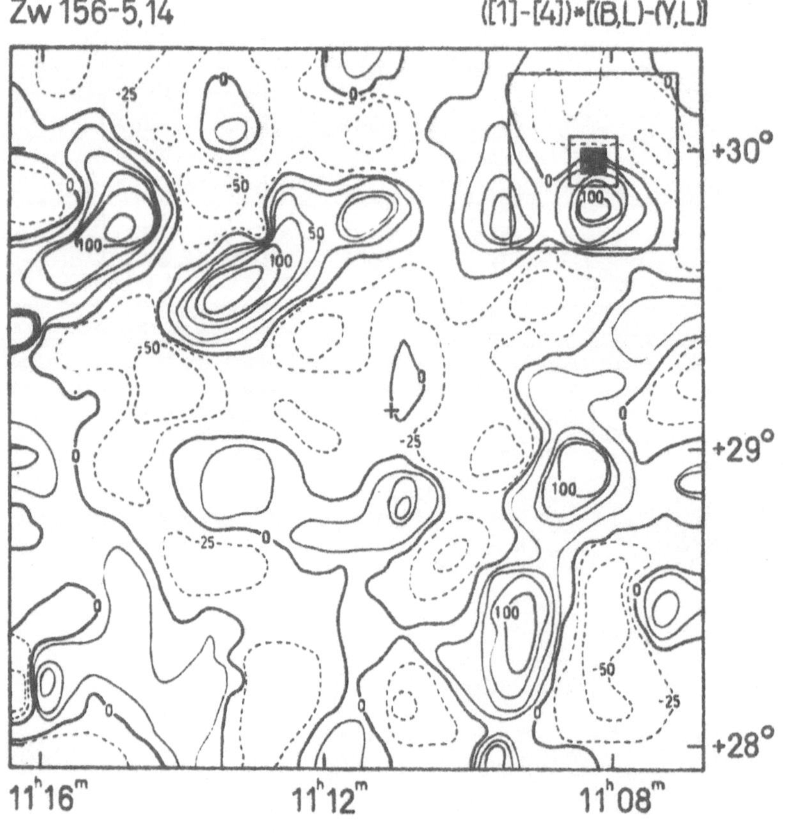
Zw 156-5,14
([1]-[4])*[(B,L)-(Y,L)]
+30°
+29°
+28°
11h16m
11h12m
11h08m

Fig. 16.

REFERENCES

ABELL G.O., 1958, Astrophys.J. Suppl. 3, 211

KALINKOV M., 1973, Compt. Rend. Acad. Bulg.Sci. 26, 855

KALINKOV M., 1974, Proc. First Europ. Astron. Meet. Springer Verlag, Berlin, Vol. 3, 142

KALINKOV M., KANEVA Il., STAVREY K., TOMOV B., VLAHOVA K., JANEV K., 1976, Compt. Rend. Acad, Bulg. Sci. (in press)

KALINKOV M., STAVREV K., KANEVA Il., 1975, Lett. Astr. Zu. 1, Nr. 2, 7

RUDNICKI K., 1963, Acta Astron. 13, 230

RUDNICKI K., BARANOWSKA M., 1966, Acta Astron. 16, 65

SHANE C.D., WIRTANEN C.A., 1967, Publ. Lick Obs. XXII, Part 1

ZWICKY F., HERZOG E., 1963, Catalogue of Galaxies and of Clusters of Galaxies. California Institute of Technology, Pasadena. Vol. 2

OCCULTATION OF THREE MARKARIAN GALAXIES AT 327 MHz

V.A. SANAMIAN, GOPAL-KRISHNA

BYURAKAN ASTROPHYSICAL OBSERVATORY U.S.S.R.

TATA INSTITUTE OF FUNDAMENTAL RESEARCH, BOMBAY, INDIA

PRESENTED BY V.A. SANAMIAN

ПОКРЫТИЕ ТРЕХ ГАЛАКТИК МАРКАРЯНА

The occultation (by Moon) of Markarian galaxies 369, 370 and 384 were observed with the Ooty radio telescope. Two of these, Markarian 369 and 384 were detected at 327 MHz. The galaxy Markarian 370 was not detected at this frequency.

Confusion limit for telescope is about 0.04 f.u. for such observations. No source in the 4 C Catalogue was to lie within 0.5 of the observed galaxies. Moreover, the observed times of occultation are in agreement with the accurate optical positions. It, therefore, becomes very unlikely that the observations were appreciably confused.

The observational and the derived radio parameters are given in Table 1. The physical parameters are computed for Hubble constant = 50 km $sec^{-1}M_{pc}$-1 and Red-Shifts 0.013, 0.003 and 0.016 for Mark. 369, 370 and 384 correspondingly.

Table 1

Markarian galaxy	Flux Density (f.u.)	Angular size (arc sec)	Linear size (kpc)	Radio Luminosity (WH_zST_r)	Brightness Temperature (oK)
369	0.1 ± 0.05	< 5	< 2	5,4 x 10^{21}	5 x 10^4
370	< 0.2	–	–	<5,8 x 10^{20}	–
384 384	0.3 + 0.1	< 5	< 2.4	2,5 x 10^{22}	1.6 x 10^5

Both the galaxies detected at 327 MHz have radio luminosities within the range known for Markarian galaxies and both are stronger than typical normal galaxies. From this occultation observation it seems that radio emission arises in their central regions, not larger than a few kpc. The observed high brightness temperatures suggest possibility of non-thermal origin of radio emission.These results are in agreement with earlier results.

Moreover, the fact that from three arbitrary occultated galaxies two turn out to be a high luminosity radio sources at 327 MHz is in favour of that in Markarian galaxies the radio emission at lower frequency band occurs more frequently than at high frequency band.

RADIO BRIGHTNESS DISTRIBUTION OF SOME MARKARIAN OBJECTS AND GALAXIES OF M 82 TYPE

R. SRAMEK, H. M. TOVMASSIAN
BYURAKAN ASTROPHYSICAL OBSERVATORY, U.S.S.R.
PRESENTED BY H.M.TOVMASSIAN

РАСПРЕДЕЛЕНИЕ РАДИОЯРКОСТИ НЕКОТОРЫХ ОБЪЕКТОВ МАРКАРЯНА И ГАЛАКТИК ТИПА М 82

In this paper we present the results of aperture syntheses observations of eight Markarian galaxies and two galaxies of M 82 type, made with the Green Bank interferometer at 3.7 and 11.1 cm.

For some of those Markarian galaxies radio emission had earlier been detected by us at 6 cm (Sramek and Tovmassian, 1975). Five of them (MRK 1, 3, 231, 273 and 348) are of Seyfert type with broad hydrogen and forbidden lines. Sixth galaxy, MRK 501, is one of the four objects without emission lines that were detected in our 6 cm survey and thus is similar to BL Lac type objects. The other two Markarian galaxies observed are MRK 171 and MRK 297. Both of them have narrow emission lines in their spectra, have very irregular appearance and are included in the Arp's Atlas of Peculiar Galaxies.

Thus Markarian galaxies of all three types which relatively often have radio emission (Sramek and Tovmassian, 1975) were included in the list of present observations.

Since Markarian galaxies possess prominent optical nuclei it was interesting to see whether the radio emission originates mainly in their nuclei or if there could be extended components as well. As we showed earlier (Sramek and Tovmassian, 1974) MRK 6, a Seyfert type galaxy, consists of an unresolved component with diameter < 0''.3, which coincides with the optical nucleus of galaxy an extended component with a diameter of ~ 10'' which is displaced from the nucleus.

Two irregular galaxies NGC 520 and 5363 were also included in the program. Both of them were considered as galaxies of M 82 type by Markarian (1963) and it was interesting to find out whether their radio structure have traces of past nuclear explosions. Their radio emission was first detected by Tovmassian (1967) at 21 and 11 cm.

The results of observations are the following:

a)none of Markarian-Seyfert galaxies was resolved. Radio brightness distribution of all of them are dominated by a point source, with diameter less than 1 arc sec which coincides with the nucleus of the corresponding galaxy to within the 6'' errors of the optical position (Peterson, 1973). Only in the case of MRK 3 a possible large component is suspected. For the other galaxies, an upper limit of 10% of the point source flux can be set for the flux density of any additional extended component smaller than 3'.

Thus, remote Seyfert type galaxies, found among Markarian objects, have radio structure similar to that of close-by bright Seyfert-galaxies, with exception of NGC 1275, which is a powerful radio galaxy at the same time.

b) MRK 501, the galaxy without emission lines, was not resolved either. As in the case of Markarian-Seyfert galaxies its radio core with diameter < 1''.0 , from where all observed radio flux is coming, coincides with the optical center of the galaxy.

c) Radio sources in two irregular galaxies NGC 520 and NGC 5363 were not resolved as well. In both of them the sizes of radio sources are less than 1''. Radio sources are situated in the central parts of both galaxies.

d) The situation is different in two peculiar galaxies MRK 171 (Arp 299) and MRK 297 (Arp 209 NGC 6052).

In the case of MRK 171 radio source is double (Fig.1b). The sizes of componens are of about 2'' and 10'' at 11.1 cm and the separation between them is ~ 20''. These radio sources coinside apprieciably well with two of a few bright condensations seen at short exposure photograph of the galaxy (Fig.1a). These condensations are not situated in the centres of the two major components of MKR 171 which appears as a double galaxy at long exposure photographs (see Fig.1c which is a reproduction from Palomar Sky Survey Red Print).

The radio source is resolved also in the case of MRK 297. Its half-power diameter is about 16'' x 26'' with PA≈75°. Radio source coincides with the bright elongated central region of the galaxy and has roughly the same shape.

Thus in most of the observed galaxies, in all five Seyferts and in one BL Lac type object, radio emission is highly localized in their central regions, confirming the activity of their nuclei.

In two other galaxies radio emitting regions are appreciably large which is in accordance with the suggestion made earlier (Sramek and Tovmassian, 1975) that these galaxies are at a later stage of evolution in comparison with Seyfert and BL Lac type objects.

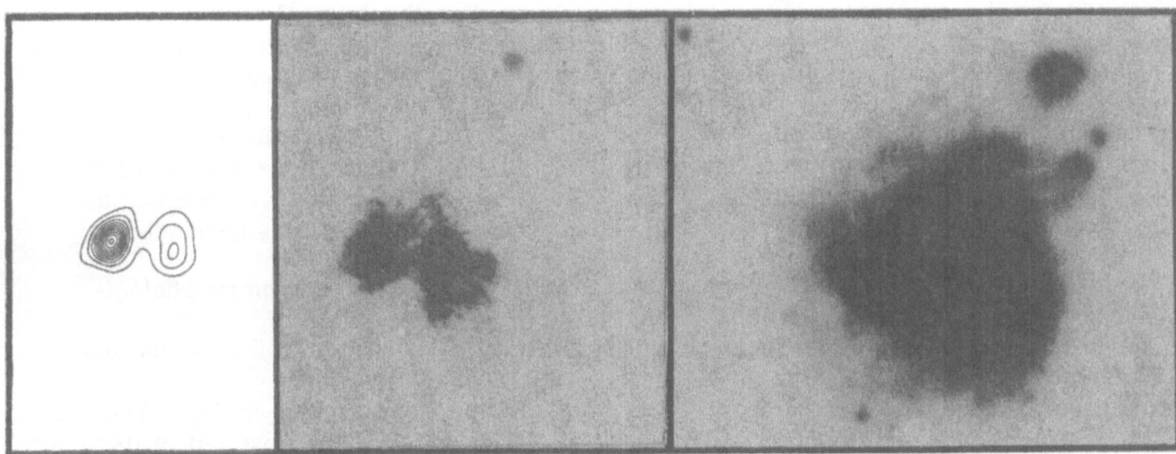

Fig. 1a. Fig. 1b. Fig. 1c.

REFERENCES

MARKARIAN B.E., 1963. Soob. Byurakan Obs., 34, 19.

PETERSON S.D., 1973, A.J., 78, 811

SRAMEK R.A. and TOVMASSIAN H.M., 1974, Ap. J., 191, 633

SRAMEK R.A. and TOVMASSIAN H.M., 1975, Ap.J., 196, 339

TOVMASSIAN H.M., 1967, Astrofizika, 3, 427

DISCUSSION

GOUGUENHEIM: MRK 297 has been studied in high dispersion at the Haute Provence Observatory by R. Duflot. She found that the nucleus is made up of two parts. MRK 201 was studied by Botinelli et al. in the 21 cm line. A secondary line was found, suggesting the existence of large motions.

TOVMASSIAN : Thank you for this interesting information. The data you mentioned also show that the nuclei of Markarian galaxies with detected radio emission are really in an active state.

MULTICOLOR PHOTOMETRY OF SEYFERT GALAXIES MARKARIAN 290 AND NGC 1275

I.I.PRONIK and I.P.METIK
CRIMEAN ASTROPHYSICAL OBSERVATORY, U.S.S.R.
PRESENTED BY I.I.PRONIK

МНОГОЦВЕТНАЯ ФОТОМЕТРИЯ СЕЙФЕРТОВСКИХ ГАЛАКТИК МАРКАРЯН 290 и NGC 1275

Numerous investigations of sample objects from the lists of Markarian show that among them there are giant and dwarf galaxies (Khachikian and Weedman, 1974), superassociations (Khachikian and Saakian, 1975) and so on. It is very interesting to know the morphological types of different Markarian objects and especially of nonstable ones.

Multicolor observations of Markarian objects have been carried out at the prime focus of 2.6-m Schajn telescope using an image converter and 9 color filters. The effective wavelengths used equal approximately 3600, 3730, 4400, 4680, 5090, 5280, 6090, 6600 and 7400 Å. The filters for 3730, 5090 and 6600 Å are centered on emission lines. For absolute calibration extra-focal star images were used. The results for Markarian galaxies 34, 42, 69 and 205 have been published (Metik, Pronik, 1974). Now we discuss the multicolor photometry of Seyfert galaxies Markarian 290 and NGC 1275.

Observations of Markarian 290 were made at 25-26.6, 26-27.6, 27-28.7 and 28-29.7.1971. 44 negatives were obtained in all 9 filters with exposures $16^s - 7^m$. Observations of NGC 1275 were made on 28.3 1968 in 7 filters with the exposure times 10^s-5^m. Photometrical recording has been applied and energy distribution for central and other regions of galaxies has been determined.

The results of this multicolor photometry show that both galaxies are common morphologically.

1. R.Minkowski showed (1968), that the structure of NGC 1275 is like that of E type galaxies but its spectral type is like that of A stars.

Photometrical perpendicular recording directions of Markarian 290 are given in Fig.1. They show that the structure of Markarian 290 is like that of an E type galaxy. The most part of its radiation comes from the 8" region (4.5 kpc if H= 75 km/sec/Mpc). The bulge of this galaxy is about 12" (6.5 kpc). Fig.2 gives the spectral energy distribution of the central parts of galaxies investigated. Markarian 290 is bluer than NGC 1275 and may consist of stars earlier than type A.

2. Direct photographs of galaxies in filter 7400 A are given in Fig.3. They show that both galaxies have satellites. In Fig.4 spectral energy distribution of satellites is given. Since the interstellar absorption in Milky Way is small in the directions of the galaxies under consideration (Weedman, 1973; Sharov, 1963), it is possible to conclude that the satellites are really red. Spectral energy distribution of NGC 1275 satellite has a gap near 5100 Å. That is the absorption band of MgH which shows that the satellite contains a great deal of dK type stars (Spinrad, Wood, 1965). It was shown that satellite of Markarian 290 consists of late type stars too. Visible and absolute magnitudes of satellites are given in the table. One can see that giant galaxies under consideration have satellites which absolute magnitudes are like those of dwarf galaxies or nuclei of giant galaxies,

Table 1

Visual and absolute magnitudes of galaxies and their satellites

Galaxy	m_V (sat.)	M_V (sat.)	M_V (total)
NGC 1275	$17.^m8 \pm 1^m$	-17^m	$-20.^m5$ (by E.Dibaj)
Markarian 290	$20.^m0 \pm 1^m$	-16^m	$-20.^m73$ (by D.Weedman)

3.A remarkable feature of NGC 1275 is the blue knot 3'' near the nucleus. This knot was found by Burbidges (Burbidge and Burbidge, 1965) in blue light. Photometrical recordings of NGC 1275 in 3 filters passing across the nucleus and blue knot are given in Fig.5. The knot is on the left of the nucleus. Its maximum radiation is at 4680 Å, while in blue and red it weakens. It is very probable that the knot has spectral energy distribution like that of A type stars or well known jet of NGC 4486 (Pronik et al.,1967). An excess radiation at 7400 Å is seen to the right of the nucleus in Fig. 5.

Fig.1 shows that the nucleus and the bulge of Markarian 290 are a little larger in ultraviolet than in infrared. Fig.6 shows photometrical recording of Markarian 290 taken across its nucleus and the satellite. It gives the evidence that there is an excess radiation in blue light on the left side of the red nucleus of Markarian 290. This excess resembles the blue knot near the nucleus of NGC 1275.

4. To all photometrical likenesses of galaxies investigated one can add that they have Seyfert type nucleus (Markarian, 1969; Sargent, 1972; Seyfert, 1943).

The detailed results will be published in Izv. Krymsk. Astrofiz. Obs.55.

We express our sincere thanks to Dr. K.K.Chuvaev for placing the image convertor for observations and remarks, V.I.Pronik for stimulating discussions and T.Korkina for drawing the figures.

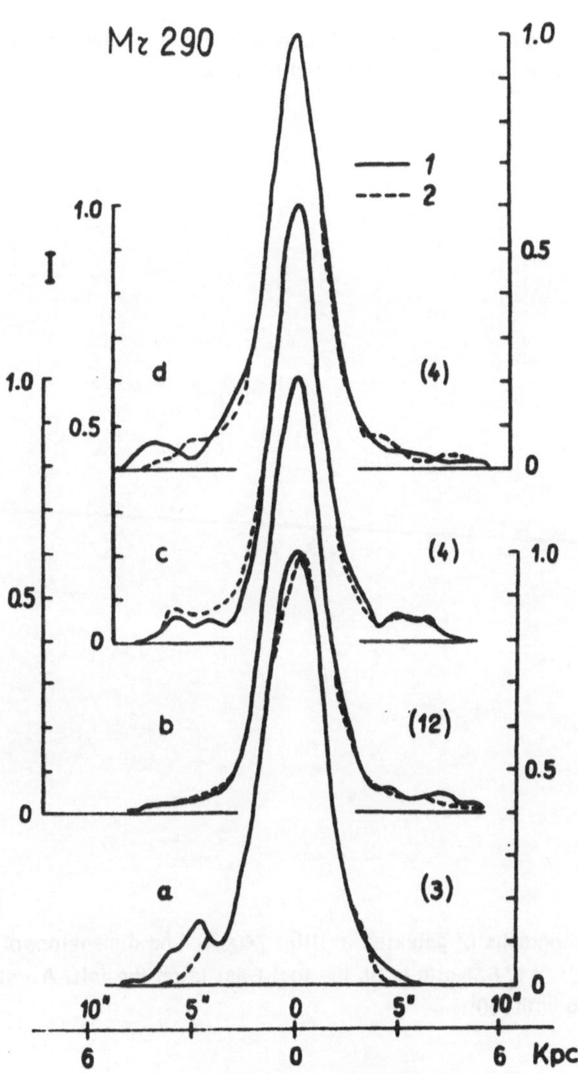

Fig. 1. The intensity distributions in two perpendicular photometric records of Markarian 290:
a - in filter 3600 Å,
b - in filter 4400 Å,
c - in filter 5200 Å and
d - in filter 7400 Å.
The numbers in brackets are equal to negatives used.

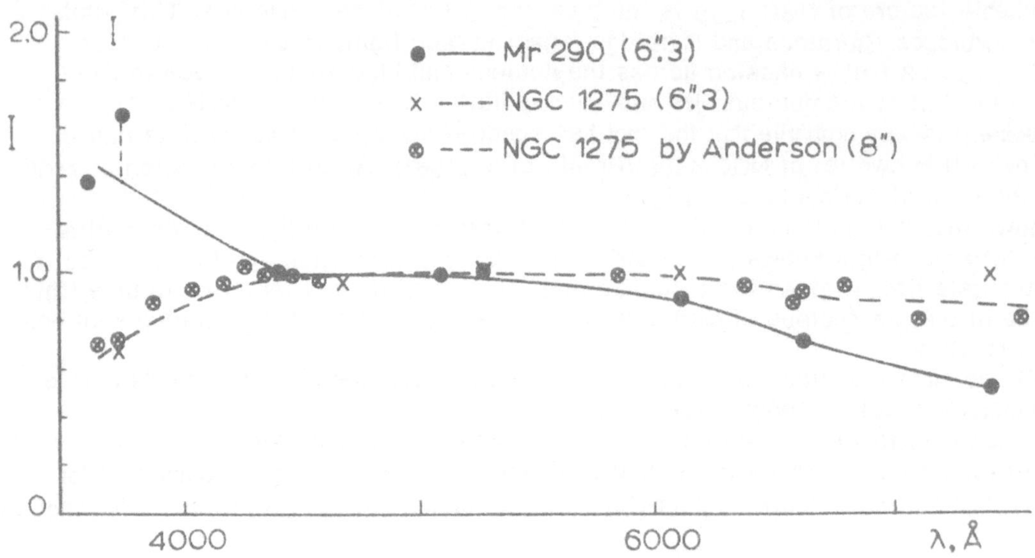

Fig. 2. Relative spectral energy distributions of galaxies. Mean square error of relative intensities are shown on the top left.

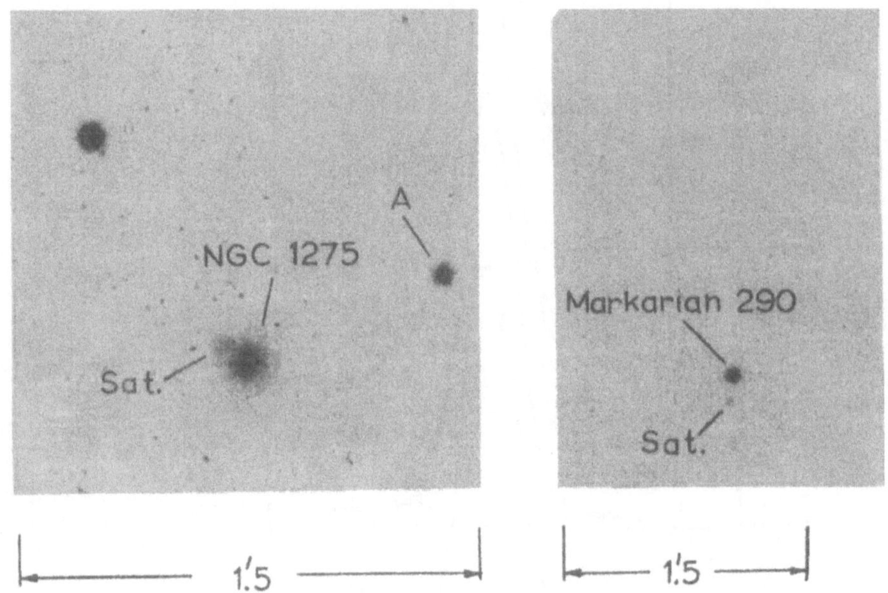

Fig. 3. Direct photographs of galaxies in filter 7400Å. The dimensions of each photo is 1'.5 x 1'.5. North is on the top, East is on the left. A - star for identifying with [10].

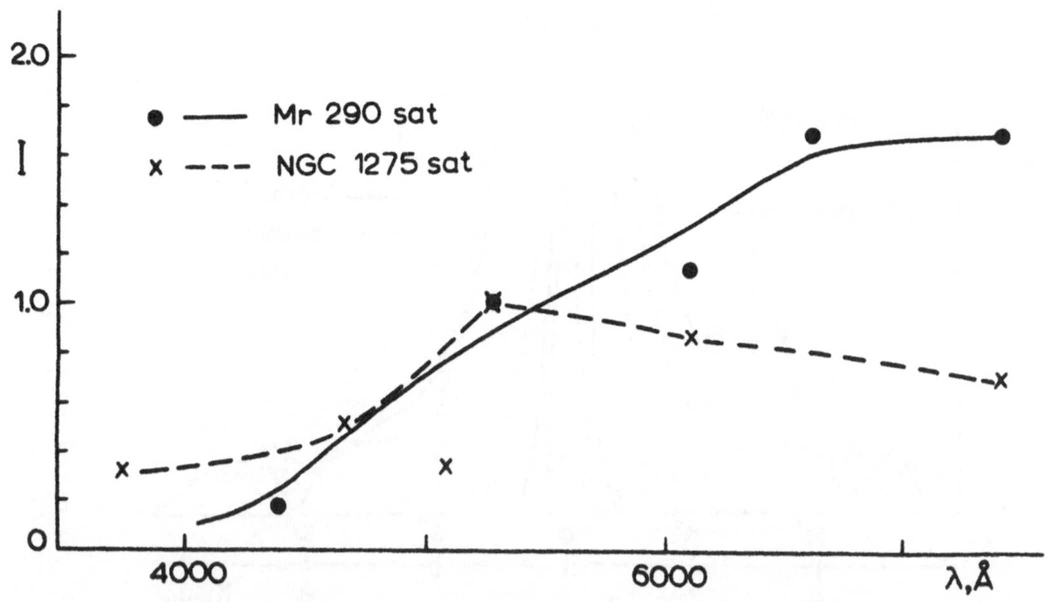

Fig. 4. Relative spectral energy distribution in galaxies' satellites.

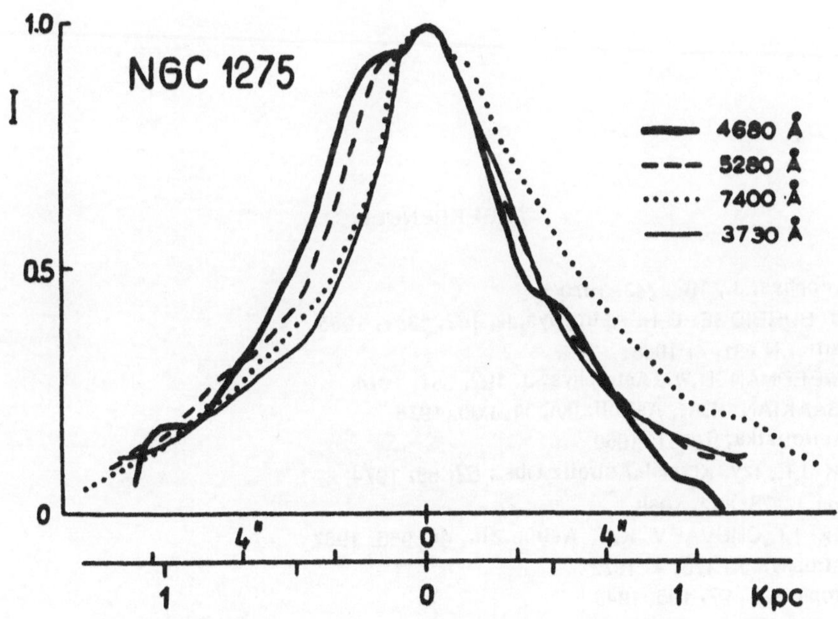

Fig. 5. Photometric record across the blue knot and nucleus of NGC 1275.

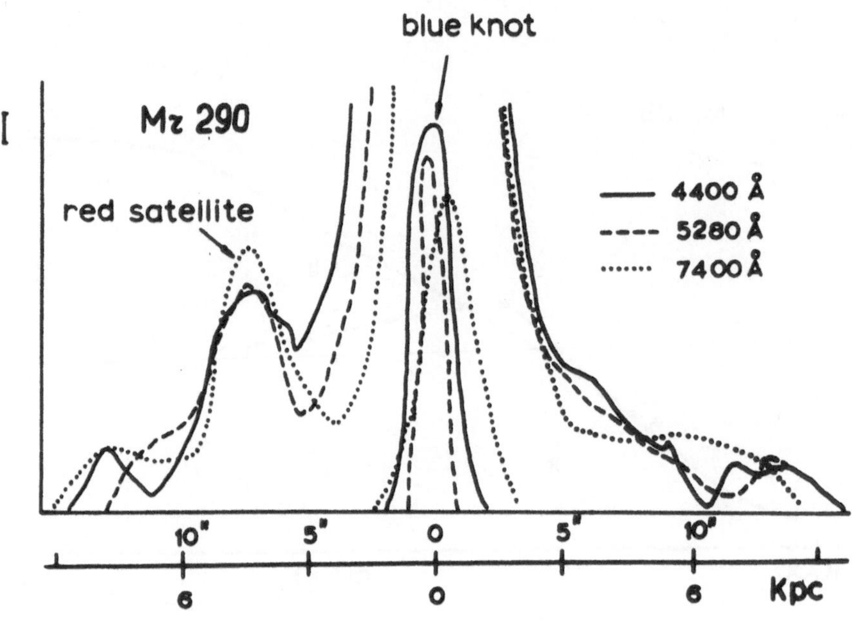

Fig. 6. Photometric record across the nucleus and red satellite of Markarian 290. The scales of intensities for nucleus and bulge are not similar.

REFERENCES

ANDERSON K.S., Astrophys. J., 162, 743, 1970

BURBIDGE E.M. and BURBIDGE G.R. Astrophys.J., 142, 1351, 1965

DIBAJ E. Astron. Tsirk., N 481, 4, 1968

KHACHIKIAN E.Y., WEEDMAN D.W., Astrophys. J. 192, 581, 1974

KHACHIKIAN E.Y., SAAKIAN K.A., Astrofizika, 11, 000, 1975

MARKARIAN B.E., Astrofizika, 5, 581, 1969

METIK L.P., PRONIK I.I., Izv. Krymsk.Astrofiz.Obs., 52, 65, 1974

MINKOWSKI R., Astron.J., 73, 842, 1968

PRONIK V.I., PRONIK I.I.,CHUVAEV K.K. Astron.Zh., 44, 965, 1967

SARGENT W.L.W. Astrophys.J., 173, 7, 1972

SEYFERT C.K., Astrophys.J., 97, 195, 1943

SHAROV A.S., Astron.Zh., 40, 900, 1963

SPINRAD H., WOOD D. Astrophys.J., 141, 109, 1965

WEEDMAN D.W., Astrophys.J., 183, 29, 1973

A KINEMATIC AND PHOTOMETRIC STUDY OF EARLY-TYPE BARRED SPIRALS

M.F. DUVAL-CHERIGUENE
OBSERVATOIRE DE MARSEILLE, FRANCE

КИНЕМАТИЧЕСКОЕ И ФОТОМЕТРИЧЕСКОЕ ИССЛЕДОВАНИЕ СПИРАЛЕЙ РАННЕГО ТИПА С ПЕРЕМЫЧКОЙ

I n t r o d u c t i o n. It is becoming more and more evident that bars play an important role in the dynamics and evolution of spiral galaxies. Unfortunately these systems are difficult to study spectroscopically, especially the earlier morphological types, because of the presence of large quantities of dust evident in lanes along the bar and the weakness of the ionized hydrogen emission.

We give here some observational results derived from plates taken with the nebular spectrograph on the Observatoire de Haute Provence 2 meter telescope. The dispersion is 35 Å/mm at H_α and the resolution is 85 arc seconds per mm.

Last year at the International Colloquium on Dynamics of Spiral Galaxies at Bures-sur-Yvette we gave rotation curves for 12 galaxies ranging from SBa to IBm. We present here results concerning 2 galaxies reobserved this winter. We can draw several supplementary conclusions regarding movements in these bars and we give some preliminary results on the gas conditions through the ratio of intensities between H_α and [NII] 6583.

N G C 2146 SBabp

This galaxy contains two well-defined but weak arms, a very brilliant nucleus and a bar, these latter two partly hidden by a dark lane which could be a third arm (de Vaucouleurs, 1950). Two plates were obtained along the direction of the bar, taken to be coincident with the major axis of the galaxy, and a third, perpendicular to the bar and centered on the nucleus. This latter plate gives a constant radial velocity of 880 km s^{-1} over 1 kpc, the velocity later accepted as the systematic velocity. Thus, this orientation (p.a. = 41°) seems to correspond to the minor axis.

On the major axis the slope of the rotation curve is very considerable 450km s^{-1} kpc^{-1}, implying a mass of 10^{10} $M_\odot$ within a diameter of 0,8 kpc. In the observed part of the bar, out to about 1 kpc on each side of the nuclear bulb, the velocity gradient is very small. There being a reasonably pronounced discontinuity with the central region, there is evidence that the bar does not rotate in a simple solid body fashion.

Using the services of the Astronomical Plate Reduction Center of Nice we have obtained iso-intensity curves in the five following emission lines [NII] 6553, H_α, [NII] 6583, [SII] 6716, [SII] 6730. The ratio H_α/[NII] 6583 varies from three to six from the northwest to the southeast so the HII regions are "classical". An exception is a strong region in the southeast bar where the nitrogen emission is up by a factor of 2. The [SII] doublet is easily visible and indicates that the electron density is low.

The central region is composed of two distinct emission zones, the first superimposed on the peak of stellar continuum, and the other 10 arc seconds removed.

N G C 5383 SB (rs) b

We have extended the study of the velocity field in this typical barred spiral. The very brilliant nucleus is composed of two elements of 15'' and 5'' diameters. The gradient in the rotation curve is sharply different following the major axis determined previously (120 km s^{-1} kpc^{-1}) (Burbidge et al., 1962, Allen et al., 1973) and the minor axis (25 km s^{-1} kpc^{-1} - note quite different from zero).

Along the bar the centers of these two regions have deviant radial velocities of -60 km s^{-1} and + 50 km s^{-1} with respect to the center defined by the stellar continuum. In the perpendicular direction the results are essentially the same.

Four spectra have been obtained in the bar, two bisect the nucleus, one passes 6'' north and the last passes 9'' south. There is good agreement with velocities where the spiral arms begin. In spite of the weak emission we have been able to obtain velocities along the entire length of the bar, velocities being reported as a function of distance from the stellar nucleus. On the two plates of the central region, there is seen a sharp discontinuity of 50 km s^{-1} between the nucleus and the bar. In the bar the velocity gradient is very low, going up where the arm joins in. A simple explanation would be that the stellar density in the bar is very low.

With respect to solid body rotation defined by the velocity differential between the extreme ends of the bar ($\omega \approx 10$ km s^{-1} kpc^{-1}) the velocity deviations reach 40 km s^{-1} at 8 kpc from the center.

The two slit positions toward the exterior provide evidence for transverse motions in the bar. Every 10'' (2, 3 kpc) in the plane of the galaxy we show differences in radial velocities between points toward the exterior and those near the center. The difference is maximum at 20'' radius on each side of the nucleus, on the northwest side on the southern edge of the bar and on the southeast side on the northern edge.

The iso-intensity tracings clearly show the presence of two nuclear components. In the bar the absence of [NII] emission on the northwest side and of H_α on the southeast side is confirmed. The ratio H_α/[NII] at several spots indicates that the HII regions at the end of the bar are "classical". In the nucleus we confirm that one extremely big region is highly excited while the second shows much weaker excitation (hydrogen-nitrogen ratios respectively 6 and 2). In the bar this ratio is less than equal to unity, a condition similar to that found in the centers of galaxies, particularly ellipticals, and suggests either high excitation or overabundance of nitrogen.

C o n c l u s i o n s . The problems presented by bars are numerous. According to morphological type, the rotation curves show differences: a nucleus of greater or lesser importance is related to rotation gradient more or less rapid.

Thus for two galaxies NGC 5383 and NGC 3351 we have observed marked differences in radial velocities between two bright, distinct elements of the nucleus (110 km s^{-1} for NGC 5383 and 140 km s^{-1} for NGC 3351). Then for NGC 5383 there is a significant discontinuity in velocity between centers of these elements and the bar.

For the third galaxy, NGC 3319 (SB(rs)cd), with a very well-defined bar but less important nucleus, the rotation curve does not deviate in passing from the nucleus to the bar but rather over the full lenght of the bar the angular velocity is roughly constant ($\simeq 25$ km s^{-1}). There is a velocity deviation at the extreme southwest, however, a phenomenon similar to that seen in NGC 7479 (Burbidge et al.,1960).

Peculiar motions in bars, that is motions not following solid body rotation, certainly will necessitate a great deal more observations. Our results on hydrogen to nitrogen flux, although spotly, show that in the bars of the two galaxies studied the conditions of ionized gas cannot be very different from "classical" HII regions.

REFERENCES

ALLEN R.J., GOSS W.M., SANCISI R., SULLIVAN W.T., van WOERDEN H. 1973. IAU Symposium, Canberra

BURBIDGE E.M., BURBIDGE G.R., PRENDERGAST K.H. 1960. Astrophys. J. 132,654

BURBIDGE E.M., BURBIDGE G.R., PRENDERGAST K.H. 1962. Astrophys. J. 136,704

VAUCOULEURS G. de 1950. Ann. d'ap. 13,362

MORPHOLOGY, KINEMATICS AND DYNAMICS OF A SPIRAL GALAXY : M 33

J.BOULESTEIX
OBSERVATOIRE DE MARSEILLE, FRANCE

МОРФОЛОГИЯ, КИНЕМАТИКА И ДИНАМИКА СПИРАЛЬНОЙ ГАЛАКТИКИ: М 33

1. Morphology of the gas : It is well known that the ionized gas is a very good indicator of the spiral structure of galaxies. More generally, the knowledge of the relative distribution of H_{II} regions, H_{I} clouds, dust and stars is absolutely necessary to understand the morphology of spiral galaxies.

A general catalogue of 369 distinct H_{II} regions has been compiled in M33 from wide field photographs obtained with a focal reducer and narrow-band interference filters, giving position intensity and size of the regions (Boulesteix et al., 1974). Narrow band interference H_{α} filters (4Å to 25Å) were necessary in order to lower the continuum flux to acceptable levels. A very fast camera f-ratio (F/1) was then needed to record the faint extended emission. The sensitivity limit obtained on this large field plates was about 150 cm^{-6} pc, i.e. just 3 times more than the general emission of the disk which can be observed by more sensitive techniques (Monnet, 1971).

Fig. 1 presents a mosaic of these plates. H_{II} regions are seen out as far as 35' from the nucleus, extending to the limit of detailed neutral observations by Wright et al. (1972). Thirteen ring-like regions have been noted till 30' from the nucleus.

Their shape is generally circular and their apparent diameter obviously increases with the distance from the nucleus (Fig. 2).

Such hollow regions, which are very similar to the ring-like regions in the irregular galaxy NGC 6822 or in our Galaxy, seen. not to be isolated cases. Probably their detection in external Sc galaxies is just limited by spatial resolution and sensitivity. Because the observations of nonthermal continuum and of H_{α} - N_{II} - S_{II} line ratios, they are certainly not supernovae remnants. Rather these rings are probably a late stage in the life of an expanding ionized region.

For the 369 regions we plotted general histograms related to the intensity diameters, positions. The number surface density of faint H_{II} regions is rather constant over the whole galaxy, while for the stronger regions, it is distinctly greater in the central part.

Fig 3, shows that the histogram of the frequency distribution of H_{II} region diameters, leads to a "most-probable" value of about 13' .

Fig.4 shows a comparative radial distribution of the principal observations in M33, all reduced to projected surface area units.

It was found that the rate of formation of massive stars is proportional to $\rho_{H_I}^{2.2}$ in agreement with the stellar rate inside galaxy.

Concerning the general spiral structure, there is in M33 a strong asymmetry in the distribution of Population I component (blue stars, H_{II} regions, H_{I} clouds), although there is no asymetry of the faint-star background. This asymmetry can be also well seen on H_{I} observation by Wright et al (1972)

In fact, density computations, show that there is a strong correlation between H_{I} and H_{II} spiral structures, the H_{II} arms and H_{I} clouds being in very good agreement (Fig.5). Consequently, neutral gas density fluctuations play an important role in so far as the existence of the spiral structure is concerned and therefore in the basic hydrodynamical equations of spiral structure theories.

At the end, the detailed distribution shows that the position of H_{II} regions with respect to H_{I} clouds leads to a general law, that in M33, most of the H_{II} regions are not situated in the densest part of H_{I} clouds but on their edges. This is particularly true in the regions situated along the southern arm, where the arm-interarm H_{I} density gradient is stronger and where a precise shock front can be seen.

II. Kinematics of the gas : More than 4000 velocity measurements were obtained from Perot Fabry observations made by Courtès, Monnet and Georgelin at the Palomar 200 inch-telescope. The mean dispersion of 20Å/mm, allows easily a measuring precision of 2 km.s $^{-1}$.

A kinematic asymmetry, correlated with the previous morphological asymmetry has been confirmed. In terms of rotational velocities, one can distinguish two curves meeting themselves in the outer parts of the galaxy, and a central detailed study shows clearly the different gradients between Northern and Southern curves, related to the dissymmetry of the galaxy. This central velocity asymmetry is certainly strongly correlated with a clear asymmetry of population I in the central part, the photometric data showing that there is no bar of Population II in M33.

Velocity variations in and between the arms appear to be clearly associated with the gravitational perturbations due to the spiral arms.

III. Dynamics : In M33, from Wright 's H_I distribution, it is easy to compute the gravitational perturbation due to the gas itself. The iso-potential curves are given in Fig.6. They are highly asymmetrical as is the distribution of the gas density, of about 4% of the total potential of gas plus stars. In M33, we are thus rather far from popular linear Lin 's theory and fully non linear computations are needed.

From these gas-potential data we tried to compute the orbits of test particles, to see if the rotational curves could be closely connected to the asymmetry of the gas distribution. It has been found that, in a potential field constituted by this gas field and the general star field of the galaxy, the closed orbits are close to a pear-shape: the distance to the center being larger in the southern part, where the rational velocities are lower the motion is more circular. This can explain two facts:

-the dissymmetry of the rotational curves due to combined effects of change of rotational velocity and distortion of the orbits (ϖ);

- the different dispersion of the observed velocities on the two curves due to the missing circularity in the northern part.

The general dynamical data of M33 from observational velocities were computed. Several reasonings lead to a position of the corotation near 5 kpc:

- the observation of the dispersion of the associations in the southern arm (Courtes , Dubout, 1971).

- the nature of the arms which are breaking up at this distance,

- the presence at 5 kpc of a very massive H_I cloud (10% H_I mass), contributing very strongly to the asymmetry of gas mass,

- the applications of Lindblad's inegalities.

This corotation corresponds to a density wave angular velocity of 20 km s^{-1} kpc^{-1}.

In conclusion, it can be said, that the complexity of morphology and kinematics in this galaxy makes difficult the satisfactory test of Lin's Q SS hypothesis. We are now leading more complex studies, especially orbits and simulation.

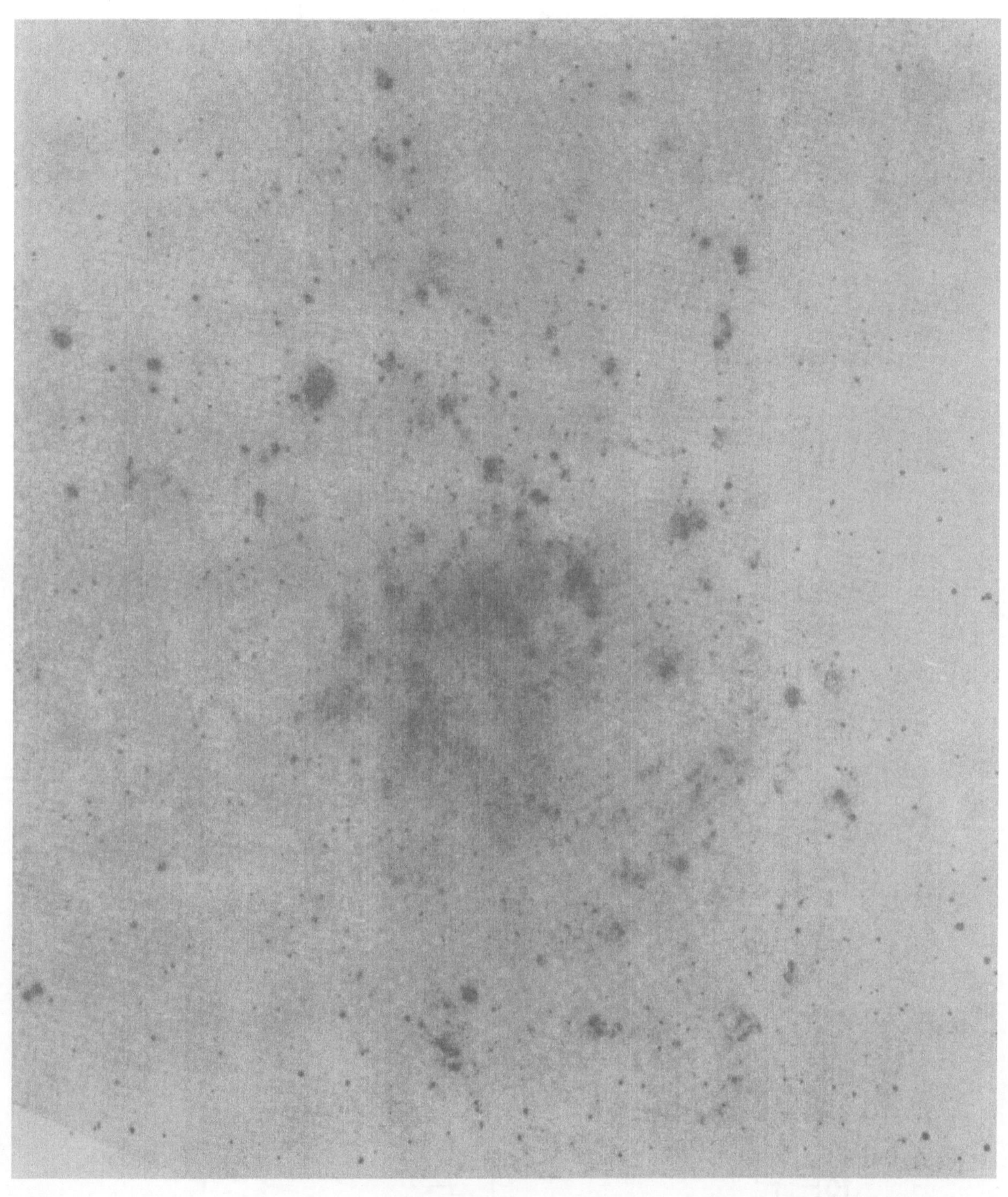

Fig. 1. Large field Hα photograph of M33 (filter band pass of 25Å exposure times 3 hours for each plate).

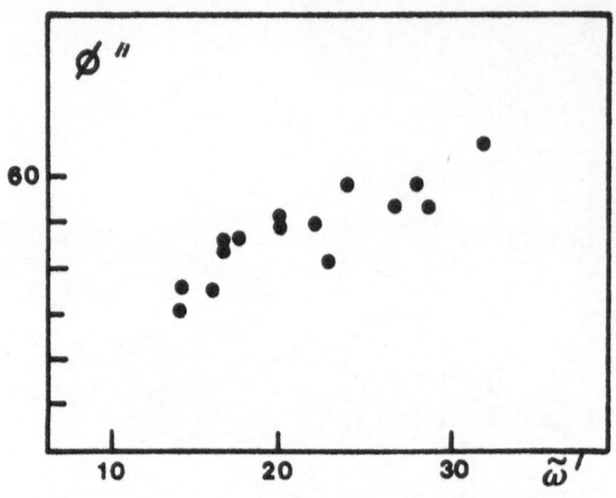

Fig. 2. Apparent diameter of ring-like H_{II} regions with respect to the distance from the nucleus

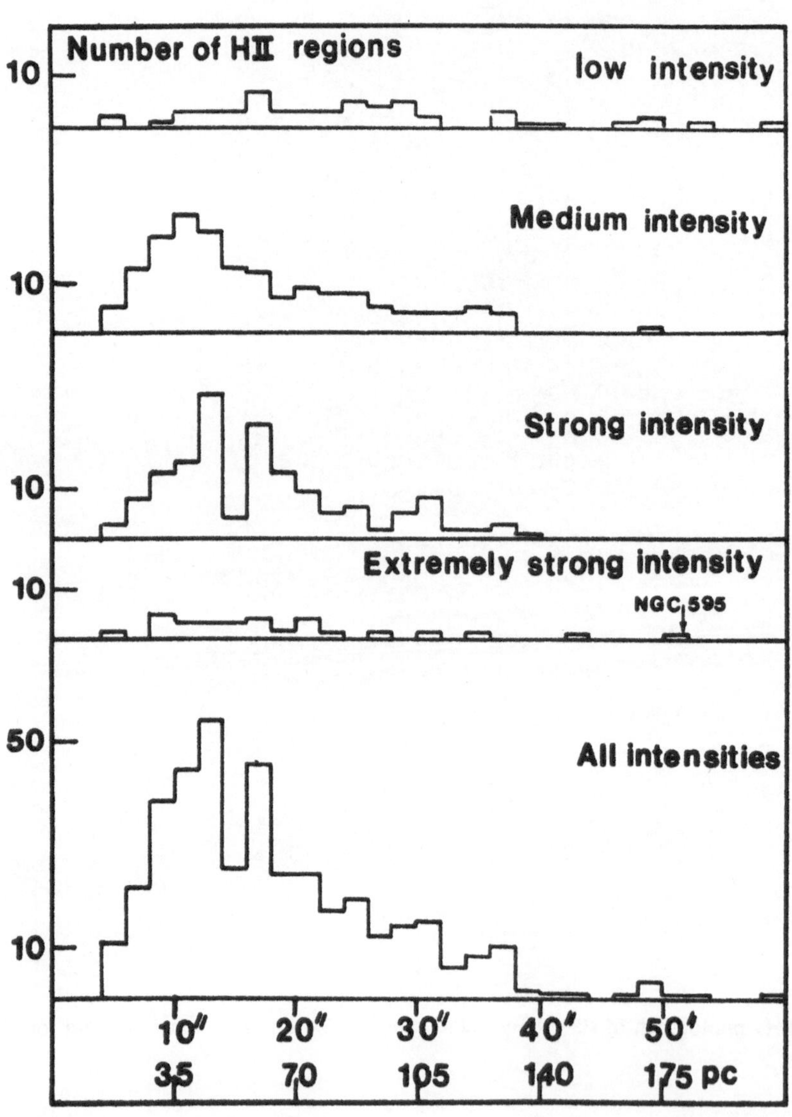

Fig. 3. Histogram of the frequency distributions of diameters of H_{II} regions.

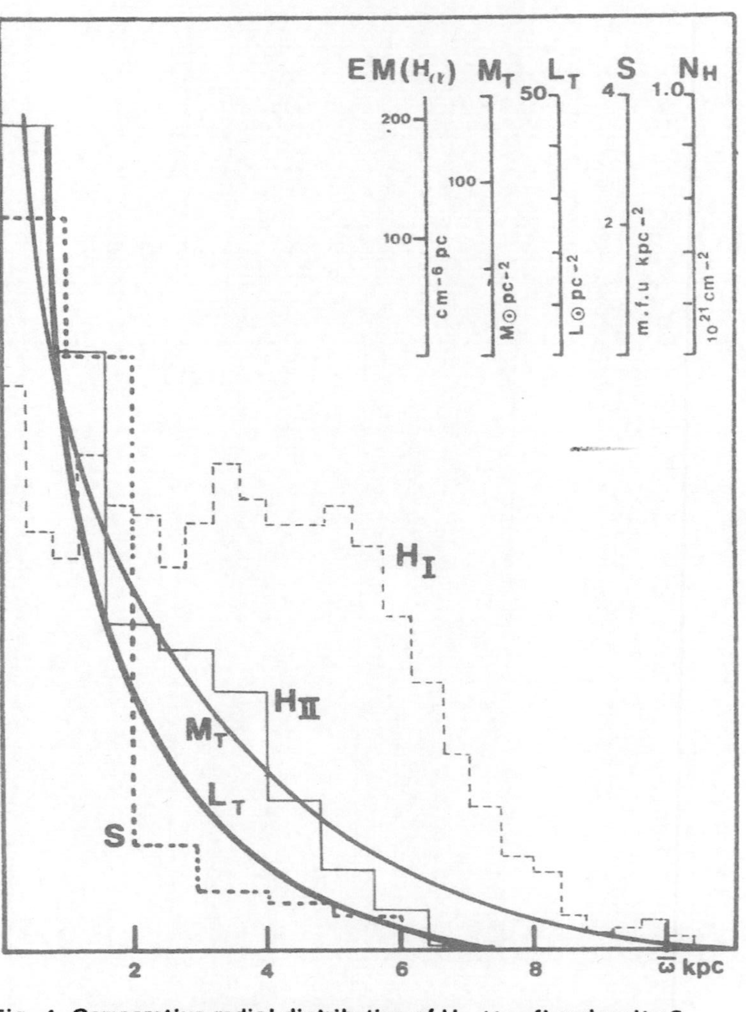

Fig. 4. Comparative radial distribution of H_I, H_{II}, flux density S, total projected mass, and luminosity per unit surface area, plotted as a function of the distance from the center of the galaxy.

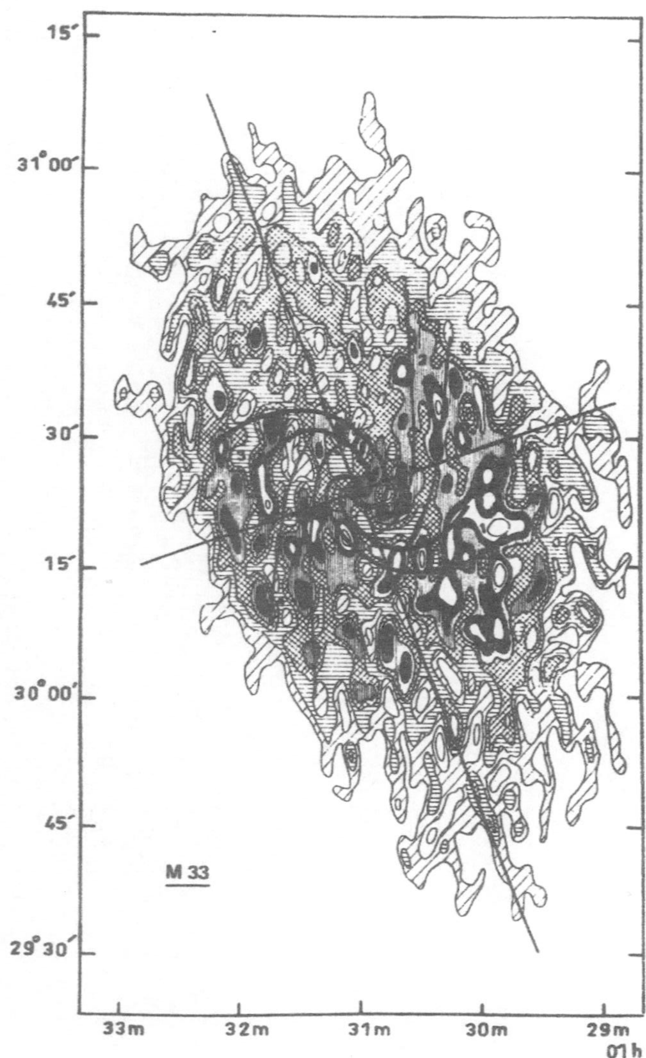

Fig. 5. Position of H_I arms represented on an H_I sky map from Wright et al (1972).

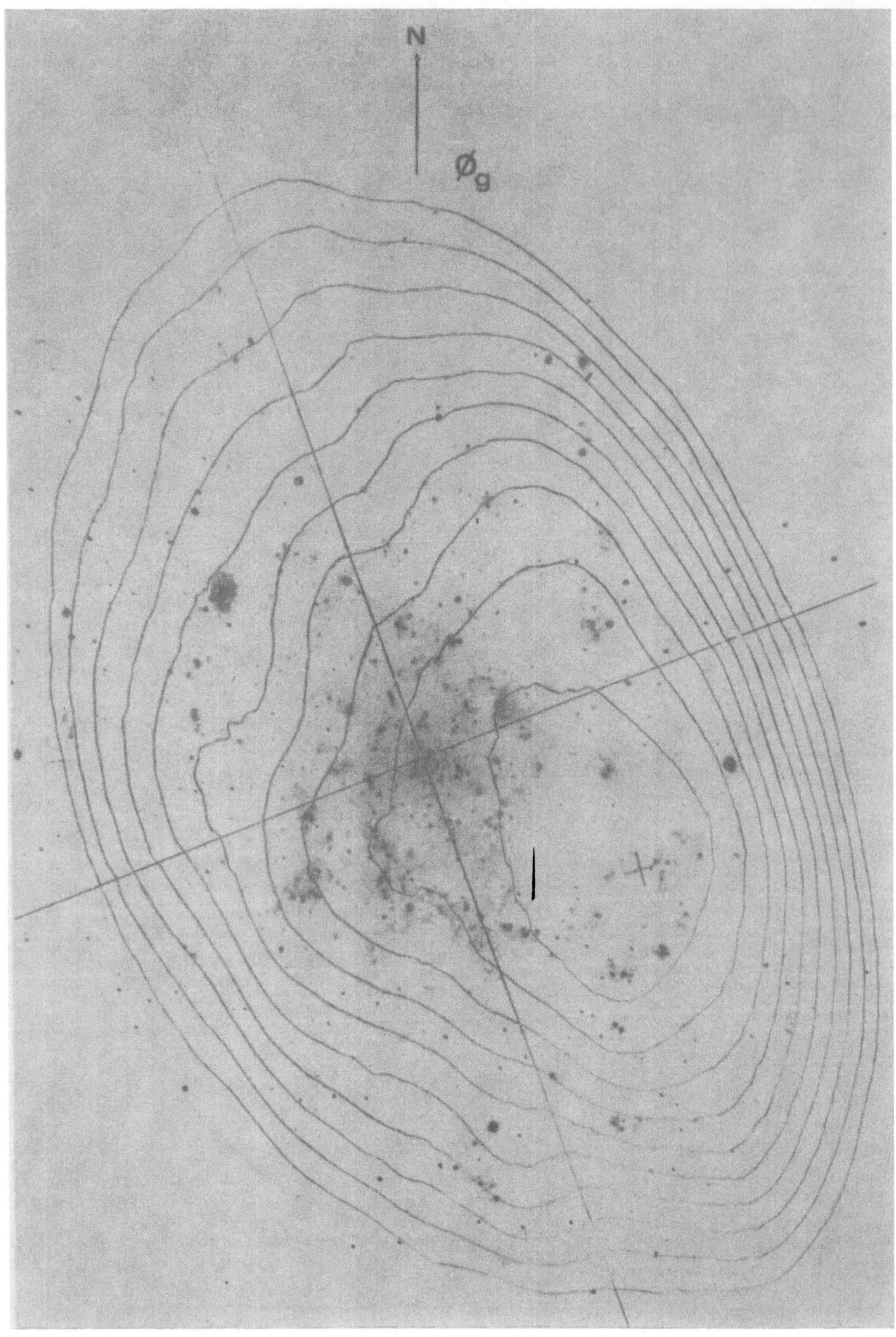

Fig. 6. Iso-gravitational potential curves of the neutral hydrogen mass distribution.

REFERENCES

BOULESTEIX J., COURTÈS G., LAVAL A., MONNET G., PETIT H., 1974, Astr. and Astroph. 37, 33
CARRANZA G., COURTÈS G., GEORGELIN Y., MONNET G., POURCELOT A., 1968, Ann. Astrophys. 31, 63
COURTÈS G., DUBOUT R., 1971, Astr. and Astroph. 11, 488
MONNET G., 1971, Astr and Astroph. 12, 379
WRIGHT M.C.H., WARNER P.J., BALDWIN J.E., 1972 Monthly Notices Roy Astr. Soc. 155, 337

STRUCTURE OF COMPACT RADIO SOURCES IN GALACTIC NUCLEI ON THE MILLI-ARC SECOND SCALE

I. PAULINY-TOTH, E. PREUSS and A. WITZEL
MAX-PLANK-INSTITUT FÜR RADIOASTRONOMIE, F.R.G.
PRESENTED BY E. PREUSS

СТРУКТУРА КОМПАКТНЫХ РАДИО ИСТОЧНИКОВ В ГАЛАКТИЧЕСКИХ ЯДРАХ В ШКАЛЕ ДУГОВЫХ МИЛЛИСЕКУНД

Two major interferometric experiments including transatlantic baselines have been carried out in 1974 between the 100-m telescope in W.Germany and stations in Sweden, West Virginia (USA), Texas (USA) and California (USA) at 6-cm and 2.8-cm wavelength. There were 10 and 6 simultaneous baselines in the 6-cm and 2.8-cm experiments, respectively. The longest baseline in this "array" was the West-Germany-California baseline providing an angular resolution of about 2×10^{-4} arc seconds.

Sources observed were 3C84, 3C120, 3C274, 4C 39.25, 3C273, 3C279, 1633+38, 3C345, 3C454.3, 2134+004, BL LAC, OJ287, and NRAO 150.

All objects showed interference fringes and are at least partially resolved.

The smallest object found is the compact source in the nucleus of M 87 (Virgo A).

The object with the most complicated visibility function coincides with the nucleus of the Seyfert galaxy NGC 1275. In particular for this source it is fair to say that no experiment before yielded so much information from such a small area of sky (diameter ≈ 6×10^{-3} arc seconds). For the first time a consistent model for the brightness distribution is presented which fits all 2.8-cm data with an r.m.s. deviation better than 1% of the total flux. The details of these results will be reported elsewhere.

JOINT OPTICAL AND 21-CM LINE STUDY OF THE PECULIAR GALAXY NGC 3448

L.BOTTINELLI, R.DUFLOT, L.GOUGUENHEIM
OBSERVATOIRE DE MEUDON AND OBSERVATOIRE DE MARSEILLE, FRANCE
PRESENTED BY L.GOUGUENHEIM

СОВМЕСТНОЕ ОПТИЧЕСКОЕ И В ЛИНИИ 21 СМ ИССЛЕДОВАНИЕ ПЕКУЛЯРНОЙ ГАЛАКТИКИ NGC 3448

1. I n t r o d u c t i o n. The peculiar galaxy NGC 3448 is classified by Arp (1966) as a galaxy (n° 205) with material ejected from the nucleus. De Vaucouleurs and de Vaucouleurs (B.G.C., 1964) consider it as an interacting pair, with a very faint dwarf spiral, lying at 3'8 to the West; Nilson (1973) and Vorontsov-Vel'yaminov and Krasnogorskaya(M. C.G., 1962)note the existence of a bridge between these two galaxies.

Its morphological type is under dispute. It is given in the B.G.C. as an irregular of the non-magellanic 10 type, but the classification is doubtful and a SOa type is suggested. Van den Bergh (1960) gives it as a spiral galaxy with a bright nucleus and distorted arms; according to Krienke and Hodge (1974) it is an irregular II galaxy, belonging to the subclass of tidally interacting objects.

In order to elucidate the true nature of this object, a joint 21-cm line and optical spectroscopic study has been undertaken at Nançay and at Haute-Provence Observatories.

II. O b s e r v a t i o n a l   d a t a. Two spectra have been obtained with an image - tube RCA 33011 with the 193 cm telescope of the Haute-Provence Observatory, in the wavelength range 6500-6570 Å, with a dispersion of 30 Å mm^{-1}, the slit being along the major axis of the galaxy. The emission lines [NII] 6548.1 Hα, [NII] 6583.6, [SII] 6717.0 and [SII] 6731.3 show a "crux-like" structure which suggests either the existence of a double system or that of important ejection motions.

The 21-cm line of neutral hydrogen has been observed with the Nançay radiotelescope at several positions along the major axis of the galaxy giving the HI emission of NGC 3448 and that of its dwarf companion.

III. R e s u l t s . The systematic radial velocity of NGC 3448 obtained from the optical spectra is (1370±10) km s^{-1}, in agreement with the HI line determination (1350±20) km s^{-1}.

All the data obtained here:

- the high excitation value (I(Hα)/I([NII]) = 6),
- the comparison of the integral parameters: HI mass to luminosity ratio, mean HI projected density, mean quasi-volumic HI density, mean quasi-volumic indicative total mass density and HI diameter to optical one ratio, with the mean value obtained for each morphological type (Durand, 1975) are consistent with a late morphological type (Sd or Sm) and our results exclude a SOa type.

It appears then that irregular II galaxies may exhibit very different integral properties; indeed, those of NGC 5253 are consistent with a So type (Bottinelli *et al.*, 1972).

The rotation curve is very unusual, exhibiting a strong asymmetry which might be due to an interaction with the dwarf companion. Moreover, the strong non-circular motions confirm the suggestion (Arp, 1966) that an explosive event has taken place in the central part of the galaxy.

The integral properties of the dwarf companion (small luminosity, small diameter, large value of the HI mass-to- luminosity ratio, small internal velocity spread) are in good agreement with the mean values found by Balkowski *et. al.*, (1974) for dwarf galaxies. Its systemic radial velocity exceeds by 130 km s^{-1} that of the main galaxy.

A c k n o w l e d g e m e n t. We are indebted to the Centre de Calcul de Nice for elaborating the computation programs allowing to obtain intensities from microphotometric recordings.

REFERENCES

ARP H., 1966, Atlas of Peculiar Galaxies, p.35, California Institute of Technology, Pasadena

BALKOWSKI C., BOTTINELLI L., CHAMARAUX P., GOUGUENHEIM L., HEIDMANN J. 1974, Astron. and Astrophys. 34, 43

BERGH S. VAN DEN 1960, Publ. David Dunlap Obs. 6, 159

BOTTINELLI L., GOUGUENHEIM L., HEIDMANN J. 1972, Astron. and Astrophys. 17, 445

DURAND N. 1975, Thesis, University of Paris VII

KRIENKE O.K. Jr, HODGE P.W. 1974, Astron. J. 79, 1242

NILSON P. 1973, Uppsala General Catalogue of Galaxies, Uppsala Astron. Obs. Ann. 6

VAUCOULEURS G. de, VAUCOULEURS A. de.1964, Reference Catalogue of Bright Galaxies, University Texas Press, Austin

VORONTSOV-VEL'YAMINOV B.A., KRASNOGORSKAYA A. 1962, Morphological Catalogue of Galaxies, Trudy Sternberg State Astron. Inst. vol. 32

THURSDAY, JULY, 3 SEVENTH SCIENTIFIC SESSION."GALAXIES".(LECTURE)
CHAIRMAN : PROF. G.COURTÉS

GALACTIC X - RAY SOURCES

I.S.SHKLOVSKY
INSTITUTE FOR SPACE RESEARCH, MOSCOW, U.S.S.R.

ГАЛАКТИЧЕСКИЕ ИСТОЧНИКИ РЕНТГЕНОВСКИХ ЛУЧЕЙ

(The author has not sent the text of his lecture).

SPACE DISTRIBUTION AND KINEMATICS OF F-STARS

A.BLAAUW, C.D.GARMANY
LEIDEN OBSERVATORY, THE NETHERLANDS
McCORMICK OBSERVATORY, BOULDER, U.S.A.

ПРОСТРАНСТВЕННОЕ РАСПРЕДЕЛЕНИЕ И КИНЕМАТИКА F-ЗВЕЗД

РЕЗЮМЕ

Представлено предварительное сообщение о состоянии и анализе наблюдательной работы по определению характеристик населенности Галактики звездами среднего возраста в зависимости от расстояний от галактической плоскости. Ведутся наблюдения звезд класса F в северных и южных полярных областях в Площадках собственных движений Мак Кормик для определения собственных движений, спектральных типов, uvby и H-бэта фотометрий. Предварительные результаты показывают, что для этих звезд имеет место постепенное изменение [M/H] от + 0.20 при Z = 0 до -0.35 около Z= 700 пс. Разброс значений достигает 0.22 при Z= 0, оставаясь повидимому постоянным до Z= 700 пс. Кинематические свойства постепенно изменяются в зависимости от Z и от обилия металлов, как и можно было ожидать : для каждой группы расстояний дисперсия скоростей больше для звезд с низким содержанием металлов, чем для звезд с более высоким содержанием ; для каждой группы звезд данного содержания металлов дисперсия скоростей увеличивается с расстоянием Z.

Данные, которые надеемся получить, будут необходимы для изучения некоторых аспектов галактической эволюции, особенно, касающихся более молодой составляющей диска. Можно надеяться, что они позволят выявить, например, более поздние стадии галактического "коллапса" для областей, сравнимых, по расстояниям от галактического центра, с Солнцем; или протяженность, до которой эффекты релаксации в движениях звезд, обусловленные столкновениями с массивными телами (комплексы облаков, волны плотности), могут быть ответственны за рассеяние рассматриваемых объектов в направлении Z .

Выбор звезд типа F для исследования этой "локальной" галактической эволюции на поздней стадии предпочтителен ввиду следующих обстоятельств:

а) диапазон их возраста от 10^9 до 8×10^9 лет;

б) диапазон содержания металлов, как признано, определяется содержанием, характерным для Гиад, и меньшим на порядок величины.

в) их истинная светимость достаточна, чтобы было возможно наблюдать фотометрически за пределами 1 000 пс с помощью телескопов средних размеров.

An interim report is presented on the status of observational work and on a provisional analysis in a project aiming at determining population characteristics of stars of intermediate age as a function of distance from the galactic plane. F-type stars in the McCormick proper motion fields in the north and south galactic polar caps are observed for proper motions spectral types and uvby, H-beta photometry. Provisional results indicate, for these stars, a gradual change of [M/H] from +.20 at z =0 to -.35 around z ≐ 700 pc. The spread around these values, amounting to 0.022 at z=0 appears to remain approximately constant up to z = 700 pc. The kinematical properties gradually change with z and with metal abundance in the sense expected : for each distance group the velocity dispersion is greater for the low metal abundance stars than for those with high abundances; and for each abundance group the velocity dispersion increases with distance z.

1. I n t r o d u c t i o n . This paper reports on progress made in a project comprising photometric, spectral classification and proper motion measurements for the purpose of studying population characteristics in the solar neighbourhood (i.e. up to distances of about 2 kiloparsec). Of the rich observational material obtained so far, we select the F stars for the first analysis.

Aim of this analysis is, to find the distribution as a function of distance from the galactic plane (the z distribution), and as a function of age and of metal abundances. This data will eventually be required for the study of certain aspects of the galactic evolution, especially those concerning the younger disk population. This population may be expected to reveal, for instance, the later stages of the galactic "collapse" for the domains at distances from the galactic centre comparable to that of the sun; it also may reveal the extent to which relaxation effects in stellar motions due to encounters with massive bodies (cloud complexes, density waves) may be responsible for a spread of the objects concerned in the z direction, i.e. a thickening of the layer as compared to the layer in which the stars were formed.

For studies of this "local" late-stage galactic evolution, the F stars are well suited for a variety of reasons:

a) they cover a range of ages from 10^9 years to about 8×10^9 years;

b) they are known to cover a range of metal abundance from about Hyades abundance to an order of magnitude less;

c) they are of sufficient intrinsic luminosity so as to lend themselves for photometric observation up to distances beyond 1000 pc with intermediate-size telescopes;

d) A photometric system has been developed (uvby, H-beta) which allows analysis of these stars in terms of the parameters required (luminosity, colour, metal abundance, age).

From the kinematic properties of the F stars it can be easily deduced that the majority of them must have spent all their life in a narrow zone between 8 and 13 kpc from the galactic centre, and these must also be the confines within which their place of origin must be located. The majority of the stars now in or near the galactic plane do not reach distances above or below the plane beyond 200 pc, and only a small fraction, about 5 percent reaches distances beyond 500pc; F stars beyond these distances from the plane are still sufficiently abundant for study of age and metal abundance characteristics, but they have such high velocities perpendicular to the plane at $z = 0$ (>40 km/sec) that they spend only a short fraction of their lives around $z = 0$.

Investigations of aspects of galactic evolution based on the bright F stars observed in the uvby system have been made by several authors of which we mention specifically Clegg and Bell (1973) and Mayor (1974). These, however, could not but be based on the observed situation in the galactic plane; the addition of information on the run of certain parameters with distance z obviously, is an essential element for more complete studies.

2. Status of the observational programmes. Observed stars are those in the McCormick proper motion fields studied earlier by van de Kamp and Vyssotsky (1937) and Vyssotsky and Williams (1948). These fields are centered on bright stars, originally observed at the McCormick Observatory for trigonometric parallaxes to be measured with respect to the faint field stars which reach down to about 12.0 or 12.5 photographic magnitude. There are about 700 such fields between Dec +90° and -30°. In our project we have given priority to the 69 fields between galactic latitudes 60° and 90°, of which 44 are in the north galactic polar cap (NGP) and 25 in the south polar cap (SGP).

The following summarizes the present status of the observations.

a) Proper motions. About 1800 stars have been measured in 99 regions which include 30 regions at intermediate and low galactic latitudes. All of this work was done at the McCormick Observatory under the supervision of Dr. Ph.I. Ianna and Dr. C.D. Garmany. For each field an average of 8 early epoch and 8 modern plates were used with an average time interval of about 50 years. The resulting relative proper motions have an exceedingly high measuring accuracy, corresponding with probably errors of about 0."001 per year. Furthermore, most of the fields are centered on FK4 stars which makes possible a reduction to the FK4 fundamental system. There is, however, still left an inherent uncertainty in the proper motions due to the limited number of stars defining the reference system per field which does not sufficiently exclude a residual expansion or contraction term in the proper motions. This difficulty requires further study. Publication of these 1800 proper motions is in preparation (Garmany, Ianna and Blaauw, 1976).

b) Objective prism spectral classification. This has been done for all NGP fields by Dr. R.A. Bartaya of the Abastumani Observatory. These classifications serve two purposes: the first choice of the stars for photometric measures and, once this has been done, a comparison of the classifications with the photometric results, especially b-y. For the SGP fields the spectra given by the early McCormick workers are checked by means of objective prism plates taken at the ESO Observatory (La Silla).

c) The photometry. This is done in the uvby, H-β system introduced by Strömgren (see f.i. Strömgren, 1966) and worked out by Crawford (1975). For the F stars, the system allows the determination of the parameters Δc_1, a measure of the star's luminosity with respect to the zero age main sequence (ZAMS), and Δm_1, a measure of the star's metal abundance with respect to the Hyades as a standard sequence. Furthermore, it provides b-y as a measure of colour. All three quantities have to be corrected for interstellar reddening.

Calibration of the luminosity parameters is based mainly on stellar model calculations and the resulting evolution tracks, see f.i. Clegg and Bell (1973) and Hejlesen et al (1972). On the other hand, calibration of the abundance parameter Δm_1 is based empirically on the observed correlation between this quantity and spectrophotometric abundance analysis. For the present, exploratory, purposes we use the relation

$$[M/H] = -12.5\,\Delta m_1 + 0.30$$

given by Crawford (1975), where [M/H] is the logarithmic abundance with respect to that of the sun. Thus, for the range $\Delta m_1 = 0$ to $+0.080$, relevant to our stars, the corresponding abundance ratios with respect to the sun range from 2.0 (Hyades) to 0.20.

Photometry has been completed for most of the stars for which proper motions were measured. A variety of telescopes was used at Kitt Peak National Observatory and at the ESO Observatory at La Silla. All NGP and SGP fields are virtually completed as well as some at intermediate latitudes (Blaauw et al, 1976; Garmany and Ianna, 1976). For stars down to photographic magnitudes 11.5 we have the following r.m.s. errors: in b-y, ±0.010; in m_p ±0.015; in c_1, ±0.026. This accuracy is not sufficient for the discussion of individual properties of the stars, but it does serve well for statistical treatments as will be shown below.

d) Radial velocities. For only few stars have radial velocities been measured; this part of the project awaits the completion of a photoelectric radial velocity device.

3. Provisional Analysis. We have explored the relation between distance from the galactic plane, z, and the photometric and kinematic properties. For this purpose the stars have been subdivided into intervals of distance modulus, m-M = < 6.0, 6.0 to 6.9, 7.0 to 7.9, 8.0 to 8.9 and 9.0 to 9.9.

Also, a subdivision according to b-y was made, but it turned out that no marked differences appeared between the subgroups b-y = 0.25 to 0.34 (F2-F8) at b-y = 0.35 to 0.39 (F8-G2). We therefore communicate here only results pertaining to the combined sample b-y = 0.25 to 0.39; F2 to G2. On the other hand, we do show separately the results for the northern and the southern polar caps.

a) Relation m-M, $\langle\Delta M\rangle$. A gradual increase of the mean excess luminosity of the stars with respect to the ZAMS, $\langle\Delta M\rangle$, appears for both hemispheres, from zero for m-M = 5 to about -2.0 for m-M = 9.5. See Table 1, which also gives the numbers of stars used. This result does not necessarily indicate a real change of mean luminosity with distance z (although such a dependence may occur as a result of increasing age with mean z).In fact,the dependence found can entirely be accounted for as due to observational selection: the limiting magnitude of the basic star lists corresponds with an increasingly luminous lower cut-off of the stars occurring in our sample.It is of some interest to see that the absolute magnitudes obtained via the ΔC_1 measures do, indeed, reveal this selection effect.

b) Relation m-M, $\langle\Delta m_1\rangle$. An interesting dependence of the mean values of Δm_1 on distance from the plane appears for both hemispheres, see Table 1 and Figure 1. This cannot be due to selection effect but must be real, except perhaps for a small contribution due to reddening in the southern hemisphere. The data presented have not been corrected for reddening effects, however for the NGP we have good evidence from H-β measures that these effects must be irtually negligible. The adopted mean relation shown by the dotted line in Fig.1 corresponds to a change from $\langle\Delta m_1\rangle$ = about +0.01 at z = 100 pc to about + 0.50 at z = 700 pc, with a corresponding change of [M/H] from about +20 to -0.35.

c) Spread in Δm_1 at a given distance z. From uvby photometry of bright stars it appears that there is a spread in the observed values of [M/H] which can be expressed as $\langle|D|\rangle = 0.22$, where we have designated [M/H] - <[M/H]> by D. This may be derived for instance, from the data presented by Clegg and Bell (1973). In order to estimate this spread with respect to the mean [M/H] at larger distance z, a special observing programme of the stars in the SGP was conducted in 1975 by Degewij (1976). Two groups of stars were selected, one around m-M=7.0, and one around m-M =9.0. These observations doubled the weight of those available from earlier observing sessions and now render the material suitable for the intended purpose. For the nearer and the more distant groups we find from the old as well as from the new observations an observed scatter $\langle|\Delta m_1 - \langle\Delta m_1\rangle|\rangle = 0.024$ and 0.026 from 25 and 18 stars, respectively. Correcting these for the spread due to observational error, amounting to about 0.014 for the old as well as for the new observations,we find in a good agreement between the two independent sets of data, that the real spread must be close to 0.020 for each of the distance groups. This corresponds to a spread of about 0.25 in [M/H], which thus appears to differ very little from the value, 0.22,found for small z.

d) Kinematic properties. Pending a more complete analysis the following provisional results, based on part of the proper motion material, are of interest. We consider the spread in the velocity components after elimination of the solar motion in the components U and V. U is directed away from the galactic centre, V in the direction of galactic rotation. As the NGP and SGP areas are close to the pole, the proper motions in these areas can be reliably analyzed for these components. Table 2 summarizes the values of $<|U - <U>|>$ and $<|V - <V>|>$ for three distance groups, and subdivided according to Δm_1. The results for the lowest z group are derived from a selection of the stars in the Strömgren-Perry catalogue. The following trends are noted notwithstanding the limited numbers of stars used so far:

A) For each distance group, the velocity spread is always larger for the low metal abundance group than for the high metal abundances.

B) The velocity spread increases with increasing distance from the plane.

The project reported here is a joint one in collaboration with Dr. R.A. Bartaya at Abastumani Observatory, Drs.Ph. A. Ianna and C.R. Tolbert of McCormick Observatory, and Dr. R. West of the European Southern Observatory.

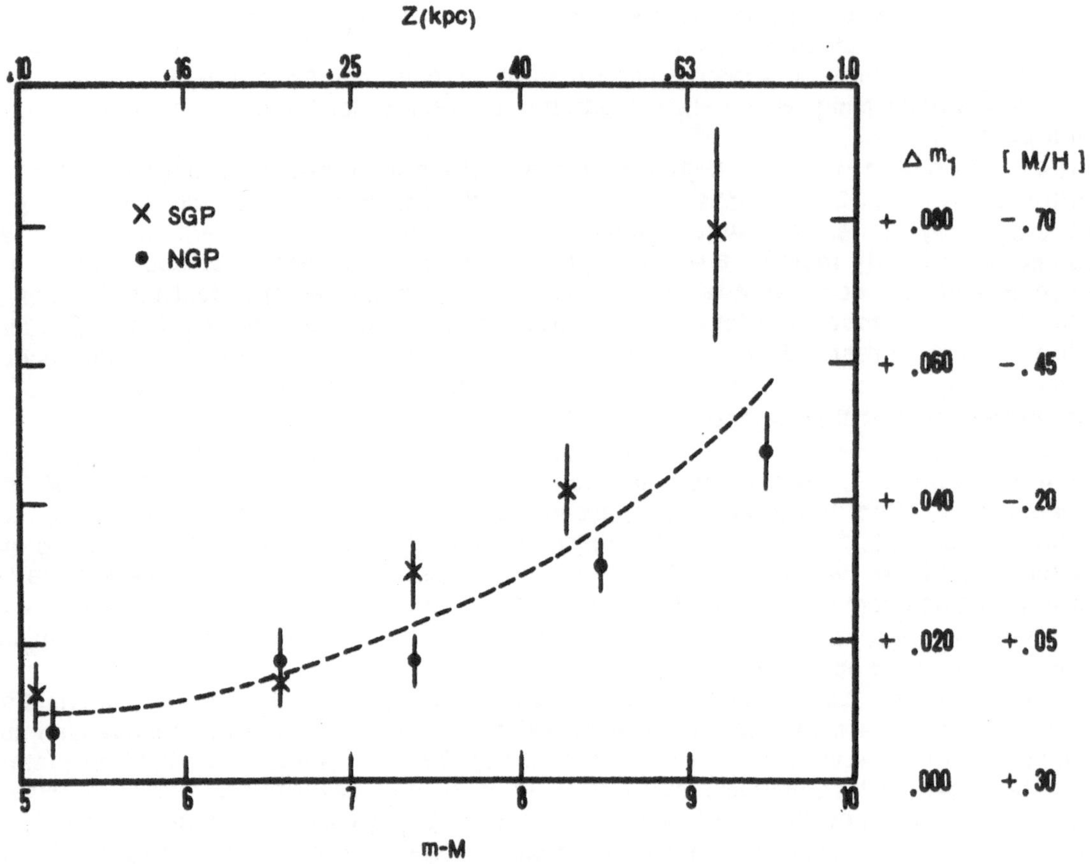

Fig. 1. Observed relation between mean Δm_1 and distance z from the galactic plane, with the corresponding variation in [M/H].

Table 1

Change of mean ΔM with distance z from the galactic plane, probably due to selection effect inherent to the choice of stars; and change of mean Δm_2 with z as a consequence of the varying mean metal abundance.

		South Galactic Pole				North Galactic Pole			
m - M	z(pc)	n	$<\Delta M>$	$<\Delta m_1>$	m.e.	n	$<\Delta M>$	$<\Delta m_1>$	m.e.
< 6.0	< 160	21	-0.1	+ .012	± .005	30	+ 0.2	+ .007	± .004
6.0 - 6.9	160 - 250	26	-0.2	+ 14	4	42	- 0.3	+ 17	4
7.0 - 7.9	250 - 400	27	-0.6	+ 30	5	70	- 0.5	+ 17	3
8.0 - 8.9	400 - 630	23	-1.1	+ 42	6	41	- 0.9	+ 31	4
9.0 - 9.9	630 - 1000	9	-2.2	+ 78	15	23	- 1.6	+ 48	5

Table 2

Kinematic properties derived from part of the $F_2 - G_2$ stars (b-y = 0.25 - 0.40

	z (pc)	Δm_1	n	$<\|U - <U>\|>$ km / sec	$<\|V - <V>\|>$ km / sec
McCormick fields	250 - 600	⩾.040	23	37	36
		<.040	45	31	26
	<250	⩾.040	9	32	33
		<.040	53	17	14
Strömgren - Perry catalogue	< 100	⩾.040	14	33	20
		<.040	133	24	10

REFERENCES

BLAAUW A., TOLBERT C.R., WEST R.M., BARTAYA R.A. 1976, Astron. and Astrophys. Supp. (in press)

CLEGG R.E.S., BELL R.A. 1973, Monthly Notices Roy. Astron. Soc. 163, 13

CRAWFORD D.L. 1975, (in press)

DEGEWIJ J. 1976, in preparation, Leiden Observatory

GARMANY C.D., IANNA Ph.A., 1976 in preparation, McCormick Observatory

GARMANY C.D., IANNA Ph.A., BLAAUW A. 1976, in preparation, McCormick Observatory

HEJLESEN P.M., JORGENSEN H.E., PETERSEN, J.O., KOMCKE L., 1972, L'age des étoiles, Coll. I.A.U. No.17, p.XVII-1. Ed. S.Cayrel de Strobel and Z A.M.Delplace

MAYOR M. 1974 , Astron. and Astrophys. 32, 321

STRÖMGREN B. 1966, Annual Review of Astron. and Astrophys. 4, 433

VAN DE KAMP P., VYSSOTSKY A.N. 1937, Publ. McCormick Obs. 7, 1

VYSSOTSKY A.N., WILLIAMS E.T.R. 1948, Publ. McCormick Obs. 10, 1

DISCUSSION

DODD: Do you consider that the systematic difference between the results of the north and south galactic polar caps is real?

BLAAUW: No! There may be an effect of reddening here. Reddening tends to simulate a higher Δm_1, and hence lower metal abundance. We have checked with the H-beta measures that this is negligible for the NGP, but for the SGP a small effect may be present.

LINDSAY SMITH: Your implicit assumption is that stars at high z distance are, on the average, older. Can you say how much older? And what is the range of ages involved?

BLAAUW: Pending the final analysis of the photometry, especially of the c_1 which depends on the u-band (that is the most uncertain) and measures the luminosity, I would rather not yet give definite figures. However, there is certainly evidence that the mean luminosity increases with distance from the plane, and this is also born out in the statistics shown earlier at this conference by Dr. Kharadze. Concerning the range in ages: from, say, 10^8 years to about 6×10^9 years.

KINEMATICS AND STRUCTURE OF SUBSYSTEMS OF DIFFERENT AGES AND CONTENTS

WILHELMINA IWANOWSKA
INSTITUTE OF ASTRONOMY N.COPERNICUS UNIVERSITY, TORUN, POLAND

КИНЕМАТИКА И СТРУКТУРА ПОДСИСТЕМ РАЗЛИЧНЫХ ВОЗРАСТОВ И СОСТАВОВ

РЕЗЮМЕ

Рассмотрены кинематические характеристики галактических подсистем различных возрастов и составов. Показано наличие глубокой связи подсистемного строения Галактики с динамической и физической эволюцией последней.

1. Historical Introduction. Galactic rotation was discovered by Lindblad and Oort in 1925-1927 through differential effects: the rotational velocity is depending on the distance to the galactic center (Oort's formulae), but at the same distance different kinds of stars revolve with different velocities. Lindblad called them subsystems and stated after Strömberg (1924) that lower rotational velocity of a subsystem is accompanied by greater dispersion of random velocities and weaker concentration towards the galactic plane. As more and better determinations of radial velocities, proper motions and stellar parallaxes appeared, the statistical characteristics of motions were redetermined again and again for different kinds of stars by several investigators.

One such major revision of the characteristics of distribution and motion of a variety of subsystems was performed by Parenago, Kukarkin and their collaborators (see e.g. Parenago, 1954) during the first post-war decade. They have determined the values of β and m - the constants of exponential concentration of subsystems towards the galactic plane and the galactic center respectively - the group motion and the velocity dispersions along the principal axes of the velocity ellipsoids. They regarded as subsystems different physical groups of stars confined to definite spectrum-luminosity or period-spectrum ranges. They distinguished three kinds of subsystems: flat, intermediate and spherical ones, intermediate being the most common kind. Some physical groups appeared soon to be non-homogeneous or mixed subsystems, e.g. Mira variables, cepheids, short-period variables. In fact, not one physical group is kinematically homogeneous, each contains some excess of high or low velocity stars compared to a normal ellipsoidal distribution.

With this in mind, I have proposed (Iwanowska,1965) to consider each physical group of stars as being composed of two components (population types), confined to the galactic disk or the galactic halo, mixed in different proportions. Kinematical and distribution characteristics of each component are different in different physical groups. Thus, both components of cepheids - δ Cep and W Vir type variables - form more flattened subsystems than is the case of short-period variables. Also the relative abundances of the two components are drastically different in these two cases, with population type I prevailing in cepheids, population type II - in short-period variables. Following this scheme, some 16 physical groups of stars, containing in total about 4000 stars with known velocities and /or positions, have been represented with the two-term frequency-functions:

$$N = N_1 + N_2 = N_1^0 \exp\left(-\frac{|z|}{\beta_1}\right) \exp\left(-\frac{R}{m_1}\right) f_1(V) + N_2^0 \exp\left(-\frac{|z|}{\beta_2}\right) \exp\left(-\frac{R}{m_2}\right) f_2(V).$$

where z and R are star distances to the galactic plane and center respectively, β_1, β_2 and m_1, m_2 are respective concentration constants, to be determined from the observational data together with the local component frequencies N_1^0 and N_2^0. $f_1(V)$ and $f_2(V)$ are exponential ellipsoidal velocity distributions, containing three dispersion values and rotational velocities as characteristic constants in each component, to be determined from the observational material. These two-term frequency-functions were used to determine statistical probabilities (statistical population indices) of each star's belonging to one of the two components according to the position and motion of the star. As an example, Table 1 contains kinematical characteristics for F, G, K giants and dwarfs in one-term (Parenago,1954) or two-term (Głębocki,1965) representation.

Table 1

Kinematical characteristics of F,G,K stars for one-term and two-term frequency-functions

Stars	$\theta_c - \theta_m$	σ_π	σ_θ	σ_z	n	Ref.
gF, gG, gK	3	30	20	16	1226	Parenago,1954
	0	29	18	15	964	Głębocki,1965
	50	80	58	45	100	
dF, dG, dK	9	37	24	20	879	Parenago,1954
	0	37	23	15	992	Głębocki,1965
	100	116	86	59	204	

2. Dynamical aspect of subsystems. Assuming the coexistence of different subsystems in our Galaxy as an observational fact, it is possible to build a composite mass distribution model of the Galaxy in equilibrium and to derive its potential (see e.g. Vandervoort,1970). Composite models, though not always connected with Lindblad's subsystems have been proposed by Oort (1932), Perek (1962), Schmidt (1956, 1965), Innanen(1966) a.o. With given mass distribution model it is possible to determine the motion of a mass-point or a star in the Galaxy, when star's initial position and velocity are known. Grids of individual stellar orbits have been calculated by Contopoulos and Strömgren (1965) in the galactic plane for their own model. The orbits are in general not closed curves. Ollongren (1962) has calculated a grid of three-dimensional orbits using Schmidt's (1956) model. They can be represented as some quasi-oscillatory non-closed meridional curves revolving around the galactic axis with the periods of rotation and oscillatory motions being different. Many new numerical orbit calculations were performed for different mass models of the Galaxy. One can choose from these data families of orbits suitable for stars of flat, intermediate and spherical subsystems getting a satisfactory representation of the state of motions in the Galaxy with a stationary present-day potential. The meaning of subsystems consists in different ranges of initial or present-day positions and motions of groups of stars. The next step is to calculate the evolutionary models of the Galaxy following numerically the changes of positions of a sample of stars with time. This has been attempted in several modern investigations for limited but increasing numbers of stars (e.g. Gott,1973; House and Innanen,1975). One might dream that in the future it will be possible to follow numerically the process of evolution of a whole galaxy with its formidable complexity, taking into account regular and irregular gravitational forces, the presence of the interstellar medium with magnetic, radiative and other fields acting, with the process of star formation and destruction going on in evolutionary or explosive way.

A statistical approach to the structure and dynamics of a system composed of stars and/or gas is possible through the averaged hydrodynamical Boltzmann equations. It is also possible to introduce into these equations, even for a steady state system, a sum of subsystems with different initial distributions and states of motions. What is the reason for such a composite structure to come into being, remains a fundamental problem of the formation and evolution of galaxies. Beginning with the works of Poincaré and ending on the present-day numerical integrations, systems similar to real galaxies, composed of a nucleus, disc and halo can be obtained as a result of gravitation, rotational momentum and random motions with given initial conditions (see e.g. Larson,1969, 1972, 1974). Expanding halo appears as a counterpart of a collapsing nucleus. Some authors invoke gravitational collapse (e.g. Eggen, Lynden Bell, Sandage,1962), others - explosive events (Burbidge and Hoyle,1963; Ambartsumian , 1965; Unsöld,1969) in order to explain the existence of the galactic halo. For this aim Innanen and Keenan (1973) introduce small dwarf galaxies infalling to the galactic center in very elongated orbits and destroyed there through collisions giving rise to the galactic nucleus and halo with the surviving remnants left as globular clusters.

3. Physical properties of subsystems. It was early recognized that different subsystems differ not only in distribution and kinematics, but also in physical properties of stars contained in them. This was clearly stated with Baade's finding that different regions of galaxies display different H-R diagrams. Baade (1944) distinguished two main population types: type I with a classical H-R diagram found in discs of spiral galaxies and type II with H-R diagrams similar to those of globular clusters appearing in elliptical galaxies, as well as in central parts and halos of spirals. Different shapes of H-R diagrams mean different evolutionary stages of systems and/or different chemical composition. Spectroscopic investigations indicate that atmospheres of population II stars are deficient in heavy elements and are in general more advanced in their evolution as compared to population I stars. Baade considered originally central bulge and halo jointly as sites of population II with low heavy element content. Later on it became clear from spectroscopic observations that there exists also a radial gradient of this content in our and other galaxies, increasing towards the center. This fact disproves to some extent the concept of a subsystem as a chemically homogeneous sample of stars. Besides significant differences in the average chemical composition between different subsystems, there is in each subsystem a systematic increase of heavy element content toward the galactic plane or center.

If Baade considered population types as local phenomena, another view, regarding subsystems and population types as age phenomena, gained general acceptance afterwards (see e.g. O'Connel,1958). It seemed that the sequence of kinematical and distribution characteristics running from flat over intermediate to spherical subsystems is a pure age sequence. Similarly, the correlated run of chemical composition from metal-rich to metal-poor objects seemed to be a pure age sequence going from young to old generations. There are, however, some other factors affecting population characteristics (Iwanowska,1958). The heavy element content depends on the time and the place of birth of a star. The kinematics and distribution of a subsystem depend on the age and the mass of stars. More massive stars form more flattened subsystems than low mass stars of comparable age. This is not quite obvious since mass and age are interrelated for the majority of stars, massive stars being in average younger. Yet, there are some kinds of stars of similar mass but different age and vice-versa. One such example are Me and M dwarfs which probably are of similar mass but different age, Me dwarfs being younger and forming a more flattened subsystem than M dwarfs. But Me dwarfs form a less flattened subsystem than M giants which are of comparable age but greater mass. When estimating the age of a stellar species from its kinematical and distribution characteristics, it is necessary to introduce a correction for the mass effect.

Thus, old and low mass stars, e.g. subdwarfs form nearly spherical subsystems - extreme population II, whereas young and massive stars, e.g. supergiants form flat subsystems - extreme population I. All other combinations of age and mass produce intermediate subsystems. The dependence of the distribution and kinematical characteristics on the mass is what should be expected for a system in statistical equilibrium. Certainly, our Galaxy has not yet reached such a state, however, it seems to proceed towards it.

The initial chemical composition of stars, usually observed in their spectra, depends on the time and place of the star formation. The theory of chemical evolution of the Galaxy underwent itself a process of evolution. Some twenty years ago it was believed (Burbidge, Burbidge, Fowler and Hoyle,1957) that originally our and other galaxies consisted of pure hydrogen, that helium and all heavy elements were formed in the interiors of stars through nuclear processes and were injected into interstellar medium mostly through supernova explosions. Subsequent generations of stars were formed out of interstellar matter enriched more and more in heavy elements. It was stated soon, however, that the lifetime of the Galaxy is too short to produce all presently existing heavy elements out of pure hydrogen through this mechanism. This appeared particularly critical for helium whose abundance presently observed amounts to about 10% by number of atoms, much too high for its creation wholly in stars. Therefore it is presently assumed and proved again to be possible that light nuclei , namely, the isotopes of hydrogen, helium and lithium were formed earlier than the Galaxy itself, probably in the big-bang of the Universe. Another fact conflicting with the time-scale of the mentioned theory of chemical evolution of the Galaxy is the insufficient number of metal-poor stars presently observed. In order to overcome the time-scale difficulty and to accelerate the "astration process" several ad hoc assumptions have been proposed. Very big stellar masses in the first generations were admitted (Schmidt,1963; Truran and Cameron,1971), much higher than those observed now. Also, much higher than presently observed frequency of occurence of supernovae was assumed. The idea of metal enhanced star formation (Talbot and Arnett,1973) is being advocated and proved to be effective in accelerating the stellar supply of heavy elements. The process of nucleosynthesis in stars in the course of their evolution has been reexamined with new nuclear data and modern theory of stellar evolution (see e.g. Clayton,1968). Thus we dispose of a theory of chemical evolution of the Galaxy with the lightest elements inherited from the big-bang and the heavier ones synthesized in stellar interiors and dispersed afterwards through mass loss or supernova explosions. Pushing the relevant processes and quantities to their extreme rates and values it is possible to reconcile the time-scale of this theory with the supposed age of the Galaxy of some 12.10^9 years. Otherwise, we have to look for heavy element formation prior to the formation of the Galaxy or to extend still more its age. In fact, the Hubble constant shows a definite inflation tendency nowadays, falling from 100 over 75 down to 50 km $s^{-1}Mpc^{-1}$.

Any theory of galactic evolution has also to explain the fact of inhomogeneous distribution of elements with the heavier ones concentrated more towards the galactic plane and the galactic center. The theory considered above is able to do so in several ways. If one assumes that star formation rate is proportional to some power of gas density higher than one, then heavy element production is running faster in denser regions with the maximum rate in the galactic plane and center. Another possibility of generating composition gradients exists when the initial mass function of stars being formed depends on the density of interstellar medium favouring bigger stellar masses in denser regions, as proposed by Madore, Sydney van den Bergh and Rogstad (1974). In this case again the heavy element production is fastest in the galactic plane and center. A similar result is attained if massive stars spend all their life in the vicinity of the galactic plane or center owing to low velocity dispersion. The accretion of intergalactic gas composed almost of pure hydrogen and helium, admitted by Quirk and Tinsley (1973), may also produce composition gradients, though its other effect of producing a maximum of heavy element content at some time moment within the interstellar gas does not seem to find a support in observational data. As a further possibility of separation of elements within the interstellar gas the process of gravitational diffusion is being considered by Ewa Basińska-Grzesik at the Toruń Observatory with the preliminary result that the effect is small but not quite negligible.

During the last few years more and more attention is paid to the role of the dust in the interstellar medium. Though it contains about one percent of the total mass of the interstellar matter, its importance for the process of star formation seems to be essential (see e.g. Reddish,1975). Another important property of the dust consists in sucking out heavy elements, particularly those easy for condensation, as e.g. Ca, Ti, Al. This process is inferred from the analysis of interstellar lines, hundreds of which are presently observed in ultraviolet, especially from the "Copernicus" satellite. A depletion of heavy elements correlated with their condensation temperature from interstellar gas has been stated by Field (1974) and confirmed by several other investigators.

It is usuallly accepted for the simplicity sake that the relative abundances among the heavy elements themselves are in average equal everywhere. This is not true for the elements sucked out by dust, but also an increase of the concentration of heavy elements towards the galactic plane and center with increasing atomic weight seems to be present (Iwanowska,1968; Kuroczkin,1974) pointing again towards statistical equilibrium.

Started by Bertil Lindblad fifty years ago as a kinematical problem the phenomenon of stellar subsystems in our Galaxy appears to be deeply connected with the dynamical and physical evolution of the Galaxy in the variety of its component stellar species.

REFERENCES

AMBARTSUMIAN V. A., 1965, Proc. 13-th Solvay Conference

BAADE W., 1944, Astrophys. J.100, 137

BURBIDGE E.M., BURBIDGE G.R., FOWLER W.A. and HOYLE F., 1957, Rev. Mod. Phys.29 547

BURBIDGE G.R. and HOYLE F., 1963, Astrophys. J.138, 57

CLAYTON D.D., 1968, Principles of Stellar Evolution and Nucleosynthesis, McGraw Hill, New York

CONTOPOULOS G. and STRÖMGREN B., 1965, Tables of Plane Galactic Orbits, Inst. of Space Studies, New York.

EGGEN O.J., LYNDEN BELL D., SANDAGE A.R., 1962, Astrophys. J.136, 748

FIELD G.B., 1974, Astrophys. J.187, 453

GLĘBOCKI R., 1965, Bull. Astr. Obs. Toruń 41

GOTT J.R., 1973, Astrophys. J.186, 481

HOUSE F.C., INNANEN K.A., 1975, Astrophys. Space Sci.32, 138

INNANEN K.A., 1966, Z.Astrophys. 64,158, 457

INNANEN K.A. and KEENAN P.C., 1973, J.Astron. Soc. Canada 67, 248

IWANOWSKA W., 1958, Bull. Astr. Obs. Toruń. No. 18

IWANOWSKA W., 1965, Vistas in Astronomy (ed. A.Beer), 7, 133

IWANOWSKA W., 1968, Astrophys. Space Sci. 2, 128

KUROCZKIN D., 1974, Bull. Astr. Obs. Toruń No. 54

LARSON R.B., 1969, Monthly Notices Roy. Astron.Soc.145, 405

LARSON R.B., 1972, Monthly Notices Roy. Astron. Soc.156, 467

LARSON R.B., 1974, Monthly Notices Roy. Astron. Soc.166, 585

LINDBLAD B., 1925-7, Arkiv Mat., Astr., Fys. 19A, Nos. 21,27,35; 19B, No. 7; 20A, No. 17

MADORE B.F., van den BERGH S., ROGSTAD D.H., 1974, Astrophys. J.191, 317

O'CONNELL D.J.K., 1958, Stellar Populations, Vatican Conference

OLLONGREN A., 1962, Bull. Astron. Inst. Neth.16, 241

OORT J.H., 1926, Groningen Publ. No. 40

OORT J.H., 1932, Bull. Astron. Inst. Neth.6, 249

PARENAGO P.P., 1954, Kurs zvezdnoj astronomii, Moskva

PEREK L., 1962, Advances in Astronomy and Astrophysics, 1, 165, Acad. Press, London and New York

QUIRK W.J. and TINSLEY B.M., 1973, Astrophys. J.179, 69

REDDISH V.C., 1975, Monthly Notices Roy. Astron. Soc.170, 261

SANDAGE A.R. and TAMMANN G.A., 1975, Astrophys. J.197, 265

SCHMIDT M. 1956, Bull. Astron. Inst. Neth.13, 15

SCHMIDT M., 1963, Astrophys.J.137, 758

SCHMIDT M., 1965, Galactic Structure (ed. R.A.Blaauw and M.Schmidt) 527

STRÖMBERG G., 1924, Astrophys.J.59, 228

TALBOT R.J. and ARNETT W.D., 1973, Astrophys.J.186, 69

TRURAN J.W. and CAMERON A.G.W., 1971, Astrophys. Space Sci.14, 179

UNSÖLD A.O., 1969, Science 163, 1015

VANDERVOORT P.O., 1970, Astrophys. J.162, 453

DISCUSSION

SHATSOVA : It is considered by many that the true velocity distribution of stars is not normal and that the density law is not exponential. If this is so, then your subsystems may be seriously changed or even not correspond to reality. Would it not be better to consider the kinematic criteria as secondary, after the physical criteria, when establishing the subsystems?

IWANOVSKA : The one-term exponential-ellipsoidal distribution is a first approximation, the two-term a second, to the real distribution. One could think of higher-term distributions as further approximations. With the two-term frequency-function, we are on the second step with twice as many fitting parameters as the first approximation, liable for physical interpretation in terms of the disk-halo model and enabling us to determine the statistical probability of each star belonging to the one or other of these components. I think that simultaneous investigations of the physical and kinematical properties of the Galaxy are very useful, both methods supporting each other.

THE KINEMATICAL PROPERTIES OF SOME STAR SUBSYSTEMS OF GALAXY

D.K.KARIMOVA AND E.D.PAVLOVSKAYA
STERNBERG ASTRONOMICAL INSTITUTE, MOSCOW, U.S.S.R.
PRESENTED BY E.D.PAVLOVSKAYA

КИНЕМАТИЧЕСКИЕ СВОЙСТВА НЕКОТОРЫХ ЗВЕЗДНЫХ ПОДСИСТЕМ

РЕЗЮМЕ

Получены кинематические параметры для нескольких групп звезд: цефеид, сверхгигантов B, переменных типа RR Лиры, A$\bar{V}$,Am и Ap звезд. В исследование включены лишь звезды с известными Vr и UBV фотометрией из числа тех, для которых определены собственные движения в Институте имени Штернберга (~800 звезд).
Изучение групп звезд, занимающих различные объемы пространства имеют свои специфические трудности. Выборка звезд в ближайших окрестностях Солнца (r $\leqslant$ 25 пс) с известными тригонометрическими параллаксами искажена так называемой кинематической селекцией. Предложен метод учета этого эффекта. Собственные движения звезд в более широких окрестностях Солнца (r < 2 кпс) не могут использоваться без введения поправки на случайные ошибки определения μ
Для исправления использован метод Эддингтона.
Для всех рассмотренных групп звезд определены параметры галактического вращения и эллипсоида скоростей. Результаты приведены в табл. 1 и 2. Выделена группа сверхгигантов B в области l ≈ 135° (на расстоянии 2- 3 кпс), имеющих общее движение. Среди переменных типа RR Лиры выделены звезды, относящиеся к населению диска Галактики.

ABSTRACT

The kinematic parameters of some stellar groups (Cepheids, B - supergiants, RR Lyrae variables, A$\bar{V}$, Ap and Am stars) were studied. The fundamental proper motions obtained in Sternberg Institute for about 800 stars were used for the investigation. For all these stars Vr and UBV magnitudes were known.
The investigation of the various groups of stars occupying different volumes of space has some difficulties. The stellar sample in the immediate neighbourhood of the Sun (r$\leqslant$ 25 pc) is deficient in stars with small tangential components of space velocity. The method of correcting for this effect (it is called the kinematical selection effect) was suggested.
The proper motions of the stars in wider vicinity of the Sun (r< 2 kpc) cannot be used without the correction for accidental errors. For this purpose Eddington's method was used.
For all considered stellar groups the differences from solar motion relative to the circular velocity in the neighbourhood of the Sun, the parameters of differential galactic rotation and the parameters of velocity ellipsoid were determined. The results are given in Tables 1 and 2. There is a group of B-supergiants located at l ≈ 135° and the distance 2 - 3 kpc. Almost all stars of this group move in the same direction. The stellar group of disk population was selected among the RR Lyrae variables.

Kinematical research in the recent years has made much progress. Advances in this field are connected with extensive use of electronic computers which made it possible to improve the methods of the analysis of observational data (Clube, Jones, 1971) and to begin the calculation of the orbital elements for many stars (Woolley, Savage, 1971; Woolley et al. 1970a). These elements are more stable properties of the stellar motion than the components of the space velocity. These elements combined with physical properties of stars yielded the improved understanding of the origin and evolution of the Galaxy. It is very important to compare the orbital elements with an age (de Strobel, 1973; Wielen, 1973) and metal abundance of stars (Epstein,1969; Woolley, 1970b). Advances in the theory of stellar evolution make it possible to estimate stellar ages from the positions of stars in the color- magnitude diagram.

Some basic galactic parameters: period and angular velocity of the galactic rotation ω (R), its derivatives ω' and ω'' and mass of the Galaxy, can be determined from the statistical investigation of stellar motions. One can also improve the distance scale and obtain some other kinematical parameters which are necessary for the construction of a dynamical model of the Galaxy.

The determination of the kinematical parameters for the stellar groups of different ages is of great interest. However these groups usually occupy different volumes of space. The kinematical investigation for each of these groups has some difficulties and peculiarities and permits to obtain only some of the above-mentioned parameters. If we consider the stars in the immediate neighbourhood of the Sun ($r \leqslant 25$ pc) we may calculate their velocity components and orbits with high precision. Consequently we may study the velocity distribution and velocity ellipsoid reliably. But it is impossible to determine the parameters of the galactic rotation.

This sample is deficient in stars with small tangential components of space velocity because the stars with high proper motions are preferential in the trigonometric parallax program. It is known as the kinematical selection effect. The method of correcting for this effect by means of electronic computers was suggested in Moscow. If the kinematical selection is not taken into consideration, the kinematical parameters are misrepresented. For instance, the obtained dispersion will be overestimated by about 10% for M dwarfs. For other stars this effect is less.

If we consider more distant stars ($r \leqslant 150$-200 pc) for which the trigonometric parallaxes are unknown and the photometric distances are used there is need to improve the distance scale by means of statistical method. Differential galactic rotation cannot be investigated in this volume.

For the determination of the parameters of differential galactic rotation it is necessary to use stars in wider vicinity of the Sun ($r<2$ kpc). These are the young stars of high luminosity with small velocity dispersion. Their proper motions are so small that the errors approach or even exceed the values of proper motion themselves. This is the principal difficulty of the kinematical investigation in this volume of space. But the study of such objects is very important.

At present almost all kinematical study of young objects is based only on their radial velocities. In this case it is impossible to study space velocities of individual stars.

Prof. Blaauw (1968) pointed out that if an absolute accuracy of proper motions of these stars of about $\pm 0\overset{''}{.}002$ can be obtained, important information on regional kinematics may be obtained up to 2 or 3 kpc.

At present we think that it is possible to determine the fundamental proper motions with such precision for many stars because in the recent years a great number of position catalogues have been published. Fundamental proper motions in the FK4 system have been obtained in Moscow for about 800 stars (Karimova, Pavlovskaya, 1971, 1972a, 1972b, 1973, 1974). Among these are O-stars, B-supergiants, Ap and Am stars, Cepheids. The precision of the determination of proper motions is shown in Fig. 1. For the determination of absolute proper motions it is very important to establish the correct system of weights, especially for stars with small proper motions Such work was initiated in Moscow; we calculated the weights based upon external errors for some new position catalogues.

Much observational material is available for deriving accurate relative proper motions. After the publication of the AGK 3 catalogue we may select the reference stars among the stars of this catalogue. In this case there is no need to reduce the derived proper motions. Such proper motions of fainter stars may be used for the determination of kinematical parameters, such as fundamental proper motions. In particular, it is very important to determine proper motions of Cepheids. Their proper motions are required to improve the distance scale in the Galaxy.

Many investigators obtain different values of solar motion with respect to different groups of objects. In our opinion, the solar motion relative to the circular velocity in the neighbourhood of the Sun should always be used. Deviations from this motion are due to different group motions of different classes of objects. To establish the frame of reference the completion of the proper motion determination for stars with respect to galaxies is very important. After establishing the frame of reference the above mentioned differences may be determined in the following way.

The equations for the determination of the kinematical parameters are:

$$V_r + V_0 \cos\lambda = R_0 \sin l \cos b \, \Delta\omega + R_0(R-R_0) \sin l \cos b \, \omega' + \frac{R_0}{2}(R-R_0)^2 \sin l \cos b \, \omega'' - \cos l \cos b \, \Delta V_R + r \cos^2 b \, \epsilon_0 , \quad (1)$$

$$4.74r\,\mu_l + V_\odot \sin\lambda\sin\psi = R_0\cos l\Delta\omega + (R-R_0)(R_0\cos l - r\cos b)\,\omega' - r\cos b\,\omega_0 + $$
$$+ \tfrac{1}{2}(R-R_0)^2(R_0\cos l - r\cos b)\,\omega'' + \sin l\Delta V_R\,, \qquad (2)$$

$$4.74r\mu_b + V_\odot \sin\lambda\cos\psi = -R_0\sin l\sin b\,\Delta\omega - R_0(R-R_0)\sin l\sin b\,\omega' - \frac{R_0}{2}(R-R_0)^2\sin l\sin b\,\omega'' +$$
$$+ \cos l\sin b\,\Delta V_R - r\sin b\cos b\,\epsilon_0\,, \qquad (3)$$

where $\Delta\omega$ is the difference in angular velocity at the distance from the Sun R_0, ΔV_R the difference in radial motions with respect to the galactic center also at the distance $R=R_0$, ΔZ the difference in motions in z-direction of the centroid under consideration and the centroid of solar neighbourhood stars regarded as the frame of reference.

Usually parameters μ_α and μ_δ are used for the determination of the kinematical model. However, the results from proper motions, as a rule, differ significantly from those obtained from radial velocity, The systematic errors of the FK 4 proper motion system appear to be the cause of this. Possibly the kinematical investigation will help to establish the character of errors of the FK 4 proper motion system.

It appears preferable to obtain the kinematical parameters from the proper motion and radial velocity together, that is to say equations (1), (2), (3) are being solved together .For the objects with strong galactic concentration only equations (1) and (2) should be regarded. It is important to receive the correct weight for each equation of condition. The following system of weights may be suggested. All the equations for V_r may be taken with the weights equal to unity. If the error of proper motion component is $\leqslant\pm0''.002$ corresponding equation should be also taken with unity weight. For other equations the weights may be calculated from the errors of proper motions. In this system all weights are $\leqslant 1$. For the statistical determination of kinematical parameters it is very important that the weights are limited.

In conclusion we should like to point out the results which have been obtained for some stellar subsystems of different age. These are cepheids and B-supergiants belonging to extreme flat population, RR Lyrae variables which belong to halo population and some other groups.

Although the proper motions of young stars have been improved we cannot use them without correction for accidental errors. The method for such correction was suggested by Eddington (Dyson 1926; Eddington 1940). The corrections $\Delta\mu_\alpha$ and $\Delta\mu_\delta$ may be calculated by means of the following formula

$$\Delta\mu_i = -\frac{\epsilon_\mu^2}{\sigma^2}(\mu_i - \bar{\mu})\,,$$

where ϵ_μ - the mean error of the μ determination, σ - dispersion of the peculiar proper motions, μ_i - peculiar proper motions.

For most RR Lyrae stars only the relative proper motions reduced to absolute frame of reference by statistical method are known Proper motions obtained in such a way already imply the dependence on the parameters of the kinematical model. But these statistical reductions are small in comparison with the proper motions themselves. Hence one can hope that the difference between the motion of RR Lyrae variables and the motion of stars in solar neighbourhood will be established even from their relative proper motions reduced to absolute motions. The proper motions of RR Lyrae variables can be used without correction for accidental errors because their errors are small in comparison with the proper motions themselves.

The stellar group of disc population was selected among the RR Lyrae variables by comparison of their osculating orbital elements with metal abundance. The dependence of elements e and i upon value $(k\text{-}b)_2$ characterizing the metal abundance is shown in Fig. 2.

The parameters which have been determined by the solution of equations (1) and (2) or (1), (2) and (3) together are given in Table 1 for all the investigated groups.

It was established that B-supergiants move to the south galactic pole with the velocity of about 2 km/sec. This velocity may be probably caused by the errors of the FK 4 proper motion system.

The semiprincipal axes of velocity ellipsoid are given in Table 2 for all the investigated groups. The motions of B-supergiants in the galactic plane are shown in Fig.3. These motions are random. But there is a group of these stars located at l = 135° and the distance 2 - 3 kpc. Almost all stars of this group move in the same direction (Fig. 4). The authors of many previous papers pointed out the peculiarity of stellar motions in this region. But this was done only on the basis of radial velocities.

The relation between box orbit elements and the age of some open clusters was established (Barchatova, Pavlovskaya, 1974).

Table 1

	n	ω_o	$\Delta\omega$	ω'	ω''	ΔV_R	ϵ_o
BI, II	162	28.8±2.9	-0.84±0.35	-2.69±0.37	+0.13±0.49	-6.3±2.9	-0.0±1.5
Cep	82	23.9 4.6	-0.07 0.29	-3.56 0.60	-4.12 1.38	-2.0 2.2	-0.5 2.6
B + Cep	244	28.8 2.4	-0.58 0.24	-2.82 0.31	-0.12 0.40	-4.6 2.0	-0.2 1.3
A$\bar{V}$+Ap+Am	819	21.2 16.8	+0.07 0.08	-1.25 2.15		+0.4 0.8	-19.0 9.6
RRLyr disc	63	30.7 21.1	-2.18 1.09	-4.80 2.20		+21.4 11.5	-20.2 8.9
RRLyr halo	216	22.9 26.6	-17.5 1.6	-2.43 4.17		-11.0 17.3	-4.7 17.4

Table 2

	B + Cep	A$\bar{V}$+Ap+Am	RRLyr disc	RRLyr halo
n	121	273	22	72
σ_1	11.3 ± 1.0	17.0 ± 1.3	37.6 ± 3.7	165.5 ± 12.3
L_1	13° 16°	28° 11°	109° 24°	358° 8°
B_1	+7° 9°	-3° 4°	+23° 72°	-2° 8°
σ_2	8.7 ± 0.7	9.4 ± 1.6	27.9 ± 5.7	124.9 ± 13.4
L_2	102° 16°	118° 11°	25° 24°	87° 8°
B_2	-2° 16°	0° 12°	-14° 17°	+13° 43°
σ_3	7.2 ± 0.8	7.3 ± 1.0	22.3 ± 6.7	111.7 ± 10.3
L_3	356° 133°	22° 212°	224° 86°	95° 39°
B_3	+83° 83°	+87° 41°	+62° 45°	+77° 43°
σ_2/σ_1	0.77 ± 0.09	0.55 ± 0.10	0.74 ± 0.17	0.75 ± 0.10
σ_3/σ_1	0.64 0.09	0.43 0.07	0.59 0.19	0.67 0.08

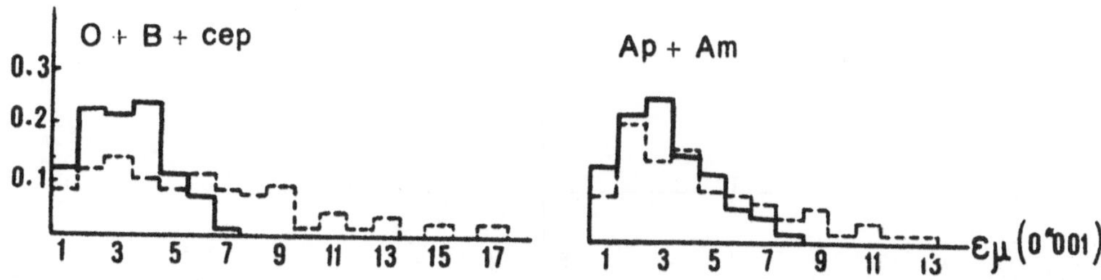

		$\varepsilon_\mu \leq 0''002$	$\varepsilon_\mu \leq 0''005$
GC stars	Ap + Am	104	258
	O + B + cep	90	230
Other stars	Ap + Am	5	92
	O + B + cep	6	94

The 1. The freguency distribution function of proper motion errors

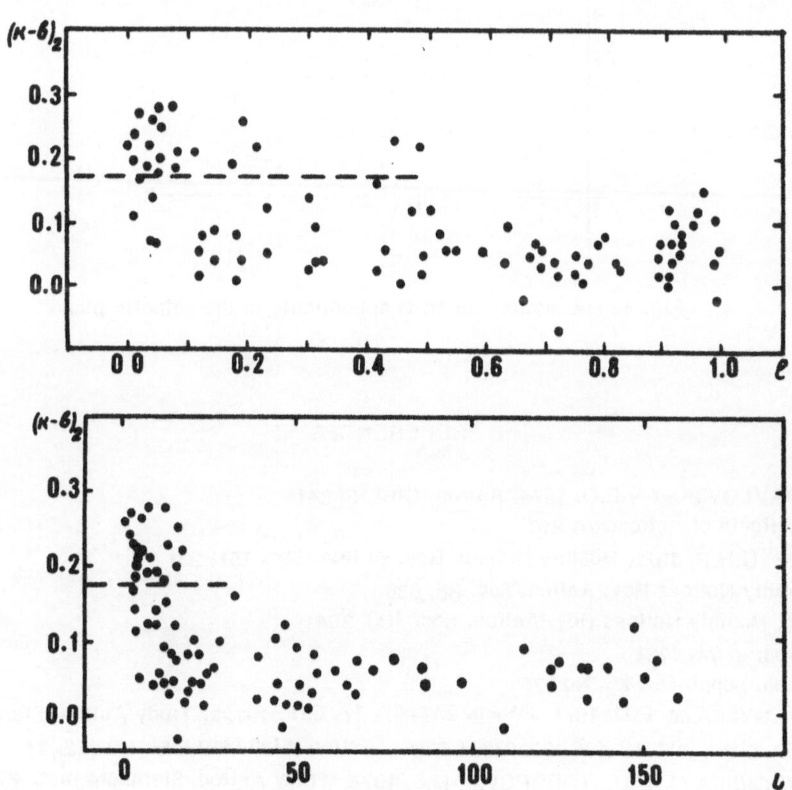

Fig. 2. The relation between osculating orbital elements and metal abundance $(k\text{-}b)_2$ for RR Lyrae stars

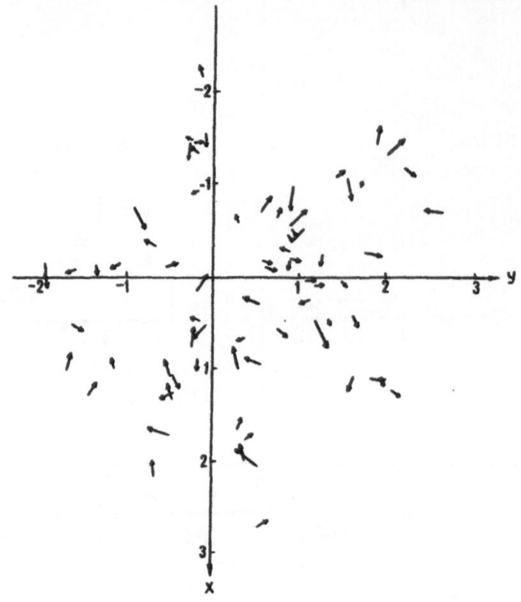

Fig. 3. The motions of B supergiants in the galactic plane

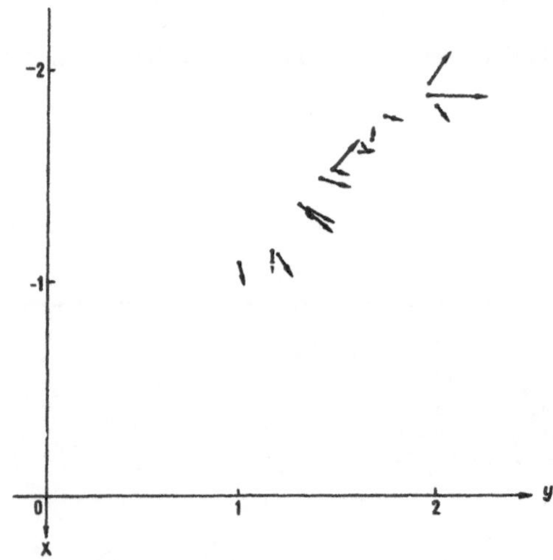

Fig. 4. The motions of 16 B supergiants in the galactic plane

REFERENCES

BARCHATOVA K.A., PAVLOVSKAYA E.D. 1974, Astron. Circ No 841

BLAAUW A. 1968, Highlights of Astronomy. 316

CLUBE S.V.M., JONES D.H.P. 1971, Monthly Notices Roy. Astron. Soc. 151, 231

DYSON F.W. 1926, Monthly Notices Roy. Astron. Soc. 86, 686

EDDINGTON A.S. 1940, Monthly Notices Roy. Astron. Soc. 100, 354

EPSTEIN I. 1969, Astron. J. 74, 1131

JONES D.H.P. 1973, Aph. Suppl. Ser. 25, No 225

KARIMOVA D.K., PAVLOVSKAYA E.D. 1971, Perem. Zvezdy. 17, 591 ; 1972a, Trudy Astron. Sternberg Inst.42,65; 1972b, Comm. Astron. Sternberg Inst. No 176, 3; 1973, Comm. Astron. Sternberg Inst. No 172, 14

KARIMOVA D.K., PAVLOVSKAYA E.D., TOROPOVA M.S. 1974, Trudy Astron, Sternberg Inst. 45, 87

de STROBEL G.C. 1973, Highlights of Astron. 3, 369

WIELEN R. 1973, Highlights of Astron. 3, 395

WOOLLEY R. et al. 1970a, Roy. Obs. Ann. No 5; 1970b, Monthly Notices Astron. Soc. 148, 463

WOOLLEY R., SAVAGE A. 1971, Roy. Obs. Bull. No 170.

FORMATION OF O-STARS AND THE RATE OF STAR FORMATION IN THE GALAXY

P.G.MEZGER AND LINDSEY F.SMITH
MAX - PLANCK - INSTITUT
FÜR RADIOASTRONOMIE
BONN, F.R.G.
PRESENTED BY P.G. MEZGER

ОБРАЗОВАНИЕ О-ЗВЕЗД И ТЕМПЫ ЗВЕЗДООБРАЗОВАНИЯ В ГАЛАКТИКЕ

РЕЗЮМЕ

Рассматриваются наблюдательные свидетельства,касающиеся образования О-звезд, а так же менее массивных звезд. Дается оценка темпов формирования звезд в Галактике на основе радионаблюдений областей Н II. Предположено, что около 71% рассматриваемых звезд образовываются в основных спиральных рукавах (волнах плотности), 17% в конгломератах облаков между основными рукавами и 12% внутри объема в пределах 200 пс от ядра Галактики. Оценено общее количество фотонов Лаймановского континуума в Галактике, которое в четыре раза выше ранее принимавшегося. Наблюдаемый темп звездообразования сравнивается в предсказываемым. Найдено, что линейная зависимость темпов звездообразования от поверхностной плотности газа лучше согласуется с наблюдениями,чем обычно принимаемая зависимость от квадрата поверхностной плотности.

Оценена вариация не в зависимости от расстояния от галактического центра и сравнивается предсказываемая вариация распространенности Не с наблюдаемым обилием He^+.

ABSTRACT

We review the observational evidence pertaining to the formation of O-stars and lower mass stars. We estimate a present star formation rate in the Galaxy of 4.2 $M_\odot yr^{-1}$ based on radio observations of H II regions and their estimated age of 6.4 E5 yr. About 71% of these stars appear to be formed in main (density wave) spiral arms, 17% in cloud complexes between the main arms and 12% within the central 200 pc. The total number of Lyc-photons in the Galaxy is estimated to 3.3 E53 s^{-1}, a value which is about four times higher than previous estimates. The observed star formation rate is compared to a predicted star formation rate. It is found that a linear dependence of the star formation rate on the surface density of gas fits the observations better than the usually adopted dependence on the square of the surface density.

We estimate the variation of He as a function of distance from the galactic center and compare the predicted variation of the He-abundance with the observad He^+- abundance.

1. I n t r o d u c t i o n. Masses of observed stars range from 0.08 $M_\odot$ to about 100 $M_\odot$. The lower mass limit appears to be well understood. Protostars with masses < 0.08 $M_\odot$ never attain central temperatures high enough for the initiation of hydrogen burning. The upper mass limit appears to be determined by the high luminosity of massive stars. The radiation pressure acts on the dust particles and through friction on the neutral gas and at a certain critical luminosity and stellar mass the infall of material is stopped. (Larson and Starrfield, 1971; Kahn,1974).

The MS lifetime of stars depends critically on their mass and varies from some 10^6 yr for O-stars to lifetimes comparable to the age of the Galaxy for stars with masses $M < 1 M_\odot$. The presence of a large number of O-stars in the Galaxy was for a long time the strongest argument that today stars still form out of the interstellar matter.

The Jeans criterion for gravitational instability requires gas densities ranging from 10^3 to 10^6 atoms cm^{-3} and more for the initiation of gravitational collapse of single protostars. The average density of the interstellar matter in the Galaxy ranges from 0.5 to 5 atoms cm^{-3}. The formation of dense clouds in the interstellar gas thus must be the first step in a sequence of events which leads to the formation of stars out of the interstellar matter.

O-stars form out of dense clouds of interstellar gas. Having reached the MS the O-stars ionize the surrounding gas and form H II regions which - as long as their average plasma densities are $n_e \gtrsim 10^2\ cm^{-3}$ - can easily be detected in the radio range throughout the Galaxy both by their free-free continuum and recombination line radiation. By surveying the Galaxy for H II regions we identify three major regions of star formation:

1) The main spiral arms, which are supposed to be maintained by density waves (DWA);2)The material arms in the region between the main spiral arms (MA); 3) The inner part of the nuclear disk, a rapidly rotating gaseous disk which extends from the nucleus of our Galaxy out to a distance of about 800 pc. If the ratio of low-mass stars to O-stars is the same everywhere in the Galaxy and if the rate of O-star formation is proportional to the number of Lyman continuum (Lyc/) photons emitted by the ionizing stars of H II regions - assumptions which will be justified later - then we find that of all stars formed during the past 6 10^5 yr in our Galaxy about 70 % have been formed in DWA, 20% in MA and 10% in the inner 200 pc surrounding the nucleus of the Galaxy.

This review consists of three parts. In Section II we review observational evidence pertaining to the formation of O-stars and of lower-mass stars and show, how the stellar mass associated with an H II region can be estimated. In Section III we estimate the present star formation rate in different parts of the Galaxy and compare it to the star formation rate predicted on the basis of the observed gas-to-mass ratio. In Section IV we discuss recent observations pertaining to the existence of gradients in the abundances of elements in the Galaxy and relate these observations to the fraction of interstellar matter which went through stars during the lifetime of the Galaxy.

II. Observational evidence pertaining to the formation of O-stars and lower-mass stars. The sun is located in the Orion arm which is a material arm. The closest DWA are the Perseus and Sagittarius arms. From observations in the solar vicinity, we obtain a rather coherent picture of star formation in MA and a more fragmentary picture of star formation in DWA. There is practically no information available on star formation in the nuclear disk and one therefore has to be extremely cautious

Most O-stars reach the main sequence (MS) embedded in a shell of neutral gas and dust. First, all stellar radiation is absorbed by dust which reradiates in the IR. After the gas density in the surrounding shell has decreased to values $\lesssim 10^5$ cm^{-3} a very compact H II region is formed. Figure 1 shows as an example the compact H II regions and IR sources W3(A) and (OH). From the integrated IR spectrum, we can estimate the total luminosity, and from the radio flux density the Lyc-photon luminosity of the ionizing star(s). In the further evolution, the surrounding shell expands and more of its mass gets ionized until the shell is fully ionized and eventually is dispersed. An H II region is observable as a strong thermal radio source until its average electron density falls below ~ 100 cm^{-3}. In this stage its optical radiation is often heavily obscured by dust and the ionizing star(s) in most cases are not observable in the optical range. We estimate that between 15 and 25% of all O-stars are hidden by dust.

It has been first shown by Blaauw (1964) that in MA OB-associations the units of star formation are subgroups which contain some 1000 $M_\odot$ of stars. In these subgroups the O-stars form last (Iben and Talbot,1966). The ages of the subgroups are spread over more than 10^7 years. Radio observations with high angular resolution of giant H II regions in DWA indicate that star formation there occurs in similar subgroups. However, the evolutionary ages of subgroups in DWA differ probably by less than 10^5 yr. This difference suggests that, in DWA, star formation is triggered by one global event, for example compression of the interstellar gas in a shock. In MA, the formation, at least of O-stars, is a much more random effect probably caused by thermal instability of the gas.

As an example of a very young OB-association in a DWA, let us consider W3, which is the fourth and youngest in a sequence of giant H II regions in the Perseus arm. Figure 2 shows an overlay of the radio continuum contours on a Mt. Palomar red print. The H II region seen in Hα is IC 1795, W3 with its various components lies in an area which in the optical range is nearly completely obscured by dust. The radio observations have an angular resolution of 2 arc min (Schraml and Mezger, 1969).

Figure 3 shows the same region, but this time the radio observations have an angular resolution of about 30 arc sec (Sullivan and Downes,1973). One sees how the radio sources are resolved into a number of compact components, each of which could be ionized by a single O-star. These compact components represent subgroups of the OB-association W3, which may become visible in the optical range in some 10^5 yr. W3(OH) and (A), whose radio and IR spectra are shown in Figure 1, are two of these compact components of W3. By fitting models to the radio and IR spectrum one can deduce detailed information on the distribution of dust and ionized and neutral gas which gives us insight into the "cocoon stage" of OB-stars (Krügel and Mezger,1975).

Is there additional observational evidence supporting this picture of star formation in MA and DWA? From observations in M51 it has been found that the time span between the compression of the interstellar matter and the appearance of the first O-stars in the spiral arm is $6 \cdot 10^6$ yr (Mathewson et al.,1972). In MA, on the other hand, T-Tauri associations have been known for a long time as places where stars of medium mass form. Recently observations in the IR and observations of the carbon recombination line have demonstrated the presence of B-stars in T-Tauri associations, which are not yet visible in the optical range (Brown et al.,1974;Gatley et al.,1974; Grasdalen et al.,1973). On the other hand, T-Tauri stars are known to coexist in H II regions such as the Orion nebula or the Rosette nebula.

Considering the whole mass spectrum of MS stars, one finds that most of the luminosity, especially of the Lyc-photon luminosity, resides in the massive O-stars while most of the stellar mass resides in lowmass stars. From radio observations we obtained the number of Lyc-photons which are required for the ionization of an H II region. If the stellar mass function is known we can relate the number of Lyc-photons to the total mass of all stars which have been formed together with the O-stars. Using Salpeter's (1955) initial mass function (IMF) we find that 2.23 E46 Lyc-photons are associated with 1 $M_\odot$ of stars (Mezger et al.,1974).

Is Salpeter's IMF valid for star formation in both MA and DWA? The mass functions of two young star clusters have been determined (Walker,1957). NGC 2264 is located in the Orion arm and thus should be typical for star formation in MA. NGC 6530 is located in the Sagittarius arm which is one of the main spiral arms in the Galaxy and thus may represent star formation in DWA. In both cases the mass functions at least for stars >1.5 $M_\odot$ agree with each other and with Salpeter's IMF. This means that present observations do not contradict the assumption of a constant IMF for star formation in MA and DWA. Nothing, however, is known about the IMF in the nuclear disk close to the galactic center.

Most H II regions are associated with dense molecular clouds, whose mass can be estimated from radio molecular lines. Within the subgroups of an OB-association, we assume that all remaining gas has been ionized. The mass of ionized gas can be estimated from the total radio flux density. The total stellar mass associated with the H II region can be estimated from the Lyc-luminosity-to-mass ratio. Applied to giant H II regions this method yields usually ratios

$$M_{cloud} : M_* : M_{H\,II} = 10:1 : 0.1 .$$

To obtain the efficiency of star formation, we compare the mass of stars to the mass of gas left over in the immediate neighbourhood - i.e. in the H II region - and get an efficiency of 90%, which is very high. On the other hand, comparing the mass of stars to the mass of the cloud, one gets 10%. In fact, applying a similar method to investigate the efficiency of star formation in DWA,one finds that only about 1 % of the total mass of gas which flows through a DWA is transformed into stars.

III. Star formation rate in our Galaxy. This section is the summary of a paper by Mezger and Smith (in preparation) in which we determine the present star formation rate based on radio observations of H II regions. This observed star formation rate is compared to the star formation rate predicted on the assumption that it is proportional to the amount of gas available.

1) The observed present star formation rate. We can investigate regions of star formation in our Galaxy by means of radio continuum and recombination line surveys of H II regions. Note that H II regions observed as radio sources are only those with fairly high average density, $n_e \gtrsim 100$ cm^{-3}, i.e. young H II regions, which, at optical wavelengths, are usually obscured by dust. Optically observable H II regions, on the other hand, are often weak undetectable at radio wavelengths.

Figure 4 shows in the upper diagram (from Mezger,1970) the distribution on the galactic plane of all H II regions detected in the radio surveys of the northern and southern parts of the Galaxy (Reifenstein et al.,1970; Wilson et al.,1970). The density of the H II region is highest in the solar vicinity and falls off with increasing distance. This is to be expected since there are a large number of relatively weak H II regions such as the Orion nebula or NGC 2024, which would not be detectable at large distances. In the lower diagram we therefore removed all small H II regions and retained only giant H II regions, whose intrinsic flux density is more than four times that of the Orion nebula. For one galactic half plane, we find a distribution of giant H II regions which shows some symmetry with respect to the galactic center. Since surveys are incomplete for distances farther than 10 kpc we see only a few giant H II regions beyond the galactic center.

The last diagram shows the distribution of all giant H II regions together with an indication of the possible large scale spiral structure of the Galaxy. This diagram comes from Yvonne Georgelin (1975a,b) who combined our radio data with optical distance determination for optically observable giant H II regions. The picture of a four-armed spiral emerges, where the arms have pitch angles between 10° and 15° and wind about ¾ of a revolution around the galactic center.

Giant H II regions occur only in the main (DWA) spiral arms and in the galactic center. After correction for absorption of Lyc-photons by dust inside the H II regions - for this we assume an absorption cross section which is a function of galactic radius (see Smith, 1975) and correction for incompleteness of the radio surveys, we obtain the number of all Lyc-photons emitted by the ionizing stars of giant H II regions in DWA

$$\Sigma N_C(\text{giant H II, DWA}) = 4.2 \text{ E52 s}^{-1} \quad (1)$$

The λ6cm continuum radiation from an area l x b ≈ 1.5° x 0.5° around the galactic center yields $\Sigma N'_C \approx 0.7$ E52 s^{-1}. This number would have to be corrected for absorption of Lyc-photons by dust which, in the central region, appears to be especially strong. On the other hand, recent observations with the 100-m telescope (unpublished) have shown that a relatively large fraction of the λ6cm radiation is non-thermal. Until better quantitative results are available, we assume that the two effects cancel and put $\Sigma N_C \approx \Sigma N'_C$ as determined from the total λ6cm radiation. This yields

$$\Sigma N_C(\text{gal. center}) = 0.7 \text{ E52 s}^{-1} \quad (2)$$

Within a radius of 2 kpc around the sun, radio observations demonstrate the existence of many small H II regions with $S_5 D^2 < 400$ (f.u.) kpc^2. We assume that the solar neighbourhood is representative of the space between main (DWA) spiral arms. We estimate the stellar Lyc-photon flux of the ionizing stars of these small H II regions in the usual way and compare it to the corresponding Lyc-photon flux of the stars ionizing giant H II regions between galactic radii l and 12kpc. Allowing for the difference in the areas we considered, we find ΣN_C(small H II) /ΣN_C(giant H II) = 0.25 We assume that this ratio holds everywhere between galactic radii 4 and 13kpc. As shown in Fig.4, giant H II regions occur only in DWA; this appears to be a consequence of the mechanism of star formation in DWA. Small H II regions are found predominantly in interarm regions and MA. Although small H II regions probably also occur in DWA their integrated Lyc-photon flux

$$\Sigma N_C(\text{small H II, interarm}) = 1.0 \text{ E52} \quad (3)$$

represents primarily star formation in regions between DWA. (How much of this star formation occurs in MA and how much in genuine interarm regions is still an open question.)

Provided that all Lyc-photons come from O-stars and that the IMF is constant over the Galaxy and is represented by Salpeter's IMF (see previous section), the Lyc-photon flux per solar mass of stars is $\langle N_C \rangle / \langle M_* \rangle = 2.23$ E 46 $s^{-1} M_\odot^{-1}$. This assumption is probably reasonable for the region between galactic radii 4 and 13 kpc but could be wrong for the central region of the Galaxy. The total number of Lyc-photons emitted in the Galaxy by stars which ionize H II regions is the sum of the values, Eq. (1) - (3) : $\Sigma N_C = 5.9$ E52 s^{-1}. With the above assumptions, it follows that these radio H II regions are associated with 2.7 E 6 $M_\odot$ of newly formed stars of all masses.

The lifetime of radio H II regions is estimated as follows: Stars of spectral type O5 - O7 contribute most to the ionization. Their lifetimes range from 3.1 to 3.6 E6 yr. From a comparison of radio H II regions and optically visible O-stars in the vicinity of the sun we find that the early O-stars that ionize radio H II regions, are nearly always hidden by dust (see Schraml and Mezger, 1969) and comprise about 20%±5 of those stars. Adopting plausible mean values of 3.2 E6 yr and 20%, we derive an average lifetime for a radio H II region of 6.4 E5 yr. The total present star formation rate in the Galaxy is obtained by dividing the total mass of stars formed together with the ionizing stars by the lifetime of radio H II regions.

$$\frac{dM_*}{dt} = \frac{\Sigma N_C}{t\,(\text{radio H II})} \cdot \frac{\langle M \rangle}{\langle N_C \rangle} = 4.2\, M_\odot \text{ yr}^{-1} \quad (4)$$

Of these, 12% are formed within the central 200 pc, 71% are formed in DWA and 17% are formed in the interarm region (i.e. probably predominantly in MA).

The number of Lyc-photons emitted by all O-stars in our Galaxy has been estimated previously by Terzian (1974) to be $\Sigma N_C = 7.5$ E52 s^{-1}. This estimate is based on the observed O-star density in the vicinity of the sun, multiplied by the volume of the Galaxy where star formation occurs, which, according to Terzian, is limited by the galactic radii 4 and 15kpc and has a width in z-direction of 200 pc. We estimate that 20% of all O-stars are in radio H II regions. Thus, from our estimates of Lyc-photon fluxes in radio H II regions, above, it follows that the number of Lyc-photons emitted by O-stars in the interarm region is 5.0 E52 s^{-1} and in DWA is 2.1 E53 s^{-1}.

Terzian's estimate refers primarily to the interarm region and agrees fairly well with our value for the interarm region alone. The total number of Lyc-photons produced by *all* O-stars per sec in the Galaxy is ΣN_c(total) = 3.3 E53, or more than four times the number estimated by Terzian.

2). Comparison of observed and predicted star formation rate. Next, we want to relate the observed present star formation rate to the ratio of gas to total mass. In Figure 5, lower diagram, we plot the distribution of the number of giant H II regions as a function of radial distance, together with the distribution of atomic hydrogen While the surface density of atomic hydrogen shows a very flat maximum which extends from 5 to 15 kpc, the distribution of giant H II regions peaks sharply between 4 and 8 kpc. The total mass surface density (not shown in the diagram) increases very sharply towards the galactic center.

In the last years, it became clear that, already at moderate gas densities such as exist in cloudlets, most of the hydrogen tends to form molecules and thus is no longer detectable in the λ21cm line. It was suggested that the difference in the observed distributions of giant H II regions and atomic hydrogen was due to this effect. Several surveys of the galactic plane in the $J=1\to0$ λ2.6mm transition of CO confirmed this suggestion. (Scoville and Solomon, as quoted by Scoville, 1975; Burton et al., 1975). The distribution of the average CO line temperature obtained by Scoville and Solomon (which is roughly proportional to the column density of H_2) is shown in the upper part of Figure 5. It follows the distribution of giant H II regions, even in details. This result stresses the generic relation between clouds and star formation. Moreover, these observations allow us to estimate the fraction of hydrogen which is in molecular form. Between 4 and 8 kpc, the observed column density of atomic hydrogen has to be multiplied by a factor of 5 to account for the molecular hydrogen. Outside 13 kpc, the fraction of molecular hydrogen appears to be very low. With these corrections, and accounting for a He-abundance of 10%, we derive a total mass of gas of 1.4 $10^{10} M_\odot$. The corresponding average ratio of atomic to total H in the Galaxy ~ 0.4. Since most material in interstellar "clouds" is in molecular form, this implies that ~60 % of the interstellar material is in such clouds.

In a very simple model, the rate of star formation is proportional to the amount of gas available in the Galaxy. We express the gas mass as a fraction of the total mass of the disk component of the Galaxy. We use Sanders' and Lowinger's (1972) value of 11 E9 $M_\odot$ for the mass of the central cluster ($R_G < 0.8$ kpc) and Contopolous (1973) tabulation of the Schmidt model from 0.8 pkc out to 17 kpc. The total mass is then M_{tot} = 1.5 E11 $M_\odot$, and the fraction of gas is $\mu = M_{gas}/M_{tot} = .09$.

We define S as the ratio of the mass of all stars formed in the life of the Galaxy, M_*, to the total mass. The fraction of matter locked in stars at present is, however, smaller since a fraction "r" (which ranges from 0.2 to 0.4; see Tinsley, 1974) is returned to the interstellar matter. We make the usual approximation that mass return occurs instantaneously. Then,

$$(1 - r)\, S + \mu = 1$$

and at the present time, in the Galaxy,

$$S = (1 - \mu)/(1 - r) = 0.91 / 0.7 = 1.3$$

for $\mu = 0.09$ and $r = 0.3$.

The assumption that the star formation rate is proportional to the amount of gas then takes the form

$$\frac{dS}{dt} = \frac{1}{M_{tot}} \frac{dM_*}{dt} = \frac{1}{(1-r)\,\tau}\, \mu$$

and it follows that

$$\mu = \exp\{-t/\tau\},$$

where τ is the characteristic time of gas evolution. Applied to the Galaxy, where μ = 0.09, this theory yields (adopting t = 10^{10} yr as the age of the Galaxy) τ = 4.2 E9 years, dS/dt = 3.1 E-11yr^{-1} and $dM_*/dt = 4.6\ M_\odot \cdot \mathrm{yr}^{-1}$.

If dS/dt is proportional to μ^2, instead of μ, then $\mu = (t/\tau + 1)^{-1}$ and, for the Galaxy at the present time, τ = 9.9 E8 yr, dS/dt = 1.2 E-11 and dM_*/dt = 1.8 $M_\odot \mathrm{yr}^{-1}$. Comparison with the rate 4.2 $M_\odot \mathrm{yr}^{-1}$ derived from counting Lyc-photons, indicates that dS/dt $\propto \mu$ is the better solution.

A more detailed comparison appears to be warranted. For this detailed comparison, we subdivide the Galaxy in concentrical rings (Figure 6) and determine the gas fraction for each of these rings (lower part of the diagram) and the predicted star formation rate (upper part of the diagram). Also shown in the upper diagram is the star formation rate estimated from the distribution of giant H II regions. The result of this comparison can be stated as follows:

1) There is agreement between theory and observations in the region of galactic radii from 4 to 12 kpc. where most of the stars of the Galaxy are formed.

2) A dependence of dS/dt on μ^2 rather than μ produces predicted values of dS/dt shown by the dotted lines in Figure 6: they are significantly lower than the values derived from countting Lyman photons.

3) In the galactic center, the observed star formation rate is 60 times the predicted formation rate. There are several plausible explanations for this discrepancy which reflect primarily our ignorance about the physical conditions of the interstellar gas close to the galactic nucleus and their possible implication for star formation.

4) For galactic radii >12 kpc the predicted star formation rate is an order of magnitude or larger than the observed formation rate. Again, there are several plausible explanations for this discrepancy. For example, we put the star formation rate proportional to the surface density rather than to the volume density. This would not alter the results inside the solar circle where the scale height of the interstellar gas is rather constant, but would decrease the star formation rate outside the solar circle where the scale increases rapidly. Furthermore, if Lin's density wave correct, the star formation rate in DWA would not only be determined by the gas density but also by the compression of the gas in the shock front which precedes a DWA. The strength of this compression decreases from the inner Lindblad resonance at 4 kpc to the co-rotation point at 13 kpc. Outside the co-rotation point the main spiral arms would gradually become MA and one would therefore expect a lower star formation rate.

IV. **Element abundance in the Galaxy and its relation to star formation.** In the big bang, the light elements H, D, ^{3}He, ^{4}He were produced. We adopt mass fractions

$$\begin{aligned} x(H) &= 0.76 \\ X(D) &= 6.2\ E\text{-}5 \\ Y(^3He) &= 2.3\ E\text{-}5 \\ Y(^4He) &= 2.4\ E\text{-}1 \end{aligned}$$

As a consequence of star formation and subsequent return of processed material, the interstellar gas was enriched by elements heavier than ^{4}He. The simple theory yields for the mass fraction of heavy elements

$$Z = \frac{P}{1-r} \ln \frac{1}{\mu} \approx 10^{-2} \ln \frac{1}{\mu}$$

with P the mass fraction of the gas returned in the form of heavy elements and P/(1-r) the yield. The numerical value of the yield has been adjusted to give $Z \approx 0.02$ for the solar vicinity.

The interstellar gas is also enriched by ^{4}He. The present mass fraction is therefore the sum of primordial He and enrichment

$$\begin{aligned} Y(^4He) &= Y_{prim} + \Delta Y \\ &= 0.24 + 2Z\ . \end{aligned}$$

The relation $\Delta Y \approx 2Z$ has been empirically determined for the solar neighbourhood.(Churchwell, et al.,1974).

^{3}He and ^{2}H are destroyed when passing through a star. The present ^{2}H-abundance is connected with its primordial value by

$$X(^2H) = X_{prim}\, \mu^{\frac{r}{1-r}} = X_{prim}\, \mu^{0.43}.$$

Evaluating these equations, we predict that :

1) The metal abundance increases from 0.02 at the distance of the sun to 0.07 at the galactic center.
2) The Deuterium at the galactic center is depleted by a factor of 7.5 compared to the solar vicinity.
3) The He-abundance should increase from 0.28 (.10 by number) at the distance of the sun to 0.37 (0.16 by number) at the galactic center.

We compare these predictions with observations.

1) An increase of heavy element abundances towards the center is well established in external galaxies. (H.E.Smith,1975). There are a number of observations which qualitatively suggest a similar effect in our own Galaxy. However, no quantitative results are available.

2) Presence of Deuterium can be observed through radio lines of isotopic molecules, such as HCN and DCN. For a gas fraction $\mu = 10^{-3}$ the Deuterium in the center region should be depleted by a factor of 7 compared to the solar vicinity. DCN has been observed in the Orion Molecular Cloud, but has not been detected in the Sgr B2 cloud (Penzias, priv. comm.).

3) He-abundances in H II regions can be determined through observations of adjacent radio recombination lines, provided He is ionized throughout the H II region. Figure 7 shows the predicted He-abundance (by number) based on a primeval He-abundance of 0.08, plus He produced in stars and returned to the interstellar matter. The combined He-abundance is rather constant about 10% between $D_G = 5$ to 11kpc. It increases towards the galactic center and decreases to a minimum value of 9% at $D_G = 14$ kpc. Also shown are our and Churchwell's (Churchwell et al.,1975, unpublished) measurements of the ionized He-abundance in H II regions which is related to the total He-abundance by

$$y^+ = Ry$$

with R the ratio of He^+- to H^+ - Strömgren sphere, weighted with the square of the proton density. While the He^+- abundance between 9 and 15 kpc behaves approximately as expected, it falls to values of 3% and less when approaching the galactic center. This behaviour is believed to be the result of selective absorption of Lyc-photons by dust (Churchwell et al.,1974). He-ionizing photons are absorbed more strongly than H-ionizing photons and with increasing absorption depth the He^+ - Strömgren sphere shrinks more rapidly than the H^+ - Strömgren sphere. The fact that the He^+ - abundance, and hence R, decrease towards the galactic center is interpreted as increasing absorption cross section per H-atom. Since the dust is made primarily of heavy elements, this suggests that Z increases towards the galactic center.

There is also evidence that suggests high He-abundance. We have recently detected He-recombination lines from several H II regions close to the galactic center and especially from Sgr A West, a compact H II region which surrounds the nucleus of our Galaxy. In Sgr A West the ionized He-abundance is already 10% and extrapolating this value to the total He-abundance suggests a value considerably higher than 10% (Mezger and Smith, subm. for publ.).

Accurate determinations of He-abundances as a function of galactic radius in our Galaxy will yield important astrophysical information. Outside 13 kpc, we can expect to determine the "big bang" abundance which is directly related to the mass density of the universe. The increase of the He-abundance above the primordial value at smaller galactic distances is directly related to the chemical evolution of the Galaxy. The He-abundance in the central region of the Galaxy may yield information about the physical processes in galactic nuclei.

With the assumptions made here, it follows that the present enrichment of the interstellar matter by He and metals depends only on μ (the ratio of gas-to-total mass) and not on the relation between star formation rate and gas-to-total mass ratio, $dS/dt \propto \mu^n$. However, the time dependence of the enrichment does depend on this relation. The solar system isolated from the interstellar matter ~ 5 10^9 yr ago. Adopting an age of the Galaxy of 10^{10} yr, the metal abundance at the time of the birth of the solar system would have been 50% of its present value for $n=1$ and 75% for $n=2$. It is often implied that the solar system abundance is identical with the present abundance in the interstellar matter; this implication cannot be corrected if recycling of material processed in stars in the process which dominates the enrichment of interstellar matter by Y and Z. In fact, the distinct and confirmed differences in the isotopic ratios of C^{12}/C^{13} and O^{16}/O^{18} between solar system and surrounding interstellar matter is a convincing prove of continuing processing of material since the isolation of the solar system.

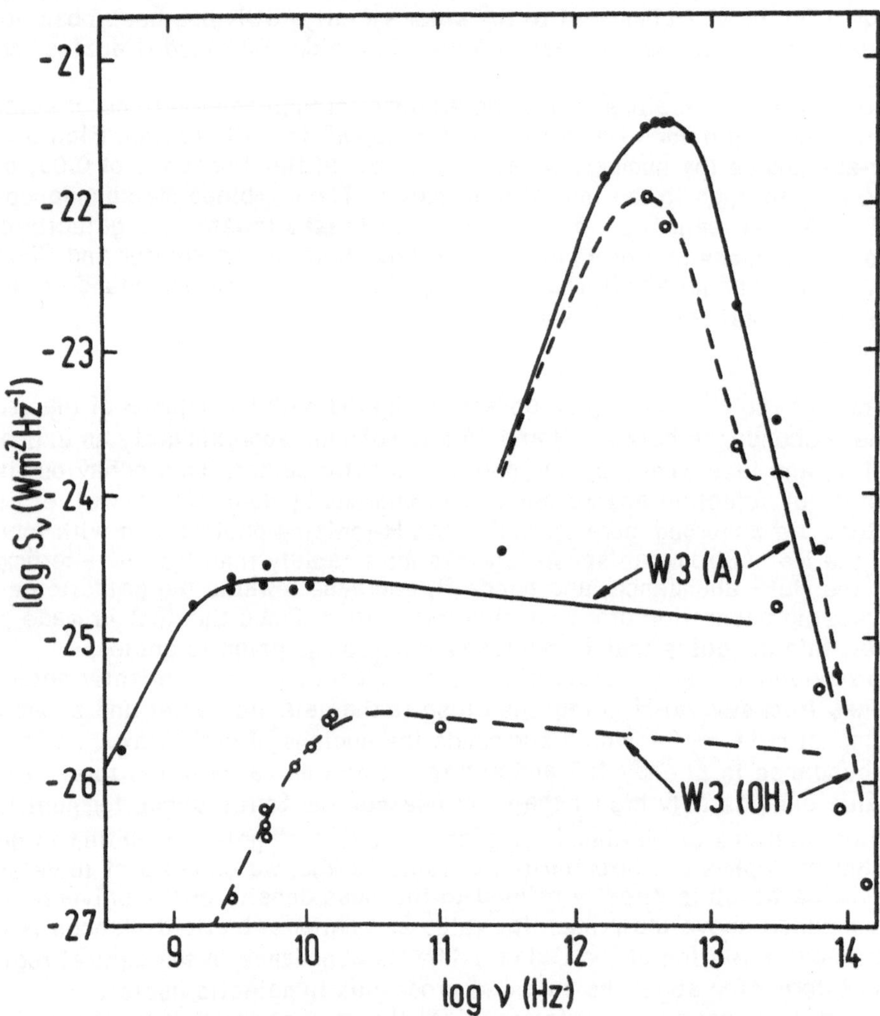

Fig. 1. Radio and IR spectra of W3(A) and W3(OH), two compact H II regions in the W3 complex. Full and dashed curves are from model fits by Krügel and Mezger (1975).

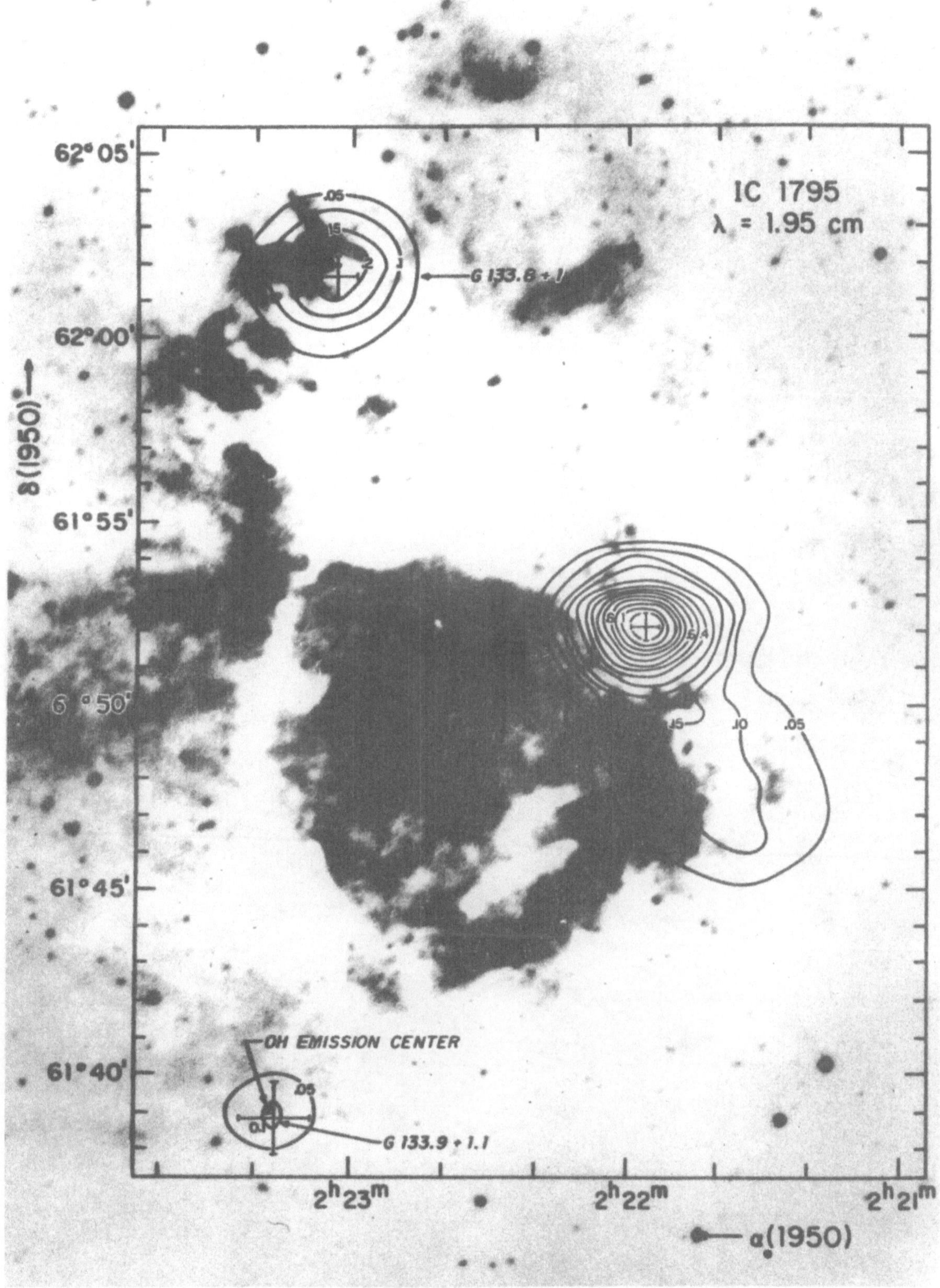

Fig. 2. Overlay of contour lines of constant radio free-free emission (angular resolution 2') on a Mount Palomar red print (Schraml and Mezger 1969). Note that the regions of strongest radio emission are completely obscured by dust in the optical picture.

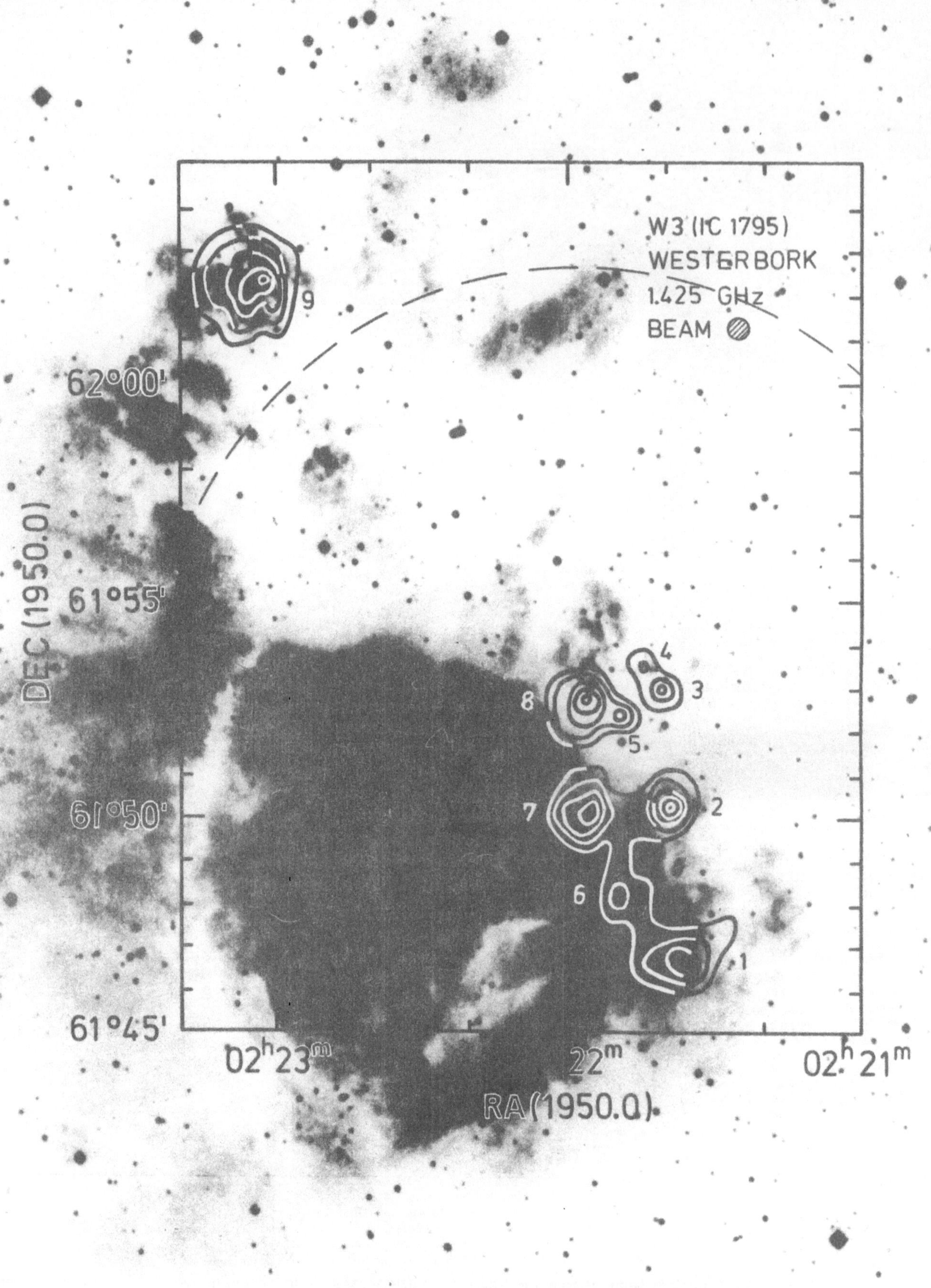

Fig. 3. Overlay similar to Figure 2, but now observed with an angular resolution of ~30" (Sullivan and Downes, 1973).

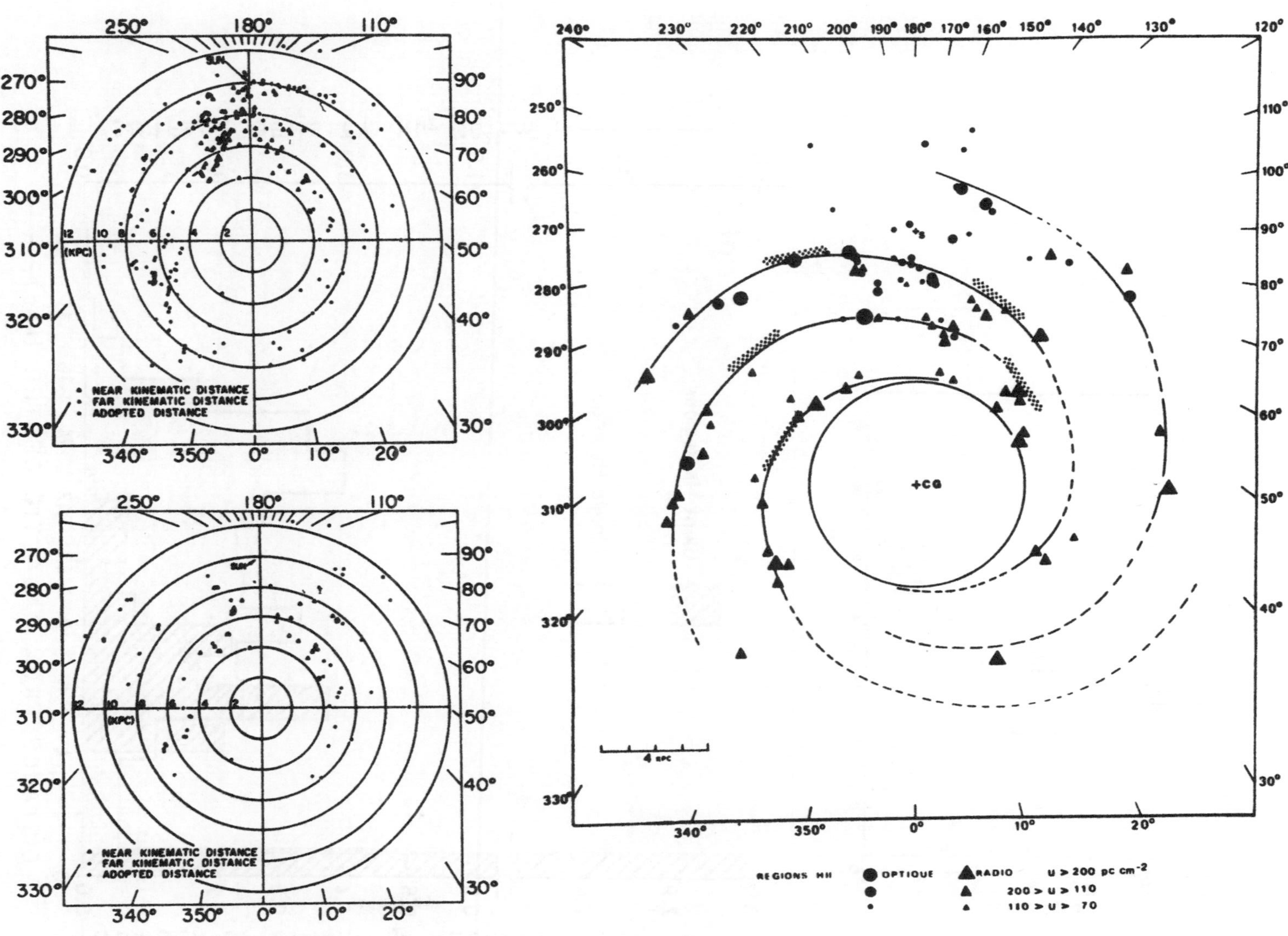

Fig. 4. The distribution of H II regions and the spiral structure of the Galaxy. Upper diagram left: All H II regions observed in radio continuum and recombination line surveys at λ6cm. Lower left diagram; Only giant H II regions are retained (Mezger, 1970). Right diagram; The main spiral arms of the Galaxy, based on optical and radio observations of giant H II regions (Georgelin, 1975 a ,b).

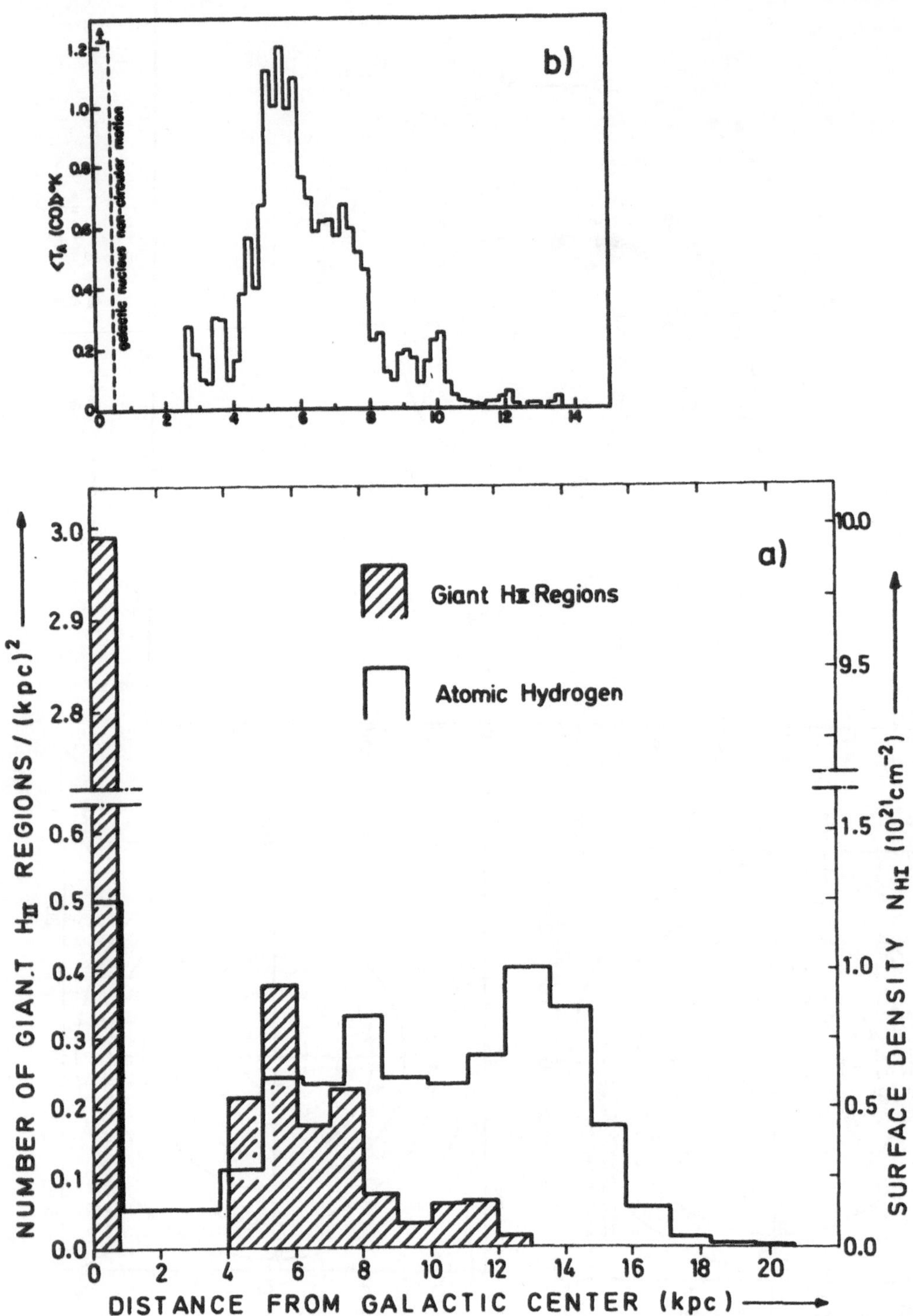

Fig. 5. Distribution of neutral and ionized gas as a function of distance from the galactic center: a) The distribution of atomic hydrogen and giant H II regions (Mezger, 1970). b) The average CO(J= 1→ 0) line temperature (Scoville and Solomon, 1975), which is roughly proportional to the column density of H_2.

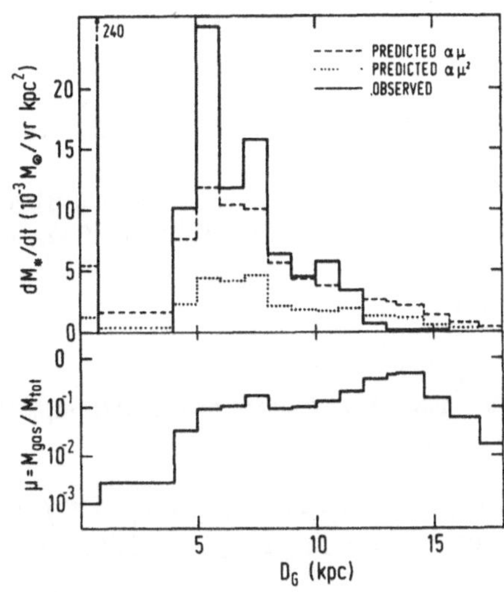

Fig. 6. Lower diagram: Ratio of gas/total mass in the Galaxy. Upper diagram; observed and predicted (for two different models) star formation rate in the Galaxy.

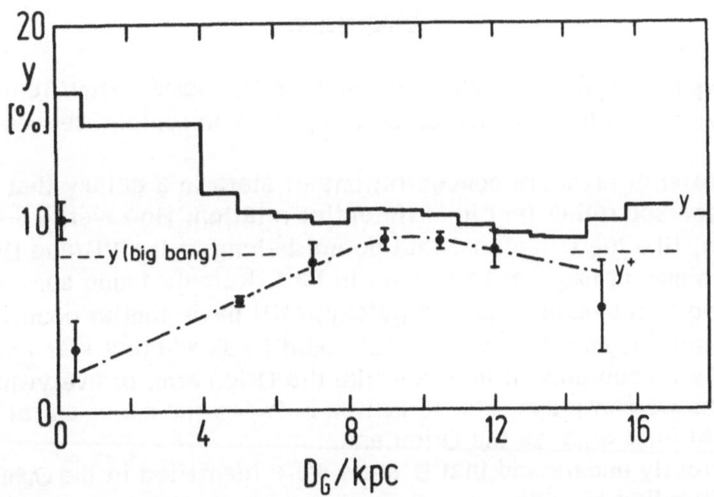

Fig. 7. He-abundance (by number) as a function of the distance from the galactic center. Heavy curve: He-abundance predicted on the basis of the gas/total mass ratio shown in Figure 6. Dashed-dotted line; Observed abundance of ionized He in galactic H II regions.

REFERENCES

BLAAUW A. 1974, Ann. Rev. Astr. Astrophys. 2, 213

BROWN R.L., GAMMON R.H., KNAPP G.R., BALICK B. 1974, Ap. J.192, 607

BURTON W.B., GORDON M.A., BANIA T.M., LOCKMAN F.J. 1975, Ap.J., subm. for publ.

CHURCHWELL E., MEZGER P.G., HUCHTMEIER W. 1974, Atron.a.Astrophys. 32, 283

CONTOPOULOS G. 1973, in "Dynamical Structure and Evolution of Stellar Systems", Geneva Observatory ed., p.1

GATLEY I., BECKLIN E.E., MATTHEWS K., NEUGEBAUER G., PENSTON M.V., SCOVILLE N.1974, Ap. J. 191, L121

GEORGELIN Y. 1975a, unpub;. thesis, Universite de Provence, Observatoire de Marseille; 1975b C.R.Acad. Sci. Paris, 280, series B, p.349

GRASDALEN G.L., STROM R.M., STROM S.E. Ap. J. 184, L53

IBEN I., and TALBOT R.J. 1966, Ap. J. 144, 968

KAHN F.D. 1974, Astron.a.Astrophys. 31, 149

KRÜGEL E., MEZGER P.G. 1975, Astron.a. Astrophys. 31, 149
LARSON R.B., STARRFIELD S. 1971, Astron. a.Astrophys. 13, 190
MATTHEWSON D.S., van der KRUIT P.C., BROUW W.N. 1972, Astron. a.Astrophys. 17, 468
MEZGER P.G. 1970, "The spiral structure of Our Galaxy", IAU symp. No. 38 (Becker and Contopoulos ed.) Reidel publ. Co., Doordrecht, pp. 107 - 121
MEZGER P.G., SMITH L.E., CHURCHWELL E. 1974, Astron. a.Astrophys. 32, 269
REIFENSTEIN E.C. III, WILSON T.L., BURKE B.F., MEZGER P.G., ALTENHOFF W.J. 1970, Astron. and Astrophys. 4, 357
SALPETER E.E. 1955, Ap. J. 121, 161
SANDERS R.H., LOWINGER Th. 1972, Astr. J. 177, 292
SCHRAML J., MEZGER P.G. 1969, Ap.J. 156, 269
SCOVILLE N.Z. 1975, in Proc. EPS Symp. "H II regions and related topics", D. Downes and T.L. Wilson ed (in press)
SMITH H.E. 1975, Ap. J. 199, 591
SMITH L.E. 1975, Proc. EPS Symp. "H II regions and related topics", D.Downes and T.L.Wilson ed.(in press)
SULLIVAN III, W.T., DOWNES D. 1973, Astron. a.Astrophys. 29, 369
TERZIAN, 1974, Ap.J. 193, 93
TINSLEY B.M. 1974, Ap. J. 192, 629
WALKER M.F. 1957, Ap. J. 125, 636
WILSON T.K., MEZGER P.G., GARDNER F.F., MILNE D.K. 1970, Astron. a.Astrophys. 6, 364

DISCUSSION

BLAAUW: With regard to the term "material arm", is the name "arm" really justified?

MEZGER: Material arms refer to dense cloud complexes in regions between the main spiral arms.

CONTOPOULOS: Material arms are concentrations of stars in a galaxy that are not density waves, but are dispersed rather fast by differential rotation. However, I think that Dr. Mezger means features, like the Orion arm, that do not belong to the "Grand Design" of the Galaxy. Such features may or may not be waves. In fact, Kalnajs found some cases where features, not belonging to the main arms of a galaxy, still have similar counterparts 180° away, and they are therefore bisymmetric waves. But could I ask whether star formation between the spiral arms takes place only in features like the Orion arm, or everywhere?

MEZGER: As far as we can judge, star formation between the major spiral arms occurs predominantly in material arms such as the Orion arm.

BLAAUW: Do I correctly understand that B stars were identified in the Ophiuchus clouds that were not known optically?

MEZGER: B stars are not hot enough to ionize a substantial part of the surrounding H but they emit enough photons $\lambda < 1100$ Å to form a CII region, which can be detected through the radio recombination lines. They also heat dust in the surroundings which give raise to IR radiation. The B stars in the Ophiuchus cloud were not known optically and were detected from the 10 micron emission and their CII radio recombination line emission.

SANDQVIST: In your discussion of the star formation rates, did you take into consideration that there is a great increase in the contribution to the mass of the gas by massive molecular clouds as you approach the galactic center?

MEZGER: Yes, we did. The total mass of the nuclear disk including H_2 is $\approx 10^7\ M_\odot$, the mass of the central stellar cluster is $\approx 10^{10}\ M_\odot$.

SANDQVIST: Recent analysis of high resolution observations of SgrA West has shown that a part appears to have a non-thermal spectrum. If, so, how would the resulting decrease in size of and flux from the thermal component affect your model for the nucleus of the Galaxy?

MEZGER: It would mean that the total He-abundance would be about 0.10 by numbers and the central mass could be higher than $10^6 - 10^7\ M_\odot$.

LONGAIR: You made the important statement that you belive there is no evidence for variations in the Salpeter initial mass function. Could you describe your evidence for this statement?

MEZGER: Walker has observed M8 and NGC 2264, the first in the Sagittarius (density wave) arm, the other in the Orion (material) arm. He finds that the luminosity functions agree with one another and with the Salpeter initial luminosity function down to $M_V \sim +3.5$ or $M = 1.5\ M_\odot$. For the region of the galactic center, however, we have no information about the initial mass function.

BIRTH-PLACES OF STARS AND GALACTIC SHOCKS

PER OLOF LINDBLAD
STOCKHOLM OBSERVATORY, SWEDEN

МЕСТА ЗВЕЗДООБРАЗОВАНИЙ И ГАЛАКТИЧЕСКИЕ УПЛОТНЕНИЯ

РЕЗЮМЕ

Если допустим, что звезды образуются, по крайней мере в определенных границах, повсеместно в Галактике, наличие галактического уплотнения может обусловить разрыв в пространственном распределении мест звездообразования для звезд находящихся в настоящее время в окрестности Солнца. Это происходит оттого, что поток газа, из которого образуются звезды, разрывается от волны плотности и звезды, образовавшиеся до и после уплотнений, но в обоих случаях проходящие вблизи нынешнего положения Солнца, будут следовать вдоль довольно различных орбит. Такой прерывный скачок прослеживается на данных Гросбола. В случае подтверждения, это может явиться наиболее непосредственным свидетельством наличия ударных волн в нашей Галактике.

I have been asked to discuss at this meeting the observational evidence of stellar evolution in different parts of the Galaxy. I will have to limit myself to the galactic distribution of one important phase of stellar evolution- the very early phase, or stellar birth.

In 1961 our knowledge about ages and chemical parameters for the stars had advanced so far that Strömgren (1962) at a conference in Princeton could formulate the ideal program for "the past history of the interstellar medium of our Galaxy" in the following way: "Ideally, one would determine on the basis of the space velocity and the age of the star its place of formation, and then one would know that at a certain time and location the star formed out of the interstellar medium, and had such and such chemical composition".

Strömgren added "In very few, if any, cases can we carry out this ideal program".

Nevertheless, Strömgren (1967) could at the symposium in Nordwijk report on three major steps towards the realization of that program.

1. The calibration of narrow band (uvby and Hβ) photometry measurements into spectral class, luminosity and metal content of stars (Strömgren,1966).

2. The calibration of the HR-diagram in terms of mass and age for stars of various chemical composition (Kelsall and Strömgren , 1966).

3. Computations of galactic orbits of stars with various velocities at the present time, with the results presented in tables (Contopoulos and Strömgren,1965).

Birth-places of stars, based on the principles drawn up by Strömgren, have been derived by Yuan (1969) and recently by Grosbøl(1975).

Grosbøl's material consists of 328 stars of spectral type BO-AO and luminosity class III-V all brighter than $m_V = 6.5$. The stars selected have been observed photoelectrically in the uvby and Hβ bands and have known radial velocities and proper motions.

The ages of the stars were determined by Hejlesen of the Copenhagen Observatory from the photometric data and computations of stellar models. All stars selected are younger than $700 \cdot 10^6$ years.

The birth-places as computed by Grosbøl in a galactic field of rotational symmetry are plotted in Figure 1.R_b is the distance from the galactic centre of the star at birth and Θ_b the corresponding galacto-centric position angle in a non-rotating frame of reference. At the top of the diagram this angle is transformed into τ which is the time it takes for a star to cover the angle Θ_b in circular motion at the distance R=10kpc. Thus τ is approximately equal to the age of the star.

The most obvious feature in Grosbøl's diagram is what seems to be a periodic variation between strong concentration to the distance R_b= 10 kpc and scatter up to larger R_b . The period is almost $200 \cdot 10^6$ years. This general behaviour may be readily explained in terms of the epicyclic representation of stellar orbits. According to this the stars can be considered, to the first order, to perform an epicyclic oscillation around a circular orbit at its mean distance from the centre, the epicyclic frequency κ being independent of the amplitude of the oscillation. κ is related to the circular angular velocity ω_c and Oort's constant A of galactic rotation through

$$\kappa^2 = 4\omega_c (\omega_c - A).$$

Let us make the reasonable assumption that the number of stars decreases with increasing deviation from circular motion at birth, i.e. stars born with low random energy are more abundant. This is still more emphasized in the present material, as stars reaching longer distances from the galactic plane are left out.

For a star from an outer or inner region of the Galaxy to reach the neighbourhood of the sun with lowest possible energy it takes half a revolution around the epicycle, which to the first approximation should be independent of the distance. With commonly adopted values of ω_c and A this time would be close to $100 \cdot 10^6$ years. Stars born in more distant regions would thus appear in our neighbourhood with a preferential age of about $100 \cdot 10^6$ years. Such stars would return a second time after 1 1/2 epicyclic revolution, and we would have another group of stars with ages around $300 \cdot 10^6$ years that also would be born at larger distances from the sun. For an age equal to the epicyclic revolution, or about $200 \cdot 10^6$ years, stars would have returned to the distance R_b of their birth-places and we would have in our neighbourhood only those with $R_b \approx 10$ kpc. All this can be seen in Figure 1. Unexplained is the lack of small R_b, which may be due to galactic structure effects. The quantitative agreement is remarkable, because it means an agreement between the galactic rotation parameter κ and the estimate of stellar ages from stellar evolutionary models. Alternatively, one could get an estimate of κ from stellar ages and kinematic distributions. Also, the figure illustrates the very strong selection effects imposed on the birth-places due to the requirement that the stars must now be in the solar neighbourhood.

Another effect which can be seen in Figure 1 is the apparently discontinuous jump in the string of birth-places close to $R_b = 10$ kpc around $\tau = 80 \cdot 10^6$ years. The argument in what follows is that the variation from somewhat low to somewhat high R_b over that age limit may be due to streaming motions on the two sides of a spiral arm and the discontinuity in that variation due to a galactic shock.

Grosbøl has plotted the birth-places in a polar co-ordinate system that rotates with the pattern velocity 13.5 km/sec/kpc assumed by Lin and his coworkers for a density wave as shown in Figure 2. We get an impression of the strong adherence to the circle $R = 10$ kpc with periodic samples from the outer structure brought here through 1/2 and 1 1/2 epicyclic orbit.

There is an observational selection effect, besides the kinematical one, to be kept in mind. Because early type stars down to a certain limiting apparent magnitude have been selected, there is a strong preference for young stars as can be seen in the figures.

If we assume that stars are born *exclusively* in spiral arms, then the birth-places ought to trace out such spiral arms, or rather selected parts of spiral arms due to the selection effects discussed. In particular one spiral arm has to pass through the present position of the sun. Such an interpretation of the observation leads to certain difficulties as has been pointed out by Grosbøl.

If, on the other hand, stars are formed to a *non-negligible* extent everywhere in the Galaxy, and thus also between the spiral arms, due to delayed triggering or secondary triggering of star formation, then the selection effects discussed above might be of primary importance for the appearance of Figure 2. As, however, the stars would be born with the average motion of the gas out of which they are formed the presence of a galactic shock (Roberts, 1969) would cause a discontinuous break in the string of low-velocity birth-places around $R_b = 10$kpc. This is so because the gas flow is broken at the shock, but the stars already born out of the gas are not. Thus, stars now reaching the position of the sun must be born in different gas stream lines depending on whether they were born before or after the gas reached the shock.

In terms of the linear density wave theory the situation might be described as follows. Stars born at the outside of a spiral arm would be born in a "high velocity stream" and would reach us preferentially from the inner region of the Galaxy, while stars born on the inside of an arm in a "low velocity stream" would reach us preferentially from the outer region of the Galaxy. Thus, for $\Omega_p < \omega_c$ and a trailing spiral pattern, when the string of birth places crosses the spiral arm for increasing ages of the stars there should be a shift from somewhat low to somewhat high R_b. The observed sense of the variation would thus be in agreement with $\Omega_p < \omega_c$ and a trailing spiral pattern, or $\Omega_p > \omega_c$ and a leading spiral pattern, according to the linear theory. In the presence of a shock the situation may be somewhat more complicated.

Figure 2 would then not trace out spiral arms but indicate the intersection between a spiral arm and the circle $R = R_{\odot}$. The present position of this intersection depends on the pattern velocity, but it is such, according to Figure 1 that a star now close to the sun and travelling in a circular orbit, would have passed through the arm about $80 \cdot 10^6$ years ago. If the apparent discontinuity around $\theta_b = -2$ radians in Figure 1 is confirmed, this may be the most direct evidence for the existence of galactic shocks in our Galaxy.

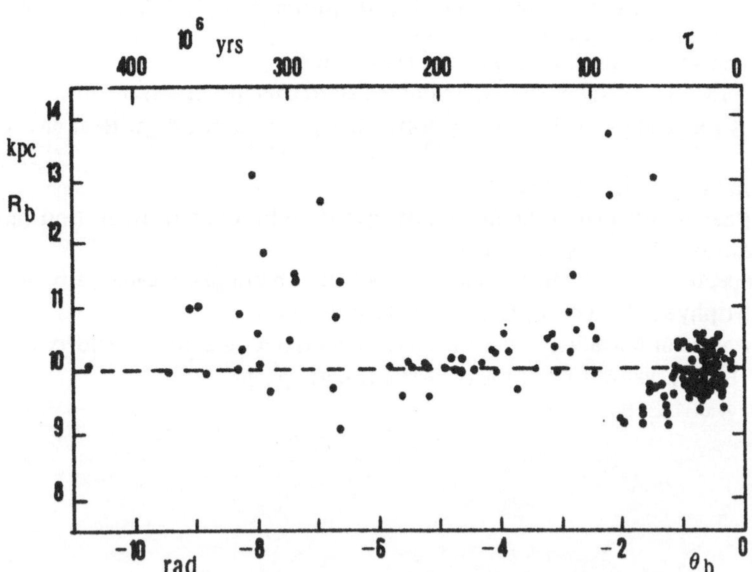

Fig. 1. The observed (m_λ -V) plotted against $\frac{1}{\lambda}$ (μm^{-1}) for Per (HD 24398) is shown by filled circles; the corrected (m_λ- V) is denoted by open circles. The model computations are indicated by a solid line.

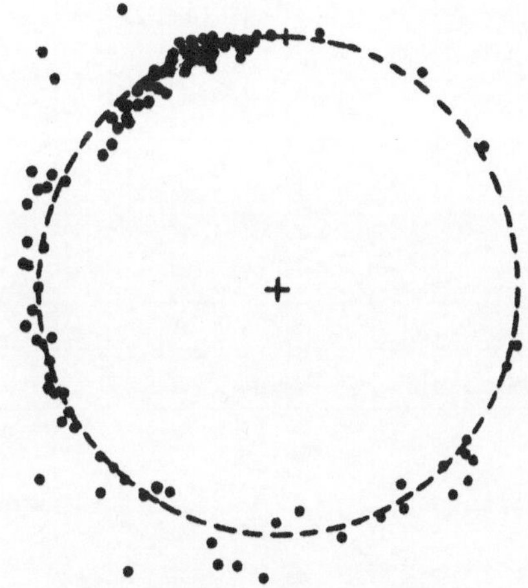

Fig. 2. a) log Te vs Q; b) log Te vs MK spectral type. Open circle are supergiants and filled circles are main sequence stars.

REFERENCES

CONTOPOULOS G. and STRÖMGREN B. 1965. Tables of plane galactic orbits. Publ. Inst. Space Studies, Goddard Space Flight Centre, NASA

GROSBØL P. 1975. manuscript (in danish, Copenhagen Observatory)

KELSALL T. and STRÖMGREN B. 1966. Calibration of the Hertzsprung-Russell diagram in terms of age and mass for main-sequence B and A stars in Vistas in astronomy, Vol. 8, Ed. A. Beer and K.Aa. Strand, p 159

ROBERTS W.W. 1969. Astrophys. J. 158, 123

STRÖMGREN B. 1962. Past distribution of the interstellar gas in The distribution and motion of interstellar matter in galaxies, Ed. L. Woltjer, p.274

STRÖMGREN B. 1966. Spectral classification through photoelectric narrowband photometry in Annual review of astronomy and astrophysics. Vol. 4, Ed. L. Goldberg, p.433

STRÖMGREN B. 1967. Places of formation of young and moderately young stars in Radio astronomy and the galactic system, Ed. H. van Woerden; I.A.U. Symp. No. 31, p.323

YUAN C. 1969. Astrophys. J. 158, 889

THE PAST POSITIONS OF THE SPIRAL ARMS OF OUR GALAXY

G.CONTOPOULOS AND P.GROSBØL
UNIVERSITY OF THESSALONIKI, THESSALONIKI, GREECE
PRESENTED BY G.CONTOPOULOS

ПОЛОЖЕНИЕ СПИРАЛЬНЫХ РУКАВОВ НАШЕЙ ГАЛАКТИКИ В ПРОШЛОМ

РЕЗЮМЕ

Рассчитаны и определены места образования 176 избранных B8-A0 звезд. Критерием наилучшей модели спиральной системы, вращающейся с некоторой угловой скоростью, служило расположение места звездообразования в четко очерченной структуре, параллельной спиральным рукавам. Наилучшее согласие к настоящему времени являет компактное спиральное поле, при угловой скорости ≈ 15 км$^{-1}$кпс$^{-1}$ и 3%-ой амплитуде осесимметричного поля.

ABSTRACT

The birthplaces of 176 selected B8 - A0 stars have been calculated in various spiral models. The criterion for the best model was that the birthplaces in a frame rotating with the angular velocity, Ω_s, of the spiral pattern should be located in well defined structures, parallel to the imposed spiral arms. The best agreement found up to now was with a rather tight spiral field with $\Omega_s \approx 15$ km s^{-1}kpc^{-1} and an amplitude 3% of the axisymmetric field.

I was surprised, on my arrival to Tbilisi, to realize that Dr. Lindblad was to speak on exactly the same topic as myself. However, our conclusions are rather different. Therefore, I will emphasize some of the differences between our points of view.

I had the opportunity to collaborate with Dr. Strömgren in the project of finding the birthplaces of stars in our Galaxy. We constructed an axisymmetric galactic model, which was used in calculating the orbits of stars backwards to their places of origin (Contopoulos and Strömgren, 1965). We expected that in this way we would find the past positions of the spiral arms of our Galaxy, assuming that stars are born basically only in spiral arms. In fact the main tracers of spiral arms in galaxies are very young objects, like O-associations, HII regions, etc.

If we know the past positions of the spiral arms, we can immediately derive the angular velocity of the spiral pattern, Ω_s. This is a very important quantity in density wave theory, but up to now it is very poorly known. Lin, Yuan and Shu (1969) give a value around $\Omega_s = 13$ km s^{-1} kpc^{-1}, while Kalnajs (1970) gives a value around $\Omega_s = 30$ km s^{-1}kpc^{-1}.

Strömgren (1967) calculated the birthplaces of 52 stars and derived a preliminary value of the angular velocity of the spiral pattern equal to $\Omega_s = 20$ km s^{-1}kpc^{-1}. In these calculations the effects of the spiral field itself were ignored. Yuan (1969) included the spiral field of Lin and Shu, and found the birthplaces of stars in the theoretical spiral arms. This result was considered as evidence in favor of the value of Ω_s adopted by Lin, Yuan and Shu. However, the total number of stars used (25) was rather small and his conclusions were criticized by Contopoulos (1972) and Kalnajs (1973).

The most complete study of this problem up to now has been made by Mr.Grosbøl from Copenhagen. Dr. Lindblad mentioned the earlier work of Mr.Grosbøl, contained in an unpublished report in Danish. Since that time Mr.Grosbøl has been in Thessaloniki for several months completing and improving his work. The main differences between his recent study and his earlier work described by Dr. Lindblad are the following.

1. He included many B8 - B9 stars and southern A0 stars, that were missing in the previous study. At the same time he made his criteria more rigorous. Thus out of about 700 stars with proper motions from FK4 and its supplement (Fricke and Kopff, 1963; Fricke, 1963), radial velocities from Wilson's (1953) General Catalogue, and spectral types between B8 - A0, he selected only 176 stars. These are expected to have an error in age less than 25 x 10^6 years.

2. The ages of these stars range between 60 and 300 x 10^6 years. Thus all young stars were excluded from this study. The reason for this exclusion was that the very young stars must have been born in local condensations, probably inside the so-called Orion arm, which does not seem to belong to the two-armed "grand design" of our Galaxy, but is believed to be an interarm branch between the main arms.

Such branches appear often in other galaxies and star formation does take place in them. In order to avoid such "local" effects, due to the strong bias of young stars, we thought that we would find more reliable results by dealing with somewhat older stars.

This is an important difference between our point of view and that of Dr. Lindblad who assumes that star formation takes place to a non-negligible extent everywhere between spiral arms.

3. Finally, the calculations were made with an axisymmetric field on which a spiral field was superimposed, while Dr. Lindblad mentioned only calculations without a spiral field.

As axisymmetric field the Contopoulos-Strömgren (1965) model was used together with a spiral field of the form

$$- a \frac{tg i}{2} \omega_0^2 \Omega_0^2 \cos \left[2(\theta - \Omega_s t) + \frac{2}{tg i} \ln \left(\frac{\omega}{\omega_1}\right) \right].$$

This spiral field depends on the following parameters:

a. the distance, ω_1, of the potential minimum nearest to the Sun from the galactic center in the direction of the Sun (Sagittarius arm $\theta = t = 0$),

b. the ratio, a, of the (maximum) radial spiral force over the axisymmetric force at the distance of the Sun, ω_0, which is equal to $\omega_0 \Omega_0^2$,

c. the pitch angle i, and

d. the spiral pattern speed, Ω_s.

A large number of models were explored with the parameters ranging in the following intervals:

ω_1	from	8.1	to	8.9 kpc
a	from	3%	to	5%
i	from	-4°	to	-14° and
Ω_s	from	7.5	to	42.5 km s^{-1} kpc^{-1}

Furthermore, two chemical compositions were assumed, with (X, Z) = (0.7, 0.03) and (X, Z) = (0.7, 0.02). The ages were calculated using the isochrones of Hejlesen *et. al.* (1972).

The criterion for selecting the best model was that the brithplaces in a frame rotating with the angular velocity of the spiral pattern should be located in well defined structures, parallel to the spiral arms.

Figure 1 (a, b, c, d, e) shows some of the cases that were studied. In the case (1a) we have only the effect of the axisymmetric field. If we compare Fig. 1a with the first figure of Dr. Lindblad, we notice, first, the lack of young stars. However, we see several stars in the region $\omega < 10$ kpc and ϕ between -50° and -100°, which are missing in his figure (notice that he uses a nonrotating frame in which this region is between -2 and -4 radians). Thus the discontinuity around the angle -2 radians, noticed by Dr. Lindblad, does not seem to appear if we include the new material of B8 - B9 stars.

Dr. Lindblad's conclusions were based, further, on calculations without a spiral field. This would be justified if the effect of the spiral field were unimportant. However, although the amplitude of the assumed spiral field is, in general, only 3% of the axisymmetric field, its effects, over the periods of time considered in this study, are significant. This can be seen by comparing Fig. 1 (b, c, d, e) with Fig. 1a. The most sensitive parameters that affect the pattern of birthplaces are i and Ω_s.

Figure 1b gives one of the best models with chemical composition z = 0.03. The parameters used are $\omega_1 = 8.4$ kpc, a = 3%, i = -9°.5, and $\Omega_s = 13.5$ km s^{-1} kpc^{-1}. In this model most of the birthplaces form a rather well defined spiral arm, almost parallel to the imposed spiral field. However, the birthplaces are rather far from the potential minimum, namely about 80° before it in azimuth.

On the other hand, in most other cases no well defined feature is apparent. Taking z = 0.02 the best agreement was found for values near those of Fig. 1e, namely $\omega_1 = 8.4$ kpc, a = 3%, i = -9°, and $\Omega_s = 15$ km s^{-1} kpc^{-1}. In this case most stars have bitrhplaces on the average 50° inside the spiral minimum. This deviation seems still rather large and better models are looked for. It should be noted, however that no adjustment of ages was made.

A more detailed paper on this subject is prepared by Mr.Grosbøl. This contains also information on the velocities of the stars at birth, the z-components at the birthplaces, and a detailed discussion of the material, models, and assumptions used in this study. A discussion of various selection effects will also be made. The above calculations were made at the 1106 Univac Computer of the University of Thessaloniki.

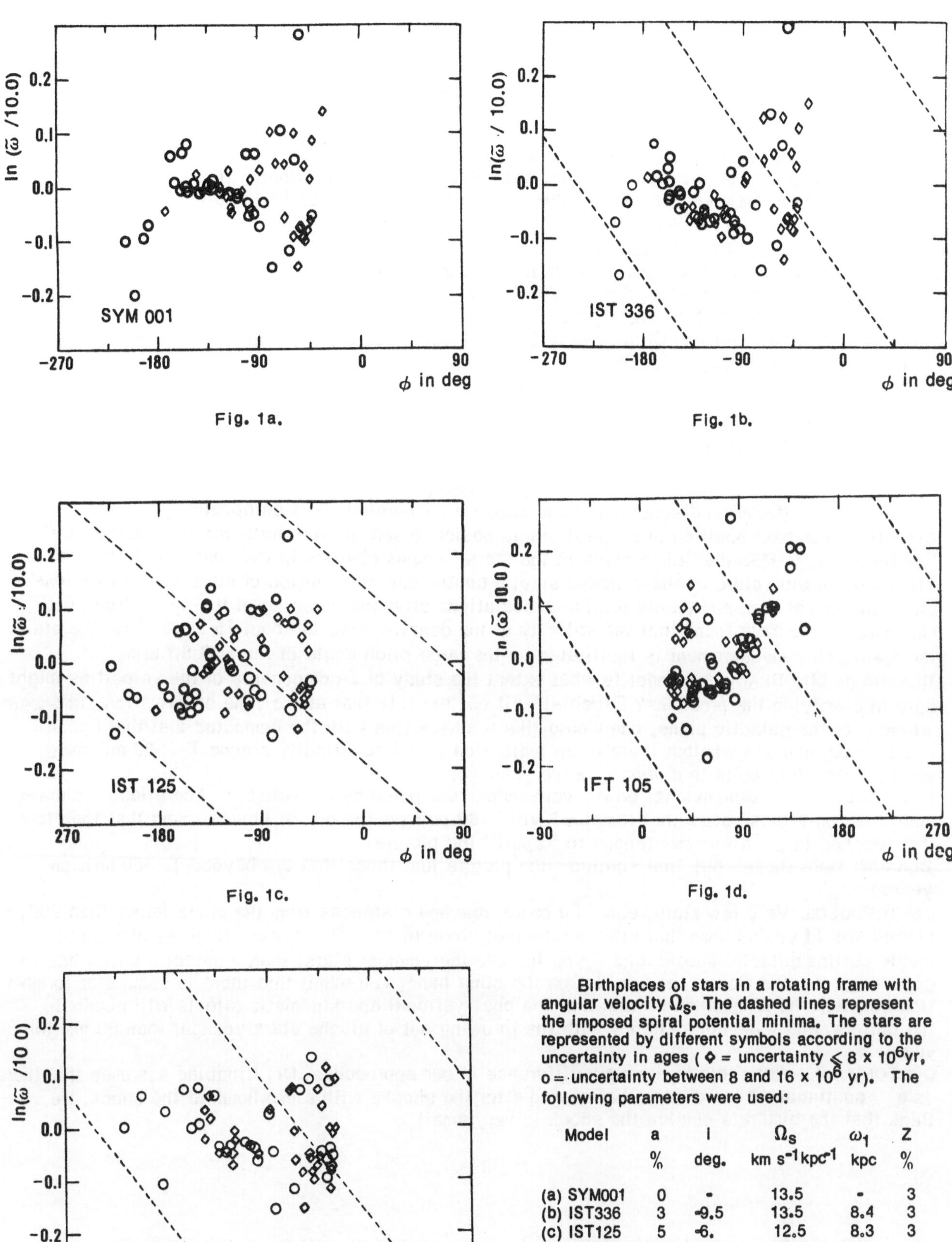

Fig. 1a.

Fig. 1b.

Fig. 1c.

Fig. 1d.

Fig. 1e.

Birthplaces of stars in a rotating frame with angular velocity Ω_s. The dashed lines represent the imposed spiral potential minima. The stars are represented by different symbols according to the uncertainty in ages (◊ = uncertainty $\leqslant 8 \times 10^6$ yr, o = uncertainty between 8 and 16 x 10^6 yr). The following parameters were used:

Model	a	i	Ω_s	ω_1	Z
	%	deg.	km s^{-1}kpc^{-1}	kpc	%
(a) SYM001	0	-	13.5	-	3
(b) IST336	3	-9.5	13.5	8.4	3
(c) IST125	5	-6.	12.5	8.3	3
(d) IFT105	5	-10.	32.5	8.5	3
(e) IST043	3	-9.	15.0	8.4	2

REFERENCES

CONTOPOULOS G., 1972, The *Dynamics of Spiral Structure,* Lecture Notes, Univ. Maryland

CONTOPOULOS G. and STRÖMGREN B., 1965, *Tables of Plane Galactic Orbits,* Institute for Space Studies, N. York

FRICKE W., 1963, *Voroeff. Astr. Rechen Inst. Heidelberg, 11.*

FRICKE W. and KOPPF A., 1963, *Voroeff. Astr. Rechen Inst. Heidelberg, 10*

HEJLESEN P.M.JØRGENSEN H.E., PETERSEN J.O. and ROMCKE L., 1972, *IAU Coll. 17* paper XVII.

KALNAJS A.J., 1970, *IAU Symp. 38,* 318

KALNAJS A.J., 1973, *Observatory 93,* 39

LIN C.C., YUAN C. and SHU F.H., 1969, *Astrophys. J. 155,* 721

STRÖMGREN B., 1966, *Ann. Rev. Astron. Astrophys. 4,* 433

STRÖMGREN B., 1967, *IAU Symp. 31,* 323

WILSON R.E., 1953, *General Catalogue of Stellar Radial Velocities,* Carnegie Inst. of Washington.

YUAN., C., 1969, *Astrophys. J. 158,* 889

DISCUSSION

(Common discussion of the papers by Lindblad and Contopoulos)

COURTÉS: The past position of a spiral arm is easier to see in other galaxies than in our own. For instance, in M33 the "birth place of the stars" seems obvious in the south west spiral arm. There are no blue stars on the concave side, a continuous distribution of stars and HII regions along the spiral arm, and plenty of blue associations after the crossing of the spiral front. With Mrs. Dubout we have found that the velocity of the density wave is about 15 km/s (from the stellar ages). This measurement is facilitated by the large pitch angle of the M33 SW arm.

BLAAUW (to LINDBLAD): I wonder to what extent the study of z-components of the velocities might help in clarifying the problems? Especially, if one expects that at the time of formation stars were close to the galactic plane, one would like to check this with the computed z at that epoch, and for instance see whether there is an indication of a discontinuity around T = -60 million years. That might point to a density wave passage.

LINDBLAD: Three-dimensional orbits were indeed computed by Dr. Grosbøl. The slides I showed contained only those stars that remained within 80 pc from the plane. It was found that the stars that reached larger distances tended to "spoil" the picture.

BLAAUW: Were those stars that spoiled the picture just those that are beyond T= -60 million years?

CONTOPOULOS: Very few stars, about 20 or so, reached distances from the plane larger than 200 pc.

LINDBLAD: If you assume that stars exclusively form in galactic shocks, then the birthplaces should outline galactic shocks and if you include the younger stars, then a galactic shock has to go through the position of the Sun. If, on the other hand, you admit that there is some star formation between the arms, then I think that the observational and kinematic effects will dominate the picture, and I am not so happy with the throwing out of all the stars younger than 60 million years.

CONTOPOULOS: Well, this is the main difference in our approaches. Dr. Lindblad assumes that there is a continuous birth of stars before and after the shocks with a maximum at the shock. We think that the birthrate outside the shock is very small.

Note added in Proof. In the calculations above the perturbation of the spiral potential on the basic solar motion has not been taken into account. Later calculations, including this effect, bring the birthplaces quite close to the theoretical spiral arms in the best models.

NEW RESULTS ON M 31 ANDROMEDA GALAXY

IONIZED HYDROGEN SYSTEMATIC DETECTION AND FIRST U.V.1400 AND 2500 Å IMAGES OF THE NUCLEUS

G.COURTES. P.CRUVELLIER, J.M.DEHARVENG, J.MAUCHERAT, G.MONNET, M.PELLET, M.SIMIEN

OBSERVATOIRE DE MARSEILLE
OBSERVATOIRE DE HAUTE-PROVENCE
LABORATOIRE D'ASTRONOMIE SPATIALE DU C.N.R.S., FRANCE
PRESENTED BY G.COURTÈS

НОВЫЕ РЕЗУЛЬТАТЫ ПО ГАЛАКТИКЕ АНДРОМЕДЫ М-31. СИСТЕМАТИЧЕСКАЯ РЕГИСТРАЦИЯ ИОНИЗИРОВАННОГО ВОДОРОДА И ПЕРВЫЕ УЛЬТРАФИОЛЕТОВЫЕ 1400 И 2500 Å ИЗОБРАЖЕНИЯ ЯДРА

1. Extended survey of ionized hydrogen. The HII region of the M 31 galaxy has been observed for the first time by Baade and Mayall (1950) twenty five years ago, using the 100 inch Mt-Wilson telescope. The filters were a red and a near infrared Shott coloured filter combination, one to select H_α and [N II] lines, the other for comparison with the continuum.

The newly designed red sensitive 103 aE Kodak plates made possible this first survey which was playing a very important role in the understanding of the spiral structure of our own Galaxy (see Courtès 1972, Plate IX). Since this survey, HII regions are considered as the best indicators of the spiral structure.

Arp has recently given an accurate catalogue of this first survey followed by his own new observations (Arp,1964 and 1973).

A first test of the interference filter-reduction focal method (Courtès, 1960, plate IV) on M 31 showed that the power of detection of faint HII regions of this method with the 120 cm telescope of the Haute-Provence Observatory was equivalent to the 100 inch telescope-coloured filter method.

More recent investigation (Courtès,1972 p. 104) showed that interference filter can reach a much higher detection power : $\Gamma = 62$ ($\Gamma = 1$ for the Palomar sky survey or the Baade survey). The telescope was the 193 cm of the Haute-Provence Observatory equipped with the F/1, 72' field, large "reducteur focal".

The range of radial velocities of M 31 (Deharveng and Pellet, 1969) lead to use a relatively large bandwidth of 20 Å. Two filters differently peaked were used for the both ends of the major axis of the Galaxy (then, $\Gamma \approx 12$).

This survey detects .

1 - New H II regions, especially the arm-emission interconnecting the H II regions already detected by Baade and Arp.

2 - Large faint H II regions and, in some part, a faint patchy repartition of the ionized hydrogen.

3 - A series of ring-like and spherical H II region very characteristic of the Strömgren spheres.

4 - The ionized hydrogen of this general survey fits exactly with the most recent neutral hydrogen ring repartition of Baldwin (1975)*.

5 - H II regions are often considered as individual H II regions in the Baade and Arp catalogue, in fact they are only the brightest points of some more extended regions.

A general Atlas is being prepared and will be published soon. The catalogue will refer systematically to the Arp and Baade catalogue whenever possible.

---

* Baldwin, private communication.

2. Detection of the nuclei-image of M 31 in $\lambda = 1400$ and 2500 Å. The Laboratoire d'Astronomie Spatiale S 183 experiment in Skylab made two photographs of the nucleus of the Andromeda nebula (Courtès - Laget - Vuillemin - 1974). A recent rocket flight (November 1974) provided by the Centre National d'Etudes Spatiales in French Guyana, detected with a F/1.2, 160 mm aperture Wynne telescope and multichannel plate detector, the nucleus of M 31 at the wavelength $\lambda = 1400$ Å.

The two experiments had a resolution of 1', and lead to the following results:

1 - The flux at 1400 Å is of the same order that the one obtained at 2574 Å ($\Phi = 10^{-11}$ erg. cm^{-2} sec^{-1} for a bandwidth of 358 Å) and confirms at least a flat spectrum below 2500 Å *.

2 - The last observation of the OAO telescope was obtained with spatial resolution of 10' with an ambiguity on the bulge or nucleus origin of the UV radiation. Our experiment shows, owing to its resolution of 1', that the nucleus is likely the main origin of this unexpected UV radiation.

---

* This result is interesting to compare with the Netherland Astronomical satellite measurements on globular clusters (this congress). It may correspond to a general property of population II.

REFERENCES

ARP H. and BAADE W., 1964. Astrophysical Journal, p. 1027 n° 139

ARP H. and BRUECKEL F., 1973. Astrophysical Journal, p.445 n° 179

BAADE W. and MAYALL N.U., 1950. Problems of Cosmological Aerodynamics USAF Central Documents Office Dayton (Ohio)

BOULESTEIX J., COURTÈS G., MELLE LAVAL A., MONNET G., 1974. ESO/SRC/CERN - Geneva, Proceedings. Ed. A. Reiz - p.221

COURTÈS G., 1972. Vistas in Astronomy, Tome XIV

COURTÈS G. - 1960, Ann. Astrophys. 23 - p. 115

COURTÈS G., LAGET M. and VUILLEMIN A., 1974. AIAA/AGU Conference on Scientific Experiments of Skylab - AIAA Paper n° 74-1250

DEHARVENG J.M. and PELLET A. 1969. Astronomy and Astrophysics I, p- 208

DYNAMICAL STUDY OF THE DOUBLE GALAXIES NGC 1253 AND NGC 7752-53*

P.BENVENUTI, S.D'ODORICO, P.VETTOLANI
OSSERVATORIO ASTROFISICO DI ASIAGO; LABORATORIO DI RADIOASTRONOMIA, BOLOGNA, ITALY
PRESENTED BY P. BENVENUTI

ДИНАМИЧЕСКОЕ ИССЛЕДОВАНИЕ ДВОЙНЫХ ГАЛАКТИК NGC 1253 и NGC 7752-53

I n t r o d u c t i o n   a n d   o b s e r v a t i o n s. Investigations on the dynamics of double systems of galaxies have gained interest since detailed computer-made models of interacting galaxies have been produced. The comparison between the observed velocity field of a double galaxy and its model can provide a better understanding of "normal" features like spiral patterns and of peculiarities such as bridges and tails. In particular, one can estimate which of the observed features are plausibly tidal and which are not.

With the aim of increasing available data on particularly interesting objects, a list of double and multiple peculiar galaxy systems was prepared and included in the program of extragalactic spectroscopic observations at the Asiago Observatory. The observations were carried on with the standard nebular spectrograph attached to the newtonian focus of the 122 cm reflector. The capability of this instrument for radial velocity measurements, together with the standard reduction technique is described in a previous work (Benvenuti et al.,1975): the spectrograph is followed by a WL-30677 image tube and it is particularly suited for observations of low surface brightness objects. After correction for line curvature and zero velocity the residual internal mean error is estimated 13 km s^{-1}. Typical exposure time was 60 minutes. The spectral region 5800 - 8000 Å was observed at 125 Å mm^{-1}. The scale perpendicular to the dispersion is 127 arcsec mm^{-1}. About 40 spectra are now available concerning 8 peculiar systems. We report here the preliminary results on the double galaxies NGC 1253 and 7752-53.

N G C 1253 (A r p 279). NGC 1253 is a normal appearing Sc galaxy with an irregular S-shaped companion. A blue photograph is reproduced in fig. 1 (courtesy of H.Arp). This object has been studied at 21-cm by Peterson and Shostak (1974). They obtained a heliocentric velocity of 1710 ± 10 km s^{-1}, with a velocity width of 380 km s^{-1}. The computed H I mass is $M_{HI} = 7.8 \cdot 10^9 M_\odot$ for an assumed distance of 17 Mpc. We have obtained 4 spectra of NGC 1253, in which H_α emission from the galaxy was easily detected. The slit of the spectrograph was set close to the major axis at P.A. 68°, 71°, 95° and 102°. In the spectrum Sp 1796 (reproduced in fig. 2), at P.A. 71°, the emission from the companion was also recorded. From the four spectra we derived a mean heliocentric velocity $V_0 = 1742 \pm 37$ km s^{-1} for NGC 1253 and $V_0 = 1837 \pm 64$ km s^{-1} for inclination of the galaxy to the line of sight $\xi = 61°$ and a P.A. for the line of nodes $\theta = 84°$. A plot of rotational velocities was finally produced (fig. 3). The curve shows a central part corresponding to rigid body rotation, with the turning point at about 30 arcsec from the nucleus. In the external regions of the galaxy some bumps are observed in the rotation curve, but their interpretation is not possible with the material of this preliminary study. The rotation curve was folded and fitted with standard Brandt (1970) functions. We obtain an estimate of the total mass $M_T = 5.3 \cdot 10^{10} M_\odot$. This value is consistent with the H I mass given by Peterson and Shostak, however the new mass to luminosity ratio, $M_T/L_{pg} = 10.7$, is considerably lower, and seems more sensible for a Sc galaxy (Vorontsov-Vel'yaminov, 1970). In the spectrum Sp 1796 the H_α emission of the companion extends about 30 arcsec from the apparent nucleus with a maximum velocity difference $\Delta V = 50$km s^{-1}. The rotation curve was used also in this case to estimate the mass. The data were corrected assuming that the companion is lying on the same plane of the principal galaxy. This assumption seems sensible because of the appearance of the two galaxies and the similarity of their axial ratios. With this hypothesis the sense of rotation is also the same for the two objects. The curve was fitted as for NGC 1253 with Brandt functions giving a total mass $M_T = 4.3 \cdot 10^8 M_\odot$.

In order to check the stability of the system, the escape velocity from the principal galaxy at the distance of the companion (20.3 kpc) was computed. With the assumption of the complanarity of the two galaxies, the escape velocity is V = 150 km s^{-1}, while the velocity difference between the two nuclei is 95 km s^{-1}. Apart from unknown projection effects, which seem unlikely in this case, we can conclude that the system is stable.

N G C 7752-53 (A r p 86). This M 51 - system is composed of a large Sc spiral (NGC 7753) with a high surface brightness, elliptical shaped companion lying at the end of its southern arm. This system was previously investigated by Arp (1969), by Bertola and D'Odorico (1973), and had received particular attention by theoretical modellists (Toomre, A. and J., 1972 . Toomre, A.,1974) who tried to understand what is tidal provoked and what is "selfmade" in its well developed spiral structure. NGC 7752, the companion of the main galaxy,is intrinsically interesting. It has a very high surface brightness and in spite of its elliptical appearance,Arp noted that in short exposure plate a knotty feature is present in the inner part of the galaxy. Moreover a spectrum along the minor axis reveals a remarkable velocity gradient that can be interpreted as an expansion motion.

Finally Bertola and D'Odorico, from the velocity field in NGC 7753 and the velocity difference of the two nuclei concluded that the system is not bound.

We have obtained three spectra of this system. In one case the slit was set at P.A. 46° and the H_α emissions of the two galaxies were recorded in the same spectrum. The main results are the following:

a) The redshifts of the nuclei are in good agreement with previous values. We obtain a heliocentric velocity $V_0 = 5200 \pm 25$ km s^{-1} for NGC 7753 and $V_0 = 4902 \pm 16$ km s^{-1} for the companion. The confirmation of a velocity difference between the two nuclei of $\sim$ 300 km s^{-1} is particularly relevant since obtained from the same spectrum.

b) In a spectrum of NGC 7752 taken at P.A. 43° a velocity gradient of 13 km s^{-1} per arcsec was measured up to 5 arcsec from the center. We tried to obtain a rotational velocity from our data and those by Arp, assuming an expanding velocity component. Two cases were considered: i) the two galaxies are lying on the same plane and the inclination to the line of sight $\xi = 38°$, defined by the axial ratio of NGC 7753, is assumed; ii) the planes of the two objects are different, and, considering NGC 7752 a flat system, an inclination $\xi = 66°$ is then assumed. The results are very close in both cases: the rotational velocity is of the order of 115 km s^{-1} at 5 arcsec from the centre and the expansion velocity, at the same distance, is about 60 km s^{-1}. A rough estimate of the mass can then be made assuming a homogeneous spheroid as a model. We obtain a value $M = 3 \cdot 10^9\ M_\odot$ within 5 arcsec.

c) A rotation curve derived from one single spectrum shows a higher turning point velocity than that observed previously by Bertola and D'Odorico.The discrepancy cannot be due to the difference in the position angles, unless extremely peculiar motions are present, The assumption of the new rotation curve would lead to an increasing in mass for NGC 7753 up to $7.8 \cdot 10^{11} M_\odot$, and in that case the instability of the double system would be checked anew.

The conclusion is that the large velocity difference between the two galaxies is confirmed but this peculiar system still deserves special care in future observations. In particular spectra at high spatial resolution are needed to give a better value for the expansion velocity in NGC 7752. High dispersion observations with good spatial resolution at various position angles will be of great interest in order to determine with higher accuracy the mass of NGC 7753 and to decide about instability.

A c k n o w l e d g e m e n t. Two of us (P.B. and S.D.) acknowledge partial support from the Italian National Research Council (C.N.R.).

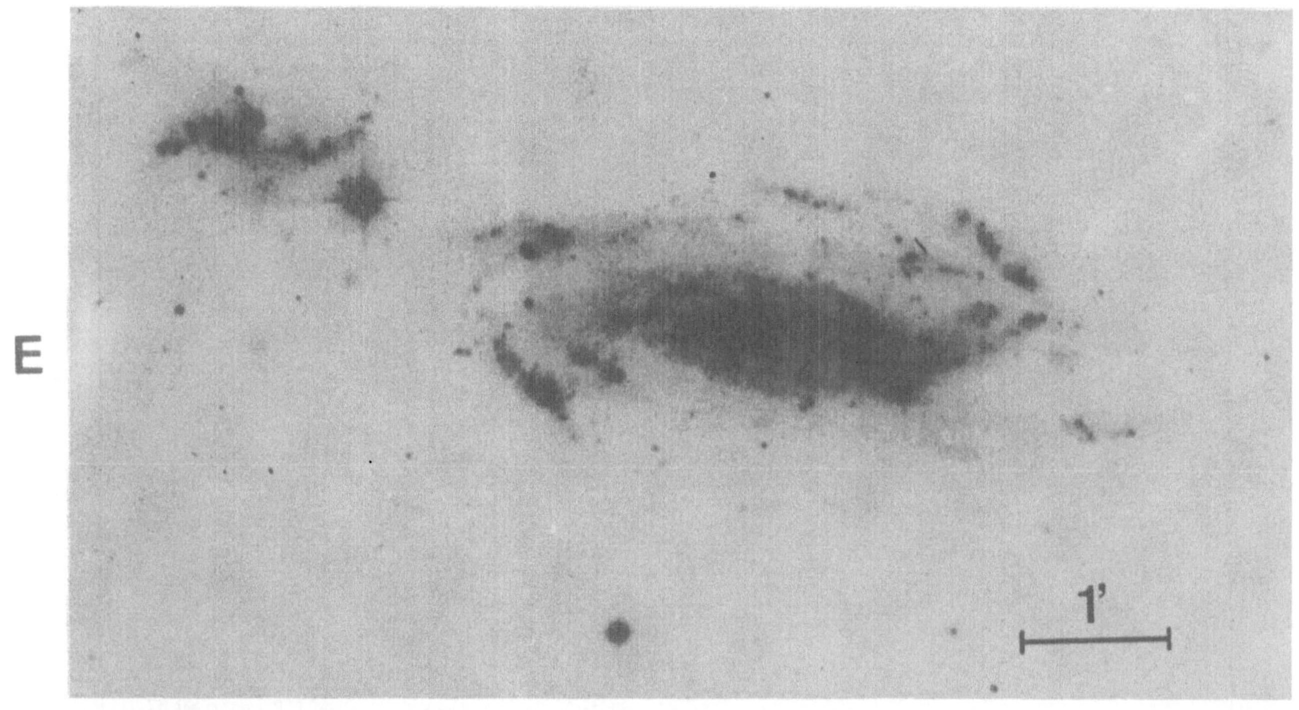

Fig. 1. Blue photograph of NGC 1253 taken with the 200' telescope (kindly made available by H.Arp)

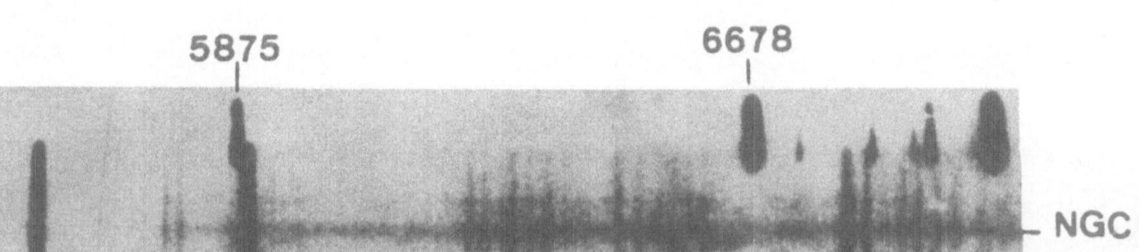

Fig. 2. NGC 1253. Spectrum N° 1796 at P.A.71°, Hα emission and positions of the two nuclei are indicated.

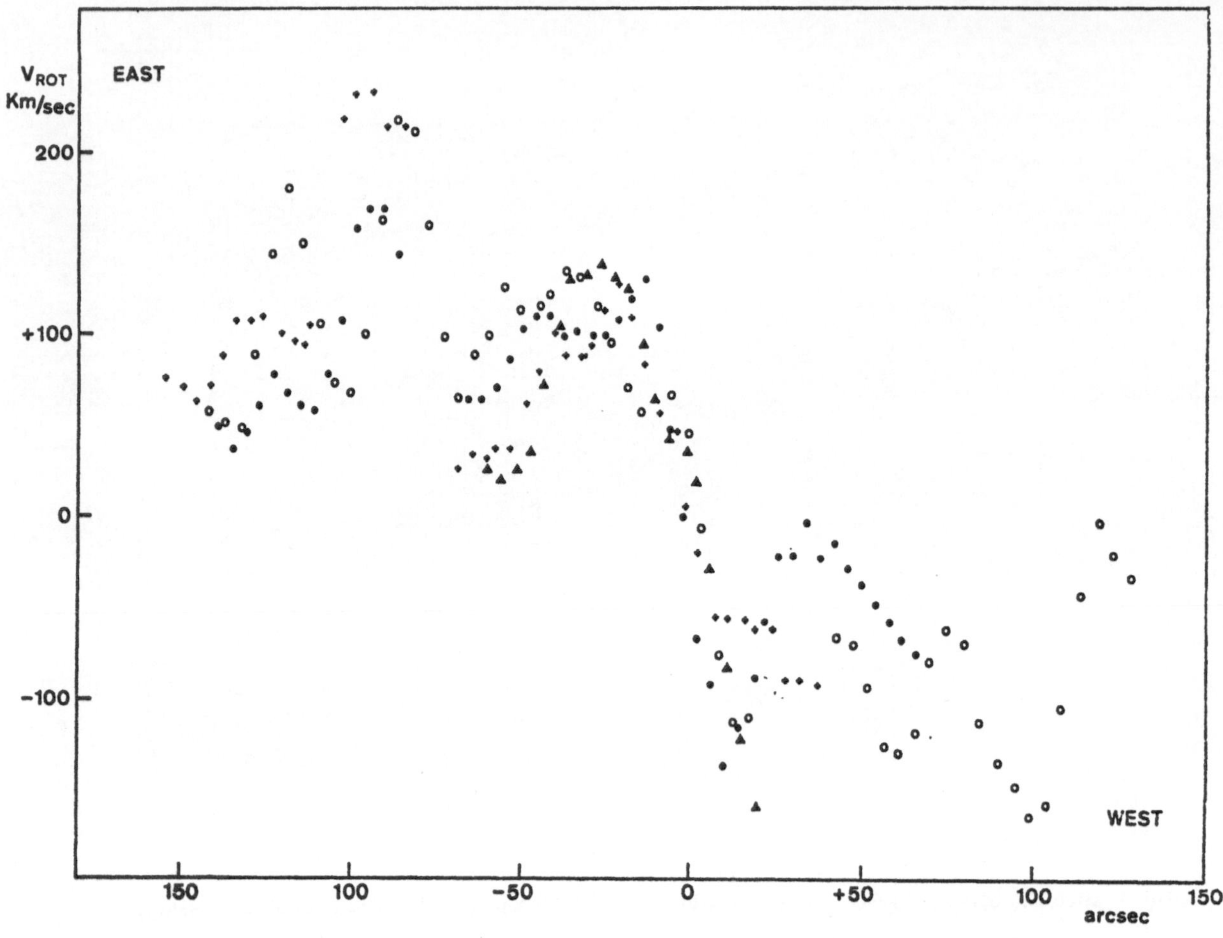

Fig. 3. Rotational velocities of NGC 1253 versus the distance from the center on the plane of the galaxy. Different symbols indicate measurements of different spectra. Assumed inclination of the galaxy to the line of sight $\xi = 61°$, P.A. of the major axis $\Theta = 84°$.

REFERENCES

ARP H. 1966, Atlas of Peculiar Galaxies (Calif. Inst. Technol., Pasadena)

ARP H. 1969, Astron. Astrophys., 3, 418

BENVENUTI P., CAPACCIOLI M., D'ODORICO S. 1975, Mem. Soc. Astron. Ital. (in press)

BERTOLA F., D'ODORICO S. 1973, Astrophys. Letters, 13, 161

BRANDT J.C. 1960, Astrophys. J., 131, 293

BRANDT J.C., BELTON M.J.S. 1962, Astrophys. J., 136, 352

PETERSON S.D., SHOSTAK G.S., 1974, Astron. J., 79, 767

TOOMRE A., TOOMRE J. 1972, Astrophys. J., 178, 623

TOOMRE A. 1974, "The formation and dynamics of galaxies", Proceedings of the IAU Symposium NO 58, D.Reidel Publ.

VORONTSOV - VEL'YAMINOV B.A. 1970, Sov. Astron.J., 14, 222

THE HIGH VELOCITIES OF EXPANSION OF COMPACT HII REGIONS

Yu.I.GLUSHKOV, E.K.DENISYUK, Z.V.KARYAGINA
ALMA-ATA ASTROPHYSICAL INSTITUTE, U.S.S.R.
PRESENTED BY E.K.DENISYUK

ВЫСОКИЕ СКОРОСТИ РАСШИРЕНИЯ КОМПАКТНЫХ HII ОБЛАСТЕЙ

Spectra of compact HII region Mil-19 were obtained in December, 1974 with a diffraction spectrograph and an UM-92 image-convertor-tube. One may see on the spectrograms with dispersions of 11 and 18 Å/mm (Fig.1) that Hα and [NII] lines have wide blue wings. The width of the wings (from the maximum of intensity of lines) is larger than 4Å. This corresponds to the velocity of expanding $\geqslant$-200 km/s. The absence of red wings might be explained by strong absorption (for Mil-19 the Hα extinction A(Hα) is ~ $8^{m}_{.}0$ - $11^{m}_{.}0$).

The blue asymmetry of Hα line was found by us in the new compact HII regions: Sh149, Sh307.For these nebulae A(Hα)is $\geqslant 7^{m}_{.}0$.Its spectra were obtained with a dispersion 42Å/mm and it seems difficult to give the estimate of the expansion velocities.

The observations are interpreted as the existence of fast expanding shells around compact HII regions with a strong absorption and high electron densities ($N_e \geqslant 10000$ cm^{-3}).

The centres of expansion might be presented by a few cocoon stars which can be inside compact HII region. Our previous observations (for example Sh235B, Mil-15, He2-446) confirm the existence of high-velocity (V>500 km/s) ionized gas around ultra-compact HII regions(the late stage of cocoon stars). Fig.1 shows the spectrum of MI -19; dispersion 18Å/mm.

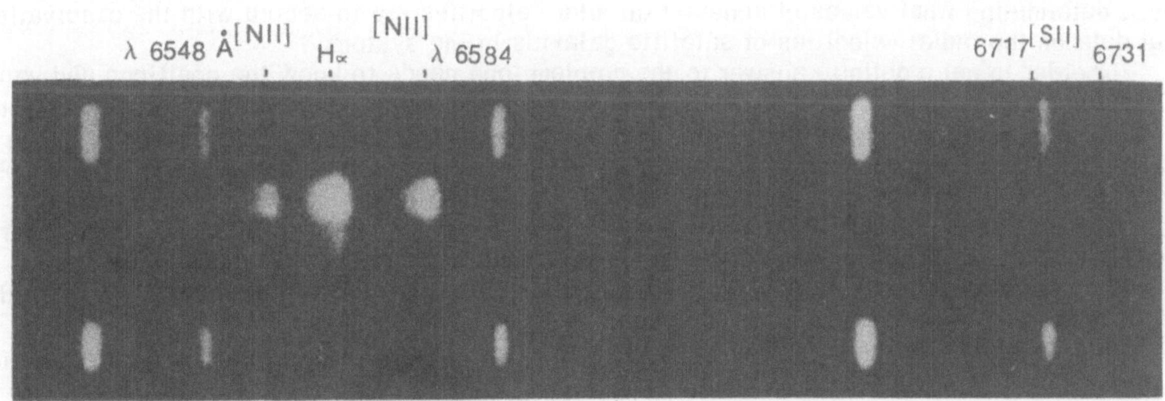

Fig. 1.

A NEW MODEL OF THE GALAXY

J.EINASTO, M.JÔEVEER, A.KAASIK
TARTU INSTITUTE OF ASTROPHYSICS AND ATMOSPHERE PHYSICS, U.S.S.R.
PRESENTED BY J.EINASTO

НОВАЯ МОДЕЛЬ ГАЛАКТИКИ

In last few years our knowledge of the structure, kinematics and stellar content of various populations in galaxies have considerably improved (Spinrad and Peimbert,1975; Blaauw, 1975). There exist strong arguments favoring the presence of large and massive invisible coronas around giant galaxies, including our own Galaxy (Gunn,1974; Einasto *et al.*,1974a;Ostriker *et al.*,1974). The system of local galactic constants is also improving. In particular various independent methods indicate that the Sun's distance from the galactic centre as well as the density of matter in the solar vicinity are considerably smaller than adopted earlier (van den Bergh,1974; Balona and Feast,1974; Plaut and Oort 1975; Jôeveer.1974).

We have constructed a new model of the mass distribution in the Galaxy. All principal new data on the structure of the Galaxy have been taken into consideration.

In Table 1 the new system of galactic constants is given. In the first column the best observed and mutually adjusted system with rms errors of all individual constants using the method proposed by Einasto and Kutuzov (1974). In the last column we present the system of constants adopted in the present model.

The mass distribution of stellar population is given by a superposition of five heterogeneous ellipsoids, representing the nucleus, the bulge, the halo, the old disc and extreme I populations respectively. On these visible stellar populations a corona of invisible matter is superimposed. The parameters of all components are given in Table 2. The density distribution of visible populations is represented by a modified exponential model (Einasto 1970), the corona is represented by a truncated isothermal model (Einasto et al. 1974b).

The mass of the corona of the Galaxy was estimated as follows.

As demonstrated by Ostriker et. al. (1974) and Einasto et.al.(1974b) the density of the corona decreases with increasing distance according to the inverse square law and the circular velocity remains constant over a wide distance interval. The mass of the corona can be found by... determining what values of constant circular velocities are in accord with the observational data on the radial velocities of satellite galaxies in the system.

In order to get a definite answer to the problem, one needs to know the positions and velocities of satellite galaxies at two different epochs, say at the present time and at the moment of the formation of the whole system. We assume, following Kahn and Woltjer and others, that satellite galaxies were formed at the perigalaxies of their orbits. The corresponding distances from the central galaxy can be estimated from the tidal radii of the satellite galaxies.

The results of our calculations are presented in Table 3. Here our Galaxy has been regarded as a satellite of the Andromeda galaxy. The estimated uncertainty of the resulting value of the constant circular velocity is approximately 50%, accounting for the possible errors in the initial data.

The constant circular velocity of 175 km s^{-1} in M31 is in good agreement with direct optical and radio data of the velocity run at the periphery of this galaxy (Roberts 1975; Emerson and Baldwin 1973). In the case of our Galaxy agreement with other data is also satisfactory.

The distribution of various quantities characterizing the gravitational field and the structure of the Galaxy in the plane of the system is presented in Tables 4 and 5.

Table 1

Constant	Unit	Observational estimate	Smoothed value	Adopted value
R_0	kpc	8.15 ± 0.71	8.0 ± 0.4	8.5
V	km s^{-1}	220 ± 20	214 ± 8	225
$W{=}AR_0$	km s^{-1}	120 ± 15	130 ± 5	136
A	km s^{-1} kpc^{-1}	16.9 ± 0.9	16,4 ± 0.5	16.0
ω	km s^{-1} kpc^{-1}	25.4 ± 2.2	26.8 ± 0.8	26.5
C	km s^{-1} kpc^{-1}	70		70
k_z		0.282±0.010	0.280±0.005	0.284

Table 2

Population		a_0 kpc	$\mathfrak{M}$ $10^{10}M_\odot$
Nucleus	0.6	0.004	0.01
Bulge	0.8	0.6	3
Halo	0.3	2	1
Disc	0.08	5.22	4.47
Young	0.0185	6	0.7
Corona	1	100	100

Table 3

Main galaxy	Satellite galaxy	V_r km s^{-1}	R_{obs} kpc	$R_{perigal}$ kpc	R_{apogal} kpc	V_{corona} km s^{-1}
Galaxy	Fornax	-40	230	62	250	112
				125	250	140
Galaxy	NGC 6822	+60	500	270	675	108
				360	720	125
M31	Galaxy	-120	690	100	900	175
				370	925	225

Table 4

R kpc	V km s^{-1}	V_k km s^{-1}	A	-B km s^{-1}	C kpc^{-1}	k_θ	k_z
0.0	0	1181	0	489000	659000	1.000	0.500
0.1	189	821	687	1205	2125	0.637	0.389
0.2	224	787	448	672	1302	0.600	0.375
0.4	248	738	281	338	759	0.546	0.353
0.6	254	702	205	219	546	0.517	0.341
0.8	255	670	160	159	437	0.498	0.332
1	254	650	131	123	373	0.486	0.327
2	246	581	65	58	245	0.475	0.322
3	242	540	42	39	194	0.486	0.327
4	241	510	31	29	159	0.488	0.328
5	239	489	25.0	22.8	132	0.477	0.323
6	236	467	21.4	18.0	110	0.458	0.314
7	233	443	18.8	14.4	92	0.434	0.302
8.5	225	417	16.0	10.5	70	0.396	0.284
10	216	396	13.8	7.8	54	0.361	0.265
20	167	319	5.9	2.4	11.9	0.294	0.227
30	143	291	3.2	1.6	5.1	0.329	0.248
40	131	271	2.1	1.2	3.3	0.354	0.262
50	123	256	1.5	0.9	2.5	0.370	0.270
100	106	219	0.6	0.4	1.1	0.412	0.292
200	96	185	0.27	0.21	0.49	0.439	0.305
300	92	164	0.17	0.14	0.31	0.460	0.315
400	90	148	0.12	0.11	0.22	0.471	0.320
500	89	136	0.09	0.08	0.17	0.458	0.314
600	87	125	0.09	0.06	0.14	0.409	0.290
1000	69	97	0.05	0.02	0.07	0.236	0.191

Table 5

$\log \frac{R}{R_o}$	R(kpc)	log ρ (R) ($\mathfrak{M}_\odot$ / pc^3) bulge	halo	disc	flat	total	P (R) ($\mathfrak{M}_\odot$ / pc^2) bulge	halo	disc	flat	total
- ∞	0.000	3.146	1.528	-0.055	-0.399	6.728	8.8 10^4	2700	490	58.4	7.42 10
-3.0	0.008	3.057	1 518	-0.055	-0.399	3.477	8.26 10^4	2630	488	58.4	9.62 10^4
-2.8	0.014	2.965	1.504	-0.055	-0.399	3.015	7.68 10^4	2600	488	58.4	8.06 10^4
-2.6	0.021	2.822	1.472	-0.055	-0.399	2.843	6.81 10^4	2540	488	58.4	7.12 10^4
-2.4	0.034	2.632	1.407	-0.055	-0.399	2.658	5.71 10^4	2420	488	58.4	6.00 10^4
-2.2	0.054	2.400	1.296	-0.055	-0.399	2.435	4.51 10^4	2210	488	58.4	4.78 10^4
-2.0	0.085	2.132	1.134	-0.055	-0.400	2.177	3.34 10^4	1920	488	58.4	3.58 10^4
-1.8	0.135	1.826	0.927	-0.056	-0.400	1.885	2.30 10^4	1570	487	58.3	2.51 10^4
-1.6	0.214	1.481	0.681	-0.058	-0.401	1.560	1.47 10^4	1210	486	58.2	1.64 10^4
-1.4	0.339	1.092	0.398	-0.063	-0.405	1.208	0.85 10^4	875	483	58.0	0.99 10^4
-1.2	0.537	0.656	0.078	-0.074	-0.414	0.843	4470	586	476	57.3	5590
-1.0	0.850	0.167	-0.284	-0.099	-0.434	0.499	2080	361	460	55.8	3000
-0.8	1.35	-0.382	-0.690	-0.155	-0.479	0.218	851	202	425	52.4	1530
-0.6	2.14	-0.998	-1.145	-0.264	-0.570	-0.006	299	101	359	45.7	806
-0.4	3.39	-1.689	-1.657	-0.459	-0.735	-0.241	88.7	45.0	259	34.8	428
-0.2	5.37	-2.465	-2.231	-0.785	-1.017	-0.569	21.8	17.4	142	21.0	202
0.0	8.50	- 3.335	-2.875	-1.315	-1.479	-1.079	4.30	5.74	50.1	8.61	68.7
0.2	13.5	- 4.312	-3.598	-2.164	-2.220	-1.880	0.67	1.59	8.62	1.89	12.8
0.4	21.4	- 5.407	-4.408	-3.514	-3.401	-3.127	0 08	0.36	0.47	0.15	1.06
0.6	33.9	- 6.637	-5.318	-5.659	-5.276	-4.902	0.01	0.06	0.00	0.00	0.08

ρ (R) is the space density in the plane of the Galaxy, P(R) is the surface density.

REFERENCES

BALONA L.A. and FEAST M.W. 1974, Monthly Not. Roy. Astr. Soc. 167, 621
BLAAUW A. 1975, report at the Third Europ. Astron. Meeting, Tbilisi
EINASTO J. 1970, Tartu Astron. Obs. Teated.No. 26, 1
EINASTO J., KAASIK A., SAAR E. 1974a, Nature, 250, 309
EINASTO J., JAANISTE J., JÕEVEER M., KAASIK A., KALAMEES P., SAAR E., TAGO E., TRAAT P., VENNIK J., and CHERNIN A.D. 1974b, Tartu Astron. Obs. Teated No. 48, 3
EINASTO J. and KUTUZOV S. 1964, Tartu Astron. Obs. Teated No. 11, 11
EMERSON D.T. and BALDWIN J.E. 1973, Mon. Not. R. Astr. Soc. 166, 9p
GUNN J.E. 1974, Comments Astrophys. Space Phys. 6, 7
JOEVEER M. 1974, Tartu Astron. Obs. Teated No. 46, 3, 18, 35
KAHN F.D. and WOLTJER L. 1959, Astrophys. J. 130, 705
OSTRIKER J.P., PEEBLES P.J.E. and YAHIL A. 1974, Astrophys. Journ. (Lett.) 193, L1
PLAUT L., OORT J.H. 1974, Preprint
ROBERTS M.S. 1975, Astroph. J.
SPINRAD H., PEIMBERT M. 1975, in "Stars and Stellar Systems", 9, ed. A.Sandage and J.Kristian
VAN DEN BERGH S. 1974, Astrophys. Journ. (Lett.) 188, L9

A GALACTIC MASS MODEL COMPOSED OF A SPHEROIDAL HALO AND AN EXPONENTIAL DISK

ROLAND WIELEN
ASTRONOMISCHES RECHEN-INSTITUT, HEIDELBERG, F.R.G.

МОДЕЛЬ МАССЫ ГАЛАКТИКИ, СОСТОЯЩЕЙ ИЗ СФЕРИЧЕСКОГО ГАЛО И ЭКСПОНЕНЦИАЛЬНОГО ДИСКА

Surface photometry of galaxies suggests that galaxies are mainly composed of two components: a spheroidal halo with a projected density distribution similar to that of giant elliptical galaxies, and a disk with an exponential decrease of the projected density. Assuming a constant mass-to-light ratio within each component and ellipsoidal surfaces of equal space densities, we derive the space densities and the contribution to the rotation curve for each component as a function of its mass, scale length and flattening. By applying this model to our Galaxy, we obtain the mass of the halo and of the disk from density determinations only, and we compare these results with those based on the rotation curve. The space density of halo stars in the solar neighbourhood is at least $0.94 \cdot 10^{-3}\ \mathfrak{M}_{\odot}/\mathrm{pc}^3$ (Jahreiss and Wielen, 1975). Using this local density, a flattening of $c/a = 0.8$ and an effective radius of 5 kpc for the halo, we derive for the total mass of the galactic halo $3.3 \cdot 10^{10}\ \mathfrak{M}_{\odot}$. The mass of the halo is larger than this value either if the actual local halo density is higher or if the effective radius of the halo is significantly larger or smaller than about 5 kpc.

INTERPRETATION OF LINE INTENSITY GRADIENTS IN SPIRAL GALAXIES

S.COLLIN - SOUFFRIN and M.JOLY
OBSERVATOIRE DE PARIS - MEUDON, FRANCE
PRESENTED BY D.KUNTH

ИНТЕРПРЕТАЦИЯ ГРАДИЕНТОВ ИНТЕНСИВНОСТИ В СПИРАЛЬНЫХ ГАЛАКТИКАХ

We study the influence of some parameters in the line emission spectrum of HII regions : stellar temperatures and gravities, gas density, number of ionizing stars and element abundances. We apply the results to the emission line intensities in the HII regions of M 101 and M 33. In particular, we show that the classical homogeneous models (with a He^{+} Strömgren sphere and an external He^{o} - H^{+} sphere) can account the best for the observations. The only parameters that need to vary across the disc are the element abundances. The gas temperature of the He^{+} region is strongly dependent on the oxygen abundance while the temperature of the He^{o} - H^{+} region depends essentially on the nitrogen abundance.

In summary :

- in the centre, we find a very low temperature of the He^{+} region ($\sim$ 3000°K) and an over-abundance of N and O (by a factor 10);
- in the outer regions, the temperature is high (12000 - 14000°K) and all the elements are underabundant by at least a factor 3.

INTERSTELLAR ABSORPTION AND SPATIAL DISTRIBUTION OF STARS IN THE TAURUS DARK NEBULA

M.D.METREVELI
ABASTUMANI ASTROPHYSICAL OBSERVATORY, U.S.S.R.

МЕЖЗВЕЗДНОЕ ПОГЛОЩЕНИЕ СВЕТА И ПРОСТРАНСТВЕННОЕ РАСПРЕДЕЛЕНИЕ ЗВЕЗД В ТЕМНОЙ ТУМАННОСТИ СОЗВЕЗДИЯ ТЕЛЬЦА

The dark nebula in Taurus is one of the most interesting objects in the Northern sky.

P.P.Parenago included this region in 1956 into the Complex Plan of the Milky Way investigation as the most *conspicuous* one.

On the basis of our Catalogue of Magnitudes, Colour-Indices, Spectral and Luminosity Classes of about 3000 Stars interstellar absorption and spatial distribution of stars in the Taurus dark nebula region covering 76 sq. degrees have been studied.

The region under consideration is clearly non-homogeneous. The densest dark clouds are present here. Fig. 1 shows density distribution of dust matter in the region under consideration in galactic longitude and latitude. The densest dust clouds, almost for all longitudes and latitudes, are concentrated within 10-280 parsecs from the Sun. Then follow the clouds of comparatively low density filling the whole cone in question up to 550 parsecs. The second dense dust cloud is situated within the longitudes of 172° - 176° and latitudes of -17° - -20° between 660 - 700 parsecs. The figure also shows that after 700 parsecs absorbing matter is present mainly in the range of 1^{II}- 168° - 177° and b^{II}- -15° -20°. The other directions are more or less free from absorption.

Of special interest is the comparison of absorbing matter distribution with that of neutral hydrogen in the direction given. It shows a positive correlation. In so doing gas distribution turned out to be in a good correlation with that of dust in two regions of Gould's Belt -in the direction of the galactic centre at positive latitudes and anticentre at negative latitudes, the latter being coincident with our region. The dark dense clouds adhere to the local spiral arm with their front side as it has been noted by others. These clouds constitute dense foreground of absorbing matter.

The study of spatial distribution of stars of different spectral classes shows the following picture: the earliest spectral classes in the region in question begin from B7.

B7-B9 stars are concentrated within 300-800 parsecs. Density maximum for them is at 400-500 parsecs. A maximum density for A stars comes at 200-400 parsecs. FO-F8 stars show their maximum densities comparatively at small distances - 200-250 parsecs. After 500 parsecs the densities fall rapidly.

gG-K stars are concentrated at 200-500 parsecs. After this they decrease and at a distance of 900 parsecs they get negligible.

Almost all types of stars coincide with the local galactic spiral arm.

Luminosity function has been constructed for the region too. Fig.2 displays that it is in a good agreement with other determinations for $M_V = +3 - +7$ and $M_V = -1 + 2$ is a little shifted downward. This points to deficiency of stars of a given luminosity in the Taurus dark nebula region which is situated between $b^{II} = -9° - -20°$ latitudes.

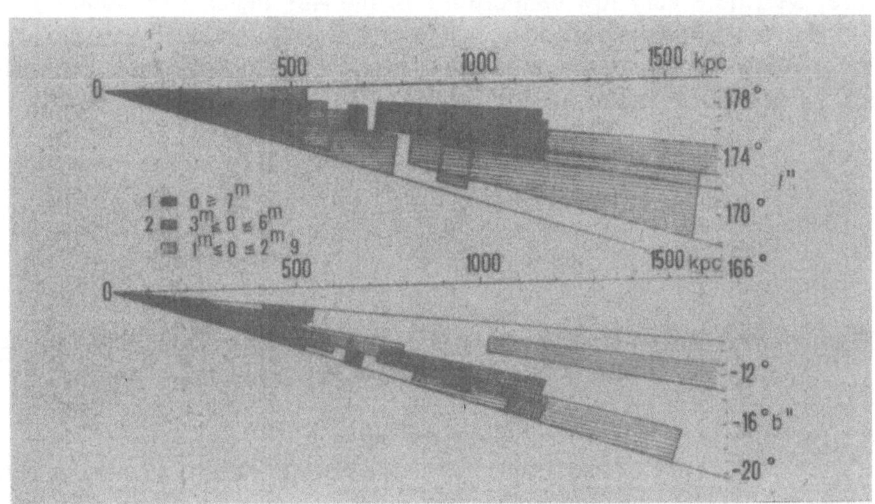

Fig. 1.

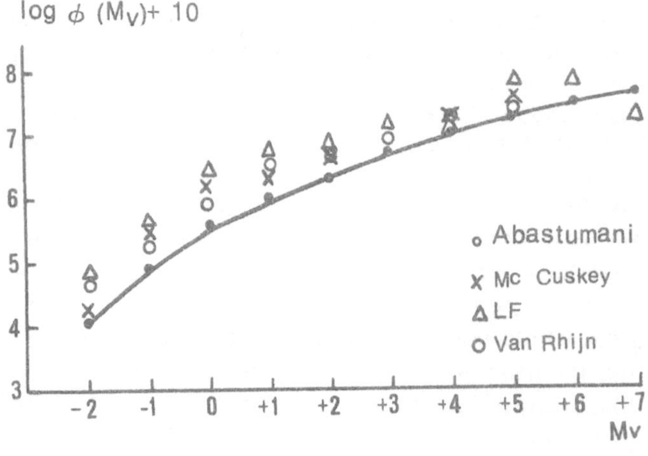

Fig. 2.

CLASSICAL CEPHEIDS AND THE STRUCTURE OF OUR GALAXY

G.R.IVANOV, N.S.NIKOLOV
UNIVERSITY OF SOFIA, BULGARIA
PRESENTED BY G.R.IVANOV

КЛАССИЧЕСКИЕ ЦЕФЕИДЫ И СТРУКТУРА НАШЕЙ ГАЛАКТИКИ

I n t r o d u c t i o n a n d b a s i c d a t a. Classical cepheids are among those objects , which are very often used for the galactic structure. Most recently the distribution of the cepheids in the Galaxy was studied by Kraft and Schmidt (1963) and Fernie (1968). Kraft's and Schmidt's main conclusion is that only the long period cepheids are located in the spiral arms of our Galaxy, as they are traced by associations.

In our present paper we have started from formula (Allen 1955)

$$M_V = 96{,}67 - 5 \log R - 10 \log Te + B.C., \qquad (1)$$

with

$$\log Te = 3{,}886 - 0{,}175\,(B-V)_0 \qquad (2)$$

and

$$M_b - M_V = 0{,}146 + 0{,}322\,(B-V)_0 \qquad (3)$$

taken from Kraft (1961). There are some suggestions, mainly by Schmidt (1971, 1972b, 1973), that Oke-Kraft's temperature scale (2) needs some small corrections, but this problem is still under consideration.

The radius of each cepheid was calculated as a mean one from the radius determination by Fernie (1965), Kurochkin (1966), Latischev (1966), Opolski (1968) and Parsons (1972). We believe that such a mean radius is better than the one obtained by a single author. It is necessary to emphasize that our comparison of radii, determined by the above cited authors, did not show systematic differences, although there are differences in the radius determination methods or in their modifications used.

The mean radii of 151 cepheids were obtained. The $(B-V)_0$ colours in (1), after (2) and (3) have been substituted in (1), were determined on using Nikolov's and Ivanov's (1975) mean excesses E_{B-V} . The latter are obtained as the weighted mean by Mianes (1963) (who takes into account Kron's and Svolopoulos' (1959) determination), by Williams (1966), Schmidt (1972a), Kelsall (1972) and the derived on the basis of infrared photometry from Fernie's (1970) E_{B-V} determinations reduced into the standard system, defined by Tsarevsky and Yakimova (1970) excesses, obtained after Kraft's method (1961) on using a new calibrated "spectrum-$(B-V)_0$"-relation and values of observed B-V, derived on the basis of Nikolov's (1968), Tsarevsky's and Yakimova's (1970) and Schaltenbrand's and Tammann's (1971) catalogues. In cases when a cepheid did not have at least an individual E_{B-V} determination, its intrinsic $(B-V)_0$ colour has been derived by means of Ivanov's statistical formula (Ivanov,1973)

$$(B-V)_0 = 0{,}298 + 0{,}206 \log P + 0{,}492 A_{B-V}, \qquad (4)$$

A_{B-V} denoting the colour amplitude.

By least squares, using 151 cepheids with known radii, we have found the following period-radius relations:

$$\log R = 0{,}710 \log P - 12{,}036, \qquad (5)$$

which is quite similar to that by Woolley and Carter (1973). From (1) with (2), (3) and (4) we have

$$M_V = 2{,}61 - 3{,}55 \log P + 2{,}07\,(B-V)_0 \qquad (5)$$

which relation has been used for the absolute magnitude determination when there isn't any radius determination of a given cepheid.

After obtaining M_V in such a way we have calculated the distances of all cepheids with a photoelectrically obtained V-magnitude from Nikolov's (1968) catalogue, as well as with the E_{B-V} from Nikolov's and Ivanov's (1975) paper on using as usually $A_V = 3E_{B-V}$. The distances and their coordinates

$$x = r \cos l'' \cos b$$
$$y = r \sin l'' \cos b$$
$$z = r \sin b$$

were obtained (the galactic coordinates l'' and b have been taken from the third edition of GCVS (Kukarkin et al. 1969) for 252 cepheids).

A study of galactic interstellar matter absorption with the help of classical cepheids

In Fig.1 we give the mean cepheid absorption A_V/kpc (A_V=3E_{B-V}, E_{B-V} from the list by Nikolov and Ivanov (1975) or from (4) against the galactic longitude. The main tendency of this dependence is the lowest absorption in the direction near the galactic anticentre (l''≈180-240°), the largest one - in the direction near the galactic centre (l''≈0 - 50°) and can be represented roughly as a sinusoidal function, as it was done by Fernie (1968). But a more detailed examination of Fig.1 shows besides one highest maximum towards the galactic centre (spiral arm in Sagittarius), two maxima towards Cygnus and Perseus-Cassiopeia spiral arms; the third maximum may appear at l''≈270°.

An additional investigation of interstellar absorption in different directions around the Sun was made as follows. We have divided the galactic cepheids in 8 sectors: I (0°<l''<45°); II (45°<l''<90°); III (90°<l''<135°); IV (135°<l''<180°); V (180<l''<225°); VI (225°<l'' <270°); VII (270°<l''<315°) and VIII (315°<l''<360°). Fig.2 presents the rough dependences of the A_V/kpc of the distance from the Sun for the eight sectors. It is evident that towards the galactic centre (sectors VIII, I and II) the mean absorption increases more rapidly than in all other directions and practically no cepheids are observed at distances larger than about 3 kpc from the Sun, because of the great absorption. In all other directions the character of the absorption-distance relationships are qualitatively identical - to a certain distance the absorption increases, after which it remains practically constant. A minimum absorption is seen approximately towards the anticentre - between l'' 180 and 270°.

The amount of the absorption in different directions is in connection with the distribution of the observed cepheids: in sectors III - VII their mean distance is 3,3 kpc, but towards the galactic centre (sectors VIII, I and II) it is 1,8 kpc. Fig.3 presents the normalized cepheid space densities as a function of the distance from the Sun.

Cepheid z-coordinate distribution

It is well known that the classical cepheids are typical representatives of the I type population and form a slim formation around the galactic plane. We have investigated in detail their z-coordinates. In Fig.4 the mean |z| are plotted against the distance from the Sun. Up to about 2-3 kpc the mean cepheid |z| coordinates are constant, after which they increase. Evidently the absorption matter in the vicinity of the galactic plane is the cause for the observation of such a small number of stars with small |z|, the greater their distances are.

The distribution of the cepheid z-coordinates is shown in Fig.5 and can be well represented by

$$N\,|z| = No\;0,188\,e^{-(z+0,042)^2}$$

Fig.5 together with the fact that the cepheids towards the galactic centre have a mean |z| close to the general mean |z| = 65 pc, while towards the anticentre they are located predominantly under the galactic plane, confirm Fernie's result, that classical cepheids form a system whose plane is inclined towards the galactic plane (Fernie 1968).

Classical cepheids in the galactic plane

The position of all 252 classical cepheids, whose distances and coordinates have been determined, are shown in Fig.6 (the Sun is in the centre, the direction towards the galactic centre is point O,O below; points denote cepheids with period log P < 0.9; crosses denote cepheids with periods log P>0,9). Concerning the reality of the picture in this Figure we must have in mind at least two points: the difference in the density of interstellar matter in different directions and the selectivity of our observations (we are better acquainted with the sky in summer!).

Fig. 6 shows clearly a higher concentration cepheid band below the Sun in direction approximately l''≈270-280°. Quite probably this band denotes the Carina spiral arm, pointed out already by Walraven et al.(1958). The Sun is at the outer edge of this arm. The space density in this arm is not uniform - it is divided at least into two parts.

A higher concentration of cepheids is observed towards the Perseus-Cassiopeia galactic arm too.

It is interesting to mention here a "cluster" with many long period cepheids approximately towards the galactic centre and at a distance of 2-4 kpc from the Sun outside a nearer group, which appears as a prolongation of the Carina arm to the right of the Sun.

If our conclusion about spiral arms traced by cepheids is correct, it is necessary to emphasize that not only long period cepheids trace these arms. We shall mention in this connection in the Carina arm one sees a "cluster" of cepheids within about 1 to 2 kpc, in which predominate stars with log P≈0,8-0,9. One can also see a second group within about 2 to 3 kpc, in which cepheids with short periods log P<0,8 predominate. And only far away to about 3 to 4 kpc in this direction stars with long periods predominate. It is quite unlikely to assume that such a distribution of cepheids with different periods is due merely to a selective effect.

Apart from the particular arrangement of cepheids with different periods towards Sagittarius and in the Carina spiral arm, there are several suggestions, that there is a dependence between the period of cepheids and their distance from the galactic centre (Shapley and McKibben 1940, Kukarkin 1949, van den Berg 1958, Fernie 1968). We return to this point once more, investigating galactic cepheids only, within 3 kpc from the Sun (the distance from the Sun to the galactic centre is accepted to be 8 kpc) in order to diminish the absorption influence and our incompleteness of knowledge of distant cepheids. The following Table shows the mean period change with the distance from the galactic centre R.

R	$\overline{\log P}$
6.0	0.96
7.5	0.82
9.0	0.80

If we limit the area to only 1 kpc from the Sun, in which one can expect that the investigated cepheid sample is full, we obtain the result:

R	$\overline{\log P}$
7.5	0.84
8.5	0.68

Moreover, not only the period, but quite probably the light amplitude A_V is also dependent on the cepheid location in the Galaxy.

We have given a special attention to the s-cepheids from Efremov's work (Efremov 1968). These stars show a spotty distribution in the Galaxy. While we find s-cepheids in the group which is nearer to the Sun towards Sagittarius, they are absent in the second more distant group in the same direction; we find the opposite situation towards l''≈290° : s-cepheids exist only in the second more distant group of stars.

A comparison of cepheid distribution in the Galaxy with other galactic objects

Fig.7 presents the cepheid distribution on the galactic plane together with that of other young population I objects as associations (▲), open clusters (●), H II regions (◆) and supergiants (■). The distribution of these objects is taken from Burton (1973). It is evident that in general cepheids are located in places occupied by other objects.

But a more detailed examination of Fig.7 shows several evident differences between cepheid distribution and the distribution of the objects:

1) an excess of cepheids is seen in the vicinity of the Sun, where there are no other objects;

2) in the Carina arm, as traced by cepheids, there are many other young objects, but no associations;

3) the other objects trace better than cepheids an arm towards l''≈140°;

4) the situation is more drastic in an arm, traced by other objects, which passes about 1 kpc from the Sun and has a direction from 1''≈40° towards 1''≈200°; in this arm cepheids are very rare.

The best coincidence is that of cepheids with supergiants. The coincidence of cepheids with open clusters is not good. Especially long period cepheids (log P>1,0) avoid the young (earlier spectral type ≤ B2) clusters, as it is seen in Fig.8, in which the clusters are taken from Moffat and Vogt (1973). Probably it is not accidental, that no long period cepheids belong to open clusters (the cepheid with the longest period is the 10.7 day one, namely TW Nor, a probable member of Linga 6 cluster (van den Berg and Hagen 1975)), which is not the case with associations.

Conclusion

In this paper a new investigation of the distribution of cepheid z-coordinates and the cepheids on the galactic plane is reported. The difference between this investigation and those by Kraft and Schmidt (1963) and Fernie (1968) is the use of mean radii and mean excesses E_{B-V} (Nikolov and Ivanov 1975).

The main conclusion of the picture of cepheid distribution on the galactic plane is the discernment of spiral arms with the exception of one, coinciding probably with the Carina arm. Another conclusion drawn is the unlikeness of the distribution of cepheids (and also of the long period cepheids) and the distribution of young open clusters. Some results about the "diversity" of the period and the light amplitude and especially of small period s-cepheids in different places of the galactic plane may be of some importance for star formation in connection with chemical composition.

Acknowledgement

The authors express their gratitude to Mrs. S. Pomenova for the indispensable help in the final stages of preparation of this paper.

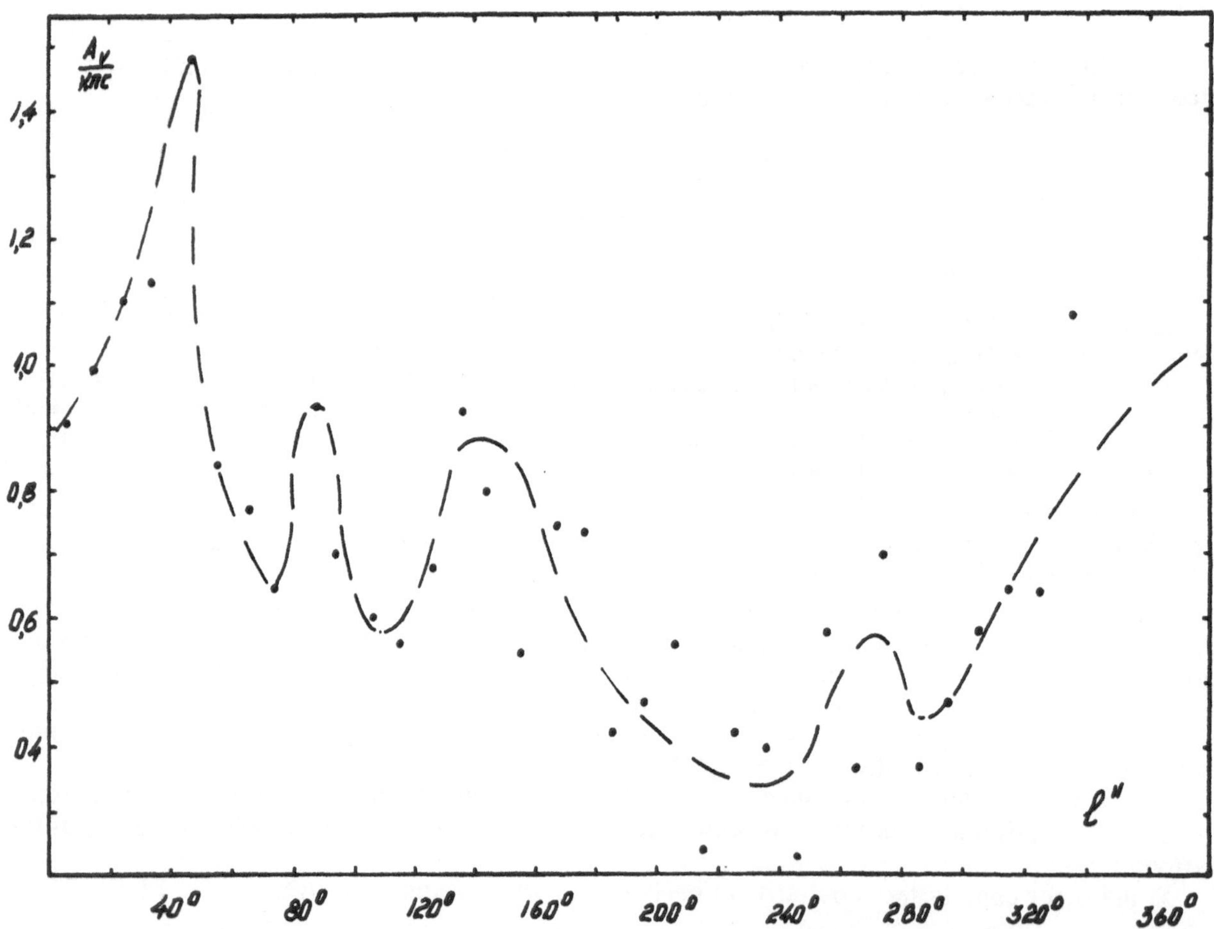

Fig. 1. Mean absorption of cepheids for 1 kpc as a function of galactic longitude.

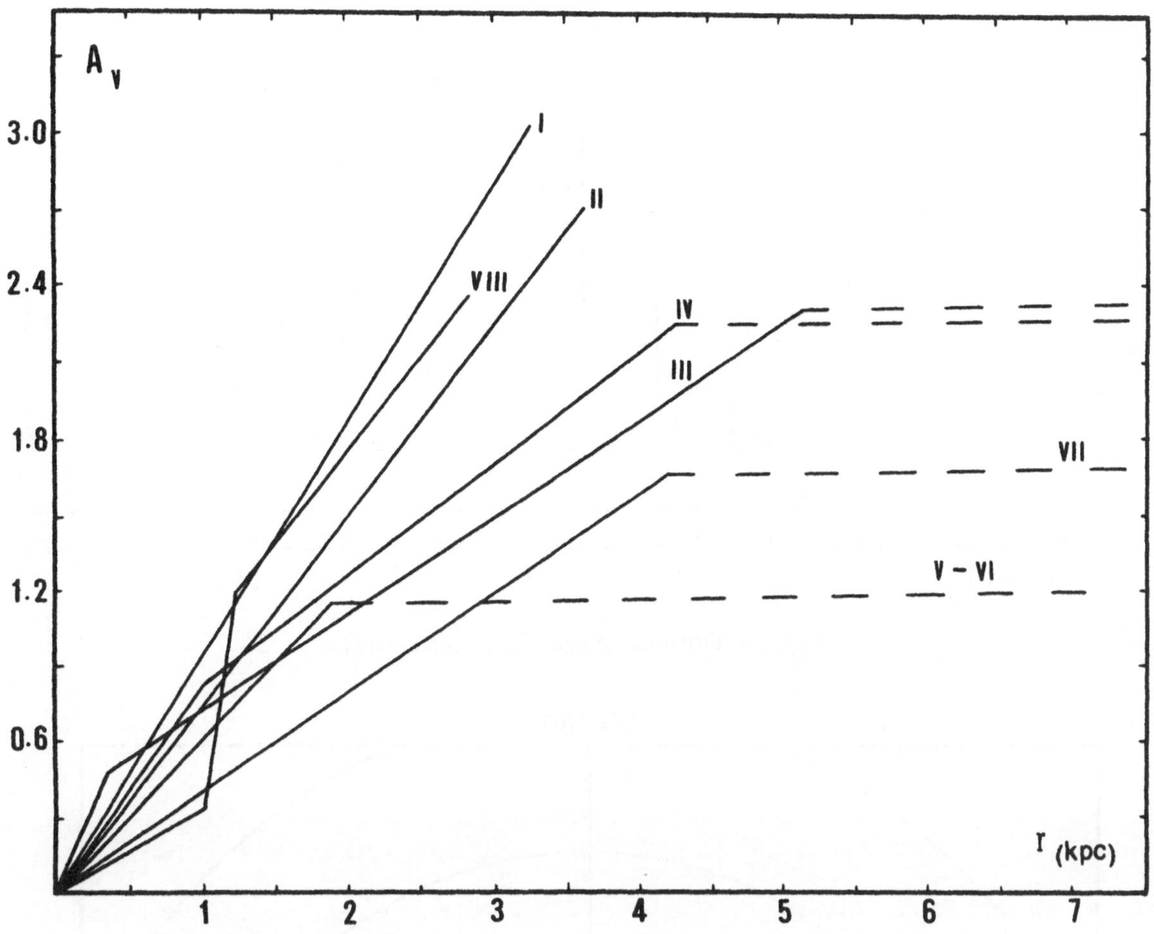

Fig. 2. Dependences of absorption A_V/kpc on the distance from the Sun for cepheids divided in 8 sectors, according to their longitudes. The character of the absorption-distance relations are qualitatively identical for all sectors: the absorption increases to a certain distance and further on (for longer distances) it is practically constant.

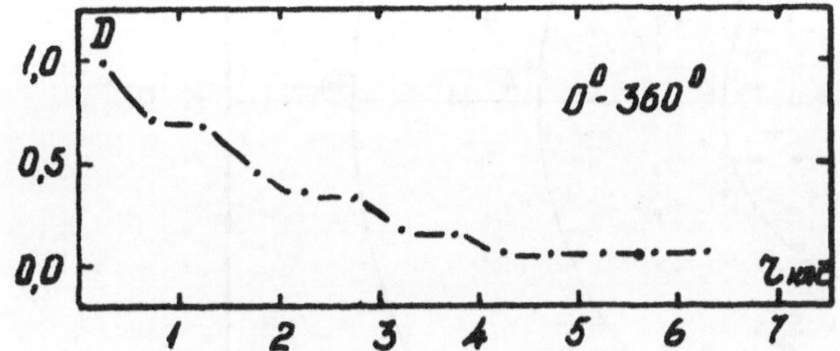

Fig. 3. The normalized cepheid space densities as a function of the distance from the Sun.

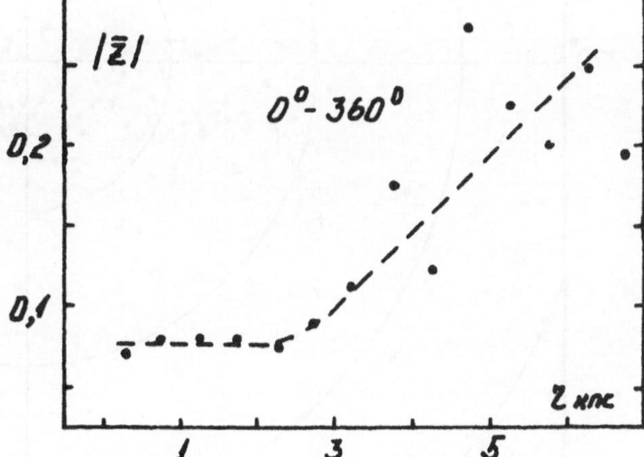

Fig. 4. The mean |z| coordinates of cepheids against the distance from the Sun. The absorption causes the increase of |z| coordinates of known cepheids for distances greater than about 2-3 kpc.

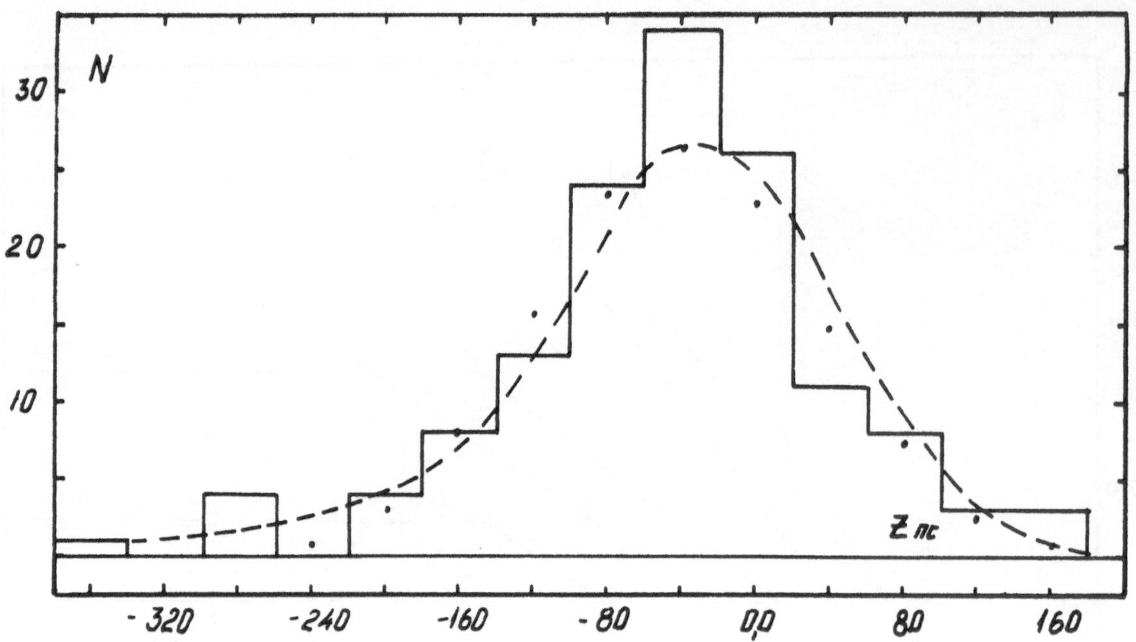

Fig. 5. Distribution of cepheid z coordinates.

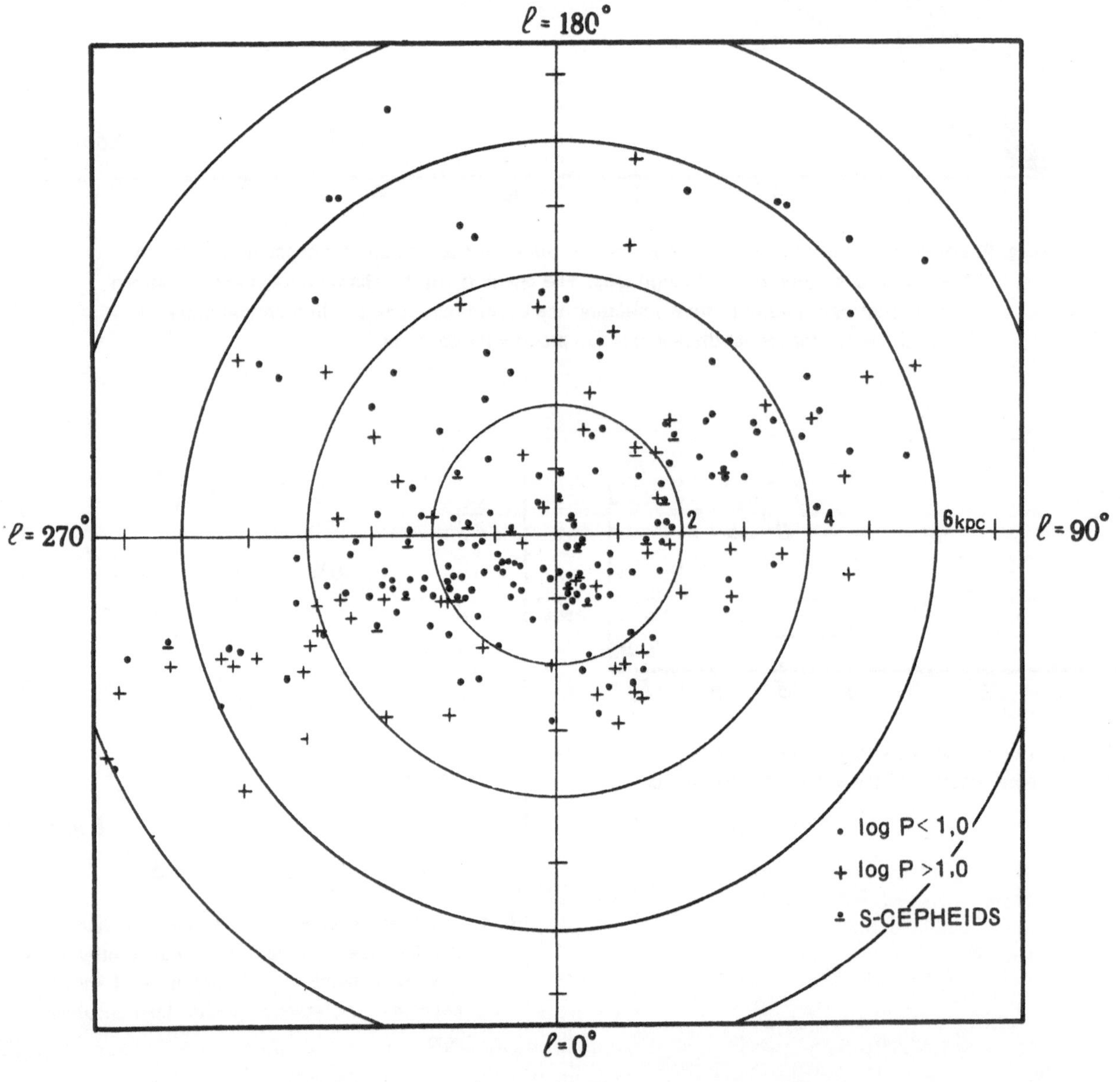

Fig. 6. Distribution of 252 cepheids on the galactic plane.

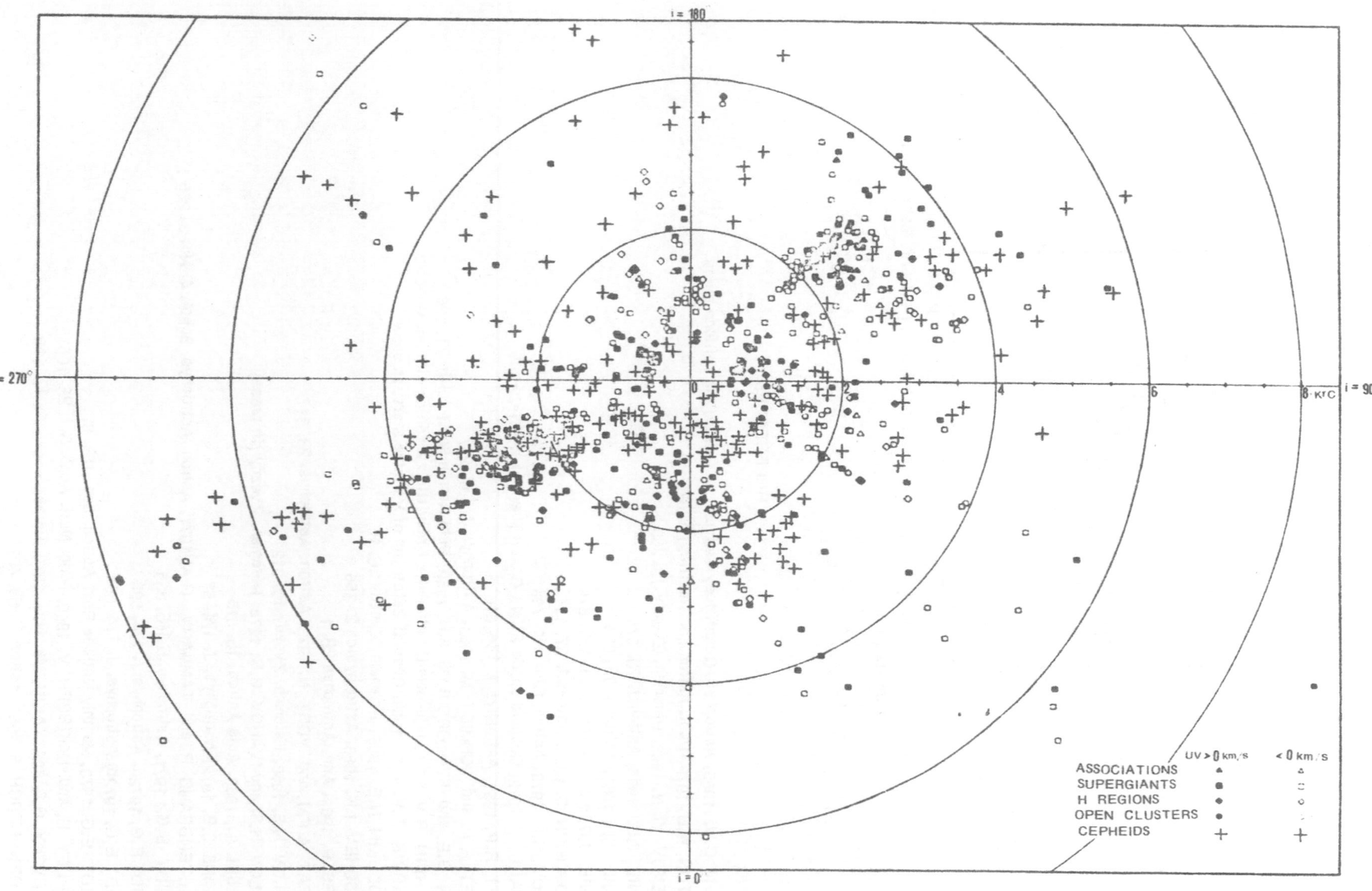

Fig. 7. Cepheids on the galactic plane together with associations (▲), open clusters (●), H II regions (◆) and supergiants (■).

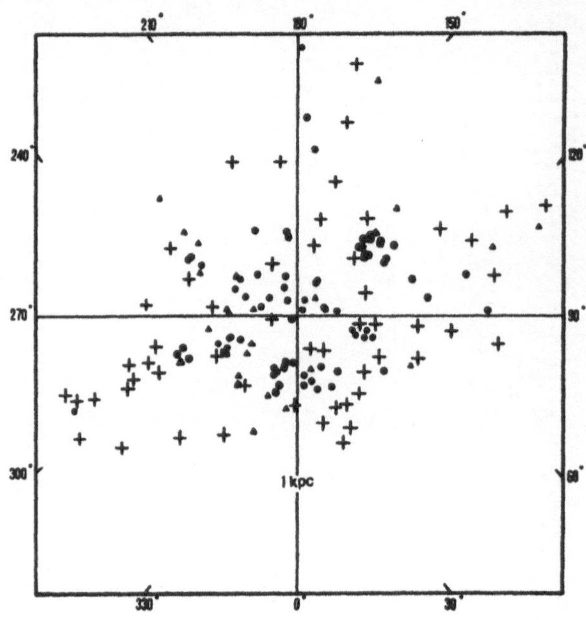

Fig. 8. A comparison of long period cepheid distribution (+) with that of open clusters (●).

REFERENCES

ALLEN C.W. 1955, Astrophys. Quantities, -Univers. of London, the Athlone press

BURTON W.B. 1973, Publ. Astron.Soc.Pacific, 85, 679

EFREMOV Yu.N. 1968, Peremen. Zvezdy 16, 365

FERNIE J.D. 1965, Astron.J. 70, 575

FERNIE J.D. 1968, Astron.J. 73, 995

FERNIE J.D. 1970, Astrophys.J. 161, 679

GAPOSCHKIN S. 1962, Astron.J. 67, 334

IVANOV G.R. 1973, Astron. Circ. No. 798

KELSALL T. 1972, Goddard Space Flight Center Preprint X-641-72-365

KRAFT R.P. 1961, Astrophys.J. 134, 616

KRAFT R.P. and SCHMIDT M. 1963, Astrophys.J. 137, 247

KRON G.E. and SVOLOPOULOS S.N. 1959, Publ.Astron.Soc.Pacific 71, 126

KUKARKIN B.V. 1949, Isledovanije stroenija i razvitija zvezdnich sistem, Moscow

KUKARKIN B.V. et al. 1969, General Catalogue of Variable Stars, Moscow

KUROCHKIN N.E. 1966, Peremen. Zvezdy 16, 10

LATISCHEV I.N. 1966, Astrophysica 2, 355

MIANES P. 1963, Ann, Astrophys 26, 1

MOFFAT A.F.J. and VOGT N. 1973, Astron. Astrophys. 23, 317

NIKOLOV N.S. 1968, Peremen. Zvezdy 16, 312

NIKOLOV N.S. and IVANOV G.R. 1975, Peremen. Zvezdy (in press)

OPOLSKI A., 1968, Acta Astron. 18, 515

PARSONS S.B. 1972, Astrophys. J. 174, 57

SCHALTENBRAND R. and TAMMANN G.A. 1971, Astron. Astrophys. Suppl. 4, No. 3, 265

SCHMIDT E.G.: 1971, Astrophys.J. 165, 335

SCHMIDT E.G. 1972a, Astrophys.J. 174, 595

SCHMIDT E.G. 1972b, Astrophys. J. 174, 605

SCHMIDT E.G. 1973, Monthly Notices Roy. Astron. Soc. 163, 67

SCHAPLEY H. and McKIBBEN V. 1940, Proc. Natl. Acad. Sci. 26, 105

TSAREVSKY G.S. and YAKIMOVA N.N. 1970, Peremen. Zvezdy 17, 120

VAN DEN BERGH S. 1958, Astron. J. 63, 492

WALRAVEN T., MÜLLER A.B. and OOSTERHOFF P.T. 1958. Bull. Astron. Inst. Neth. 14, 81

WILLIAMS J.A. 1966, Astron.J. 71, 615

WOOLLEY R. and CARTER B. 1973, Monthly Notices Roy. Astron. Soc. 162, 379

PERIOD-AGE RELATION OF CEPHEIDS AND THE HISTORY OF STAR FORMATION IN M31 and THE LMC

Yu.N.EFREMOV
STERNBERG ASTRONOMICAL INSTITUTE, MOSCOW, U.S.S.R.

ОТНОШЕНИЕ ПЕРИОД-ВОЗРАСТ ДЛЯ ЦЕФЕИД И ИСТОРИЯ ОБРАЗОВАНИЯ ЗВЕЗД В М31 И БМО

The origin of cepheids from massive main sequence stars implies that a relation must exist between cepheid periods and their ages. Such a relation was found (Efremov, 1964) from data on ages of few galactic clusters and periods of their cepheid members. Now the period-age relation is established from data on about 70 cepheids and 30 clusters, mainly from the Magellanic Clouds. This relation has the following approximate form: $\lg t = 8.23 - 0.74 \lg P$, and nearly coincides with the theoretical relation of Kippenhahn and Smith (1969) for the abundance X=0.602 and Z=0.044(the dashed line in Fig.1). The same abundance was taken by Dixon et al. (1972) whose data I used for age calibration of the Magellanic Cloud cluster integral colours and colour-magnitude diagrams.

The period-age relation suggests that the difference between average cepheid periods in some areas of galaxies is due mainly to different average ages of young stars in these areas rather than to difference in the initial mass function or abundance.

A gradient of cepheid periods across the Baade spiral branch S4 in M31 was found some years ago (Efremov, 1971; Fig.2). The age gradient across a branch is in accordance with the predictions of the density wave theory of spiral pattern. Fig.2 shows that stars near the inner part of branch are, on the average, younger than stars near the outer part. There is also the period gradient along the same branch: the average period increases with increasing distance from the nucleus along the branch (Fig.3). This gradient may be connected with some observational selection of cepheid discoveries. The distribution of cepheid periods in the van den Bergh''OB associations'' shows both effects: the average periods increase to the inner side across the branch and from the nucleus along the branch.(Fig.4).

Investigations of cepheids outside Baade's fields in M31 would be most valuable for understanding the nature of spiral pattern.

In the LMC the period gradient is found only along its bar. Large period cepheids predominate in the western part of the bar (Fig. 5, 6). Dixon and Ford (1972) found an age gradient along the major spiral branch of the LMC in the same direction. Probably the density wave theory cannot be applied to the galaxies of LMC type.

There is a number of areas in the Magellanic Clouds, M31 and our Galaxy where cepheids have rather similar periods. The LMC data show that the average cepheid periods on such areas usually correspond to the ages of clusters on the same areas. The average diameter of such an area is about 500 pc as well as average size of HI concentrations in the LMC (575 pc), van den Bergh's OB associations (star cloud) in M31 (480 pc), superassociations, and dimension of the Local system in our Galaxy. Tendency of the LMC clusters to group together in space and time was pointed out by Hodge (1973).

We conclude that space-period distribution of cepheids confirm the conception of burst of star formation in the cells with diameters of few hundred parsecs. The dispersion of the cepheid periods in these cells allows to conclude that star formation in a given cell occurs during few ten millions of years.

The most active cell of contemporary star formation (which is too young for cepheid occurrence) in the LMC is the well known region 30 Dor (NGC 2070). In radio it looks very similar to the nucleus of our Galaxy (Mathewson, 1971). This has led to a suggestion that active nuclei of galaxies are those in which the star formation is going now at a very high rate.

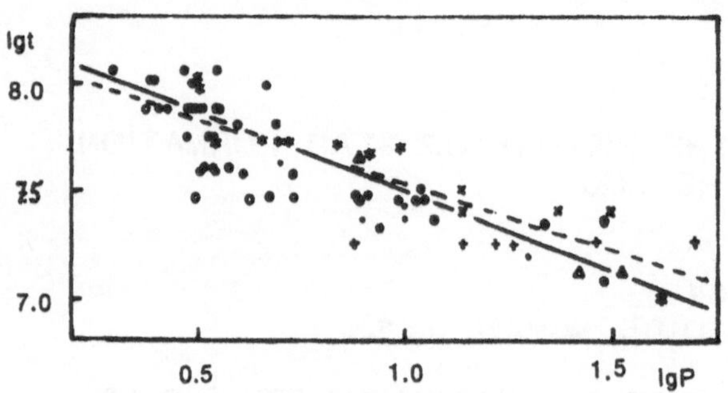

Fig.1. The period - age relation
• galactic cluster cepheids
●○• magellanic clouds cluster cepheids
x + magellanic clouds association cepheids
▲△ M 31 cluster cepheids

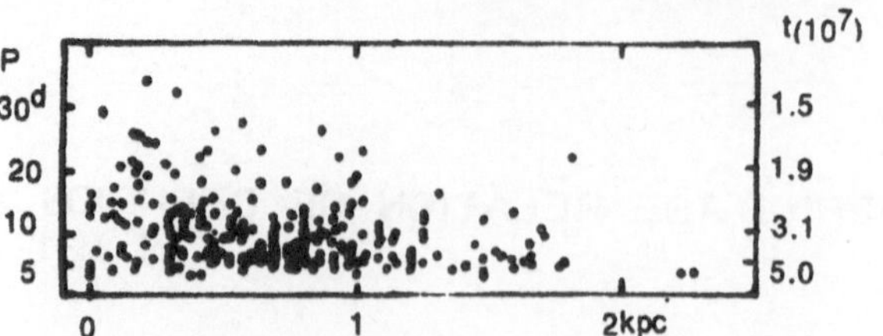

Fig. 2. The period (age) gradient across spiral branch S4 IN M 31

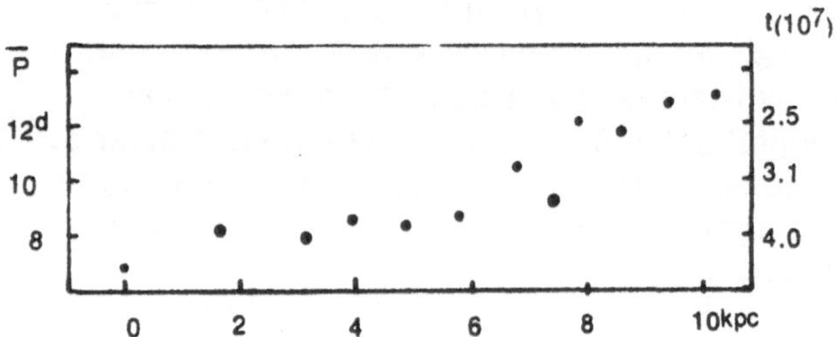

Fig. 3. The period (age) gradient along the spiral branch S4 IN M31

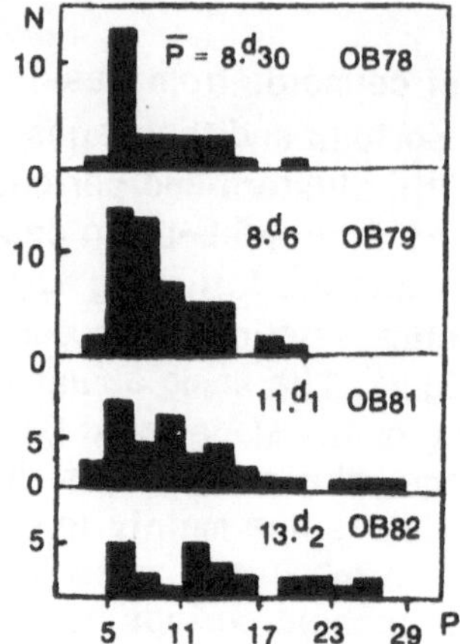

Fig. 4. The distribution of cepheids periods in the van den Berg's OB associations in M31

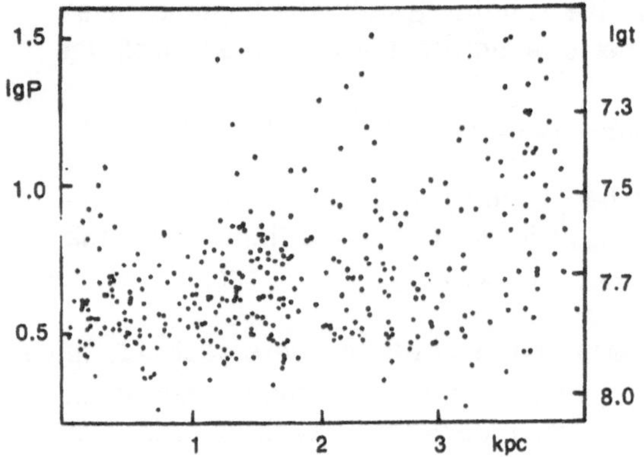

Fig. 5. The period (age) gradient along the bar of the LMC

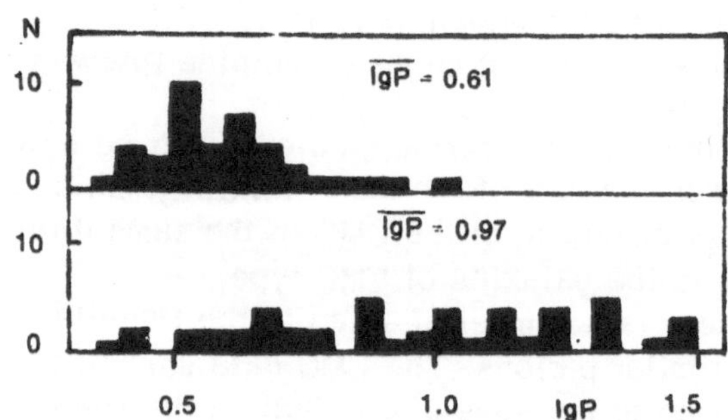

Fig. 6. The distribution of the cepheid periods in eastern and western part of the LMC bar

REFERENCES

DIXON M.E., and FORD V.L. 1972, Astrophys. J., 173,35
DIXON M.E., FORD V.L., ROBERTSON J.W. Ast 1972, Astroph.J., 174, 14
EFREMOV Yu.N. 1964, Variable Stars, 15, 242; Astron. Circ. No. 311
EFREMOV Yu.N.1971, Astron. Circ. No. 639
HODGE P.W. 1973, Astron.J., 78, 807
KIPPENHAHN R. and SMITH L. 1969, Astron.and Astroph.1, 142
MATHEWSON D.S., 1971, The Magellanic Clouds (ed. Müller), 98

THE SPIRAL STRUCTURE OF THE LARGE MAGELLANIC CLOUD

Th. SCHMIDT-KALER
ASTRONOMISCHES INSTITUT DER RUHR-UNIVERSITÄT, BOCHUM, F.R.G.

СПИРАЛЬНАЯ СТРУКТУРА БОЛЬШОГО МАГЕЛЛАНОВОГО ОБЛАКА

The optical structure of the LMC is subject of a longstanding discussion (de Vaucouleurs, 1974; Westerlund, 1974). This is also true for the structure in the neutral hydrogen from the 21 cm-line (McGee and Milton, 1966, Kerr, 1971). However, recent observations of supergiants (Ardeberg et al., 1972, Isserstedt, 1975) suggest spiral features and lead together with other excellent spiral tracers to a rather clear picture of the over-all spiral structure of the LMC (Schmidt-Kaler and Isserstedt, 1975).

The best spiral indicators are the HII-regions and the heavy dark lanes. We selected from Henize's (1956) H_α survey all regions with diameters $\geq 25'' \cong 6_{pc}$. The smaller H_α-regions have lower detection probabilities in the vicinity of the larger ones, and are (in the average) excited by stars of spectral types B2 and later which are less reliable spiral tracers. The HII- regions (or the O-B1 stars) and the blue supergiants with $(U-B)_0 \leq -0.60$ (approximately O-B8 Ia-O to Iab) show a clear spiral pattern with two main arms (Fig.1) Both arms have complex structure, the northern arm is split up in two filaments. The 10 known supernova remnants (Mathewson and Clarke, 1973) follow exactly the spiral filaments. The distribution of the OB-associations, young open clusters, Wolf-Rayet stars, X-ray sources and M-supergiants is also concentrated to the spiral features but spread out in increasing degrees. Thus, the quality as a spiral tracer correlates well to the age of the type of object.

The spiral filaments start from the enormous HII-complex 30 Doradus. Like in other galaxies, as for instance M31, the spiral structure in the inner parts is best seen in the distribution of the dark clouds. These have been taken from the surveys of van den Bergh (1974) and Hodge (1972).

Since the interpretation of the 21 cm-line profiles is burdened with the difficulty to split up the velocity components and with the uncertainties of the rotation curve we analyzed the total intensities integrated over all velocities. Although the distribution appears at first sight very irregular the ridge lines are well defined and show, apart from many differences in details, the same basic pattern as the HII-regions and blue supergiants. (Fig.2).

It is most remarkable that the spiral structure of the LMC is both, extremely asymmetric and completely unrelated to the central Bar. The centre of mass and the centre of spiral arms do not coincide. In normal spiral galaxies a dominating massive nuclear region and - probably due to it - a high degree of symmetry is observed. Disturbed systems are experiments of nature which help us like in biology to understand together with the disturbed function the normal function as well. The LMC as such a case yields some hints on the origin of spiral structure. Three facts are hard to explain by a purely gravitational theory:

1) The density-wave theory starts from an axisymmetric stationary ground state. In the LMC the centres of mass, spirals and rotation of population I are all different. (The tidal perturbation by our Galaxy is in the inner parts of the LMC of no importance.) Spiral structure develops without *a* dominating central mass.

2) The massive Bar is crossed by a very sharp filament without any recognizable perturbation. If this filament is not in a plane far from the Bar the gravitational action of the masses of the Bar in the spiral is small.

3) The supergiant HII-complex 30 Doradus is starting-point of the spiral filaments. It is, regarding the density and mass of ionized hydrogen, the velocity dispersion and the peculiar radially filamentary structure unique among all the HII-regions of the LMC and our Galaxy (except for Sgr A). A detailed examination (Schmidt-Kaler and Feitzinger, 1975) shows that 30 Dor presents all the general properties of a galactic nucleus except, perhaps, for the high mass concentration. The semistellar object R136 of type O+WN and extremely high variable luminosity $M_V \cong -11$ is at the centre. It also displays evidence of activity, especially the strong non-thermal component and the splitting of the [OIII] 5007 profile into two components. This corresponds to an expansion velocity of 50 kms^{-1} and a mass loss of about 0.05 $M_\odot/yr$. Relative to the total mass this equals

about the mass loss of our Galaxy and is about 100 times that of M31.

On account of these observations an ejection theory (Schmidt-Kaler, 1975; Schmidt-Kaler and Feitzinger, 1975a) appears more promising than a density-wave theory to explain the structure of the LMC.

LMC and SMC form a double system interconnected by a massive bridge of neutral hydrogen. It is tempting to speculate that the isolated dark cloud at the centre of the Bar might represent the remains of an earlier location of the nucleus, and the spiral structure described by de Vaucouleurs the remains of an earliar spiral phase. In a galaxy of low mass the nucleus may be wandering (and even be ejected) due to anisotropic explosions, and so induce the asymmetry of low-mass galaxies in contrast to the symmetry of massive galaxies whose active centre is fixed by a deep potential well.

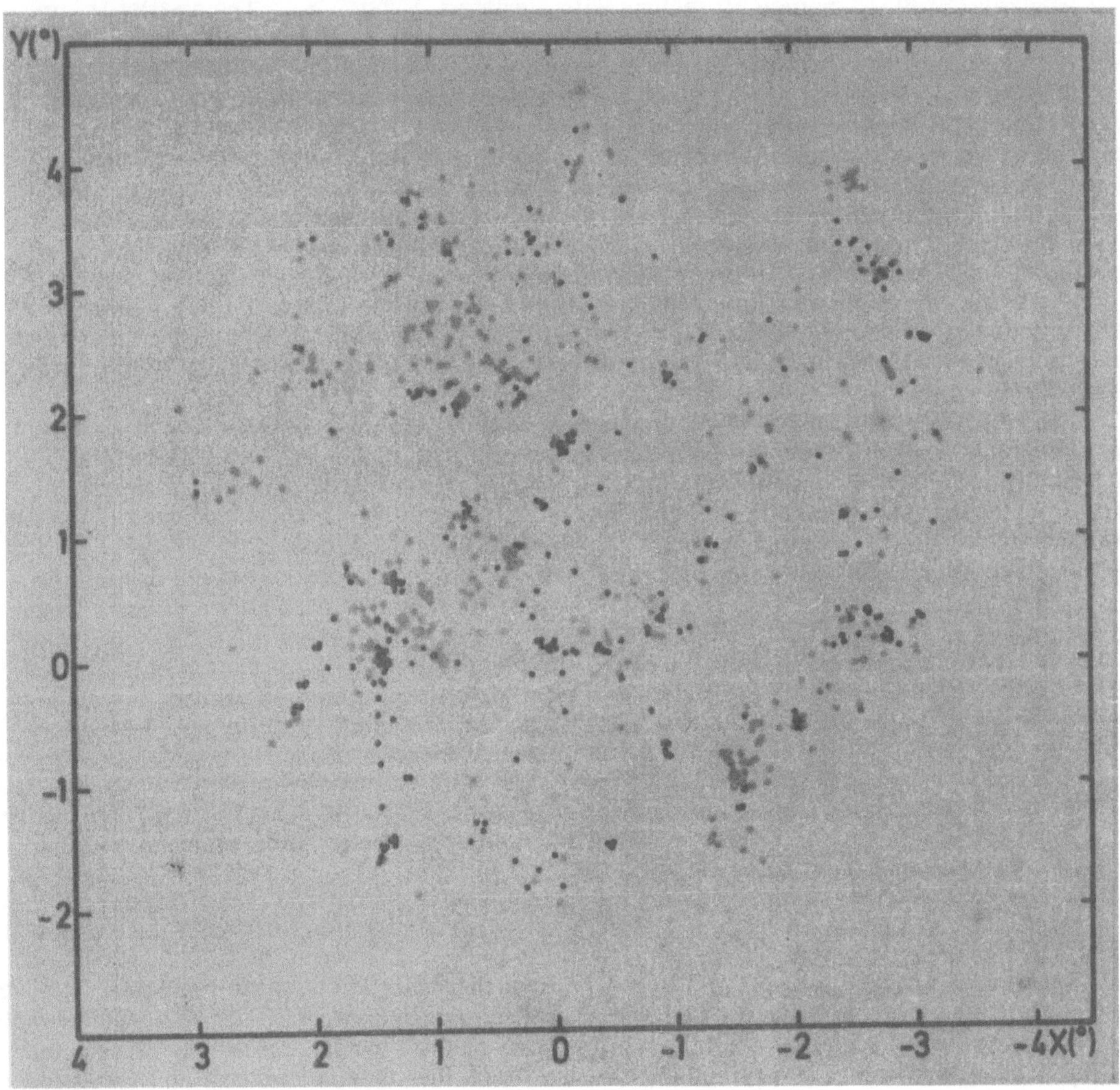

Fig. 1. The Spiral Structure of the LMC.
Black dots: HII regions with diameter $D \geqq 25''$.
Grey dots: Supergiants with $(U-B)_0 \leqq -0\overset{m}{.}6$.

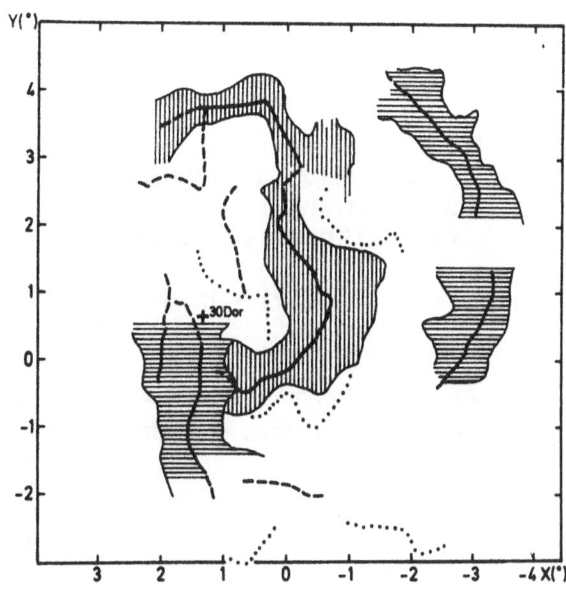

Fig. 2. The distribution of neutral hydrogen in the LMC. The lines indicate the ridge-lines of the intensities (Summed up over the line of sight); in case of low contrast the ridge-lines are dashed.
The half-intensity widths of the spiral arms are hatched (arm I ≡ , arm II ⦀).

REFERENCES

ARDEBERG A., MAURICE E., BRUNET J.P., PRÉVOT L. 1972. Astron. Astrophys. Supp. 6, 249

HENIZE K.G. 1956. Astrophys.J. Suppl. 2, 315

HODGE P.W. 1972. Publ. Astron. Soc. Pac. 84, 365

ISSERSTEDT J. 1975. Astron. Astrophys. Supp. 19, 259; Astron. Astrophys. 41, 21

KERR F.J. 1971. in: "The Magellanic Clouds" (ed. Muller), p.50, Dordrecht, Holland

MATHEWSON D.S., CLARKE J.N., 1973. Astrophys. J. 180, 725

McGEE R.X., MILTON J.A. 1966. Austr. J. Phys. 19, 343

SCHMIDT-KALER Th., 1975 in: Optische Beobachtungs Programme zur galaktischen Struktur und Dynamik (ed.Th Schmidt-Kaler), p.75, Bochum 1975

SCHMIDT-KALER, Th., ISSERSTEDT J.1975. Astrophys. Space Sci. in press

SCHMIDT-KALER Th., FEITZINGER J. 1975. Astrophys. Space Sci. in press; 1975a, in preparation

VAN DEN BERGH S. 1974. Astrophys. J. 193, 63

VAUCOULEURS G. DE 1974. IAU-Symp. 58, 1 (ed. J.B.Shakeshaft)

WESTERLUND B.E. 1974. Proceed. 1st Europ. Astr. Meet. Vol. 3, 39. Springer, Berlin

DISCUSSION

KOMBERG : Why do we not observe spiral structure extending from the superassociations, for instance in Markarian galaxies?

SCHMIDT-KALER : I have not looked at Markarian galaxies this far; it may be worthwhile.

ARDEBERG : Would you comment on the far-ultraviolet photograph by Watts?

SCHMIDT-KALER : The EUV (1050 - 1300 Å) photographs in my opinion clearly show most of the structure in question except for regions with strong dust concentrations.

ARDEBERG : But you do not presume that all the features seen are in the same plane?

SCHMIDT-KALER : My statement was that the gravitational action of the Bar is small if all features lie in the plane. But should we not be happy that a reasonable distribution of HI similar to the optical features results, *without* the necessity to introduce more than one plane?

KOLESNIK (L.N): Do your stellar spirals coincide with the neutral hydrogen?

SCHMIDT-KALER : The same basic structure is observed for both. The HI ridge of arm I is just outside the axis of arm I as defined by optical tracers, for arm II the situation is more complicated.

GALACTIC STRUCTURE AT $l = 66°, 77°, 113°$ and $187°$

N.B.KALANDADZE, L.N.KOLESNIK and V.I.VOROSHILOV
ABASTUMANI ASTROPHYSICAL OBSERVATORY; GOLOSEEVO ASTRONOMICAL OBSERVATORY, KIEV, U.S.S.R.
PRESENTED BY L.N.KOLESNIK

ГАЛАКТИЧЕСКАЯ СТРУКТУРА В ДОЛГОТАХ $l = 66°, 77°, 113°$ и $187°$

I n t r o d u c t i o n . Photographic photometry and spectral classification of about 16768 stars in four regions (Table 1) of 18 square degrees near the galactic plane have been used to study the distribution of OB stars and dust.

Table 1

Region	l	b	n
NGC 6834	65°.7	+1°.2	4463
NGC 6913	76°.9	+0°.6	3550
NGC 7654	112°.8	+0°.5	4286
NGC 2129	186°.6	+0°.1	2469
			16768 stars

The spectra (166 and 666 Å/mm at H_γ.) were taken by N.B.Kalandadze with the 8° and 4° objective prisms attached to the 70cm telescope of the Abastumani observatory. Photographic BV magnitudes to the limited magnitude $V = 13^m.2$ were determined by L.N.Kolesnik and V.I.Voroshilov at the Main Astronomical Observatory of the Academy of Sciences of the Ukrainian SSR in Kiev.

The interstellar absorption was obtained adopting $A_V = 3E_{B-V}$. The run of absorption with distance is summarized in Table 2. The interstellar dust has the highest density in the direction $l = 77°$. The absorption is also heavy in the direction $l = 113°$ (toward the Perseus spiral arm).

Fig. 1 shows the variation of interstellar absorption and stellar densities of OB stars with distance on the studied areas. There is a large cloud complex in the direction $l = 77°$ between 1-2 kpc from the sun associated with the local spiral arm. A region of high reddening is also in the direction $l = 113°$ between 300-800 pc (local arm) and in the distance range 2.4 - 3.0 kpc (Perseus arm).

S t e l l a r   d e n s i t i e s. Fig.2 shows the space density of O - B2 stars (well known spiral arm tracers) as a function of distance at four galactic longitudes. The greatest concentration of OB stars is in the direction at $l = 77°$. There is a rather strong maximum in the distance range 1.2 - 2.0 kpc. This density maximum is in the local spiral arm. The density peak is due to the presence of the association Cyg OB 1.7 Wolf-Rayet stars were found in this region.

The maximum frequency of occurence of the OB stars in the direction $l = 113°$ is in the range 2.6 - 3.8 kpc. This density maximum is in the Perseus spiral arm. The region appears to contain many faint OB stars at very great distances from the sun. There is an indication of two concentrations of these stars at the distances 5.5 and 7.5 kpc, but the location of these concentrations is uncertain because of the errors in photometry, mean absolute magnitudes and intrinsic colours amount to about 15% for a star.

In the anticenter direction the space density of the OB stars is lower than in other regions. If we interpret two weak density peaks in the distance ranges 2.5 - 3.3 and 4.0 - 4.8 kpc as an evidence of their association with the spiral structure of the Galaxy, then the extension of the Perseus arm in the anticenter direction is at about 3 kpc and the extension of the outer arm at about 4.5 kpc.

On the direction at $l = 66°$ the space density of the OB stars reaches a maximum at 2.6 kpc. The majority of the OB stars lies in the distance range 2.2 - 4.3 kpc. It seems to be an extension of the local spiral arm in this direction.

Table 2

Region	l	A_V r = 0.5 kpc	r = 1 kpc	r = 2 kpc	r = 3 kpc	A max
NGC 6834	66°	$0\overset{m}{.}5$	$1\overset{m}{.}1$	$1\overset{m}{.}75$	$2\overset{m}{.}4$	$3\overset{m}{.}5$
NGC 6913	77°	$0\overset{m}{.}6$	$1\overset{m}{.}5$	$3\overset{m}{.}3$	-	$4\overset{m}{.}8$
NGC 7654	113°	$0\overset{m}{.}4$	$1\overset{m}{.}1$	$2\overset{m}{.}0$	$2\overset{m}{.}2$	$4\overset{m}{.}8$
MGC 2129	187°	$0\overset{m}{.}4$	$0\overset{m}{.}75$	$1\overset{m}{.}25$	$1\overset{m}{.}6$	$2\overset{m}{.}4$

Fig.3 shows that the OB stars are asymmetrically distributed with respect to the galactic plane. In the directions l = 77° and l = 187° most of the OB stars are situated above the galactic plane. The tilt of the "OB star plane" against the galactic plane is about 1°.

The B3 - B5 stars show association with the spiral structure but the increase in their densities on the inner sides of the spiral arms begins earlier, than for the O-B2 stars.

In the direction along the spiral arm toward Cygnus the OB stars appear to be embedded in dust associated with the arm: large amount of absorbing material coincides with the maximum frequency of OB stars in the distance range 1 - 2 kpc. At l = 113° the dust also shows association with the OB stars at the distances 2.4 - 3.0 kpc (Perseus spiral arm.). There is a large amount of absorbing material in this direction near the sun, where OB stars are scarce.This may indicate that the dust occupies areas larger than those occupied by OB stars.

The O - B2 star distribution shows little similarity with the distribution of neutral hydrogen shown in Verschuur's new map (1973). As a rule the stellar arms lie behind the neutral hydrogen arms.

T h e   l u m i n o s i t y   f u n c t i o n. The luminosity function is determined for $-5^m \leqslant M_V \leqslant +8^m$ in four Milky Way regions (Fig.4). For stars within $0^m \leqslant M_V \leqslant +3\overset{m}{.}5$ the over-all agreement between the luminosity function for the four regions and luminosity function derived by Luyten (1968) is good. Fig.4 illustrates the longitude variation in log φ(M). There appears to be an excess in log φ(M) for the directions l = 66° and l = 77° as compared with the directions l = 113° and l = 187°. The increase in log φ(M) is more prominent for the intrinsically bright stars ($M_V < 0^m$) and for the direction at l = 77°. This is probably due to the contribution of the local spiral arm and to association Cygnus OB 1 in this direction. This difference is certainly real because the results are based on quite homogeneous data.

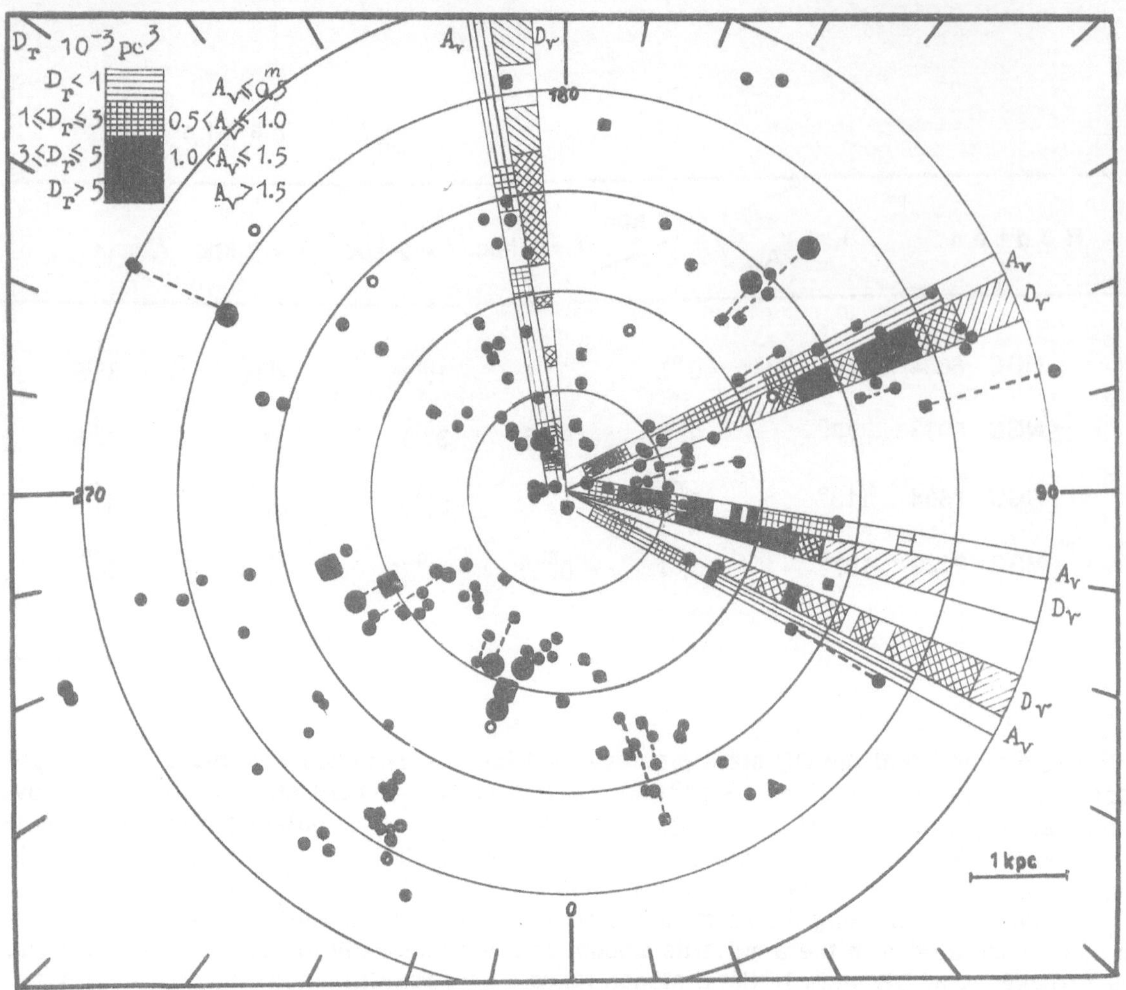

Fig. 1. Space distribution of the O - B2 stars (D(r) - stars per 10^3pc^3) and dust (A_V per kpc^3).

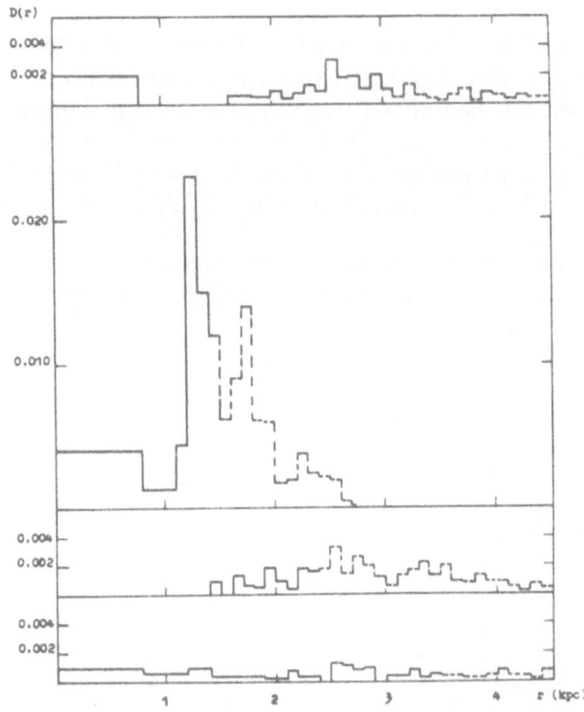

Fig. 2. Space densities (stars per 10^3pc^3) of the O - B2 stars as a function of distance.

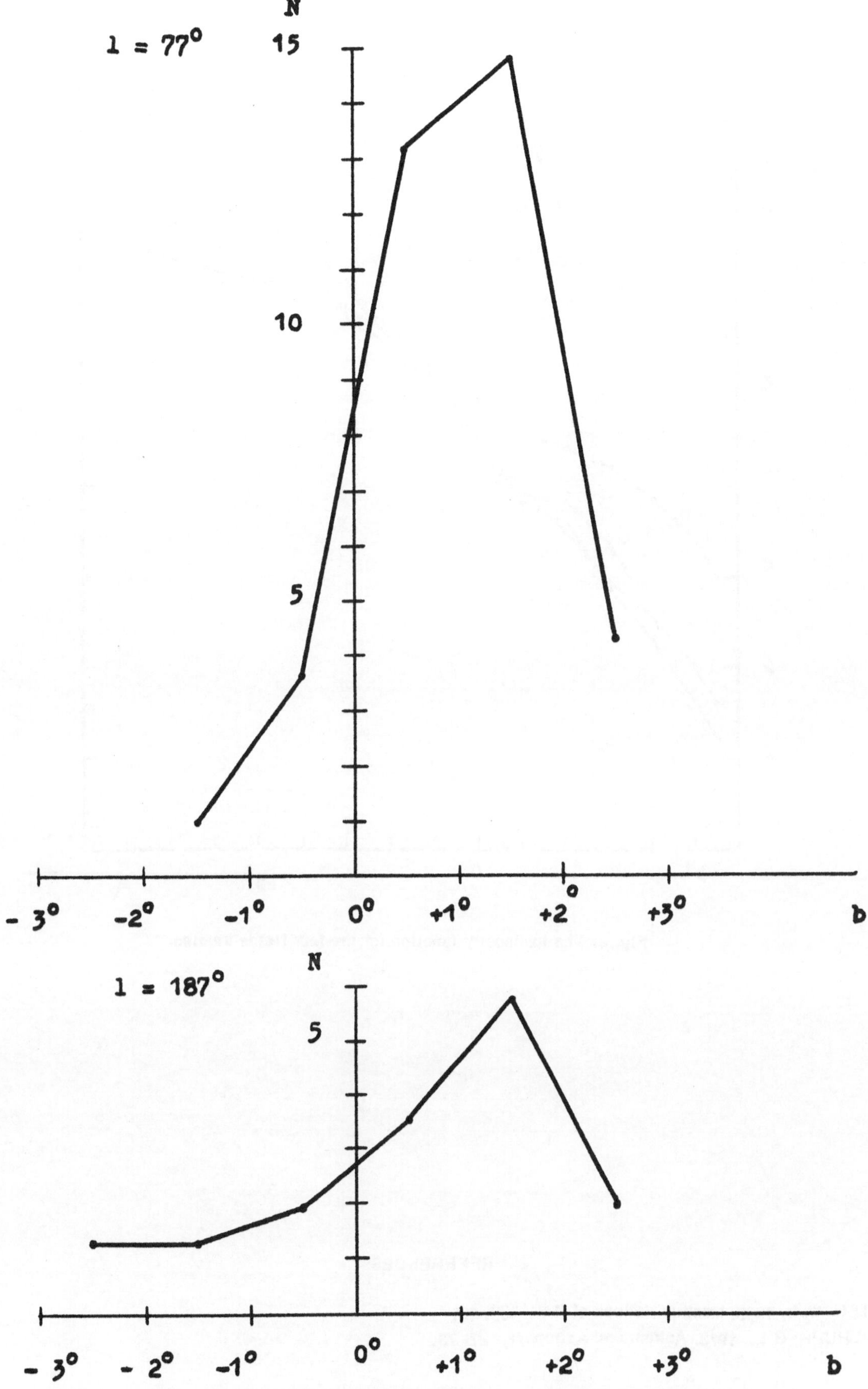

Fig. 3. Distribution of the O - B2 stars with respect to the galactic plane.

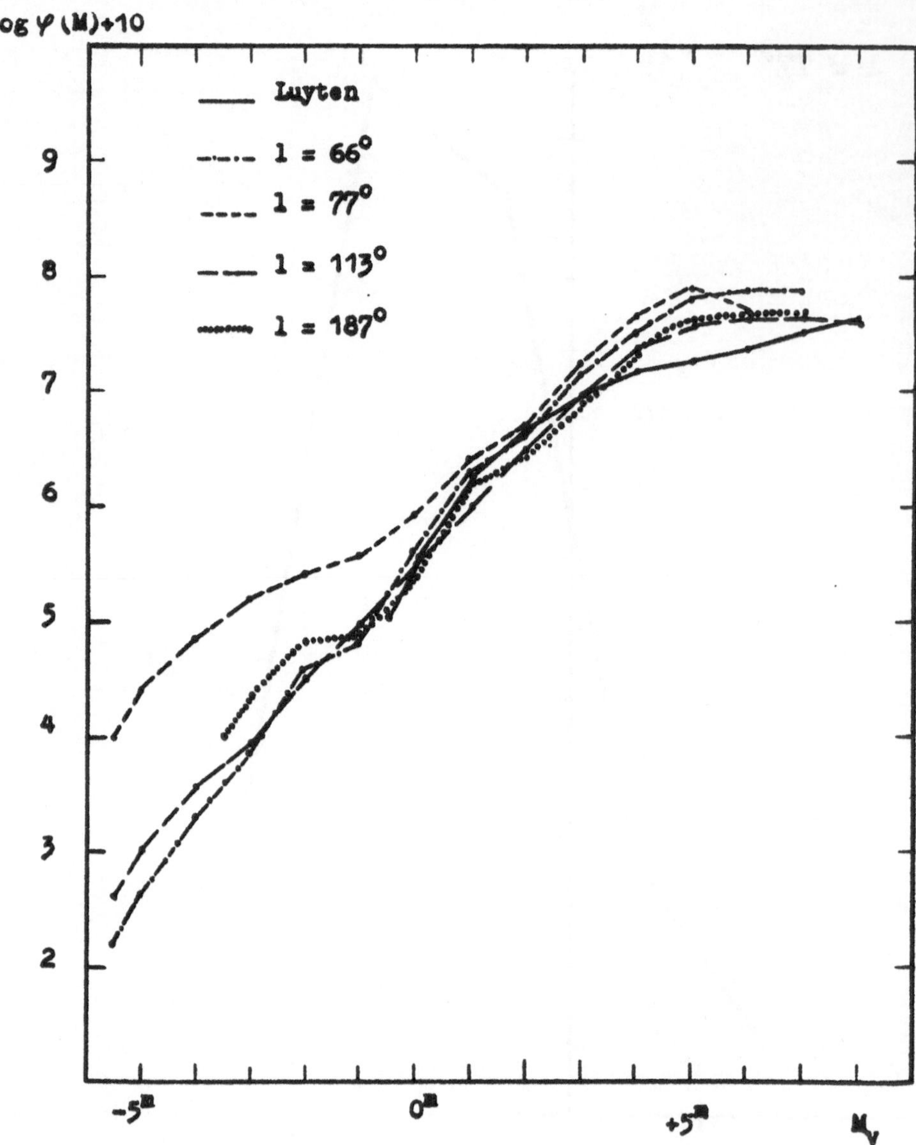

Fig. 4. The luminosity function for the four fields studied.

REFERENCES

LUYTEN W.J., 1968. Monthly Notices of RAS 139, 221
VERSCHUUR G.L., 1973. Astron. and Astrophys. 27, 73

MISSING MASS IN GALAXIES

F.BERTOLA AND G.DI TULLIO
ASIAGO ASTROPHYSICAL OBSERVATORY, UNIVERSITY OF PADOVA, ITALY
PRESENTED BY F.BERTOLA

СКРЫТАЯ МАССА В ГАЛАКТИКАХ

РЕЗЮМЕ

Обсуждаются некоторые аспекты проблемы скрытой массы. Наблюдения диффузных источников X-лучей, инфракрасных гало вокруг спиральных галактик и обширных гало, окружающих эллиптические галактики, заставляют считать преуменьшенными прежние определения массы и средней плотности Вселенной. Однако, они недостаточны для объяснения известных расхождений, обнаруженных при определении масс скоплений на основе теоремы о вириале.

ABSTRACT

Some observational aspects of the problem of the missing mass are discussed. Particularly, observations of the X-ray diffuse sources, infrared halos around spirals and extended halos around ellipticals, lead to an increase of the previous mass determination and of the mean density of the Universe. However this is not sufficient to explain the well known discrepancy found in determining the virial mass of clusters.

The problem of the missing mass in clusters of galaxies, groups of galaxies and in galaxies or, if we want to consider the problem the other way around, the possible evidence of their instability is clearly expressed in the diagram by Karachentsev (1966). Although derived almost ten years ago it has still a validity in illustrating an overall phenomenon. It shows that the mass to light ratio, when the mass is derived by the virial theorem, increases with the total luminosity (or size) of the cluster. In addition one can see that the value of the mass to light ratio ranges from 100 to 1000, i.e. quite well outside the range established for the single galaxies. Summing up the mass of each galaxy in the clusters one finds a discrepancy with the mass derived from the virial theorem, in the sense that the latter results are higher. There are two possibilities of explaining this discrepancy. Around the galaxies or spread throughout the cluster, there is a lot of missing mass, which has not yet been detected even with the most modern and sophisticated techniques. The second possibility is that we are not allowed to make use of the virial theorem, which can be applied only to bound systems, since the galaxies, the systems of galaxies and the universe all are characterized by different degrees of instability, as proposed several years ago by V.A.Ambartsumian(1958). The above mentioned discrepancy has been reduced in the last years by several times. For instance in the Coma Cluster it has decreased from a factor of several hundreds (Zwicky 1959) to a factor eight (Rood et al. 1972) without introducing additional mass in some previously unknown form, by just using the classical procedure with improved observational data.

Among the factors which can contribute to explain the above discrepancy special attention has to be paid to the intergalactic medium, indicated by the diffuse X-ray sources in clusters of galaxies, if they are interpreted in terms of thermal bremsstrahlung from hot plasma. In fact the X-ray spectrum of Coma does fit well a thermal bremsstrahlung emission at $T=10^8$K; in Perseus a more concentrate emission from the very active NGC 1275 may superimpose on a thermal bremsstrahlung diffuse emission. According to very recent computation by Cavaliere (1975, private communication), the intergalactic medium can contribute to a large fraction of the mass of a cluster. Proper considerations of the equilibrium of the hot gas in a cluster potential well, as traced by galaxy counts, lead to gas distribution depending on the parameter $<V^2_{rad}>/2.KTm$. Independently observed values of this ratio produce a remarkable fit of the X-ray scans out to 120' in Coma. These models by Cavaliere entail somewhat more mass hidden in the cluster than derived in the previous models. The mass of the gas results in the order of 150% - 200% times the mass of the galaxy content still short by a factor five of the virial mass.

The second important point related to the problem of the missing mass concerns the fact, which has been recently pointed out by some authors, that we have probably underestimated the masses of the single galaxies.

Ostriker, Peebles and Yahil (1974) stressed the point that the spiral galaxies have been underestimated by a factor 10 or more. In addition Ostriker and Peebles (1973) suggested, on pure theoretical considerations based on stability criteria of a cold galactic disk, that extended massive halos are present around spirals. Other evidences of halos around galaxies have been discussed by Einasto, Kaasik and Saar (1974). On the contrary in a recent paper G.Burbidge (1975) tries to show that there is no unambigous dynamical evidence which demonstrates that galaxies have very massive halos.

The above discussions have stimulated new observations. In order to check the presence of halos around spiral galaxies, H.Spinrad (1975, private communication), has observed photographically and photoelectrically low surface brightness regions rather far from the disk of edge-on galaxies, especially in the red and infrared part of the spectrum. Spinrad finds a halo around NGC 4565 which looks like an E4 or E5 galaxy, rounder than the disk of the system, and with a color resembling that of M stars. A similar large elliptical halo, red in colour,is found also in NGC 253. On the assumption of $M/L = 100$ Spinrad estimates that the total mass increment would be a factor 2 or 3.

A different approach in order to derive how massive is the halo of our own galaxy has been attempted by M.Schmidt (1975) using the luminosity function of a complete sample of high velocity stars. The conclusion of Schmidt is quite definite: the local mass density of halo stars is about 1.7×10^{-4} solar masses per cubic parsec, or an order of magnitude lower than that corresponding to the massive halo proposed by Ostriker and Peebles.

The fact that the elliptical galaxies possess very extended faint envelopes, up to 0.2 Mpc in diameter, is a well established phenomenon, which has been discovered by Arp and Bertola (1969, 1971) and confirmed by de Vaucouleurs (1970), who (1972) estimated that the B light emitted by M 87 is factor 4 larger than prior to the discovery of the halo.

By means of the virial theorem, the mass of M 87 has been estimated by Bertola and Capaccioli (1970) $3.10^{13} M_{\odot}$, an order of magnitude larger than previous determinations. The conclusions by Arp and Bertola (1971) were that all ellipticals, regardless of their size,tend to increase in diameter with the same rate when passing from a standard isophote to the limiting one. The same phenomenon was not present in the spirals, which seem to be sharply bounded. The latter statement has not been confirmed in a successive paper by Kormendy and Bachall (1974), who claim that spiral galaxies are not substantially smaller than ellipticals of the same luminosity.

We present here some results of an investigation, which should be considered a continuation of the work by Arp and Bertola (1971), on the Virgo and Coma clusters and on two edge on spirals.

T h e V i r g o C l u s t e r. One 10 x10 inch 111a-J plate (plus Wr 4 filter) exposed 140^m, taken by one of us (F.B.) in April 1970 with the Palomar 48 inch Schmidt telescope, centered on M87, has been traced with the Padova Joyce-Loebl Isodencitraced (scanning spot 250μ square and density stepsize of 0.01). We estimated that using the above plate-filter combination and exposure time, the limiting B magnitude of about 27 mag/□ '' has been reached. We considered the galaxies with $m_p \leqslant 14$ in the Catalogue by Zwicky et al.(1961-68),morphologically classified also in the Reference Catalogue (G. and A. de Vaucouleurs, 1964). Ten ellipticals, fourteen lenticulars and twenty spirals were traced and the maximum detectable diameter was measured. In addition we measured for all the galaxies an inner diameter, related to a fixed density level in our tracings, which corresponds to the B surface brightness 24 mag/□ '', after a calibration with Liller's (1960) photometry.

From our analysis we can establish the following results:

a) the giant elliptical galaxies are characterized by very large size, as already established in the previous paper (Arp and Bertola, 1971). On the plate used in this investigation M 87 extends 42' while NGC 4406 extends 29'.

b) the diameters of giant spirals never reach that of the giant ellipticals. We got 13' for NGC 4569 and 12' for NGC 4501.

c) the increase of the outermost dimensions in the ellipticals with respect to the inner diameter at 24 mag/□ '' level is not dependent on their luminosity or size. The corresponding increment for spirals is much less, leading to the conclusion that they tend to be sharply bounded. Lenticular galaxies behave in an intermediate way. All this is well illustrated in Fig.1 where for each galaxy, represented by different symbols according to the morphological type, the magnitude from Zwicky et al. (1961-68) is plotted against the ratio of the outermost to inner diameter measured. The right hand side is populated by ellipticals, while in the opposite side the spirals are clumped together.

T h e C o m a C l u s t e r. We had available two deep 48-inch Schmidt plates taken under the same conditions that the previous Virgo plate. We traced all the galaxies brighter than $m_p = 15$ (Zwicky et al., 1961-68). The morphological type is mainly from the Reference Catalog. The inner diameter measured is estimated from the de Vaucouleurs' (1970) photometry to be at the level of 25 B mag/□". In Fig.2, which is similar to Fig. 1, we plotted points representative of 18 ellipticals, 4 lenticulars and 5 spirals. It has to be noted that the range in luminosity covered in Coma is much less extended than in Virgo. Fig.2 shows that also in Coma elliptical galaxies behave differently from spirals as far as the relative increase of diameter is concerned. Although the inner diameter is at a lower surface brightness level than in Virgo, we notice a tendency toward higher values. This would indicate for the Coma galaxies a more pronounced tendency to larger extensions when possible observational errors are completely ruled out. We intend to go further on this point with new observational material.

Another point we would like to mention briefly here and which is related to the problem of the missing mass, concerns the optical presence of intergalactic matter in Coma. The appearance in the tracings of some features on the SW side of the two bright central galaxies has been interpreted by Sastry and Welch (1972) as due to intergalactic matter. The result has been confirmed also by Kormendy and Bahcall (1974). Similar features are presented also in our tracings, but it is our opinion that they might be mainly explained as the effect of the overlapping of the faint envelopes belonging to a concentration of faint galaxies on the SW side. Therefore we believe that there is still no clear evidence of optical intergalactic matter in clusters.

N G C 891 - N G C 4565. The best way to detect faint halos around spiral galaxies is to look at edge on systems.
We had available from Dr.Arp. two deep 111a-J plates of NGC 891 and NGC 4565, whose tracings are reproduced in Fig.3 and 4. In table 1 we report our measurements of the major and minor axes together with the Holmberg's (1958) values.

Table 1

Comparison with Holmberg's (1958) diameters for NGC 891 and NGC 4565

	NGC 891		NGC 4565	
	Holmberg	Present paper	Holmberg	Present paper
major axis	15'	16'	20'	18'
minor axis	3'8	7'	3'6	5'6

It is very interesting to see that, while the dimensions of the major axis are not getting bigger on the deep plates, the extension of the minor axis is almost twice as much for NGC 891 and one and half time for NGC 4565. This fact is in agreement with the above findings in cluster. The disk tends to be sharply bounded, with the spheroidal component, which photometrically behaves like the elliptical galaxies, increases in diameter. From our material it is difficult to estimate the amount of light added to the spheroidal component, by using the 111a-J technique. It is our opinion that such light, even if characterized by high M/L, has difficulty in accounting for massive halos around spirals.

C o n c l u d i n g R e m a r k s. There is one point concerning the mass discrepancy as derived by the virial theorem and the mass resulting from the sum of single galaxies,which is quite clear. Such a discrepancy has decreased from the large amounts of several years ago to the present small values. This reduction is mainly due to the improved observational data available. Nevertheless the discrepancy (a factor 8 in Coma according to Rood et al (1972))is still large. Hidden material is certainly present in the clusters and the diffuse X-ray sources are an indication of it.

There is a tendency, we have mentioned above in criticizing the previous mass determinations, in the sense that they were underestimated. The final aim of this tendency is a cosmological one, i.e. an approach to get close to the critical density in the universe.

There are new observations which add some material to that previously known. Our results show that elliptical galaxies, regardless of their size, tend to increase their luminosity on deeper plates. This statement is true only for the ellipticals and does not hold so conspicuously for spirals. Spinrad has found some evidence of infrared halos around edge-on spiral galaxies. He estimates a total mass increment of a factor of 2 or 3. In any case we are not close to reduce completely the discrepancy.

At this point we would still come back to Karachentsev diagram and his interpretation saying that the possibility of instability in systems, from binary upward, has not been ruled out by the continuous search for the missing mass.

A c k n o w l e d g e m e n t We are grateful to Dr.H.Arp for kindly making available his plates of NGC 891 and NGC 4565.

The plates of the Virgo and Coma Clusters were taken by one of us (F.B.) during a period of guest investigator at the Hale Observatories, for which privilege thanks are due to Dr. Babcock.

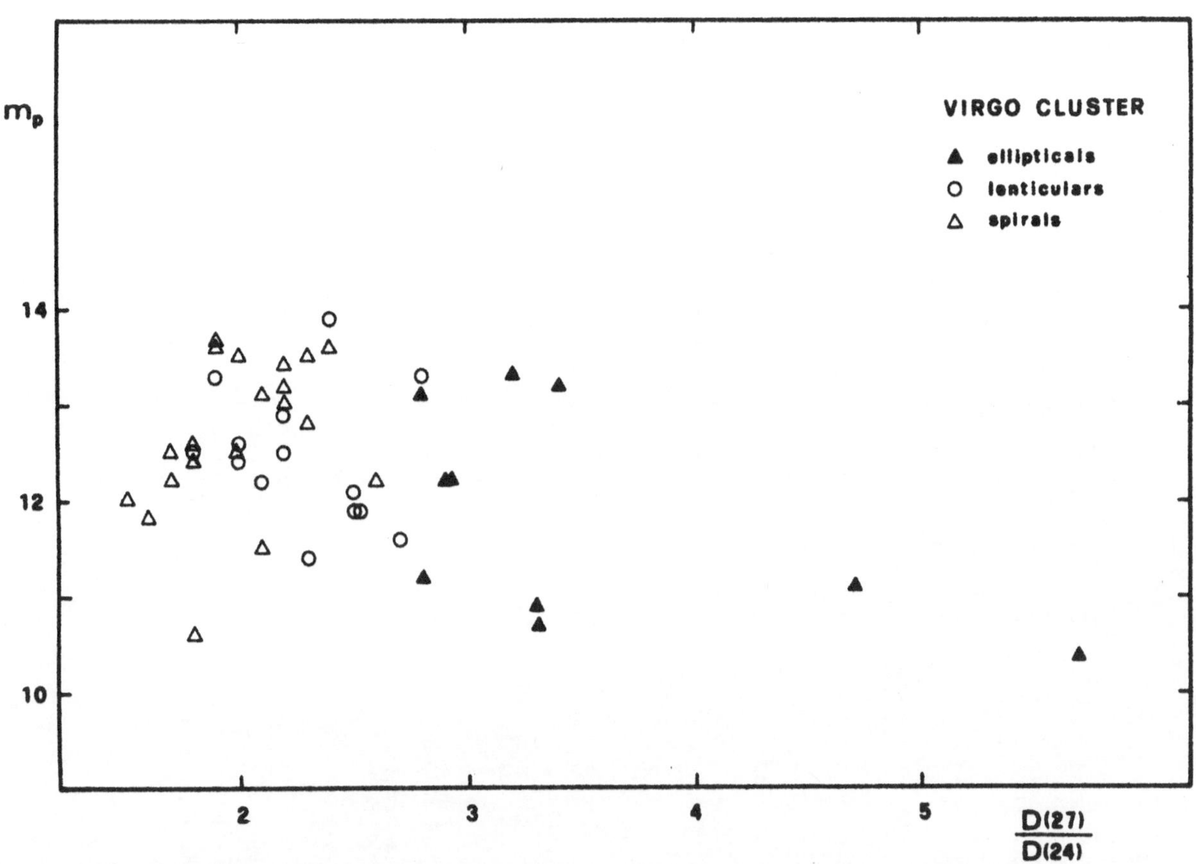

Fig. 1. Ratio of the outermost measured diameter, estimated at a level of surface brightness of about 27 Bmag/ " to the diameter at 24 B mag/ ", as a function of the apparent magnitude.

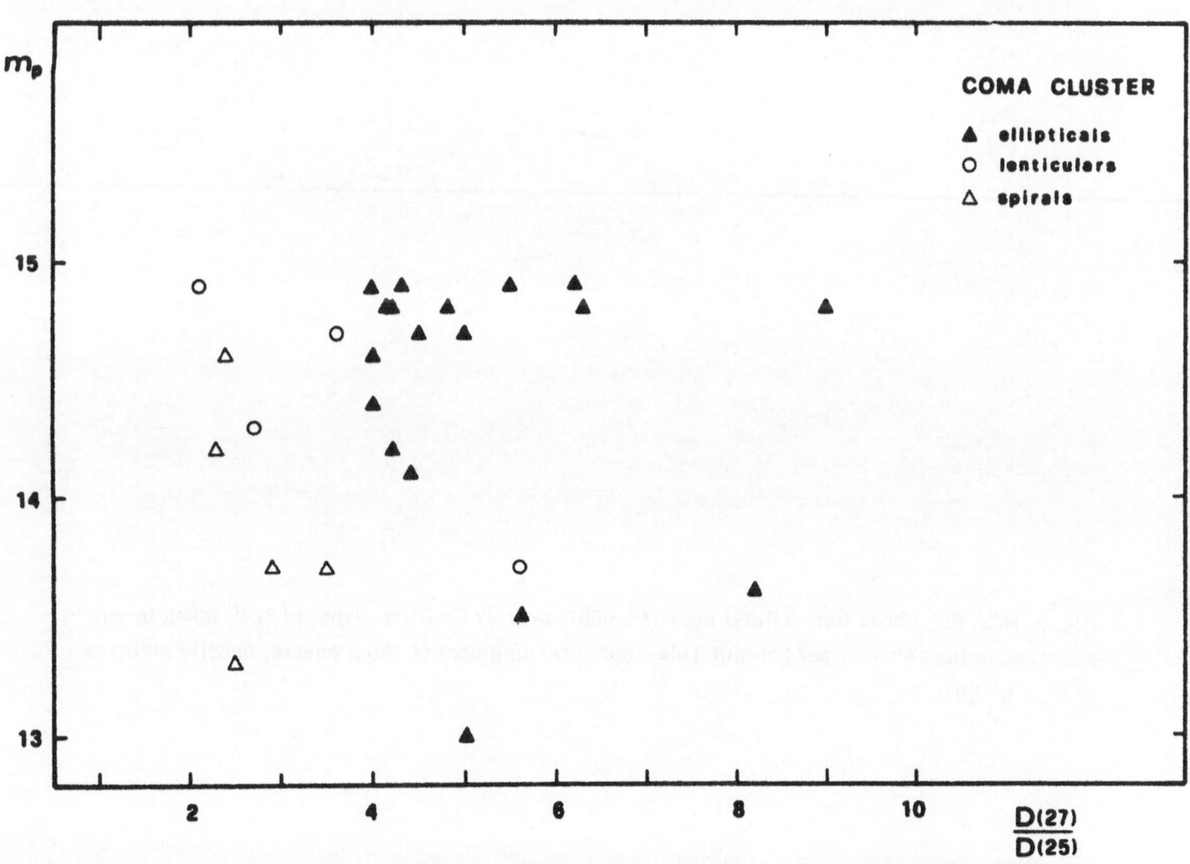

Fig. 2. Ratio of the outermost measured diameter, estimated at a level of surface brightness of about 27 B mag/ " to the diameter at 25 Bmag/ ", as a function of the apparent magnitude.

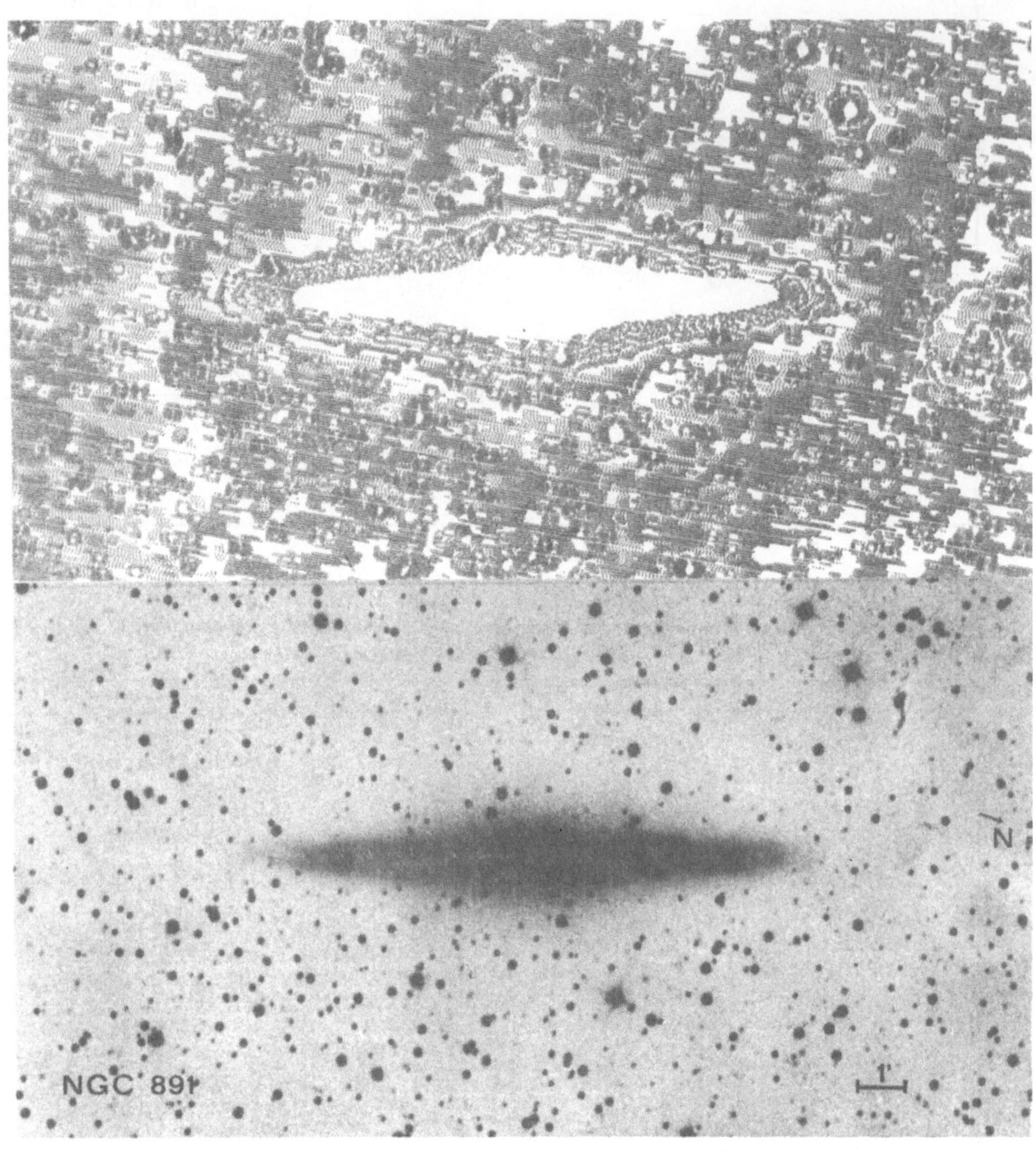

Fig. 3. NGC 891 traced from a IIIa-J plate (PS 9607) plus Wr 2c filter, exposed 60^m, taken by Arp with the 48" Palomar Schmidt Telescope. Scanning spot of 250 μ square, density stepsize of 0.01.

Fig. 4. NGC 4565 traced from a IIIa-J plate (PS 8861) plus Wr 2c filter, exposed 70^m, taken by Arp with the 48" Palomar Schmidt Telescope. Scanning spot of 250 μ square, density stepsize of 0.01.

REFERENCES

AMBARTSUMIAN V.A. 1958, La Structure et l'evolution de l'univers. Solvay Conference (Brussels: R.Stoops). p.241.

ARP H.C. AND BERTOLA F. 1969, Astrophys. Letters. 4, 23.

ARP H.C. AND BERTOLA F. 1971, Astrophys. J. 163, 195.

BERTOLA F. AND CAPACCIOLI M. 1970, Mem. S.A. It. vol. XLI, 57.

BURBIDGE G. 1975, Astrophys. J. (Letters). 196, L7.

EINASTO J., KAASIK A. and SAAR E. 1974, Nature. 250, 309.

HOLMBERG E. 1958. Medd. Lunds Astr. Obs. Ser. 11, No. 136.

KARACHENTSEV I.D. 1966, Astrofizika. 2, 81.

KORMENDY J. and BAHCALL J.N. 1974, Astron. J. 79, 671.

LILLER M.H. 1960, Astrophys. J. 132, 306.

OSTRIKER J.P. and PEEBLES P.J.E. 1973, Astrophys. J. 186, 467.

OSTRIKER J.P., PEEBLES P.J.E. and YAHIL A. 1974, Astrophys. J.(Letters). 193, 1

ROOD H.J., PAGE T.L., KINTER E.C. and KING I.R. 1972, Astrophys. J. 175, 627.

SCHMIDT M. 1975, preprint.

VAUCOULEURS G. de and de VAUCOULEURS A. de 1964, Reference Catalogue of Bright Galaxies.

VAUCOULEURS G. de 1970, In Stars and Stellar System. vol 9.

VAUCOULEURS G. de 1972, Astrophys. Letters. 10, 145.

WELCH G.A. and SASTRY G.N. 1971, Astrophys. J. (Letters). 169, L3.

ZWICKY F. 1959, Hanb. Phys. 53, 590.

ZWICKY F., et al. 1961-1968. Catalogue of Galaxies and of Clusters of Galaxies.

DISCUSSION

KOMBERG : Peebles and Ostriker suggested that the disk of a flat galaxy does not develop a bar if it has a massive corona and that barred spiral galaxies do not have coronas. An attempt to verify this statement by the use of the mass-luminosity ratic for isolated pairs in the Karachentsev catalogue failed, since there was only one pair in which both components were barred spirals. Have you any information about the luminosity of the faint external regions of barred spirals?

BERTOLA : I did not observe barred spirals in particular. Your suggestion is interesting and I think that barred spirals deserve more attention.

J.HEIDMANN: Is the brightness contour for the spirals sharper than corresponding to the exponential law?

BERTOLA : My results are not based on quantitative photometry, so I cannot answer this question properly.

THE MISSING MASS AROUND GALAXIES

J. EINASTO, M. JÔEVEER, A. KAASIK AND J. VENNIK

TARTU INSTITUTE OF ASTROPHYSICS AND ATMOSPHERE PHYSICS, U.S.S.R.
PRESENTED BY EINASTO

СКРЫТАЯ МАССА ВОКРУГ ГАЛАКТИК

РЕЗЮМЕ

Исходя из устойчивости систем галактик (пар, групп, скоплений) и основываясь на радионаблюдениях нейтрального водорода ряда галактик и теоретических моделях, делается вывод о наличии обширных массивных корон у гигантских галактик. Массы таких корон могут на порядок превосходить массы оптически наблюдаемых галактик. Физически реальные карликовые спутники, окружающие гигантские галактики погружены в упомянутые короны и образуют с основными галактиками единые динамические системы, называемые гипергалактиками. Морфологические особенности строения карликовых спутников объясняются их движениями в короне в зависимости от близости к основной галактике. Рассматриваются наблюдательные и теоретические аргументы в пользу высказанных предположений.

1. Introduction. About 40 years ago Zwicky (1933) discovered mass discrepancy in clusters of galaxies. The mass of a cluster per galaxy, calculated from the velocities of galaxies, is by 1 to 2 orders of magnitudes greater than the mean mass of galaxies derived from their inner motions. Karachentsev (1966) and Page (1966) have shown that mass discrepancy occurs in double galaxies as well as in groups of galaxies.

In recent years evidence has been accumulated that mass discrepancy is present also in single giant galaxies (Einasto 1972, Einasto *et al.* 1974 a, b, c, Ostriker and Peebles 1973, Ostriker *et al.* 1974, Ozernoy 1974 a, b).

It is well known that giant galaxies, like our own Galaxy, are surrounded by a cloud of dwarf companion galaxies. Both our Galaxy and the Andromeda galaxy have 10 to 12 companions, other nearby giant galaxies also have on an average about 10 companions each. The luminosities and masses of the companion galaxies are much smaller than the luminosity and mass of the central giant galaxy, therefore the visible mass is strongly concentrated towards the centre of the systems of galaxies considered. If no other masses are present in the system, the circular velocity should follow the Keplerian law, as it is the case in the Solar system where mass is also strongly concentrated towards the centre. Radio observations of hydrogen clouds on the periphery of galaxies as well as the dynamics of companion galaxies indicate that the circular velocity is approximately constant throughout the whole system up to a distance of ~ 1 Mpc from the central galaxy (Fig. 1).

The constancy of the circular velocity can be accounted for if we assume that the giant galaxies are surrounded by huge coronas made up of the hidden or missing mass, which penetrate the whole systems of companion galaxies. The mass of these invisible coronas is about 10 times the mass of the visible stellar populations in galaxies.

The concept of the presence of massive invisible coronas around giant galaxies has been suggested on the basis of dynamical arguments. These dynamical arguments are based on two assumptions: (a) that the systems of galaxies studied are physical systems and (b) that they are bound (Einasto 1974, Burbidge 1975). Principal objections against the missing mass concept are connected with the questioning of these assumptions (Burbidge 1975).

When discussing the missing mass problem in galaxies we shall first study the validity of these two assumptions. To check the validity of the first assumption we compare two alternative hypotheses: that the systems of galaxies studied are either physical systems or that they represent random samples of field galaxies. We have come to the conclusion that a variety of existing morphological, dynamical and structural data strongly support the first alternative. There exists also observational evidence that the systems of galaxies studied are bound systems. The dynamical data indicate that in most of these systems of galaxies there exists a discrepancy between the virial-theorem masses and the conventional masses of visible galaxies.

II. Physical and dynamical properties of the systems of companion galaxies. If the galaxies located in the vicinity of giant galaxies are not their physical companions but constitute random samples of field galaxies, then the statistical properties of these galaxies must correspond to the mean properties of field galaxies.

According to the general field hypothesis, the spatial density of galaxies should be approximately the same everywhere. This, however, is not the case. The systems of the companion galaxies surrounding giant galaxies possess characteristic radii of about 1 Mpc. The spatial density of galaxies in these systems is by 1 to 2 orders of magnitude greater than the mean density of galaxies. The space between these systems is almost void of galaxies. According to de Vaucouleurs (1971), about 90% of galaxies are members of various systems of galaxies. This estimate is a conservative one, the actual percentage of isolated field galaxies seems to be even smaller.

The general luminosity function of galaxies is a decreasing one: the number of galaxies per unit magnitude is the smaller the higher is the luminosity of the galaxy. Hence, according to the general field hypothesis, the number of giant galaxies in any given volume should be smaller than the number of dwarf ones. This, however, is not the case. Bright companions are much more closely concentrated round the central giant galaxy than the companions of lower brightness (Einasto *et al.* 1974a).

The galaxies have a certain distribution of their morphological types. According to the general field hypothesis, this distribution should be the same for any given volume. This is not the case. Elliptical and non-elliptical companions of giant galaxies are separated from each other: elliptical companions are strongly concentrated round the central galaxy, non – elliptical companions populate the peripheral regions of satellite systems. This separation was predicted theoretically by Chernin and discovered by the Tartu team (Einasto *et al.*1974c).

According to the general field hypothesis, the mean luminosity of the companion galaxies should be constant.This is not the case.The mean luminosity of the companion galaxies is the higher the more luminous are the central galaxies (Einasto *et al.* 1974a).

And, finally, according to the general field hypothesis, the mean velocity dispersion of the companion galaxies should be the same for all systems. This is not the case either. As seen from Fig. 2 and Table 1, the velocity dispersion is the larger the higher is the luminosity of the central galaxy. This result is based on a study of all available kinematical data. In the next section we shall discuss these kinematical data in details.

During this study we have re-analysed all principal lists of groups of galaxies and pairs of galaxies (de Vaucouleurs 1966, Karachentsev 1966, 1970). We have come to the conclusion that in many cases the listed groups of galaxies contain either foreground or background objects or consist of several subcondensations. For this reason in classical groups the properties mentioned above are only weakly expressed. But after a careful re-arrangement of galaxies we were able to separate systems of galaxies where the above-mentioned properties are clearly expressed. In order to avoid confusion with classical groups of galaxies, defined as density enhancements in space (de Vaucouleurs 1966), and following a proposal by Chernin, we call systems of galaxies with these properties *hypergalaxies*.

To check our sample of galaxies in hypergalaxies for possible foreground or background objects, we derived the distribution of relative velocities ΔV_r of galaxies with large absolute values $|\Delta V_r|$. This method was used earlier by Fesenko (1975) to separate possible background galaxies. The distribution shown in Fig. 3 is rather symmetrical, and we conclude that our sample is practically free of foreground and background objects.

We shall now consider the second assumption that the systems of galaxies studied are bound. This assumption is confirmed by the following three observational arguments.

Firstly, the expansion ages of clusters and rich groups of galaxies are of the order of 10^9 years or smaller, the ages of individual galaxies derived from the nucleochronology and by other methods are of the order of 10^{10} years.

Secondly, as demonstrated by Einasto et *al.* (1974 a) and Ambartsumian *et al.* (1974),compact clusters of compact galaxies have all the properties of the nuclei of hypergalaxies. In particular,the mean harmonic radii of compact clusters of compact galaxies are practically identical with the corresponding radii of the nuclei of the nearby hypergalaxies. All compact clusters of compact galaxies studied so far are located at a large distance of ca 2 x 10^9 lightyears (Mirzoyan *et al.* 1975). If these clusters were expanding objects with an expanding age of 10^9 years, then they must have been formed about 3 x 10^9 years ago in contrast with the nearby hypergalaxies, which must have been formed not earlier than 10^9 years ago. This geocentric conclusion from the expansion hypothesis is apparently unrealistic.

Absence of expansion can be seen directly in the case of the Local Group (Einasto 1974)- about half of the satellites of our Galaxy have *negative* radial velocities.

Statistical arguments against the expansion hypothesis were given by Ozernoy (1974 a,b).

III. The missing mass around galaxies. In this section we shall discuss the possible presence of the missing mass around galaxies. The arguments given in the previous section clearly indicate that the dynamics of companion galaxies can be used for determining the masses of the corresponding systems.

First of all, we note that there exist no separate missing mass problems in galaxies on the one hand and in systems of galaxies on the other. Companions surround practically all giant galaxies and the invisible coronas found to be existing around these giant galaxies belong, in fact, to the whole system (Einasto *et al.* 1974a). In this respect the parent giant galaxies with their coronas can be considered as inseparable constituents of the corresponding hypergalaxies.

The most powerful argument demonstrating the dynamical link between the parent galaxies and their companions comes from the comparison of their kinematics. As demonstrated by Burbidge (1975), the scale of internal velocities in galaxies and relative velocities between companion galaxies is practically identical. A more detailed study (Einasto *et al.* 1975 b, c.) demonstrates that the equality of internal velocities in galaxies with relative velocities of companion galaxies is not a statistical property of mean motions but is valid in all the particular indiv dual systems studied so far. The relative velocities of galaxies in groups and clusters of galaxies are the larger the higher is the inner velocity dispersion of stars in the respective parent galaxies (Fig. 4). This remarkable dynamical property of parent galaxies and of their companion galaxies has far-reaching cosmogonic consequences (Einasto *et al.* 1975b). In the present context we use this property to study the mass distribution around giant galaxies.

The possible presence of massive coronas or halos around giant spiral galaxies was emphasized by Sizikov (1969) and Einasto (1972) on the basis of flat rotational curves of galaxies. Detailed model calculations showed that visible stellar populations with the mass-to-luminosity ratio calibrated dynamically cannot represent observed rotational curves: the calculated rotational curves have much steeper radial gradients than the observed ones (Fig. 5). A formal agreement of calculated rotational curves with observations was established by Emerson and Baldwin (1973) for M31. However, in the corresponding model the mass-to-luminosity ratio for the disk of M31 is unbelievably high (approximately 20) in comparison with the ratio for the bulge (about 5). On the other hand, Roberts (1975) has shown that in M31 as well as in some other galaxies (M101, IC 342) the rotational velocity is not a decreasing function of the distance but remains constant on the periphery or even increases with the increasing distance. This behaviour cannot be represented by any combination of mass-to-luminosity ratios of visible stellar populations. In all models which represent well the observed rotational curves an outer shell or a population with a very high mass-to-luminosity ratio is present directly or indirectly (Fig.5). Among models where such an invisible population is present indirectly we note the Schmidt (1965) model of the Galaxy as well as all models based on the Bottlinger circular velocity law (Roberts 1966). In these models the mass density decreases on the periphery according to the power law $\varrho \sim R^{-4}$, whereas the luminosity density decreases much more rapidly, according to the exponential law $\varrho \sim \exp(-R/R_0)$(de Vaucouleurs 1974, Sandage *et al.* 1970). As demonstrated by Einasto (1972) in these "conventional" models about a half of the mass is located in the invisible population. In particular, the mass of visible populations of M31 is about $1.8 \times 10^{11} M_\odot$ the total mass of the "conventional" model of M31 being $3 \times 10^{11} M_\odot$.

In principle the discrepancy between the distribution of the mass and that of the luminosity can be explained if non-circular motions are present in the outer parts of spiral galaxies(Einasto 1969, 1972, Burbidge 1975). Large scale deviations from the circular motion occur indeed in spiral galaxies, as can be seen by a comparison of the rotational velocities of different sides of galaxies. This phenomenon, however, cannot explain the discrepancy between the mass and luminosity distributions. If the motions are non-circular and the circular velocity on the periphery of galaxies follows the Keplerian law, then at large distances from the galaxy the circular velocities, calculated from the relative motions of companion galaxies, should follow the same Keplerian law, too. This is not the case. As seen from Table 1 and Fig.1, the circular velocity remains approximately constant throughout the whole system of companion galaxies up to a mean distance of about 600 kpc. The dynamics of companion galaxies can be accounted for only by the presence of an extensive and massive corona of invisible matter. The estimated masses and mean harmonic radii of the coronas of galaxies of different luminosity and type are indicated in Table 1.

The number of companion galaxies in individual groups and hypergalaxies with known radial velocities is usually small (2-5). Hence the velocity dispersion and the corresponding total virial mass cannot be calculated with sufficient accuracy. We have found no large mass discrepancy (the ratio of the virial mass to the estimated visible mass): it lies between 1 and 30. In this respect our results coincide well with the data given by Tully and Fisher (1975) and by Materne and Tammann (1974, 1975).

Of special interest are the systems of galaxies that have no appreciable mass discrepancy. Examples of such groups are given by Sandage and Tammann (1974) and Materne and Tammann (1974). In this case we have to do with two alternative possibilities: (a) that these groups are really stable without any missing mass or (b) that absence of the mass discrepancy is due to the small number of galaxies with known radial velocities. The second possibility has been discussed in detail by Aarseth and Saslaw (1972). They demonstrated by numerical experiments that the virial masses calculated from "observed" relative radial velocities are, as a rule, about twice smaller than the real masses, if the number of galaxies with known velocities is small (4-8). A look at the data presented by Materne and Tammann (1974, 1975) indicates that in most cases the number of galaxies with known velocities does not exceed 7. We conclude that available data are insufficient to solve this problem and additional accurate velocities are urgently needed.

Additional velocities are available for the NGC 1023 hypergalaxy (Tully, personal communication). New data confirm the conclusion drawn by Materne and Tammann that this system is bound without any hidden matter. We note that this hypergalaxy differs from the others by its very large dimensions: its radius exceeds the radius of normal hypergalaxies approximately twice.

Materne and Tammann (1974) include the Local Group also among bound groups without any hidden mass. This conclusion has been drawn on the basis of a study of the Local Group by Herbst (1969). In this study an unrealistically low absolute value of 267 km s^{-1} was used for the heliocentric radial velocity of M31. At the mean value of optical and radio data -301 ± 3 km s^{-1} (Rubin and D'Odorico 1969) the Local Group of Galaxies without missing mass is definitely unstable, as seen from Table 2. For a number of high velocity objects in the Local Group this Table presents their distances from the parent galaxy R and their galactocentric radial velocities V_{GSR}. The reduction of heliocentric velocities V_{LSR} was performed according to the formula $V_{GSR} = V_{LSR} + 225 \sin l^{II} \cos b^{II}$ (Mathewson *et al* 1974). These galactocentric velocities are compared with the escape velocities V_k calculated under the assumption that only visible stellar populations are present in the Local System. According to our new model of the Galaxy (Einasto *et al.* 1975a) we take $1.0 \times 10^{11} M_{\odot}$ for the mass of visible populations of our Galaxy. For the mass of visible populations of the Andromeda galaxy we use the value $1.8 \times 10^{11} M_{\odot}$ quoted above. In most cases the observed velocities greatly exceed the calculated escape velocities (in the case of the Magellanic Stream we have used not the observed maximum galactocentric radial velocity but the total velocity calculated from a dynamical model of the Stream).

As an argument against the presence of large quantities of hidden matter around our Galaxy Burbidge (1975) pointed to the tidal radii of distant elliptical dwarf galaxies, which should correspond to perigalactic distances of ~80 kpc. We have recently shown (Einasto and Jõeveer 1975) that when using the perigalactic distances of 80 kpc (Hodge 1966) the tidal radii of these dwarf ellipticals are not in conflict with the proposed model of our Galaxy with a corona of mass $10^{12} M_{\odot}$ and the mean harmonic radius of 100 kpc. According to this model the mass effective at the distance of 80 kpc is just $2 \times 10^{11} M_{\odot}$, which is in good agreement with the estimate of Burbidge (1975).

Concluding this report, we note that there exists good dynamical evidence that most giant galaxies are like our own Galaxy, surrounded by massive invisible coronas. Apparently there exist galaxies like NGC 1023, which have no such coronas. The data available at present are insufficient to make a final decision on all particular cases and new accurate radial velocities are needed, especially for distant members of physical groups of galaxies.

A c k n o w l e d g e m e n t s. We are deeply grateful to Dr. B. Tully for giving us new radial velocities for the NGC 1023 group prior to publication, and to Prof. I.S.Shklovsky, Dr. A.G. Tammann and Dr. I.D. Karachentsev for fruitful conversations.

Table 1*

Type	⟨R⟩ kpc	⟨σ^2⟩ $(100\ km/s)^2$	n	N	a_0 kpc	⟨M_{VT}⟩ $10^{12} M_\odot$	⟨L_0⟩ $10^{10} L_\odot$	⟨L⟩ $10^{10} L_\odot$	⟨M_{VT}⟩/⟨L⟩
S V	20	0.60							
	100	0.40							
	600	0.04							
		0.34 ± 0.06	5	1	75	1.1	1.1	1.4	80
S IV	18	1.7 ± 0.4	19						
	150	0.8 0.2	18						
	770	0.5 0.2	19						
		1.0 ± 0.2	56	17	75	3.2	1.4	3.6	90
S III	21	1.6 ± 0.8	21						
	125	1.2 0.4	20						
	750	0.7 0.3	21						
		1.2 ± 0.3	62	32	100	5.1	3.5	6.7	75
S II	21	4.5 ± 2.4	21						
	70	1.8 1.0	17						
	450	2.8 0.7	21						
		3.0 ± 1.0	59	28	110	14.2	8.9	15.9	90
E III	16	3.5 ± 0.9	18						
	150	2.7 1.4	18						
		3.1 ± 0.8	36	20	100	13.2	3.2	6.0	220
E II	14	17.7 ± 4.5	16						
	75	14.9 5.8	16						
	320	8.0 3,0	16						
		13.5 ± 2.7	48	22	110	63	6.5	20	320

* The designations used in Table 1 are as follows: ⟨R⟩ is the mean projected distance from the centre of parent galaxy; ⟨σ^2⟩ denotes the mean square relative velocity of galaxies with respect to the mean velocity with its estimated rms error; n is the number of galaxies used in this determination and N - the number of hypergalaxies of the corresponding type. The types and luminosity classes of hypergalaxies are given according to Einasto et al. (1974a, 1975c), a_0 is the estimated harmonic mean radius of hypergalaxies. M_{VT} is the mean virial-theorem mass derived from the formula $M_{VT} = 4.25 \times 10^{10}\ \sigma^2 a_0$ (Einasto et al. 1974a). In this formula masses are expressed in solar units, velocities in 100 km s^{-1} and mean radii in kpc. ⟨L_0⟩ is the mean blue luminosity of the primary galaxy. ⟨L⟩ the mean blue total luminosity of the hypergalaxy.

Table 2

Primary galaxy	Companion	R kpc	V km s^{-1}	V_k km s^{-1}
Galaxy	Fornax	230	-60	61
	NGC 6822	500	60	60
	NGC 5694	25	-275	186
	Magellanic Stream	17	-310	225
M31	Galaxy	690	-122	61

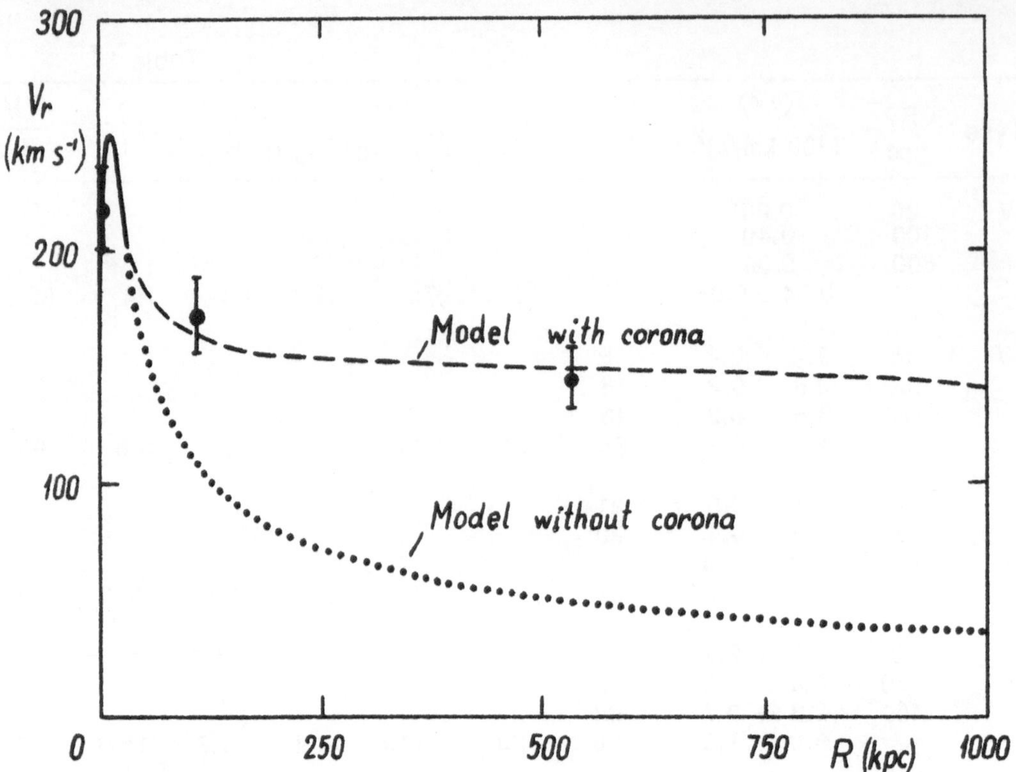

Fig. 1. Circular velocity V_c versus distance R in the hypergalaxy M31. The solid line represents radio observations (Emerson and Baldvin 1973); the dashed line denotes circular velocities calculated from a model with a corona having a mass of $4 \times 10^{12}\ \mathfrak{M}_\odot$, the dotted line signifies the circular velocity calculated from a model without a corona (Kepler's law). The dots represent values of circular velocities calculated from relative radial velocities of 266 galaxies in 120 hypergalaxies (Table 1); the velocity dispersion of hypergalaxies of different types has been reduced to the same mean dispersion 100 km s^{-1}.

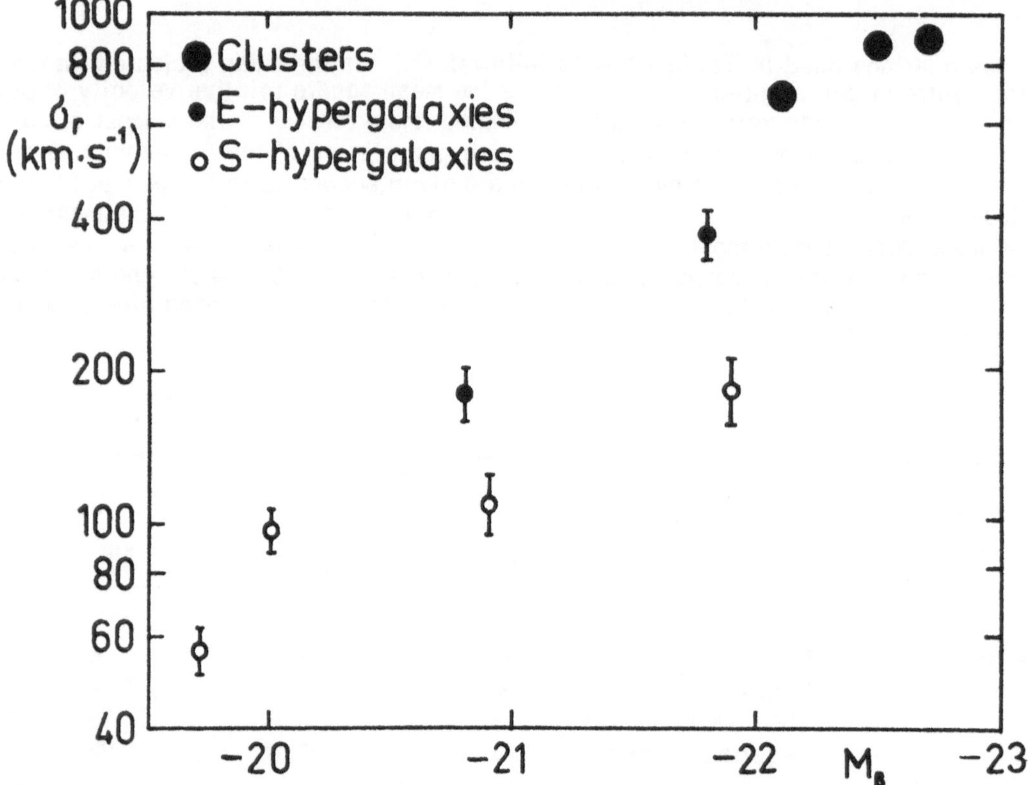

Fig. 2. Velocity dispersion σ_r versus blue absolute magnitude of the parent galaxy M_B for clusters of galaxies and for E- and S-type hypergalaxies. For hypergalaxies the data have been drawn from Table 1, for clusters - from the paper by Einasto et al. (1975b).

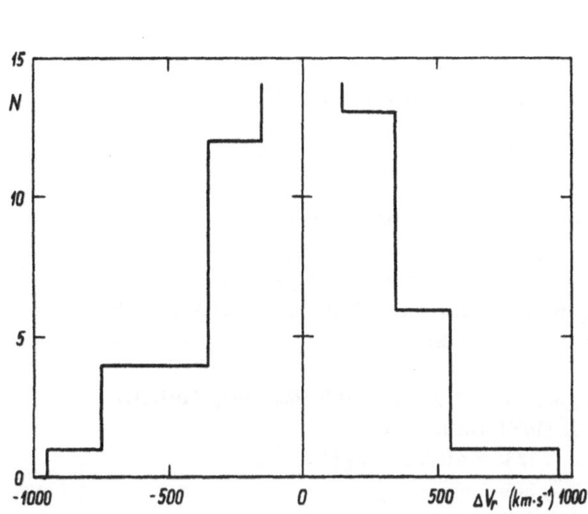

Fig. 3. Distribution of companion galaxies according to relative velocities.

Fig. 4. Dispersion of internal velocities in parent galaxies, σ_{nucl}, versus dispersion of relative velocities of companion galaxies, σ_{comp} (Einasto et al. 1975b).

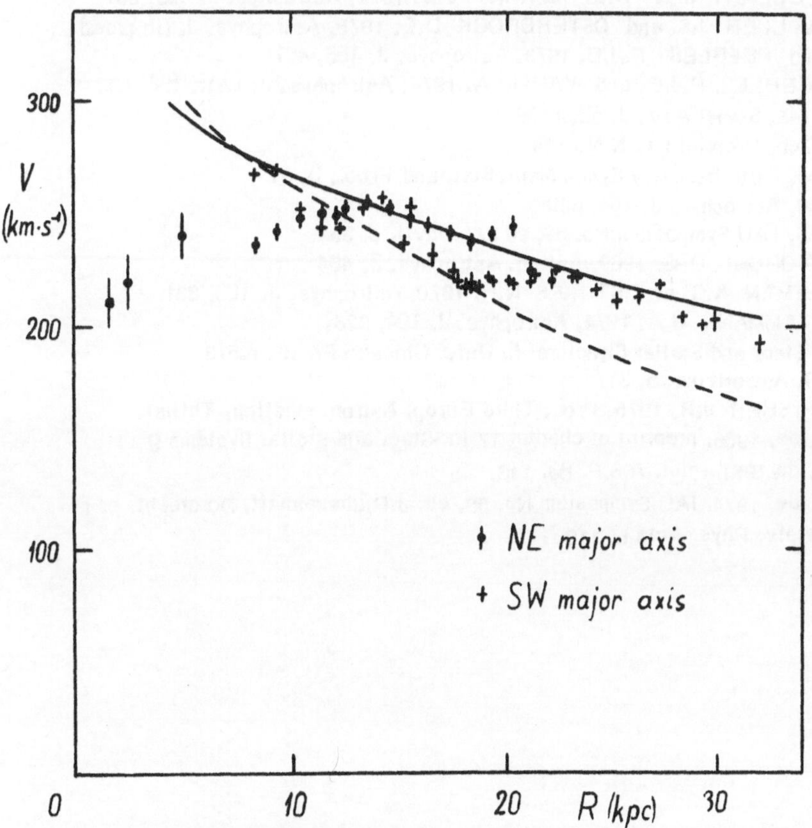

Fig. 5. Rotational curve of M31. Dots and crosses represent radio observations carried out by Emerson and Baldwin (1973), solid line denotes the model with a corona and the dashed line signifies the model without a corona.

REFERENCES

AARSETH S.J. and SASLAW W.C. 1972, Astrophys. J. 172 17

AMBARTSUMIAN V.A., MIRZOYAN L.V., PETROSIAN M.B. and SHAHBASIAN R.K. 1975, Third European Astron. Meeting, Tbilisi.

BURBIDGE G. 1975, Astrophys. J. Lett. 196, L7

EINASTO J. 1969, Astrofizika 5, 137

EINASTO J. 1972, Tartu Astron. Obs. Teated 40 (Proc. First Europ. Astr. Meet. Ed. L.N. Mavrǿdǿs, Springer-Verl. Berlin-Heidelberg-New York, 2, 291, 1974

EINASTO J. 1974, Astr. Circ. USSR No. 840, 1

EINASTO J., JAANISTE J., JÔEVEER M., KAASIK A., KALAMEES P., SAAR E., TAGO E., TRAAT P., VENNIK J., and CHERNIN A.D. 1974a, Tartu Astron. Obs. Teated 48

EINASTO J. and JÔEVEER M. 1975, Pis'ma v Astron. Zh. (in press).

EINASTO J., JÔEVEER M. and KAASIK A. 1975a, Proc. Third European Astr. Meeting, Tbilisi.

EINASTO J., JÔEVEER M., KAASIK A. and VENNIK J. 1975b (in press)

EINASTO J., KAASIK A., KALAMEES P. and VENNIK J. 1975c, Astron. Astrophys. 40, 161

EINASTO J., KAASIK A. and SAAR E. 1974b, Nature, 250, 309.

EINASTO J., SAAR E., KAASIK A. and CHERNIN A.D. 1974c, Nature, 252, 111

EMERSON D.T. and BALDWIN J.E. 1973, Mon. Not. R. Astr. Soc. 165, 9

FESENKO B.I. 1975, Astron. Zh. (in press)

HERBST W. 1969, Publ. A.S.P. 81, 619

HODGE P.W. 1966, Astrophys J. 144, 869

KARACHENTSEV I.D. 1966, Astrofizika 2, 81

KARACHENTSEV I.D. 1970, Problemy kosm. fiz. 5, 201

MATERNE J. and TAMMANN G.A. 1974, Astron. Astrophys. 37, 383

MATERNE J. and TAMMANN G.N. 1975, Proc. Third Europ. Astron. Meeting, Tbilisi.

MATHEWSON D.S., CLEARY M.N AND MURRAY J.D. 1974, Astrophys. J. 190, 291

MIRZOYAN L.V., MILLER J.S. and OSTERBROCK D.E. 1975, Astrophys. J. (in press)

OSTRIKER J.P., and PEEBLES P.J.E. 1973, Astrophys. J. 186, 467

OSTRIKER J.P., PEEBLES, P.J.E. and YAHIL A. 1974, Astrophys. J. Lett. 193, L1

OZERNOY L.M. 1974a, Soviet Astr. J. 52, 1108

OZERNOY L.M. 1974b, Preprint FIAN No 124

PAGE T. 1966, Proc. Fifth Berkeley Symp. Math. Stat. and Prob., 3, 31

ROBERTS M.S. 1966, Astrophys. J. 144, 639

ROBERTS M.S. 1975, IAU Symposium No. 69, ed. A. Hayli, p. 331

RUBIN V.C. and D'ODORICO S. 1969, Astron. Astrophys. 2, 484

SANDAGE A., FREEMAN K.C. and STOKES N.R. 1970, Astrophys. J. 160, 831

SANDAGE A. and TAMMANN G.A. 1974, Astrophys. J. 194, 223.

SCHMIDT M. 1965, Stars and Stellar Structure 5, Univ. Chicago Press, p.513

SIZIKOV V.S. 1969, Astrofizika 5, 317

TULLY R.B. and FISHER J.R. 1975, Proc. Third Europ. Astron. Meeting, Tbilisi.

VAUCOULEURS G. de, 1966, preprint of chapter 17 in -Stars and Stellar Systems 9

VAUCOULEURS G. de, 1971, Publ. A.S.P. 83, 113

VAUCOULEURS G. de, 1974, IAU Symposium No. 58, ed. J.R.Shakeshaft, Dordrecht, p. 1

ZWICKY F. 1933, Helv. Phys. Acta 6, 110

HIDDEN MASS PROBLEM FOR DOUBLE GALAXIES

I.D.KARACHENTSEV
SPECIAL ASTROPHYSICAL OBSERVATORY, ZELENCHUK, U.S.S.R.

ПРОБЛЕМА СКРЫТОЙ МАССЫ В ДВОЙНЫХ ГАЛАКТИКАХ

РЕЗЮМЕ

Представлены результаты определения отношения вириальной массы к светимости f для 134 двойных галактик. Рассмотрено влияние на оценку f фактора проекции и ошибок измерения лучевых скоростей. Вычислен закон распределения отношения вириальной массы к светимости у двойных галактик для трех типов движения: круговых движений, осцилляций и радиальных несю вязанных движений компонентов. Среднее отношение вириальной массы к светимости существенно зависит от типа взаимодействия компонентов пары, от морфологического типа двойных галактик и от степени изолированности пары относителнно соседних галактик. Подчеркнуты трудности объяснения наблюдательных данных с точки зрения существования в парах галактик скрытой массы в виде газа и звездных корон.

ABSTRACT

Results of determination of virial mass-to-luminosity ratio, f, for 134 double galaxies are presented. The effect of the projection factor and radial velocity measurement errors on the estimate of f is considered. A distribution law of virial mass-to-luminosity ratio in double galaxies for three types of motion: circular motions, oscillations, and radial unconnected motions of components is calculated. Mean virial mass-to-luminosity ratio is essentially dependent on the type of interaction between the pair components, the morphological type of double galaxies, and on the degree of a pair isolation relative to neighbouring galaxies. Difficulties in explaining the observational data from the pointof existence in double galaxies of hidden mass in the form of gas and stellar coronas are emphasized.

I n t r o d u c t i o n. As is well known the discrepancy between estimates of mass of a system of galaxies by mutual motions of galaxies and by total luminosity of the system members occurs for different scale systems of galaxies ; pairs, groups, clusters and superclusters. Despite the 40 - year history of this "virial paradox" existence, no satisfactory explanation has been found for it so far. The estimate of the mean density of the universe and the characteristic time scale of evolutionary phenomena in the realm of galaxies depend essentially on one or another solution of the "virial paradox" namely the presence of hidden mass or instability of systems of galaxies. That is why the problem of masses and motions of galaxies in systems has become the central problem of astrophysics for the past years. It may be proved by the "informational outburst" of the number of publications on the problem of hidden mass of galaxies.

It is natural to assume that the reason of discrepancy in mass estimates is common to all systems of galaxies irrespective of their scale. In this case one can readily establish the type of a system of galaxies, where the situation is the most favourable for investigation of the "virial paradox" reason.

In Table 1 we present the basic mean parameters of systems of galaxies which explain the dynamical situation in systems. The examination of these data shows that in view of very large virial densities, double galaxies seem to be the most appropriate for the analysis of different hypotheses concerning the virial problem. Therefore, we shall further focus our attention just upon this simplest type of systems of galaxies.

(1) - type of systems of galaxies, (2) - relative number of galaxies belonging to systems of the present type, (3) - effective radius of system in Mpc, (4) - root mean square radial velocity motions relative to system center in km/s, (5) - virial mass-to-luminosity ratio in solar units, (6) - mean virial density of system in critical density units, (7) - mean expansion time of system in units of reverse value of Hubble's constant, (8)- number of systems with virial mass determination.

Table 1. Mean Parameters of Systems of Galaxies

Type of system	q	< R >I	σ_V	<$\mathfrak{M}$/L>I	< ρ / ρ_c >I	< T/h^{-1}>I	n
1	2	2	4	5	6	7	8
Pairs	0.15	0.03	100	27	2.10^6	0.03	134
Groups	0.35	0.30	250	200	2.10^3	0.10	70
Clusters	0.40	2.0	700	600	2.10^2	0.30	15
Superclusters	0.60	12.0	1000:	1200:	5:	1:	1

G e n e r a l   c h a r a c t e r i s t i c s   o f   d o u b l e   g a l a x i e s. A definition of a pair of galaxies can be performed in a great number of modes. In this procedure, it is desirable that the criterion of a pair should be quantitative, expressed through directly measurable values, and should allow for the condition of isolation of the pair components with respect to surrounding galaxies.

Our further statement we shall base upon the data on double galaxies from the Catalogue (Karachentsev, 1972) composed in the process of examining the Palomar Sky Survey using one of such criteria. The Catalogue includes 603 pairs of galaxies whose components are brighter than $m = 15.7$ in the sky region $\delta > -3°$. More than a half of them show features of interaction between components. We have divided evidence of interaction into three types; "LIN"- linear structures ("tails", "bridges"), "ATM" - presence of "atmosphere" enveloping the components, "DIS" - distortion of the structure of components. Examples of types of interaction in pairs of galaxies are shown in Fig.1.

Comparing the number of double galaxies, 2x603, with the total number of galaxies in the Catalogue of Zwicky et al. (1961-1968), N = 31 350, we have the relative "catalogic" number of double galaxies, $q_c = 0.039$. As it follows from the statistical analysis of distribution of galaxies, the relative number of double galaxies in space is $q = 0.15$ (Karachentsev, 1970). The difference between q_c and q is caused by two reasons; a) due to selection by apparent magnitudes, about a half of pairs will be represented in the Catalogue by only one (bright)component; b) the criterion of isolation we have chosen cuts off half of the physical pairs (mainly broad pairs). Both types of selection may affect essentially the estimate of the mean virial mass of double galaxies.

V i r i a l   m a s s - t o - l u m i n o s i t y   r a t i o. For a pair of galaxies with the measured radial velocity difference between the components, y, and the mutual separation in projection, x, it is possible to calculate the estimate of the total mass

$$\mu = \langle \eta \rangle^{-1} \quad \gamma^{-1} \, x.y^2 , \tag{1}$$

where γ is the gravitational constant, and $\langle \eta \rangle$ is the mean value of the projection factor, which depends on the type of motion of components. The true total mass of a pair

$$\mathfrak{M} = \eta^{-1} \quad \gamma^{-1} \quad x.y_0^2 \tag{2}$$

differs from μ for two reasons: a) we are ignorant of the concrete orientation of the pair,and therefore of the factor η; b) the radial velocity difference between the components, y, is determined with the error, so that

$$y = y_0 + u , \tag{3}$$

where y_0 is the true radial velocity difference, and u is the accidental measurement error.

In conformity with the summary of data on 101 pairs of galaxies with measured radial velocity (Karachentsev, 1974), we have determined estimates of mass and standard confidence intervals [$\mathfrak{M}_*$, $\mathfrak{M}^*$] which contain, with the probability $p = 0.6827$, the value of the true mass for each pair. At any type of motion of components, the projection factor η has distribution with large dispersion and asymmetry. This circumstance in combination with velocity measurement errors do not allow, as a rule, to draw a conclusion for an individual pair, on the agreement or contradiction between the virial and "photometric" masses of the pair, The ratio, $\varphi = \mu / L$, of the measured mass of a pair to its total luminosity may be considered as a random value determined over the sample of pairs of galaxies. By sampled distribution of this quantity, $p_\varphi\{\varphi\}$ it is possible to restore the density distribution of the unknown quantity, $f \equiv \mathfrak{M}/L$, which is the true virial mass-to-luminosity ratio. It is most convenient to determine the distribution $p_f\{f\}$ by the method of moments. From relations (1)-(3) we have

$$\varphi^{1/2} = (\mu/L)^{1/2} = \langle\eta\rangle^{-1/2}\eta^{1/2}(\mathfrak{M}/L)^{1/2} + \gamma^{-1/2}\langle\eta\rangle^{-1/2} x^{1/2} u L^{-1/2} \tag{4}$$

Raising both sides of (4) to the power 2k, and allowing for the evident mutual independence of random values $\{\eta, u, \mathfrak{M}/L\}$, after averaging we obtain

$$\langle\varphi^k\rangle = \sum_{i=0}^{2k} C_{2k}^{i}\, \gamma^{-(2k-i)/2}\langle\eta\rangle^{-k}\langle u^{2k-i}\rangle\langle x^{(2k-i)/2}\cdot\eta^{1/2} f^{i/2} L^{-(2k-i)/2}\rangle \tag{5}$$

$$k = 1, 2, \ldots$$

In particular, since the velocity measurement error is distributed, as a rule, symmetrically, ($\langle u\rangle = 0$), then

$$\langle f\rangle = \langle\varphi\rangle - \gamma^{-1}\langle\eta\rangle^{-1}\langle x\cdot u^2\cdot L^{-1}\rangle \equiv \langle\varphi\rangle - \langle\Delta\varphi_u\rangle. \tag{6}$$

In the general case, determination of the unknown moments, $\langle f^k\rangle$, from the sampled moments, $\langle\varphi^k\rangle$, requires the knowledge of correlation between the dependent random values $\{\mathfrak{M}, L, x, \eta\}$. If radial velocity measurement errors are negligibly small, equation (5) has a very simple form:

$$\langle\varphi^k\rangle = \langle\eta\rangle^{-k}\langle\eta^k\rangle\langle f^k\rangle, \qquad k = 1, 2, \ldots \tag{7}$$

We have considered three types of possible motions of components: 1) circular motions, 2) radial oscillations, 3) radial unconnected motions corresponding to disruption of double galaxies. At the supposition of chaotic orientation of pairs relative to the line of sight, we have the following expressions for the moments of the projection factor (Karachentsev, Shcherbanovsky, 1970)

$$\langle\eta^k\rangle = \frac{\Gamma(k+1/2)\,\Gamma(3k/2+1)}{2\,\Gamma(k+1)\,\Gamma(3k/2+3/2)} \quad \text{- circular motions}$$

$$\langle\eta^k\rangle = \frac{2^{k-1}\Gamma(k/2+1)\,\Gamma(k+1/2)\,\Gamma(k+1/2)}{\Gamma(3k/2+3/2)\,\Gamma(k+2)} \quad \text{- oscillations} \tag{8}$$

$$\langle\eta^k\rangle = \frac{\Gamma(k/2+1)\,\Gamma(k+1/2)}{2\Gamma(3k/2+3/2)} \quad \text{- radial unconnected motions}$$

where $\Gamma(x)$ implies a gamma-function.

Calculation of the first four moments $\langle f^k\rangle$ from (7) allows to find the analytical kind of distribution, $p_f\{f\}$, say, by the method of Pearson diagrams (Hahn, Shapiro, 1967).

O b s e r v a t i o n a l   d a t a   a n a l y s i s . In addition to the summary of 101 pairs of galaxies with the measured radial velocities of both components (Karachentsev, 1974) we have made use of new data on radial velocities of double galaxies (Afanasjev et al., 1975, Karachentsev et al., 1975a, Karachentsev at al., 1975b, Sargent, 1973, Jenner, 1974, Denisyuk et al.,1974). The list of 134 double galaxies is presented in the Appendix. Estimates of virial mass-to-luminosity ratios, $\varphi = \mu/L$, and their unbiased values, $f = \varphi - \Delta\varphi_u$ have been calculated at the supposition of the circular motion of components ($\langle\eta\rangle = 3\pi/32$), and Hubble 's constant, h= 75km/s. Mpc.

Distribution of 134 double galaxies by the values of $\lg(\mu/\mathfrak{M}_\odot)$ and $\lg(L/L_\odot)$ is presented in Fig.2. As can be seen, most of the pairs consist of galaxies of high luminosity, exceeding that of our Galaxy. The correlation between the virial mass and luminosity of pairs is weakly pronounced, which is apparently due to the effect of projection and velocity measurement errors. Essentially the correlation of masses and luminosities is revealed only in existence of an upper limit to the ratio μ/L.

In calculation of the mean unbiased virial mass-to-luminosity ratio we have divided the sample of pairs into groups according to different properties: morphological type of components, type of interaction, degree of isolation of a pair with respect to neighbouring galaxies. A representative enough sample is made up of the pairs having one or both Markarian galaxies, or compact galaxies from Zwicky Catalogue (1971). Some pairs which satisfy the criterion of isolation, are, nevertheless, the members of clusters and groups of galaxies with the measured radial velocities. We have united them in a separate sample. Mean values of $f = \mathfrak{M}/L$ and standard errors of the mean for different samples are presented in Table 2. Since the distribution of double galaxies by the value of f has quite large dispersion and asymmetry we provide also the medians of the mass-to-luminosity ratio, $f_{0.50}$, and the values of $f_{0.90}$, within which there are 90 per cent of f-values. Analysis of the data presented permits to draw the following conclusions:

a) Judging by median values, $f_{0.50}$, most of double galaxies have normal mass-to-luminosity ratio, $\mathfrak{M}/L \lesssim 10\ \mathfrak{M}_\odot/L_\odot$, which agrees with the estimates of mass by the internal motions in galaxies.

b) Pairs of galaxies with spiral components have on the average smaller $\mathfrak{M}/L$ ratios than pairs with elliptical or mixed (ES, SE) components.

c) Pairs which contain "active" galaxies (blue Markarian galaxies, Zwicky compact objects) do not stand out among others by large virial mass-to-luminosity ratio. Note that estimates of mass for these galaxies by intrinsic motions are extremely scarce. Therefore the mean values of $\mathfrak{M}/L$ for 41 pairs with Zwicky galaxies and 21 pairs with Markarian galaxies presented in Table 2 are still the most reliable data for these objects.

Table 2. Mean virial mass-to-luminosity ratio for different samples of double galaxies

Sample	n	S(%)	$\langle f \rangle \pm \sigma_{\langle f \rangle}$	$f_{0.50}$	$f_{0.90}$
1	2	3	4	5	6
All pairs	134	69	27 ± 6	3.4	75
EE - galaxies	25	0	42 ± 18	11.5	133
ES - galaxies	33	50	47 ± 16	8.7	190
SS - galaxies	76	100	13 ± 6	0.5	68
Compact	41	56	14 ± 5	2.8	60
Markarian	21	78	10 ± 7	0.4	69
"LIN" (all)	38	70	7 ± 3	0.5	39
with "tail"	12	62	5 ± 6	0.0	50
with "bridge"	15	77	8 ± 6	0.8	50
with "tail" and "bridge"	11	68	7 ± 4	2.6	20
"DIS" (all)	39	86	38 ± 14	5.3	130
one distorted component	22	77	57 ± 22	6.9	300
both distorted components	17	100	12 ± 13	0.1	100
"ATM" (all)	34	40	40 ± 16	7.4	133
with amorphous "atmosphere"	11	0	32 ± 17	4.5	60
with shredded "atmosphere"	23	59	45 ± 22	7.7	200
noninteracting	23	80	21 ± 12	-1.0	120
strong isolation	56	75	11 ± 6	0.7	42
moderate isolation	54	64	31 ± 11	2.3	96
slight isolation	24	65	53 ± 17	26.3	194
in groups and clusters	28	70	36 ± 16	1.5	130
with $\Delta \varphi_u < 5\ f_\odot$	76	60	22 ± 4	6.5	72

(1) type of the sample of double galaxies, (2) - number of objects in the sample, (3) - relative number of spiral galaxies in (%), (4) - mean virial mass-to-luminosity ratio and standard deviation of the mean in solar units, (5) - median of distribution of f, (6) - 90 per cent value for distribution of f.

d) The mean virial mass-to-luminosity ratio is dependent essentially on the type of interaction between components. Pairs with linear properties of interaction ("tails", "bridges") and those the spiral structure in both components of which is markedly distorted have the least value of $\langle f \rangle$. The components of pairs enveloped by a common "atmosphere" have on the average larger mass-to-luminosity ratios. Features of interaction correlate with $\langle f \rangle$ to a greater extent than the morphological type of pair components. This circumstance may indicate that there is connection between features of interaction and a concrete phase of mutual motion and orientation of galaxies in a pair.

e) The mean value $\langle f \rangle$ decreases with increasing degree of isolation of a pair. This may be caused by different effects, in particular by the presence of accidental optical pairs.

As it has been noted, the mean value of virial mass-to luminosity ratio and shape of distribution, $p_f\{f\}$, itself depend on the assumed type of motion of components of a pair. To determine the function $p_f\{f\}$ we have applied equation (8) to the sample of 76 double galaxies with the most reliable radial velocity measurements. Selection of these pairs was carried out according to the condition that the systematic error in f due to radial velocity measurement errors should not be larger than $\Delta\varphi_u = 5 f_\odot$. The data for this sample of pairs are in the last line of Table 2.

From calculations it follows that for each of the considered types of motions, the distribution density can be represented by beta-function

$$p_f\{f\} = \frac{\Gamma(m+n+2)}{\Gamma(n+1)\ \Gamma(m+1)}\ s^m (1-s)^n, \qquad [0 \leqslant s \leqslant 1] \tag{9}$$

where $s = f/f_{max}$, and m and n are the distribution parameters. The values for the mean $\langle f \rangle$, standard deviation $\sigma(f)$, and also for the parameters f_{max}, m and n are given in Table 3. Distribution densities of the virial mass-to-luminosity ratio for the three types of motions are presented in Fig.3.

Table 3. Parameters of the distribution function, $p_f\{f\}$, of double galaxies for three types of motion of components.

Motion type	$\langle f \rangle / f_\odot$	$\sigma(f)/f_\odot$	$f_{max}/f_\odot$	m	n
Circular	21.9	20.8	72.2	-0.5277	+0.0861
Oscillations	65.7	34.0	96.6	-0.4803	-0.7559
Radial unconnected	32.8	40.6	199.4	-0.6178	+0.9410

The feature common to the three distributions is a sharp maximum at $f = 0$. In case of oscillations the distribution density has a second maximum at $f_{max} = 96.6 f_\odot$. For such U-shaped distribution it is difficult to find an appropriate physical interpretation. Probably, the model of pure oscillations does not conform to real motions in double galaxies.

P r o b l e m o f o p t i c a l p a i r s . Analytical estimate of the relative number of optical pairs, q_{opt}, satisfying the criterion of isolation we have chosen, is rather complicated. Preliminary calculations show that in the sample considered, optical pairs do not exceed 10 per cent of the total number of pairs. Let us provide several arguments in favour of the assertion that q_{opt} is not large.

The presence of traces of interaction between components show strong evidence for them as belonging to a common system. If many double galaxies with large virial mass-to-luminosity ratio were optical, the percentage of interacting galaxies among them would then be lower. As it follows from the list of pairs (Appendix) such an effect is not expressed.

The presence of many optical pairs may become evident in the distribution of the number of pairs by the difference in radial velocities of components, N(y). The monotonously exponential distribution, N(y), in Fig. 4 does not point to the marked number of optical pairs either.

As an example, take pair N 584 whose both components are Markarian galaxies (N 325+326). A common rare feature evidences for the physical connection of these galaxies. Their radial velocities are measured quite reliably. Nevertheless, at any type of motions and at the most favourable orientation of the pair, its virial mass-to-luminosity ratio cannot be less than $f_{min} = 20 f_\odot$. For spiral galaxies this quantity is much too large.

In view of great importance of the problem of optical pairs, we undertake calculation of q_{opt} simulating spatial distribution of systems of galaxies and conditions of selection of isolated pairs with a computer.

D o e s   h i d d e n   m a s s   e x i s t   i n   d o u b l e   g a l a x i e s ? As it follows from Table 2, attributing 10 per cent of the largest values of f to the contribution of optical pairs, we still confront with the necessity to account for the large values of $f_{0.90} = (50 \div 200) f_{\odot}$. Many authors suggested the existence of some hidden matter in pairs. It can be readily shown that at the density $\langle \rho \rangle = 2 \cdot 10^{-23} g/cm^3$, needed to explain mutual velocities of galaxies in pairs, hidden mass in the form of gas would give an anomalously large flux of optical-, radio-, or X-ray radiation. Einasto et al. (1974) have suggested that many galaxies, including double ones, are surrounded with massive coronas of faintly luminous stars. In the opinion of these authors, stellar coronas extend to distances of $(0.1 \div 1.0)$ Mpc and have masses by an order of magnitude larger than the apparent mass of a galaxy. This hypothesis seems doubtful for the following reasons:

a) According to the data of Table 2 most double galaxies have normal values of $f \leqslant 10 f_{\odot}$. Morphology, dimensions, and luminosities of double galaxies with large virial mass-to-luminosity ratios do not differ much from those for double galaxies with $f \leqslant 10 f_{\odot}$. Structural differences should be expected in view of unequal conditions of formation and evolution of galaxies with large and small masses.

b) The weak point of the corona hypothesis is revealed in our opinion in the features of virial mass-to-luminosity dependence on the distance between the components of a pair. This dependence has been discussed by Einasto et al. (1974). We have calculated the cumulative mean of $\langle f \rangle$ for double galaxies whose projections of distance between the components are not larger than the fixed value, x_{kpc}. For pairs with {SS} - and {EE, SE, ES} - components these dependences are shown in Fig.5. Attention is attracted by uneven variations of $\langle f \rangle$ as a function of the distance x similar for both samples of double galaxies. It is noteworthy that in the region $x = (10 \div 25)$ kpc there is a plateau on both curves corresponding to normal "photometric" values, $\langle f_S \rangle 5 f_{\odot}$ and $\langle f_E \rangle = 18 f_{\odot}$. A sharp increase of the virial ratio $\langle f \rangle$ by a factor of ~ 3 occurs in both cases at $x = (25 \div 35)$ kpc. At greater distances there is practically no rise in the virial mass-to-luminosity ratio. If this effect is verified by further observations, to explain it within the framework of the stellar corona hypothesis one should consider that coronas have the shape of a thin spherical layer with sharp boundaries. Dynamical stability of such formations is rather doubtful. The noted peculiarities of the dependence $\langle f / x \rangle$ are difficult to account for from other points of view. It is possible that their cause is due to a play upon effects of orientation and selection of double galaxies.

In the light of the virial problem the following data are of interest for systems of galaxies. For 22 galaxies included in the pairs considered, the mass-to-luminosity ratio calculated from rotational curves has a normal value, $\langle f \rangle = 7.2 f_{\odot}$. On the other hand, 28 pairs located in groups and clusters (Table 2) do not differ essentially in $\langle f \rangle = (36 \pm 16) f_{\odot}$ from the rest of double galaxies. For 16 groups and 5 clusters, where these pairs are, the mean virial mass-to-luminosity ratio is $(405 \pm 90) f_{\odot}$. Therefore, at each structural level the total virial mass of subsystems does not conform to the virial mass of the system as a whole.

In our opinion the data considered do not exclude Ambartsumian's (1956) hypothesis according to which large virial masses are caused by disruption of a certain part of double galaxies.

To promote the soonest solution of the virial problem, two ways seem to be most actual: a) systematic high accuracy measurement of radial velocities for isolated double galaxies which make up a statistically uniform sample; b) investigation of the selection effects and the role of optical pairs in such a uniform sample.

The author is indebted to Dr.B.I. Fesenko for his most valuable remarks and advice.

APPENDIX

List of 134 double galaxies with virial mass-to-luminosity ratio.

N°	Type	Inter-action	$\varphi/f_{\odot}$	$f/f_{\odot}$	Note
1	2	3	4	5	6
11	ES	LIN	0.1	-1.1	Zw, Mr
13	SS	DIS	54.8	54.5	Mr
23	EE	ATM	8.9	8.3	
31	SS	LIN	0.3	0.2	
32	EE	ATM	0.5	-3.4	
40	SS	-	5.0	-11.9	
46	EE	ATM	0.0	0.0	Zw
47	SS	ATM	57.2	50.5	Mr
64	SS	LIN	5.1	-4.5	Zw
67	SS	-	1.2	-0.2	Zw
73	SE	ATM	34.2	30.2	
84	EE	ATM	31.6	31.3	Zw
99	EE	DIS	81.1	78.5	Zw, Mr
103	SE	-	82.3	42.1	Zw
125	SS	-	9.8	8.8	
127	SE	DIS	131.0	128.7	Zw
133	SS	-	8.6	0.6	Mr
135	SS	ATM	0.0	-0.2	Zw, Mr
144	SE	-	34.0	26.3	
156	SS	LIN	0.0	-4.2	
161	SS	LIN	0.9	-13.1	
175	EE	ATM	32.0	29.6	
181	SS	DIS	7.9	4.6	
202	ES	DIS	134.2	131.0	
203	SS	DIS	2.3	-0.6	Mr
210	SS	-	127.8	114.0	
218	SS	-	124.9	124.3	
228	SS	ATM	19.5	-11.8	
234	ES	LIN	35.2	34.5	
236	SS	LIN	12.1	4.7	
240	SE	ATM	10.2	7.1	
249	SS	ATM	0.7	-0.7	
255	SS	DIS	14.9	-56.0	
257	SS	LIN	11.3	-37.9	
268	ES	ATM	12.8	9.7	
271	SS	DIS	29.6	24.2	
278	EE	-	39.8	37.8	
281	SS	-	15.1	-16.3	
288	SS	DIS	0.7	0.6	Mr
294	SS	DIS	0.8	-3.1	
295	SS	DIS	88.5	77.0	
296	SS	DIS	0.9	-1.3	
311	SS	LIN	24.0	22.6	
324	SS	-	10.8	-47.5	Seyf
327	EE	LIN	6.2	-12.4	
334	ES	DIS	0.0	-27.5	
340	EE	LIN	12.6	6.8	Zw, Mr
341	SS	DIS	22.1	15.3	
343	SS	-	9.4	3.6	
347	SS	ATM	0.2	-0.9	

1	2	3	4	5	6
349	SS	DIS	236.0	160.0	
350	SS	DIS	1.3	-0.7	
352	SS	DIS	48.2	31.4	
353	SE	DIS	3.3	2.3	
354	ES	LIN	0.4	-2.4	Zw, Mr
355	ES	LIN	3.8	0.4	
356	ES	DIS	390.0	380.8	
358	SS	DIS	13.4	-79.6	
362	SS	DIS	42.6	6.0	Zw, Mr
368	SS	LIN	33.5	27.4	
369	SE	LIN	15.1	12.5	Ze
372	EE	ATM	1.9	1.5	
379	SS	LIN	1.9	0.9	
388	SS	ATM	3.0	2.2	Zw, Mr
389	SS	LIN	7.7	-9.1	
390	SS	LIN	0.0	-2.8	Zw, Mr
396	SS	DIS	342.0	306.7	
416	SE	-	97.5	71.5	
419	SE	-	244.0	198.8	
422	SS	-	0.0	-15.5	
438	SS	DIS	56.8	-1.2	Zw
439	ES	LIN	72.5	55.3	
440	SS	-	19.0	-40.1	
444	SS	-	10.6	-43.1	Mr
454	EE	ATM	0.6	-1.3	Zw, Mr
455	SS	-	0.6	-10.3	
466	SS	LIN	8.2	7.2	Zw
468	ES	LIN	0.6	-0.2	
471	SS	DIS	12.8	0.3	Mr
472	SS	LIN	43.9	39.7	Mr
476	ES	-	2.7	-2.9	
504	SS	DIS	33.0	30.8	
508	SS	LIN	10.0	0.8	
534	SS	LIN	13.5	11.9	Zw
536	SS	DIS	9.1	6.3	Zw
548	SE	DIS	55.2	48.2	
549	SS	ATM	33.3	27.4	Zw
551	SS	LIN	0.1	-22.4	Zw
552	SE	LIN	9.2	7.7	Zw
560	SE	LIN	1.0	0.5	Zw
564	EE	ATM	0.0	-0.4	Zw
567	SS	DIS	0.9	0.5	Zw
570	ES	-	88.6	-20.9	
571	SS	DIS	20.7	6.9	Zw
575	SS	DIS	1.3	-1.7	
578	SS	-	0.1	-9.1	
584	SS	-	73.3	69.9	Zw, Mr
587	SS	LIN	0.6	-4.6	Mr
588	EE	ATM	76.8	76.7	
590	SS	DIS	15.1	3.5	Zw
591	ES	LIN	66.5	63.4	Zw
592	SS	DIS	3.5	-3.6	
603	SS	DIS	0.0	-0.2	Zw
(0016+4603)	EE	LIN	12.0	5.8	Zw
(0017+0351)	EE	ATM	2.2	-3.0	Zw
(0048-0720)	SS	ATM	4.8	3.1	
(0056+2636)	EE	ATM	16.7	16.6	Zw

1	2	3	4	5	6
(0117+3155)	SE	ATM	327.0	310.3	
(0121+1346)	EE	ATM	11.1	7.4	Zw
(0159+2918)	EE	ATM	0.0	-0.2	Zw
Maffei 1+2	ES	-	0.2	-1.7	
(0303-2741)	SS	DIS	9.6	8.6	
(0318+1546)	EE	ATM	59.3	58.8	Zw
(0321-3720)	ES	DIS	9.9	6.7	
(0520-1132)	SE	DIS	0.3	-0.2	
(0924+7437)	EE	ATM	191.0	187.8	
(0930+5527)	EE	ATM	28.9	24.4	Zw
(0956+2906)	EE	LIN	18.4	17.3	
(1016+5740)	SS	ATM	15.2	-4.4	
(1115+5402)	SS	DIS	5.5	-2.3	
(1159-1835)	SS	LIN	0.0	0.0	
(1204-2930)	EE	LIN	15.7	14.7	
(1252-1219)	EE	ATM	47.1	46.9	
(1255-4834)	SS	DIS	71.0	69.6	
(1340+2638)	EE	ATM	430.0	423.8	
(1355+2902)	SE	ATM	18.0	15.0	Mr
(1401-0548)	SS	LIN	3.6	0.7	
(1429+2727)	SS	ATM	4.4	3.3	
(1603+1744)	ES	LIN	46.2	39.4	
(1749+5641)	SE	LIN	3.1	-0.1	Zw
(2213+3943)	ES	LIN	1.2	-2.0	Zw
(2223+3835)	SS	LIN	0.1	-2.9	Zw
(2255-0403)	SS	ATM	51.5	28.5	
(2337-1254)	SS	DIS	0.0	-0.2	Zw

(1) - the number of the pair according to the Catalogue (Karachentsev, 1972), bracketed are α, δ - coordinates of pairs not included in the Catalogue (southern or faint double galaxies), (2) - morphological type of components, (3) - type of interaction between components, (4) - estimate of virial mass-to-luminosity ratio in solar units, (5) - unbiased value of virial mass-to-luminosity ratio in solar units, (6) - presence of Markarian (Mr), Seyfert (Seyf) or compact Zwicky (Zw) galaxies.

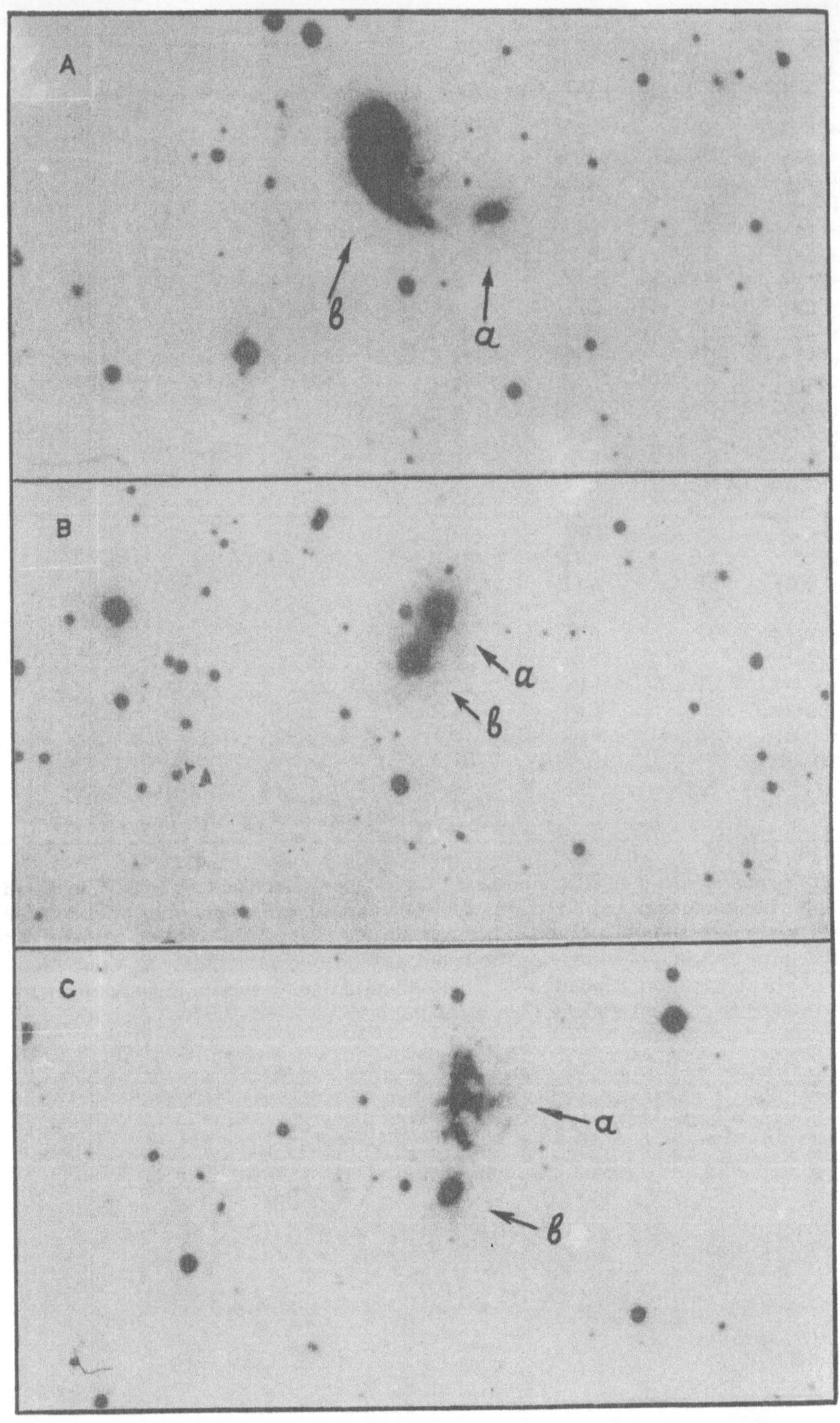

Fig. 1 . Examples of types of interaction between components of double galaxies. a)pair № 472 with "tail" and "bridge" ("LIN"). b) pair N 493 whose both components are enveloped in an "atmosphere" ("ATM"). c) pair N 414, whose components have structure distortion ("DIS").

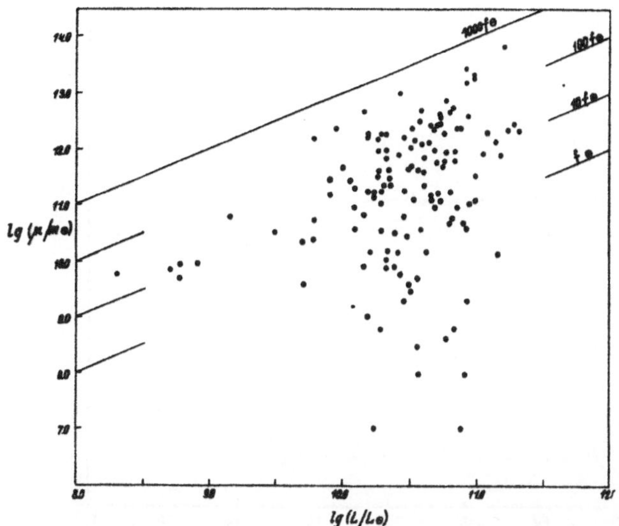

Fig. 2. Distribution of 134 double galaxies by virial masses and total luminosities. The diagonal straight lines indicate the value of mass-to-luminosity ratio in solar units.

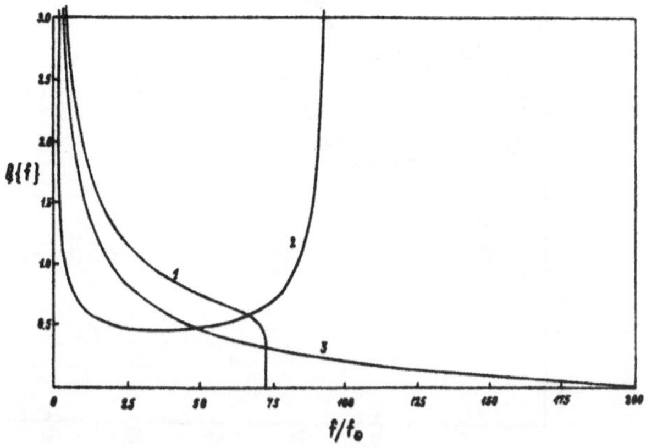

Fig. 3. Distribution density of the virial mass-to-luminosity ratio for: (1) -radial motion of components. (2) -oscillations. (3) - radial unconnected motion of galaxies.

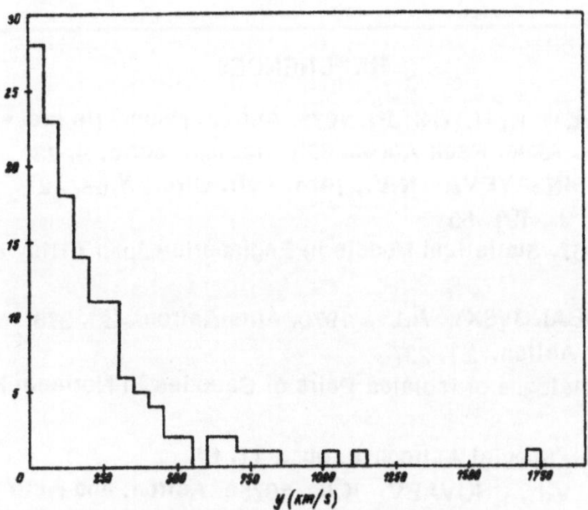

Fig. 4. Distribution of number of double galaxies by the absolute value of radial velocity difference of components.

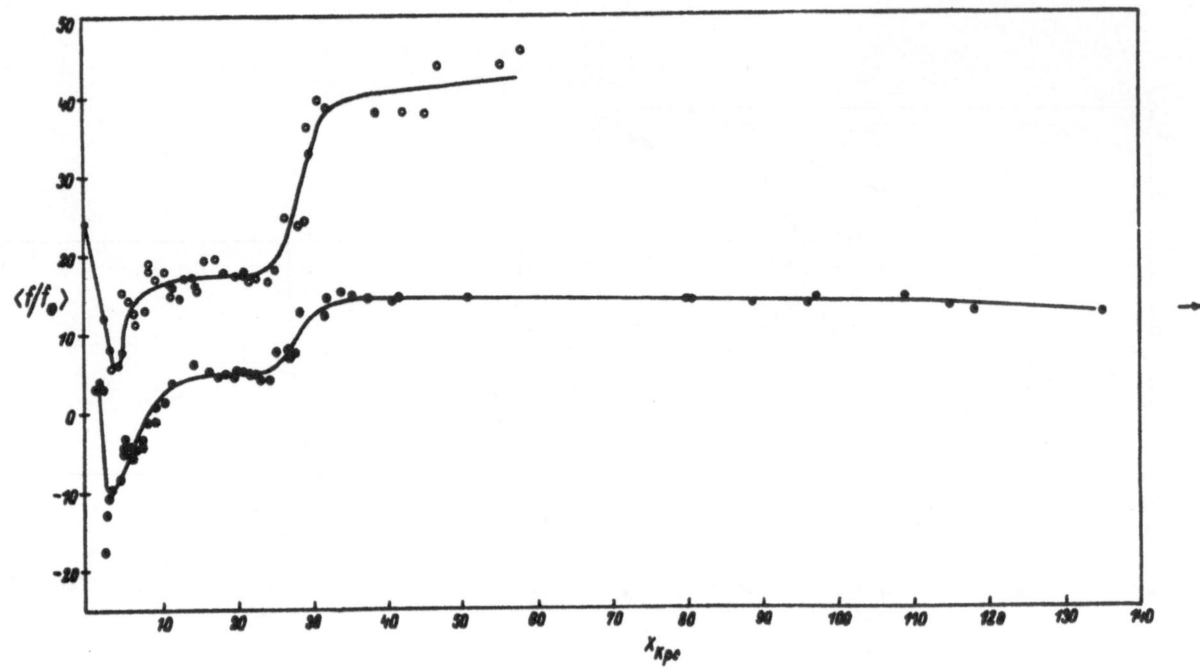

Fig. 5. Mean virial mass-to-luminosity ratio for pairs in which the projection of distance between components does not exceed x_{Kpc}. The dots are for pairs of spiral galaxies, the circles are for pairs of elliptical and mixed types of galaxies.

REFERENCES

AFANASJEV V., KARACHENTSEV I., NOTNI P., 1975, Astron. Nachr. (in press)

AMBARTSUMIAN V.A., 1956, Izv. Akad. Nauk Armen. SSR, fiz. mat. seria, 9, 23

DENISYUK E.K., BABKIN I.G., SINJAYEVA N.V., 1974, Astr. Circ., №.837, 2

JENNER D.C., 1974, Astrophys., J., 191, 55

HAHN G.J., SHAPIRO S.S., 1967, Statistical Models in Engineering,John Willey and Sons, Inc., New-York.-London-Sydney

KARACHENTSEV I.D., SHCHERBANOVSKY A.L., 1970, Acta Astron., 20, 373

KARACHENTSEV I., 1970, Acta Astron., 21, 237

KARACHENTSEV I.D., 1972, Catalogue of Isolated Pairs of Galaxies in Northern Hemisphere, Publ. Special Astrophys. obs., 7, 3

KARACHENTSEV I.D., 1974, Publ. Special Astrophys. obs., 11, 51

KARACHENTSEV I.D., PRONIK V.I., CHUVAEV K.K., 1975a, Astron. and Astrophys., (in press)

KARACHENTSEV I.D., PRONIK V.I., CHUVAEV K.K., 1975b, (to be published)

SARGENT W.L.W., 1973, Astrophys. J., 182, L13

ZWICKY F., HERZOG E., WILD P., KARPOWICZ M., KOWAL C., 1961-1968, Catalogue of Galaxies and of Clusters of Galaxies, 1 - 6, California Institute of Technology, Pasadena

IS THERE ANY MISSING MASS ?

H. OLEAK
POTSDAM-BABELSBERG ZENTRALINSTITUT FÜR ASTROPHYSIK DER AKADEMIE DER WISSENSCHAFTEN DER D.D.R.

СУЩЕСТВУЕТ ЛИ СКРЫТАЯ МАССА?

РЕЗЮМЕ

Сравнением функции распределения M/L для изолированных спиральных и иррегулярных галактик с таковыми для двойных галактик ((S,S и E,E)) из списка Караченцева (1974) показано, что несоответствие массы может быть приписано сильно уклоненной форме этой частотной функции, и, следовательно, большой зависимости среднего значения M/L от дисперсии. Неизбежное увеличение дисперсии вследствие ошибок измерений и неправильное включение объектов в выборку, всегда завышает среднее значение M/L.
Этот эффект может быть принят за несоответствие массы.

ABSTRACT

By comparing the M/L-distribution function of single spiral and irregular galaxies with that of double galaxies (S,S and E,E) from the list of Karachentsev (1974) it is shown that the mass discrepancy can be ascribed to the strongly skewed form of this frequency function and thus to the strong dependence of the mean value of M/L on the dispersion. An unavoidable enlargement of the dispersion by errors of measurement and wrong attachment of objects to the sample always raises the mean of M/L. This effect could be mistaken for a mass discrepancy.

I n t r o d u c t i o n. The problem of the missing mass or the mass discrepancy dates back to 1933 when Zwicky found that the mass of the Coma cluster derived with the virial theorem from the velocity dispersion was much larger than the sum of the masses of the cluster members inferred from their luminosities. This appeared to be a general property of systems of galaxies. Different hypotheses have been proposed to solve this problem: a) the systems are not stable, b) the missing mass exists in a form that has prevented observational detection up to now, c) single galaxies are more massive than hitherto known. In the last years the tendency for a gradual decrease of the derived amount of the missing mass in some well investigated systems can be noticed, for instance in the Coma cluster from factors of several hundreds to about four or five (Tarter and Silk 1974) and in bound groups to about 10(Turner and Sargent 1974). The main reason for this decrease is the increase of observational material. It is the aim of this paper to show that the mass discrepancy could be probably further reduced to a factor equal or less than two. Let us assume that systems of galaxies are dynamically stable or that at least the instability inferred from Ambartsumian's(1956) cosmogonic conception can be neglected to a first order.

T h e M/L - d i s t r i b u t i o n a n d i t s i m p l i c a t i o n. Chiao and Reinhardt (1973) showed that the mass-to luminosity ratios (M/L in solar units) of 104 spiral and irregular galaxies for which Roberts (1972) determined masses from rotation curves are not normally distributed but belong to a very wide, extremely skewed and longtailed frequency function. They could further demonstrate that the logarithm of the M/L ratios obey a gaussian distribution.

This behaviour obviously reflects some cosmogonic reasons and should be inherent in every sample of spiral and irregular galaxies. This form of the frequency function has some far reaching consequences.

Let σ be the dispersion of the ln (M/L)- frequency function

$$f(\ln M/L) = (2\pi\sigma^2)^{-1/2} e^{-\frac{1}{2\sigma^2}\{\ln M/L - \langle \ln M/L \rangle\}^2}, \tag{1}$$

then the mean <M/L> strongly depends on the dispersion of the sample (see Chiao and Reinhardt)

$$\langle M/L \rangle = e^{\langle \ln M/L \rangle}\, e^{\frac{1}{2}\sigma^2}. \tag{2}$$

There is only the median independent of σ :

$$\text{median } (M/L) = e^{\langle \ln M/L \rangle} \tag{3}$$

Therefore, *either* the mean <ln M/L> or the median and not the mean <M/L> of different samples should be compared if we want to prove whether they belong to some extent to the same parent population. Only if the dispersion is taken into account the mean value of M/L can be used. Any contamination of the sample, for instance by objects not really belonging to it, will enlarge the dispersion and hence raise the value of the mean exponentially! Thus a mass discrepancy is easily concluded. (If σ is twice the real value, the mean $\langle M/L \rangle$ is shifted by a factor 4; if σ is 3 times larger, $\langle M/L \rangle$ becomes already 30 times larger.)

Comparison with double galaxies. This consideration can be best verified for double galaxies. Each pair represents two members of such a sample. If we consider only pairs with components of the same morphological type we can expect that the M/L ratios derived from the virial mass and the total luminosity should be distributed in a first approximation like single galaxies provided that there is no mass discrepancy.

Figure 1 shows the ln(M/L) distribution of 42 spiral pairs (S,S) and of 16 pairs with elliptical components (E,E) from the list of Karachentsev (1974). (Double galaxies for which the velocity difference is smaller than half the rms error caused by the errors of the radial-velocity measurements have been omitted.) The histograms can be fitted by gaussian frequency functions. The parameters are shown in the table. Both the means $\langle \ln M/L \rangle$ for S,S-pairs and single galaxies lie inside the 95% confidence limit. If we take the difference serious, the mass discrepancy is about 2. Regarding the E,E-pairs we find that there is apparently a real difference between spirals and ellipticals. The higher standard deviations in both samples (S,S and E,E) contribute essentially to the large values of the mean $\langle M/L \rangle$. This can be easily tested because the values of $\langle M/L \rangle$, $\langle \ln M/L \rangle$ and σ should be connected by relation (2) if the samples have common properties. This is demonstrated in Figure 2. All samples which consist of galaxies of the same morphological type lie near the line for the theoretical relation (2). This is not the case for the whole sample of double galaxies. Here the ln M/L cannot be expected to belong to a gaussian distribution because they consist of a mixture of S,S-and E,E-pairs and mixed S,E-pairs.

It is evident that the mass discrepancy for double galaxies can be reduced when the frequency functions are compared. This is the correct way of comparison if a strongly skewed distribution function might be broadened by unavoidable errors of measurement and attachment. The same mechanism might work in groups and clusters of galaxies. An investigation in this direction is under way. The relation (2) could then explain why the mass discrepancy turns out to be positive in most cases.

Table

	N	$\langle \ln M/L \rangle$	σ	median (M/L)	$\langle M/L \rangle$
Single galaxies +)					
S + I	104	1 62	0.93	5.1	7.8
Double galaxies ++)					
S,S - pairs	42	2.35	1.79	10.5	42.1
E,E -pairs	16	3.27	1.30	26.3	62.6
all pairs	100	1.64	2.84	5.2	44.2

+) CHIAO and REINHARDT (1973)

++) from the list of KARACHENTSEV (1974)

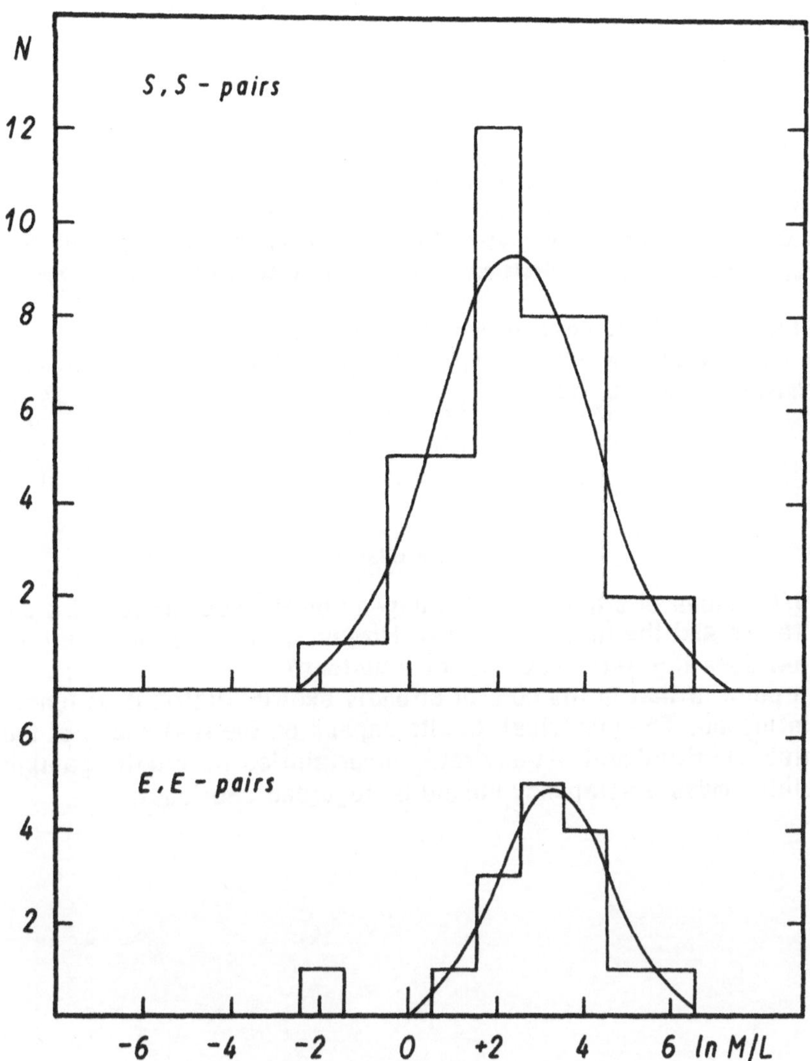

Fig. 1. The frequency histograms of ln(M/L) for S,S-pairs (above) and E,E-pairs (below) from the list of Karachentsev (1974). The curves are analytic fits by normal frequency functions.

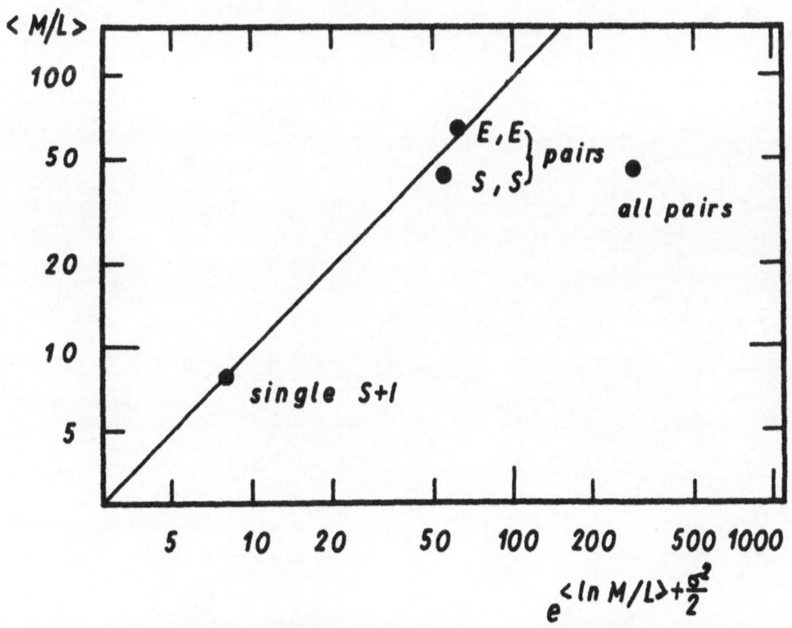

Fig. 2. Relation between the mean values <M/L>, <ln M/L> and the standard deviation σ for the samples of single S + J galaxies, pure S,S - and E,E-pairs and all double galaxies. The line represents equation (2).

REFERENCES

AMBARTSUMIAN V.A. 1956, Izv. Acad. Nauk Arm SSR Ser. Fiz. Mat. 9, 23.

CHIAO R.Y. and REINHARDT M. 1973, Astron.and Astrophys. 22, 257.

KARACHENTSEV I.D. 1974, Communications Spec. Astroph. Observ. Acad. Sciences USSR, N.11

ROBERTS M.S. 1972, in "Stars and Stellar Systems" Vol. IX, Ed. A. Sandage and M.Sandage, University of Chicago press (in print)

TARTER J. and SILK J. 1974, Quart. Journ. R. Astr. Soc. 15, 122.

TURNER E.L. and SARGENT, W.L.W. 1974, Astrophys. Journ. 194, 587.

ZWICKY F. 1933, Helv. Phys. Acta 6, 110.

DISCUSSION

BROSCHE : The M/L values of single galaxies depend on the assumption of a constant ratio between the Holmberg radius and the turn-over radius. However, they depend certainly also on the morphological type. How does this influence your conclusions?

OLEAK : The main point is that in the core of strongly skewed distribution functions the dispersion law has a large influence. The numerical results depend on the real shape of the distribution function derived by Chiao and Reinhardt. At any rate, uncertainties for single galaxies only support the idea that a calculated mass discrepancy should be regarded cautiously.

ON THE STABILITY OF GROUPS OF GALAXJES AND THE QUESTION OF HIDDEN MATTER

J.MATERNE and G.A.TAMMANN
HAMBURGER STERNWARTE, HAMBURG, F.R.G. EUROPEAN SOUTHERN OBSERVATORY, GENEVA, SWITZERLAND; ASTRONOMICAL INSTITUTE OF THE UNIVERSITY OF BASEL, BINNINGEN, SWITZERLAND
PRESENTED BY J.MATERNE and G.A.TAMMANN

СТАБИЛЬНОСТЬ ГРУПП ГАЛАКТИК И ВОПРОС СКРЫТОЙ МАТЕРИИ

During the last 25 years our knowledge on groups of galaxies has greatly increased.Since the pioneering work of Holmberg (1950) several authors have isolated new groups. The total number of proposed groups with 5 to 20 members is now in the order of 100. However, detailed information on most groups and their members is still lacking. The reality of some suggested groups is still questionable, the membership assignments are uncertain for outlying and incomplete for faint galaxies, redshifts are still lacking for many possible members , and the accuracy of the redshifts is often insufficient. These points must be kept in mind when an investigation of the properties of groups is attempted.

The first applications of the virial theorem to groups of galaxies seemed to indicate that they are highly unstable. This has been taken as additional evidence for "missing matter" in the universe. However, during the last two or three years groups of galaxies have been found which are bound. Their number is still increasing. There is a tendency that groups of galaxies are found to be stable as more data and better redshifts become available for their members.Thus it is questionable whether the existence of unbound groups of galaxies can presently be considered to be significant. Of course, there is still the possibility that some groups may only be chance concentrations of field galaxies, and that there may exist groups or clouds of galaxies which, analogous to stellar associations, have positive total energy and are expanding on time scales comparable to the expansion of the universe (cf. Turner and Sargent, 1974, Chincarini et al.,1975).

In section I of this paper we shall give a brief account of groups of galaxies, - especially nearby groups of galaxies. - for which there is little room left to postulate any large amounts of hidden matter. Additional argument shall be given in section II that the masses of many galaxies are probably not much higher than hitherto assumed.

1. Bound Groups of Galaxies. A group of galaxies is stable if its total energy $E = T + \Omega$ is negative, i.e. if $T/|\Omega| < 1$. It has often been assumed that stable groups have also to fulfill the virial theorem, i.e. $T/|\Omega| < 1/2$. de Vaucouleurs (1965) and Turner and Sargent (1974) have stressed the possible pitfalls of the latter assumption. Not only holds the virial theorem only for *stable, relaxed* systems, but also the necessary time-averaged parameters must be replaced by the observed instantaneous values. In addition stable, dynamically evolved systems may be surrounded by "evaporated"(escaped) members,which in general cannot be distinguished from bound members and which hence enter the calculation. (The latter problem can affect, of course, not only the virial mass of a group, but also its total energy.).

As in a previous paper (Materne and Tammann, 1974b) we prefer here to discuss the total energy of a group instead of its virial mass. We also adopt here the conventional ("visible") masses of galaxies as in the previous paper. These mass values imply mass-to-light ratios of $\mathfrak{M}/L=4$ for spirals, $\mathfrak{M}/L = 20$ for SO galaxies, and $\mathfrak{M}/L = 30$ for elliptical galaxies. The $\mathfrak{M}/L$ - value for spirals is taken from Roberts (1969) and reduced to $H_0 = 55$ km s^{-1} Mpc^{-1} (this value is used throughout this paper; cf. Sandage and Tammann, 1975b), the value for ellipticals is taken from Smart (1973), and the value for SO galaxies is adopted as a reasonable guess. The luminosities used are based on Holmberg's(1958) m_{pg}-system; the magnitudes are corrected for the full amount of the internal absorption following Holmberg (1958) and for galactic absorption according to Sandage(1973). In the following we shall refer to the masses resulting from these luminosities and mass-to-light ratios as to the "conventional mass" of a galaxy.

The potential energy of a group is given by

$$\Omega = -\alpha G \sum_{pairs} \frac{m_i \, m_j}{r_{ij}} \qquad (1)$$

and the kinetic energy by

$$T = \frac{1}{2} \ \beta \Sigma m_i \hat{v}_i^2$$

Here m_i, m_j are the masses of individual member galaxies and r_{ij} their separation. $\hat{v}_i$ is the corrected radial velocity of galaxy i in respect to the barycenter of the group; the procedure to correct these velocities for systematic effects from observational errors is explained elsewhere (Materne, 1974). It must be stressed that the random errors of $\hat{v}_i^2$ are sometimes so large that the formal kinetic energies are actually indeterminate. The values α and β are projection factors; following Limber and Mathews (1960) we have assumed that $\alpha = \frac{2}{\pi}$ and $\beta = 3$. The most probable values of α and β depend on the assumptions made; in the case of random orientation the statistical averaging in different coordinate systems leads to different results for α and β; variations of about ±30 percent are quite reasonable (Spaenhauer, 1975). In addition α and β are influenced by statistical fluctuations; in groups with few members these fluctuations can amount to 60 to 70 percent (Limber and Mathews, 1960; Smart, 1973).

The conventional masses used here are only first-order approximations because they are determined by the gross method of adopting mean mass-to-light ratios.If the masses of individual galaxies are randomly over- or under- estimated by factors of 2, which seems possible,the resulting uncertainty of T and Ω is considerable. This is mainly because different galaxies enter with very different weight into their determination; the effect has been discussed elsewhere (Materne and Tammann, 1974b). An additional error source for T and Ω is our incomplete knowledge of the fainter group members, which leads to a systematic overestimate of $T/|\Omega|$. In the same direction works our neglecting the intergalactic matter, which is known to exist in some groups (cf. Mathewson et al., 1974, 1975; Davies, 1974; Huchtmeier, 1975b).

These remarks should be taken as a warning to overrate the accuracy with which the kinetic and potential energy can be determined. Errors of factors 2-3 of the ratio $T/|\Omega|$ are probably a conservative guess.

In table I we have compiled data for the twelve groups of galaxies within 25 Mpc which have been found by different authors to have small (or indeterminate) total energy. Column 1, 2, and 3 give the designation(s) of the group, the number of group members with known radial velocities, and the radial velocity of the barycenter, respectively. The types of galaxies involved are shown in column 4 (E = ellipticals; SO = SO's and IrII's; S = spirals and IrI's). In column 5 the corrected velocity dispersion $< \hat{v}_i^2 >^{1/2}$ is given. The crossing times in column 6 are calculated by taking the apparent linear diameter (in most cases the maximum separation between member galaxies) and by dividing them by the velocity dispersion in column 5. The mass of a group is shown in column 7 as the sum of the conventional galaxy masses; the total masses are lower limits because only members with known radial velocities are used. The ratio $T/|\Omega|$ in column 8 is taken from the sources shown in column 9; in cases where T is indeterminate due to the mean errors of the available radial velocities, the ratio is shown in brackets. The groups in table I are arranged in order of increasing total mass.

Except for two groups (G 7 and G 10) the crossing times of the groups in table I are shorter than or of the order of the age of the universe ($\sim 16 \cdot 10^9$ ys). These groups could be relaxed. However, not all groups seem to be in dynamical equilibrium: the strongly negative total energy of the NGC 3190 group (G 47) suggests that the group is in a phase of rapid dynamical evolution (possibly after the evaporation of a formerly bound member); the negative energy of the M 51 group is marginally significant. Alternatively, the strong negative total energies could be explained by the assumption that the group members move in a preferential plane; this would, of course, invalidate the adopted projection factors in equations (1) and (2).

Seven of the twelve groups in table I have $T/|\Omega|$ - values between 1.5 and 0. There is no reason to doubt that these groups are bound. (For the Local Group there is, however, the well-known Kahn-Woltjer paradox (1959) that the Galaxy and M31 may approach each other; the significance of this paradox has been judged differently by different authors, cf. Burbidge, 1975). The kinetic energies of the remaining five groups are indeterminate and only upper limits can be derived. For these groups (G7, G10, Leo) these upper limits are so low that it is very likely that they are bound, too. We believe that the remaining two groups (G8, G14) shall eventually also be found to have $T/|\Omega| \lesssim 1$, once the redshifts of their members shall be better determined.

Outside the volume considered here there are many more distant groups which are bound.A few of these groups are compiled by Sandage and Tammann (1974). A systematic search for such groups was made by Smart (1973); he has found more than 20 external groups which are bound. These groups include so well known aggregations as Stefan's Quartet (cf. Limber and Mathews, 1960), VV116(cf. Burbidge and Burbidge, 1961), the NGC 3395 group (G 43) (cf. Karachentsev, 1966), the NGC 833 group (cf. Burbidge and Sargent, 1971) and the Fornax cluster (cf. Welch et al., 1975). A particularly interesting case of an apparently "overstable" galaxy aggregation, which seems to have a large negative total energy, is Shakhbazian I (Robinson and Wampler, 1973).

Of course the question arises why nearly all groups of galaxies originally seemed to require large amounts of missing mass, and why now more and more groups turn out to be bound by their conventional mass. The answer is simply that the known data for groups and their members have become more complete and more accurate. The most important, single factor is that many more reliable radial velocities have become available during recent years, mainly from 21 cm observations. It can be seen in table I that the typical velocity dispersion of a bound group is in the order of ~ 100 km s^{-1}, and in several open groups it is as low as 50 km s^{-1}. Since the older, optical radial velocities had errors of this order, they were simply unsuitable for the determination of the kinetic energies of groups of galaxies.

The small velocity scatter in many groups sets stringent upper limits to the size of any non-Doppler component of the observed redshifts. The conclusion that such anomalous redshift components are for galaxies in groups is consistent with a recent investigation by Harrison (1975).

If groups of galaxies are bound by their conventional mass there is little room left for any large amounts of hidden matter. In particular, it is not likely that the true total masses of the galaxies involved are much larger than their conventional masses. Indeed the six stable groups in table I, the kinetic energies of which are well determined, give a mean ratio of $T/|\Omega| = 0.6$. Since these groups are probably also relaxed, they should obey the virial theorem. The ratio of their total virial mass to their total conventional (visible) mass then is $\mathfrak{M}_{VT}/\mathfrak{M}_{vis} = 1.2$. This suggests that the true mean mass-to-light ratios are only insignificantly higher than adopted: $\mathfrak{M}/L = 4.8$ for spirals and $\mathfrak{M}/L = 36$ for ellipticals. These values can be compared with Smart's (1973) virial masses for seventeen more distant groups, for which he finds on the average $\mathfrak{M}/L = 7$ for spirals and $\mathfrak{M}/L = 39$ for ellipticals. The evidence from galaxy groups argues therefore against the assumption that the true mass-to-light ratios of group members could be larger than 10 for S galaxies and 50 for normal E-galaxies.

Although this result holds strictly only for galaxies which are members of investigated, bound groups, it is important to stress that these galaxies represent a considerable fraction by number and by mass of all nearby galaxies. The majority of proposed nearby groups of galaxies are contained in table I; nine of these groups, containing 59 known members, are unquestionably bound and lie north of $\delta = -36°$ and within 22 Mpc ($v_0 < 1200$ km s^{-1}). Excluding the Virgo cluster and its extensions the Shapley-Ames cataloque (1932) lists 227 galaxies within the same volume. Although a few group members included are not Shapley-Ames galaxies the two samples may be used for a rough comparison. This comparison suqqests that at least 26 percent by number of the nearby galaxies are members of groups. A more detailed comparison is shown in table II. Here the galaxies are subdivided in three groups of morphological type. The number of Shapley-Ames galaxies within the stated limits are shown in column 2. Their conventional masses have been calculated exactly as for the galaxies in groups (column 3). The number and fraction (in percent) of galaxies which are members of bound groups are given in column 4 and 5. Their mass and mass fraction (in percent) are shown in column 6 and 7. The figures in the last two columns imply that roughly 30 percent of the total mass of nearby Shapley-Ames galaxies lie in bound groups. The evidence for the correctness of the conventional masses therefore comprises a sizable fraction of all nearby galaxies.

It has been suspected that spirals do not occur in bound groups. The data in table I contradict this notion: there are bound groups which include only spirals. Table II shows that about 25 percent of the nearby spirals are in known bound groups. SO galaxies may be slightly underrepresentated in bound groups, but the numbers involved are too small to be significant. The numbers for E-galaxies are also quite low; this corresponds to the known fact that this type of galaxies is relatively scarce within the volume considered. Taken at face value the data in table II suggest, however, that the relative frequency of E galaxies bound in groups is high. - half of the nearby E galaxies being in bound groups. Whereas the E galaxies have about average mass (56 percent of the E galaxies carrying 41 percent of the mass), there is a hint that spirals in bound groups are overmassive: they represent 25 percent by number but carry 40 percent of the mass. These remarks on the distribution of group members over types are very tentative and possibly biased by observational selection effects; they should be checked with a much larger set of data.

The important point here is that the existence of bound groups sets upper limits for the mass of elliptical galaxies as well as of spiral galaxies out to volumes of the order of 1 Mpc.

II. Independent Mass Determinations of Galaxies. The above results are in contradiction with theoretical considerations which require very large halo masses for spiral galaxies (Ostriker and Peebles, 1973; Einasto et al., 1974; Gunn, 1974). Extended, very faint outer regions have been detected in spiral galaxies by Kormendy and Bahcall (1974), - however, these envelopes could represent the extension of the exponential disk, which would not be surprising in view of the large HI diameters of spirals (e.g. Roberts,1972). Direct observational evidence for large halos is lacking in spite of high-sensitivity searches in the near infrared (Schild et al., 1975; Simkin, 1975). Ostriker et al. (1974) have argued that there is observational evidence for the mass of spirals to increase linearly with the radius, - as in isothermal gas spheres, - out to distances of ~1 Mpc! However, from the present results on groups of galaxies it is obvious that their dynamics would become a major puzzle with the corresponding masses (masses of $10^{12-13}\mathfrak{M}_\odot$ and mass-to-light ratios of 100-200). The derivation of such high masses depends entirely on the *assumption*, as Burbidge (1975) has emphasized, that the systems considered are *physical, bound* systems.

In the following we shall briefly review the most recent evidence from independent mass determinations. The different methods are arranged in such an order as to reflect the mass within increasing volumes around a galaxy.

1) Rotation curves. The typical, optically observed rotation curves of spiral galaxies extend normally only slightly beyond the turn-over point. The total mass of these spirals depends therefore quite strongly on the adopted extrapolation of the rotation curve. Masses derived from conventionally extrapolated rotation curves lead to a mean mass-to-light ratio of $\mathfrak{M}/L = 4$ (Roberts,1969); this value was used above. New 21 cm observations show that the rotation curve of some spirals does not approach a Keplerian curve after passing its peak rotational velocity, but that it remains nearly constant,- particularly in the case of M 31 out to 30 kpc from the center (Roberts, 1974). This has cast considerable doubt on the conventional masses derived from rotation curves. However, the evidence of this very interesting observation against the conventional masses should not be exaggerated. First, it is not proved that the flat rotation curves are caused by *circular* motions of the gas in these outer regions; in contrary, the existence of some galaxies with asymmetric rotation curves suggests the influence of non-circular motions. Secondly, the flat rotation curve of M 31, as far as it is now observed, requires only a rather modest increase of the conventional mass (~31 instead of $25 \cdot 10^{10}\mathfrak{M}_\odot$; Roberts, 1974). Finally, it is very important that Huchtmeier (1975) has recently found a number of spirals with clearly descending rotation curves. These galaxies, outnumbering those with flat curves by 8:5, comprise giant spirals (M 101) as well as moderate-size spirals (M 33). At least in these latter cases the 21 cm rotation curves do seem to confirm the conventional masses.

2) Globular clusters. The motion of globular clusters does not only reflect the mass of the flattened stellar population, but the total mass out to their typical distances. In particular, the method accounts for the halo mass within the reach of globular clusters. So far the method has been applied to two galaxies: Lohmann (1961) has derived a mass of $2.1 \cdot 10^{11}\mathfrak{M}_\odot$ for the Galaxy, and Hartwick and Sargent (1974) a mass of $(3.4 \pm 1.4)\ 10^{11}\mathfrak{M}_\odot$ for M 31. These values agree very well with the masses determined from rotation curves. The globular clusters used for the solutions fill volumes with typical radii of 12-16 kpc. Of course, there is the possibility that outside these volumes there is still a considerable (or even the largest) fraction of the halo mass. But out to the distances involved the agreement between mass determinations from rotation curves and from globular clusters contradicts the assumption of any significant excess halo mass.

3) According to Hodge (1966, 1971) the three nearest elliptical dwarf galaxies (Sculptor,Draco,and Ursa Major) are probably bound too and tidally limited by the Galaxy. This conclusion implies a galactic mass of $2 \cdot 10^{11}\mathfrak{M}_\odot$ as seen from these companions. Hence, the conventional galactic mass seems to hold out to the distances of these companions, viz. 70-80 kpc. Applying the same method to three more distant elliptical dwarf galaxies (Leo I, Leo II, and Fornax) Ostriker et al. (1974) have found a very high galactic mass of $(3.9 \pm 2.2) \cdot 10^{12}\mathfrak{M}_\odot$. As Burbidge (1974) has stressed, this assumes, however, that these galaxies were tidally limited at their *present* distances of about 200 kpc, whereas Hodge (1966) has shown that perigalacticons of < 80 kpc are very reasonable also for these objects.

4) Pairs of galaxies. Page (1972) and Smart (1973) have determined average masses and mass-to-light ratios for galaxies in pairs. The latter found for spirals $\mathfrak{M}/L = 1 \pm 1$. This value is surprisingly low, but in view of the large error the deviation from the adopted value $\mathfrak{M}/L = 4$ may not be judged to be significant. In any case the value is *lower*, not larger than 4. Since the separations of double galaxies are $\lesssim$ 50-80 kpc there is direct evidence against any excess mass out to this distance.

Our Galaxy has a Holmberg radius of R ~ 18 kpc, a luminosity of ~ $4 \cdot 10^{10}\ L_\odot$ and a conventional mass of $1.8 \cdot 10^{11}\ \mathfrak{M}_\odot$ (Tammann, 1975); If its mass were to increase as $\mathfrak{M}(R) \propto R$ it would have at R = 80 kpc a mass of $8 \cdot 10^{11}\mathfrak{M}_\odot$ and $\mathfrak{M}/L = 20$. Such a high mass-to-light ratio for a spiral is in open conflict with the evidence from pairs.

5) Clusters of galaxies. Since the times of Zwicky (1933) the stability of rich clusters of galaxies, like the Coma cluster, seemed to be a major problem. The application of the virial theorem has required consistently mass-to-light ratios of ~ 175, thus making the postulation of a "missing mass" attractive. However, the derivation of a virial mass of a cluster is a very intricate problem. Performing N-body calculations Wielen (1974) has shown that clusters cannot be expected to be isothermal; they are hotter in the center. This is confirmed by observations, e.g. for the Virgo cluster (Sandage and Tammann, 1975c). There are additional problems: in the outer zones of a cluster the velocities are mainly in the radial direction, the density profile of the cluster is difficult to determine, the radius of the core is uncertain, and the total luminosity is not precisely known.

A new development has been initiated by Smart (1973), who has shown that the inner galaxies of the Coma cluster form a subsystem and are apparently bound to NGC 4874, the next to brightest cD galaxy in the cluster. With this assumption the virial theorem leads to a mass of "only" $3.4 \cdot 10^{13}\ \mathfrak{M}_\odot$ for this supergiant galaxy. This value, which seems very high in comparison with normal ellipticals, is supported in several ways: for seven cD galaxies in clusters Smart derived an average virial mass of $2.3 \cdot 10^{13}\ \mathfrak{M}_\odot$ with a surprisingly small scatter. Bertola and Capaccioli (1970) have derived the mass of M 87, a cD member of the Virgo cluster, from the line width on spectra of the innermost region; their result is $(5.0 \pm 4.4) \cdot 10^{13}\ \mathfrak{M}_\odot$. From various considerations Wolf and Bahcall (1972) favor a mass range of $10^{13-14}\ \mathfrak{M}_\odot$ for cD's, and Jenner's (1973) result on the cD or db galaxies with double nuclei imply an average mass of $>(1.1 \pm 0.7) \cdot 10^{13}\ \mathfrak{M}_\odot$. Hence, there is observational evidence for cD galaxies, *and only for these galaxies,* to have very large masses. These masses of ~ $3 \cdot 10^{13}\ \mathfrak{M}_\odot$ combined with the standard photographic magnitudes of cD galaxies correspond to a mass-to-light ratio of $\mathfrak{M}/L \sim 160$. However, cD galaxies have faint, very extended halos which carry large fraction of the light (de Vaucouleurs and de Vaucouleurs, 1970; de Vaucouleurs, 1972). Taking this additional luminosity into account leads to $\mathfrak{M}/L \sim 50$ (Bertola and Capaccioli, 1972; Smart, 1973), which is only about a factor of 2 higher than the preferred value for normal ellipticals.

It is tempting to conclude with Smart (1973): "It would therefore appear that ... the 'missing mass' must be *inside* the [cD] galaxy, i.e. there is no 'missing mass'."

7) The mean mass density. There is increasing evidence that the universe is open. According to an argument by Sandage et al. (1972), which was put into an explicit form by Silk (1974a), the isotropy of the Hubble flow requires q_0 to be small (Sandage and Tammann, 1975a; Sandage, 1975), - although this isotropy in itself is still controversial (Peebles, 1975; Rubin and Ford, 1975; de Vaucouleurs, 1975); the final answer on the isotropy is to be expected in the near future from multicolor photometry of elliptical galaxies by Sandage and Visvanathan. A complementary method to measure q_0 shall become available once the fluctuations of the 3 K background radiation can be measured reliably (Silk, 1974b; Zel'dovich, 1975). Independent evidence for small values of q_0 comes from the synthesis of light elements in the big bang (Wagoner, 1973; Peimbert and Torres-Peimbert, 1974, Reeves, 1974 from the age of globular clusters (Sandage and Tammann, 1975b) and other arguments (Gott et.al., 1974). A value of $q_0 \simeq 0.025$ is compatible with all these results, although they are not free of assumptions. In a Friedmann universe with $H_0 = 55$ km s^{-1} Mpc^{-1} and zero cosmological constant the corresponding mean mass density is $3 \cdot 10^{-31} g\ cm^{-3}$ or $4 \cdot 10^9 \mathfrak{M}_\odot\ Mpc^{-3}$.

The mean luminosity density in the universe is about $2 \cdot 10^8\ L_\odot\ Mpc^{-3}$ (Tammann, 1974). In a more recent determination Christensen (1975) has found $3 \cdot 10^8\ L_\odot\ Mpc^{-3}$, which is in very good agreement with the previous value after reducing the latter value to the space outside the Local Supercluster, subtracting 25 percent for the questionable contribution of hypothetical galaxies fainter than $M_B = -8^m$, and increasing it by the internal absorption in spirals.

A mean mass density of $4 \cdot 10^9 \mathfrak{M}_\odot Mpc^{-3}$ and a luminosity density of $2 \cdot 10^8\ L_\odot\ Mpc^{-3}$ imply a mean mass-to-light ratio of $\mathfrak{M}/L \sim 20$. Admittedly the uncertainties are considerable, but the result is in most reasonable agreement with the mean of the conventional mass-to-light ratio of 4 for spirals and 30 for ellipticals. - The result can be stated in a different form: there is evidence that we see most of the total mass of the universe in form of galaxies with conventional masses.

In this paper we have not excluded the possibility that there may exist a few individual spiral galaxies with very massive halos. But the present arguments do not support the view, that our Galaxy or any spirals in bound pairs and groups belong to this species.

Acknowledgement: We have greatly profited together or separately from discussions with Dres. W.K.Huchtmeier, D.Lynden-Bell, N.C. Smart, R.Wielen, and L.Woltjer; we thank them as well as those which let us use pre-publication data, and which are listed in the references. This paper has been made possible by a grant of the Swiss National Science Foundation.

Table I: Bound Groups of Galaxies Within 25 Mpc ($v_o < 1400$ km s^{-1})

Group (1)	Number of vel. (2)	Types (3)	$\langle v_o \rangle$ (km s^{-1}) (4)	$\langle v_i^2 \rangle^{1/2}$ (km s^{-1}) (5)	Crossing time (10^9 yrs) (6)	$\mathfrak{M}_{vis}$ ($10^{10} \mathfrak{M}_\odot$) (7)	T/|Ω| (8)	Source (9)	Remarks (10)
G8 (N 2997)	3	2 S, 1 S0	536	138	7.4	26	[230]	1	kinetic energy indeterminate
M 101	7	7 S	402	58	8.8	33	0.8	1, 6	stable
N 7331	4	4 S	1102	40	8.5	43	0.7	2	stable
Local Gr.	10	7S, 1S0, 2E	-	136	6.2	53	0.2	3	stable, but difficulties with dynamical history
M 51	3	3 S	588	42	15	54	0.2	1	stable
G 14 (N 6300)	3	3 S	1243	95	12	57	[11]	1	kinetic energy indeterminate; strong galactic absorption
G 10 (CVn II)	5	5 S	710	51	42	64	[2]	1	probably stable; kin. energy indet., but small; not relaxed
G 7 (N 1023)	5	4 S, 1 S0	724	29	31	79	[0.7]	4	stable; kin. energy indeterm., but small; not relaxed
G 49	7	3S, 2S0, 2E	1057	276	5	120	1.3	5	stable
G 47 (N 3190)	7	5 S, 2 E	1198	107	8	130	0.04	5	bound, highly negative total energy
Leo	14	9 S, 2 E	667	103	16	230	[0.6]	1	stable; kinetic energy indeterminate but small
G 16	5	1S, 3S0, 1E	999	205	4	340	0.6	5	stable

Sources: 1. Materne and Tammann, 1974b. 2. Materne and Tammann, 1974a. 3. Herbst, 1969. 4. Materne, 1974. 5. Smart, 1973. 6. Sandage and Tammann, 1974.

Table II: A Comparison of Nearby Galaxies in the Shapley-Ames and in Bound Groups of Galaxies ($\delta > -36^o$, $v_o < 1200$ km s^{-1}, the Virgo Cluster is excluded)

Types of galaxies	All Galaxies		Galaxies in Bound Groups			
	Number	Mass ($10^{10} \mathfrak{M}_\odot$)	Number		Mass ($10^{10} \mathfrak{M}_\odot$)	
(1)	(2)	(3)	(4)	(5)	(6)	(7)
E	16	440	9	56%	180	41%
S0	32	800	6	19	130	16
S	179	1260	44	25	500	40
Σ	227	2500	59	26	810	32

REFERENCES

BERTOLA F., and CAPACCIOLI M. 1970. Mem. Soc. Astron. Italiana, N.S. 41, 57

BERTOLA F., and CAPACCIOLI M. 1972. Mem. Soc. Astron. Italiana, N.S. 43, 539

BURBIDGE E.M., and BURBIDGE G.R. 1961. Ap. J. 134, 248

BURBIDGE E.M., and BURBIDGE G.R. 1969. *Galaxies and the Universe,* ed.A. and M.SANDAGE and J.KRISTIAN (=Stars and Stellar Systems, vol. IX), preprint

BURBIDGE E.M., and SARGENT W.L.W. 1971. Semaine d'Etude on the Nuclei of Galaxies (Pontificae Academiae Scientiarum Scripta Varia 35) 351

BURBIDGE G. 1975. Ap. J. (Letters) 196, L 7.

CHINCARINI G., ROOD H.J., and WELCH G.A. 1975. Mon. Not. R. astr. Soc. 170, 441

CHRISTENSEN C,G, 1975. A.J. 80, 282

DAVIES R.D. 1974. *Formation and Dynamics of Galaxies,* ed J.R. SHAKESHAFT (=I.A.U.Symp. No. 58), p.119

EINASTO J., KAASIK A., and SAAR E. 1974. Nature 250, 309

GOTT J.R., GUNN J.E., SCHRAMM, D.N., and TINSLEY B.M. 1974. Ap. J. 194, 543

GUNN J.E. 1974. Comm. Astrophys. Space Phys. 6, 7

HARRISON E.R. 1975. Ap. J. (Letters) 195, L61

HARTWICK F.D.A., and SARGENT W.L.W. 1974. Ap.J. 190, 283

HERBST W. 1969. Publ. Astron. Soc. Pacific 81, 819

HODGE P. 1966. Ap. J. 144, 869

HODGE P. 1971. Ann. Rev. Astron. Astrophys. 10, 227

KORMENDY J., and BAHCALL J.N. 1974. A.J. 79, 671

HOLMBERG E. 1950. Medd. Lund Obs. Ser. II, No. 128

HOLMBERG E., 1958. Medd. Lund Obs. Ser.II.No. 158

HUCHTMEIER W.K. 1975. preprint

JENNER D.C. 1973. Publ. Astr. Soc. Pacific 85, 533

KAHN F.D., and WOLTJER L. 1959. Ap. J. 130, 705

KARACHENTSEV I.D. 1966. Astrofisika 2, 81

LIMBER D.N., and MATHEWS W.G. 1960. Ap. J. 132, 286

LOHMANN W. 1961. Sitzungsber. Oesterr. Akad. Wiss. Math. - naturw. Kl. Abt. II 169, 171

MATERNE J. 1974. Astron. Astrophys. 33, 451

MATERNE J., and TAMMANN, G.A. 1974a. Astron. Astrophys. 35, 441

MATERNE J., and TAMMANN G.A. 1974b. Astron. Astrophys. 37, 383

MATHEWSON D.S., CLEARY M.N., and MURRAY J.D. 1974. Ap. J. 190, 291

MATHEWSON D.S., CLEARY M.N., and MURRAY J.D. 1975. Ap.J. (Letters) 195, L97

OSTRIKER J.P., and PEEBLES P.J.E. 1973. Ap. J. 186, 467

PAGE Th. 1972. *Galaxies and the Universe,* ed. A. and M.SANDAGE and J.KRISTIAN (= Stars and Stellar Systems, vol. IX), preprint

PEEBLES P.J.E. 1975. preprint

PEIMBERT M., and TORRES-PEIMBERT S., 1974. Ap. J. 193, 327

REEVES H. 1974, Ann. Rev. Astron. Astrophys. 12, 437

ROBERTS M.S. 1969. A.J. 74, 859

ROBERTS M.S. 1972 *External Galaxies and Quasi-Stellar Sources,* ed. D.E. EVANS (= I.A.U. Symp. No.44). p.12

ROBERTS M.S. 1974. preprint

ROBINSON L.B. and WAMPLER E.J. 1973. Ap. J. (Letters). 179, L135

RUBIN V.C. and FORD W.K. 1975. Bull A.A.S. 7, 253

SANDAGE A. 1973. Ap. J. 183, 711

SANDAGE A. 1975. preprint

SANDAGE A., and TAMMANN G.A. 1974. Ap.J. 194, 223

SANDAGE A., and TAMMANN G.A. 1975b, Ap. J. 197, 265

SANDAGE A., and TAMMANN, G.A. 1975c, in preparation

SANDAGE A., TAMMANN G.A. and HARDY E. 1972. Ap. J. 172, 253

SHILD R., FRANKSTON M., McCORD T.B., and BERGH S. van den 1975. preprint.

SHAPLEY H., and AMES A. 1932. Harvard Annals 88, No. 2

SILK J. 1974a, Ap. J. 193, 525

SILK J. 1974b, Ap. J. 194, 215

SIMKIN S.M. 1975. A.J. 80, 415

SMART N.C. 1973. Thesis, University of Cambridge.
SPAENHAUER A. 1975. private communication
TAMMANN G.A. 1974. *Confrontation of Cosmological Theories with Observational Data*, ed. M.S..LONGAIR (=I.A.U.Symp. No.63), p.47
TAMMANN G.A. 1975. *Optische Beobachtungsprogramme zur galaktischen Struktur und Dynamik der Milchstrasse* Meeting in Bochum Febr. 1975. preprint.
TURNER E.L., and SARGENT W.L.W. 1974, Ap.J. 194.587
VAUCOULEURS G. de 1965. *Galaxies and the Universe*, ed. A. and M.SANDAGE and J.KRISTIAN (= Stars and Stellar Systems, vol. IX), preprint
VAUCOULEURS G. de 1972. Astrophys. Letters 10. 145
VAUCOULEURS G. de and VAUCOULEURS A. de 1970. Astrophys. Letters 5, 219
WAGONER R.V. 1973, Ap. J. 179. 343
WELCH G.A., CHINCARINI G., and ROOD H.J. 1975. A.J. 80,.77
WIELEN R. 1974. *Stars and the Milky Way System*, ed. L.N.MAVRIDIS. p. 326
WOLF R.A., and BAHCALL J.N. 1972. Ap.J. 176. 559
ZEL'DOVICH Ya.B. 1975. private communication
ZWICKY F. 1933. Helv. Phys. Acta 6. 110

DISCUSSION

LONGAIR: Emerson and Baldwin have shown that there is a maximum in the rotation curve of M31, and that there are differences in this curve as determined in opposite directions along the major axis, and some indication for non-circular motions. Freeman has made an important point: if one takes the luminosity distribution determined optically and assumes a constant M/L ratio, one can work out what the velocity curve should be. The maximum lies further out than the Holmberg radius. Such models provide a satisfactory explanation of radio and optical data on M31. This argument does not indicate that there cannot be more mass outside; it indicates that a constant M/L ratio can explain the velocity curves and that there is no necessity to introduce additional unseen mass.

EINASTO: We have carried out model calculation of the rotation curve of M31 and compared it with radio observations from Great Britain and optical observations. We still find it necessary to add a large corona in order to explain the discrepancies.

TAMMANN: Just a weak reply. To the best of our knowledge, M31 is a member of the Local Group. which is a bound system already with "Conventional" masses:

KARACHENTSEV: What can you say about the dynamical conditions in the M81 group of galaxies?

MATERNE: I do not remember the exact figures by heart , but the group cannot be terrible stable, or it would have appeared in our list. The value of $T/|\Omega|$ may be about 10. But there is a hydrogen cloud in the group.

LUMINOSITY FUNCTION OF THE STARS AND THE PROBLEM OF THE "MISSING MASS" IN THE SOLAR NEIGHBOURHOOD

W. GLIESE
ASTRONOMISCHES RECHEN-INSTITUT, HEIDELBERG, F. R. G.

ФУНКЦИЯ СВЕТИМОСТИ ЗВЕЗД И ПРОБЛЕМА "СКРЫТОЙ МАССЫ" В ОКРЕСТНОСТИ СОЛНЦА

1. Introduction. The problem of the "missing mass" in the solar neighbourhood has been well known for many years as the discrepancy between the density resulting from the masses of known objects or matter and the total mass density derived from dynamical investigations. But where to search for the missing matter?

The space density of stars given by the luminosity function seems to be fairly well determined for stars intrinsically brighter than $M_V = +10$. Beyond that luminosity, our knowledge is still uncertain and the supposed existence of a very large number of low velocity red dwarfs in the vicinity of the sun (Weistrop, 1972) has reactivated the discussion about this problem.

The total mass density as determined by Oort (1965) is $0.15 M_\odot$ pc^{-3} or, if the mass is concentrated to the galactic plane in a thin layer, it may be even $0.21 M_\odot$ pc^{-3}.

From the number of proper motion stars on Schmidt-Palomar plates, Luyten (1968a) has derived a luminosity function which has a maximum between $M_{pg} = +15$ and $+16$. The increasing branch of this curve up to $M_{pg} = 15$ is virtually confirmed by the luminosity function (Jahreiss and Wielen, in: Wielen, 1974) based on the data of the "Catalogue of Nearby Stars" (Gliese, 1969). Both curves are shown in Fig.1 together with the luminosity function derived by Weistrop (1972) between $M_V = +10$ to $+13$, extrapolated to $M_V = +15$ which corresponds to $M_{pg} = +17$. The Weistrop curve includes only red main sequence stars but it lies remarkably above the other two curves which include all stars (white dwarfs!).

From these data, combined with assumptions on white dwarfs, and on the frequency of companions,the conflicting values given in Table I have been derived.

Table I. Mass density in the solar neighbourhood ($M_\odot$ pc^{-3})

luminosity function	Luyten 1968	Jahreiss, Wielen 1974	Weistrop 1972
density from stars	0.064 (all stars)	0.046 (all stars)	M_V <13.5 : 0.075 >13.5 : 0.088 (main-sequence stars) 0.02 (white dwarfs)
density of interstellar matter	0.02 to 0.03		0.03
Sum	0.09	0.07	0.21

According to this result, Weistrop (1972) and Veeder (1974) believe to have found the missing mass and to have solved the problem. But it is striking that more than 40 per cent of their total mass comes from objects of the extrapolated run of the luminosity function. Therefore, the problem remains open until the existence of these low-velocity red dwarfs has been confirmed.

In this paper we investigate whether or not the observed numbers of faint red dwarfs with large proper motions are in agreement with one of the luminosity functions and in contradiction to the other. As long as reliable distance determinations are not available for a representative number of main-sequence stars intrinsically fainter than $M_{pg} = +12$, such an investigation seems to be justified. The research will concentrate mainly upon proper motion distributions in areas near the galactic poles where the number of distant giants is minimum.

A report on a very preliminary comparison between computed numbers of proper motion stars and the stars observed by Luyten (1971) in the vicinity of the South Galactic Pole (SGP) was given in 1973 (Gliese). The present discussion is restricted to the same area of about 285 square degrees or 0.69 per cent of the sky:
RA: 23h49m to 1h36m; Dec: -20°45' to -32°45' (1950.0), but the data published in the volume "The South Galactic Pole" (Luyten and La Bonte, 1973) have been used. The data for the opposite area (11h49m to 13h36m; +20°33' to +32°33') are published in several papers (Luyten, 1961; 1964b; 1968b; 1973; 1974a).

The northern plates have been hand-blinked. At first, plates in the vicinity of the SGP (as used in the preliminary discussion 1973), were hand-blinked too, but then were also investigated with the automated-computerized plate scanner and measuring machine (Luyten and La Bonte, 1973). The authors emphasized that the lists will not be complete for stars brighter than $m_{pg} = 12$ or 13. The following investigation is restricted to faint stars with $m_{pg} \geq 13.0$. The material has been subdivided into four magnitude groups:

From $m_{pg} =$ 13.0 to 14.9
15.0 to 16.9
17.0 to 18.9
19.0 to 21.2

Each of them contains about hundred or even more objects with proper motions exceeding 0.″2 annually ($\mu \geq 0.''20$).

II. Observed Data

a) Proper motions

In the lists of the "Proper motion survey with the forty-eight inch Schmidt telescope" Luyten has tried to record the stars with $\mu \geq 0.''20$ ann as completely as possible. In the first years the plates were hand-blinked only. Luyten (1975) supposes that, in the vicinity of the NGP, the percentage of missed proper motion objects may well be 30 or 40. The plates near the SGP were also machine-processed, which makes the situation much more favourable. Luyten guesses that the listed motions larger than 0.″2 ann. are at least 90% complete for stars fainter than 14pg. Therefore, one expects to find a larger number of proper motion stars in the lists of the SGP areas than in the catalogues of the NGP plates.

There is no possibility of checking the proper motion system of the faintest LP-stars. But in the range $m_{pg} = 12$ to 17, comparisons between the Lowell Proper Motion Survey (by Giclas et al., 1959 - 1975) and the 48-inch data show that, on the average, the Lowell proper motions are larger by 0.″02 or 0.″03 ann. - obviously in all parts of the sky. In the worst case if this difference is entirely due to errors in the LP-proper motions, their number will be too small by a few per cent. But, according to several investigations by Luyten (see 1974b), the 48-inch proper motion system seems to be virtually correct.

The mean errors of hand-blinked proper motions derived from an epoch difference of 11 years amount to about ±0.″020 in each coordinate (Luyten, 1964a). The errors of the machine-processed proper motions are of the order ± 0.″016 (Luyten and La Bonte, 1973). "All motions are relative and require corrections for the average motion of the comparison stars to render them absolute. We estimate that for this region these average roughly +0.″010 and -0.″007 respectively" (Luyten and La Bonte, 1973). In the following investigation these corrections have been applied to the motions near the SGP and corresponding corrections have been applied for the proper motions in the NGP-area. All objects with $\mu \geq 0.''200$ annually have been counted. Double stars and moving pairs are counted as one object, the primary only.

b) Apparent photographic magnitudes

The apparent magnitudes m_{pg} of the LP-stars are given to the tenth of a magnitude class. Luyten (1968a) says "these are all estimates made at the eye-piece of the blink microscope and are based on a simple eye interpolation between the known magnitudes of some B.D. stars on the plates, and the assumption that the limit of the blue survey plates is 21 pg". In the SGP catalogue the plate limit has been always assumed 21^{m}2 pg.

The main problem in this discussion is not the inevitably fairly large accidental errors (Luyten guesses the mean errors to be not larger than $\pm 0^m_.6$ (1964a) but the effects of systematic errors. As the system of the "Durchmusterung magnitudes" varies over the sky, no uniform LP-magnitude system can exist. Additional, slight variations from plate to plate are observed. More serious are extinction effects on the Palomar plates near the SGP, and the question whether or not there exist systematic scale errors varying with magnitude in the range considered. They would shift the limits between the subdivided magnitude regions and alter their widths.

The photoelectric U, B, V series of southern LFT stars (Rodgers and Eggen, 1974) and of the proper motion stars near the SGP (Eggen, 1975) allow an examination of the m_{pg} from 10 to 17 (see Table II).

Table II. Comparison of the LFT- and LP-magnitudes with photoelectric B-magnitudes

m_{pg}	LFT: Rodgers, Eggen				LP: Eggen, 1975			
	n	m(LFT) - B		m.e.$_1$	n	m(LP) - B		m.e.$_1$
10.0 - 12.9	41	$-0^m_.08$	$\pm 0^m_.09$	$\pm 0^m_.60$	42	$-0^m_.01$	$\pm 0^m_.07$	$\pm 0^m_.47$
13.0 - 14.9	66	-0.17	0.045	0.37	61	-0.20	0.055	0.43
15.0 - 15.9	20	-0.05	0.12	0.54	34	-0.54	0.10	0.61
16.0 - 17	4	+0.45	0.19	0.38				

n = number of stars; m.e.$_1$ = mean error of an estimated magnitude

From m_{pg} = 10 to 16 the LFT data from the whole southern sky are, on the average, too bright by a nearly constant amount of $-0^m_.12 \pm 0^m_.04$. The LP-magnitudes in the vicinity of the SGP, however, differ from the LFT system and they show a scale difference: Δm_{pg} = 4.0 corresponds to ΔB = 4.5 in the range 12^m to 16^m which means, that m_{pg} = 13.0 to 14.9 (resp. 12.95 to 14.95) corresponds about to B = 13.00 to 15.25. Therefore, the observed number of proper motion stars with m_{pg} = 13.0 to 14.9 may be too large by 10 or 12%. Also the next region, m_{pg} = 15.0 to 16.9, may be too large. But as the faintest objects cannot be fainter than 21.2 pg (perhaps they are even brighter near the SGP by an extinction effect), the two faintest regions, m_{pg} = 17.0 to 18.9 and 19.0 to 21.2, should be somewhat smaller than $2^m_.0$ resp. $2^m_.3$. No similar comparisons are available for the LP-magnitudes in the NGP areas.

c) Colour classes

The colour classes b to m estimated by Luyten do not coincide exactly with the spectral classes B to M. It is an open question, whether or not this colour system varies slightly from one pair of plates to another, or, whether or not it has slightly changed over the years. Again, the estimates in the LFT catalogue and the LP-colours in the vicinity of the SGP have been compared with the photoelectric data from Rodgers and Eggen (1974) and from Eggen (1975). For the stars with m_{pg} = 10 to 17 the mean B-V of each colour class are given in Table III, together with the extreme values (lowest and highest values) and the standard deviation of a colour class from the mean B-V. As each colour class includes a certain range of B-V values the mean B-V should increase with m_{pg} since lower luminosities with greater B-V are contributing to the proper motion stars of fainter apparent magnitudes. This phenomenon is really existent in the area near the SGP but not among the LFT data which come from all parts of the southern hemisphere. From the difference in Table III it seems advisable to restrict further conclusions to the comparison with Eggen's data.

The scatter of the individual B-V in one colour class allows an estimate of how many objects of this class will be red dwarfs (B-V $\geq$+1.15). The very rough results are shown in Table IV.

For the colours in the northern areas, no direct comparison between the classes b to m and photoelectric B-V is available. For the Giclas colour classes 0 to +4 comparison with B-V is possible near the SGP (Eggen, 1975) and at the northern hemisphere (Priser, 1970; Routly, 1972) which show that the percentage of red dwarfs is larger in the northern colour classes than in the classes near the SGP. Similarities between the relations Luyten-colours/B-V and Giclas-colours/B-V near the SGP let us draw the conclusion that also in the NGP areas the Luyten-classes include somewhat more red dwarfs. In this way the percentages are estimated and given in the last column of Table IV.

Table III. Mean B-V of Luyten's colour classes

Colour class	LFT - colours / Rodgers - Eggen						LP - colours / Eggen, SGP					
	n	<B - V>		$s.d._1$	limits		n	<B - V>		$s.d_1$	limits	
m	57	$+1^{m}.52$	$\pm0^{m}.02$	$\pm0^{m}.14$	$1^{m}.22$	$1^{m}.93$	42	$+1^{m}.44$	$\pm0^{m}.03$	$\pm0^{m}.20$	$0^{m}.84$,	$1^{m}.80$
k - m	6	1.50	0.12	0.30	0.93,	1.82	19	1.29	0.05	0.23	0.88,	1.56
k	24	1.46	0.055	0.27	0.45	1.73	45	1.06	0.04	0.29	0.45,	1.625
g - k							13	0.77	0.065	0.23	0.41,	1.205
g	6	1.30	0.12	0.29	0.88	1.57	14	0.61	0.04	0.135	0.44,	0.895

Table IV. Percentage of red dwarfs for different colour classes near the SGP and near the NGP

Colour class	SGP m_{pg} = 13.0 - 14.9	SGP m_{pg} = 15.0 - 16.9	NGP
m	95%	100%	100%
k - m	90	100	100
k	75	65	90
g - k	15:	-	50
g	-	-	20:

Table V. Observed numbers of proper motion stars in areas of 285 square degrees near the galactic poles

$\mu = 0.''200-0.''299$ annually

m_{pg}	SGP								NGP							
	m	k-m	k	g-k	g	f-g ..a	red stars	O_c	m	k-m	k	g-k	g	f-g ..a	red stars	O_c
1	2	3	4	5	6	7	8	9	10	11	12	13	14	15	16	17
13.0 - 14.9	21	21	19	6	4	1	54	47	15	5	20	1	2	1	39	51
15.0 - 16.9	80	22	15	3	1	6	112	110	50	16	23	3		1	88	125
17.0 - 18.9	62	5	8		1	3	72	78	77	5	5	1		5	87	113
19.0 - 21.2	74		1	1	1		75	89	71	1	1	1			73	98

$\mu > 0.''300$

m_{pg}	m	k-m	k	g-k	g	f-g ..a	red stars	O_c	m	k-m	k	g-k	g	f-g ..a	red stars	O_c
13.0 - 14.9	13	7	6		1	2	23	19	6		14	1			19	20
15.0 - 16.9	38	10	7	2		4	53	50	36	8	10			1	53	62
17.0 - 18.9	26	2				1	28	29	27		1		1		28	32
19.0 - 21.2	22		2				23	26	20		1	1			21	25

The columns 1 - 17 are explained in the text.

d) Accidental errors

The chapter on observed quantities cannot be closed without some comments on the systematic effects of the accidental errors. It is well known (e.g. Dyson, 1926) that in statistical investigations, observed data x need a correction Δx depending on their average mean error and on their frequency distribution:

$$\Delta x = \sigma^2 \frac{\nu'(x)}{\nu(x)} ,$$ where $\nu(x)$ is the

smoothed frequency curve of the measured values x and $\nu'(x)$ is its derivative.
In this investigation, the application of the formula is somewhat risky as, strictly spoken, various frequency functions should be used (for different magnitude intervals and for different proper motion groups). But then the number of objects is too small for accurate statistical corrections. The effects can only be estimated.

Proper motions: With $\sigma(\mu) = \pm 0\rlap{.}''016$ of the machine-measured proper motions, a noticeable effect will occur only at the lower limit $\mu = 0\rlap{.}''200$ where the proper motions in the four magnitude groups need, on the average, small negative corrections of the order - 0.005. Their application would diminish the numbers of observed proper motion stars by about 7 per cent. Near the NGP only hand-blinked proper motions have been published. Their mean errors are significantly larger. Therefore, the observed numbers of stars with $\mu \geq 0\rlap{.}''200$ near the NGP may be too large by even 15%.

Magnitudes: The magnitude errors are of the order $\pm 0\rlap{.}^m4$ to $\pm 0\rlap{.}^m6$. As the frequency curves of the northern and of the southern proper motion stars show their maximum between 15 and 17pg, the effect of these errors, probably, has diminished the number of stars between 15 and 17 by about 7 % and, compensating for this, the observed numbers in the three other regions may be too large by a few per cent (≤ 5%).

e) Observed number of stars with proper motions exceeding $0\rlap{.}''20$ annually

Table V gives the numbers of observed proper motion stars for the different colour classes in four magnitude groups (columns 2-7, 10-15). An estimate of the percentages of red dwarfs in the classes m, k-m, k, g-k, and g is possible by comparisons between these classes and photoelectric B-V (see Table IV). The numbers of red objects resulting from these percentages are given in columns 8 and 16. But these figures are not yet the true numbers of red proper motion stars in both areas. The final values can be estimated only by taking into account all the disturbing effects quoted below:

1) Incompleteness corrections are positive corrections. They introduce the largest uncertainties into the final results.

2) Systematic errors of the proper motions are unknown and probably insignificant.

3) Corrections due to the systematic effects of the accidental errors in the proper motions are negative corrections.

4) Systematic errors of the photographic magnitudes m_{pg} in the vicinity of the SGP probably produce negative corrections from 13 to 17 pg, and positive corrections from m_{pg} = 17 to 21. Near the NGP the corrections are unknown.

5) Corrections due to the systematic effects of the accidental errors in the m_{pg} are negative in the groups 13 - 15 pg, 17 - 19 pg, 19 - 21 pg (small) and positive in the range m_{pg} = 15 to 17.

6) Knowledge of the percentages of red dwarfs in the colour classes k-m, k, g-k, and g is uncertain.

7) Knowledge of the true differences between relative and absolute proper motions is uncertain.

8) Knowledge of the statistical (cosmic) dispersion is uncertain as each area includes only 0.69 per cent of the sphere.

Columns 9 and 17 give the "corrected observed numbers" O_c of red dwarfs with large proper motions. Due to the uncertainties cited above, errors up to ±30% (SGP) or even ±35% (NGP) may be possible. The figures O_c will be compared with computed numbers which are based on reasonable assumptions concerning the luminosity function and the velocity dispersion of the red dwarfs in the solar neighbourhood.

III. Computed numbers of faint proper motion stars. In the following, the numbers of stars with proper motions $\mu \geq 0\rlap{.}''20$ annually have been computed in models based on plausible suppositions concerning the luminosity function and the velocity dispersion of red dwarfs. The investigations are restricted to the motions in the galactic U,V plane which is the tangential plane at the galactic poles. As both areas have latitudes $|b|>75°$, the deviations between the planes of the proper motions and the U,V plane are negligible.

From the velocity distributions, the percentage of those stars are read whose $TV = (U^2 + V^2)^{1/2}$ exceed certain values. The formula $TV = 4.74\ \mu/\pi$ allows a derivation of the percentage P of stars with $\mu > \mu_0$ as a function of the parallax π, or as a function of the distance modulus m - M. In practice, these relations have been graphically slightly smoothed. The space around the sun is subdivided into thin shells. The i-th shell lies between the two spheres with the radii which are defined by the distance moduli $(m-M)_i - 0.^m1$ and $(m-M)_i + 0.^m1$. The number n_{ij} of stars with the absolute magnitude M_j (from $M_j - 0.^m1$ to $M_j + 0.^m1$) is given by the luminosity function and the volume of the i-th shell. A homogeneous distribution of the stars in the vicinity of the sun has been assumed.

The product $0.0069\ n_{ij}\ P_i = N_{ij}$ is the number of objects of luminosity M_j observed as $m_i = M_j + (m-M)_i$ with $\mu > \mu_0$ where $\mu_0 = 0.''200$ resp. $0.''300$ annually. Finally, the numbers N_{ij} from all shells which contribute to the stars in the magnitude interval $m_1 < m_i < m_2$ have been summed up. The results are the computed numbers of proper motion stars in Table VI which have to be compared with the "O_c" in Table V.

The method of deriving the numbers of proper motion stars from the models is fairly exact except some uncertainty of the products "percentage P times volume of the shell" in larger distances where P rapidly decreases and where the volume from shell to shell grows significantly. It is estimated that such uncertainties do not exceed 10 per cent of the final computed numbers.

The computations are based on the following six models:

A) The luminosity function of red stars intrinsically fainter than $M_{pg} = +8$ is that by Luyten (1968a) diminished by the luminosity function of the white dwarfs between $M_{pg} = 10$ and 17. The latter has been derived from the data given by Allen (1973).

The velocity distribution of red dwarfs from $M_{pg} = 8$ to 20 is that of the McCormick stars (Vyssotsky et al., 1943, 1946, 1952, 1956) with $M_{pg} = 8$ to 12. This is a very uncertain assumption as nothing is known about the velocities of samples of fainter stars free from selection effects. But this supposition is supported by the observations that the velocity dispersion does not vary significantly from dG to dM2 stars (Gliese, 1956; Wielen, 1974).

On the other hand, M dwarfs do not form a homogeneous group. About one third of the McCormick stars nearer than 22 pc (Gliese, 1969) are designated as emission-line objects;their velocity dispersion is significantly smaller than that of the dM stars (Delhaye, 1953;Gliese, 1958). The observed velocity distribution seems to depend on the ratio of dMe to dM stars.

A further assumption here is that the fraction of emission-line objects is virtually constant for M dwarfs of all luminosities, namely about 30 or 35 per cent.

To demonstrate the effects of different velocity dispersions the computations have been made for three samples of Vyssotsky stars:

1) dM stars only,
2) dMe stars only
3) group "C", namely stars with the velocity distribution of the 292 McCormick stars with known U, V, W components in the "Catalogue of Nearby Stars" (Gliese, 1969). Case "C" is the mixture of case dM and case dMe, namely the McCormick star population as observed near the sun.

Recent observations (Wilson and Woolley, 1970) and investigations (Wielen, 1974) have shown that such a simple subdivision into two classes cannot be more than a first rough approximation. Probably, with a sufficiently large spectral dispersion, Ca II emission will be seen in the spectra of most of the late red dwarfs. However the emissions are of very different intensity. This intensity is strongly correlated with the age of the stars and, combining stars of equal ages, the age is correlated with the velocity dispersion. Wielen's Table VI shows an obvious correlation between the Ca II emission intensity and the velocity dispersion among the McCormick dwarfs. The objects classed as dMe in the "Catalogue of Nearby Stars" are compiled from various sources regardless of the emission-line intensity and thus they form no homogeneous group. Their velocities have a mean dispersion of the dMe star population. It can be estimated that the stars with strongest emission lines have a velocity dispersion nearly 20 per cent smaller than that of the dMe star sample used here. But the observed data are not yet sufficient for statistical investigations of various Me dwarf groups subdivided according to different emission-line intensities.

The velocity distribution of the McCormick stars used here is disturbed by observational errors which

a) slightly increase the observed velocity dispersion, and, but also b) which cause, on the average, a decrease in the tangential velocities as, due to the restriction to parallaxes exceeding 0.″044, there is an excess of positive parallax errors. Both these effects are smaller than 4 per cent and have been neglected.

B) Weistrop (1972) assumes a rich population of low velocity red dwarfs in addition to the known M dwarf population in the solar neighbourhood. The corresponding luminosity function is shown in Fig. 1. It has been derived for the stars with $M_V = +10$ to 13 and then extrapolated down to $M_V = 15$. The velocity dispersion of these objects is still unknown, but it is supposed to be of the order ±10 km s^{-1} in one coordinate. Also the relative solar motion should be smaller than for other star groups.

The Weistrop function is combined with a velocity dispersion $\sigma_{UV} = \pm 10$ km s^{-1} in the U, V plane and with three assumed values for the solar velocity $s_\odot = (U_\odot^2 + V_\odot^2)^{1/2}$:

B1) $s_\odot = 0$ km s^{-1}. This model, certainly, is unrealistic as the observed asymmetry in the proper motions contradicts a star group moving with the sun's velocity.

B2) $s_\odot = 10$ km s^{-1}

B3) $s_\odot = 20$ km s^{-1}

The results of these six computations are shown in Table VI. They are compared with the observed numbers of red proper motion stars from Table V "O_C" and with the uncorrected numbers of the observed stars.

IV. Comparison between observed and computed numbers of proper motion stars. In general, observed and computed numbers are of the same order. But a careful examination shows some irregularities of the differences "observed minus computed". A more detailed analysis is in progress. This paper confines itself to the main features. Nearly all red dwarfs with large proper motions are at distances nearer than 100 pc, the stars of the low velocity population are even nearer than 40 pc. The decrease of the star density with distance z from the galactic plane cannot appear in these numbers.

Table VI. Computed numbers of faint red proper motion stars in areas of 285 square degrees near the galactic poles

Luminosity function	Case	S_0 [km s^{-1}]	σUV	μ = 0.″200 - 0.″299 m_{pg}: 13	15	17	19 – 21	μ > 0.″300 13	15	17	19	21
Luyten 1968	dM	22	±40	64	88	128	185	33	46	64	62
	dMe	12	30	23	31	47	47	12	17	20	12
	"C"			37	51	77	98	26	37	50	46
Weistrop 1972	B_1	0	±10	3	12	24	7	3	8	7	0
	B_2	10	10	5	24	56	30	7	19	22	4
	B_3	20	10	10	65	225	221	14	48	80	17
Luyten "C" + Weistrop	B_2			42	75	133	128	33	56	72	50
Luyten "C" + Weistrop	B_3			47	116	302	319	40	85	130	63
Observed data =0.5 (north + south) red stars, uncorrected				47	100	79	74	21	53	28	22
O_C				49	117	95	93	19	56	30	26

The different models are explained in the text. The computed star numbers are given for the four magnitude intervals between $m_{pg} = 13$ and 21. For comparison, the mean numbers of the proper motion stars discovered near the NGP and of those observed near the SGP (Table 5) are added.

In nearly all groups, the observed differences between the numbers of northern and southern proper motion stars are insignificant. The excess of the northern "O_C" over the southern "O_C" in the group m_{pg} = 17-19 may be partly due to uncertain incompleteness factors.

With variations in the ratio dMe: dM numbers Luyten's luminosity function can explain the observed quantities in three magnitude groups for $0''.2 \leq \mu < 0''.3$. However the large number of proper motion stars of m_{pg} = 15-17 which is observed near the SGP as well as in the vicinity of the NGP seems to be unexplicable by a Luyten/McCormick model alone. Additional stars from a low velocity population would bring computed and observed numbers into better accordance.

Table VI shows that only a few members of the supposed low-velocity dwarfs can be observed in the proper motion surveys of stars brighter than m_{pg} = 17.

Therefore, down to this limit Luyten's observations cannot decide in favour of or against the Weistrop population from M_V = 10 to 13 (or M_{pg} = 11.5 to 15). Among the fainter stars, however, the Luyten/McCormick model alone is in sufficient agreement with the observed data. Additional objects from the low-velocity models would yield computed quantities considerably higher than the observed numbers. Therefore, the existence of the low velocity stars in the extrapolated range $M_V > 13$ ($M_{pg} > 15$) is not confirmed by the proper motion surveys between $0''.20$ and $0''.30$ ann.

The discussion and interpretation of the data of stars with large proper motions ($\mu \geq 0''.300$ ann.) seem to be more complicated. A study in greater detail has been postponed. But the conclusion that the extrapolation of the luminosity function by Weistrop is not justified is strongly confirmed by the low frequency of these objects among the stars fainter than m_{pg}= 17.

The essential results for the problem of the "missing mass" are the following:

The existence of many low-velocity red dwarfs with luminosities between M_V = 10 and 13 seems to be possible, but on the basis of only these data this question cannot be solved definitely. However, among the intrinsically fainter stars with $M_V > 13.5$ ($M_{pg} > 15$) no extraordinary large number of low-velocity objects can be present in the solar neighbourhood.

Just these suspected red dwarfs yielded the principal contribution of 0.09 $M_\odot$ pc^{-3} to the total mass density as shown in Table I. Without this quantity, only a mass density of at the most 0.12$M_\odot$ pc^{-3} has been discovered up to now. Therefore, at least a small amount of matter of 0.03 $M_\odot$ pc^{-3} is still unknown as "missing mass" in the solar neighbourhood.

The problem of the "missing mass" has not yet been solved completely by the suspected low-velocity red dwarfs even if they exist with luminosities between M_V = 10 and 13.

The possibility of solving the problem by correcting Oort's values of the total mass density in the solar neighbourhood is not discussed in this paper.

A c k n o w l e d g e m e n t s : The author is greatly indebted to Prof. O.J.Eggen for sending a preprint of his paper containing photoelectric measurements of magnitudes and colours of proper motion stars near the South Galactic Pole. I also wish to thank Prof. W.J.Luyten for detailed information on his observed data. The English version of the text was graciously corrected by Dr. R.Scholl.

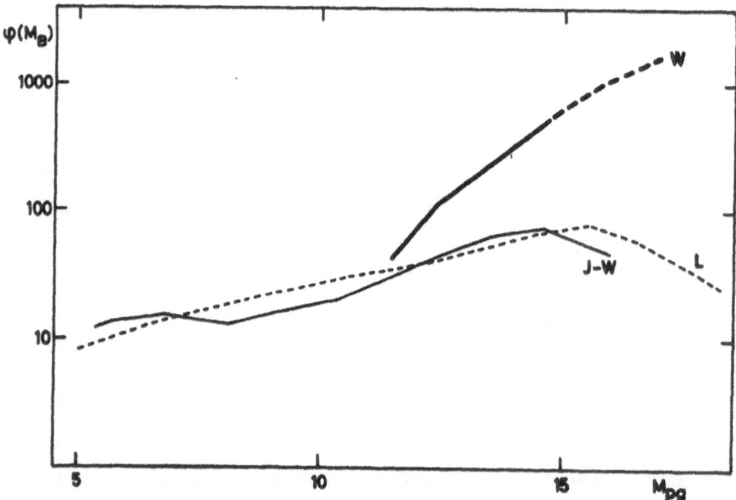

Fig. 1. Luminosity functions $\varphi(M_B)$ by Luyten (1968), Jahreiss-Wielen (Wielen, 1974), and by Weistrop (1972). Stars per unit magnitude interval in a sphere of radius, r = 10 pc.

REFERENCES

ALLEN C.W., 1973. Astrophysical Quantities, 3rd ed., The Athlone Press, Univ. of London, p.226, 249

DELHAYE J., 1953, C.R.Acad. Sci. Paris 237, 294

DYSON F., 1926, Monthly Notices Roy. Astron. Soc. 86, 686

EGGEN O.J., 1975, Astrophys. J., in press

GICLAS H.L., SLAUGHTER C.D. and BURNHAM Jr., R., 1959, Lowell Obs. Bull. No. 102

GICLAS H.L., BURNHAM Jr.,R. and THOMAS N.G., 1961, Lowell Obs. Bull. No. 112; 1963, ibid. No. 120;1964, ibid. Nos. 122 and 124; 1965, ibid. No. 129; 1966, ibid. Nos.132 and 136; 1967, ibid. Nos. 138 and 140; 1968,ibid. No. 144; 1969, ibid. Nos. 150 and 151; 1970, ibid. No. 152; 1972, ibid. No. 158; 1973, ibid. No. 160; 1975, ibid.No. 162.

GLIESE W., 1956, Z. Astrophys. 39, 1

GLIESE W., 1958, Z.Astrophys. 45, 293

GLIESE W., 1969, Veröffentl. Astron. Rechen-Inst. Heidelberg No.22

GLIESE W., 1973, Mitt. Astron. Ges. 34, 100

LUYTEN W.J., 1961, Publ. Astr. Obs. Univ. Minnesota 3, No. 10

LUYTEN W.J., 1964a, Proper motion survey with the forty-eight inch Schmidt telescope, Univ. of Minnesota, Minneapolis, No. 2.

LUYTEN W.J., 1964b, Ibid. No. 3

LUYTEN W.J., 1968a, Monthly Notices Royal Astron Soc. 139, 221

LUYTEN W.J., 1968b, Proper motion survey with the forty-eight inch Schmidt telescope, Univ. of Minnesota, Minneapolis, No. 14.

LUYTEN W.J., 1971, Ibid. No. 27

LUYTEN W.J., 1973, Ibid. No. 35

LUYTEN W.J., 1974a, Ibid. No. 36

LUYTEN W.J., 1974b, Ibid. No. 38, 16

LUYTEN W.J., 1975, private communication

LUYTEN W.J., and La BONTE A.E., 1973, The South Galactic Pole, Minneapolis, Minnesota

OORT J.H., 1965, in A.BLAAUW and M.SCHMIDT (eds), 'Galactic Structure', Univ. of Chicago Press, Chicago, p.455

PRISER J.B., 1970, Publ. U.S.Nav. Obs. (2) 20, pt. 3, 27

PODGERS A.W. and EGGEN O.J., 1974, Publ. Astron. Soc. Pacific 86, 742

ROUTLY P.M., 1972, Publ. U.S. Nav. Obs. (2) 20 pt. 6

VEEDER G.J., 1974, Astrophys. J. Letters 191, L 57

VYSSOTSKY A.N., 1943, Astrophys. J. 97, 381

VYSSOTSKY A.N., 1956, Astron. J. 61, 201

VYSSOTSKY A.N., JANSSEN E.M., MILLER W.J. and WALTHER M.E., 1946, Astrophys. J. 104, 234

VYSSOTSKY A.N. and MATEER B.A., 1952, Astrophys. J. 116, 117

WEISTROP D., 1972, Astron. J. 77, 849

WILSON O. and WOOLLEY R., 1970, Monthly Notices Roy. Astron. Soc. 148, 463

WIELEN R., 1974, in G.Contopoulos (ed.), 'Highlights of Astronomy' 3, 395

STRINGENT UPPER LIMIT OF THE BLACK HOLE MASS IN THE CENTRE OF OUR GALAXY

L.M. OZERNOY
P.N.LEBEDEV PHYSICAL INSTITUTE
ACADEMY OF SCIENCES, U.S.S.R.

ТОЧНЫЙ ВЕРХНИЙ ПРЕДЕЛ МАССЫ ЧЕРНОЙ ДЫРЫ В ЦЕНТРЕ НАШЕЙ ГАЛАКТИКИ

During last years many theorists following the outstanding Lynden-Bell (1969) paper, think that a source of activity in the nuclei of galaxies including our own may be an accreting supermassive black hole. Considering the nuclei of galaxies as dead quasars many authors suppose that the mass of a "corpse" in the form of a black hole is as large as $10^{7-11}\,M_\odot$ (e.g. Novikov and Thorne, 1973 and literature cited there). I should like to suggest that the available observational data may limit the mass of a black hole in the centre of our Galaxy to a much smaller value.

The present uncertainty for the mass and luminosity of the central black hole is caused by the unknown value of the accretion rate of interstellar gas because we do not know the rate of its angular momentum loss. However, we can calculate from the observational data the lower limit of the accretion rate due to the *inevitable* accretion onto the black hole of the stars surrounding it and disrupted by its tidal forces. The stars from the galactic nucleus getting into a sphere of the Roche radius

$$R_R = \left(\frac{6}{\pi}\,\frac{M_h}{\rho_s}\right)^{1/3} = 1.4\cdot 10^{11}\left(\frac{M_h\,/\,M_\odot}{\rho_s\,/\,\rho_\odot}\right)^{1/3}\text{cm} \qquad (1)$$

where M_h is the black hole mass and ρ_s - the mean density of a star, are disrupted by the tidal forces and are transformed into the gas. Being within the sphere of capture by the black hole

$$R_h = \frac{2GM_h}{V_*^2} = 6.7\cdot 10^{11}\,\frac{M_h\,/\,M_\odot}{(V_*\,/2\cdot 10^7\ \text{km}\cdot\text{sec}^{-1})^2}\ \text{cm}, \qquad (2)$$

this gas is accreted onto the black hole. Due to the low angular momentum of the gas as compared with the angular momentum at the last stable orbit near the black hole, the accretion proceeds in a spherically-symmetric manner. The energetic efficiency of the turbulent and randomly magnetized gas from the dispersed stars may be estimated as large as 0.1 (Schwarzman,1971), i.e. the electromagnetic luminosity of the black hole due to accretion is

$$L = 0.1c^2\,\frac{dM_a}{dt} = 0.57\cdot 10^{46}\,\frac{dM_a\,/\,dt}{M_\odot\ /\text{year}}\ \text{erg}\cdot\text{sec}^{-1} \qquad (3)$$

The rate of accretion of gas released is equal to

$$\frac{dM_a}{dt} = \langle \sigma\,\rho_*\,V_*\rangle\ , \qquad \sigma = \pi r^2\,(1 + 2GM_h\,/\,rV_*^2\,) \qquad (4)$$

where ρ_* and V_* are the concentration and velocity dispersion of the surrounding stars respectively. Since $V_p \approx (2GM_h\,/\,R_R\,)^{1/2} = 2\cdot 10^7 (M_h/M_\odot\,)^{1/3}(\rho_s\,/\rho_\odot)^{1/3}$cm/sec is expected to be greater than V_*, the cross-section $\sigma \simeq \pi R_R R_h$, and it is large enough to make the capture of stars and their disruption an effective energy source.

If we neglect the dependence of ρ_* and V_* on r due to the influence of the black hole, the rate of accretion is estimated to be (cf.Hills, 1975)

$$\frac{dM_a}{dt} = 6.3\cdot 10^{-16}\,\frac{M_h}{M_\odot}\,\frac{\rho_*}{M_\odot\cdot pc^{-3}}\,\frac{R_R}{R_\odot}\left(\frac{V}{1\text{km}\cdot\text{sec}^{-1}}\right)^{-1}\frac{M_\odot}{\text{year}} =$$

$$= 1.4\cdot 10^{-6}\left(\frac{M_h}{10^4\,M_\odot}\right)^{4/3}\left(\frac{\rho_*}{10^6\,M_\odot pc^{-3}}\right)\left(\frac{\rho_s}{\rho_\odot}\right)^{-1/3}\left(\frac{V_*}{200\ \text{km}\cdot\text{sec}^{-1}}\right)^{-1}\frac{M_\odot}{\text{year}} \qquad (5)$$

In fact, it is necessary to take into account the re-distribution of stars by the black hole at $r \leqslant R_n$ whose influence leads to the following distributions (Peebles, 1972):

$$\rho_* \propto r^{9/4}, \quad V_* \propto r^{-1/2} \quad \text{at} \quad r \leqslant R_h , \tag{6}$$

and we obtain instead of Eq. (5)

$$\frac{dM_a}{dt} \approx 1\left(\frac{M_h}{10^4 M_\odot}\right)^{5/2} \frac{\rho_*}{10^6 M_\odot \ pc^{-3}} \left(\frac{V_*}{200 \ km \cdot sec^{-1}}\right)^{-9/2} \cdot \left(\frac{\rho_s}{\rho_\odot}\right)^{1/4} \frac{M_\odot}{year} . \tag{7}$$

Inserting Eq.(7) into Eq.(3) and solving the latter with respect to M_h, we obtain the upper limit of the black hole mass depending only on the observable quantities:

$$M_h = 3.2 \cdot 10^2 \left(\frac{L}{10^{42} erg \cdot sec^{-1}}\right)^{2/5} \left(\frac{\rho_*}{10^6 M_\odot \ pc^{-3}}\right)^{-2/5} \left(\frac{V_*}{200 \ km \cdot sec^{-1}}\right)^{9/5} \left(\frac{\rho_s}{\rho_\odot}\right)^{-1/10} M_\odot \tag{7}$$

As is known, for the nucleus of our Galaxy we have $\rho_* \approx 7 \cdot 10^6$ stars/pc^3 and $V_* \approx 200$km/sec. The total non-thermal luminosity of the nucleus is estimated usually as 10^{42}erg/sec (e.g.Lynden-Bell and Rees, 1971), but it may be as much as 10^2 times smaller, taking into account the recent high-resolution observations at 10μ (Becklin and Neugebauer, 1975). Inserting these values into Eq.(7) we obtain

$$M_h \leqslant 150\text{-}23 \ M_\odot \quad \text{at} \quad L = 10^{42} - 10^{40} \ erg/sec \tag{8}$$

This extremely low upper limit of the black hole mass in the galactic centre leads to three important conclusions.

1. The critical (Eddington) luminosity corresponding to the limit (8) is not greater than $L_E = 10^{40}$ erg/sec. This value is much smaller than the total luminosity of Seyfert galaxy nuclei not to mention that of quasars. Because of the fact that Seyfert galaxies differ from normal giant spirals only by the excited state of their nuclei, it is impossible to explain the non-thermal radiation of Seyfert nuclei in the framework of the black hole concept.

2. The Lynden- Bell's (1969) hypothesis that nuclei of galaxies are dead quasars is also in contradiction with the upper limit (8) for nuclei of spiral galaxies, at least. The source of activity in a quasar nucleus is not transformed completely into a black hole, and somehow (apparently, by an explosion) throws off the most part of its mass.

3. The upper limit (8) is much smaller than the value $\sim 10^7 M_\odot$ up to which the mass of a relict black hole must increase if it serves as a centre of the formation of the Galaxy by an accretion (Ryan, 1972; Gribbin 1974; Carr, 1975). Consequently, the assumption widely discussed at present, that the relict black holes may be the main factor of galaxy formation is in contradiction with the observational data.

REFERENCES

BECKLIN E.E. and NEUGEBAUER G. 1975, Preprint

CARR B.J. 1975, Preprint OAP-389

GRIBBIN J. 1974, Nature 252, 445

HILLS J.G. 1975, Nature 254, 295

LYNDEN-BELL D. 1969, Nature 223, 690

LYNDEN-BELL D. and REES N. 1971, Monthly Notices of R.A.S. 152, 461

NOVIKOV I.D. and THORNE K.S. 1973, in "Black Holes", eds. C.DeWitt and B.DeWitt (New York: Cordon and Breach pp. 343-450

PEEBLES P.J.E. 1972, Astrophys. J. 178, 371

RYAN M.P. 1972, Astrophys. J. 177, L 79

SCHWARZMAN V.F. 1971, Astron. Zh. 48, 471

A NOTE ON THE COLLECTIVE EVOLUTION OF RICH CLUSTERS OF GALAXIES

G. PAÁL
KONKOLY OBSERVATORY, BUDAPEST, HUNGARY

ЗАМЕТКА О ГРУППОВОЙ ЭВОЛЮЦИИ БОГАТЫХ СКОПЛЕНИЙ ГАЛАКТИК

I n t r o d u c t i o n. Recently stability considerations have led Ostriker and Peebles (1974) and Einasto et al. (1974) to the idea that hidden mass may typically be present in massive but subluminous halos (or coronas) around some giant members of systems of galaxies*. This concept has since been widely developed and discussed; both supported by independent arguments and criticized**. It is however rather obvious that even if true, this concept cannot be extended to rich clusters of galaxies without any modification ***. It is therefore important to search for further possible effects the massive halos might produce and to compare theoretical predictions with empirical data. The present note aims at proposing and applying empirical tests sensitive to the distribution and amount of hidden mass in halos of members of rich clusters of galaxies.

Let us first approach the problem from the point of view of the theory of collisional relaxation in gravitationally bound systems, and add massive halos to the giant members of rich galaxy clusters. In this case the number of dynamically important "individual" objects(giant galaxies with halos and -perhaps - companions) may turn out smaller roughly by an order of magnitude than the total number of member galaxies in the system. "Rich" clusters may be dynamically "poor": they may well be built up from, say, a few scores of important objects with typical mass close to 10^{13} solar masses each. If rich clusters are bound (with dark halos of giant members containing the missing mass required by the Virial theorem), the time scale of their expected dynamical evolution may be very short or very long depending on the *extent* and stability of halos. Distributing the hypothetical hidden mass in the form of large overlapping halos (essentially one common halo for the whole cluster,or its main condensation) would make the individual encounters ("collisions") ineffective and slow down the collisional relaxation process. On the other hand distributing the same dark matter in an essentially discrete, clumpy form (partially isolated and autonomous massive halos) would speed up the dynamical evolution by collisional relaxation. The reason for the latter lies in the fact that the number of dynamically significant objects would decrease and the total mass increase by about an order of magnitude. The relaxation time t_r is related to the crossing time t_c, the total mass M, the total number N and harmonic mean radius R of the cluster by the following relations $t_r/t_c \sim N/\lg N$ and $t_c \sim (R^3/M)^{1/2}$ (Chandrasekhar, 1942; Lecar, 1970; Hennon, 1970; Aarseth and Saslaw, 1972). In our case the range of reasonable relaxation times for "rich" clusters of galaxies turns out to be very wide. A typical value t_r may be everything from 10^9 to 10^{12} years depending on the extent and number of assumed massive halos (Paál, 1973). Moreover it is known (Zel'dovich and Podurets, 1965) that the evolution of collisionally dominated systems "runs away" (as N and $R^3/M \approx 1/\rho$ become smaller) and if t_c and t_r are of the order of 10^9 years, then rich clusters can be "old enough" to have reached their age of "rapid"evolution (before the unification of individual halos into a single coalesced halo and subsequent slowing down of further relaxation). Thus any upper or lower observational limit of the speed or effectiveness of relaxation processes in rich clusters of galaxies could in principle be converted into a corresponding limit of the distribution and amount of hidden masses in them. - Note that if large halos (or "coronas") were becoming coalesced soon after the formation of rich clusters of galaxies, the fact of

* For the sake of generality of our considerations no distinction is made here between the names "halo" and "corona".

** See e.g. Burbidge (1974) and Materne and Tammann (1974).

*** The mass to luminosity ratio, size and autonomy of typical giant galaxies with coronas in loose systems seem to be typical in dense rich clusters.

certain degree of partial segregation of bright and faint member galaxies in some typical clusters would be very hard to explain. - Evolutionary tests of clusters of galaxies are however important also outside the frame of collisional relaxation theory of gravitationally bound systems and without regard to speculations on hidden masses : any cosmogonical or cosmological theory is in need of data on evolution. Therefore we are going to discuss briefly the evidences of evolution independently of the previous suggestions and other theoretical predictions. We should however bear in mind that in the light of new independent results concerning massive halos an appreciable evolution of *rich* galaxy clusters has lost its aspect of being theoretically unacceptable or implausible, it finds a natural explanation in the frame of conventional physics.

Outline of the evolutionary tests. If a definite theory of evolution of clusters of galaxies were adopted, an observed "final" product of a long evolution could serve as an integrated measure of all past evolutionary changes. This is the case when we judge on evolutionary stage by the mere existence of partial segregation in some systems. If, however, "direct empirical data" are needed to see how the evolution proceeds - whatever its correct interpretation, - then some distant ("younger") extragalactic samples are to be compared with the corresponding nearby ("older") ones in order to find a measure of systematic differences between two samples of different cosmological epoch*. Such a comparison were straightforward, if our data referred to two or more different parameters of the same physical nature in all distances considered (e.g. two different characteristic angular sizes of clusters in all distances). If however, our data for each object are of different physical nature (e.g. brightness, size, spectrum), then far-reaching cosmological presuppositions are generally invoked when comparing distant and local samples. In this case observations are compared with cosmological predictions - not with each other - and almost any deviation between theory and observation may be interpreted either as an unrecognized cosmological effect, or the evolutionary effect looked for. The independence of theory of clusters is paid for by a dependence on theory of the universe. In what follows a maximum possible separation of collective evolution from cosmological effects will be attempted applying two specific methods - and their combination - to limit or exclude much of the undesirable dependence of evolutionary tests on the world model adopted.

Method 1. A homogeneous sample of characteristic angular sizes, ϑ_i describing some structural feature of clusters of galaxies at different distances and times (e.g. Paál, 1971, 1973), together with a homogeneous apparent brightness - red shift relation, (l_i, z_i) constructed for the same sample permit us to compare "surface brightnesses", $B = \frac{l}{\pi \vartheta^2}$, near and far**. Hubble and Tolman (1935), Kristian and Sachs (1966), Penrose (1966), Sachs (1968) and others have already pointed out in several independent ways that in an empty space populated by equal objects B is remarkably independent of the world model: it only depends on red shift z according to a unique formula valid in all relativistic cosmological models including inhomogeneous and anisotropic ones /the geometric and dynamical effects vanish/. It is perhaps interesting to note that using a result of Whitrow (1954) an analogous formula can be derived for "all" static world models with some progressive reddening of "tired light". A formula valid for both classes of models reads

$$B = \frac{L}{\pi \Delta^2} = \frac{l}{\pi \vartheta^2} \cdot (1 + z)^n \qquad (1)$$

where l and ϑ are the apparent brightness and size, L and Δ the corresponding absolute quantities seen from unit distance, z is the red shift, while n is equal to -4 or -1 in expanding or static world models respectively. Identity (1) is valid in an absolutely transparent space, unless the properties of "standard" objects studied are functions of spatial position or time. Let us consider briefly the possible causes of deviation from identity (1)- i.e. deviation of observed value of n from -4 or -1.

Neglecting first extinction and luminosity evolution the following non-geocentrical cases are to be distinguished from the point of view of evolution.

---

* An evolution of the corresponding samples is termed "collective" in the title of present note.

** These are only fictitious "surface brightnesses" obtained by dividing apparent brightnesses of some member galaxies by $\pi \vartheta_i^2$, i.e. a characteristic solid angle of the clusters themselves.

$n > -1$: the clusters are "expanding". (More exactly their specified characteristic size is increasing, whatever the true nature of extragalactic red shifts.)

$n = -1$: the clusters can be non-evolving in a non-expanding universe. (In some respect they can "partake" in the general rest. Some of their observed structural features show no distinguishing kinematical properties when contrasted with the general substratum in the range observed.)

$n = -2$: the clusters can be co-expanding with the universe. (They can "partake" in the general expansion showing no distinguishing kinematical properties again.)

$n \quad -2$: the metric evolution of clusters differs from that of the universe; some structural features of clusters do not "partake" in the general cosmological motion (whatever it may be); the clusters are subjected to their own law (e.g. the influence of their own gravity).

$n = -4$: the clusters can be non-evolving in an expanding universe.

$n < -4$: the clusters are "contracting". (Their specified size is decreasing, whatever the true world model, the nature of red shifts and the cause of evolution of structure in clusters).

According to the usual relaxation theory in an expanding universe the values $n > -4$ and $n < -4$ are expected for characteristic sizes of outer parts (halo) and inner parts (core) of clusters respectively, if there is a detectable dynamical evolution in clusters (e.g. in case of less extended massive halos around giant members), and the value $n = -4$ for *all* sizes, if there is practically no evolution (e.g. hidden mass in too extended coronas and clusters without an appreciable partial segregation of light and heavy members).

Admitting extinction would lead to an overestimation of n. Any significant nonlinearity of the relation between lg B and lg (1 + z) would mean some kind of temporal change in large samples of the universe, unless geocentrical inhomogeneities are accepted. Serious deviation from $n = -4$ as a result of strong extinction or rapid luminosity evolution is most improbable, because of the practically complete absence of spectral differences between the nearest and farthest known giant elliptical galaxies (Oke, 1971). Disregarding highly artificial cases evolution or extinction should be accompanied by a change of color index. Nevertheless in order to extract all information from the existing empirical data we shall present another independent way of comparison between angular sizes and other observables in an effort to exclude many of the uncertainties inherent in Method I without loosing too much of its independence on cosmological presuppositions.

M e t h o d II. From the basic cosmological formulae one can easily derive expressions (cf. Paál, 1965; Heckmann, 1942) for comparing the number count - distance relation and the angular diameter - distance relation. We have

$$N \quad \vartheta^3 (1+z)^{-3} \cong \text{const} \begin{cases} \dfrac{\chi - \sin\chi \cos\chi}{\sin^3\chi} & \text{if } k = +1 \\ \dfrac{2}{3} & \text{if } k = 0 \\ \dfrac{\text{sh}\,\chi\, \text{ch}\,\chi - \chi}{\text{sh}^3\chi} & \text{if } k = -1 \end{cases} \qquad (2)$$

where N is the total number of objects (clusters) counted, χ is the co-moving radial coordinate of the Robertson-Walker metric of the form $ds^2 = dt^2 - R^2 /t/ . /d^2 - .../$, k is the sign of the curvature of the three space t = constant, t is the universal cosmic time, while ϑ and z are some characteristic angular size and red shift of the cluster. Of course expression /2/ is valid only in case of no evolutionary change in co-moving number density and intrinsic size of objects, but its validity is practically unaffected by luminosity evolution and extinction - i.e. the factors influencing the test of evolution by Method I-! Unless the cosmological "equator" is reached - a possibility that can easily be ruled out empirically - the increment of the expression lg / $N \vartheta^3 (1+z)^{-3}$/ cannot exceed 0.37 up to the limit of the sample (without evolution). The actual values of expression /2/ can be calculated for each cluster with known z, ϑ and N / the total number of clusters not fainter than the one chosen). A cosmologically "forbidden" increment implies evolution!

According to relaxation theory N is expected to be too large at large distances, i.e. earlier epochs, (because more galaxies should be found within the contour of the "counting area" and more clusters catalogued as reaching a given "richness"/, while ϑ may be too large or small depending on its definition (inner or outer size)*. Method II is therefore most sensitive to evolution by relaxation, if inner sizes (core diameters) are used in evaluating the increment of expression (2). No effect is however expected above the numerical cosmological threshold and the limit of detectability, if the invisible massive halos of giant member galaxies mentioned in Section I are too large and constitute a coalesced unique halo for the whole cluster.

I n f o r m a t i o n f r o m e m p i r i c a l d a t a. Angular sizes for rich clusters of galaxies at widely different distances can be taken from the following sources:

a/ for the subsystem of bright members and only for the "core" from counts and measurements by G.Paal (1971, 1973, and unpublished data); the corresponding sample of clusters can be subjected to reliable check of homogeneity only up to a red shift of about 0.2;

b/ for the bright and fainter parts together from works of Bahcall (1973) and Noonan (1974). The later is sensitive to the distribution of galaxies in the outer parts of clusters, the former to both the inner and outer distribution. Although to a different extent, both depend on estimates of background density, changes of spectral region due to red shift, and changes of magnitude interval between the brightest and faintest members counted.

Brightnesses for 1st ranked cluster galaxies of rich clusters are found in the work of Sandage (1972). They might in principle be affected by luminosity evolution and extinction.

Counts of rich clusters of galaxies are taken from Abell's Catalogue (1958). They are sensitive to several possible systematic effects: change of the magnitude interval used at counting the galaxies, change of the linear diameter corresponding to the "counting circle" used at determining the membership,. change of background estimate, etc.

Method I applied to Paál's diameters and Sandage's magnitudes gives a lgB versus lg/1+z/ relation with a slope $n = -9.1 \pm \begin{cases} 2.2 \\ 2.9 \\ 3.9 \end{cases}$ with probabilities $P = \begin{cases} 0.95 \\ 0.99 \\ 0.999 \end{cases}$ and an empirical coefficient of correlation of $r = -0.85$ (see Figure 1). As we have seen above this value of n corresponds to a contracting core of the subsystem of bright members, unless Sandage's data are distorted by luminosity evolution or extinction by more than half a magnitude (!) without detectable spectral changes (Oke, 1971). Method II applied to the same type of diameters and the homogeneous part of Abell's Catalogue gives Figure 2. - Apparent local irregularities have been smoothed out in function N/m_{Abell} / when constructing this diagram. - Obviously there appears in Figure 2 an excess of value $N.\vartheta^3$ of about 5 at least over and above the very improbable cosmological threshold corresponding to the strongest positive space curvature not immediately excluded. The possibility of having reached the "equator" of world can be ruled out directly,e.g. by the $\vartheta(z)$ relation (cp. Paál, 1965): the quantity $\vartheta/1+z$ ought to show a minimum, which is definitely not the case. - We note that in the rather "pathological" case of a static world model with "tired light effect" the above discrepancy would be even more prominent. - The observed excess can easily be explained by evolution of ϑ or N, but not by that of l : the decrease of brightness due to evolution or extinction can hardly account for it. Some systematic errors in Abell's sample may well be present, but it seems absurd to accept a relative error of N exceeding a factor of 5! A careful examination of possible systematic errors of the values ϑ leads to the same negative result.

The combination of Method I and II strongly suggests an explanation in terms of changes of diameters: physical changes of linear sizes and corresponding "anomalies" of apparent sizes when compared to those expected on the basis of other observables (l, z and N). The distant clusters - observed in an earlier cosmological epoch - seem to have too large cores.

---

* Computer simulations of clusters clearly show that the subsystem of heavy members has a contracting core, while the outlying parts of clusters expand in order to keep energy balance. See e.g. Henon (1970), Aarseth (1974).

We should remember that a contraction of bright cores is just the phenomenon expected, if the massive halos of giant members (indicated by independent arguments) are not too extended. At the same time as explained in the previous Section an excess of the number of distant clusters is also a consequence of the same type of evolution. A cosmological evolution of rich clusters of galaxies detectable in a homogeneous sample covering "only" about 3.10^9 years should certainly seem surprising even after having found the above theoretical and observational evidences in favour of it. To be sure the above interpretation of data in terms of relaxation is not the only possible one. Nor can we claim that within the frame of relaxation theory this is the unique interpretation allowed by observations. Still adopting it is useful,it provides us with a productive working hypothesis, that leads to many strong predictions and serves as a guide to constructing further informative empirical test. It may look improbable, but not untestable and unproductive. Its important advantage is its "vulnerability". Proving or disproving it means commensurable contribution to our knowledge of the Universe. A simple-minded expectation might be that a number of "easy" empirical tests, distinguishing this tentative suggestion from the drastically different alternatives, can immediatelly be advanced and performed. We mention a few of the most obvious predictions to be tested.

a) Because of the dispersal of cluster members the number of clusters themselves ought to be too large at large magnitudes or red shifts, if one applies a direct cosmological N(m) or N(z) test without recourse to angular diameters.

b) Collisional relaxation should produce a higher rate of evolution in clusters with denser central condensations, consequently the ratio of compact to medium compact clusters might be expected to decrease in the course of time.

c) Diameters characteristic of outer parts of clusters should be an increasing function of universal time.

d) In contrast with the core or halo diameters the characteristic sizes obtained by fitting inner *and* outer parts to a single model curve ought to reveal only minor cosmological evolution. In looking for the corresponding effects with a somewhat naive optimism in the existing scanty observational material of preliminary character one finds the following data.

a) The writer (Paál, 1964) points out an unexpected excess of faint clusters of galaxies in the "homogeneous" part of Abell's Catalogue. The writer was reluctant to consider it as evolution, but later Rowan-Robinson (1972) rediscovered and reinterpreted this excess as a probable sign of evolution similar to that of radio galaxies!

b) Again according to the writer's old work (Paál, 1964, and unpublished) there is a correlation between morphological type (compactness) and distance of rich clusters of galaxies just in the sense "predicted" above. (cp. also the work of Just (1959) on a correlation between richness and distance in the catalogue).

c) In Noonan's work (1974) one finds that diameters corresponding to non-central parts are too small at large distances. - Noonan himself tentatively interprets the discrepancy as a possible segregation effect, if not expansion. However segregation in itself is in general regarded as a sign of evolution/improbable in case of a continouos distribution of hidden mass in clusters

d) Bahcall's "core diameters" (1973) are obtained by fitting Emden's isothermal sphere not only to the "core" (in any sense of the word) and - as expected - they give intermediate results, i.e. no obvious evolution.

Of course the individual value of these empirical findings is rather limited. Those engaged in studying clusters of galaxies are well aware of the innumerable systematic effects obscuring real features of observational samples. No doubt each of these results can find more or less natural *ad hoc* explanations other than that tentatively suggested here, e.g. one can always speculate on unrecognized observational errors.

What is certainly more difficult to explain is the remarkable coincidence that "simply by chance" all these phenomena appear to fit spontaneously a single theoretical picture obtained in a completely independent way. Is all that the malice of Nature? Should we not take this coincidence itself as an additional - though certainly indirect - evidence? On the other hand these preliminary indications are also useful in illustrating the power of our approach;when studied in more detail, such findings may be really decisive.

Summing up we may say - as a minimum conclusion - that there seems to be no observational and theoretical support for the concept of necessarily slow evolution of dynamically "young" rich clusters of galaxies. It is difficult to see how the positive evidences of detectable evolution might all be in serious error. If they are not, the existence of suppermassive giants with essentially autonomous halos in dynamically "older" clusters would follow from the relaxation theory of gravitationally bound systems. On the other hand, true or false, this concept leads to strong and well-testable predictions thus helping us to narrow down empirically the range of theoretical possibilities. Moreover according to this suggestion appreciable changes are expected not in luminosities and spectra, but merely in visible configurations and the study of the latters is known to be much simpler task. - This note was meant for a contribution to the current debate on massive dark halos of giant galaxies. It seems to me that there are unrecognized questions and possibilities in this field of research. A proper exploitation of these possibilities might facilitate "to find the missing key to the problem of missing mass" in rich clusters of galaxies.*

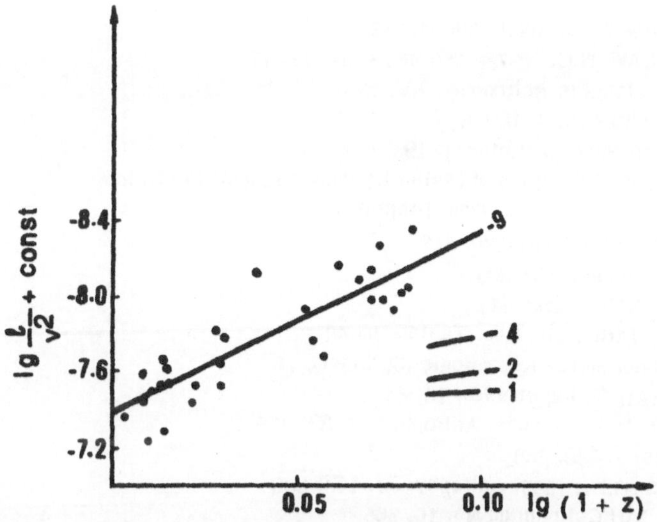

Fig. 1. Comparing the magnitude - red shift, and the angular radius - red shift diagrams: the apparent brightness of the 1st ranked cluster galaxy, l, devided by the area of the bright core of cluster, ϑ, is a quantity that can deviate from const. $/1+z/^{-4}$ only in case of evolution.

---

* In developing these ideas the writer benefited much from valuable discussion with representatives of several eminent schools of astronomy and astrophysics, especially the Moscow, Tartu, Byurakan and Cracow schools, and from discussions with my Hungarian colleagues as well as with Drs. M. Henon, M. Lecar and S.J.Aarseth and from unpublished computer simulations by the latter, kindly put at my disposal. My thanks go to all of them.

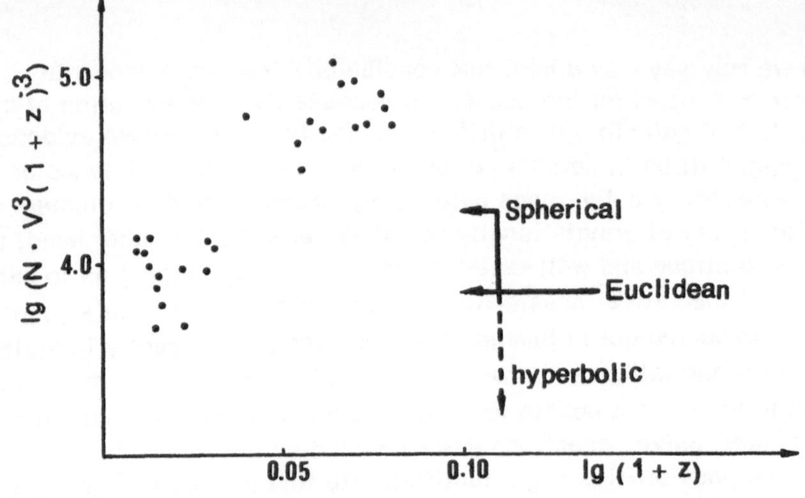

Fig. 2. Comparing the number count - redshift, /N,z/, and the angular radius - red shift, /ϑ, z /, diagrams: the increment of the function lg (N.ϑ^3. /l+z/^{3}) from z = 0 to the limit of the sample can exceed 0.37 only in case of evolution /unless the cosmological equator is reached/.

REFERENCES

ABELL G.O., 1958, Astrophys. J. Suppl. Ser. 3, 211

AARSETH S.J. and SASLAW W.C., 1972, Astrophys. J. 172, 17

AARSETH S.J., 1974, in Vistas in Astronomy, XV, 13

BAHCALL N.A., 1973, Astrophys. J. 180, 699

BURBIDGE G., 1974, Astrophys. J. (Letters), 196, L 7.

CHANDRASEKHAR S., 1942, Principles of Stellar Dynamics (Dover Publ.) 204

CHERNIN A.D. et al., 1974, Tartu Astr. Obs. Preprint 4

EINASTO J. et al., 1974, Astron. Circ. No. 811

EINASTO J. et al., 1974, Nature, 250, 309

EINASTO J. et al., 1974, Nature, 252, 111

EINASTO J. et al., 1974, Tartu Astr. Obs. Teated Nr. 48

HECKMANN O., 1942, Theorien der Kosmologie (Springer), 70

HENON M., 1970, Proc. IAU Colloquium No. 10, 44

HUBBLE E. and TOLMAN R.C., 1935, Astrophys. J. 82, 315

JUST K., 1959, Astrophys. J. 129, 268

KRISTIAN J. and SACHS R.K., 1966, Astrophys. J. 143, 380

LECAR M., 1970, Proc . IAU Colloquium No. 10, 368

MATERNE J. and TAMMANN G.A., 1974, Astron. Astrophys. 37, 383

NOONAN T.W., 1974, Astron. J. 79, 775

OKE J.B., 1971, Astrophys. J. 170, 193, and 169, 209

OSTRIKER J.P. and PEEBLES P.J.E., 1974, Astrophys. J. 186, 467

OSTRIKER J.P. et al., 1974, Astrophys. J. Lett. 193, L1

PAÁL G., 1964, Mitteilungen der Sternwarte der Ungarischen Acad. Wiss. Nr. 54

PAÁL G., 1965, Soviet Astron.- A.J. 9, 14. Астрон. Ж. 42, 19

PAÁL G., 1971, Astrofizika 7, 435. Astrophysics 7, 257

PAÁL G., 1973, Proc. IAU Symp. No. 63, 251

PENROSE R., 1966.in "Perspectives in Geometry and Relativity" (ed. Hoffmann, B. Indiana Univ. Press), 259

ROWAN- ROBINSON M., 1972 Astron. J. 77, 543

SACHS R.K. and EHLERS J., 1968, in "Astrophysics and General Relativity, Vol. 2. "(ed. Chretien, M; Cordon and Breach) 312

SANDAGE A., 1972, Astrophys. J. 178, 1.

TOLMAN R.C., 1934, Relativity. Thermodynamics and Cosmology, pp. 467, 481

ZEL'DOVICH Ja.B. and PODURETS M.A., 1965, Астрон. Ж. 42, 963

WHITROW G.J., 1954, Monthly Notices Roy.Astron. Soc. 114, 180

THE STABILITY OF "DE VAUCOULEURS' GROUPS"

R.B.TULLY and J.R.FISHER
OBSERVATOIRE DE MARSEILLE, FRANCE
PRESENTED BY R.B.TULLY

СТАБИЛЬНОСТЬ "ГРУПП ДЕ ВОКУЛЕРА"

I n t r o d u c t i o n. It is commonly stated that for many of the small groups of galaxies seen locally the discrepancy between the virial mass and the visible mass can exceed two orders of magnitude. If true, then either (i) the groups are not physical (chance projection-effects), or (ii)ninety-nine percent of the mass is unseen, or (iii) the groups are very unstable and the galaxies within them have ages of the order of 10^9 years. We will present preliminary evidence that none of these possibilities is correct; that the problem has been observational.

At present, we are reanalysing the structure of the local universe out to a redshift of 3,000 km/s, introducing three major improvements over the previous works: (i) the sample with known redshifts has been approximately tripled, (ii) for most of the sample the quality of the redshifts has been improved from an accuracy of about 100 km/s to one of about 20km/s and (iii) we have a powerful new tool for measuring distances to individual galaxies.

Although our study is not yet completed, it is illuminating to look in the light of our new data at several specific groups which previously caused great difficulties. Three such groups which are well isolated and supposedly reliably constituted were specified by Gott, Wrixon and Wannier (1973, ApJ 186, 777); the groups 6, 12 and 51 from a list by de Vaucouleurs (unpublished, Stars and Stellar Systems Vol IX, Ch 17) and a study of their stability by Rood, Rothman and Turnrose (1970, ApJ 162, 411). In a recent study of the nearest 14 of these de Vaucouleurs' groups by Materne and Tammann (1974, AandA 37, 383), only groups 6 and 12 were cited to be "clearly unstable".

Group 6

How do we constitute these same three groups? In Fig. 1 we show in galactic coordinates the region of sky containing Group 6 (NGC 2841 group). All galaxies in this region observed to have a corrected systematic velocity less than 1500 km/s are indicated. The six systems in large open circles are those identified by de Vaucouleurs as group members. Those with underlined names had measured velocities when de Vaucouleurs made his study.

In fact we propose that Group 6 is really *two groups* (circled and identified *a* and *b* in Fig. 1). A third group (*c*) now appears for which there were previously no known velocities. The reality of this division into separate groups is demonstrated in the velocity histograms of Fig.2a. The original de Vaucouleurs group 6 members are circled. There is a clear separation in velocity between the filled circles, those galaxies within the confines of annulus *a*, and the X's, within annulus *b* in Fig.1. Of the 17 galaxies in the entire region, only three are not clearly identified with one of our three groups: one is adjacent *c* and has a redshift indicating it is in the background, another is near *b* spatially but near *a* in velocity, and the third is near *b* spatially but near *c* in velocity.

It is noteworthy that the three groups in this field have distinctive morphological characteristics, although this feature was not used in identifying the groups.

Group *a* (principal galaxy: NGC 2841) is composed of moderately early spirals, group *g* (principal galaxy: NGC 2541) of late spirals and irregulars, and group *c* of nothing but dwarf irregulars.

The group virial mass M_{vt} is given by the equation:

$$M_{vt} = V^2 R / G$$

where V is the group velocity dispersion, R is a measure of the group radius and G is the universal gravitational constant. This mass can be compared with the visual mass M_L in the group, found by totaling the luminosity from the individual galaxies and assuming a mass to luminosity ratio.

If, for the present purposes, we take very crude estimates for R and M_L and we falsely pretend that each of the systems within a group has the same mass we can find the order of magnitude effect on the M_{vt} /M_L ratio. Our results for group 6 are given in Table 1 and the decrease in the above ratio is a factor of ten. A more detailed analysis following Materne and Tammann would undoubtedly further lower the M_{vt} /M_L ratio and for each of these two groups the stability could be said to be ambiguous.

Table 1

	RRT	this paper : group a	this paper : group b
V	151 km/s	46 km/s	42 km/s
R	R_{RRT}	1/2 R_{RRT}	1/2 R_{RRT}
M_L	M_{RRT}	0.8 M_{RRT}	1/3 M_{RRT}
M_{vt} /M_L	275	16	32

Group 12

The region of de Vaucouleurs' region 12 (NGC 3184 group) is shown in Fig.3; all systems with velocities between zero and 1100 km/s are indicated. Again the systems identified by de Vaucouleurs are circled and the systems that had velocities available to him are underlined. A loose group about NGC 2903 (not in de Vaucouleurs' list) is seen to lower galactic latitudes.

It is seen that we would extend the group considerably in one direction. The number of galaxies with known velocities is increased from 4 to 12 and we have radio velocities for all except Markarian 36. A histogram of the velocities is given in Fig. 4a; distinguishing between the regions associated with group 12 and the NGC 2903 group. The four original group 12 members are given as open circles. These can be compared with the situation that existed when only optical redshifts were available, shown in Fig. 4b. Table 2 summarizes the stability analysis. The ratio M_{vt} /M_L is reduced by over an order of magnitude and again the case against stability can be considered ambiguous.

Table 2

	R R T	this paper
V	257 km/s	68 km/s
R	R_{RRT}	$2R_{RRT}$
M_L	M_{RRT}	$2M_{RRT}$
M_{vt} / M_L	513	35

Group 51

Fig. 5 shows the spatial distribution of all the galaxies with velocities between 1000 and 2500 km/s in an extensive region around group 51 (NGC 6643 group). As before, the galaxies identified by de Vaucouleurs and the ones with velocities available to him are indicated. A histogram of velocities is given in Fig. 6. Circled entries indicate the optical velocities used to originally define the group but which are of lower weight. The clear indication is that the two systems at 2100 km/s are chance superpositions from the background (X's in Fig. 5). The galaxy at 1100 km/s is taken to be foreground (small dot in Fig. 5).

Analysis of this region is made difficult by incompleteness at this distance, plus the uncertainties caused by the lower weight optical velocities. Taking the region as one sole group, the dispersion in velocity of 152 km/s is still large but is much less than the 457 km/s quoted by Rood et al. The virial mass discrepancy is reduced by an order of magnitude to approximately 50.

If the region is regarded as a superposition of two groups there ceases to be any problem with the Virial mass. If the sample is cut at about b = 31° it can be noted that all those below have systematic velocities exceeding 1600 km/s (open circles in Fig. 5) while all but one above have systematic velocities less than 1600 km/s (filled circles in Fig. 5).

C o n c l u s i o n s . Much greater completeness and improved velocities suggest that even groups formerly thought to be *very* unstable may be stable. The problem in the case of group 6 was that there were really two groups adjacent each other on the plane of the sky. With group 12 the high dispersion was completely due to the low quality of the optically derived velocities. In group 51 two of the five original members were from the background.

These results are qualitatively in agreement with our experience over the whole sky: the vast majority of galaxies can be assigned to groups and the dispersions within the groups are sufficiently small that even if there is no more than the visible mass present at least they are not manifestly unstable.

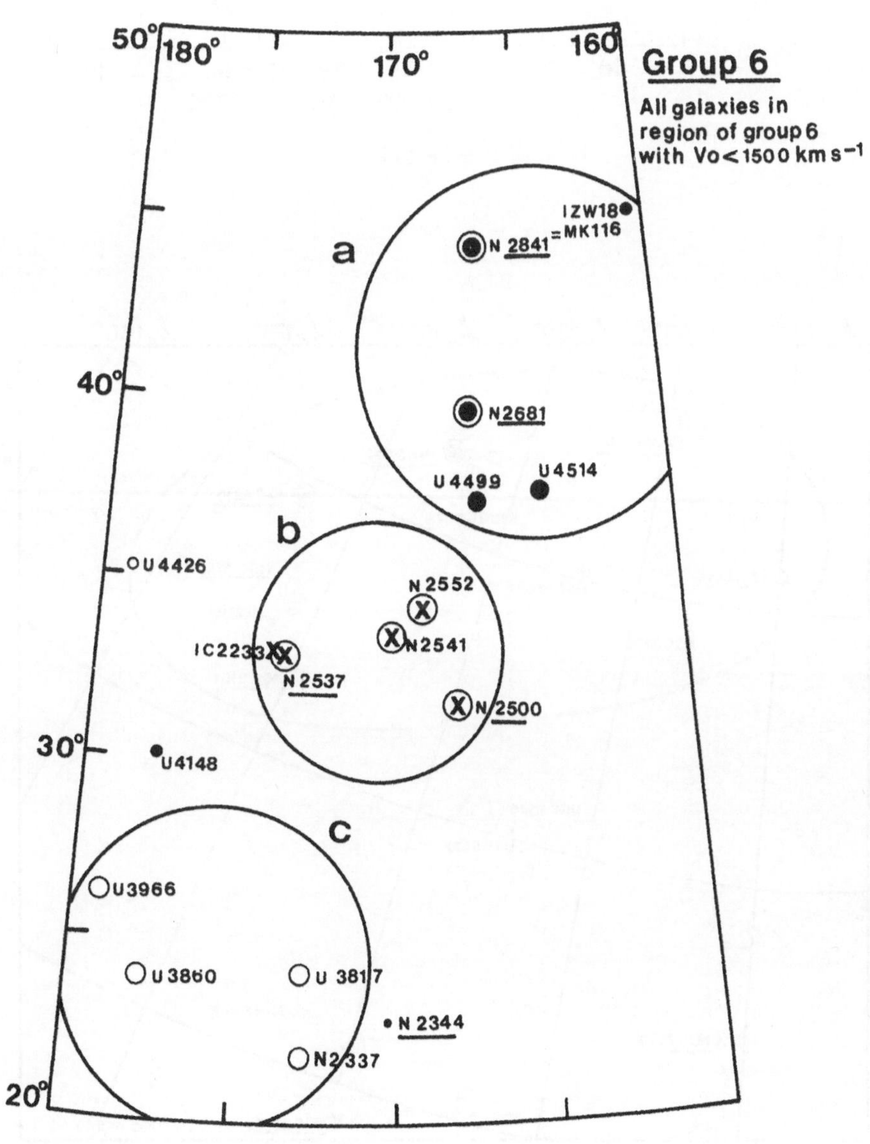

Fig. 1.

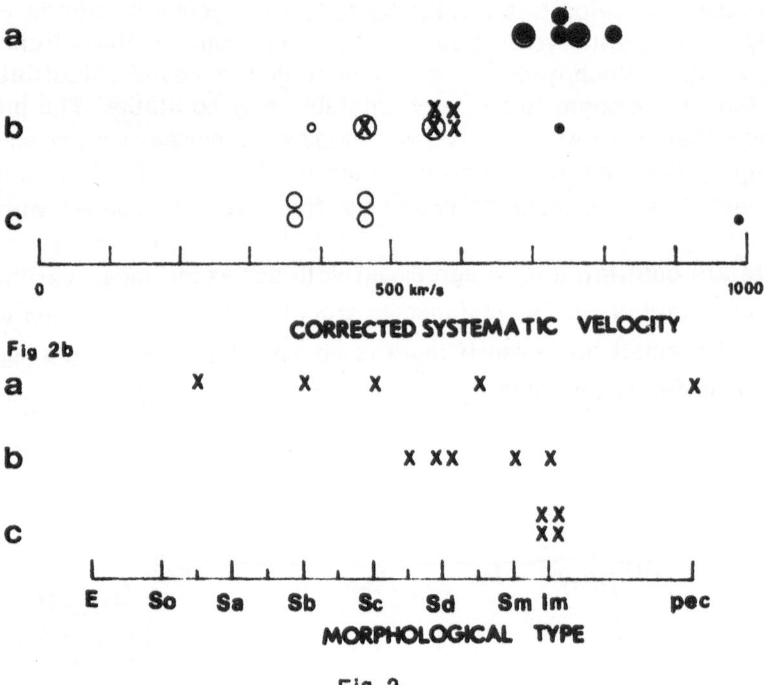

Fig 2a
a
b
c
0
500 km/s
1000
CORRECTED SYSTEMATIC VELOCITY
Fig 2b
a
b
c
E
So
Sa
Sb
Sc
Sd
Sm
Im
pec
MORPHOLOGICAL TYPE

Fig. 2.

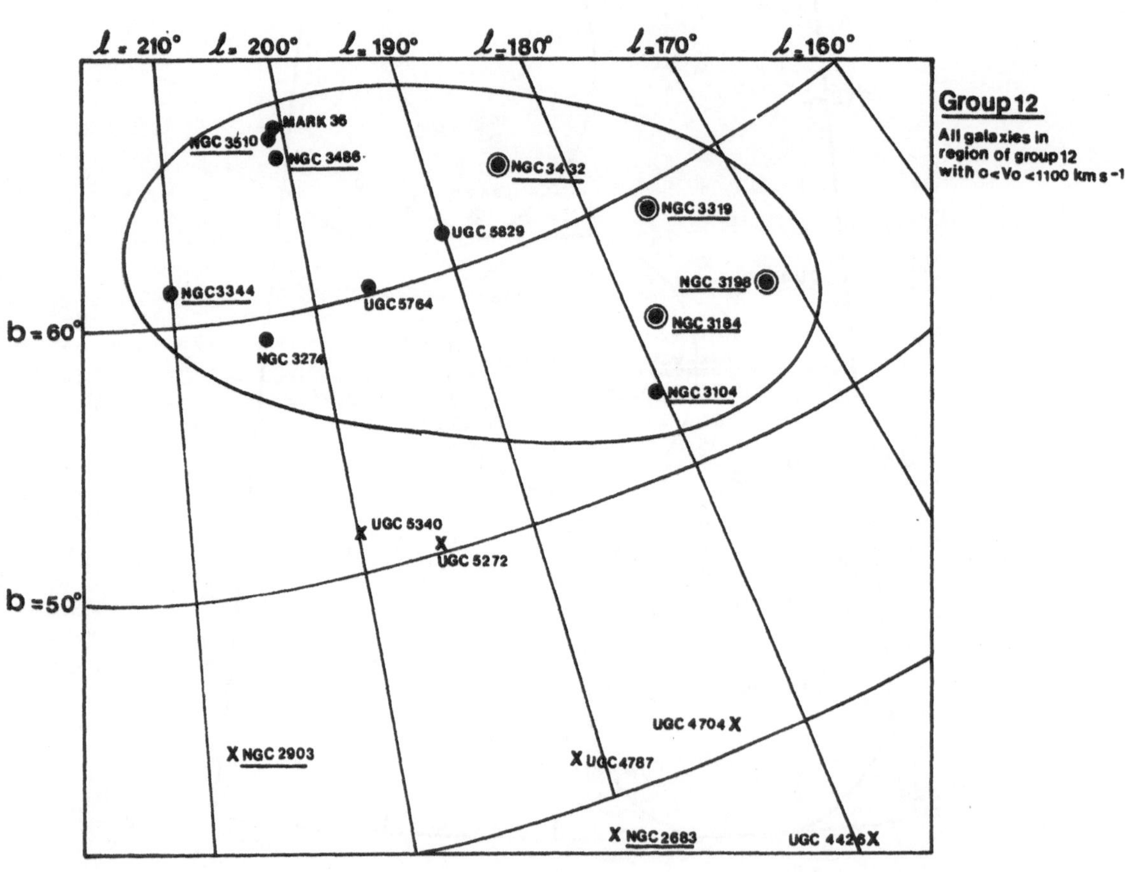

l = 210°
l = 200°
l = 190°
l = 180°
l = 170°
l = 160°
Group 12
All galaxies in
region of group 12
with 0 < Vo < 1100 km s-1
MARK 36
NGC 3510
NGC 3486
NGC 3432
NGC 3319
UGC 5829
NGC 3198
NGC3344
UGC5764
NGC 3184
b = 60°
NGC 3274
NGC 3104
UGC 5340
UGC 5272
b = 50°
UGC 4704
NGC 2903
UGC4787
NGC 2683
UGC 4426

Fig. 3.

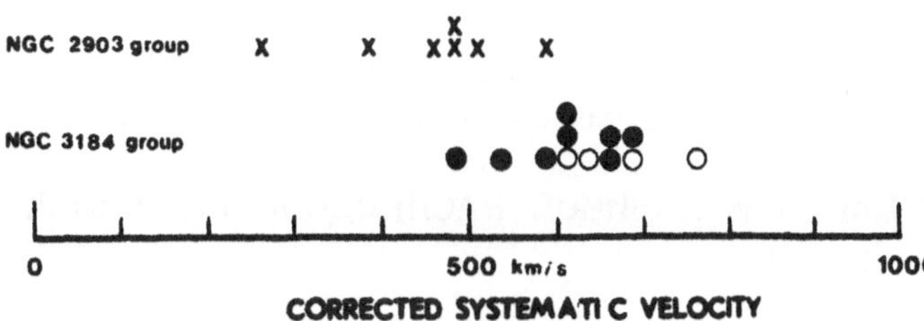
Fig 4a
NGC 2903 group
NGC 3184 group
0
500 km/s
1000
CORRECTED SYSTEMATIC VELOCITY

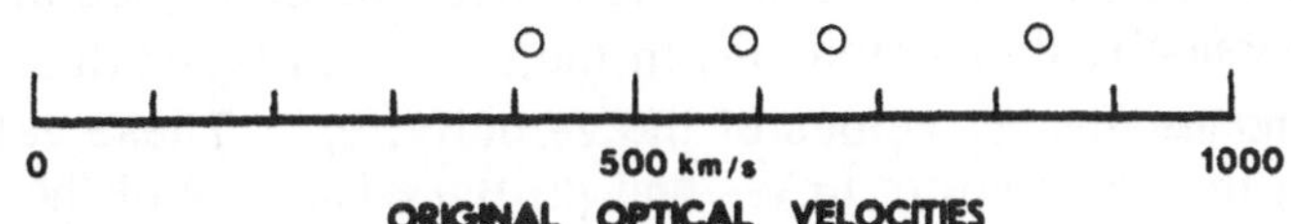
Fig 4b
0
500 km/s
1000
ORIGINAL OPTICAL VELOCITIES

Fig. 4.

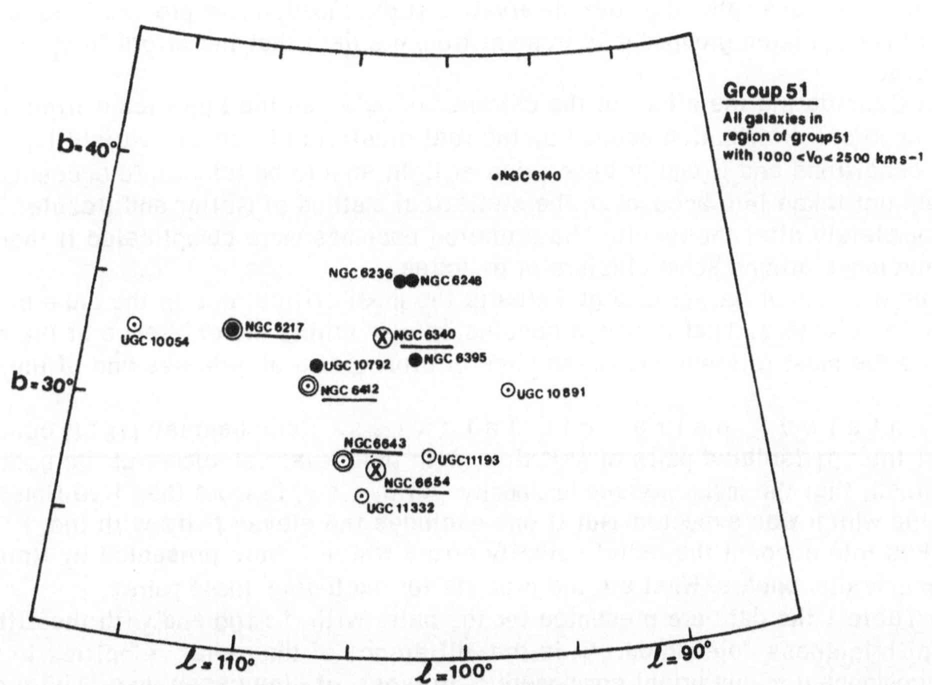
Group 51
All galaxies in region of group 51 with 1000 <V0< 2500 km s-1
b=40°
b=30°
NGC 6140
NGC 6236
NGC 6248
UGC 10054
NGC 6217
NGC 6340
UGC 10792
NGC 6395
NGC 6412
UGC 10891
NGC 6643
UGC 11193
NGC 6654
UGC 11332
l=110°
l=100°
l=90°

Fig. 5.

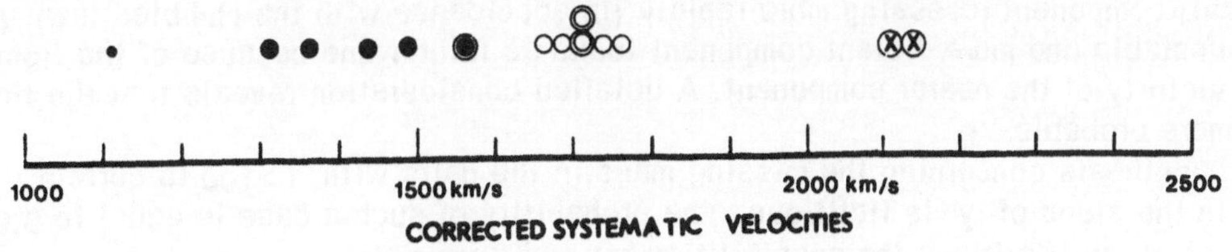
1000
1500 km/s
2000 km/s
2500
CORRECTED SYSTEMATIC VELOCITIES

Fig. 6.

THE MISSING MASS AND THE APPARENT DISTRIBUTION OF GALAXIES

B.I.FESENKO
PSKOV PEDAGOGICAL INSTITUTE, U.S.S.R.

СКРЫТАЯ МАССА И ВИДИМОЕ РАСПРЕДЕЛЕНИЕ ГАЛАКТИК

A critical discussion of the missing mass problem was given by Burbidge (1975). Here some additional arguments against the existence of the missing mass are considered.

G e n e r a l   r e m a r k s . The conclusion about the existence of the missing mass is based on the large velocity dispersion, σ^2, in the systems of galaxies. The value of σ^2 is very sensitive to the marking out values of the velocity, V_r . These values of V_r may be obtained due to (1) the crude errors in V_r and (2) the admixture of the background galaxies.

(1) The case with the radial velocities (of some galaxies) obtained in the Lick observatory tells on the possibility of large errors. According to Lewis (1974) these velocities should be diminished to 98 km/s. But the cause of such errors is not clear. Materne (1974)shows taking concrete examples that the observational data made more precise lead to the stability of the close galaxian groups. It is evident from his data that the errors in V_r sometimes reach 400 km/s.

(2) Considering the effect of the extraneous galaxies the significant irregularity of the galaxian apparent distribution caused by the real clustering in space, variability of the observational conditions and irregular absorption of light should be taken into account. The last reason was not taken into account in the statistical method of Geller and Peebles (1973) and that may completely alter the result. The situation becomes more complicated if there exist hollow dust envelopes around some clusters of galaxies.

The problem of background galaxies is the most critical one in the case of the scattered groups and clusters. That is why a conclusion concerning the existence of the missing mass looks as the most reliable one in the case of close pairs of galaxies and of the rich Coma cluster.

I s o l a t e d   p a i r s   o f   g a l a x i e s. Karachentsev (1974) concludes from the data for the 101 isolated pairs of galaxies, with the radial velocities of the both components determined, that the mean mass-to-luminosity ratio, $< f >$, is more than five times greater than the value which was expected. But if one excludes the eleven pairs with the $f > 100$ ($f_\odot = 1$) and takes into account the radial velocity errors which were presented by Karachentsev, the discrepancy disappears. What are the grounds for excluding these pairs?

In Table 1 the data are presented for the pairs with $f > 100$ and with the differences of an apparent brightness determined. y is the difference of the radial velocities in the sense of "faint component minus bright component". In eight of nine cases $y>0$. The probability of obtaining the condition $y > 0$ in eight or nine cases is equal to $(1/2)^9 + 9(1/2)^9 \approx 0.02$.

If the systems are bound the peculiarity of the signs of y indicates the significant systematical errors of V_r , errors which depend on the apparent brightness and morphological class. In that case excluding of the pairs is justified.

There are two more possibilities. (1) the pairs with $f > 100$ are optical ones with the fainter (and distant) component recessing more rapidly (in accordance with the Hubbles law). (2) These pairs are unstable and more distant component looks as fainter one because of the light absorption in the vicinity of the nearer component. A detailed consideration reveals that the first possibility is more probable.

The hypothesis concerning the missing mass in the pairs with $f > 100$ is correct if the peculiarity in the signs of y is fictitious. The probability of such a case is equal to 0.02. But if the signs of y are fictitious, the possibilities (1) and (2) remain.

C o m a   c l u s t e r. According to Rood, Page, Kintner and King (RPKK, 1972) the missing mass 7 times larger than the mass observed is required for the Coma cluster stabilization.

To consider the importance of the background galaxies two groups of galaxies were selected, the list of the probable cluster member of RPKK was used. The objects described in MCG as E and E as well as the galaxies with no MCG inclination classes and measured outer diameters had been included in the E-group (22 objects altogether). 11 galaxies with the spiral or ring structure indicated (in MCG) were included in the S group. The difference of the mean radial velocities in the sense of "E-S" is equal to 1130 + 240 km/s. The distribution of V_r is asymmetrical for other galaxies from the list of RPKK. The facts considered make the interpretation of the galaxies discussed as belonging to the same cluster more complicated. But there is another possibility: large systematical errors of V_r are errors which depend on the morphological description in MCG. In both cases the idea of the large invisible mass existing in Coma cluster is premature.

It is interesting to note that only the system of the galaxy description which is used in MCG provides the selection of two morphological groups so distinctly distinguished.

Table 1

Karachentsev's number	Δm	y	f
127	0.1	+406	131
202	0.5	+640	134
210	1.1	+248	128
218	1.1	+272	125
356	0.5	+522	390
396	0.7	+464	342
419	0.1	-311	244
584	1.2	+643	516
0924+7437	0.6	+799	244

REFERENCES

BURBIDGE G.R., 1975, Ap. J., 196, 7

GELLER M.J., PEEBLES P.J.E., 1973, Ap.J., 184, 329

KARACHENTSEV I.D., 1974, Soobshch. Spec.astrofiz. Obs. Akad.Nauk SSSR, 11, 51

LEWIS B.M., 1974, Observatory, 94, N 998, 9

MATERNE J., 1974, Astr. and Ap., 33, N 3, 451

ROOD H.G., PAGE T.L., KINTNER E.C. and KING I.R., 1972, Ap.J., 175, 627

NATURE OF REDSHIFTS AND THE PROBLEM OF MISSING MASS

T. JAAKKOLA
OBSERVATORY AND ASTROPHYSICS LABORATORY, HELSINKI, FINLAND

ПРИРОДА КРАСНЫХ СМЕЩЕНИЙ И ПРОБЛЕМА СКРЫТОЙ МАССЫ

I have studied correlations of redshift with the other empirical parameters, taking into account the possibility of non-velocity redshifts. After studying the redshift as a function of galaxian type, luminosity, diameter, colour, inclination, peculiarity, compactness, mean surface brightness and other surface brightness parameters (Jaakkola, 1971a, b, 1973; Jaakkola and Moles, 1975) it seems that redshift escapes all such tight correlations with the other parameters which would make the reduced z-dispersion sufficiently small and which would remove the missing mass problem in a direct manner. However, there are correlations, such as those of type, magnitude, inclination and surface brightness with the redshift, which establish the presence of non-velocity redshifts, and a more profound knowledge of these correlations may well solve the missing mass problem. Of course, when redshift appears to be an unreliable measure of radial velocity, there is no longer a missing mass problem in the classical sense, rather a more general question of the reasons for the redshift dispersion in the systems. As first pointed out by Holmberg (1961), these reasons would include real motions, non-velocity redshifts, errors of measurement, projected members, etc.

Fig. 1 presents the result of Jaakkola and Moles (1975) which confirms in de Vaucouleurs' groups the earlier result (Jaakkola, 1971) on the relation between redshift and morphological type concerning clusters and Karachentsev's groups. The crosses give mean residual redshifts based on the optical data, and the circles show data corrected for the type-dependent effect found by Lewis (1975), the circles meaning essentially a result for the 21-cm line redshifts. The z-difference between Sb-Sc type and the other types, $\Delta V = 116 \pm 43$ km/s for the optical data, is significant at 2.7σ level, similarly as in the earlier results. Thus, the type-redshift effect is present independent of sampling of the data and appears both for optical and 21-cm line redshifts.

Also Fig. 2 shows the necessity for taking into account non-velocity redshifts in the context of the missing mass problem. The distribution of residual redshifts is given for those galaxies in clusters; groups and pairs which are compact in the sense of Zwicky's two catalogues (Zwicky et al., 1961-1968; Zwicky, 1971). This significantly asymmetric distribution proves that excess redshifts appear in compact galaxies.

Evidently these results give direction to the partial answer to the missing mass problem; but probably only to a partial answer. E.g., it is well-known that the Virgo cluster includes galaxies with a blue-shift. This cannot be explained by non-velocity redshifts and it seems that a small fraction of galaxies really escapes from the systems as a result of non-stationary phenomena and encounters of galaxies, the groups and clusters as a whole still being not expanding but quasi - stationary.

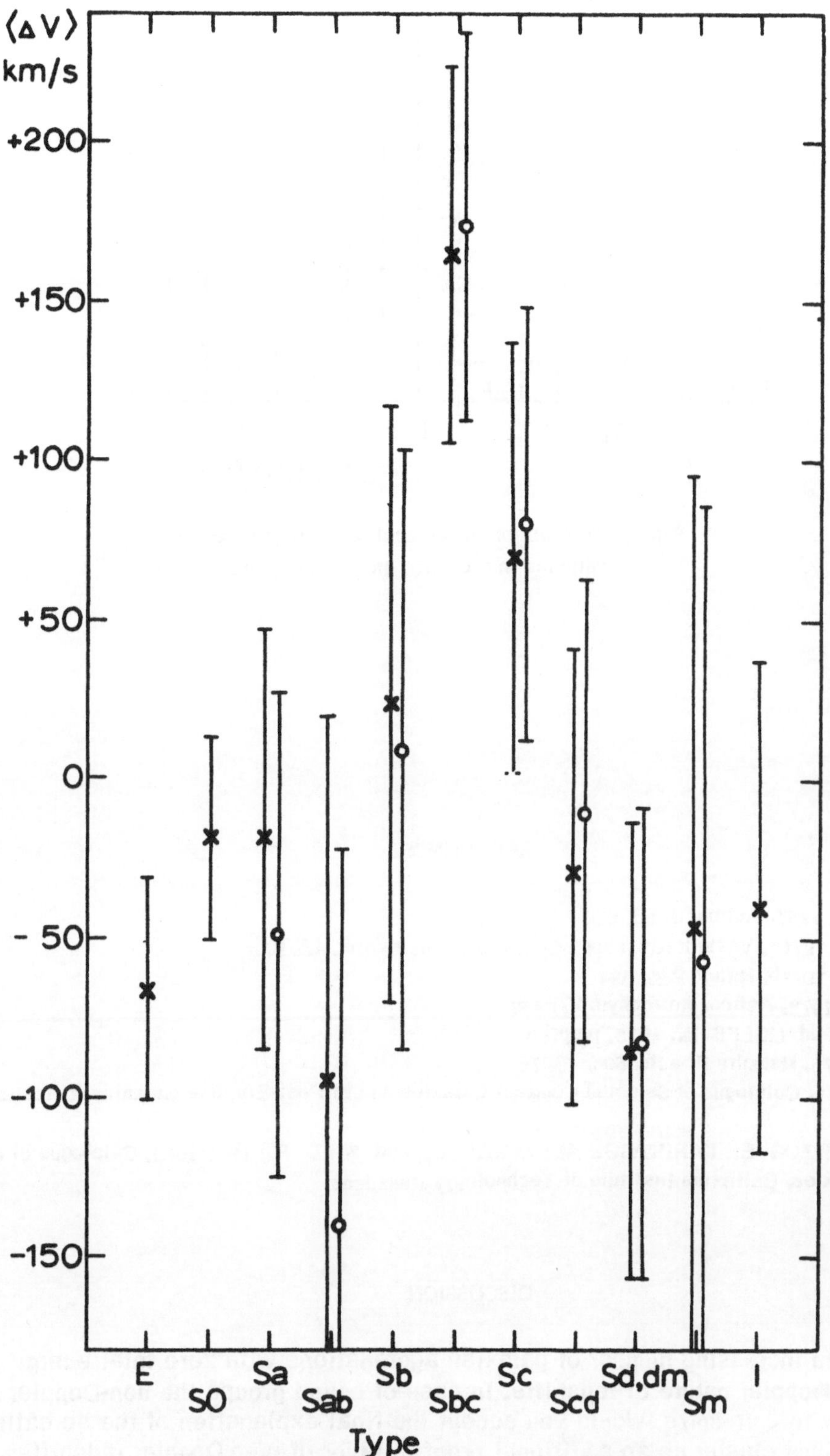

Fig. 1. Mean residual redshifts (given in km/s) as a function of morphological type for galaxies in de Vaucouleurs' groups. Crosses are for optical redshifts, circles for values corrected for the systematic effect found by Lewis (1975). Standard errors are indicated.

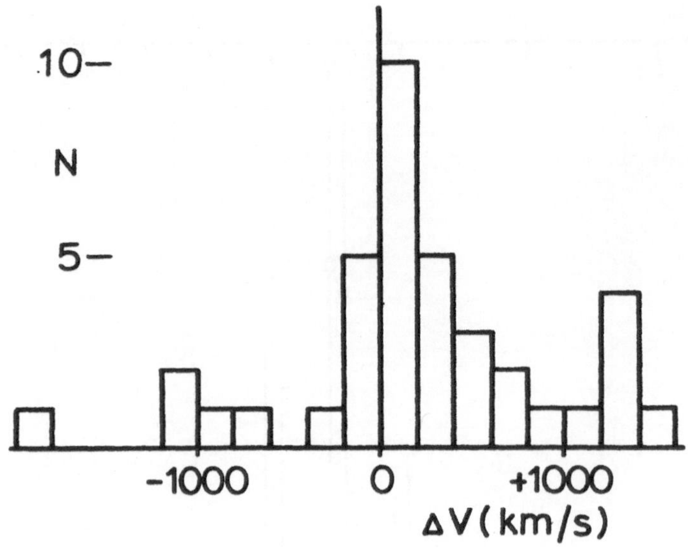

Fig. 2. Distribution of residual redshifts for compact galaxies in clusters, groups and pairs.

REFERENCES

HOLMBERG E., 1961, Astron. J. 66, 620

JAAKKOLA T., 1971a, Vestnik Kievskogo Gos. Univ. Ser. Astron. 13, 97

JAAKKOLA T., 1971b, Nature 234, 534

JAAKKOLA T., 1973, Astron. Astrophys. 27, 449

JAAKKOLA T. and MOLES M., 1975, preprint

LEWIS, B.M., 1975, Memoirs R. astr. Soc. 78, 75

ZWICKY F., 1971, Catalogue of Selected Compact Galaxies and of Post-Eruptive Galaxies, Offsettdruck L. Speich, Zürich.

ZWICKY F., HERZOG E., KARPOVICZ M., KOWAL C., and WILD P., 1961-1968, Catalogue of Galaxies and of Clusters of Galaxies, California Institute of Technology, Pasadena.

DISCUSSION

TAMMANN: The increasing number of galaxian aggregations with zero total energy proves in favor of the Doppler nature of redshifts. In case of bound groups the non-Doppler component is limited now to < 10 km/s. Would you accept the final explanation of the so called missing mass in the Coma cluster as an additional proof in favor of pure Doppler redshifts?

JAAKKOLA: Scott paid attention to the increase of the redshift dispersion in groups and clusters toward larger distances and I have found the same effect for a larger data material. Scott interpreted this effect in terms of nearly systems being selected because of similarity in the redshifts. Drs. Tammann and Materne have studied only nearby groups where the redshift dispersion is indeed small, but due to the mentioned effect they may not all be entire groups. One should not conclude too much about the question of stability and pure Doppler redshifts, before this bias has been properly reconsidered.

DAVIES : A comment should be made on M31 when discussing the missing mass in galaxies I refer to the flattening of the rotating curve at a distance of 100 arcminutes (20 kpc) from the centre. Beyond this distance the rotation velocity becomes constant; this implies the existence of significant mass at large distances from the centre, extending to at least 30 kpc from the centre. The M/L ratio is more than 20 and could be much higher.

OZERNOY : During the last years extensive halos have been discovered around many galaxies and there appeared in the literature a number of speculations that these halos may contain appreciable hidden mass and may be the internal parts of much more extensive dynamical coronae. A recent investigation by Bobrova and myself (1975, Astr, Zh., in press) shows that the observed halos do not contain any appreciable amount of hidden mass. The idea of our analysis is to compare the brightness distribution, calculated from the expected distribution of the hidden mass, with available photometric data on the brightness distribution in the faint halos. In particular, we have compared with detailed photometry of the halo of the giant elliptical galaxy M87. We find that the theoretical distributions are beyond the errors of the observed distribution, and it appears that the dynamical and photometrical distribution are incompatible when hidden mass is in the form of usual stars or gas.

An important question is what maximum mass may be contained inside a faint halo.

1. If the halo consists mainly of hot gas, its optical emission (4000 - 5000Å) is composed from free-free losses and H-beta radiation. Using the exponential character of the observed brightness distribution in M87 we conclude that the amount of hidden mass in M87 in gaseous form does not exceed $10^{12} M_{\odot}$.

2. If the halo is formed mainly from stars, the latter must be similar to that of globular clusters and dwarf galaxies with M/L=1-2. We find for M87 the mass of the stellar halo $3\text{-}6 \cdot 10^{10} M_{\odot}$. Consequently, independent of whether the halo is in the form of stars or gas, it can not contain an amount of hidden mass as large as one attributes to dynamical considerations, and the observed faint halos of galaxies have no direct relation to the supposed hidden mass.

EINASTO : There is a growing number of small systems of galaxies with very small velocity dispersions, where no mass discrepancy is seen. On the other hand, the mass discrepancy in large clusters and small groups and pairs can only be avoided by splitting the clusters into smaller systems. For instance, as Dr.Fesenko suggested, the Coma cluster may consist of two independent systems, the E-cluster and the S-cluster, the mean radial velocity of the E-cluster being about 1000 km/s larger than that of the S-cluster. A widely different result was obtained by Tifft, who found that the mean radial velocity of the E-cluster is 700 km/s smaller than that of the S-cluster. It seems to me that this difference reflects only the difficulties in classification of galaxies, and that at present there is no need to split the Coma cluster into smaller, independent clusters. The overall velocity dispersion is 900 km/s, corresponding to a mass discrepancy of about 10 times.

To obtain more exact information on the distribution of mass in small systems, radial velocities are needed for dwarf galaxies, located far from the center of the system. The program for determining such velocities from 21cm observations, carried out by Tully and Fisher, is of crucial importance for solving the mass discrepancy in small systems of galaxies.

FESENKO : Professor Einasto has expressed doubt about the reality of my results for the Coma cluster, citing the work of Tifft. I should like to stress that Tifft investigated those galaxies which have been classified as spirals (S) according to the Hubble-classification, whereas I only used galaxies with clearly established spiral structure or ring-like structure (from the Palomar Atlas). That is not the same thing.

GAS AND STARS IN THE OUTER REGIONS OF GALAXIES

R.D.DAVIES
UNIVERSITY OF MANCHESTER
NUFFIELD RADIO ASTRONOMY LABORATORIES
JODRELL BANK
CHESHIRE, U.K.

ГАЗ И ЗВЕЗДЫ ВО ВНЕШНИХ РАЙОНАХ ГАЛАКТИК

РЕЗЮМЕ

Высокочувствительные наблюдения нейтрального водорода указывают на наличие обширных образований в окрестностях нашей Галактики и нескольких близлежащих галактик. Для большинства образований оптические эквиваленты неизвестны. Обсуждаются некоторые причины этого явления.

ABSTRACT

High sensitivity neutral hydrogen observations show the existence of extensive features in the vicinity of our Galaxy and several nearby galaxies. For most of these no optical counterparts are known. Several reasons for this are discussed.

1. Introduction. The outermost regions of galaxies have received considerable attention in recent years for two main reasons. One is the discovery that a substantial proportion of galaxies are subject to tidal interactions with their neighbours; the tidal forces act preferentially on the mass lying in these outer regions. Also, measurements of the velocity field in the outer regions lead to an estimate of the mass distribution there; taken with the distribution of luminosity this gives the mass-to-luminosity ratio which characterizes the material in the outer regions of a particular galaxy.

In this short review I will concentrate on the new information which has become available from measurements with the 21 cm line of neutral hydrogen. High angular resolution aperture synthesis observations have been made of approximately 20 galaxies. These provide information on the higher brightness emission and allow a comparison with the HII regions, dust and OB associations in the clearly recognizable spiral features. Pencil beam observations provide complementary information on the lower brightness emission from the outermost regions which are inaccessible to aperture synthesis techniques. Optical data is becoming available using several techniques to obtain deep sky photographs of the star fields and using sensitive spectrograph or filter systems to detect weak line emission.

2. Our Galaxy - the High Velocity Clouds. Inside the Sun's distance from the galactic centre the neutral hydrogen lies close to the galactic plane in a layer which has a thickness that ranges from 100 to 200 pc. Outside this distance an important systematic trend occurs in the disposition of the neutral hydrogen layer. This well-known tilt is above the plane in the sector ℓ = 30° to 130° and below the plane in the sector ℓ = 260° to 340°. Neutral hydrogen observations of the lower density gas at the outer edge of the Galaxy reveal the outer spiral structure (Kepner 1970, Davies 1972 and Verschuur 1973). This gas has rotational velocities reaching 150 km s^{-1}, negative in the northern galactic plane and positive in the southern plane. Here the line integral through spiral arms is typically 5×10^{19} atoms cm^{-2} compared with 2×10^{21} atoms cm^{-2} in spiral arms near the Sun. There is no doubt about the reality of these spiral features although it is not easy to be certain about their precise distance because the galactic rotation is not well-established at large distances from the galactic centre. On the basis of the Schmidt model which is simply the extension of an analytical expression that fits the data to R ~ 14 kpc, the distances of these outer features are in the range 15 to 25kpc. In the northern sky the centre of the outermost arms can be as much as 3 kpc above the galactic plane; their thickness can be 2 kpc. Such a distortion from the plane requires a major force — most likely the tidal interaction of a close passage of the Magellanic Clouds.

The high velocity clouds near the galactic plane in the longitude range 40° to 160° probably are concentrations in the outer spiral structure whereas those clouds at higher latitudes (|b| > 30°)

require further discussion. Oort (1970) has argued that they are produced by the infall of low density intergalactic material which sweeps up neutral hydrogen in the galactic halo. Davies (1972) and Verschuur (1973) argue that the major part of this material which lies in bands emanating from near the galactic plane is closely connected with the outer spiral structure since it shows the same velocities and a similar distortion above the plane.

With the discovery of the neutral hydrogen Magellanic Streams which extend from the Magellanic Clouds and reach the galactic plane on opposite sides of the sky, a new parameter is introduced to the discussion. The existence of these Streams confirms that a strong tidal interaction has occurred between the Galaxy and the Magellanic Clouds. Indeed the passage of the Magellanic Clouds has been so close that the material at the outer ends of the Streams has been pulled from the Clouds and is now falling into the Galaxy. A number of features both above and below the plane in the galactic anti-centre region have velocities relative to the galactic standard of rest which are similar to those in the Magellanic Stream. Thus there are strong arguments that there are two sources of the high velocity clouds namely the tidal distortion of the outer parts of the Galaxy and the tidal capture from the Magellanic Clouds. Work is currently in progress to investigate the distribution and velocity structure of the southern and anticentre high velocity clouds from which it is hoped to learn more about their distance and origin.

3. T h e M a g e l l a n i c S t r e a m s. The Magellanic Streams are broad filaments of neutral hydrogen extending from the neutral hydrogen halo surrounding the Magellanic Clouds (Mathewson, Cleary and Murray 1973). The filaments show several adjacent nearly parallel branches which all lie on a celestial great circle passing through the Magellanic Clouds. One section of the Stream passes through the southern galactic pole and extends to within 30° of the galactic equator at $\ell = 90°$ (Wannier and Wrixon 1972); this appears to form a bridge between the Magellanic Clouds and the Galaxy. The outer major section passes out of the opposite side of the Clouds and can be traced to 20° above the galactic plane at $\ell = 300°$; this is presumably a galactic tail.

Recent neutral hydrogen measurements at Jodrell Bank have shown that the Magellanic stream type material is much more broadly distributed over the southern galactic hemisphere than had previously been supposed (Cohen and Davies 1975). New observations show emission covering an area at least 15° x 15° with a velocity of approximately -300 km s^{-1} relative to the lsr centred near $\ell = 165°$, b = -50°. Other high velocity clouds at $\ell = 120°$, b = -20° near M 31 (Davies 1975) and at $\ell = 133°$ b = -32° near M33 (Wright 1974) have been found with negative lsr velocities 400 km s^{-1} and greater. These latter features could be associated with their respective galaxies although the velocities are similar to that of the Stream.

The areas of the brighter neutral concentrations in the Stream are being studied optically with the UK 48" Schmidt telescope. Plates have been taken in the red and blue and star counts are being processed down to approximately magnitude 20. A deep Hα plate with a 5 hr exposure using a mosaic of filters of 100 Å width has been taken of a field centred at RA 00^h55^m Dec = -50°. No Ha emission was detected at an upper limit emission measure of about 10cm^{-6} pc. The neutral hydrogen surface integral in this concentration in the Stream is 300cm^{-3} pc.

One important unresolved problem connected with the Stream is the high negative velocity of its northern end. At first sight this velocity appears to be too high for unimpeded infall from the Magellanic Clouds. Recent model calculations performed by Dr. A. E. Wright and the author indicate that by allowing for the geometry of the sun's position and making not too extreme assumptions about the solar motion around the galactic centre, this velocity problem can be significantly diminished or even removed completely.

5. T h e M a g e l l a n i c C l o u d s. The Magellanic Clouds are sufficiently close that pencil beam observations of neutral hydrogen can be compared profitably with the optical data to give detailed information about the correlation between HII regions, the stellar populations, and neutral hydrogen. An Ha survey of the LMC has recently been completed by Davies, Meaburn and Elliott (1975). This shows many new HII regions which have a lower surface brightness than those found in previous surveys. It is found that the brighter HII regions lie at the centre of the main neutral hydrogen concentrations. Further, the extended HII regions lie within areas where the neutral hydrogen column density is greater than 10^{20} atoms cm^{-2}.

Another subject of interest in the case of such a nearby system as the Magellanic Clouds is the stellar content of its outer regions. In particular the inter-cloud region lying between the two Clouds is important since it is known to contain large amounts of neutral hydrogen and probably contains a high proportion of the outer parts of both Clouds which have been displaced by the tidal interaction. The new Southern telescopes are currently investigating these problems.

6. The Andromeda Nebula - M31. Of all the galaxies investigated so far with the aperture synthesis technique M31 allows the most rewarding comparison between the neutral hydrogen and optical data. Emerson (1974) finds that the peaks of HI emission correlate with the dust lanes. They are coincident to 100 pc at least out to 12 kpc (1°.0) from the centre beyond which distance the dust is difficult to see. As in the case of the Magellanic Clouds the strongest HI emission is associated with clusters of HII regions, although he finds that on the local scale there is an anticoincidence in the sense that the HII regions are clustered around a HI peak. The surface density of HII regions is proportional to the HI surface density to the power 2.23 ± 0.11. Similarly there is excellent agreement between the lines of OB associations and ridges of HI; 99 percent of the OB associations are in regions where the neutral hydrogen surface density is greater than $0.8 \times 10^{20} cm^{-2}$. The surface density of young OB associations is proportional to the surface density of HI to the power 1.73 ± 0.08. These data suggest that a rate of star formation proportional to the square of the neutral hydrogen density is appropriate for M31.

The HI in M31 can be traced out for beyond the listed HII regions and OB associations where the surface density is less than 10^{20} atoms cm^{-2}. The velocity field of this gas indicates the presence of significant amounts of material at distances of 25 to 32 kpc (2°.0 to 2°.5) from the centre of M31 (Davies and Davidson 1975, Roberts and Whitehurst 1975). This material has a significantly higher M/L ratio than in the main spiral arm regions.

7. The M81 group. The three major galaxies of the M81 group, M81, M82 and NGC 3077, lie in a common envelope of neutral hydrogen (Roberts 1972, Davies 1974). It is not yet clear whether this gas has been torn from these galaxies by their mutual tidal interactions or whether it is the remnant of the primordial cluster gas which has been inhibited from condensing by the complex gravitational field of the 3 galaxies.

The long exposure photograph by (Arp 1965) shows an interesting loop structure on the NE side of M81 which suggests tidal interaction or an ejection process in M81. Recent HI aperture synthesis observations by Gottesman and Weliachew (1975) show that this loop contains HI. This feature is therefore unique in showing optical emission in apparenty low density outer HI structure. The kinematics of the neutral hydrogen indicate that the loop merges into the main body of M81, although at its furthest point its velocity deviates by approximately 100 km s^{-1} from that expected from circular motion in the plane of M81.

8. Conclusion. Up to the present there is almost a total lack of any positive optical identification with the HI in the outer parts of galaxies. In some galaxies this may be because of the low neutral hydrogen density which is typically less than 10^{20} atoms cm^{-2} and probably indicates a mean spatial density of less than 10^{-1} cm^{-3}; such densities may not support star formation. In other cases where the neutral hydrogen probably acts of a tracer for tidal interactions with neighbouring galaxies, any lack of an optical counterpart is more surprising since not only might young stars form in gas condensations but also older halo objects should also be torn away from the parent galaxy along with the neutral hydrogen. One such example of the latter situation is the optical and HI loop in M81. Careful optical observations of other phenomena like the Magellanic Streams are clearly required.

REFERENCES

ARP, H., 1965. *Science*, 148, 363.

COHEN, R.J. AND DAVIES, R.D., 1975. *Mon.Not.R.astr.Soc.*, 170, 23P.

DAVIES, R.D., 1972. *Mon. Not.R. astr. Soc.*, 160, 381.

DAVIES, R.D., 1974. J.R.Shakeshaft (ed), *"The Formation and Dynamics of Galaxies" IAU Symposium*, 58, 119.

DAVIES, R.D., 1975,. *Mon. Not.R. astr. Soc.*, 170, 45P.

DAVIES, R.D., MEABURN, J. AND ELLIOTT, K.H., 1975. In preparation.

EMERSON, D.T., 1974. *Mon.Not.R. astr.Soc.*, 169, 607.

GOTTESMAN, S.T. AND WELIACHEW, L. 1975. *Astrophys. J.* in press.

KEPNER, M., 1970. *Astr. Astrophys.*, 5, 444.

MATHEWSON, D.S., CLEARY, M.N. AND MURRAY, J.D., 1973. *Astrophys. J.*, 190, 291.

OORT, J.H., 1970. *Astr. Astrophys.*, 7, 381.

ROBERTS, M.S., 1972, in D.S. Evans (ed.) *"External Galaxies and Quasi-Stellar Objects", IAU Symposium*, 44, 12.

VERSCHUUR, G.L., 1973. *Astr. Astrophys.*, 22, 139.

WANNIER, P. AND WRIXON, G.T., 1972. *Astrophys.J.Letters*, 173, L119.

WRIGHT, M.C.H., 1974. *Astr. Astrophys.*, 31, 317.

THE PRESENT STATUS OF DATA AVAILABLE AT THE STELLAR DATA CENTER

F.OCHSENBEIN
STELLAR DATA CENTER, FRANCE

СОВРЕМЕННОЕ СОСТОЯНИЕ МАТЕРИАЛА, НАКОПЛЕННОГО В ЦЕНТРЕ ЗВЕЗДНЫХ ДАННЫХ

РЕЗЮМЕ

Информация накопленная Страсбургским Центром Данных обобщается с упором на данные имеющие астрофизическое значение: спектральные типы, лучевые скорости, фотоэлектрические величины и тригонометрические параллаксы. Описаны вычислительные средства имеющиеся в Страсбурге для использования этих данных, как для статистических исследований, так и для планирования наблюдательных программ.

I would like first to present the Stellar Data Center.

The Stellar Data Center, settled at Strasbourg Observatory, in France, has been created by our National Institute for Astronomy and Geophysics in January 1972. The aims of this Center were defined by the 5 following points:

1 - To collect existing astronomical data and to put them on machine - readable form;
2 - To improve astronomical data by extensive comparison of catalogues;
3 - To make these data available to the astronomical community;
4 - To suggest useful observations to complete our knowledge on individual stars;
5 - To provide galactic researches with the collected material.

The collecting work is done in connection with the Participating Observatories, namely:

- Geneva Observatory (Switzerland)
- Astronomisches Rechen-Institut at Heidelberg (Germany)
- Marseille Observatory (France)
- Paris-Meudon Observatory (France).

The first problem the Data Center has faced is the crossidentification of catalogues; this has led us to build the General Catalogue of Stellar Identifications, abridged by CSI, which is widely described in the Information Bulletins published twice a year by our Institute. This catalogue now contains roughly 410 000 stars, and gives for each star:

- fundamental information : coordinates, magnitudes, 1-dimensional spectral type and widely used identifications.
- identifications in catalogues of observational data in astrometry, MK classification, UBV and uvby photometries, and radial velocities; these identifications being known, all subsequent information can easily and automatically be retrieved.

Fig.1 shows the histogram versus visual magnitude for the CSI stars. The full line is the number of stars in the Catalogue of Stellar Identifications per interval of one tenth of magnitude; by dotted line, we have plotted the same diagram for HD stars. It follows from this figure that the CSI is complete up to 9.1 in visual magnitude, while HD catalogue is only complete up to 8.5.

Table I gives the number of data available by means of the CSI. This table lists, in the first column, the type of data available:

μ means "existence of proper motions"
HD means "existence of HD spectral type"
(30 000 HDE stars are included in this group)
UBV means "existence of (UBV) or (U_CBV) wide-band photometry measurement"
MK means "existence of a 2-dimensional spectral type"
RV means "existence of radial velocity measurement "
uvby means "existence of intermediate-band photometry in the system defined by Stroemgren" (1963)
and π means "existence of a trigonometric parallax"

The column labelled "Number" lists the total number of stars which fall in each group; the repartition in percent between the Northern and the Southern hemisphere is given in the 3-rd column. It can be pointed out that most of the observations in the five latest groups lay in the North; the proportion of Northern stars reaches 2/3 for radial velocity and MK classification data.

Since it is not possible to speak about all data, I have selected four of them:

1 - The UBV group, which is composed by the UBV catalogue by Blanco et al. (1968), and its continuation which is kept up to-date by Mermilliod (1973).

2 - The MK group, which is composed by the catalogue published by Jaschek et al.(1964), and its continuation by Kennedy (1972).
The number of stars with known MK classification will increase in a very large amount with the survey of Michigan Observatory undertaken by Dr. N. Houk (1973).

3 - The RV group, which is composed by Wilson's General Catalogue of Radial Velocities (1953), and the Bibliography of Stellar Radial Velocities, by Abt (1972). Data in radial velocities are also compiled by Evans and Mme Barbier at Marseille Observatory (1971, 1975).

4 - The π group, which is composed by Jenkins' Catalogue (1952).

It should be noticed that most of these catalogues have been compiled by bibliographical survey; therefore, some of the available data could have been omitted.As a second point, the statistics presented are derived from the part of these data which are now connected to the CSI ; this connection was only possible when good positions could be found.

Fig. 2 shows the histogram, for each of the four groups, versus visual magnitude.

The quantity which is plotted is the number of stars falling in each interval of ½ magn., divided by the total number of stars of the group. It can be seen that faint stars are numerous in the UBV group (many clusters have been studied photoelectrically), and much less in the RV group; it should be noticed that, among the stars brighter than 6m5, which constitute the Bright Star Catalogue by Mrs Hoffleit, observational data are missing, particularly in the UBV and MK groups.

Fig. 3 shows the temperature histogram for each group. The spectral types have been splitted in 9 groups: O-B4, B5-B9, AO-A4, A5-F4, F5-F9, GO-G9, KO-K4, K5-K7, M. Here,we have plotted the relative number of stars falling in each of these intervals for the following samples of stars:

HD stars brighter than $m_V = 6.5$
HD stars " " $m_V = 7.5$
HD stars " " $m_V = 8.5$
CSI stars

and for each of the four groups UBV, MK, RV and π.

It can be seen first, in view of the upper diagrams, that, with an increasing limiting magnitude, the proportion of hot stars decreases, and the proportion of F5-G9 stars increases.

The shapes of the lower diagrams are different from the upper ones: hot stars are very numerous, particularly in the UBV group, but also in the MK and RV groups; there is a lack of late-type stars in the 3 groups UBV, MK and RV. The amount of very late type stars in the RV group is due to the survey made by Humphreys (1970). The π group is essentially composed by late-type stars; this seems quite obvious.

Now, let us suppose that these data are to be used for statistical purposes. One can say that statistical work provides good results when it fulfills two conditions at least, namely:

1st : that the data are good

2nd : that the data are representative of the sample they are taken from.

From what I have shown, it is clear that the UBV data existing now are by no means well suited for studies of all spectral types. Even if one takes only stars, let us say, of A type, if we are dealing with a distribution which is scattered over many magnitudes, we will have effects depending on reddening and on distance which have to be eliminated before doing the statistics we are interested in.

Another important effect is the heterogeneous coverage of the sky, which is very important for instance in radial velocity studies, when one wants to derive the solar motion from stars concentrated in one part of the sky.

An example of the bias has been pointed out in view of Table 1, especially for MK and RV data.

Another illustration of this bias is given in Fig. 4, where we have plotted the relative number of B stars laying within -2.5° and $+2.5^{\circ}$ of galactic latitude falling in each of the 3 groups, as a function of the galactic longitude. It can be pointed out that the shapes of the arms,in Perseus (120°), Cygnus (80°), Sagittarius (10° and 340°), and Carina (290°) vary from one group to the other.

Now to derive the interesting parameters from the observations, combined data are generally required for a given star or group of stars. Table 2 lists the number of stars available on the whole sky with another property besides the radial velocity.

It has been seen from Table 1 that roughly 22 000 stars have been observed for radial velocity. Among these 22 000 stars, it follows from the 6th line of the table that only 13755 have also a MK classification, and from the 7th line that for only 13 147 stars, radial velocity and proper motions and MK classification can be found. This means that, though 22 000 stars have a known radial velocity, it is possible to derive spatial velocities for only 13 000 stars. In fact, the interstellar absorption should be taken into account; it is possible to derive an estimation of this absorption by means of UBV measurements. Therefore, to derive spatial velocities, the following data should be combined: RV,MK,μ and UBV. The number of such stars can be found in the line of the table, labelled "μ+MK +UBV", and this number is only 9 272. And, if we are now interested in stars for which data of RV, MK, μ , UBV, uvby and trigonometric parallaxes are available, the number of such drops to 1 105, which is the number listed on the last line of the table.

I would like now to come back to the Data Center. The description of the observational material which was shown, was an illustration of the possible uses of the Data Center. To give another example of its possible applications, a list of the best MK spectral types available will soon be published by Jaschek, who assumes the directionship of the Center.

But I believe that the most important service the Stellar Data Center is able to assume is to make available to any astronomer the observational material which is now known for any star. Let me give an example: you suspect,for instance that a given star shows some peculiarity. It should be interesting to know what has been published about this star; but this task is very tedious, and time consuming, and it is not obvious that it will result after all because of the number of literature to be searched.

The Data Center is now able to list, the references of the papers published in commonly used revues since 1950 in which your star is mentioned. This bibliographical Star Index has been created by Spite and Cayrel at the Meudon Observatory (France); the information contained in the corresponding file can now be retrieved automatically (Cayrel et al. 1974).Other possibilities, such as star sampling, are described in the Information Bulletins of the Stellar Data Center. For such requirements it is possible to write or telex tp the Stellar Data Center, at Strasbourg Observatory, in France.

Very recently we have started a subcenter at Potsdam (GDR) and you should thus write to Dr. Ruben, Zentral Institut für Astrophysik. I am sure he will provide you with the details of what is being done at Potsdam.

Table 1

Distribution of CSI stars

Group	Number	Repartition	
		N	S
μ	357 660	51	49
HD	253 499	40	60
UBV	27 203	55	45
MK	22 475	68	32
RV	21 934	68	32
uvby	6 411	60	40
π	5 701	57	43

Table 2

Distribution of stars with known RV

Group	Number	%
Var	1 221	5.6
Double	4 842	22.1
μ	20 084	91.6
HD	19 656	89.6
μ + HD	18 888	86.1
MK	13 755	62.7
μ + MK	13 147	59.9
UBV	12 744	58.1
μ + HD + UBV	11 693	53.3
μ + MK + UBV	9 272	42.3
μ + HD + uvby	4 593	20.9
μ + MK + uvby	3 794	17.3
π	4 157	20.0
All	1 105	5.0

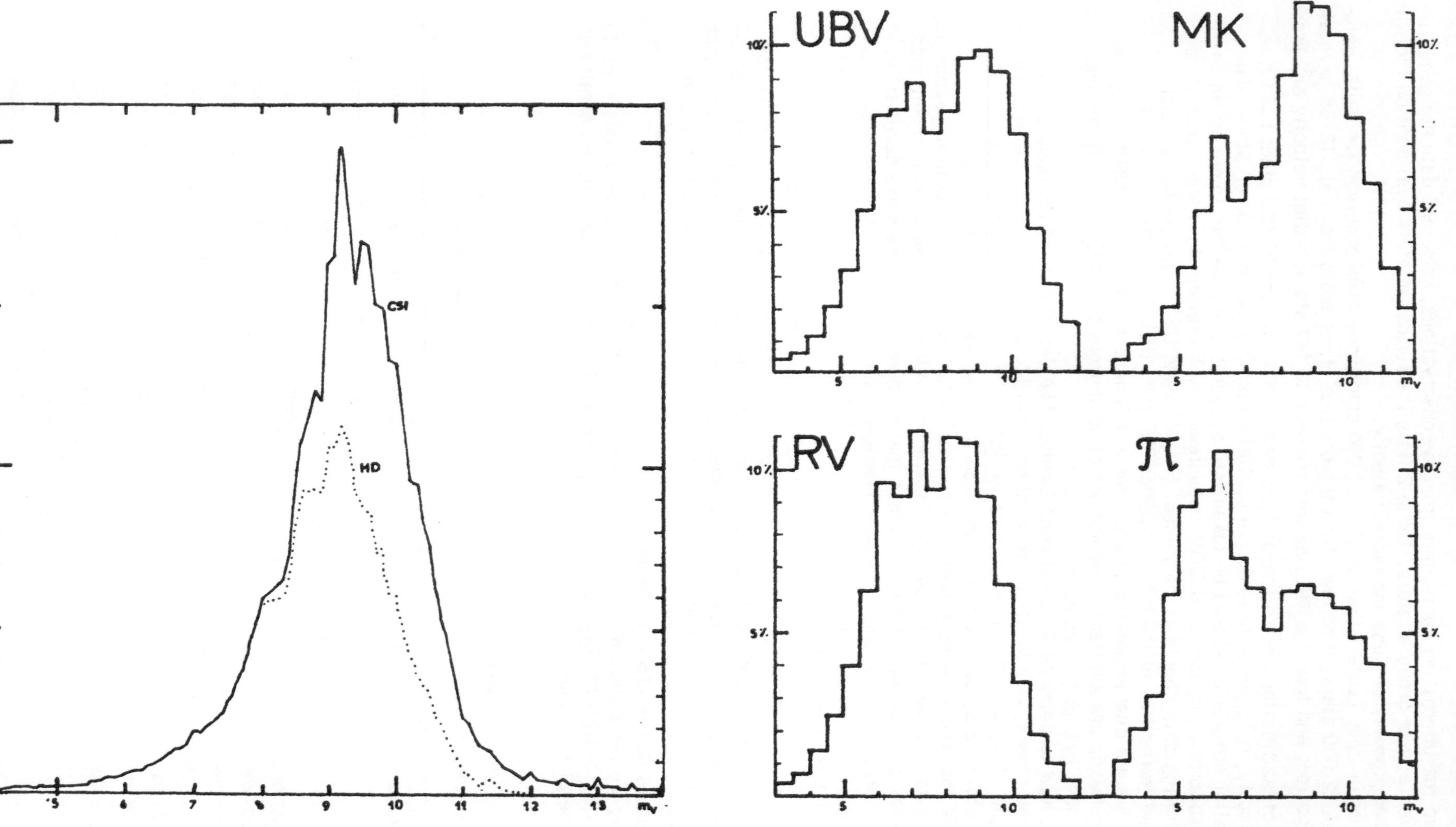

Fig. 1. Histogram, versus visual magnitude, for HD stars and stars taken from the Catalogue of Stellar Identifications.

Fig. 2. Relative distribution, versus visual magnitude, for stars with known UBV photometry (UBV), with known MK classification (MK), with known radial velocity (RV) and known triginometric parallax (π).

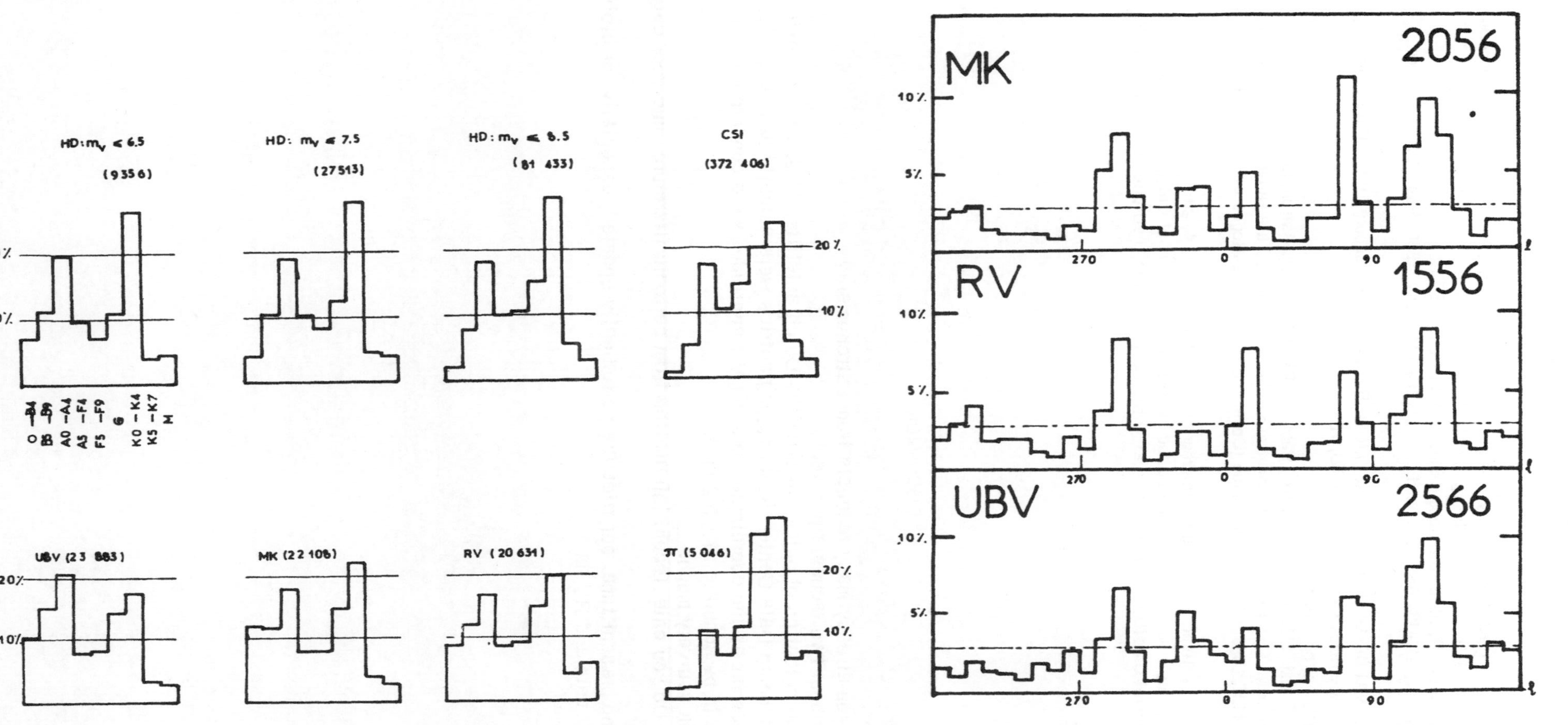

Fig. 3. Relative distribution, versus spectral type, for stars taken from HD catalogue with 3 different limiting magnitudes (top), CSI catalogue (top), and catalogues of UBV measurements, MK classification data, Radial Velocities measurements, and trigonometric parallaxes.

Fig. 4. Relative distribution of B stars near the galactic plane ($-2.^{\circ}5 \leq b \leq +2.^{\circ}5$) versus galactic longitude :
- B stars with known MK classification (top, 2 056 stars)
- B stars with known radial velocity (middle, 1 556 stars)
- B stars with known UBV photometry (bottom, 2 566 stars)

REFERENCES

ABT H.A. and BIGGS E.S., 1973, Bibliography of Stellar Radial Velocities, Kitt Peak National Observatory

BARBIER M., 1971, Inform. Bull. of the Strasbourg Stellar Data Center 1, 7

BARBIER M., 1975, Inform. Bull. of the Strasbourg Stellar Data Center 8, 12

BLANCO V.M., DEMERS S., DOUGLASS C.C. and FITZGERALD M.P., 1968, Publ. U.S. Naval Obs., 2nd Series, 21

CAYREL R., JUNG J., VALBOUSQUET A., 1974, Inform. Bull. of the Starsbourg Stellar Data Center 6, 24

HOUK N. and COWLEY A., 1973. IAU Symposium 50, 70

HUMPHREYS R.M., 1970, Astron. Journal 75, 602

JASCHEK C., CONDE H. and SIERRA A.C., 1974, Publ. La Plata Obs. Ser. Astron. 28

JENKINS L.F., 1952, General Catalogue of Trigonometric Stellar Parallaxes, Yale University Observatory.

KENNEDY P.M., 1972, M.K. Classification Extension (Complement to the Catalogue by C.Jaschek et al.), Mount Stromlo

MERMILLIOD J.Cl., 1973, Inform. Bull. of the Strasbourg Stellar Data Center, 4, 20

STROEMGREN B., 1963, Basic Astronomical Data edited by Kuiper, University of Chicago Press, 204

WILSON R.E. 1953, General Catalogue of Stellar Radial Velocities, Carnegie Institution of Washington Publ. 601, Washington D.C.

DISCUSSION

IWANOWSKA : Have you already many requests from astronomers?

OCHSENBEIN : Yes, about two requests per week.

WESSELIUS : For the selection of guide stars for the Dutch satellite, ANS, we have extensively used the Strassbourg Data Center. Some 10.000 entries were actually used and proved to be quite reliable, as far as the coordinates and the V magnitudes were concerned. The V-magnitudes proved to be accurate to $\pm$ 0.2 mag.

OCHSENBEIN : Thank you very much.

SHCHERBANOVSKY : Do you enter recent information from periodic literature into your catalogues?

OCHSENBEIN : For the present time, our files are periodically updated, especially for bibliographical data, UBV data and RV data.

OBSERVATION OF AN Sc TYPE SPIRAL IN A COMPUTER MODEL OF A GALAXY

D.R.K. BROWNRIGG
COMPUTER SCIENCE DEPARTMENT, READING UNIVERSITY, U.K.

ПОЛУЧЕНИЕ СПИРАЛИ ТИПА Sc В КОМПЮТЕРНОЙ МОДЕЛИ ГАЛАКТИКИ

РЕЗЮМЕ

Использована компютерная модель для образования устойчивого вращающегося диска с двумя спиралями, внешне напоминающего галактики типов Sc, включая NGC 309. В качестве модели диска использовалась модель из 25000 изолированных и самосогласованных звезд, представляющая поведение Населения I. Находящиеся вне плоскости галактики звезды Населения II и скопления представлены в качестве фиксированного накладывающегося гравитационного поля. Рассматривается динамика представленной системы, графически иллюстрируемая снятым с компютера 16 мм кинофильмом.

A computer model has been used to generate a stable rotating disk dominated by a two arm spiral mode. The visual appearance of the disk strongly resembles that of Sc type galaxies, including NGC 309.

A three dimensional, particle-mesh, computer model of a galaxy has been constructed. A model disk of 25000 isolated and self-consistent model stars is used to represent Population I star behaviour. Population II stars and clusters out of the galactic plane are represented by a fixed imposed gravitational field.

For a simplified physical setup, the dispersion relation behaviour matches that of a gravitationally bound, finite size mass particle system. The effective radius of each model star is three kpc and its mass is about 4×10^6 solar masses. Thus each model star behaves like a cloud of mass. For typical galactic velocity dispersions, the wave behaviour is close to that of a point mass system, especially for the long wavelengths corresponding to global spiral structure. Furthermore, for the mesh size and number of model stars used, the collisional relaxation time of the model is of the same order as, or longer than, the collision time for our galaxy. The conclusion is that the model may be used validly to simulate a thin disk galaxy.

The stability of such systems is investigated for varying fractions of Population I stars by mass and the effects of stellar motions out of the galactic plane as compared to a strictly two-dimensional model are considered.

The motions out of the galactic plane do not have a significant effect on the violent instabilities of an initially azimuthally cold, balanced, self-consistent disk. However, when a large fraction of the mass is contained in the Population II star field, the axial motions greatly delay the onset of the bar mode that seems to be the eventual fate of all dynamically stable disks.

For a halo to disk mass ratio of four to one, the three dimensional model quickly develops spiral structure which is maintained in the face of very strong differential rotation - a factor of six to one between the rotation periods of stars at the inner and outer limits of the spiral. The period of the experiments is 6×10^9 years, covering fifteen rotations for a star initially on the outermost edge of the disk (at 20 kpc radius) and over thirty rotations for stars where the angular velocity is greatest. Fourier analysis of the spiral structure shows that a two arm mode is most often dominant and is of generally global extent. Furthermore, it continues to grow slowly throughout the experiment, whereas other spiral arm numbers reach an early peak in intensity and then fade or remain roughly constant.

The propagation mechanism is basically that of a density wave. The wave also tends to break at the point of maximum shear, occasionally, and each inner part moves on to join the corresponding next outer part. The pitch angle remains constant, on average. The breaking and rejoining mechanism suggests a mode of spiral propagation mixed with that of density waves and may help to explain the broken spiral structure of many galaxies which has been taken, in the past, as an argument against global spiral theory.

A 16 mm, computer generated movie film of the 'stable spiral' case has also been made and graphically illustrates the dynamics of the system.

DISCUSSION

SHCHERBANOVSKI : How much computer time was needed to make your film?

BROWNRIGG: In all about 4 hours.

SHULMAN: Some years ago I saw a similar movie made in USA. Could you briefly summarize the differences?

BROWNRIGG: The film you saw was for a model limited to two dimensions and without lasting spiral structure. Our film is the first for a large, self-consistent, three - dimensional model, showing long-lived spiral structure.

GUIBERT: How long can your spiral pattern last?

BROWNRIGG: It is still present after 16 rotations(about 30 rotations at the radius of the Sun), even though a bar begins to form during the 14th or 15th rotation.

TOVMASSIAN: Is it possible, by choosing appropriate parameters, to get a galaxy with two definite arms?

BROWNRIGG: Yes, but these spirals are rather shortlived; usually the structure has disappeared after three to four rotations.

TOVMASSIAN: Can this model explain the difference in population of spirals and disk?

BROWNRIGG: Only in the way that the system is idealized. The model stars are taken to be bright, low-velocity dispersion, young stars that are therefore limited to the plane. All high-velocity dispersion, dim objects are considered to be members of the halo, or Population II.

DODD: Did you try initial conditions in which the Population I stars were not coplanar?

BROWNRIGG: Not yet. So far, the work has been limited to thin disks. Modelling of protogalaxy collapse is an aspect of this work which I plan to study in the near future.

IWANOWSKA: Have you taken all stellar masses equal?

BROWNRIGG: Each "star" in the model represents about $2 - 4.10^6$ $M_\odot$ in a real galaxy.

THE NUCLEI OF SEYFERT GALAXIES

E.A.DIBAY
STERNBERG ASTRONOMICAL INSTITUTE, MOSCOW, U.S.S.R.

ЯДРА СЕЙФЕРТОВСКИХ ГАЛАКТИК

РЕЗЮМЕ

Спектральные и фотометрические наблюдения позволяют определить параметры оболочек ядер сейфертовских галактик. Широкие эмиссионные линии возникают в плотной ($n_e \approx 10^8$ см$^{-3}$) газовой области. Для ряда галактик вычислены значения объемов, масс газа, кинетической энергии, характерных размеров, болометрической светимости и т.д. Рассмотрен баланс сил, действующих на оболочку, для чего проведено сравнение потоков лучистой и механической энергии. Сделан вывод о том, что ядра находятся в квазистационарном состоянии. В предположении квазистационарности проведены оценки масс центральных тел ядер, в среднем $\approx 10^8$ солнечных масс. Построена геометрическая модель ядра.

Характеристики ядер тесно связаны с характеристиками окружающих галактик. Безразмерное отношение светимостей ядра и галактики повидимому связано с градиентом плотности галактики, влияющим на концентрацию яркости. Наблюдение галактик с высоким градиентом яркости позволяет обнаруживать сейфертовские объекты в количестве, превышающем среднее число сейфертовских галактик среди объектов поля.

I n t r o d u c t i o n. There is not conventional definition of Seyfert galaxy. In this paper we will call as "Seyfert" the galaxy with a bright nucleus that has broad emission lines exceeding 1000 km/sec. Such features are confidently noted on the spectra with a small resolution.

The distribution of the effective doppler velocities (at half intensity) for 22 well determined nuclei is given in Fig 1. The low velocity end is poorly determined. The galaxies with moderate emission-line widths, similar to NGC 1068 (v_{eff} = 800 km/sec) are poor for detection. Such objects appear to be rather numerous.

G e o m e t r y o f n u c l e u s . The unhomogeneous structure of emission lines is a very marked peculiarity of Seyfert galaxy spectra. According to Dibay and Pronik (1967) two distinguished gas regions are responsible for the origin of both permitted and forbidden lines. Namely, the broad Balmer lines are emitted in dense zone (Fig.2) while the narrow forbidden ones are emitted in low density zone with small velocities. Such model permits to explain the main properties of Seyfert galaxy spectra.

1. The profiles of hydrogen and of forbidden lines are similar, density and velocity being low (NGC1068, Mark 198). In this case the only gas subsystem with low density is present.
2. There are the same features in the forbidden lines and cores of the hydrogen lines, only the hydrogen lines having wide wings (NGC 4151, 5548, 3C-120 etc). There are two emitting subsystems.
3. The profiles of the forbidden and hydrogen lines are identical, velocity and density are large(NGC1275, 3C-273). There is the only one internal dense subsystem.

In this paper we consider in some more detail some properties of dense gas moving with velocities about several thousands of kilometers per second. Numerical data being the result of photometric and spectroscopic observations of 14 selected objects are presented in Table 1. The Hubble parameter value H=75 km/sec. Mps is adopted. Column I gives NGC, 3C or Markarian numbers, column 2 - bolometric power in 0.3-10 micron domain (infrared fluxes according to Rieke and Low,1972). The mean values of UBV-magnitudes were taken for variable sources. The next point is connected with the important problem of electron density of rapid gas.

There are two approaches for determination the electron density in internal zone. The direct measurement is possible using forbidden line intensities, if this latter origin inside the dense zone. For 3C-273 (Dibay and Pronik,1964) and NGC1275 (Dibay and Pronik,1967) the value $n_e = 10^7$cm^{-3} is obtained. The second way is rapid variations of H_α - line observations of Seyfert nuclei (Cherepashchuk and Lyutyj,1973). For NGC 4151, 3516 the characteristic recombination time leads to the value $n_e = 10^7 - 10^8$ cm^{-3}. In this investigation we assume that

$$n_e = 10^8 \mathrm{cm}^{-3} \; ; \; T_e = 3.10^4 \mathrm{K}$$

are characteristic values of dense internal gas subsystem.

With these data we calculate the emission power of one cm^3 optical thin gas in H_β-line

$$E(H_\beta) = 0.96\ 10^{-10} \mathrm{erg/sec.\ cm}^3$$

Then with the power emitting in H_β - line wings $F(H_\beta)$ erg/sec the effective volume of gas can be obtained

$$V_{eff} = F(H_\beta) : E(H_\beta) \ \mathrm{cm}^3$$

The effective radius of the dense gas envelope is equal to

$$r = \sqrt[3]{3V_{eff} / 4\pi}$$

Values of r are given in the third column of Table 1. Dimensions of the envelope can be estimated independently from Cherepashchuk and Lyutyj observations ($10^{16} - 10^{17}$ cm for NGC 4151 and 3516).

The density and the volume product results in mass (column 4), and with doppler velocity taken from emission-line profile - the gas mechanical energy (column 5).

Now we consider the dynamical state of the gas envelope. Let us introduce a flux of radiative energy $L/4\pi r^2$ (column 6 in Table 1) coming from the central source of ionization, and a flux of mechanical energy ρv^3 (column 7). The comparison of these quantities is presented in Fig.3. It can be seen, that the fluxes of radiative and mechanical energies are the quantities of the same order of magnitude. Under such condition the Seyfert nuclei (Inside the sample of the considered objects) are assumed to be in a quasiequilibrium state.

On this basis the masses of Seyfert nuclei may be estimated. The observed high velocities (broad emission lines) are believed to correspond to escape velocities in the vicinity of some massive body. In this case

$$M = V^2 r / 2G$$

where G is the gravitation constant, The mass values obtained in such a way are given in column 8 of Table 1.

The masses can be associated with characteristic times of optical variability, according to Dibay and Lyutyj,1975 (Fig. 4). The distance in Fig.4 between the observed (dashed) line and the theoretical $t = R_g / c$ line determines the size of the optical variability region (in dimensionless units r/R_g).

Now a geometrical model of Seyfert nucleus can be constructed. There is some massive, probably relativistic, body in the central region of the nucleus. A typical mass is $10^7 - 10^8$ solar masses and the gravitational radius is about 10^{13} cm. On the size of the order of $10^{15} - 10^{16}$ cm the optical variability takes place, gas clouds with the common masses of $10^2 - 10^3$ solar masses are situated at the distance $10^{16} - 10^{17}$ cm.

Finally the mass-luminosity correlation for Seyfert nuclei is presented in Fig 5.

R e l a t i o n s h i p   b e t w e e n   n u c l e i   a n d   g a l a x i e s. This problem was discussed by Ambartsumian (1954). At present the connection of nuclei with surrounding galaxies appears to be established rather reliably. There is correlation between the nucleus luminosity - the galaxy luminosity (Dibay,1968; Wade,1968) and the nucleus luminosity - the galaxy mass (Zasov and Dibay,1970). In general the nucleus parameters (luminosity, mass, etc.) are increased in the sequence: spiral galaxies, N-galaxies, quasistellar objects. Let us introduce the dimensionless ratio of the nuclear luminosity to the galaxy luminosity and assume that this quantity depends on the density gradient of the galaxy

$$L_{nucl} / L_{gal} \sim f(\rho_{centr} / \rho_{gal})$$

(brightness gradient ≈ density gradient is one from the meaningful classification parameters in Morgan's scheme). Actually, Zasov and Lyutyj, (1973) showed that the classical Seyfert galaxies have greater concentration of the distribution of the matter towards the center than the ordinary spirals. Recently Arakelian (see III Tbilisi EAM, 1975) made a list of 600 galaxies with the high brightness gradient. Observations of such galaxies (Arakelian, Dibay and Yesipov, 1975) show Seyfert-type objects in the amount exceeding the mean number of Seyfert objects among the field galaxies.

The number of Seyfert objects. In the original paper of Seyfert (1943) twelve galaxies are described. Today (1975) the number of Seyfert objects is more than eighty (see for example, Khachikian and Weedman, 1974). About fifty per cent of new discovered objects belong to Markarian lists of blue continuum galaxies. More than forty galaxies with Seyfert-type nuclei were found by Arakelian, Dibay and Yesipov at the Crimean Sternberg Station as a result of the joint work of two observatories: Byurakan observatory and Sternberg Institute.

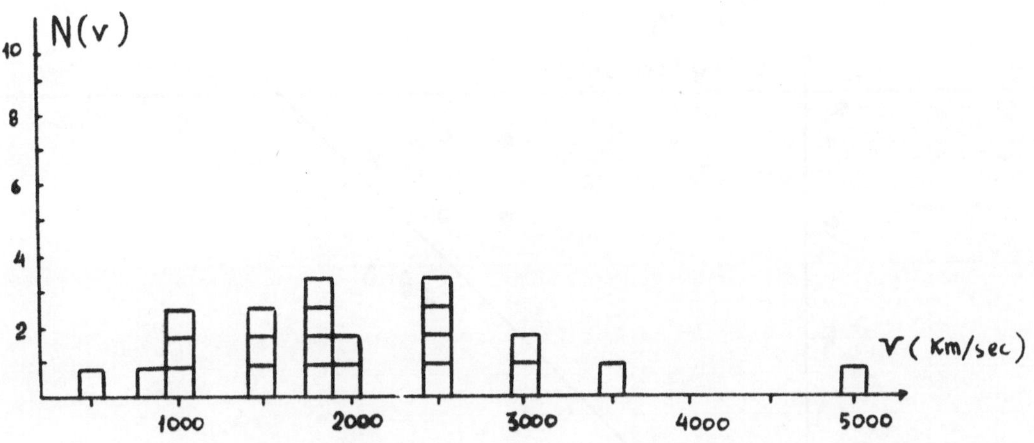

Fig. 1. Doppler velocity distribution of Seyfert nuclei

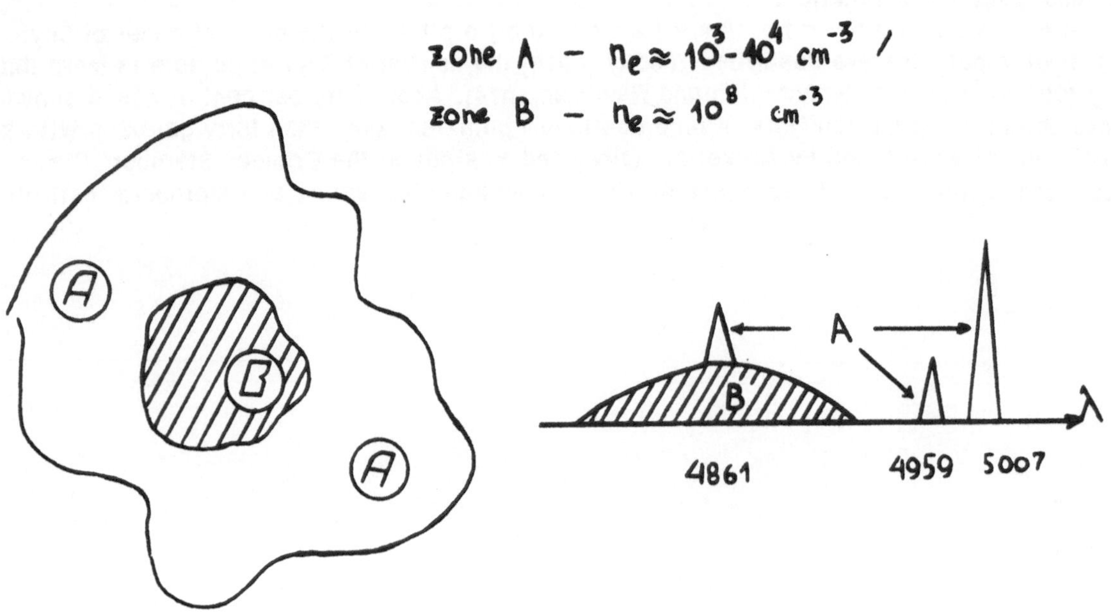

Fig. 2. Model of space geometry of emitting gas

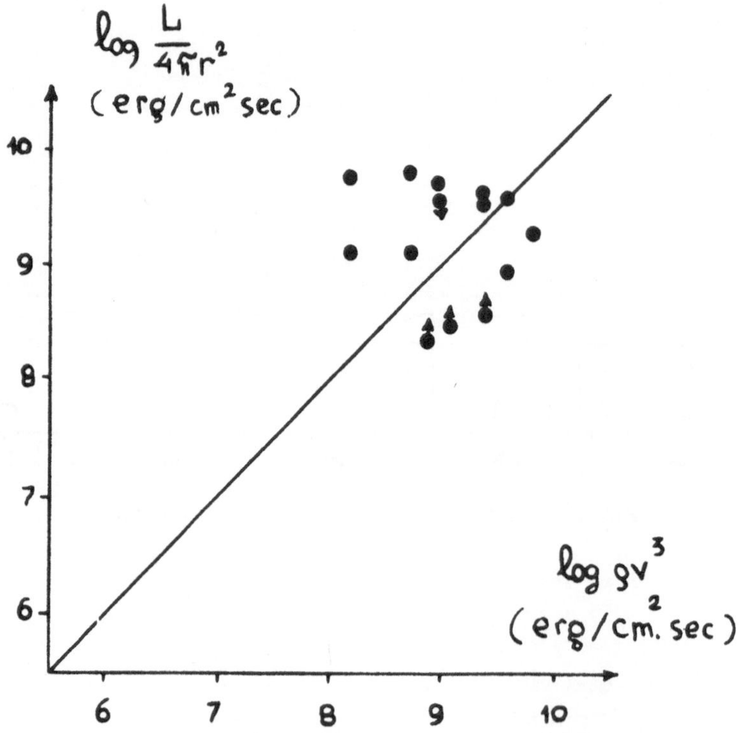

Fig. 3. Flux of radiative energy throughout the gas envelope versus a flux of mechanical energy.

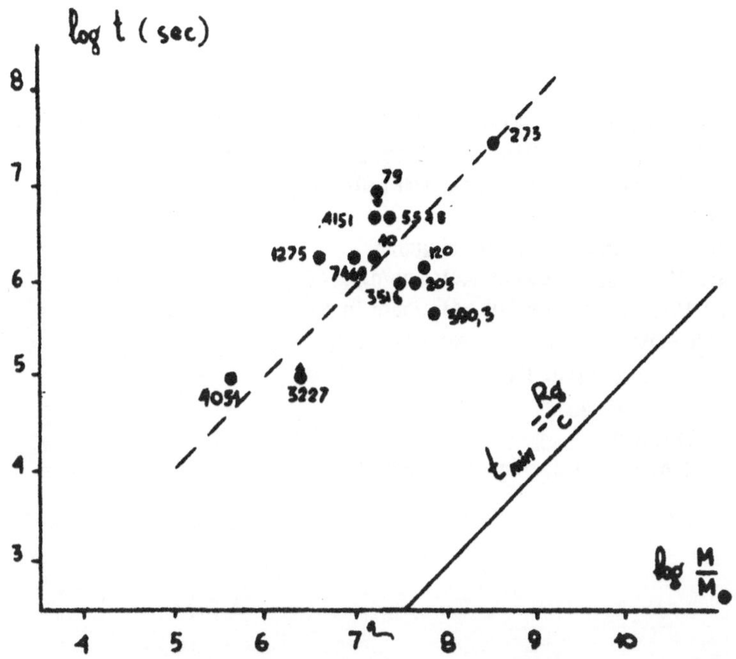

Fig. 4. Dependence: time of optical variability-mass of central body

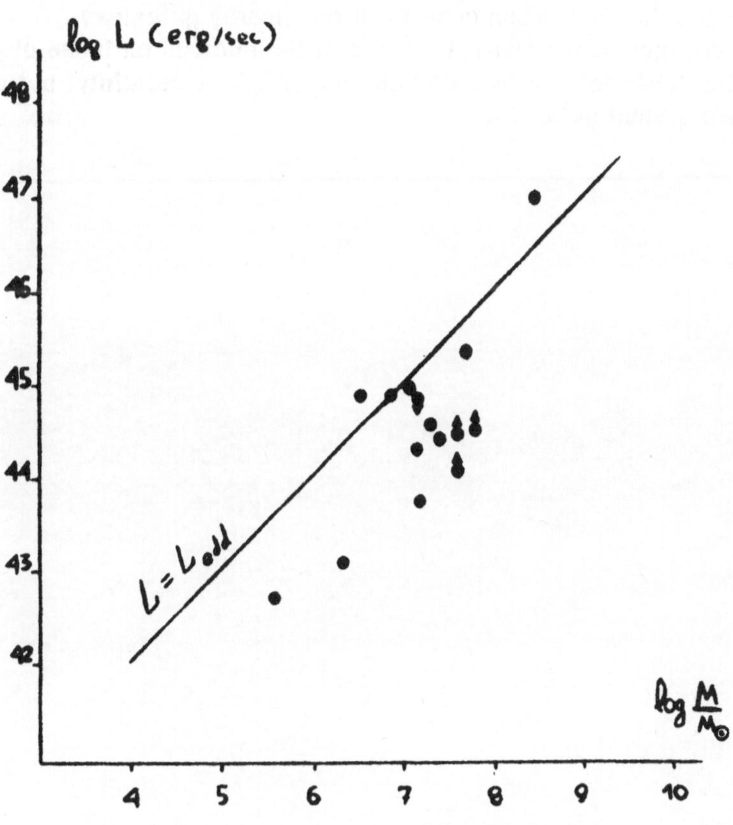

Fig. 5. Mass-luminosity relation of Seyfert nuclei

REFERENCES

AMBARTSUMIAN V,A. 1954, In book "Problems of the Universe evolution (in Russian) Erevan 1968
ARAKELIAN M.A., DIBAY E.A., YESIPOV V.F. 1975, Astrofizica, 11, 5
CHEREPASHCHUK A.M., LYUTYJ V.M., 1973, Aph.Lett.13,165
DIBAY E.A., 1968, Astr.Zirc.N 481, 2
DIBAY E.A., PRONIK V.I., 1964, Astr.Zirc. N 286, 1
DIBAY E.A., PRONIK V.I., 1967, Russian Astr.Journ.,44, 952
DIBAY E.A., LYUTYJ V.M., 1975, Russian Astr.Journ.,in press
KHACHIKIAN E.E., WEEDMAN D.W. 1974, Ap.J.192, 581
RIEKE G.H., LOW F.J., 1972, Ap.J. 176, L 95
SEYFERT C.K. 1943, Ap. J.97, 28
WADE. C.M., 1968, A.J.73, 876
ZASOV A. V., DIBAJ EA, 1970, Russian Astr.Journ.47, 23

DISCUSSION

TOVMASSIAN : When you talk about the correlation between the brightness of Seyfert galaxy nuclei and their total mass, which part of the central region does this brightness refer to?
DIBAY: For classical Seyfert galaxies we measure the brightness of the nucleus photoelectrically through a 5 arcsec diaphragm.
TOVMASSIAN : So this has only been done for a few nearby galaxies?
DIBAY: The dependence of the characteristics of the nucléus on those of the entire galaxy has been well estabilshed for nearby galaxies and with less certainty, but still rather convincingly, for the more distant galaxies.

NEUTRAL HYDROGEN AND DENSITY WAVES IN M 31

THICKNESS OF THE GAS LAYER *

J.GUIBERT and F.VIALLEFOND
OBSERVATOIRE DE MEUDON, FRANCE
PRESENTED BY J. GUIBERT

НЕЙТРАЛЬНЫЙ ВОДОРОД И ВОЛНЫ ПЛОТНОСТИ В М31; ТОЛЩИНА ГАЗОВОГО СЛОЯ

This paper is based on 21 cm line observations of M 31 made with the Nançay radiotelescope. A well defined spiral structure appears in the neutral hydrogen distribution. Most of HI arms are in good agreement with optical arms. The density wave theory developed by Lin and Shu has been applied to M 31, in order to explain the presence of a spiral pattern of nearly constant interarm spacing (4 kpc) throughout the whole nebula. It appears that solutions of the dispersion equation consistent with observations from the centre to the outer regions are strongly modified when the gas content and finite thickness of the galaxy are taken into account.

Using the observed HI density contrast and the relation : $c_r^2 / c_z^2 = 1 + c_r^2 / c_\theta^2$ between the radial, vertical and tangential stellar velocity dispersions proposed by Kuzmin, it is shown that, in M 31, the gas disk thickness increases from the central region to the outer parts. Such a prediction is consistent with several observations of the HI disk in our Galaxy and of the dust layer in the edge-on galaxy NGC 5866. Hence, in both galaxies, Kuzmin's relation seems to be valid.

Direct determination of the disk thickness for Population I objects as a function of radius thus may provide information on the stellar dynamics in spiral galaxies.

* Paper submitted to Astronomy and Astrophysics.

THE ESO / UPPSALA SURVEY

R.M. WEST
ESO SKY ATLAS LABORATORY GENEVA, SWITZERLAND

ОБЗОР ДАННЫХ ESO/UPPSALA

The ESO/Uppsala Survey of the ESO(B) Atlas was undertaken in 1974 in a collaboration between the European Southern Observatory and the Uppsala Observatory. Its aim is to identify objects of astrophysical interest on the ESO(B) plates.

The ESO(B) Survey of the Southern Sky is carried out with the ESO 1m Schmidt telescope on La Silla, Chile, on unbaked Kodak II-O plates, through a GG 385 (2mm) filter, giving standard B-colour response. Each 30x30 cm plate covers $5^{\circ}.5 \times 5^{\circ}.5$ (scale 67'',4/mm); the exposure time is 60 min. The survey will cover the Southern sky from -90° to -20° and is expected to be terminated in 1976. The limiting magnitude is $21^{m}.5$.

The ESO(B) Atlas of the Southern Sky is produced from these plates on-glass and on-film by the ESO Sky Atlas Laboratory, Geneva, Switzerland. A detailed description of this project may be found in a summary by West (1974).

The ESO/Uppsala Survey identifies the following objects on the ESO(B) Atlas:

1. All NGC and IC objects,
2. All galaxies larger than 1',0 (1mm),
3. All disturbed galaxies,
4. All stellar clusters in the Budapest list, and
5. All planetary nebulae in current lists.

The survey therefore lists known objects for easy identification as well as new objects of interest for observers.

Until now (July 1975), three lists have been compiled (Holmberg et al., 1974a, 1974b, 1975) containing about 3200 objects in 116 fields, practically all south of -45°. Most of the listed objects are new (70%). A number of IC objects were not found and have been indicated as "blank field objects" An Atlas comprising several hundred peculiar galaxies is in preparation.

REFERENCES

HOLMBERG E., LAUBERTS A., SCHUSTER H.E. and WEST R.M. 1974 a, Astron. and Astrophys, suppl. ser, 18, 463

HOLMBERG E., LAUBERTS A., SCHUSTER H.E. and WEST R.M. 1974b, Astron. and Astrophys, suppl. ser, 18, 491

HOLMBERG E., LAUBERTS A., SCHUSTER H.E. and WEST R.M. 1975, Astron. and Astrophys, suppl. ser, in press

WEST R.M. 1974, ESO Bull, 10, 25

EVOLUTIONARY CHANGES OF MAIN CHARACTERISTICS OF OPEN CLUSTERS
NONSIMULTANEOUS STAR FORMATION IN OPEN CLUSTERS

O.B.DLUZHNEVSKAYA, A.E.PISKUNOV
ASTRONOMICAL COUNCIL, ACADEMY OF SCIENCES, U.S.S.R.
PRESENTED BY O.B.DLUZHNEVSKAYA

ЭВОЛЮЦИОННЫЕ ИЗМЕНЕНИЯ ГЛАВНЫХ ХАРАКТЕРИСТИК ОТКРЫТЫХ СКОПЛЕНИЙ
НЕОДНОВРЕМЕННОЕ ОБРАЗОВАНИЕ ЗВЕЗД В ОТКРЫТЫХ СКОПЛЕНИЯХ

Investigations of stellar ages in open clusters (Iben and Talbot, 1966; Williams and Cremin, 1969; Dinescu et al. 1974), and of periods of Cepheids in Magellanic Clouds associations (Efremov, 1975) have shown that there exists an age spread in these Population I stellar systems, that reaches several tens of million years. Here we briefly present some results of our investigation of influence of nonsimultaneous star formation process in open clusters on their mass functions and integrated colours (more detailed discussion see Piskunov A. 1975). For this purpose we have constructed evolutionary sequences of cluster models using the evolutionary traces for Population I stars (Paczynski, 1970) and Salpeter-type of initial mass function (Salpeter, 1955).

Star formation was assumed to last Δt years with constant rate,($\Delta t=10^7$, 10^8, 10^9years for different models).Computations have shown that the observed mass spectra F(m) in evolved models differ from the initial one. Mass spectra for such clusters demonstrate deficiency of massive stars, that is leading to the increase of the slope of F(m) for these clusters. Evolution of mass function of the model with $\Delta t=10^7$ years is shown in Fig.1. The largest deviation from the initial form F(m) is in models with age close to Δt. In cluster models with ages considerably larger than Δt, observed mass spectra again take the initial form. Deviations of F(m) from Salpeter form in the model cluster with some fixed age increase if the period of star formation increases. Some open clusters of intermediate age have mass functions that show similar features (Muzyliev V. and Piskunov A., 1974). Comparison of the theoretical and empirical mass functions confirms the existence of considerable age spread in some open clusters.

Evolution of integral colours of clusters with continuous star formation process may be investigated in a similar manner.

Fig. 2 demonstrates evolutionary paths of cluster models with different star formation time in integral colour - lg t diagram. It shows that general features of the evolution of the cluster with simultaneous star formation are retaining for models with $\Delta t\neq 0$. However, integrated colour in the models with $\Delta t\neq 0$ grows slowly with the time. Thus, age spread of cluster members may partly explain the existence of considerable dispersion of real clusters in the I_{B-V} - lg t diagram.

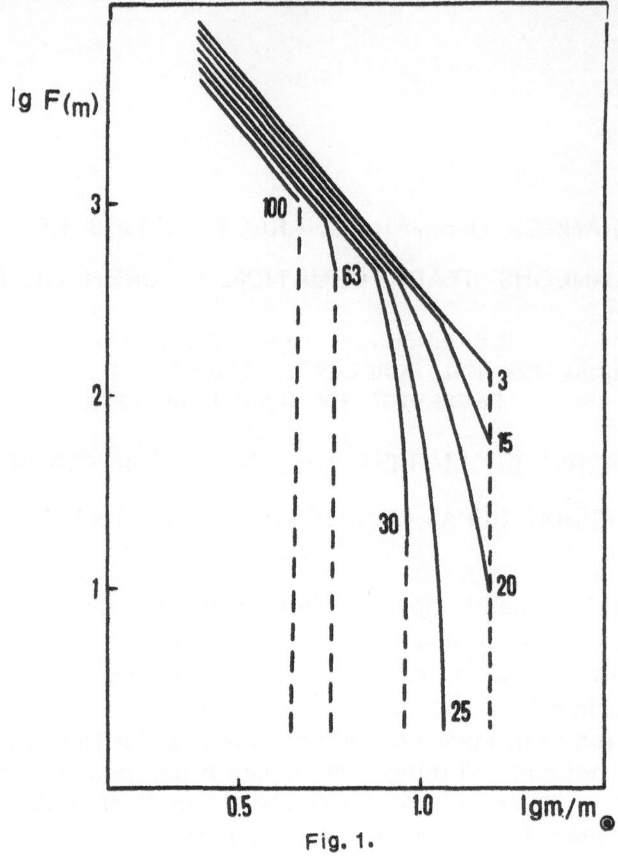

Fig. 1.

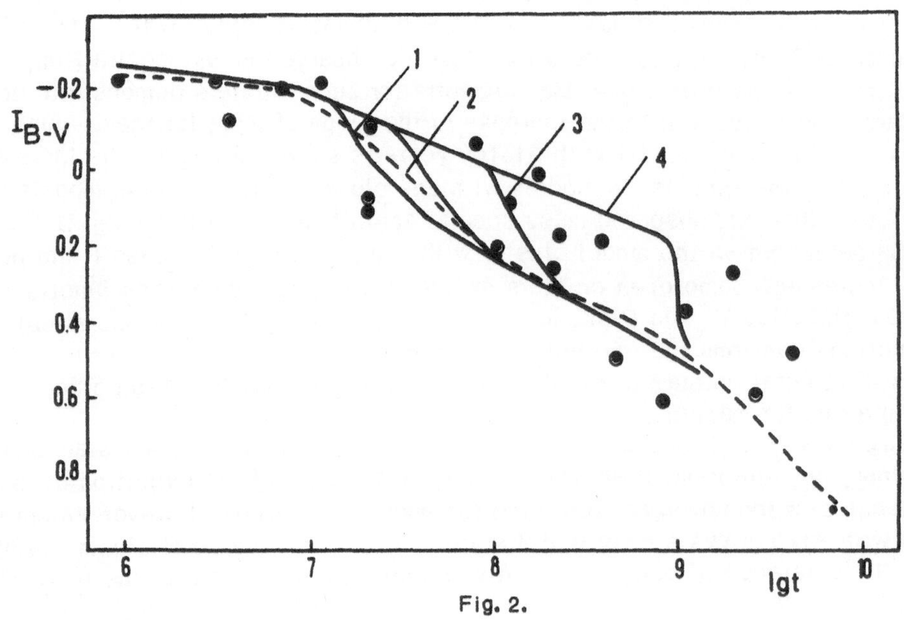

Fig. 2.

REFERENCES

DINESCU R., DLUZHNEVSKAYA O., MUZYLIEV V., PISKUNOV A., 1974, Nauch. Inf. Astron.Sov. AN SSSR, 31, 113

EFREMOV Yu., 1975, this issue

IBEN I., TALBOT R., 1966, Ap.J., 144, 968

MUZYLIEV V., PISKUNOV A. 1974, Nauch. Inf. Astron. Sov. 32, 116

PACZYNSKI B.,1970 Acta Astr., 20, 47

PISKUNOV A. 1975, Nauch. Inf. Astron. Sov. AN SSSR (in press)

SALPETER F., 1955, Ap. J., 121, 161

SEARLE L., SARGENT W.L.W., BAGNUOLO W.G., 1973, Ap.J., 179, 427

WILLIAMS I.P., CREMIN A.W., 1969, MNRAS, 144, 359

THE COMPARATIVE STUDY OF EMISSION STAR GROUPS

M.V. DOLIDZE
ABASTUMANI ASTROPHYSICAL OBSERVATORY, U.S.S.R.

СРАВНИТЕЛЬНОЕ ИССЛЕДОВАНИЕ ГРУПП ЭМИССИОННЫХ ЗВЕЗД

The aim of our paper is the comparative study of some peculiarities in distribution and evolution of groups of emission stars of different type.

Figure 1 shows a diagram in galactic coordinates of possible groups of Orion type emission stars (crosses in the diagram denote the groups known as O and T associations, crossed circles-T associations, open circles - the groups comprising Herbig's Be, Ae stars, filled circles - the places of the diagram, where crossed circles coincide with filled ones).

These groups were determined on the basis of observations made at the Abastumani Observatory as well as on the data from the literature.

A part of these groups is near the galactic plane and corresponds to the Orion and Sagittarius spiral arms.

The groups known as T and also O+T associations are seen. They belong to the Gould's belt and are in connection with the nearby complexes of dark nebulae. (Concentrations belonging to the Gould's belt are at : l=150°-230°, b>|-10°| and l = 338°-05°, b > +10°).

Other groups of longitudes near 110° and 330° to the North and South from the galactic plane are known as T - associations and are connected with isolated dark nebulae (In this respect these groups differ from those belonging to the Gould's belt). The isolated dark nebulae are in their turn connected with the reflection nebulae of different kinds, with Herbig's Be and Ae stars and with groups of very red stars.

It is possible to suggest that in this second case we observe another local detail in the Orion arm. (Corresponding galactic coordinates are l=97°-115°, b> ±10° and l= 297°-360°, b>±10°).

In addition we can note that the zones of avoidance in longitude (120°-135°, 138°-170°, 227°-300°, 308°-348°) seen in the first diagram (Fig.1) exist only in the distribution of emission line stars of the Orion type. The objects connected with them such as HII - regions, groups of reflecting nebulae as well as groups of emission stars of other than Orion type do not show the zones of avoidance seen in the first diagram (Fig.1).

Before the beginning of our observations we published a list of possible groups of emission line stars. This list was supplemented continuously and contains now 182 regions which are shown in the second diagram (Fig.2).

In 62 of these 182 areas, groups of emission stars are already found (crossed circles in the diagram)

The groups of the list were selected in those parts of the sky where dark and bright nebulae coexist (crossed circles and crosses in the diagram. Open circles in this diagram denote O associations connected with the emission stars, but not associated with nebulae). Because of such a selection criterion groups belonging to more distant spiral arms as well as the groups of other types as compared with the first diagram are presented.

Nevertheless the second diagram confirms the picture of spiral arms and partially the existence of two details of the Orion arm seen in the first diagram.

It should be noted that in discussing the character of emission star groups we come across difficulties on some areas of the southern sky. (These regions are l=250°-285° and l=310°-350° near the galactic plane). The faint emission stars are not well known there.

The objects from the second diagram could be used for some search observations in these regions of the sky.

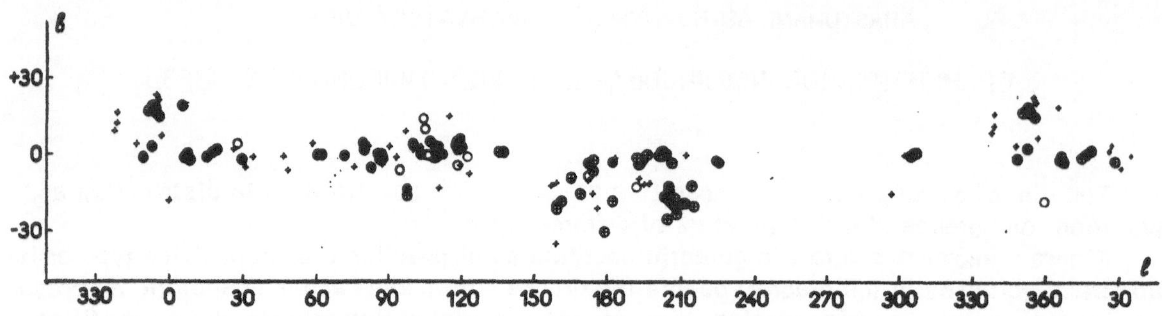

Fig. 1. Diagram of distribution of the possible groupings of Orion type emission stars in galactic coordinates. Crosses denote the groupings known as O and T associations, crossed circles- T associations, open circles - the groupings comprising Herbig's Be, Ae stars, filled circles - the places of the diagram where crossed circles coincide with filled ones, a square shows cluster An (King 10) as is also considered here as Orion type emission stars possible group.

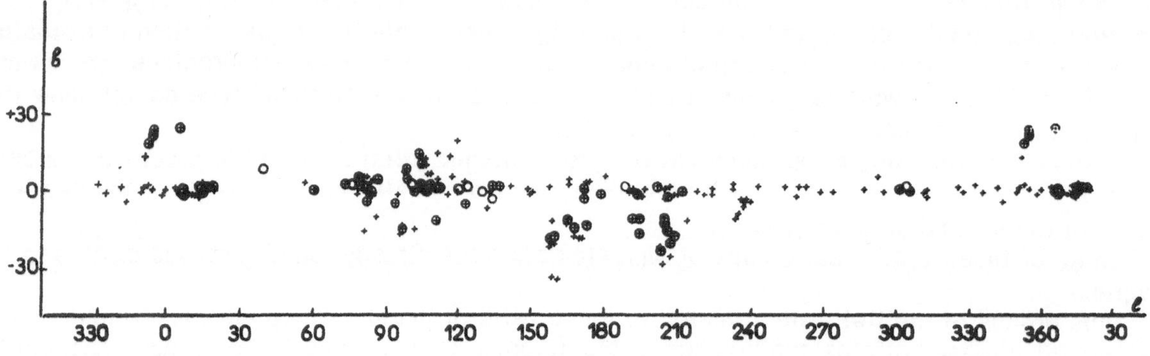

Fig. 2. Diagram of distribution of possible groupings of stars associated with bright-dark nebulae in galactic coordinates. Crosses show possible regions of emission star groupings locations associated with the complex of bright and dark nebulae, crossed circles- the same regions already studied with positive results, open circles denote O associations with the emission stars but not associated with nebulae, a square shows cluster An (King 10) as is also considered here as Orion type emission stars possible group.

THE ROTATION OF THE NEUTRAL HYDROGEN SUBSYSTEM WITHIN THE INNER REGIONS OF THE GALAXY

I.V.PETROVSKAYA
LENINGRAD STATE UNIVERSITY, U.S.S.R.

ВРАЩЕНИЕ НЕЙТРАЛЬНОЙ ВОДОРОДНОЙ ПОДСИСТЕМЫ ВО ВНУТРЕННИХ ОБЛАСТЯХ ГАЛАКТИКИ

The rotation of the neutral hydrogen subsystem in inner regions of our Galaxy was investigated at the Leningrad University in 1970 (Petrovskaya, 1970). At that time we had at our disposal the catalogue of 21-cm profiles by Muller and Westerhout (1957) for $|b^I| \leqslant 10°$. Using these data we traced the angular velocity of the neutral hydrogen subsystem up to 300 pc in both directions from the galactic plane.

Now in addition to this old catalogue we have the catalogues of the H-line profiles obtained by van Woerden, Takakubo and Braes (1962) for $10° \leqslant |b^I| \leqslant 25°$ for the northern sky and by McGee, Milton and Wolf (1966) for the southern sky. For the present investigation we have taken all the profiles of these catalogues with $|b^{II}| \leqslant 28°$.

We assume rotation to be the only motion of the neutral hydrogen subsystem and that the line of sight intersects each surface of equal angular velocity $\omega = \omega(R, z)$ at two points (Fig.1). Then for the observed optical depth of radiation of these two points which have equal angular velocities ω and hence equal radial velocities V, we have the folowing formula

$$\tau(v, l, b) = \frac{kN(R_1, z_1)}{\left|\frac{dv}{dr}\right|_{R_1, z_1}} + \frac{kN(R_2, z_2)}{\left|\frac{dv}{dr}\right|_{R_2, z_2}},$$

where R_1, z_1 and R_2, z_2 are the coordinates of these points in the galactocentric cylindrical coordinate system, r - the distance of the point from the Sun.

The surfaces of equal angular velocity are assumed to be surfaces of rotation ellipsoids in the region considered, and their parameters are determined. The following assumption was made for the dependence of neutral hydrogen density on R and z

$$N(R, z) = N(R, 0)[1 - \beta(R) z^2),$$

where the functions $N(R,0)$ and $\beta(R)$ may also be obtained together with the kinematic parameters.

Our method uses the whole profile of the line.

In Fig.2 the surfaces of the equal angular velocity are presented. Here ω_0 is the angular velocity of local standard. The oblatness of ellipsoids varies only slightly when one approaches the galactic center. The same picture takes place on the other side of galactic centre. In this figure one can also see the rotation curve at the galactic plane that was made more precise by this method. It may be seen that the mean ratio of the ellipsoid axes is about 0.13. At distances of 7-8 kpc from the galactic centre the linear velocity of the neutral hydrogen subsystem is diminished by 10km/sec, the distance from the galactic plane increasing by 100 pc.

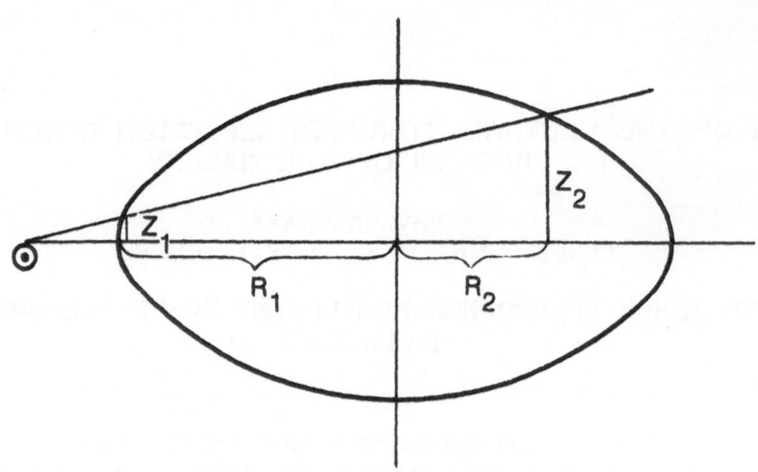

Fig. 1. The surface of equal angular velocity, scheme.

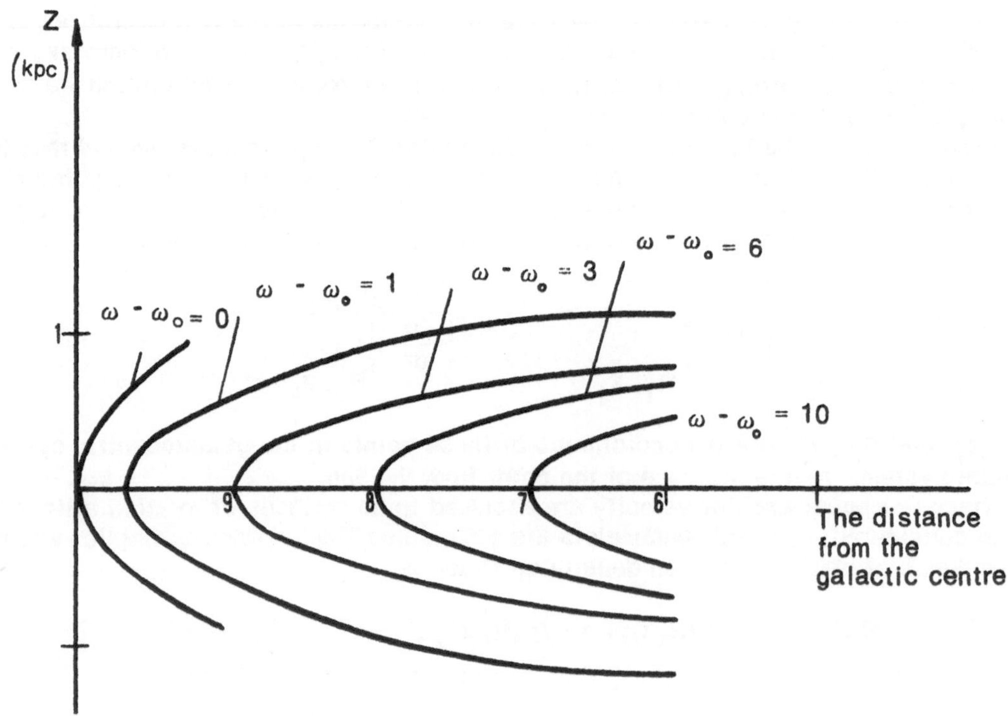

Fig. 2. The ellipsoids of equal angular velocity for the neutral hydrogen subsystem.

REFERENCES

McGEE R.X., MILTON JANICE A.and WOLF WENDY 1966,Aust.J.Phys.Astrophys.suppl., N 1,1

MULLER C.A., WESTERHOUT C., 1957, Bull.Astron.Inst. Netherl., 13, N 475, 151

PETROVSKAYA I.V. 1970, Vestnik Leningradskogo Universiteta., N 19, 129

VAN WOERDEN H., TAKAKUBO K. and BRAES L.L.E. 1962, Bull. Astron. Inst. Netherl., 16, 321

PHOTOMETRIC INVESTIGATIONS OF HD 124224

W. WEISS
UNIVERSITY OBSERVATORY, VIENNA, AUSTRIA

ФОТОМЕТРИЧЕСКИЕ ИССЛЕДОВАНИЯ HD 124224

(The author has not sent the text of his paper).

H I, H_2CO and CH IN DARK INTERSTELLAR CLOUDS ALONG THE PLANE OF GOULD'S BELT

(SHORT COMMUNICATION)

A. SANDQVIST, P.O. LINDBLAD AND K.P. LINDROOS
STOCKHOLM OBSERVATORY, SWEDEN
PRESENTED BY A.SANDQVIST

H I, H_2CO И CH В ТЕМНЫХ МЕЖЗВЕЗДНЫХ ОБЛАКАХ ВДОЛЬ ПОЯСА ГУЛДА

At the present time, a research group at the Stockholm Observatory is studying what is suspected to be a local, slowly expanding interstellar cloud, or rather cloud complex of interstellar matter and relatively young stars. The expansion age of the cloud, as derived from 21-cm observations, appears to be about 60×10^6 years and its dimensions in the plane of the Galaxy of the order of 600 by 300 pc, the Solar System passing through the outer parts of the cloud. In radio spectra of the interstellar neutral hydrogen, this structure reveals itself through a narrow emission component called "Feature A" (Lindblad 1967, Lindblad *et al.*, 1973) with a velocity dispersion of about 2.5 km s^{-1}. It is observed over a large part of the celestial sphere and has positive radial velocity with respect to the local standard of rest almost everywhere in and near the galactic plane. Concerning young stars of early spectral type, this complex may be revealing itself through the so-called Gould's Belt. The kinematics of the stars of Gould's Belt have been studied by Lesh (1972), who finds an expansion age of about 65×10^6 years.

In connection with the investigation of this complex, an extensive survey of dark clouds of small angular diameters in the region of Gould's Belt has recently been undertaken by us in the 3335-MHz CH, 4830-MHz H_2CO and 1420-MHz H I lines using the 25.6-m radio telescope at Onsala and the 42.7-m and 91.5-m radio telescopes at NRAO. Out of 60 clouds chosen from Lynds (1962) and observed at Onsala, 36 were found to contain formaldehyde and 24 of these were later detected by their self-absorption feature in the neutral hydrogen line with the 91.5-m radio telescope at NRAO. These 36 clouds were found to contain also CH through the detection of the 3335-MHz CH main line using the Onsala dish. The 3263- and 3349-MHz CH satellite lines were also detected for 11 clouds which showed strong CH main line emission. The CH observations have been published as part of a larger report containing more than one hundred positions in optically dark nebulae (Hjalmarson *et al.*, 1975). A typical and a not-so-typical set of H I, H_2CO and CH profiles are shown in Figures 1 and 2, respectively.

Figure 1 shows the observations for cloud 1535 from Lynds' catalogue. (The position observed is not that given by Lynds, as an inspection of the Palomar Sky Atlas shows that the cloud itself is actually 14' north of her position where no cloud is seen. The correct position for cloud 1535 is $\alpha(1950.0) = 4^h 32.5^m$, $\delta(1950.0) = 24°02'$ which corresponds to $l = 174°.7$, $b = -15°.5$.) At the rather high negative latitude of this cloud mainly local H I contributes to the appearance of the profile and the character of Feature A is well represented in Figure 1. The presence of cloud 1535 shows itself clearly by the pointed self-absorption feature near the maximum of the profile. The tendency for the cloud feature to have a slightly positive velocity (of the order of +1 km s^{-1}) relative to the maximum of Feature A is quite typical for most of our clouds. This may imply that any breaking mechanism, possibly slowing down the expansion of Feature A, will be less effective on the much denser dark clouds.

The asymmetric H_2CO absorption profiles indicate the presence of the hyperfine components that make up the 6-cm formaldehyde line. This line structure was used to determine excitation temperatures, optical depths and other parameters employing a least squares method for a considerable number of clouds (see below). In this analysis it was found that best fits for all the clouds were obtained with an LTE assumption regarding the relative intensities of the hyperfine components.

The CH results showed that the satellite: main: satellite line ratios were generally about 1:2:1 which is expected for optically thin and LTE cases. But, as Rydbeck *et al.*(1975) have shown, the clouds may give the appearance of being in LTE and yet still behave as weak masers in the CH line. Column densities were in fact determined under the latter assumption, as the observations showed the probable likelihood of weak masering of the CH lines in the dark clouds.

A preliminary analysis of the ratios of the column densities, N_{H_2CO}/N_{CH}, showed that, regardless of the longitudes of the clouds, the ratios predominantly scatter around the value 1, ranging from about 1/2 to 2. with a few in the range of 4 to 6. Herbst and Klemperer (1973) in a study of ion-molecule reactions in dense clouds, predict values for $[H_2CO]$ $[CH]$ of 1000, 250 and 12 for molecular hydrogen densities of 10^6, 10^5 and 10^4 cm^{-3}, respectively. Our results, which are derived from dark optical clouds generally accepted to have molecular hydrogen densities of 10^3- 10^4 cm^{-3}, therefore give some support to their theory of molecule formation.

The set of profiles shown in Figure 2 are observations of cloud 1529 from Lynd's catalogue. The appearance of the H_2CO profile, at first sight, gives the impression of severe non-LTE or high-optical depth ratios of the intensities of the hyperfine lines, specifically with regards to the hyperfine component which is at a positive velocity of +1.1 km s^{-1} relative to the main component. Heiles (1973) showed that this effect may be due to two velocity components in this cloud separated by an amount of 0.9 km s^{-1} which happens to be very close to the separation of these two hyperfine components. The H 1, and more specifically the CH observations, tend to give support to the two-velocity-component theory. The H 1 self-absorption feature has an abnormally flat bottom and the 3-kHz resolution profile of the 3335-MHz CH line., which is not inherently confused by multiple hyperfine components, shows the presence of two velocity components agreeing quite well with Heiles' derived values of + 5.7 and + 6.6 km s^{-1} for the H_2CO profile.

A southward extension of the Onsala formaldehyde survey was performed with the NRAO 42.7-m radio telescope. 39 small dark clouds were chosen from the southern extension to the Palomar Sky Atlas (declination, -36° and -42°) and accurate positions and approximate areas and opacities were determined for these clouds. 33 of these clouds contain formaldehyde in a detectable amount , and in 27 of them the absorption is sufficiently strong to permit a detailed analysis of the hyperfine components of the 6-cm H_2CO line. Physical parameters for the clouds have been determined by the method of least squares using the following expression for the observed line antenna temperature, $\Delta T(v)$, in Kelvins as a function of velocity, v, in km s^{-1},

$$\Delta T(v) = (\eta \frac{\Omega_c}{B\Omega_B}) (T_x - T_c) [1 - \exp \{ -\tau_m \sum_{i=1}^{6} \alpha_i \exp [-½ (\frac{v - v_i - V}{\sigma})^2] \}] + K$$

where $\eta_B = 0.8$ is the beam efficiency, Ω_c / Ω_B is the fraction of the main beam occupied by the cloud, T_x is the excitation temperature, T_c is the background continuum temperature assumed to be 2.7 K, τ_m is the optical depth of the main component (F : 2 → 2), α_i is the ratio of the optical depth of component i to that of the main component and v_i is the velocity separation of each component from the main one, V is the radial velocity of the cloud, and σ is the velocity dispersion which is assumed to be the same for all the components, K is a constant which accounts for small zero-level errors. The observed clouds generally were of opacity class 5 or 6 and examples of the filling of this function to the observed profiles can be seen in Figure 3. This procedure yields value for the parameters T_x, τ_m, V and σ. The optical depth. τ_t, for the total 6-cm transition is obtained by taking the negative value of the expression inside the curly brackets in the above equation.

One interesting result of this analysis is presented in Figure 4 which shows a plot of T_x vs τ_t. It is apparent that some relation exists between T_x and τ_t which is rather ill-defined for T_x at lower but becomes more stringent at higher optical depths. The uncertainty in the derived values of T_x and τ_t is found to be less than 20% and is highest in the region of low τ_t. No hyperfine analysis was performed on weak absorption lines ($|\Delta T| < 0.15$ K) although such lines were in fact detected. These weak lines would fall in the hatched area in Figure 4 and one has to bear in mind that this region covers a rather large range in T_x for small values of τ_t, but that it decreases as τ_t increases. There exists a large region, below and to the right of the plotted points,in which the values of T_x and τ_t could have been combined to produce strong absorption lines, so this region of avoidance appears to be a physical property of the clouds if our analysis is correct. It is therefore apparent from the plotted points that at low τ_t, T_x can have a fair range of values from 1 to 2 K and possibly greater, but at higher optical depths ($\tau_t > 1$) the excitation temperature is restricted to values above 2K. Evans *et al.* (1975) have shown that the collisional pumping mechanism predicts that the excitation temperature should increase with increasing density, finally resulting in a quenching of the cooling mechanism. This may be the cause of the effect shown in Figure 4 and if so, our results would tend to support the collisional pumping mechanism for the abnormal cooling of the 6-cm H_2CO line. Finally, it may be mentioned in passing that Zuckerman *et al.*(1975) have recently detected 6-cm H_2CO emission from the Orion nebula which has a deduced density of 10^5 hydrogen molecules cm^{-3} showing that, at this high density, the cooling mechanism is in fact quenched

Lindblad *et al.* (1973) suggested that some of the dark clouds, the neutral hydrogen Feature A and Gould's Belt of early type stars may be related, and that this local, now expanding, system may have had as its origin the passage of interstellar gas through the galactic shock located at the position of the Carina spiral arm, 60 x 10^6 years ago. On the other hand Quirk and Crutcher (1973) assume that the cold cloud (CC) of Riegel and Crutcher (1972) is a region of high density in the very shock of the Orion "arm". This is disputed by Rickard(1974) who places the cold cloud near the shock of the Sagittarius arm.

In Figure 5, we have plotted a velocity-longitude diagram which includes the theoretical model (dashed line) for Feature A presented by Lindblad *et al.* (1973) which has been determined for the plane of the Galaxy, and tentative velocities (open circles) for the Orion "arm" or "other local feature"(OLF) which have been obtained from observations by various authors. Furthermore, we have plotted the velocities (crosses) of the dark clouds that we have observed at Onsala and NRAO. Although some clouds have velocities that are close to OLF's, a majority of them show a distinct preference for the relation for Feature A. This preference is extremely pronounced in the region $30° > l > 335°$. This region contains the CC and its velocity components of + 4 and +7 km s^{-1} seem to rule out its association with OLF which has a velocity of about - 7 km s^{-1} at $l = 345°$ (one limit of the CC) and about +16 km s^{-1} at $l = 25°$ (the other limit of the CC). This does not, however, rule out any possible association with a shock in the Orion "arm". The added observational fact that a number of the dark clouds, which show kinematical behaviour similar to the CC and Feature A, are at distances that are of the order of 150pc would tend to argue against the placing of this complex in the Sagittarius arm. In summary, we judge that the velocity longitude relationships of the CC, most of the dark clouds, and Feature A imply a physical association of these three components of the local interstellar matter.

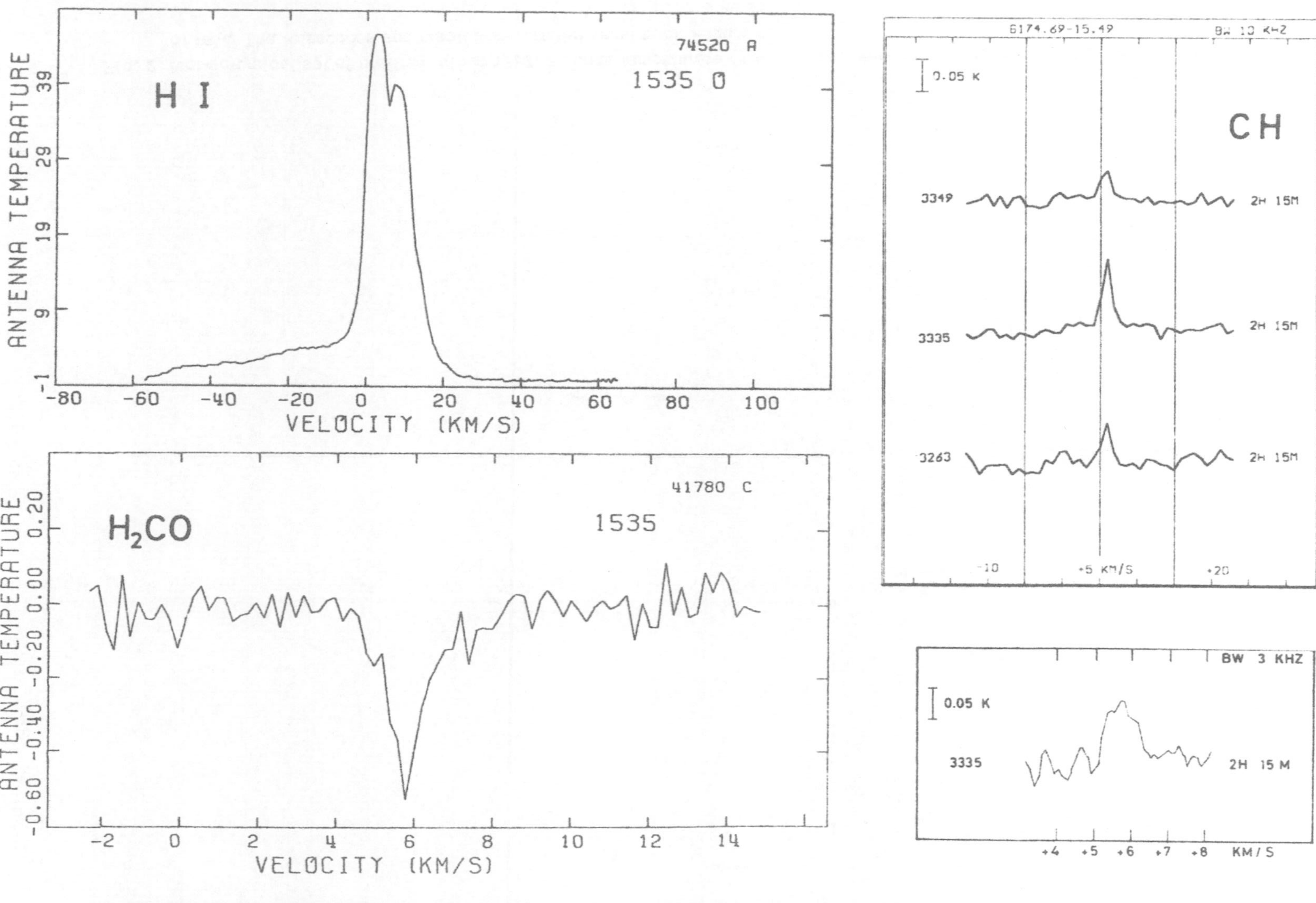

Fig. 1. Typical set of profiles, cloud 1535, antenna temperature (Kelvins) vs. velocity with respect to the local standard of rest. H I : 91.5-m telescope, NRAO, 0.7 km s^{-1} resolution. H_2CO: 42.7-m telescope, NRAO, 0.2 km s^{-1} resolution. CH: 25.6-m telescope, Onsala, 0.9 and 0.3 km s^{-1} resolution.

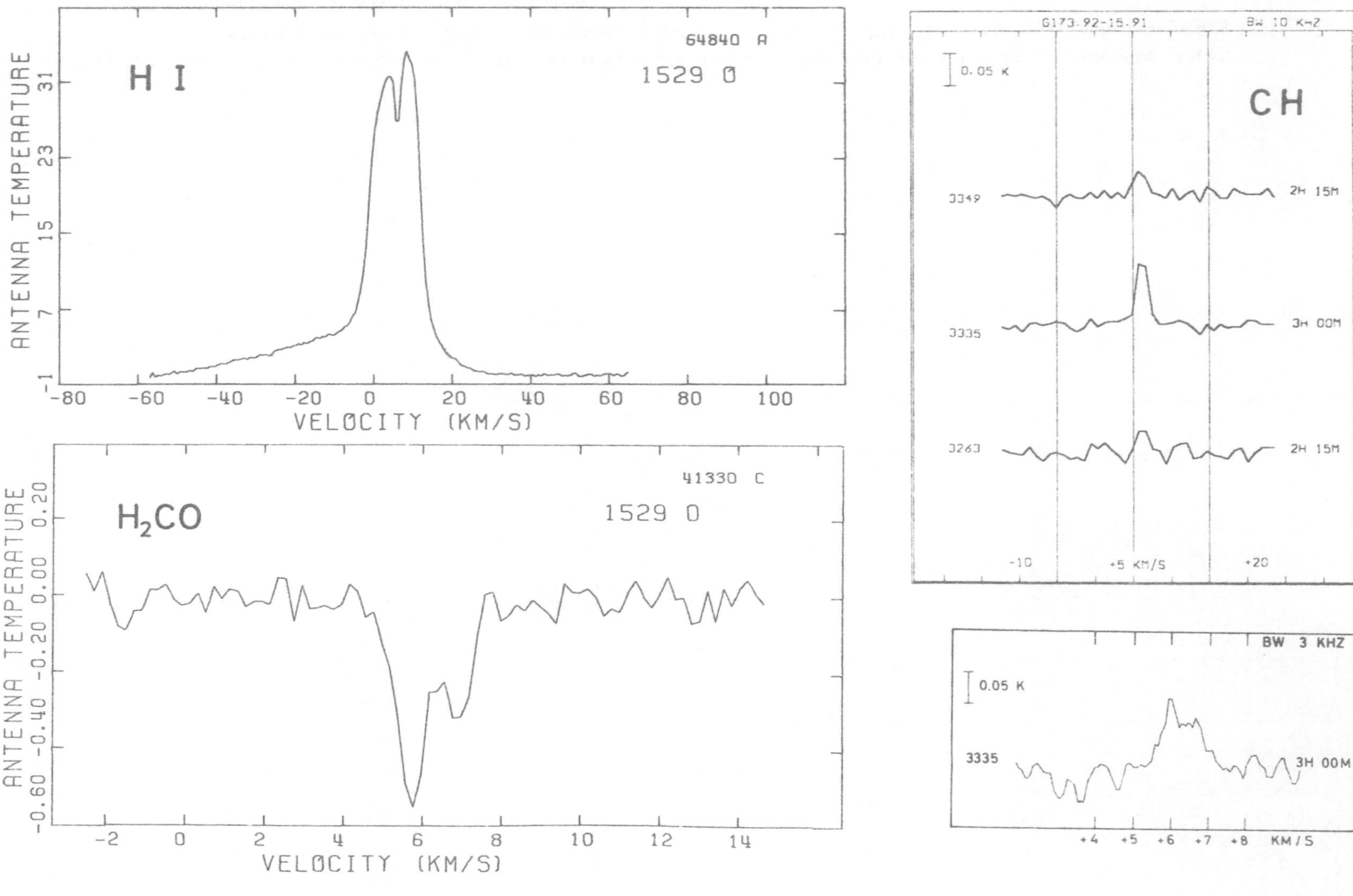

Fig. 2. Not-so-typical set of profiles, cloud 1529, antenna temperature (Kelvins) vs. velocity with respect to the local standard of rest. The telescopes and resolutions are the same as in Figure 1.

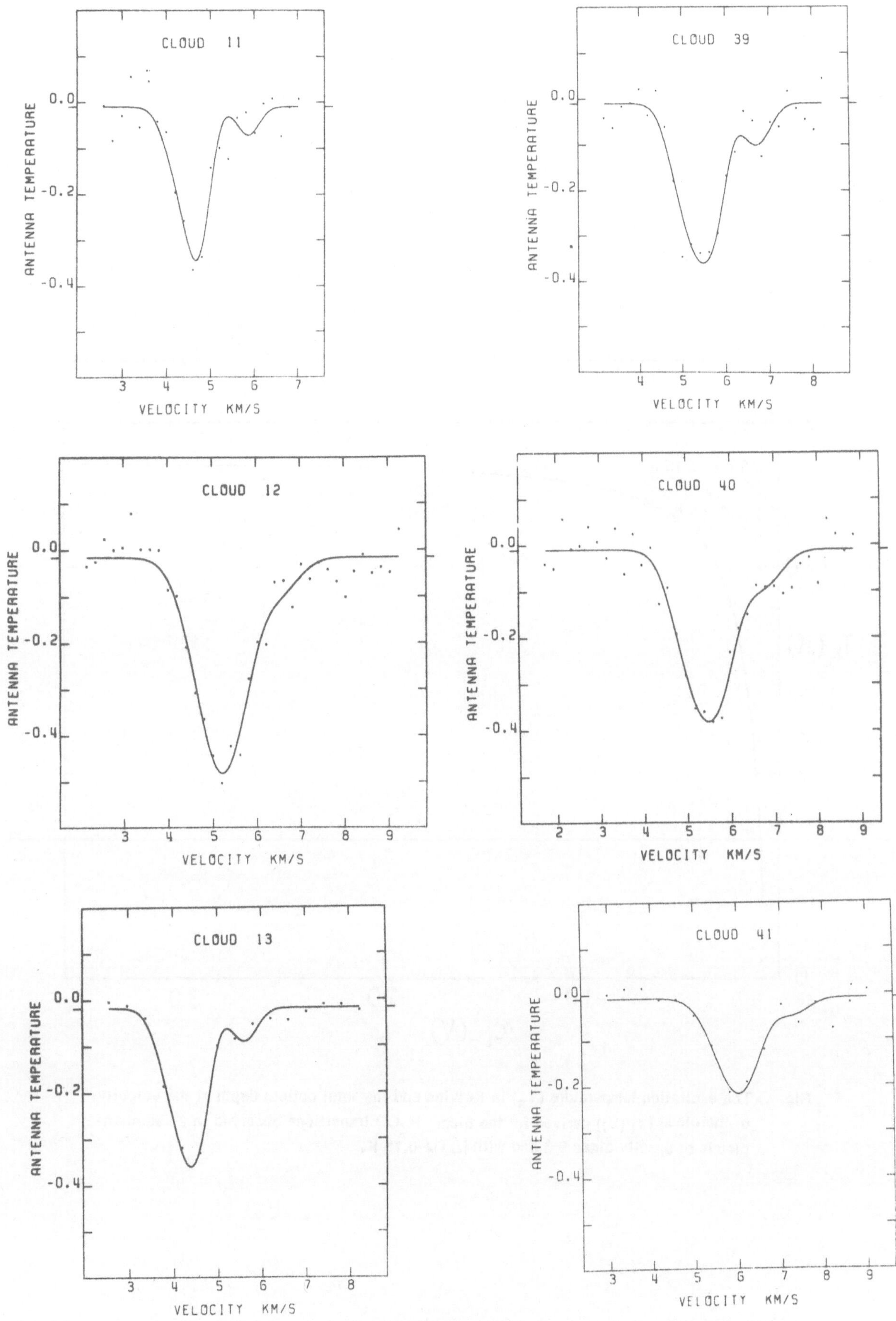

Fig. 3. The observed (dots) and the theoretical (solid lines) 6-cm H_2CO line profiles, resulting from the least squares fitting process discussed in the text, for a number of southern clouds.

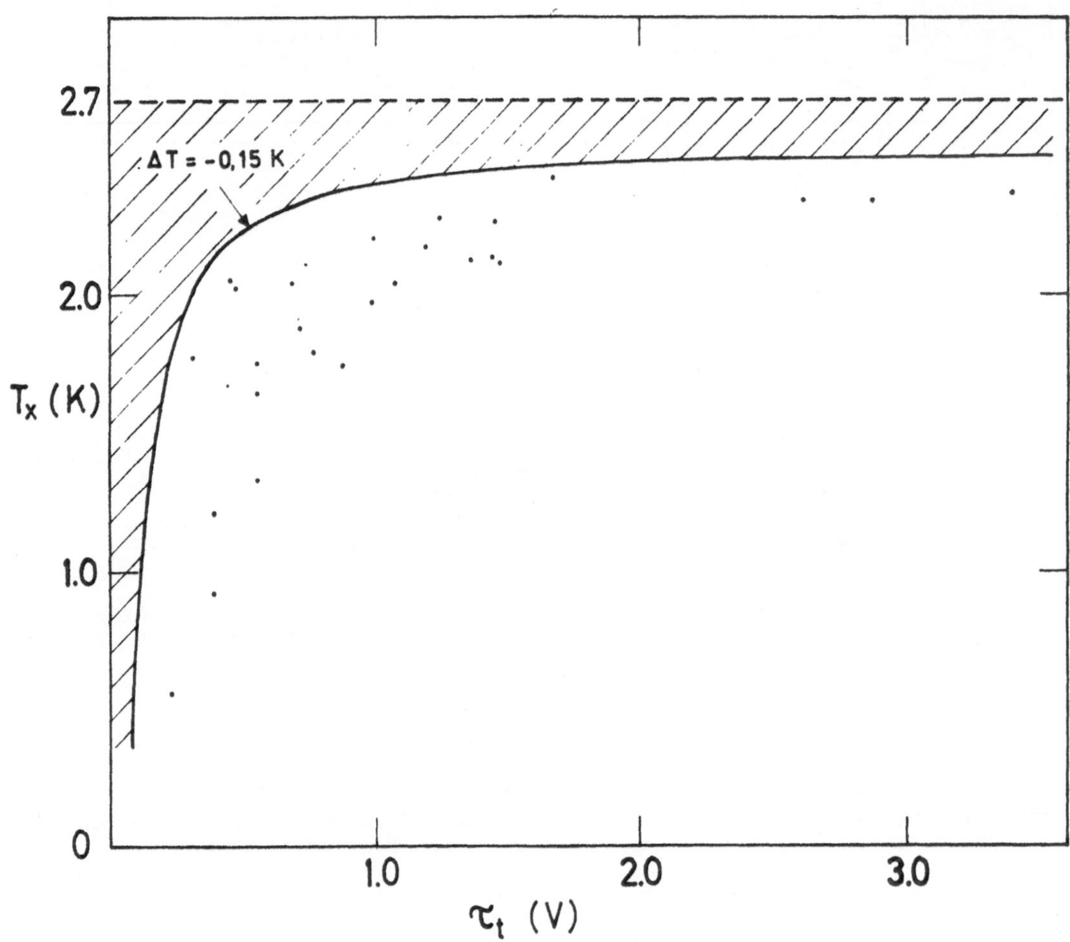

Fig. 4. The excitation temperature (T_x) in Kelvins and the total optical depth at the velocity of the cloud (τ_t (V)) derived for the 6-cm H_2CO transitions observed in 27 southern clouds of opacity class 5-6 and with $|\Delta T|> 0.15$ K.

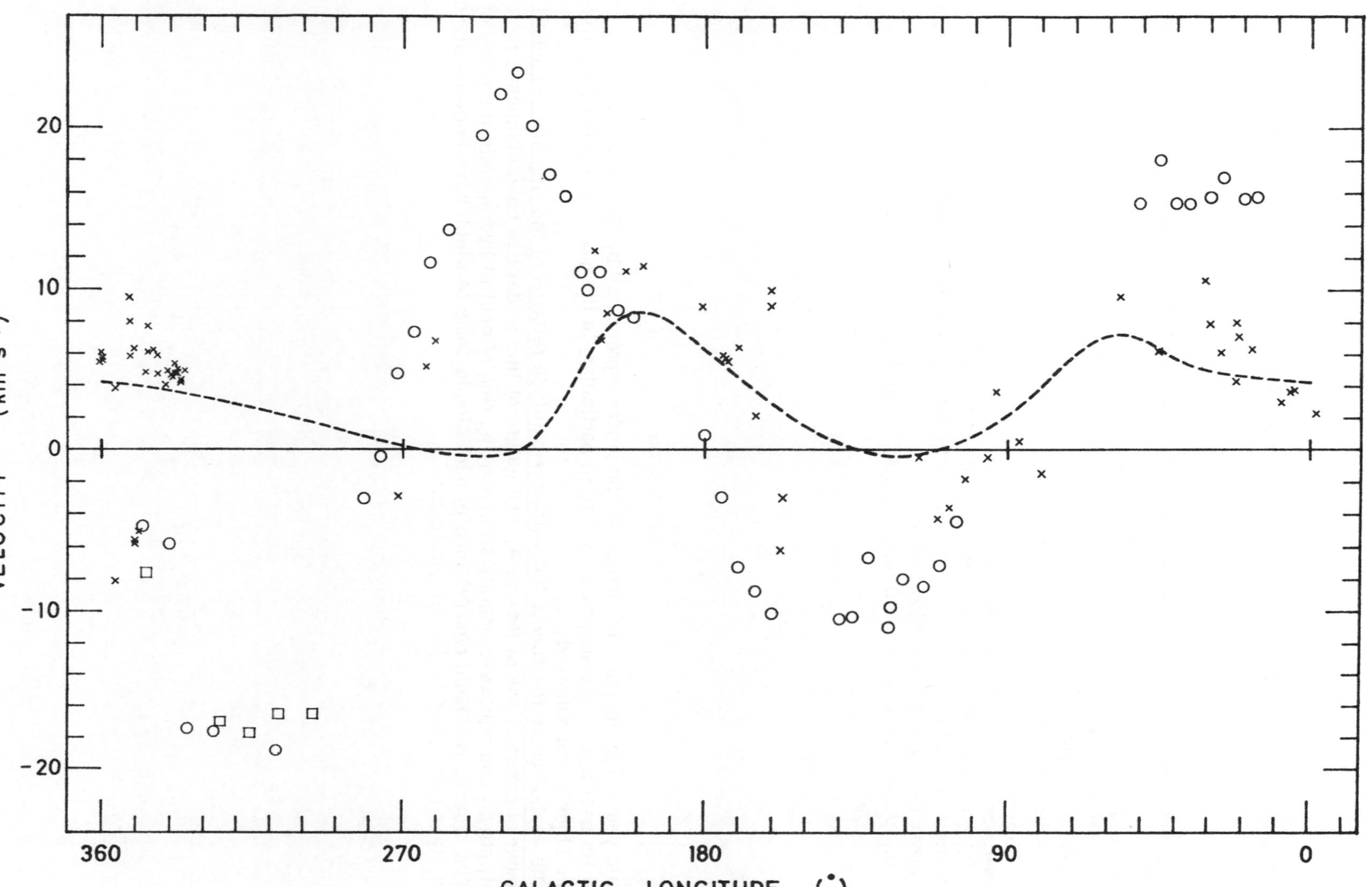

Fig. 5. Velocity-longitude diagram for the local interstellar matter. Lindblad et al. 's (1973) theoretical relation for Feature A (dashed line), tentative velocities for the "other local feature" chosen from H I observations of various authors (open circles), and observed velocities of dark clouds (crosses).

REFERENCES

EVANS N.J., ZUCKERMAN B., MORRIS G., and SATO T., 1975, Ap.J. 196, 433

HEILES C., 1973, Ap. J. 183, 441

HERBST E. and KLEMPERER W., 1973, Ap. J. 185, 505

HJALMARSON A., SUME A., ELLDÉR J., RYDBECK O.E.H., MOORE E., HUGUENIN R., SANDQVIST Aa., LINDBLAD P.O., and LINDROOS P., 1975, Radio Observations of Interstellar CH Part II, Research Report No 124, Research Laboratory of Electronics and Onsala Space Observatory, Chalmers University of Technology, Sweden

LESH J.R., 1972, In G.CAYREL DE STROBEL AND A.M.DELPLACE (eds.), 'L'age des étoiles', IAU Colloq. 17, Paper XXIII

LINDBLAD P.O., 1967, BAN 19, 34

LINDBLAD P.O., GRAPE K., SANDQVIST Aa. and SCHOBER J., 1973 A. and Ap. 24, 309

LYNDS B.T., 1962, Ap. J. Suppl. 7, 1

QUIRK W.J. and CRUTCHER R.M., 1973, Ap. J. 181, 359

RICKARD J.J., 1974, A. and Ap. 31, 47

RIEGEL K.W. and CRUTCHER R.M., 1972, A. and Ap. 18, 55

RYDBECK O.E.H., KOLLBERG E., HJALMARSON A., SUME A., ELLDÉR J., and IRVINE W.M., 1975, Radio Observations of Interstellar CH Part I, Research Report No 120, Research Laboratory of Electronics and Onsala Space Observatory, Chalmers Univeraity of Technology, Sweden.

ZUCKERMAN B., PALMER P., and RICKARD L.J., 1975, Ap. J. 197, 571

DISCUSSION

BLAAUW: Could you briefly indicate the nature of the model represented by the dashed curve in your Fig.5 and its bearings on the observed points? In particular: is the model computed for a fixed distance of the objects you observed?

SANDQVIST: The dashed curve represents the velocity-longitude relation of the ring model applied to a large number of observations of feature A. The nature of this model has been discussed by Lindblad et al, (1973) and represents mainly an expanding ring of neutral hydrogen in the plane of the Galaxy. The crosses represent observations of dark clouds some of which have known distance, but no fixed distance was used in the calculations.

FORMATION OF GALAXIES IN FRIEDMANIAN UNIVERSE *
(Lecture)

Ya.B.ZEL'DOVICH
INSTITUTE OF APPLIED MATHEMATICS, MOSCOW, U.S.S.R.

ОБРАЗОВАНИЕ ГАЛАКТИК В ИЗОТРОПНОЙ ВСЕЛЕННОЙ

РЕЗЮМЕ

Наблюдаемая структура Вселенной, хорошо описываемая однородной и изотропной космологической моделью Фридмана, возникла при гравитационном усилении первоначально малых возмущений плотности и скорости вещества. Согласно нелинейной теории гравитационной неустойчивости при возрастании первоначальных малых возмущений возникают тонкие плотные облака, ограниченные ударной волной, анизотропно сжимающей вещество под действием приливных сил внешних возмущений. Сжатые ударной волной облака под действием тепловой и гравитационной неустойчивости распадаются на отдельные образования - галактики, которые могут входить в состав более крупных объединений - скоплений галактик. При прохождении вещества через фронт ударной волны про - исходит быстрая турбулизация движения, что приводит к образованию галактик со значительным моментом вращения. Это, в дальнейшем, спиральные галактики. Эллиптические галактики возникают из газа, обладавшего малым моментом вращения.

The structure of the Universe is characterized by an enormous range of dimensions of structural elements - from stars to gigantic galaxies and clusters of galaxies. At the same time recently it has been established, while studying relict radiation, that inhomogeneities even in larger scales of matter distribution are small and the Universe evolution is well described by a homogeneous and isotropic cosmological model of Friedman.

According to the suggested theory of formation of galaxies such a structure of the Universe was created at gravitational enhancement of originally small perturbations of density and velocity. Nonlinear theory of gravitational instability predicts that as a result of increasing small perturbations thin dense clouds of gas are formed, limited from all the sides by a shock wave contracting and heating the falling in gas. As is known, dissipation of perturbations during the period of recombination leads to surviving of only large scale perturbations. The distributions of density and velocity of motion are found to be smooth functions of coordinates. The characteristic dimensions of perturbations considerably exceed the Jeans wavelength in the neutral gas. Therefore the development of perturbations in the neutral gas during the postrecombination period is practically independent on its pressure.

The nonlinear theory of gravitational instability leads to the conclusion that matter contraction is of a sharply anisotropic character. This anisotropy is already in the linear theory of gravitational perturbations and it is the result of action of tidal forces of external perturbations. As a result of perturbation development contraction takes place predominantly in one direction. There are formed flattened condensations strongly contracted in one direction.

Dense gas clouds contracted by a shock wave due to a thermal and gravitational instability decay into different gravitationally bound formations - galaxies, which can become a component of larger aggregates - clusters of galaxies.

In the suggested theory the matter motion up to the passage through the front of the shock wave is non-turbulent. But at passing through the shock wave front the non-turbulent character of the motion is violated and in the contracted matter the vortex components of the velocity are commensurable with the potential ones. Fast turbulization of the motion occurs promoting matter fragmention and formation of galaxies with a considerable rotational momentum. Such galaxies at their further evolution are transformed into spiral galaxies. Elliptical galaxies are formed from the gas with a small momentum of rotation.

The details of the described problems see in the book of Ya.B.Zel'dovich, I.D.Novikov "Structure and evolution of the Universe", Izd. Nauka, Moscow, 1975 and in the paper of A.G. Doroshkevich Ya.B.Zel'dovich, R.K.Syunyaev in the Proceedings of IAU Symposium N 63, Cracow, Poland, 1973.

---

* Editor's note: only the abstract of Prof, Zel'dovich's lecture is published according to the author's wish.

CLOSING CEREMONY

Prof. W.Iwanowska (Poland). (Speech on behalf of the IAU)
Prof. A.Blaauw (The Netherlands). (Speech on behalf of foreign participants)

After closing the Meeting two facultative communications were presented

ON THE NATURE OF STELLAR ASSOCIATIONS

P.N.KHOLOPOV
ГERNBERG ASTRONOMICAL INSTITUTE, MOSCOW, U.S.S.R.

О ПРИРОДЕ ЗВЕЗДНЫХ АССОЦИАЦИЙ

COMET MATTER IN INTERSTELLAR SPACE AND MAY BE IN OTHER GALAXIES

S.K.VSEKHSVYATSKIJ
UNIVERSITY, KIEV, U.S.S.R.

КОМЕТНОЕ ВЕЩЕСТВО В МЕЖЗВЕЗДНОМ ПРОСТРАНСТВЕ И ВОЗМОЖНО В ДРУГИХ ГАЛАКТИКАХ

INDEX
(Authors and those who took part in discussion)

CONTENTS

ТРУДЫ ТРЕТЬЕЙ ЕВРОПЕЙСКОЙ АСТРОНОМИЧЕСКОЙ КОНФЕРЕНЦИИ

ЗВЕЗДЫ И ГАЛАКТИКИ В НАБЛЮДАТЕЛЬНОМ АСПЕКТЕ

Тбилиси 1976

Напечатано по постановлению Редакционно - издательского совета
Академии наук Грузинской ССР

Сдано в набор 25.5.1976; Подписано к печати 18.5.1976;
Формат бумаги 60x84⅛; Печатных л.66,50; Уч.издат.л.55,4; Бумага офсетная

УЭ 11182; Тираж 1700; Заказ 1756

Цена 4 руб. 60 коп.

Издательство "Мецниереба", Тбилиси, 380060, ул.Кутузова,19.
Типография АН ГССР, Тбилиси, 380060, ул.Кутузова, 19.